UNIVERSITY OF BRISTOL
Food Refrigeration and
Process Engineering Research Centre
F R P E R C Churchill Building
Langford, Bristol. BS18 7DY
Tel: + 44 (0)117 928 9239   Fax: + 44 (0)117 928 9314

FOOD ENGINEERING AND MANUFACTURING SERIES

R. Paul Singh, Series Co-Editor
University of California, Davis

Dennis R. Heldman, Series Co-Editor
Weinberg Consulting Group, Inc.
Washington, D. C.

Published Titles
Advances in Food Engineering
R. Paul Singh and M. A. Wirakartakusumah

Transport Phenomena of Foods and Biological Materials
Vassilis Gekas

Forthcoming Titles
Food Engineering: Principles and Applications
B. O. Balaban

CRC Press
Boca Raton   Ann Arbor   London   Tokyo

MINIMAL PROCESSING OF FOODS AND PROCESS OPTIMIZATION
An Interface

Edited by

R. Paul Singh
Professor of Food Engineering
Department of Biological and
  Agricultural Engineering
University of California
Davis, California

Fernanda A. R. Oliveira
Assistant Professor
Escola Superior de Biotecnologia
Porto, Portugal

CRC Press
Boca Raton   Ann Arbor   London   Tokyo

Library of Congress Cataloging-in-Publication Data

Minimal procesing of foods and process optimization—an Interface /
    edited by R. Paul Singh, Fernanda A.R. Oliveira.
        p.   cm. — (Food engineering and manufacturing)
    Includes bibliographical references and index.
    ISBN 0-8493-7903-2
    1. Food industry and trade.   I. Singh, R. Paul.   II. Oliveira, Fernanda A.R.
    III. Series.
    TP370.M56 1994
    664—dc20   94-4206
                                                                                                          CIP

No claim to original U.S. Government works
International Standard Book Number 0-8493-7903-2
Library of Congress Card Number 94-4206
Printed in the United States of America   2 3 4 5 6 7 8 9 0
Printed on acid-free paper

Dedicated

to

Raj

Ana Catarina

Rui Pedro

Carlos Filipe

TABLE OF CONTENTS

PREFACE

Preservation of foods with the use of physical, chemical and biochemical means such as canning, freezing, drying and fermentation are commonly employed in the food industry. During the last several decades, scientists and engineers have made numerous advances in increasing our understanding of these processes. These advances have lead to the optimization of large scale commercial processing of foods. An overview of the published literature in this field yields several mathematical schemes that are useful in optimizing food manufacturing processes. Process optimization is vital to delivering improved quality of foods with increased assurance of food safety, while also maintaining the economic competitiveness of the food industry.

More recently, minimal processing of foods has been gaining considerable interest, partially because of the need to mitigate the undesirable effects associated with many traditional processing methods. In industrially advanced countries, the consumer is increasingly demanding foods that have "fresh-like" qualities. These demands have encouraged the development of processes that cause minimal adverse changes in a food, yet offer desirable benefits derived from increasing shelf life.

It is therefore appropriate that in our quest to develop innovative processing methods, we conduct a critical examination of notable advances made in traditional and emerging processes. With this goal in mind, a workshop was held on September 20-23, 1993, at the Escola Superior de Biotecnologia, Porto, Portugal. Leading scientists and engineers from around the world, who are actively involved in studies on process optimization and minimal processing of foods, were invited to this workshop. Forty participants from Europe, North and South America and Asia participated in the workshop.

The workshop format consisted of keynote presentations on selected themes. These overview lectures were then followed with technical presentations containing recently obtained results from laboratories with active research programs. A significant number of these programs were supported by the European Community, FLAIR and AIR research and development projects. Each topical theme was followed with extended periods allocated for discussion. The purpose of these discussions was to identify issues that must be addressed for making future advances. Constructive dialogues held between participants on engineering and microbiological problems were found to be rewarding.

All technical presentations made at the workshop are included in this book. The discussion held at the end of each presentation was recorded and is presented along with each chapter.

The workshop was supported by a number of organizations. Appreciation is extended to the European Community (Program AIR), Junta Nacional de Investigacoa Cientifica e Tecnologica, Fundacao Calouste Gulbenkian and FIMA Produtos Alimentares for their financial support.

A special word of gratitude is given to Professor Augusto Medina for his constant encouragement and support in the establishment of international collaboration, and also for his active role in creating the opportunity to convene such type of scientific exchange.

R. PAUL SINGH   F. OLIVEIRA

THE EDITORS

R. Paul Singh, Ph.D., is Professor of Food Engineering, Department of Biological and Agricultural Engineering, Department of Food Science and Technology, University of California, Davis, California.

Dr. Singh graduated in 1970 from Punjab Agricultural University, Ludhiana, India, with a degree in Agricultural Engineering; he obtained an M.S. degree from University of Wisconsin, Madison, and a Ph.D. degree from Michigan State University in 1974. Following a year of teaching at Michigan State University, he moved to University of California, Davis in 1975 as an Assistant Professor of Food Engineering. He was promoted to Associate Professor in 1979, and again to Professor in 1985.

Dr. Singh is a member of the Institute of Food Technologists, American Society of Agricultural Engineers and Sigma Xi. He received the First Place Paper Award, American Society of Agricultural Engineers in 1982; A. W. Farrall Young Educator Award, American Society of Agricultural Engineers, 1986; NATO Senior Guest Lecturer, 1987, 1993; IFT International Award, Institute of Food Technologists, 1988; and Distinguished Alumnus Award, Punjab Agricultural University, 1989.

Dr. Singh has authored and co-authored 6 books and over 150 technical papers. He is a co-editor of the Journal of Food Process Engineering. His current research interests are in transport phenomena in foods as influenced by structural changes during processing. His research is supported by grants from federal and state agencies and food industries from U.S., Japan and European Countries.

F. A. R. Oliveira, Ph.D., is Assistant Professor, Escola Superior de Biotecnologia, ESB, (College of Biotechnology), Universidade Católica Portuguesa (Portuguese Catholic University), Porto, Portugal and Head of the Graduate School of ESB.

Dr. Oliveira graduated in 1985 from Universidade do Porto (University of Porto), Portugal, with a degree in Chemical Engineering; she obtained a Ph.D. degree from Leeds University (United Kingdom) in 1989, becoming then Assistant Professor at ESB.

Dr. Oliveira is a professional member of the Institute of Food Technologists. She received the "Centenário" and "Eng° Cristiano Spratley" awards.

Dr. Oliveira has authored and co-authored over 10 technical papers in international journals and written around 30 communications in international Congresses and Workshops. Her current main research interests are mass transfer in food systems and microstructural relations with texture evolution in food processing. Her research is supported by grants from state agencies and by international collaborative research contracts funded by the European Commission.

LIST OF CONTRIBUTORS

Andres, A. Department of Food Technology, Universidad Politecnica de Valencia, Valencia, ITALY

Andrzej, Lenart Dept. of Food Engineering, Warsaw Agricultural University, UL. Nowoursynowska 166, 02-766 Warszzawa, POLAND

Balaban, M.O. Agricultural Engineering Department, University of Florida, Frazier Rogers Hall, Gainesville, FL 32611, U.S.A.

Banks, J. G. Campden Food & Drink RA, Chipping Campden, GLOS, Gl55 6LD, UNITED KINGDOM

Baptista, P.N. Escola Superior de Biotecnologia, Rua Dr. Antonio de Bernardino, 4200 Porto, POTUGAL

Brennan, J.G. University of Reading, Dept. of Food Science & Technologie, P.O. Box 226 Whiteknights, Reading, RG6 2AP, Berkshire UNITED KINGDOM

Bruhn, Christine Center for Consumer Res., University of California, , Davis, CA 95616

Busta, F. University of Minnesota, 228 Food Science & Nutr. Bldg., 1334 Eckles Ave., St. Paul, MN 55108

Campbell-Platt, G. University of Reading, Dept. of Food Science & Technologie, P.O. Box 226 Whiteknights, Reading, RG6 2AP, Berkshire UNITED KINGDOM

Cano Dolado, Maria Pilar Unidade de Congelation, Instituto del Frio, Ciudade Universitaria, 2840 Madrid, SPAIN

Capell, C.J. Escola Superior de Biotecnologia, Rua Dr. Antonio de Bernardino, 4200 Porto, POTUGAL

Chau, Khe V. Agricultural Engineering Department, University of Florida, Fraziers Rogers Hall, Gainesville, FL 32611 , U.S.A.

Chiralt, A. Department of Food Technology, Universidad Politecnica de Valencia, Valencia, ITALY

Chitarra, A.B. Department of Food Science, ESAL, Caixa Postal 37, 37200 Lavras MG, BRAZIL

Chitarra, M. Isabel F. Department of Food Science, ESAL, Caixa Postal 37, 37200 Lavras MG, BRAZIL

Cole, M.B. Dept. of Applied Biochemistry and Food Scinece, University of Nottingham, Sutton Bonington Campus Loughborough, Liecestershire LE12 5RD, UNITED KINGDOM

De Cordt, S.   Katholiec Universiteit Leuven, Faculty of Food Technology, Kardinaanal Mercierland 32, B-3001 , Heverlee BELGIUM

De Baerdemaeker, Josse   Afdl. Landelijk Genie Kul, Kardinaal Mercier, Laan 92, Heverlee, BELGIUM

Ellison, Annette   Dept. of Applied Biochemistry and Food Scinece, University of Nottingham, Sutton Bonington Campus Loughborough, Liecestershire LE12 5RD, UNITED KINGDOM

Farkas, J.   Dept. of Refrigeration & Livestock, Products TEchnology, University of Horticulture & Food Industry, H-1118 Budapest, Ménesi ut 43-45   HUNGARY

Fito, P.   Department of Food Technology, Universidad Politecnica de Valencia, Valencia, ITALY

Gekas, Vassilis   Lunds Universitet, Avd, , Fur Livsmedelsteknik, P. O. BOX 124, SWEDEN

Gormley, T. Ronnan   The National Food Center, Dunsinea, Castleknock, Dublin 15, IRELAND

Gorris, Leon G. M.   Agrotechnological Research Institute, ATO-DLO Agrotechnologie, Haagsteeg 6, Postbus 7, N-6700 AA , Wageningen   THE NETHERLANDS

Grandison, A. S.   University of Reading, Dept. of Food Science & Technologie, P.O. Box 226 Whiteknights, Reading, RG6 2AP, Berkshire   UNITED KINGDOM

Henderson, J.T.   Agricultural Engineering Department, University of Florida, Frazier Rogers Hall, Gainesville, FL 32611, U.S.A.

Hendrickx, Mark   Katholiec Universiteit Leuven, Faculty of Food Technology, Kardinaanal Mercierland 32, B-3001 , Heverlee BELGIUM

Holland, N.   Department of Food Science, ESAL, Caixa Postal 37, 37200 Lavras MG, BRAZIL

Irwe, Sten   Head of Food Technology, Aseptic and Food Technology, Tetra Pack Research & Development AB, Ruben Rausings, S-221 86, Lund   SWEDEN

Kirby, R. M.   Escola Superior de Biotecnologia, Rua Dr. Antonio de Bernardino, 4200 Porto, POTUGAL

Knorr, Dietrich   Technische Univ. Berlin, Dept. of Food Tech., Konigin-Luise St. 22, D-1000 Berlin, 333   GERMANY

Maesmans, G.   Katholiec Universiteit Leuven, Faculty of Food Technology, Kardinaanal Mercierland 32, B-3001 , Heverlee   BELGIUM

Mannapperuma, J.D.   Dept. of Biological and Agricultural Engineering, University of California, Davis, CA   95616

Lazarides, Harris N.   Department of Food Science & Technology, Aristotelean University of Thessaloniki, Box 255, , Thessaloniki 540 06, GREECE

Martens, T.   Afdl. Landelijk Genie Kul, Kardinaal Mercier, Laan 92, Heverlee, BELGIUM

Nicolai, B.M.   Afdl. Landelijk Genie Kul, Kardinaal Mercier, Laan 92, Heverlee, BELGIUM

Nihmura, M.   Central Research Laboratories, Ajinomoto General Foods, Inc., Minamitamagaki-cho 6410, Suzuka-city, Mie-prefecture   JAPAN

Noronha, J.   Katholiec Universiteit Leuven, Faculty of Food Technology, Kardinaanal Mercierland 32, B-3001 , Heverlee   BELGIUM

Nychas, G. J.   Food Microbiologist, Institute of Food Technology, S. Venizelou 1, Lycovrisi 141 23, Athens   GREECE

Oliveira, Fernanda, A.R.   Escola Superior de Biotecnologia, Rua Dr. Antonio de Bernardino, 4200 Porto, POTUGAL

Oliveira, J.C.   Escola Superior de Biotecnologia, Rua Dr. Antonio de Bernardino, 4200 Porto, POTUGAL

Olsson, I.   Head of Food Technology, Aseptic and Food Technology, Tetra Pack Research & Development AB, Ruben Rausings, S-221 86, Lund   SWEDEN

Pastor, R.   Department of Food Technology, Universidad Politecnica de Valencia, Valencia, ITALY

Pereira, P.M.   Escola Superior de Biotecnologia, Rua Dr. Antonio de Bernardino, 4200 Porto, POTUGAL

Sastry, S. K.   Dept. of Agricultural Engineering, Ohio State University, 590 Woody Hayes Drive, Columbus, Ohio   43210

Schellekens, M.   Afdl. Landelijk Genie Kul, Kardinaal Mercier, Laan 92, Heverlee, BELGIUM

Shewfelt, Robert L.   Dept. of Food Science & Technology, Experiment Station, University of Georgia, Griffin, GA   30223

Silva, C.L.M.   Escola Superior de Biotecnologia, Rua Dr. Antonio de Bernardino, 4200 Porto, POTUGAL

Singh, R. Paul   Dept. of Biological and Agricultural Engineering, University of California, Davis, CA   95616

Stewart, G.S.A.B.   Dept. of Applied Biochemistry and Food Scinece, University of Nottingham, Sutton Bonington Campus Loughborough, Liecestershire LE12 5RD, UNITED KINGDOM

Strauss, A.   Escola Superior de Biotecnologia, Rua Dr. Antonio de Bernardino, 4200 Porto, POTUGAL

Talasila, P.C.   Agricultural Engineering Department, University of Florida, Fraziers Rogers Hall, Gainesville, FL   32611 , U.S.A.

Taniguchi, R.   Central Research Laboratories, Ajinomoto General Foods, Inc., Minamitamagaki-cho 6410, Suzuka-city, Mie-prefecture   JAPAN

Teixeira, A.A.   Agricultural Engineering Department, University of Florida, Fraziers Rogers Hall, Gainesville, FL   32611 , U.S.A.

Tobback, P.   Katholiec Universiteit Leuven, Faculty of Food Technology, Kardinaanal Mercierland 32, B-3001 , Heverlee   BELGIUM

Torres, A.P.   Escola Superior de Biotecnologia, Rua Dr. Antonio de Bernardino, 4200 Porto, POTUGAL

Van Dijk, C.   ATO-DLO Agrotechnologie, Haagsteeg 6, Postbus 6, NL-6700, Wageningen   THE NETHERLANDS

Van Loey, A.   Katholiec Universiteit Leuven, Faculty of Food Technology, Kardinaanal Mercierland 32, B-3001 , Heverlee   BELGIUM

Vaz-Pires, P.   Escola Superior de Biotecnologia, Rua Dr. Antonio de Bernardino, 4200 Porto, POTUGAL

Wilocx, F.   Katholiec Universiteit Leuven, Faculty of Food Technology, Kardinaanal Mercierland 32, B-3001 , Heverlee   BELGIUM

Zottola, Edmund A.   University of Minnesota, Dept. of Food Science & Nutrition, 1334 Eckles Ave., St. Paul, MN   55108

Part I

Developments in New Technologies for Minimal Processing of Foods

NON-THERMAL PROCESSES FOR FOOD PRESERVATION

Dietrich Knorr
Department of Food Technology
Berlin University of Technology
Königin-Luise-Str. 22, D-14195 Berlin, Germany

INTRODUCTION

During the last few years non-thermal processing as well as minimally processed , freshlike products have become commonplace terms within food science and technology (Anon. 1993, King and Bolin 1989, Mertens and Knorr 1992). A long standing goal of food technology has also been to extend the shelf life of fresh products. For example, until recently activities were mainly devoted to extend shelf life of whole fruits and vegetables. Now the new challenge is considered to include peeled, cut and otherwise partially processed fruits and vegetables (King and Bolin 1989). Increasing activities are also centering on natural preservation systems such as plant metabolites or microbially derived antimicrobials (Banks et al. 1986, Gould et al 1986, Hoover and Steenson 1993).

Process developments that deal with design and application of non-thermal food processing and preservation processes include utilization of electric or magnetic fields, microwave radiation, ionizing radiation, light pulses, ultrasonics, high isobaric pressure and chemical agents such as subcritical or supercritical carbon dioxide, polycationic polymers, and enzymes with antimicrobial activities (Mertens and Knorr 1992). This chapter will concentrate on research regarding the use of biopolymers including chitosan, chitinase, lactoperoxidase, lysozyme and glucose oxidase as well as on the application of ultra high hydrostatic pressure and subcritical and supercritical carbon dioxide.

BIOPOLYMERS FOR FOOD PRESERVATION

Chitosan

The antimicrobial activity of chitosans (partially deacetylated chitin) was described by Allan and Hadwiger (1979), Hirano and Nagao (1989), Papineau et al. (1991), Stossl and Leuba (1984), Sudarshan et al. (1992). Two modes of action for microbial growth inhibition have been proposed. In one mechanis, the polycationic nature of chitosan interferes with the negatively charged residues of macromolecules at the cell surface and consequently alters the cells permeability. The other mechanisms involves the binding of chitosan with DNA to inhibit mRNA synthesis (Hadwiger et al. 1985).

Water soluble chitosan salts (chitosan lactate, chitosan hydroglutamate) proved effective against *Escherichia coli*, *Staphylococcus aureus* and *Saccharomyces cerevisiae* with *S. cerevisiae* being the most sensitive of the organisms evaluated (Papineau et al. 1992) When examining the effects to the cell envelopes of microorganisms caused by exposure to chitosans different levels of inactivation could be found (Table 1).

The release of extracellular β-galactosidase from *E. coli* V517 was affected by exposure of the microorganism (Fig. 1). The greatest leakage of β-galactosidase occured at a chitosan concentration of 0.1mg xml-1 (Fig. 2) which was also the concentration of maximum agglutination and maximum inhibitory effect of chitosan (Sudarshan et al. 1992). At higher chitosan concentrations less β-glucosidase could be identified in the supernatant (Fig. 2).

Based on these data it appears to us that at least one mechanism of antibacterial effects of chitosan was due to leakage of intracellular material. At lower concentrations, the polycationic chitosan may have bound to negatively charged constituents bacterial surface to disturb the cell membrane and to cause cell death due to leakage of intracellular components. At higher concentrations, chitosan may have additionally "coated" the bacterial surface to prevent leakage of intracellular components as well as impede mass transfer across the cell barrier (Fig. 2).

TABLE 1
Reduction of bacteria after a sixty-minute exposure to chitosans at pH 5.8 and 32 °C (after Sudarshan et al. 1992)

Viability of bacteria versus chitosan

Bacteria	Chitosan Glutamate 2 mg/ml	Chitosan Lactate 2 mg/ml	Chitosan Fungal 2 mg/ml
E. coli V517	+	+	+
S. typhimurium 7136	++	++	+
B. linens 492	++++	++++	-
Moraxella spp. 419	+++++	+++++	+++
E. aerogenes LC	++++	++++	+
S. aureus MF 31	+	+	+
Arizona spp. 318	+	+	+
P.morgani 3661	++	++	-
P.acidilactici HP	++	++	-

key: -   = Not tested;   +   = 1 log cycle reduction;   ++ = 2 log cycle reductions
     and so on

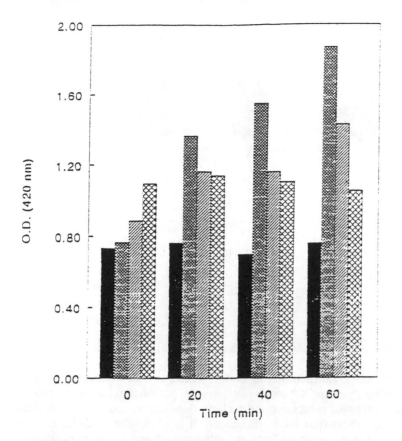

O.D. (420 nm)

Time (min)

■ 0.1 M phosphate buffer pH 5.8, control

▨ 0.1 mg/ml chitosan glutamate solution

▨ 2 mg/ml chitosan glutamate solution

▨ 5 mg/ml chitosan glutamate solution

FIGURE 1

Detection of β-galactosidase in supernatants of lactose-induced *Escherichia coli* V517 cell suspensions treated with chitosan glutamate for one hour at pH 5.8 and 32 °CC (after Sudarshan et al. 1992)

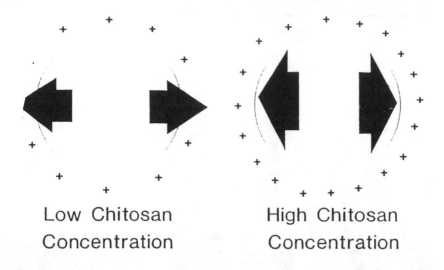

Low Chitosan
Concentration

High Chitosan
Concentration

FIGURE 2
Suggested effects of various chitosan concentrations on microbial membranes

Overall, the data obtained so far indicate that chitosans could cause leakage of intracellular components in microorganisms. Chitosan may have the potential to be added to foods and beverages as multifunctional processing aid such as clarification and concurrent control of spoilage microorganisms fruit juices (Boguslawski et al. 1990); or to be used in combination with existing unit operations such as high pressure homogenization (Popper and Knorr 1990) to create new food preservation processes.

Antimicrobial Enzymes

Endogenous plant enzymes such as amylases and proteases have been employed in traditional food processes for a long time. However most enzyme isolates used in food processing and preservation are derived from microbial sources. Since more and more conventional operations in food processing are being replaced by enzyme-catalyzed processes, we became actively involved in utilizing plant systems (eg. germinating seeds or processing wastes) as valuable, additional sources for enzyme recovery (Teichgräber et al. 1993).

It has been postulated that chitinases are among the inducible, biochemical, non-specific defense mechanisms of plants against invading chitin-containing pathogens (eg. fungi). This theory is supported by the fact that chitinase activity in plants increases following infection with pathogens. It is also interesting to note that antifungal plant chitinases and microbial chitinases differ in their mode of action. While microbial chitinases act mainly as exochitinases, plant chitinases are principally endochitinases, liberating chitin oligosaccharides. This ability to cleave any portion of a chitin polymer may be the reason for the higher antifungal activity of plant chitinases, since the attack of microbial chitinases is restricted to non-reducing termini of chitin, which are of limited availability in intact fungal cell walls (Teichgräber et al. 1993).

Chitinanse could be recovered from germinating soybean seeds by aqueous extraction followed by simple affinity chromatography with chitin (Colantuoni et al. 1992). Using these techniques yields of 20% of total chitinase activity could be obtained from the waste water of an industrial seed-germination operation. Antimicrobial activity of the plant derived chitinase has been confirmed by trials with *Aspergillus oryzae, Rhizopus stolonifer, R. japoni* and *Mucor rouxii* (Teichgräber et al. 1992).

Germinating seeds or waste waters of cereal or legume processing could become low-cost sources for chitinases which - due to their specificity for the chitin polysaccharide substrate and their shown antifungal activity - have the potential to act as antifungal agents in foods.

More complex antimicrobial enzyme systems with different modes of action have also been evaluated. Data on the inactivation of *E. coli, Bacillus subtilis* and *Lactococcus lactis* by lysozyme, lactoperoxidase or glucose oxidase or combinations of these enzymes are presented in Fig. 3.

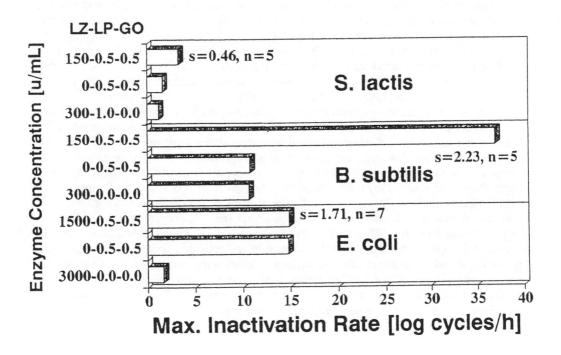

FIGURE 3
Combined action of lysozyme (LZ), lactoperoxidase (LP) and glucose oxidase (GO) during two hours at 28 °C on the survival of various microorganisms suspended in Ringer solution supplemented with 2 % glucose and 25 ppm thiocyanate (after Popper and Knorr 1993)

Maximum inactivation rates of 37 log cyles per hour could be achieved for *B. subtilis* by such an " enzyme cocktail" in a simple model system. When the model system consisted of sucrose, sodium chloride, skim milk powder, glucose and thiocyanate a reduction in the effectiveness of the enzyme system caused by the skim milk powder was found. Nevertheless inactivation rates for *B. subtilis* at 4 °C and 37 °C were 1.2 D x h^{-1} and 5.8 D x h^{-1} respectively. Although future work regarding the composition of foods and their functionality (eg. viscosity, particulate size, Aw value) is required, we are confident that such enzyme cocktails can become valuable processing tools to control microbial growth during processing or to be useful as a built-in safety device during storage (including cold storage) of highly perishable food products.

ULTRA HIGH HYDROSTATIC PRESSURE FOR FOOD PRESERVATION

The application of ultra high pressure (UHP) to food systems has been initiated at the end of the last century (Hite 1899), has again been evaluated twentyfive years ago (Sale et al. 1970) and has now reemerged as the most promising "new" non-thermal process. Key advantages of UHP application include its ability to be transferred throughout food systems instantly and uniformely (Pascal principle) and consequently, its independence of sample size and geometry. It can be applied at ambient or even lower temperatures and only non-covalent bonds are being affected by high pressure. This can result in the retention of essential food quality criteria such as color, nutrients and flavors (Hayashi 1989). Recent reviews on UHP treatment of foods and food components include Balny et al. (1992), Hayashi (1993), Mertens (1993) and Overview (1993).

The resistance of microorganisms to UHP varies. Vegetative cells in the growth phase are considered to be relatively pressure sensitive, while spores have been found to be resistant and to survive - if treated at ambient temperatures - pressures above 100 MPa (Hoover et al. 1989). However, combination processes such as elevated temperatures in conjunction with pressures of 400 MPa have been shown to be effective against spores (Seyderhelm and Knorr 1992). Effects of high pressure on foods and food constituents include modification of physico-chemical properties of water (phase transition, density), protein denaturation and modification of biopolymers (gelling properties, susceptibility to enzymatic degradation). When the activation volume of chemical reactions during pressure treatment is < 0, i.e. when the volume of the activated complex is smaller than that of the initial reactants, higher pressures and lower temperatures increase the overall reaction rate constant (Cheftel 1992). Hence, activation or inactivation of enzymes can be obtained under selected pressure/temperature conditions. In Table 1 potential food applications of UHP have been summarized from the research data currently available.

TABLE 1
Potential food application of UHP (after Cheftel 1992)

Pasteurization/sterilization at moderate temperatures
Protein modification
Changes in phase transition
Gas removal or solubilization
Extraction of food or microbial constituents
Powder agglomeration
Surface coating

We have shown that UHP pressure treatment is also suitable for process development such as high pressure blanching (Esthiaghi and Knorr 1993, Knorr 1993) or permeabilization of plant membranes for product recovery (Dörnenburg and Knorr 1993).

Subjection of apple cylinders to high pressure treatments prior to dehydration resulted in higher drying rates for the UHP treated samples (Fig. 4). This is most likely due to the permeabilization of cell walls and membranes by the pressure treatment (Dörnenburg and Knorr 1992).

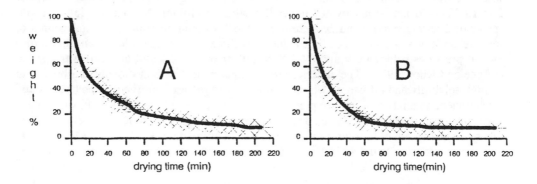

FIGURE 4
Drying rates of apple cylinders (h = 1cm, d = 2cm) without (A) and with UHP pretreatment (600 MPa, 50 °C, 15 min) during fluidized bed (air velocity 4m x sec^{-1} , air temp. 70 °C) drying (after Esthiaghi and Knorr, unpublished data)

Since enzyme inactivation is an integral part of many food preservation operations it was important to evaluate pressure effects on food enzymes.

Polyphenoloxidase (PPO) and pectinesterase have been identified as highly pressure resistant with PPO requiring up to 1,200 Mpa (10 min) at room temperature for its inactivation (Knorr 1993). We monitored inactivation kinetics of PPO at UHP between 700 and 900 MPa at different pH values (Fig. 5) and found increased barosensitivity of PPO at higher pH values. Shorter processing cycles were required at higher pressures and treatments for 30 min at 900 MPa and 45 °C were required to accomplish a 99 % inactivation of PPO at pH 7.

Inactivation of pectinesterase was achieved at lower pressures and lower pH values as for PPO (Fig. 6). These data indicated that high pH values (pH 9) resulted in increased barosensitivity, that higher pressures produced higher inactivation rates, and that pressures > 700 MPa at 45 °C for 10 min were sufficient for complete inactivation of pectinesterase. However, these data have been compiled from tests in model systems and enzyme activities have been measured immediately after pressure treatment. Further work is required to examine the effects of food constuents on the baroresistance or barosensitivity of enzymes as well as studies on possible regeneration mechanisms of enzyme activities in food systems after pressure treatment and subsequent storage.

SUBCRITICAL AND SUPERCRITICAL CARBON DIOXIDE IN FOOD PRESERVATION

Carbon dioxide has been shown to possess considerable antimicrobial activity. Carbon dioxide under pressure killed bacteria, molds and yeast with increasing effectiveness at elevated temperatures and lower pH values with and decreased effects at lowered Aw values (Haas et a. 1989). Kamihira et al. (1987) showed that various vegetative microorganism species could be inactivated with supercritical carbon dioxide (20 MPa, 35 °C) when the water content was between 70 and 90 %, while dry cells with a water content of 2 to 10 % could not be inactivated under the same conditions. No inactivation of endospores of *Bacillus subtilis* and *B. stearothermophilus* could be obtained. Total counts of fresh celery leaves and stalks could be reduced by three to four log cycles when carbon dioxide at pressures between 6.9 and 62.8 MPa (40 or 60 °C, 30 and 60 min) was applied (Kühne and Knorr 1990). The effectiveness of supercritical carbon dioxide can be due to its high solubility and diffusivity which allows CO_2 to diffuse into the microbial cells and - if sufficient amounts of water are present - to be converted into carbonic acid with subsequent inactivation of the organisms due to a reduced internal pH value. Chen et al. (1993) reported inactivation of Florida spiny lobster polyphenoloxidase (PPO) with CO_2 at atmospheric pressure (33, 38 and 43 °C). Kinetic data revealed that PPO was more labile to CO_2 and heat treatment than to heat alone and that CO_2 - in addition to pH changes - was involved in the loss of PPO activity. It appears of interest to indicate that the activity of β-glucosidase was higher and more Aw dependent under supercritical CO_2 condition than under comparable UHP conditions (Figure 7). It is most likely that the change in enzyme activity was caused by a shift of pH towards the pH optimum of the enzyme during the supercritical CO_2 treatment (Stegmann and Knorr, unpublished data).

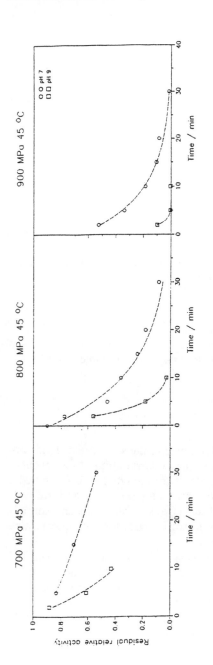

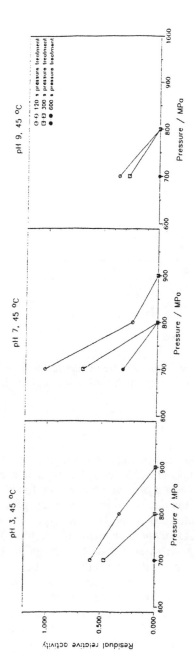

FIGURE 5
Inactivation kinetics of polyphenol
oxidase at different hydrostatic pressures
and pH values at 45 °C processing
temperature

FIGURE 6
Inactivation of pectinesterase by
hydrostatic pressure at different pH
values (9), constant processing
temperature and different times of
treatment (2, 5, 10 min)

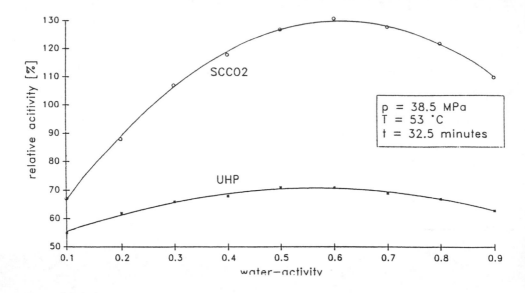

FIGURE 7

Activity of β-glucosidase in supercritical carbon dioxide and under comparable (38.5 MPa, 53 °C, 32.5 min) conditions using high hydrostatic pressure

CONCLUSIONS

Various non-thermal processes seem promising tools to extend the shelf life of foods or food components. However, kinetic data are missing and the understanding of the mechanisms of action of these processes on foods, food components or microbial cells is limited. Most of the data available have been generated immediately after treatment and are not taking into account the - at least hypothetical - possibilities of regeneration mechanisms of treated biopolymers or microorgansism. Therefore future work needs to also concentrate on storage dependent changes after non-thermal treatments.

At this point it seems unlikely that one single non-thermal unit operation will result in an effective sterilization process for complex food systems. However, combinations of mild heat with the above mentioned processes or combinations of non-thermal processes with each other, including the use of high electric field pulses (Dörnenburg and Knorr 1993, Knorr et al. 1993), or ultrasonic treatment are under investigation in our laboratory and may provide the basis for new preservation concepts and processes.

REFERENCES

Allan, C.R. and Hadwiger, L.A. The fungal effect of chitosan on fungi of varying cell wall composition. Exp. Mycol. 3, 285, 1979.

Anon. The heat is off!Enter the era of non-thermic food processing. Food Engineering Internat. 18(1), 1993.

Balny, C., Hayashi, R., Heremans, K. and Masson, P. High Pressure and Biotechnology, John Libbey & Co. Ltd., London, 1992.

Banks, J.G., Board, R.G. and Sparks, N.H. Natural antimicrobial systems and their potential in food preservation of the future. Biotechnol. Appl. Biochem. 8, 103, 1986.

Boguslawski, S., Bunzeit, M. and Knorr, D. Einfluss der Behandlung mit Chitosan auf Klarheit und Keimzahl von Apfelsaft. ZFL, Internat. J. Food Technol., Marketing, Packaging and Analysis, EFS, 41(7/8)42, 1990.

Cheftel, J.C. Effects of high hydrostatic pressure on food constituents: an overview, In, High Pressure and Biotechnology, Balny, C., Hayashi, R., Heremans, K. and Masson, P., Eds., John Libbey & Co. Ltd., London, 1992.

Chen, J.S., Balaban, M. o., Wei, C.I., Gleeson, R.A. and Marshall, M.R. Effect of carbon dioxide on the inactivation of Florida spiny lobster polyphenol oxidase. J. Food Sci. Agric. 61, 253, 1993.
Colantuoni, D., Popper, L. and Knorr, D. Purification of chitinase from soybean seed by affiity chromatography on chitin. Agro-Food-Industry Hi-Tech. , 3(2), 24, 1992.

Dörnenburg, H. and Knorr, D. Cellular permeabilization of cultured plant tissues by high electric field pulses or ultra high pressure for the recovery of secondary metabolites. Food Biotechnol. 7, 35, 1993.

Eshtiaghi, M. and Knorr, D. Comparison of effects of water blanching or ultra high pressure treatment on potato cubes. J. Food Sci. (in press).

Goulds, G.W., Rhodes-Roberts, M.E., Charnley, A.K., Cooper, R.M. and Board, R.G. Natural Antimicrobial Systems, Bath Univ. Press, Bath, UK, 1986.

Haas, G.J., Prescott, jr. H.E., Dudley, E., Dik, R., Hintlian, C. and Keane, L. Inactiation of microorganisms by carbon dioxide under pressure. J. Food Safety, 9, 253, 1989.

Hadwiger, L.A., Kendra, D.F., Fristensky, B.W. and Wagoner, W. Chitosan both actiates genes in plants and inhibits RNA synthesis in fungi, in Chitin in Nature and Technology, Muzzarelli, R.A.A., Jeuniaux, C. and Gooday, C. Eds., Plenum Press, New York, 210, 1985.

Hayashi, R. Use of High Pressure in Food: Research and Development, San-Ei Shuppan Co., Kyoto, 1989.

Hayashi, R. High Pressure Bioscience and Fodd Science, San-Ei Shuppan Co., Kyoto, 1993.

Hirano, S. and Nagao, N. Effects of chitosan, pectic acid, lysozyme and chitinase on the growth of several phytopathogens. Agric. Biol. Chem. 53, 3065, 1989.

Hite, B. H. The effect of pressure in the preservation of milk. West Virginia Univ. Agric. Exppt. Sta. Bull 58, 15, 1899.

Hoover, D.G., Metrick, C., Papineau, A.M., Farkas, D.F. and Knorr, D. Application of high hydrostatic pressure on foods to inactivate pathogenic and spilage organisms for extension of shelf life. Food Technol., 43(3), 99, 1989.

Hoover, D.G. and Steenson, L.R. Bacteriocins of Lactic Acid Bacteria, Academic Press, Inc., San Diego, 1993.

Kamihira, M., Taniguchi, M. and Kobayashi, T. Sterilization of microorganisms with supercritical carbon dioxide. Agric. Biol. Chem. 51(2), 407, 1987.

King, A.D. and Bolin, H.R. Physiological and microbiological storage stability of minimally processed fruits and vegetables. Food Technol. 43(2), 1321989.

Knorr, D. Effects of high-hydrostatic pressure processes on food safety and quality. Food Technol. 47(6)156, 1993.

Knorr, D., Geulen M., Grahl, T. and Sitzmann, W. Food application of high electric field pulses. Trends Food Sci. & Technol. (submitted), 1993.

Kühne, K. and Knorr, D. Effects of high pressure carbon dioxide on the reduction of microorganisms in fresh celery. ZFL, Internat. J. Food Technol, Marketing, Packaging and Analysis, EFS, 41(10), 55, 1990.

Mertens, B. and Knorr, D. Developments of nonthermal processes for food preservation. Food Technol. 46(5).124, 1992.

Mertens, B. Developments in high pressure food processing. ZFL, Internat. J. Food Technol, Marketing, Packaging and Analysis, (44)100 & 182, 1993.

Overview. Use of hydrostatic pressure in food processing. Food Technol. 47(6), 149, 1993.

Papineau, A.M., Hoover, D.G., Knorr, D. and Farkas D.F. Antimicrobial effect of water-soluble chitosans with high hydrostatic pressure. Food Biotechnol. 5, 45, 1991.

Popper, L. and Knorr, D. Application of high pressure homogenization for food preservation. Food Technol. 44(7), 84, 1990.

Popper, L. and Knorr, D. Nicht-thermische Inaktivierung von Mikroorganismen durch antimikrobielle Enzymsysteme. BioEngineering, 9(1), 27, 1993.

Sale, A.J.H., Gould, G.W.and Hamilton W.A. Inactivation of bacterial spores by hydrostatic pressure. J. Gen. Microbiol. 60(3), 323, 1979.

Seyderhelm, I. and Knorr, D. Reduction of *Bacillus stearothermophilus* spores by combined high pressure and temperature treatments. ZFL, Internat. J. Food Technol., Marketing, Packaging and Analysis, 43(4), 17, 1992.

Stossel, P. and Leuba, J.L. Effect of chitosan, chitin and some aminosugars on growth of various soil-borne phytopathogenic fungi. Phytopath. Z. 111, 82, 1984.

Sudarshan, N.R., Hoover, D.G. and Knorr, D. Antibacterial action of chitosan. Food Biotechnol. 6, 257, 1992.

Teichgräber, P. Zache, U. and Knorr, D. Enzymes from germinating seeds - potential applications in food processing. Trends Food Sci. & Technol. 4, 145, 1993.

ACKNOWLEDGEMENTS
Parts of this work have been supported by grants from the German Research Association (DFG), the German Industrial Research Foundation (AIF) and by CPC Europe.

OHMIC HEATING

Sudhir K. Sastry

I. INTRODUCTION

The concept of ohmic heating, or heating foods by passage of electrical currents through them, dates back to the previous century. Several patents refer to heating of flowable materials. In the early twentieth century, "electric" pasteurization of milk was achieved by pumping the fluid between parallel plates with a voltage difference between them (Anderson and Finkelstein, 1919). Six states in the United States had commercial electrical pasteurizers in operation. It was believed that extra-thermal lethal effects were caused by the presence of electricity. This technology disappeared in the succeeding years, being apparently due to lack of suitable electrode materials and controls. Its primary use has been restricted to space exploration and military applications. One notable common application has been the heating of frankfurters in foodservice operations.

In recent years, interest in ohmic heating has been revived, due at least partly to availability of imrpoved electrode materials. The limitations of conventional surface heat treatments are well recognized in the food industry, where product fouling and heat sensitivity pose special problems. Microwave heating has gained widespread application in home food preparation, however its commercial application for microbial inactivation has been limited by thermal nonuniformity. Considerable research is currently under way to extend the applicability of microwave treatment for this purpose.

Although ohmic treatment is not a true "minimal" treatment in a thermal sense, it can, if carefully designed and implemented, provide significant improvements over conventional thermal processing. Unlike conventional heating, where heating occurs from heated surface to interior locations, the ohmic process involves internal generation at controllable rates; so that high temperature short time (HTST) strategies for solid foods can be physically realized.

A. APPLICATIONS OF OHMIC HEATING

Ohmic heating is useful in a variety of situations where conventional heat treatment is difficult. Fouling is a major problem with proteinaceous foods that are exposed to hot heat transfer surfaces. Ohmic processes can be used to advantage, since heating occurs by internal energy generation, and no heated surfaces are necessary. Another emerging application is in the heat treatment of surimi and various fish products. Under conventional heat treatment, parts of these products may experience long residence times under temperatures that are optimal for protease activity. Using suitably designed ohmic treatments, it is possible to accelerate heating through this temperature range, resulting in improved texture in the finished product. Thawing is another application that has been investigated (Mizrahi et al., 1975, Henderson, 1993), and will be discussed extensively in another paper in these proceedings. Ohmic heating also shows considerable promise in continuous sterilization of

17

solid-liquid mixtures.

The design of ohmic heating equipment and processes is still in its infancy, since many of the important parameters are not yet well understood. Nevertheless, a number of interesting possibilities arise. One involves product formulation and design for ohmic treatment. Control of ionic concentrations can be used to enhance product heating rates. The physical design and configuration of equipment also lends itself to considerable ingenuity. Common configurations have included parallel plate systems, where the electrical field is perpendicular to the product flow, and designs where the electrical field is aligned with product flow. Since voltage is easily regulated, it is possible to implement sophisticated process control strategies.

II. BASIC PRINCIPLE

A food of electrical conductivity σ, placed between electrodes with a field strength ∇V across them, experiences an internal energy generation rate ($\dot{u}$) given by:

$$\dot{u} = |\nabla V|^2 \sigma \tag{1}$$

The critical property that affects energy generation rate is σ. For solid foods, the electrical conductivity depends on temperature and voltage gradient. When vegetable tissue is subjected to conventional heating, the electrical conductivity undergoes a sharp increase at about 60°C, due to breakdown of cell walls (Palaniappan and Sastry, 1991a, Fig. 1). This

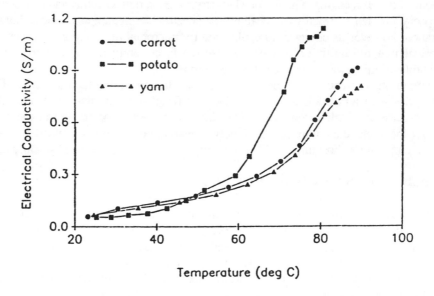

Figure 1. Electrical conductivity curves for vegetable tissue during conventional heating. (From Palaniappan and Sastry, 1991a. With permission)

phenomenon has previously been observed in relation to sugar beet tissue (Brüniche-Olsen, 1962).

When cellular tissue is heated ohmically, the electrical-conductivity temperature curve becomes more and more linear as the voltage gradient is increased. A typical set of plots for carrot tissue, shown in Fig. 2 shows that some nonlinearity persists at the lower voltage gradients (20 to 30 V/cm), but this disappears completely at the higher voltage gradients. An explanation is that electro-osmosis occurs when ohmic heating is used, the extent of which depends on the voltage field. Under high voltage gradients, electro-osmosis drives ions

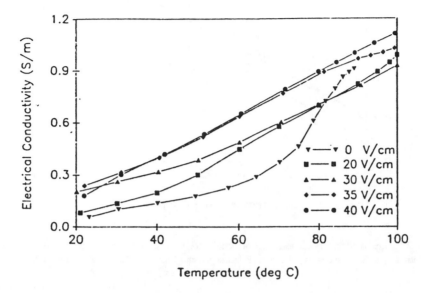

Figure 2. Electrical conductivity curves for carrot (parallel to stem axis), subjected to various voltage gradients. (From Palaniappan and Sastry, 1991a. With permission).

through cell-wall membranes even at the lower temperatures. Under sufficiently high field strengths, a linear σ-T relation may be used.

$$\sigma_T = \sigma_{ref}[1 + m(T - T_{ref}]$$ (2)

where σ_T is the electrical conductivity at temperature T, σ_{ref} is the electrical conductivity at reference temperature T_{ref}, and m is the temperature coefficient. This increase in conductivity means that ohmic heating becomes more effective at higher temperatures; indeed typical heating curves for batch ohmic heating show increasing slopes over time. Halden et al. (1990) have observed electrohydrodynamic effects in beet tissue immersed in brine. When samples of beet tissue were immersed in brine and heated ohmically, the rate of loss of betanine was far greater than with conventional heating.

Since electrical conductivity depends on ionic concentration, it is possible to change it by relatively simple treatments, such as salt infusion or leaching. Fig. 3 illustrates the effect of soaking carrot samples in water and various concentrations of sodium chloride solutions on the electrical conductivity curves. The decrease in electrical conductivity of the water soaked samples seems to be due to loss of ionic constituents to the water. The dramatic differences in heating rates of these samples can be seen in Fig. 4. Since most products that are to undergo continuous sterilization will likely need thermal pretreatment, such as blanching or precooking, it is possible to alter electrical conductivities before the ohmic process by controlling salt concentrations in the cooking medium. The diffusion of salt into potato solids has been investigated by Wang and Sastry (1993a,b). Vacuum infusion has been found useful for accelerated salt infusion into the outer layer of potato tissue, although deeper penetration is difficult for tissue of low porosity.

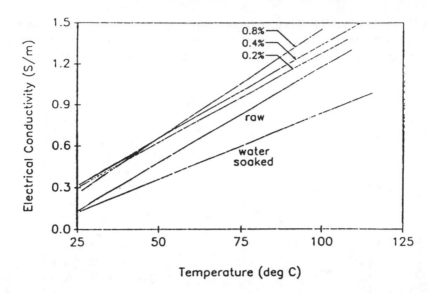

Figure 3. Electrical conductivity curves for carrot samples (raw and saturated with water, 0.2, 0.4 and 0.8% NaCl solutions, parallel to the stem axis. From Palaniappan and Sastry 1991a. With permission).

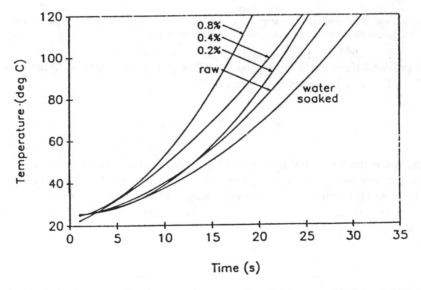

Figure 4. Ohmic heating curves for carrot samples (raw and soaked in water, 0.2, 0.4 and 0.8% NaCl solutions. From Palaniappan and Sastry, 1991a. With permission).

For liquids, the σ-T curve is linear regardless of the mode of heating used, and σ decreases with increasing pulp content (Palaniappan and Sastry, 1991b). Typical results for tomato juice are illustrated in Fig. 5. A typical relationship for orange and tomato juices is:

$$\sigma_T = \sigma_{ref}[1 + K_1(T - T_{ref})] - K_2S \tag{3}$$

where S is solids content and K_1 and K_2 are constants.

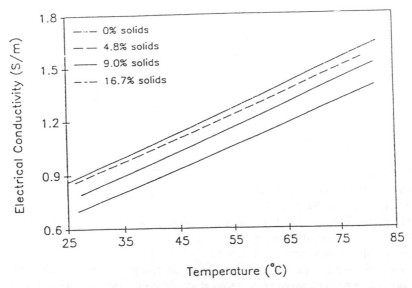

Figure 5. Electrical conductivity curves for tomato juice of various solids contents. (From Sastry and Palaniappan, 1991b. With permission.

The size of solids has been shown to affect electrical conductivity. Fig. 6 shows that electrical conductivity tends to increase as particle size decreases, although general conclusions cannot be made without accounting for particle shapes and orientations.

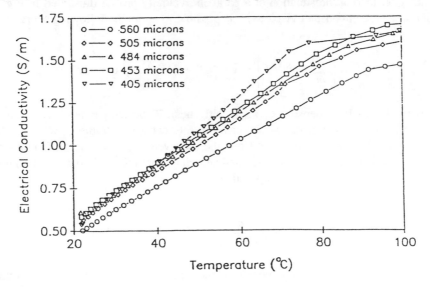

Figure 6. Electrical conductivity curves for carrot solids of various particle sizes suspended in sodium phosphate solutions. (From Palaniappan and Sastry, 1991b. With permission.)

III. HEATING OF LIQUIDS

A. ANALYSIS

The heating of liquids for pasteurization or sterilization requires that all locations within the product be sufficiently processed to destroy viable microorganisms. If flow through the heater is turbulent and well mixed, temperatures throughout the heater cross section may be considered uniform, and process design is a relatively simple matter. If flow is laminar, however, a residence time distribution is set up, and the fastest moving sections achieve the lowest temperatures. Depending on the design of the ohmic heater (longitudinal or transverse voltage gradient), the effect may be somewhat different. If the voltage is applied longitudinally (i.e. along the axis of flow), the fast-moving regions at the center of the heater receive a lower

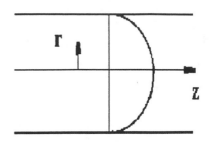

Figure 7. Laminar velocity profile in ohmic heater.

current density because of their lower temperature and greater resistance. Thus, nonuniformity can become more pronounced under these conditions. Transversely applied voltages would result in currents moving through the entire fluid, although the low-residence elements would still receive a lower thermal treatment.

Although these heating problems can and should be approached numerically for accurate solutions, it is possible, with several simplifying assumptions, to obtain approximate analytical estimates of heating nonuniformity in the case of transverse (and constant) voltage gradient. Considering the situation of a parabolic velocity profile illustrated in Fig. 7, and considering a case of constant viscosity, it is possible to apply the following energy balance:

$$\rho\, C_p\left[\frac{\partial T}{\partial t} + v\cdot\nabla T\right] = \nabla\cdot k\nabla T + |\nabla V|^2\sigma_0[1 + mT] \qquad (4)$$

where ρ represents fluid density, C_p its specific heat, T the temperature, t the time, v the velocity function, and the electrical conductivity reference temperature is 0°C. In the case of steady, fully developed, symmetric flow, the only velocity component of significance is in the axial (z) direction, and temperature and velocity variations are in the radial (r) direction only, the following simplification is possible:

$$v_z\frac{\partial T}{\partial z} = \frac{k}{r}\frac{\partial}{\partial r}\left(r\frac{\partial T}{\partial r}\right) + |\nabla V|^2\sigma_0[1 + mT] \qquad (5)$$

where v_z refers to the velocity component in the z direction. If conductive heat transfer is slow in comparison to the energy generation rate, and the tube walls are considered insulated, the conduction term on the right hand side may be eliminated, to yield:

$$v_z \frac{\partial T}{\partial z} = |\nabla V|^2 \sigma_0 [1 + mT] \tag{6}$$

This equation may be integrated, to yield:

$$\ln\left(\frac{1 + mT}{1 + mT_o}\right) = \frac{m |\nabla V|^2 \sigma_0 z}{\rho C_p v_z(r)} \tag{7}$$

where the initial condition $T(z=0) = T_0$ has been applied. The function v_z may be a commonly used expression for flow of a non-Newtonian (or Newtonian, if appropriate) fluid in a tube. For example, for a power law fluid of consistency coefficient K and flow behavior index n, the following expression may be used (Hughes and Brighton, 1967):

$$v_z(r) = v_m\left(\frac{3n + 1}{n + 1}\right)\left[1 - \left(\frac{r}{R}\right)^{\frac{n+1}{n}}\right] \tag{8}$$

where v_m is the mean fluid velocity over the tube cross-section, and R is the radius of the tube. By applying different values of the fluid velocity, it is possible to estimate the temperature achieved at a given radial location. This approach may also be extended to parallel plate heaters, for situations where simple expressions for the fluid velocity may be used.

B. OTHER PROBLEMS

Although ohmic heating is useful in reducing fouling, heat is generated within electrodes, and their surfaces may foul during ohmic heating; thus it is usually necessary to cool electrodes by some external means.

An additional consideration is the possibility of boiling within the heater. Even though the system may be pressurized, boiling may occur because there is no upper temperature limit for ohmic heating. Consequently, it is possible to achieve temperatures above the boiling point even under these elevated pressure conditions. This phenomenon has been discussed by Zhang and Fryer (1993).

IV. HEATING OF SOLID-LIQUID MIXTURES

The heating of solid-liquid mixtures poses some interesting and challenging problems, since multicomponent systems would in general, be expected to have differing properties, with complicated effects on the electrical and thermal field. One necessary item of information for sizing of ohmic heaters is the effective electrical conductivity of a solid-liquid mixture.

A. EFFECTIVE ELECTRICAL CONDUCTIVITY OF MIXTURES

A number of models have been developed over the years to determine the effective conductivity of two or three component systems consisting of a continuous and a dispersed phase. These have included Maxwell (1881), Fricke (1924), Meredith and Tobias (1961),

Kopelman (1966), Kopelman (1966), and Chaudhary and Bhandari (1968), among many others. Palaniappan and Sastry (1991c) have evaluated the effectiveness of a number of models. Many of these models have been found to be reasonable approximators of the effective conductivity of two-phase systems consisting of a liquid phase and cubic particles. However, as the system is heated ohmically, accuracy suffers because most of these methods assume isothermal conditions, and thermal nonequilibrium in distributed systems results in a complex electrical conductivity distribution. Recently, Kim and Torquato (1990) have developed approaches for prediction of the effective conductivity of multicomponent systems in random motion.

Orientation of particles plays an important role in these results. If the mixture consists of long-thin particles, their orientation relative to the electrical field has a significant influence on electrical properties as well as relative heating rates of phases (De Alwis and Fryer, 1990). If the solids are of aspect ratio close to unity, as for example with spheres, cylinders of length equal to diameter, and cubes; the effect of orientation on overall electrical conductivity is small (Sastry and Palaniappan, 1992a)

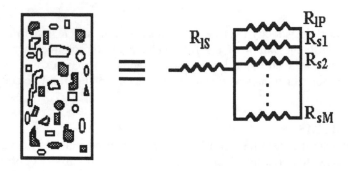

Figure 8. Equivalent resistance for a multicomponent system with M dispersed (solid) phases.

For the case of an isotropic mixture of one continuous (liquid) phase, and *M* dispersed (solid) phases, it is possible to use circuit theory in the manner of Kopelman (1966) to approximately determine the effective resistance of the mixture. The model discussed here needs further verification, will likely apply primarily under high solids concentration, and it is recognized that the morphology of individual phases will play a role; however the intent is to obtain approximations for engineering design purposes. Considering, therefore, the system illustrated in Fig. 8; we can determine the effective electrical conductivity σ_{eff} of the mixture, if we know the effective resistance of the mixture (R):

$$\sigma_{eff} = \frac{L}{AR} \tag{9}$$

where L is the length of the section tested, and A the cross-sectional area. We may determine

R as:

$$R = R_{lS} + R_P \tag{10}$$

where the subscript l refers to the liquid phase, S refers to a series resistance, and P to an equivalent parallel resistance. R_P may be determined as:

$$R_P = \frac{1}{\dfrac{1}{R_{lP}} + \displaystyle\sum_{m=1}^{M} \dfrac{1}{R_{sm}}} \tag{11}$$

where the subscript s refers to the solid phase. If the volume fraction of solid phase m is ϕ_m, the total discontinuous fraction is then:

$$\phi_s = \sum_{m=1}^{M} \phi_m \tag{12}$$

We now make the assumption that for each phase m, the *area fraction*, $\phi_{\alpha m}$ can be represented as:

$$\phi_{\alpha m} = \phi_m^{2/3} \tag{13}$$

and the *length fraction*, phi$_{\lambda m}$ is then defined as:

$$\phi_{\lambda m} = \phi_m^{1/3} \tag{14}$$

so that the volume fraction of any phase becomes:

$$\phi_m = \phi_{\alpha m} \phi_{\lambda m} \tag{15}$$

We now make the assumption that the length of all dispersed phases m in a section of length L are equal, and determined as:

$$\frac{L_m}{L} = \phi_{\lambda m} = \phi_s^{1/3} = \left(\sum_{m=1}^{M} \phi_m \right)^{1/3} \tag{16}$$

The area fraction for each phase is then calculated from the length fraction of each phase, as:

$$\phi_{\alpha m} = \frac{\phi_m}{\phi_s^{1/3}} \tag{17}$$

Then, each of the solid phase resistances, R_{sm} may be determined as:

$$R_{sm} = \frac{L\phi_s^{1/3}}{A\phi_{\alpha m}\sigma_m} = \frac{L\phi_s^{2/3}}{A\phi_m\sigma_m} \tag{18}$$

where σ_m is the electrical conductivity of phase m. The following relations then follow:

$$R_{lP} = \frac{L\phi_s^{1/3}}{A(1 - \phi_s^{2/3})\sigma_l} \tag{19}$$

where σ_l is the electrical conductivity of the liquid phase, and:

$$R_{lS} = \frac{L(1 - \phi_s^{1/3})}{A\sigma_l} \tag{20}$$

The value of R_P may be determined by substitution of eqs. (18) and (19) into eq: (11). Substituting this result and that of eq. (20) into eq. (10) we may obtain the effective resistance. Finally, the effective electrical conductivity arises from eq. (9).

B. HEATING OF SOLID-LIQUID MIXTURES

1. Mathematical Models

As with other situations involving continuous sterilization, the critical consideration is the prediction of cold-spot temperatures during the process. Since temperature measurement in continuously flowing mixtures is difficult, models are necessary. Models have been developed for a single particle in a static heater by De Alwis and Fryer (1990). For multiple particles in static and continuous flow ohmic heaters, models have been presented by Sastry and Palaniappan, (1992); Sastry, (1992), and Sastry and Li, (1993). The electrical field calculations in this approach involve the use of circuit theory approximations, that are increasingly accurate under conditions of high solids concentrations as are used in industrial applications. Zhang and Fryer (1992) have developed an analysis for heating of two-phase mixtures consisting of spherical particles under conditions of no fluid motion using a unit cell approach. Details of the models of Sastry and Palaniappan (1992) and Sastry (1992) have been discussed extensively in a number of publications, and will not be repeated here. However, some of the key results will be discussed to illustrate important points.

Solids concentration has been found to be important in ensuring more rapid heating of the solid phase. If a single, low electrical conductivity particle is placed within a high electrical conductivity fluid, the fluid tends (on average) to heat faster than the particle, since the current tends to bypass the particles (some local field variations occur, and will be discussed later). As the solids concentration is increased, the fluid fraction decreases, and the more conductive paths through the fluid become narrower and more tortuous, increasing the effective resistance of this phase. Consequently, a greater proportion of the overall current flows through the solid phase and the solids tend to heat faster than the liquid. It should be noted that for the same applied voltage, the *overall* average heating rate of the mixture decreases because of the increased low-conductivity solids concentration. This finding indicates the importance of using high solids concentrations to ensure relatively rapid heating of the solid phase.

Illustration of a situation involving a continuous flow ohmic heater with longitudinally (along the product flow) applied voltage, high solids concentration and equal electrical conductivity between fluid and particles is presented in Fig. 9.

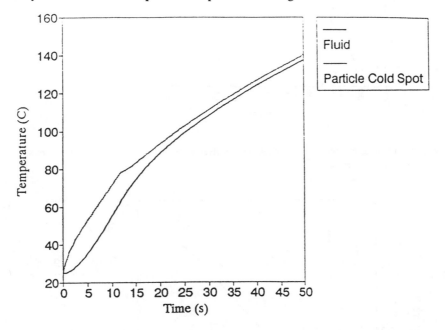

Figure 9. Particle cold-point and fluid temperatures versus time for equal electrical conductivity case. (From Sastry, 1992. With permission).

The simulation shows that cold-spots of particles in this situation heat faster than the fluid. If the solids are of low electrical conductivity, the result of Fig. 10 is observed. Here, the solid phase still heats faster than the fluid phase, but the overall heating is severely limited, because the mixture represents a circuit of extremely low electrical conductivity. Under these conditions it may be necessary either to increase the overall voltage or, if this is not feasible, to redesign the heater to ensure low electrode gaps. In the worst case, ohmic heating may not be feasible at all.

A situation conducive to underprocessing may occur if a single (isolated) low-conductivity particle enters a heater with contents at otherwise high electrical conductivity. As is shown in Fig. 11, the particle cold-spot lags significantly behind the fluid phase, and the danger of an underprocessed particle exists. This effect may be quite different if the particle flows between parallel plate electrodes, and becomes the dominant resistance in the circuit.

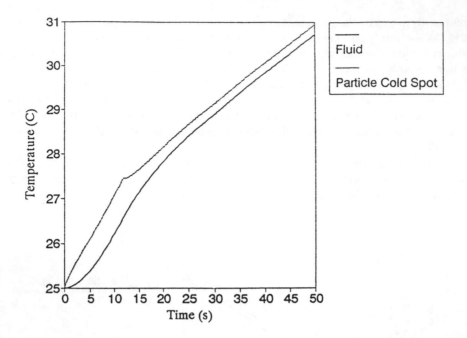

Figure 10. Particle cold-point and fluid temperatures versus time for low-conductivity particles. (From Sastry, 1992. With permission).

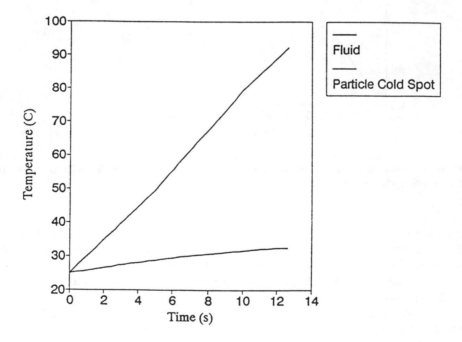

Figure 11. Particle cold-point and fluid temperatures versus time for an isolated low electrical conductivity particle (From Sastry, 1992. With permission.)

The findings illustrated here indicate that a number of factors need to be considered in designing equipment, products and processes for ohmic heating. The effects of orientation also need consideration. De Alwis and Fryer (1990) have shown that if a long thin nonconductive particle in a conductive electrical medium is oriented along the electrical field, the current tends to bypass it, and it therefore heats slower than the fluid. If such a particle is placed in an orientation across the electrical field, it tends to "block" the current path and accordingly may heat faster than the fluid. The reverse situation occurs when the particle is of higher electrical conductivity than the fluid. An important consideration controlling relative heating rate appears to be which component's resistance is dominant in the mixture.

2. Fluid Temperature Nonuniformities

It has been observed (Fryer et al., 1993) that when a low electrical conductivity particle is placed in a high electrical conductivity field in a static heater, zones of nonuniform heat generation are established in the fluid about the particle (Fig. 12). The primary reason appears to be the current density being high on the sides of the particle, and low in the front and rear regions, because of the current attempting to bypass the solid. The resulting temperature nonuniformity is not significant if fluid mixing is sufficient; however if highly

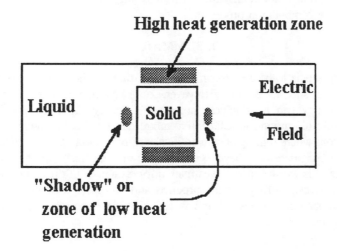

Figure 12. Zones of high and low heat generation in the fluid around a particle of low electrical conductivity.

viscous fluids are used, such nonuniformities may become important. Recent flow visualization studies in our laboratory have indicated that the extent of these effects depend on the rotation of particles, and the presence of other particles in the heater.

V. NON-THERMAL MICROBICIDAL EFFECTS

Much has been written regarding the effect of electricity on microorganisms. In particular, it has been demonstrated that high voltage discharges can destroy microorganisms by causing pores to develop in microbial cell walls. This has been used as the basis for electroporation treatment (Palaniappan et al, 1990). Low voltage treatments have been found to decrease microbial counts over long time periods, without heating to lethal temperature ranges. However, under ohmic heating conditions, neither are the voltages are not sufficiently high to cause electroporation nor are residence times sufficiently long to result in low-voltage effects. Indeed, the rapidity of ohmic heating is such that microbial death appears to occur due to thermal effects alone. Studies by Palaniappan et al. (1992) involved treatment of yeast cells by conventional and ohmic heating while maintaining identical temperature histories in each case. The results, presented in Table 1, showed no significant difference between treatments.

Table 1. D values of yeast cells (from Palaniappan et al., 1992)

Treatment	D (49.75°C) (s)	D (52.3°C) (s)	D (55.75°C) (s)	D (58.75°C) (s)
Conventional	294.6	149.7	47.2	16.9
Ohmic	274.0	113.0	43.1	17.8

Further studies on E. Coli cells by the same authors showed that if mild, thermally sublethal electrical treatments were applied before heat treatment, the D values could be significantly reduced in some, although not in all, cases. This may suggest possible alternative treatments to achieve effective combination processes.

VI. SUMMARY

Ohmic heating seems to be a promising treatment that can be used to provide thermally processed product of superior quality. However, some key unknowns need to be addressed before the technology can be routinely accepted. Among these are issues of thermal nonuniformity and fluid mixing in solid-liquid mixtures. For liquids, ohmic heating shows promise in situations where fouling of conventional equipment may pose problems. Although nonthermal microbicidal effects have been found not significant in the limited studies conducted to date, it is worth further investigation for a number of different microorganisms and food systems. Design of equipment, products and processes for ohmic heating are still in the early developmental stages. There is considerable opportunity for innovation and creativity in this emerging field.

REFERENCES

1.   **Anderson, A.K., and Finkelstein, R.** 1919. A study of the electropure process of treating milk. J. Dairy Sci. 2:374-406.

2.   **Brüniche-Olsen, H.,** 1962. Solid-liquid Extraction. NYT Nordisk Forlag, Arnold Busck, Copenhagen.

3.   **Chaudhary, D.R. and Bhandari, R.C.,** 1968, Heat-transfer through a 3-phase porous medium, Brit. J. Appl. Physics, Ser. 2, 1, 815-817.

4.   **de Alwis, A.A.P, and Fryer, P.J.,** 1990. A finite element analysis of heat generation and transfer during ohmic heating of food. Chem. Engr. Sci. 45(6):1547-1559.

5.   **Fricke, H.,** 1924, The electrical conductivity of a suspension of homogeneous spheroids, Physical Rev., 24, 575-587.

6.   **Fryer, P.J., de Alwis, A.A.P., Koury, E., Stapley, A.G.F., and Zhang, L.** 1993. Ohmic processing of solid-liquid mixtures: heat generation and convection effects. J. Food Engr. 18:101-125.

7.   **Halden, K., de Alwis, A.A.P., Fryer, P.J.,** 1990. Changes in the electrical conductivities of foods during ohmic heating. Int. J. Food Sci. Technol. 25:9-25.

8.   **Henderson, J.T.** 1993. Ohmic thawing of shrimp. M.S. Thesis, University of Florida, Gainesville, FL.

9.   **Hughes, W.F., and Brighton, J.A.,** 1967. "Theory and problems of fluid dynamics", Schaum's Outline Series, McGraw Hill Book Co., New York.

10.   **Kim, I.C., and Torquato, S.** 1990. Determination of the effective conductivity of heterogeneous media by Brownian motion simulation. J. Appl. Phys. 68(8): 3892-3903.

11.   **Kopelman, I.J.** 1966. Transient heat transfer and thermal properties in food systems. Ph.D. dissertation, Michigan State University, East Lansing, MI.

12.   **Maxwell, J.C.** 1881. "A treatise on electricity and magnetism" 2d. edn. Vol. 1, Clarendon Press, Oxford.

13.   **Meredith, R.E., and Tobias, C.W.** 1960. Resistance to potential flow through a cubical array of spheres. J. Appl. Phys. 31:1270-1273.

14.   **Mizrahi, S., Kopelman, I.J., and Perlman, J.** 1975. Blanching by electroconductive heating. J. Food Tech. (Brit.) 10:281-288.

15.   **Palaniappan, S., Sastry, S.K.,** 1991a: Electrical conductivities of selected solid foods during ohmic heating. J. Food Proc. Engr. 14(3):221-236.

16.   **Palaniappan, S., Sastry, S.K.,** 1991b: Electrical conductivity of selected juices: influences of temperature, solids content, applied voltage and particle size. J. Food Proc. Engr. 14:247-260.

17.   **Palaniappan, S., Sastry S.K.,** 1991c. Modelling of electrical conductivity of liquid-particle mixtures. Food & Bioproducts Proc., Trans. Instn. Chem. Engrs. Part C. 69:167-174.

18.   **Palaniappan, S., Sastry, S.K., Richter, E.R.,** 1990: Effects of electricity on microorganisms: a review. J. Food Proc. Pres. 14:393-414.

19.   **Palaniappan, S., Sastry, S.K., Richter, E.R.,** 1992: Effects of electroconductive heat treatment and electrical pretreatment on thermal death kinetics of selected microorganisms. Biotech. & Bioeng. 39:225-232.

20.   **Sastry, S.K.** 1992. A model for heating of liquid-particle mixtures in a continuous flow ohmic heater. J. Food Proc. Engr. 15(4):263-278.

21.   **Sastry, S.K., and Li, Q.,** 1993. Models for ohmic heating of solid-liquid food mixtures. in "Heat Transfer in Food Processing", Proceedings of the 29th. National Heat Transfer Conference, HTD-Vol 254, American Society of Mechanical Engineers, New York.

22.   **Sastry, S.K., and Palaniappan, S.** 1992. Influence of particle orientation on the effective electrical resistance and ohmic heating rate of a liquid-particle mixture. J. Food Proc. Engr. 15(3):213-227

23.   **Sastry, S.K., and Palaniappan, S.** 1992. Mathematical modeling and experimental studies on ohmic heating of liquid-particle mixtures in a static heater. J. Food Proc. Engr. 15(4):241-261.

24.   **Wang, W-C., and Sastry, S.K.** 1993a. Salt diffusion into vegetative tissue as a pretreatment for ohmic heating: determination of parameters and mathematical model verification. J. Food Engr. 20: 311-323.

25.   **Wang, W-C., and Sastry, S.K.** 1993b. Salt diffusion into vegetative tissue as a pretreatment for ohmic heating: electrical conductivity profiles and vacuum infusion studies. J. Food Engr. 20:299-309.

26.   **Zhang, L. and Fryer, P.J.** 1992. Models for the electrical heating of solid-liquid food mixtures. Chemical Engineering Science. 48(4):633-642.

27.   **Zhang, L., and Fryer, P.J.** 1993. Modelling heat generation and transfer in laminar flow of food materials. Abstract No. P2.50. Presented at the Sixth International Congress on Engineering and Food, Makuhari Messe, Chiba, Japan, May 23-27, 1993.

DISCUSSION

ZOTTOLA: Is Ohmic heating of foods being used commercially? If so, where and by whom? Who developed the processes that are being used?

SASTRY: In the U.K., one company is using Ohmic heating for producing sterile solid liquid mixtures. In Japan, surimi products are processed commercially. In the U.S., a center for Ohmic commercialization has been established in Arden Hill, MN. The recent re-emergence of the process is due to the electricity council of Great Britain with a license to APV Baker, although other small manufactures are emerging.

SINGH: What approaches do you find useful to minimize problems related to cold spots?

SASTRY: One answer lies in product formulation, if the electrical conductivities of phases are not widely different, cold spots may be minimized. If this cannot be achieved, adequate fluid mixing (which may occur under the right conditions) and suitable factors of safety in process design may be necessary.

KNORR: Would the use of static mixer help overcoming the problem of cold spots in continuous processing of particulates?

SASTRY: The particles may be large and their concentrations high, consequently it may be difficult to implement static mixing.

LAZARIDES: Would the application of turbulent flow help minimize the "cold spot" problem?

SASTRY: Certainly some means of fluid mixing is desirable. Whether shear induced turbulence can be induced in such systems is uncertain, because of the typically low flow rates and high viscosity fluids involved. Nevertheless, some particle induced turbulence may in fact be occurring.

REDUCTION OF PECTINESTERASE ACTIVITY IN ORANGE JUICE BY HIGH PRESSURE TREATMENT

S. Irwe , I. Olsson
Tetra Pak Processing Systems AB, S -221 86 Lund, Sweden

Background

In recent years there has been a growing interest for high pressure as an alternative to thermal processing of various foods. There is a wide variety of foods where high pressure treatment can be utilized to inactivate microorganisms and enzymes at levels corresponding to thermal pasteurization but without any noticeable heat induced flavor changes.

Orange juice is an example of a product where flavor, color and nutritional value of the fresh product is affected by conventional heat pasteurization. Currently there is a rapidly expanding market for fresh squeezed orange juice in the USA and some other countries. It is a high priced premium quality product, but with a very limited shelf life, even at low refrigerated temperatures. Off-flavors caused by microbial activity are usually the shelf-life limiting factors for fresh squeezed orange juice at refrigerated storage- and distribution-temperatures.

The utilization of high pressure treatment in combination with a good barrier package could result in a commercial stable orange juice with the flavor of fresh squeezed orange juice but with the shelf-life of heat pasteurized juice.

In the processing step it is necessary to achieve sufficient inactivation of microorganisms and enzymes by utilizing proper pressures, temperatures and processing time. These process parameters must however also be utilized in a cost optimal way and without changing the sensory quality of the juice.

In order to roughly define a range of possible pressure - temp - time combinations for a high pressure process for orange juice, a number of trials was done in co-operation between Tetra Pak and ABB Metallurgy AB. The trials described here were focused on the inactivation of pectinesterase.

Pectinesterase (PE)

In orange juices an opaque or cloudy appearance is considered a desirable characteristic. The cloud is composed of finely divided particulates of pectin, cellulose, protein and lipids in suspension. When the suspension collapses, the juice is converted into a two phase system of a flocculate sediment and a more or less clear serum. This separation is normally seen as a quality defect although in some markets it is now claimed as being an evidence of no thermal treatment of the juice.

Cloud loss in orange juice has been attributed primarily to the action of pectinesterase on pectin causing removal of methoxy groups from the pectin. This de-esterfication of pectin results in pectic acid complexes with calcium which precipitate as insoluble calcium pectate.

Also other cloud forming components is entrapped in this precipitation process , resulting in total cloud loss in single strength juice.

In conventional commercial orange juice processing PE is inactivated by heat in the evaporator during concentration or during the pasteurization of single strength "not from concentrate" juice. A temperature 88 - 90 °C in 15-30 seconds is generally sufficient to inactivate most of the PE although total inactivation of PE is not always considered necessary. A 90 - 100 % reduction of the PE activity is normal in commercial heat pasteurized orange juice.

There are several iso-enzymes of PE. At least five different forms of PE has been identified in Naval oranges (Versteeg). They all show different heat resistance. The most thermostable iso-enzyme require temperatures up to 90°C for inactivation. Other forms of the enzyme is affected by much lower temperatures. The heat resistant forms accounts for only a minor part of the total activity but are still the factor that determines the temperatures and times to be used in orange juice processing. It is well established that the thermal resistance of the microorganisms normally occurring in orange juice is less than that of the most heat resistant forms of PE.

Inactivation of PE with high pressure

Ogawa et al. (1992) found that a complete inactivation of PE was not possible with high pressure even at 600 MPa for 10 minutes, 23°C. A relative PE activity of approx. 15 % remained after pressure treatment of Valencia Orange juice at these conditions. The remaining activity was even higher for other citrus fruits at 800 and 1000 MPa , 20 °C, 10 minutes

According to available literature it seems that PE in citrus-juices is more pressure-resistant than micro-organisms normally occurring in citrus juice. Consequently for orange juice it is the pressure resistance of the PE that actually determines the total requirement of the processing parameters: pressure - temperature - time.

Temperature effect of high pressure treatment

When evaluating the effect of high pressure on foods, attention should be paid to the temperature increase in the food during pressurizing. The temperature increase depends on pressure and product-composition . At 400 and 800 MPa the temperature of pure water is increased 12, 22°C, respectively . For single strength orange juice the temperature effect is almost the same as for water. The inactivation of PE at high pressures could be seen as a combined effect of a high pressure and heat generated by the applied pressure.

High pressure treatment trials with orange juice

In order to get a better understanding of the pressure and temperature necessary to achieve >90 % inactivation of PE within a reasonable processing time, a number of trials were done in co-operation with ABB Metallurgy. The trials were made with an ABB-press unit with a pressure capacity of 830 MPa. Fresh squeezed orange juice were either purchased from a local processor or extracted in-house with a small kitchen unit. Different varieties of oranges were used during the different trials. These trials were limited to three different pressure levels: < 400 MPa, 400 MPa and 600 MPa.

Pressures < 400 MPa

As expected pressures below 400 MPa did not result in > 90 % PE reduction at processing times up to 2 minutes at process temperatures of 40°C and 50°C. (Fig. 1). No further trials were made as higher temperatures and/or longer processing times would be necessary to reach required PE reduction at these pressures, if possible at all.
With higher process temperatures the risk for heat induced changes of the sensory quality increases and longer processing times is not attractive from a commercial point of view.

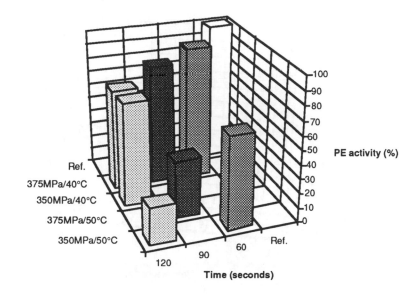

Figure 1.

400 MPa

At a pressure of 400 MPa the temperature increase in orange juice was approx. 12°C . All samples were consequently preheated to a temperature 12°C below targeted process temperature before high pressure treatment.

The preheating to 38°C and 48°C had alone a strong impact on PE activity (fig 2 and fig 4). In neither of two trials with 400 MPa at 40 °C, 40 s or 5 minutes, gave satisfactory PE reduction. (Fig. 2 - 3)

In another trial with a different variety orange, 400 MPa, 40°C for 2 min. resulted in a 90% reduction. Further reduction could not be achieved at longer processing times. (Fig. 4)

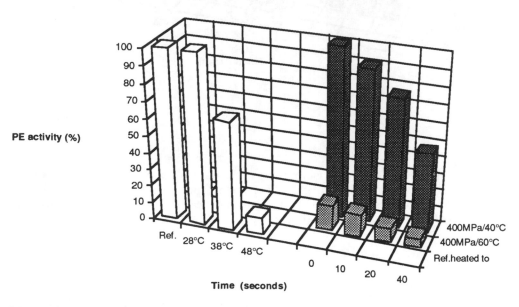

Figure 2.

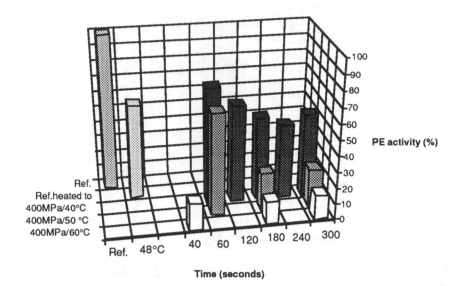

Figure 3.

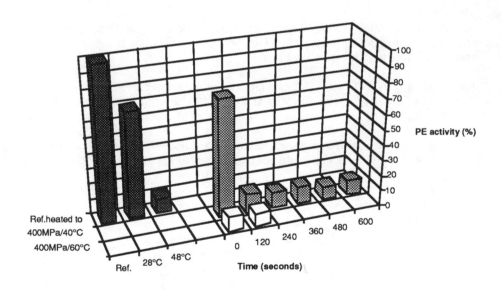

Figure 4.

600 MPa

With a pressure of 600 MPa the temperature of the juice increases 17°C. With a processing temperature of 20°C, (pre-chilled to 3°C), 1 to 2 minutes processing time was necessary to achieve a 90 % reduction of the PE activity. With an increased process temperature to 40°C this time could be reduced to 20 sec. (Fig. 5, Fig. 6)

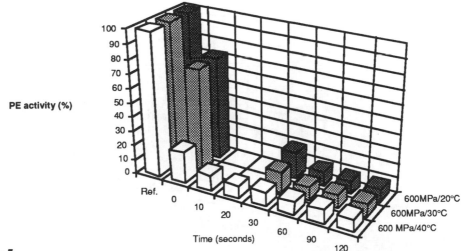

Figure 5.

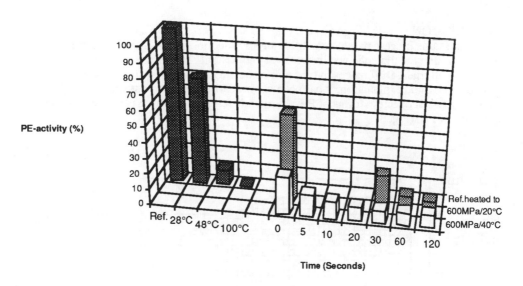

Figure 6.

Multiple treatment cycles

Multiple treatment cycles with relatively low pressures, low temperatures and short processing times was also evaluated.

The oranges used in this trial contained very heat-resistant PE. Pre-heating to 48°C resulted in only 5% reduction of PE activity. In previous tests up to 90% reduction have been achieved by the pre-heating step to 48°C alone.

No specific effect of multiple treatment cycles could be seen with low pressures. At 600 MPa without any holding time the effect of the second treatment cycle was almost 4 times compared to the first treatment cycle. (Fig. 7)

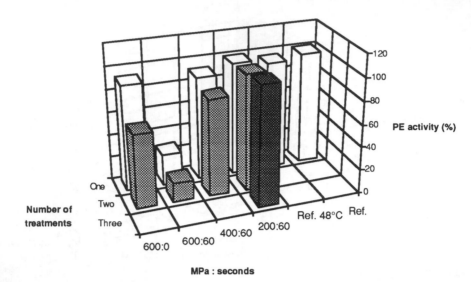

Figure 7. All treatments at 40°C

DISCUSSION.

With the same process parameters different PE reduction was achieved in different trials. However within the trials the results were very consistent. The same variety of orange were not used in all trials. Differences in type and amount of PE in the different oranges used could be a possible explanation. It has been shown earlier that the heat stability differs strongly between different forms of PE. Similar differences might exists regarding pressure resistance of PE. (Fig. 8)

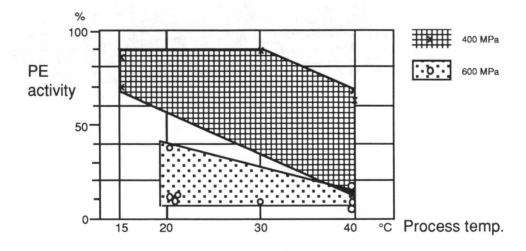

Figure 8. All treatments 120 sec. holding time.

A substantial reduction in PE activity was achieved during pre-heating of most of the samples between 38°- 48°C. In the following pressure treatment the temperature increases further. The processing temperature is thus a function of incoming product temperature and applied pressure. The total reduction of PE activity with high pressure treatment is a result of heat and pressure were part of the heat is generated by the applied pressure.

According to achieved results in these trials there are several different pressure - time -temperature combinations that have a potential to be used for > 90% reduction in PE activity in orange juice.

* 400 MPa,   short holding time   high temperature

* 600 MPa   short holding time   low temperature

* 600 MPa multiple treatment   no holding time   low temperature

To evaluate and optimize these process options from technical and commercial point of views further trials are necessary. The processing temperatures should be minimized to avoid heat induced off-flavors, and the processing times kept as short as possible to reduce total processing cost. The high pressure equipment must however have enough flexibility to process juices with various forms and amounts of heat- and pressure resistant PE. The demand for higher pressures allowing lower temperatures and shorter processing times should be balanced with the increasing cost for equipment that could be expected with increased pressure capacity.

When an optimal high pressure solution is defined it should be compared with conventional thermal processing in terms of cost and product quality. In this comparison market information from the juice industry is necessary together with technical and economical process data Only with relevant market information will it be possible to judge if the increased processing costs with high pressure can be justified by the demonstrated improved product quality.

REFERENCES

Eagerman , E. A. & Rouse A.H. Heat inactivation temperature-time relationship for PE inactivation in citrus juice. *Journal of Food sciences* . Vol. 41, 1976

Kimball, D.A. Citrus Processing, Van Nostrand Reinhold, N.Y. 1991 chap. 8.

Ogawa, H. et al., Effect of hydrostatic pressure on sterilization and preservation of citrus juice. *High pressure and biotechnology* . Eds. C. Balny, R Hayashi, K. Masson, Colloque INSERM/ John Libbey Eurotext Ltd Vol. 224. 1992

Olsson, I. High pressure pasteurization of orange juice. Internal Tetra Pak development report. 1993

Träff A. & Bergman C. High pressure equipment for food processing. *High pressure and biotechnology* . Eds. C. Balny, R Hayashi, K. Masson, Colloque INSERM/ John Libbey Eurotext Ltd Vol. 224. 1992

Versteeg et al . Thermostability and orange juice cloud destabilizing properties of multiple pectinesterases from oranges. *Journal of Food science* . Vol. 45, 1980

DISCUSSION

GORRIS: If you use a multiple treatment at high pressure (no holding time), does it extend the total processing time, especially as you may have to use more than two successive treatments?

IRWE: The total processing time is one of the issues in high pressure processing. If the build up and release of pressures prolongs the processing time substantially, this makes the multiple treatment alternative less attractive for commercial applications. We still do not have sufficient data to specifically answer the question.

ZOTTOLA: Why is pressure processing needed in the preservation of acid foods, fruit juices, jams, etc.? Is there a spoilage problem with these foods using current methods?

IRWE: The reason for using high pressure processing instead of thermal processing is to avoid the heat induced quality changes associated with thermal processing. For the orange juice, the objective is to achieve a product with the flavor of freshly squeezed juice and with the shelf-life of a pasteurized juice.

COMBINATION OF MILD PRESERVATIVE FACTORS TO IMPROVE SAFETY AND KEEPING QUALITY OF CHILLED FOODS

J. Farkas
University of Horticulture and Food Industry, Budapest,
Hungary

I. INTRODUCTION

Combinations of mild treatments and/or preservative hurdles are frequently enhancing each others' effects. Due to this synergism, combinations of individually ineffective factors and mild treatments may result in improved microbial safety and stability, and better quality of combined-treated foods.

Decreasing the energy demand of food preservation and improving the quality--retention of preserved foods are timely and generally valid requirements, therefore, exploring the possibilities of using ionising radiation in combination with preservative hurdlcs may significantly contribute to the development of new types of combination processes and new types of preserved foods. [Urbain, 1986; WHO, 1988]

In relation to the "hurdle concept" of Leistner [Leistner, 1987], and to the current interest in "minimally processed" products and extended shelf-life chilled foods, [Livingston, 1985], microbiological research is in progress in a number of laboratories on combination of antimicrobial factors in order to study survival and growth responses of microorganisms influenced by combinations of factors, and to develop models for predicting the stability and safety of such foods. It is of particular importance that extended shelf-life, minimally processed chilled foods maintain also their microbiological safety with particular reference to those food-borne pathogens, which are capable of growth at refrigeration temperatures. [Palumbo, 1986: Lechowich, 1988]

Based on these considerations, our research work to be reported here covered two fields of studies:

a) Preservation of a vacuum-packaged, ready-to-fry, chilled meat product, tenderloin rolls by combination of 2 kGy gamma radiation dose with reduction of pH and/or reduction of the water activity.

b) Model studies on combined effects of incubation temperature, water activity, pH, and nutrient media on growth of unirradiated and irradiated populations of *Listeria monocytogenes*.

II. EFFECTS OF GAMMA IRRADIATION, pH-REDUCTION AND Aw-REDUCTION ON THE SHELF-LIFE OF CHILLED 'TENDERLOIN ROLLS'

A. EXPERIMENTAL PROCEDURES

A ready-to-fry, comminuted product, "tenderloin rolls", prepared by a Hungarian company "Budapesti Hütőipari Vállalat" by comminution of fresh and frozen beef and pork as well as soaked wheat bread with small amounts of soy flour, sodium chloride, polyphosphate and a spice mixture was used as raw material. After eventual addition of acidulant and/or the humectant, small rolls of ca. 50 g-s were formed and individually vacuum-packaged into PA/PE laminated foil of 20/50 μm thickness by a MULTIVAC machine. Permeability characteristics of the packaging material:

O_2-permeability: 30-40 cm^3/m^2, day at 1 bar,
CO_2-permeability: 60-80 cm^3/m^2, day at 1 bar,
vapour-permeability: 6-8 g/m^2, day at 38 °C.

The vacuum-packaged 'tenderloin rolls' were preserved by combinations of reduction of pH from 6.1 to 5.6 by ascorbic acid, reduction of the water activity from a_w=0.975 to a_w=0.962 by sodium-lactate, and/or a radiation dose of 2 kGy. Storage of untreated and irradiated samples at +2 °C for 4 weeks was followed by one-week incubation at +10 °C. Total plate counts, counts of presumptive lactobacilli, the *Enterobacteriaceae* and sulphite-reducing clostridia were estimated at weekly intervals. pH-changes during storage were also followed. Comparative estimation of sensory qualities, thiamin contents, and TBA-values were also performed.

B. RESULTS OF SENSORY AND ANALYTICAL TESTING

Neither use of the ascorbic acid as acidulant, nor Na-lactate as a_w-reducing agent at the applied concentrations caused any statistically significant change in the sensory quality of the ready-to-eat (fried) product. Thiobarbituric acid (TBA-) values were not significantly changed neither by irradiation nor by the 5-week refrigerated storage. The thiamin level (approx. 1.2 mg/kg in the control samples) was reduced by approx. 33 % as an effect of the 2 kGy gamma radiation. However, samples containing ascorbic acid and/or Na-lactate lost only approx. 10 % of their thiamin content as an effect of the irradiation.

C. RESULTS OF MICROBIOLOGICAL TESTING

The results of microbiolgial testing and pH-measurements are summarized in Figures 1 and 2.

The initial total aerobic viable cell count of the original paste of tenderloin rolls was approx. $3x10^5$/g while its *Enterobacteriaceae* count was less than 10^2/g. Microbiological shelf-life was estimated as a number of days elapsed until the total aerobic viable cell counts reached the level of 10^8/g. According to this arbitrary definition of microbiological spoilage, the microbiological shelf-life of unirradiated samples without pH- and/or a_w-reduction was approx. 11 days at +2 °C. Both pH-reduction (addition of ascorbic acid) and a_w-reduction (addition of Na-lactate) slowed down somewhat the microbial growth and resulted in a microbiological shelf-life of approx. 16 days and 19 days, respectively. By combined application of the two additives, the level of 10^8 total aerobic viable cells/g level was reached during 31 days of storage.

In spite of these promising extensions of the shelf-life of the aforementioned experimental batches, their microbiological safety can not be considered as fully assured by the additives, because any considerable further drop of pH occurred only when the viable cell counts of lactic acid bacteria reached or exceeded the 10^8/g level, and because at the end of the above-defined shelf-life the *Enterobacteriaceae* counts amounted to the order of magnitude of 10^6/g in the additive-free samples, and to the 10^5/g level in samples containing both ascorbic acid and Na-lactate. The counts of the lactic acid bacteria reached the 10^6-10^7/g level in these samples at the end of shelf-life.

The radiation dose of 2 kGy reduced efficiently, by approx. two log cycles the initial viable cell counts, and so much the *Enterobacteriaceae* counts that the latter remained under the detection limit (approx. 10 cells/gram) during the entire storage period of the irradiated samples. The total aerobic viable cell counts of the irradiated samples did not reach the 10^6/g level during storage. The combined effect of pH-reduction plus irradiation, or a_w-reduction plus irradiation expressed itself in the extension of the lag

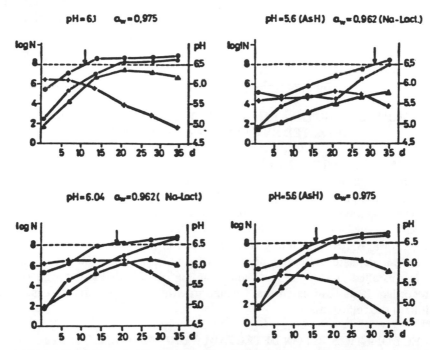

Fig.1.: Microbiological and pH_changes of unirradiated samples during storage for four weeks at +2 °C, followed by one week at +10 °C.

●——● TPC ●——● LB ▲——▲ EB ◆——◆ pH

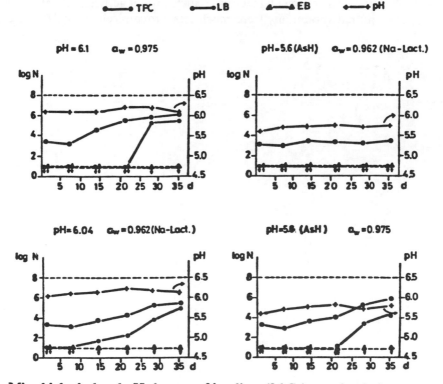

Fig.2.: Microbiological and pH-changes of irradiate (2 kGy) samples during storage for four weeks at +2 °C, followed by one week at +10 °C.

●——● TPC ●——● LB ▲——▲ EB ◆——◆ pH

phase of bacterial growth, while microbiological stability of the product was fully maintained during the five weeks of storage after the triple combination, i.e. practically no growth of the initial residual counts happened during the entire storage period. The residual microflora of such samples consisted mainly from aerobic bacterial spores. Sulphite-reducing clostridia were always under the detection level (10/g).

III. GROWTH OF UNTREATED AND RADIATION-DAMAGED *LISTERIA* AS AFFECTED BY ENVIRONMENTAL FACTORS

A. EXPERIMENTAL PROCEDDURES

The growth of untreated *Listeria monocytogenes* 4ab No. 10 and that of the surviving fraction of its population treated with 0.8 kGy gamma rays (2D-dose) was investigated in a microtitre-plate system in duplicates at various incubation temperatures between 3 °C and 35 °C in media made either from Tryptic Phosphate Broth (TPB) or Brain Heart Infusion Broth (BHIB) containing NaCl between 0.25 to 16.75 % (w/v), and acidified with citrate-phosphate buffers to pH values between 4.63 and 7.06. The initial viable count was installed to 3.10^3/ml level, and time periods for visible growth (TG, in days) were recorded. It was estimated that "visual growth" represents an increase of at least 1000-fold in the cell counts.

B. EFFECT OF STRESSES ON DETECTABLE GROWTH OF *LISTERIA*

Results have shown that radiation survivors had increased salt- and pH-sensitivities and increased minimum temperature of growth in TPB-based media (Table 1). Adverse effects of sub-optimal environmental factors (reduced water activity, -pH and -temperature) and radiation injury were much less pronounced in BHIB-based media (Table 2).

Table 1

Effect of pH and NaCl Concentration on the Growth of Untreated and Irradiated *L monocytogenes* in TPB

pH	NaCl %	\multicolumn{4}{c}{Time to visible growth (days)}						
		\multicolumn{4}{c}{Unirradiated}				\multicolumn{4}{c}{0.8 kGy}		
		35°C	30°C	10°C	5°C	35°C	30°C	10°C	5°C
7.1	0.25	2	2	20	64	2	2	20	64
	1.75	2	2	20	64	2	2	20	64
	3.25	2	2	20	64	2	2	24	64
	4.75	2	6	24	n.g	3	3	24	n.g
	6.25	n.g.	6	24	n.g.	n.g.	6	31	n.g.
	7.75	n.g.	6	24	n.g.	n.g.	6	52	n.g.
	9.25	n.g.	13	41	n.g.	n.g.	n.g.	n.g.	n.g.
	10.75	n.g.	n.g.	n.g.	n.g.	n.g.	n.g.	n.g.	n.g.
5.9	0.25	2	3	24	64	3	3	31	64
	1.75	6	3	34	n.g.	n.g.	24	41	n.g.
	3.25	8	6	34	n.g.	n.g.	24	n.g.	n.g.
	4.75	n.g.	20	64	n.g.	n.g.	n.g.	n.g.	n.g.
	6.25	n.g.	n.g.	n.g.	n.g.	n.g.	n.g.	n.g.	n.g.
5.5	0.25	6	6	64	n.g.	6	7	n.g.	n.g.
	1.75	7	6	71	n.g.	n.g.	n.g.	n.g.	n.g.
	3.25	n.g.	n.g.	n.g.	n.g.	n.g.	n.g.	n.g.	n.g.
5.3	0.25	6	6	n.g.	n.g.	7	13	n.g.	n.g.
	1.75	24	13	n.g.	n.g.	n.g.	n.g.	n.g.	n.g.
	3.25	n.g.	n.g.	n.g.	n.g.	n.g.	n.g.	n.g.	n.g.
5.2	0.25	7	7	n.g.	n.g.	n.g.	24	n.g.	n.g.
	1.75	n.g.	n.g.	n.g.	n.g.	n.g.	n.g.	n.g.	n.g.
	3.25	n.g.	n.g.	n.g.	n.g.	n.g.	n.g.	n.g.	n.g.

n.g.: no visible growth within 10 weeks

Table 2

Effect of pH and NaCl Concentration on the Growth of Untreated and Irradiated
***L. monocytogenes* in BHIB**

pH	NaCl %	Time to visible growth (days)							
		Unirradiated				0.8 kGy			
		30°C	20°C	8°C	3°C	30°C	20°C	8°C	3°C
7.1	0.25	1	2	6	27	1	2	6	31
	2.25	1	2	6	27	1	2	7	31
	4.25	1	2	7	27	2	2	8	36
	6.25	2	3	10	36	2	3	13	43
	8.25	3	6	17	n.g.	3	6	20	n.g.
	10.2	17	7	41	n.g.	n.g.	7	48	n.g.
	12.2	n.g.	n.g.	n.g.	n.g.	n.g.	n.g.	n.g.	n.g.
5.6	0.25	1	2	8	n.g.	1	2	10	n.g.
	2.25	1	2	17	n.g.	2	3	20	n.g.
	4.25	2	3	n.g.	n.g.	2	6	24	n.g.
	6.25	3	6	n.g.	n.g.	3	7	n.g.	n.g.
	8.25	n.g.	n.g.	n.g.	n.g.	n.g.	n.g.	n.g.	n.g.
5.3	0.25	1	2	27	n.g.	2	3	24	n.g.
	2.25	2	2	n.g.	n.g.	2	6	45	n.g.
	4.25	3	6	n.g.	n.g.	3	7	n.g.	n.g.
	6.25	n.g.	22	n.g.	n.g.	7	15	n.g.	n.g.
	8.25	n.g.	n.g.	n.g.	n.g.	n.g.	n.g.	n.g.	n.g.
5.1	0.25	2	3	44	n.g.	2	6	n.g.	n.g.
	2.25	2	6	n.g.	n.g.	2	6	n.g.	n.g.
	4.25	3	8	n.g.	n.g.	6	10	n.g.	n.g.
	6.25	n.g.	n.g.	n.g.	n.g.	n.g.	n.g.	n.g.	n.g.
4.9	0.25	3	n.g.	n.g.	n.g.	6	n.g.	n.g.	n.g.
	2.25	3	22	n.g.	n.g.	6	31	n.g.	n.g.
	4.25	n.g.	29	n.g.	n.g.	n.g.	34	n.g.	n.g.
	6.25	n.g.	n.g.	n.g.	n.g.	n.g.	n.g.	n.g.	n.g.
4.7	0.25	8	n.g.	n.g.	n.g.	n.g.	n.g.	n.g.	n.g.
	2.25	22	55	n.g.	n.g.	22	n.g.	n.g.	n.g.
	4.25	n.g.	n.g.	n.g.	n.g.	n.g.	n.g.	n.g.	n.g.
4.6	0.25	17	n.g.	n.g.	n.g.	n.g.	n.g.	n.g.	n.g.
	2.25	n.g.	n.g.	n.g.	n.g.	n.g.	n.g.	n.g.	n.g.

n.g.: no visible growth within 10 weeks

C. GROWTH MODELLING

On the initiative of a paper by Cole and co-workers [Cole et al., 1990], for quantitative analysis of interaction of environmental factors and the effect of radiation injury on growth, the data obtained at 30 °C, as most abundant informations, were analysed statistically for a least squares fit to a quadratic model and stepwise variable selection by a Statgraphics software. Thus, polynomial equations linking the salt [S] (%) and hydrogen ion concentrations $[H^+]$ (μmol/1) to the natural logarithms of the "apparent relative growth rates" (100/TG, day^{-1}) were calculated. After omitting terms insignificant at 95 % probability level, the following quadratic equations and determination coefficients (r^2) were obtained for the four combinations of radiation dose and media:

TPB, 0 kGy:

$$Ln(100/TG)=3.91-0.166[H^+]-0.020[S]^2-0.157[H^+][S]; \ r^2=0.751$$

0.8 kGy:

$$Ln(100/TG)=3.99-0.295[H^+]-0.013[S]^2-0.577[H^+][S]; \ r^2=0.890$$

BHIB, 0 kGy:

$$Ln(100/TG)=5.05-0.133[H^+]-0.028[S]^2-0.010[H^+][S]; \ r^2=0.907$$

0.8 kGy:

$$Ln(100/TG)=4.81-0.157[H^+]-0.027[S]^2-0.006[H^+][S]; \ r^2=0.914$$

The equations show that the apparent growth rate was significantly higher in BHIB-based media than in TPB-based media. The logarithm of the apparent growth rate was a linear function of the hydrogen ion concentration and a quadratic function of the salt concentration. In TPB-based media a significant interaction (synergism) was found between the effect of hydrogen ion concentration and salt concentration. The sensitizing effect of radiation injury appeared in the considerably increased constants of the $[H^+]$ and $[H^+][S]$-terms as compared to those obtained with unirradiated inoculum. However, no statistically significant radiation sensitization effects could be proved in the more rich (BHIB-based) growth media.

The polynomial functions calculated were used to construct three-dimensional response surface plots in order to visualise the relationship of the growth variable to the environmental parameters (Fig. 3).

The roles of pH, salt concentration and radiation injury as well as their combined effects on the growth kinetics at 30 °C are also illustrated by the border-lines of growth/no growth during 3, 7, 14 and 21 days of incubation, respectively (Fig. 4).

III. DISCUSSION AND CONCLUSIONS

Summarizing the results on combination preservation of "tenderloin rolls", one can conclude that radiation treatment with 2 kGy gamma radiation dose could reduce the viable cell counts by approx. 99 %, and extended the microbiological shelf-life of the vacuum-packaged, chilled meat product tested by a factor of at least three, without

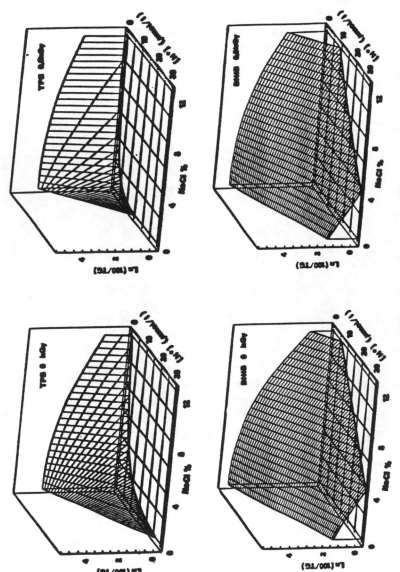

Fig.3.: Response surfacess corresponding to the polynomial equations describing the combined effect of hydrogen ion and salt concetrations on the apparent growth of untreated (0 kGy) and irradiated (0.8 kGy) populations of *Listeria monocytogenes* at 30 °C.

TG= time to visually observable growth in days.
TPB= Tryptic Phosphate Broth; BHIB= Brain Heart Infusion Broth.

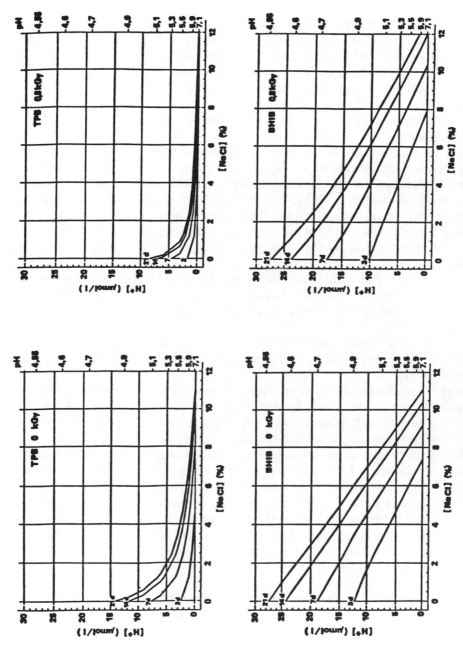

Fig.4.: Border-lines of growth/no growth of untreated (0 kGy) and irradiated (0.8 kGy) populations during 3, 7, 14, or 21 days of incubation at 30 °C.

significantly lowering the sensory quality of the 'tenderloin rolls'. Comnibining the radiation treatment with a slight reduction of the pH and the water activity may result in a microbiologically stabile product of acceptable sensory quality for at least 5 weeks at +2 °C. It would be worthwhile to perform investigations to determine whether these shelf-life extensions were due to the pH- and a_w-reduction only, or specific antimicrobial effects of these additives could also be demonstrated.

Our results are in agreement with the literature showing the radiation sensitivity of Enterobacteriaceae [Ingram and Farkas, 1977; Tarkowski et al., 1984a,b] and a flora--shift favouring lactic acid bacteria in irradiated, vacuum-packaged pork products [Ehioba, 1987; Lebepe et al., 1990; Grant and Patterson, 1991a]. Extension of shelf-life of vacuum-packaged refrigerated ground beef by addition of 0.5 % ascorbic acid was reported by Von Holy and Holzapfel [1983].

Concerning the microbiological safety of the extended shelf-life chilled products our studies are aiming, too, it is well established in the literature of radiation microbiology of foods that radiiation dose of 1-2 kGy is capable to decrease the viable cell count of non-sporogenic, psychrotrophic pathogenic bacteria such as *Yersinia enterocolitica*, enterotoxigenic *Echerichia coli*, *Aeromonas hydrophila*, and *Listeria monocytogenes*, by at least 2-4 orders of magnitude [ICMSF, 1980; Tarkowski et al., 1984b; Patterson et al., 1989]. Therefore, the radiation treatment and further antimicrobial factors introduced by reduction of pH and/or water activity may extend not only the shelf-life of combined treated refigerated products, sutch as studied here, but they can improve their microbiological safety, for instance, in case of a temperature abuse [Grant and Patterson, 1991b]. Flora shift in irradiated and combined-treated batches favouring lactic acid bacteria may be considered as a built-in safety factor against the growth of less acid-tolerant pathogenic bacteria [Mattila-Sandholm and Skytta, 1991; Holzapfel, 1989; Gombas, 1989].

Our resuslts on the increased susceptibility of an irradiated population of *L. monocytogenes* surviving a $2D_{10}$ radiation dose indicate a sublethal injury of the radiation survivors.

Although the simple model based on visual determination of growth (representing at least 1000-fold increase in numbers) helps to visualize the basic relationship between the variables investigated, it allows only very rough estimates of the impact of stress factors on the growth of the bacterium studied.

Nevertheless, these model studies demonstrate also the efficiency of combining antimicrobial hurdles, and illustrate that growth ranges of *L. monocytogenes* cells are reduced when surviving a radiation dose foreseen for radurizing minimally processed products such as described in the first part of this report. Our observations are consistent wiith the findings of other authors who reported an increase of the lag phase of growth of *L. monocytogenes* which survived an irradiation process [Patterson et al., 1993]. Therefore, low-dose irradiation in combination with mild antimicrobial stress factors may contriibute to the microbiological safety of those extended shelf-life chilled products where *L. monocytogenes* presents a potential hazard. Storage of the product at temperatures of ≤ 3.3 °C will remain, however, a major safety factor to reduce the risk of botulism from spores of non-proteolytic strains of *C. botulinum*, which easily survive radiation doses applied in our experiments. Thus, dependable, controlled-temperature distribution system is a critical requirement for safety, product stability and shelf-life. [Notermans et al., 1990] Nevertheless, selecting a model product which requires heating, thereby inactivation of botulinum toxins, before consumption of the food was also considered when the test material was chosen for our studies.

Acknowledgement
This study was supported by the Hungarian National Science Foundation (OTKA) Project No.611 and by the International Atomic Energy Agency Research Contract No. HUN-6272/RB. Éva Andrássy, Diána Bánáti and László Mészáros are thanked for their capable technical assistance.

REFERENCES

Cole, M.B., Jones, M.V., and Holyoak, C.(1990): The effect of pH, salt concentration and temperature on the survival and growth of *Listeria monocytogenes, Journal of Applied Bacteriology*, 69, 63-72.

Ehioba, R.M.(1987): Effect of low dose (1 kGy) gamma radiation and selected phosphates on the microflora of vacuum-packaged ground pork, Abstract of a PhD Thesis, Iowa State University, USA. Publication No. AAC87 16760.

Gombas, D.E.(1989): Biological competition as a preserving mechanism, *J. Food Safety*, 10, 107-117.

Grant, I.R. and Patterson, M.F.(1991a): A numerical taxonomic study of lactic acid bacteria isolated from irradiated pork and chicken packaged under vaarious gas atmospheres, *J. Applied Bacteriol.*, 70, 302-307.

Grant, I.R. and Patterson, M.F.(1991b): Effect of irradiation and modified atmosphere packaging on the microbiological safety of minced pork stored under temperature abuse conditions, *Internat. J. Food Sci. Technol.*, 26, 521-533.

Holzapfel, W.H.(1989): Lactobacillus ecology of meat and meat products, with special reference to radurisation, pp.317-321. in *Trends in Food Product Development* Ghee, A.H. (Chief ed.) Singapore Institute of Food Sci. Technol., Singapore.

ICMSF(1980): Ionizing irradiation, in *Microbial Ecology of Foods*, Vol.1, pp.46-69, The International Commission on Microbiological Specifications for Foods. Academic Press, New York.

Ingram, M. and Farkas, J.(1977): Microbiology of foods pasterurised by ionising radiation, *Acta Alimentaria*, 6, 123-185.

Lebepe, S., Molins, R.A., Charden, S.P., Farrar, H. and Skowronski, R.P.(1990): Changes in microflora and other characteristics of vacuum-packaged pork loins irradiated at 3.0 kGy, *J. Food Sci.*, 55, 918-924.

Lechowich, R.V.(1988): Microbiological challenges of refrigerated foods. *Food Technology*, 42, 84-85 and 89.

Leistner, L.(1987): Shelf-stable products and intermediate moisture foods based on meat, pp.295-327, in *Water Activity: Theory and Applications to Food*, Rockland, L.B. and Beuchat, L.R., Eds. Marcel Dekker, Inc.

Livingston, B.E.(1985): Extended shelf-life chilled prepared foods, *J. Food Service System*, 221-230.

Mattila-Sandholm, T. and Skyttä, E.(1991): The effect of spoilage flora on the growth of food pathogens in minced meat stored at chilled temperature, *Lebensm. Wiss.-u. Technol.*, 24, 116-120.

Notermans, S., Dufrenne, J. and Lund, B.M.(1990): Botulism risk of refrigerated, processed foods of extended durability. *J. Food Protection*, 53, 1020-1024.

Palumbo, S.A.(1986): Is refrigeration enough to restrain foodborne pathogens? *J. Food Protection*, 49, 1003-1009.

Patterson, M.(1989): Sensitivity of *Listeria monocytogenes* to irradiation on poultry meat and in phosphate buffered saline. *Lett. Appl. Microbiol.*, 8, 181-184.

Patterson, M.F., Damoglou, A.P. and Buick, R.K.(1993): Effects of irradiation dose and storage temperature on the growth of *Listeria monocytogenes* on poultry meat. *Food Microbiology*, 10, 197-203.

Tarkowski, A., Beumer, R.R., Kampelmacher, E.H.(1984a): Low dose irradiation of raw meat. II. Bacteriological effects on samples from butcheries. *Int. J. Food Microbiol.*, 1, 25-31.

Tarkowski, J.A., Stoffer, S.C.C., Beumer, R.R. and Kampelmacher, E.H.(1984b): Low dose gamma irradiation of raw meat. I. Bacteriological and sensory quality effects in artificially contaminated samples. *Int. J. Food Microbiol.*, 1, 13-23.

Urbain, W.M.(1986): *Food Irradiation*, Academic Press, Inc., Orlando, USA.

Von Holy, A. and Holzapfel, W.H.(1983): The influence of extrinsic factors on the microbiological spoilage pattern of ground beef. *Int. J. Food Microbiol.*, 6, 269-280.

WHO(1988): *Food Irradiation: A Technique for Preserving and Improving the Safety of Food*. World Health Organization, Geneva.

DISCUSSION

BUSTA: What do you suggest is the mechanism of irradiation damage.

FARKAS: We did not study this question in the frame of the present project. I guess, however, that radiation damage (increased sensitivity of radiation survivors) may be related to increased permeability of cell membranes reducing the cells capability to keep their homeostasis. Due to this effect, systems involved in DNA repair may become also more sensitive to environmental stresses.

GORRIS: In a recent congress on microbiological foods issues, a representative of the World Health Organization (Mr. Kaferstein), strongly advocated the wide application of irradiation. If I recall well, his irradiation level was ±4 times lower. Do you have information on that, your studies did not indicate sensory effects?

FARKAS: Dr. Kaferstein at the Food Micro' 93 conference in Bingen referred to the Report of the Joint FAO/IAEA/ WHO Expert Committee on Wholesomeness of Irradiated Foods convened in 1980. The JECFI concluded after evaluation of a tremendous amount of available data, that food irradiated up to an overall average dose level of 10 kGy (1 Mrad) can be considered as safe from the toxicological point of view, and proper uses of food irradiation do not create specific microbiological or nutritional problems. This "legal" dose limit is not lower but 5 to 20 times higher than doses required in consideration with mild preservative factors/environmental stresses to improve microbiological safety and/or microbiological stability of minimally processed products. This low-dose irradiation does not induce sensory changes, as I mentioned during my presentation, in the product we have investigated, or in many other fresh or minimally processed commodities for combined preservation.

IMPROVEMENT OF THE SAFETY AND QUALITY OF REFRIGERATED READY-TO-EAT FOODS USING NOVEL MILD PRESERVATION TECHNIQUES[*]

L.G.M. Gorris

**Agrotechnological Research Institute, Haagsteeg 6,
P.O. Box 17, NL-6700 AA Wageningen, The Netherlands**

I. SUMMARY

Refrigeration is most commonly used to extend the durability of vegetable based ready-to-eat foods, *i.e.* fresh or minimally processed preparations. Because of the difficulty of maintaining sufficiently low temperatures throughout the food chain from production and processing to retail, additional barriers to the growth of spoilage and pathogenic microorganisms are required. The present study was initiated to develop these barriers based on novel mild preservation techniques, *i.e.* biopreservation, modified atmosphere packaging (MAP) or coating (MAC), and coatings containing food-grade antimicrobial agents (active MAC). These techniques were selected because they comply with the high quality product standards demanded by consumers nowadays. In the course of the research, basic knowledge on microbiology, product physiology and preservation techniques are combined with practical evaluation of microbiological safety and product quality obtained with novel or optimised techniques. Computer modelling will eventually be used to integrate the various data on product quality, microbiology and storage c.q. packaging conditions.

II. INTRODUCTION

The range of foods which may be classified as ready-to-eat is quite extensive, including (mixtures of) raw vegetables, minimally processed (washed, trimmed, sliced) vegetables with or without dressings, combinations of raw vegetables and cooked food items, and Sous Vide preparations (cooked vegetable and potato based dishes). When the processing includes a heat treatment, the resulting preparations may be grouped as Refrigerated Processed Foods of Extended Durability (REPFEDs) The physiological and microbiological characteristics of products within this range varies substantially. Heat processed products are generally not metabolically active, while fresh and minimally processed vegetables are active and may create a modified gas atmosphere when they are

---

[*] This research is conducted under EC Contract AIR1-CT92-0125. The credits for the work described are due to the projectleaders and their teams: F.M. Rombouts (Wageningen Agric. Univ.), C. Nguyen-The (INRA, Montfavet), S. Guilbert (CIRAD-SAR, Montpellier), G.J. Nychas (NAgReF-ITAP, Athens), M.W. Peck (AFRC-IFR, Norwich), D. O'Beirne (Univ. Limerick), P. Callaghan (Nature's Best Ltd, Duleek), S. Le Hesran (Fruidor S.A., Cavaillon) and L.G.M. Gorris (projectcoordinator).

Table 1
Food categories and model foods characteristics

Category	Model food	Characteristics
Fresh	lettuce, carrot	low metabolic activity, moderate epiphyte level
	mungo bean sprouts	high metabolic activity, high epiphyte level
Minimally processed	cut lettuce, sliced carrot	low metabolic activity
	salad with dressing	low pH-selective environment
REPFEDs*, incl. Sous Vide	vegetable/egg product	model to study *Clostridium botulinum*

* REPFED: refrigerated, processed food of extended durability

are packaged. The metabolic activity of the fresh produce will depend on the type of product, the type of processing and the storage conditions. Also, these products possess, part of, their natural epiphytic microbial flora. Pathogens may form part of this flora, posing a potential safety problem, even under the low-oxygen conditions generally prevailing in these Modified Atmosphere Packaging (MAP) systems. The pathogens encountered under refrigerated storage conditions are mainly psychrotrophic bacteria (*Listeria monocytogenes, Yersinia enterocolitica, Aeromonas hydrophila*). Some mesophyllic pathogens are able to proliferate at abuse temperatures (*Salmonella typhimurium, Staphylococcus aureus*). Several of these pathogens reportedly survive well or even proliferate to high numbers on produce stored under low-oxygen conditions. [Berrang *et al.*, 1989a,b; Hotchkiss and Banco, 1992; Kallander *et al.*, 1991] With Sous Vide preparations and other REPFEDs, the mild heat treatment included in the processing diminishes most of the microbial flora but not spores of sporeforming bacteria, like *Clostridium botulinum* and *Bacillus cereus*. Non-proteolytic strains of *C. botulinum* and some strains of *B. cereus* are able to grow at refrigeration temperatures, and conceivably better in the absence of a competitive microflora.

The current project aims at optimizing existing techniques for mild preservation, *i.e.* refrigeration and MAP, for application with vegetable based convenience foods. In addition, novel biopreservation techniques are integrated into these existing techniques to establish an extra 'hurdle' towards the psychrotrophic, low-oxygen tolerant pathogens. Because of the substantial variation in ready-to-eat food products, the study is restricted to a number of model food products, varying mainly in the degree of processing and the size of the initial epiphytic microflora (Table 1). The research tasks comprise both fundamental and applied topics (Figure 1). Fundamental studies are devoted to investigating the basic physiology of a number of relevant foodborne pathogens in relation to key parameters (temperature, pH, a_w), their interactions with epiphytic microorganisms and with the absolute and relative levels of O_2 and CO_2 in the gas atmosphere. A specific objective is to assess the effect of heat processing and subsequent mild preservation factors on the safety of REPFEDs in relation to non-proteolytic *C. botulinum*, a low temperature pathogen. Based on studies of the interaction of products with the atmosphere and on packaging film characteristics, gas compositions and packaging materials are deduced which ensure optimal product quality during storage. Using natural polymers such as waxes and proteins, coatings are developed which have gas and vapour transmission characteristics comparable to plastic films but which are fully biodegradable. The various studies are designed to allow integration of the data obtained in a mathematical computer model. This model should be able to link information on product quality, microbiology and preservation technique and may be used to predict the effects of changes in a specific parameter on the shelf life of the product.

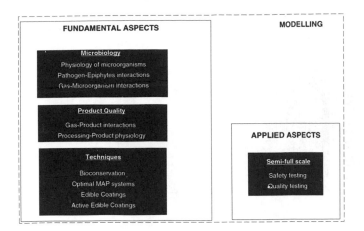

Figure 1. Outline of the various research topics included in the project.

III. RESEARCH TOPICS & RESULTS

The research reported on here is carried out by 7 institutions (two universities, five government/mixed research institutes) and two SMEs involved in preparation of ready-to-eat foods. In the following, a description of a number of research topics and a discussion of the preliminary results obtained within the first year of the project are given.

A. PHYSIOLOGY OF PATHOGENS AND SPOILAGE BACTERIA

Growth of psychrotrophic foodborne pathogens such as *L. monocytogenes* and non-proteolytic *C. botulinum* are of particular concern in minimally processed chilled foods. These bacteria are able to grow at refrigeration temperatures and heat-resistant spores of non-proteolytic *C. botulinum* would survive a minimal heat treatment and possibly germinate, grow and produce toxin in foods held at temperatures in excess of 3.3°C. The growth of the pathogens and of psychrotrophic spoilage bacteria needs to be investigated in relation to the characteristics of the food matrix, because several important factors (pH, a_w, preservatives, osmoprotectants, protective compounds) therein, interact with product quality and safety. The major topics studied are regulation of intracellular pH and osmotic strength and their respective links to the bioenergetic status of the cell, and the utilization of various sugars (carbon- and energy source) and nitrogen sources (amino acids and peptides) under different environmental and food conditions. Two research themes are highlighted below.

1. Osmoregulation in *Listeria monocytogenes*

The property to adapt to changes in the osmotic strength of the environment is inherent to most living cells. Cellular adaption to osmotic stress is a cardinal process, because organisms require a relatively constant cytoplasmic composition for optimal activity. Osmoregulation has been most extensively studied in Gram(-)-bacteria, *i.e. E. coli* and *S. typhimurium*. [Csonka, 1989] By accumulating compatible solutes (osmoprotectants) these organisms may counteract water stress conditions. The primary response after an osmotic upshock involves the accumulation of K^+ and glutamate which are replaced after or during trehalose synthesis. Proline and betaine, compounds which occur in many foods of animal and plant origin, are preferentially used as osmoprotectants.

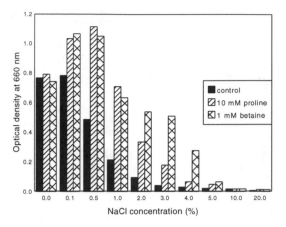

Figure 2. Stationary phase levels of growth for *Listeria monocytogenes* strain 311 in mineral medium (control), and in mineral medium with 10 mM proline or 1 mM betaine.

The ability of *L. monocytogenes* to grow in foods with a low a_w (up to 10 % NaCl) suggests that it can adapt well to an adverse osmotic environment, an ability which may contributes to its important role as a food poisoning agent. The mechanism through which Gram(+)-bacteria such as *L. monocytogenes* counteract osmotic stress conditions are not yet elucidated. Proline and betaine have been identified as osmoprotectants for *Streptomyces* spp., *Staphylococcus aureus* and *Lactobacillus acidophilus*. [Killham and Firestone, 1984; Bae and Miller, 1992; Jewell and Kashket, 1991] In the present study, therefore, the influence of low a_w on the growth of *L. monocytogenes* and the effect of the proline and betaine was studied in more detail. Part of this work has been submitted for publication. [Beumer *et al.*, 1993]

Growth experiments were performed in mineral medium in which the osmotic strength was varied by the addition of NaCl. Growth with or without osmoprotectants was assessed turbidimetrically during incubation at 37°C. The effect of betaine on the lag time and growth rate in the mineral medium was examined by plate count experiments at 10 and 37°C.

It was found that, while *L. monocytogenes* grew poorly in mineral medium at high NaCl concentrations, the presence of 1 mM betaine or 10 mM proline allowed growth up to 5% NaCl (Figure 2). Betaine was much more effective than proline at 1 mM, a concentration which is comparable with levels in foods. At 1 mM, the osmoprotective effect of proline was negligible small, which is in keeping with findings of Patchett *et al.* (1992). In foods, proline will occur mainly bound in proteins. However, it is conceivable that the liberation of proline from proteins by proteases of other bacteria in foods could result in its availability to *Listeria* resulting in growth stimulation at low a_w.

Betaine influenced both the lag time and the growth rate of *L. monocytogenes* in the mineral medium with salt at 37°C. The levels reached in the stationary phase for growth of *L. monocytogenes* in mineral medium with osmoprotectants were comparable with the levels reached in the mineral medium in absence of salt (about 10^9 CFU/ml). At 10°C, betaine reduced the lag time and doubling time of *L. monocytogenes* cells to a lesser extent. The combined effect of low temperature and high salt concentrations is probably more difficult for the organism to overcome.

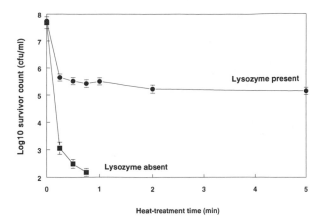

Figure 3. Effect of 10 µg/ml hen egg white lysozyme in the plating medium on the recovery of *C. botulinum* from heat-treated spores (data from Peck *et al.*, 1992a)

2. Growth from heat-treated spores non-proteolytic *C. botulinum*

Non-proteolytic strains of *C. botulinum* are able to grow and produce neurotoxin at temperatures as low as 3.3°C, and therefore pose a safety risk in refrigerated, ready-to-eat foods. Concern is particularly great with products receiving a mild heat-treatment as this treatment will reduce the vegetative cells but not the spores of the pathogen. This project will develop the use of mild preservation techniques to reduce the risk of growth and toxin production from spores of non-proteolytic *C. botulinum* in such foods. In order to design optimal techniques, initially, the interaction of the food matrix and the environmental conditions is studied in laboratory media. As for the composition of the food matrix, the role of factors affecting the ability of non-proteolytic *C. botulinum* to survive in pasteurized foods is investigated.

In previous studies it was found that lysozyme was capable of increasing the survival and outgrowth from sub-lethally heat damaged spores which were naturally permeable to it (Figure 3), or which had been made permeable to it by alkaline thioglycolate treatment (Peck *et al.*, 1992a,b, 1993). Lysozyme may be present in many foods, is relatively heat-resistant and may thus remain active after mild heating. Other lytic enzymes may be present in foods which similarly to lysozyme cause germination of heat-damaged spores and affect their survival. [Lund and Peck, 1993]

Also in this project, recovery of sub-lethally heat damaged spores of non-proteolytic *C. botulinum* type B strain 17B was improved by the presence of lysozyme. The heat-treatment involved five time periods exposure to 75°C and recovery in the absence of lysozyme and to 90°C and recovery in the presence of 10 µg lysozyme/ml. Recovery was assessed at 30°C, pH 6.8 in the presence of 0.5% NaCl. These spores are being used to study the effect of the gas atmosphere composition, incubation temperature, salt concentration, presence of lysozyme, and combinations of these on growth and toxin production in a model, vegetable and egg-containing food. Growth is assessed visually at regular time intervals. Samples are tested for *C. botulinum* toxin at the start and end of the experiment.

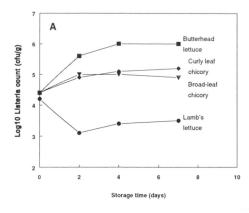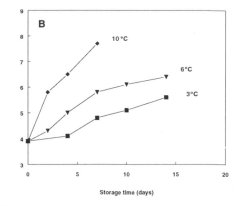

Figure 4. Growth of *Listeria monocytogenes* strain Scott A as assessed A) on different types of leafy vegetables at 10°C and B) on broad leaf chicory at three different temperatures.

B. INTERACTIONS BETWEEN PATHOGENS AND EPIPHYTIC FLORA

Interactions of psychrotrophic food pathogens with microorganisms of the normal epiphytic flora of fresh or minimally processed vegetables is studied in order to evaluate the potential of the pathogens to grow to concern levels under conditions of mild preservation, especially under modified atmosphere, despite the possible suppressive action of this natural microflora.

1. Growth of *L. monocytogenes* on leafy vegetables

The ability of foodborne pathogens, especially *L. monocytogenes*, to proliferate on ready-to-eat vegetable preparations is studied in dependence of biological factors and storage conditions. Biological factors considered are: species of salad, age of the salad leaves, amount of epiphytic microflora on the leaves, and strain of *L. monocytogenes*. The vegetables used in the experiments were checked for the occurrence of natural infections of *L. monocytogenes*. A preliminary account of the work has been given elsewhere. [Carlin and Nguyen-the, 1993]

L. monocytogenes strain Scott A and two strains of *L. monocytogenes* isolated from infected coleslaw were used in this work. [Nguyen-the and Lund, 1991] All strains tested grew on chicory leaves at 10°C, although strain Scott A grew best. Growth of the Scott A strain on broad leaf chicory, curly leaf chicory, butterhead lettuce and leaves of lamb lettuce is shown in Figure 4a. The highest growth rate was obtained on butter head lettuce and the lowest on lamb lettuce. Growth was comparable on the two chicory salads. Growth of aerobic mesophyllic microflora and spoilage development were not significantly different on leaves of the four different salads (data not shown).

On broad leaf chicory, *L. monocytogenes* strain Scott A was found to grow at 2, 6 and 10°C (Figure 4b), although it grew at a lower rate than the epiphytic microflora at the lower temperatures. Strain Scott A grew faster on young, yellow leaves than on older green leaves, with higher spoilage after 7 days at 10°C for young leaves than for old leaves. A positive correlation existed between the extent of spoilage of the leaf and the number of listeria as well as of the number of epiphytes after storage (data not shown).

Disinfection of leaves of broad leaf chicory with H_2O_2 (10%) reduced the number of epiphytic bacteria by 1 to 2 log units as compared to leaves rinsed only with water and did not cause a significant additional spoilage after storage at 10°C for 7 days. *L. monocytogenes* inoculated on disinfected leaves grew faster and reached higher counts after 7 days at 10°C than on leaves rinsed with water only (data not shown). This finding could indicate a competition between the epiphytic microflora and the pathogen. The various experiments are being duplicated to take into account the variability of the plant material.

In practice, the results obtained show that the leaves which are the more likely to carry a high load of *L. monocytogenes* at the end of storage are leaves which are more susceptible to decay and are therefore less likely to be consumed after a long period of storage. In addition, the level of *L. monocytogenes* naturally contaminating vegetables has always been found to be very low (below 50 cfu/g) and this presumably reduces the maximum level the pathogen could reach during storage. The fact that disinfection enables an infection of *L. monocytogenes* to grow faster to higher numbers than without, has important implications for the processing of ready-to-use salads.

2. Survival of pathogens on vegetable products under MAP conditions

Modified atmosphere packaging of produce prolongs the shelf life of respiring products through reduction of physiological and microbiological deterioration processes. In general, aerobic pathogens do not survive the low-oxygen conditions in MAP systems, although some pathogens, such as *L. monocytogenes*, survive and even may proliferate to concern levels. [Brackett, 1987]. Little is known about the impact of the natural epiphytic microflora on the survival of pathogens under MAP conditions. To obtain more insight in this aspect, the survival of *B. cereus* and *L. monocytogenes* was determined here on two vegetable products which differ strongly in the size of the initial epiphytic microflora, namely mungo bean sprouts and chicory endive. Products were stored while packed under moderate vacuum (MVP), which is a novel type of MAP storage. [Gorris and Peppelenbos, 1992]

On both produce, *B. cereus* was found to lose viability quickly during storage at 4 and 7°C, at both ambient (air, atmospheric pressure) and MVP conditions (400 mB, reduced oxygen tension), whereas the pathogen proliferated during storage at 10°C under ambient conditions but diminished in the MVP system. [Gorris *et al.*, 1993] Apparently, the ability of this aerobic pathogen to grow under ambient conditions is not influenced by the presence of a high or a low natural epiphytic microflora.

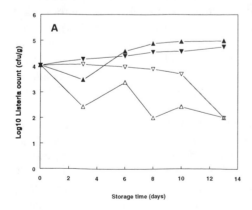 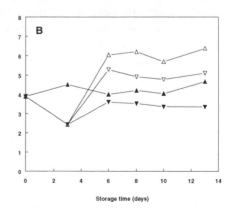

Figure 5. Survival of *Listeria monocytogenes* challenged to A) mungo bean sprouts and B) chicory endive, stored at 4°C (symbol down) and 7°C (symbol up) under ambient (air, atmospheric pressure) conditions (open symbols) and moderate vacuum (reduced oxygen tension, 400 mB pressure) conditions (closed symbols).

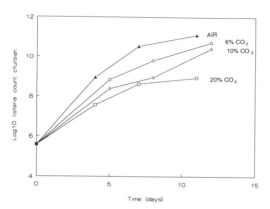

Figure 6. Viable count of *Listeria monocytogenes* on Brain Heart Infusion at 8°C under air or under different levels of carbon dioxide.

With *L. monocytogenes*, growth was observed during ambient storage at 4°, 7°C and 10°C. In the MVP system, *L. monocytogenes* expired on mungo bean sprouts (Figure 5a) but grew to concern levels on chicory endive (Figure 5b). The different behaviour of *L. monocytogenes* on the two products may well be related to the larger initial size of the epiphytic microflora on mungo bean sprouts as compared to chicory endive. By the mechanism of competitive exclusion, the growth of the pathogen may be successfully suppressed on the former. However, it may alternatively be related to the presence and size of a specific subpopulation on the produce, more specifically the lactic acid bacteria (LABs). On mungo bean sprouts, just under 10^6 cfu/g LABs occurred initially. During MVP storage, this number increased sharply to 10^8-10^9 cfu/g. With chicory endive, about 100 cfu/g of LABs were present initially, while the number increased to 10^7-10^8 cfu/g during storage. LABs produce several antimicrobial compounds (lactic and acetic acid, bacteriocins) which may hamper the growth of *L. monocytogenes*, an effect that will be more pronounced on bean sprouts considering the higher amount of LABs present there.

C. INTERACTIONS BETWEEN GAS ATMOSPHERE AND MICROORGANISMS

The packaging of food under modified atmospheres in general, has been the subject of some controversy over the last decade, mainly because there is considerable concern about the lack of knowledge about the effects of particular gas atmosphere compositions on growth and proliferation of food pathogens, like for instance *L. monocytogenes* and *A. hydrophila* under low oxygen conditions in modified or controlled atmosphere storage systems. The use of a particular gas mixture may select for a microbial association different from that of the unpreserved product. In the case of the psychrotrophic pathogens mentioned, the low oxygen conditions apparently favour their undesirable growth. However, manipulation of the gas atmosphere composition may result in proliferation of a desired subpopulation, for instance the lactic acid bacteria which are regarded as natural antimicrobial agents.

In this part of the project, the effect of different combinations and concentrations of O_2, CO_2 and N_2 on the growth of *L. monocytogenes*, *Staph. aureus*, *Salmonella* spp. and *B. cereus* are studied in laboratory media and on the model produces. Special attention is paid to the antimicrobial effect of CO_2. In Figure 6 the effect of increased levels of CO_2 on the growth of *L. monocytogenes* is given. With the *in situ* experiments, food products inoculated or not with a specific pathogen are packaged in different mixtures of gases using high barrier packaging materials and stored at a range of temperatures. At regular time intervals packs are removed and the level of relevant microbial associations determined using methods described previously. [Nychas & Arkoudelos, 1990] Results of the *in situ* experiments will be reported elsewhere.

D. BIOPRESERVATION

The objective is to suppress pathogens relevant to fresh and minimally processed ready-to-eat foods *in situ* using natural antagonistic microorganisms. With vegetable based food products, application of this "biopreservation" principle at present is restricted to the preparation of fermented vegetable preparations through natural or controlled lactic acid fermentation. In this process, the conservation effect resides mainly with organic acids produced by lactic acid bacteria. Although this type of bioconservation renders a safe and durable end-product, the sensory characteristics of the fresh product are lost. Milder biopreservation procedures need to be designed to comply with the consumer demand for fresh, natural produce. These milder techniques well be integrated with refrigeration and modified atmosphere systems to form an extra safety factor against the psychrotrophic, facultative aerobic, anaerobic or fermentative pathogens which may be able to survive or even grow under these conditions. Using lactic acid bacteria as antimicrobial agents, there may be two distinct possibilities for application, namely from the endogenous or an exogenous source.

1. Endogenous lactic acid bacteria

Agricultural produce harbour lactic acid bacteria (LABs), although usually at low numbers. Under micro-aerophyllic conditions, as in MVP and MAP systems, LABs grow faster that the aerobic, spoilage microflora. [Gorris *et al.*, 1993a] In addition to the effect of refrigeration, LABs exert an antimicrobial effect due to the production of lactic and acetic acid), and possibly also bacteriocins they produce. Bacteriocins are low-molecular weight proteins produced by certain species of genera such as *Lactobacillus, Lactococcus, Streptococcus* and *Pediococcus*. For example, the bacteriocin nisin produced by *Lactococcus lactis* has a broad-spectrum activity towards Gram(+)bacteria.

To bring the natural microflora of LABs to expression, manipulation of the gas atmosphere composition will be used. Presently, therefore the influence of gas phase concentrations of O_2 and CO_2 on LABs isolated from the endogenous microflora of the model foods is studied. This will identify gas phase compositions suitable for a controlled, selective promotion of LABs to be included in MVP or MAP as a general safety factor.

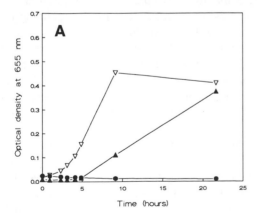 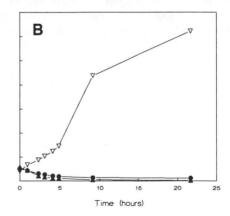

Figure 7. (A) Growth of *L. monocytogenes* at 30°C (▽) is inhibited strongly by cell-free supernatant of a nisin-producing strain of *Lactococcus lactis* (•), but only weakly by dialysed supernatant (▲); conclusion: *L. monocytogenes* is only weakly inhibited by nisin alone, but strongly by the mixture of organic acids and nisin. (B) T3, a LAB isolated from mungo bean sprouts, is very sensitive to both supernatants, thus it is inhibited strongly by the nisin, while the organic acids present in the complete supernatant do not affect it.

2. Exogenous lactic acid bacteria

By employing exogenous sources of LABs, better control of the action of the LABs may be obtained. Also, it may be possible to take benefit of the rather specific antimicrobial effect that bacteriocins from LABs have on Gram(+)-bacteria, *a.o.* the potentially hazardous pathogens in MVP or MAP such as *L. monocytogenes* and *C. botulinum*. In this respect, appropriate bacteriocin producing isolates or the respective bacteriocins in purified form may be applied to the product. In the former case, isolates which are ecologically adapted to the vegetable product are considered to be advantageous.

For this reason, LABs were isolated from fresh vegetables and from vegetables which had been stored under MVP for 7 days. [Gorris *et al.*, 1993b] Currently, these isolates are screened for suitable bacteriocin producing (so called bacteriocinogenic) strains, using a liquid medium assay system in which the effect of cell-free supernatant of LABs on growth of indicator strains is judged by measuring OD_{655} of inoculated broths in microtiter plates. Assays of dialysed (MWCO 3500) supernatants, enable discrimination between antimicrobial effects of organic acids and bacteriocins (Figure 7). In addition, bacteriocinogenic strains obtained from other research groups are included in the studies. These comprise both vegetable derived isolates, as well as isolated obtained from other foodstuffs. While the application of purified bacteriocins is subject to the approval of legislative authorities, natural bacteriocinogenic strains may be applicable freely in most countries. Therefore, the latter option is the main subject of this study.

E. GAS ATMOSPHERE COMPOSITION AND PRODUCT QUALITY

Currently, MAP of fresh and minimally processed vegetable-based food items makes use of plastic packaging films which are commercially available but which may, in many cases, not yield an optimal gas atmosphere composition around the packaged product. This is partly due to the fact that packaging films are available in a rather limited range of gas permeabilities, but also because of processing and packaging companies tend to limit the range of films they employ because of practicle reasons. Current practice is to package products under a specific gasmixture in a high barrier film or under air in a low barrier (semi-permeable) type of film. Because the packaged products continue to respire actively, an equilibrium MA develops which is mainly determined by the respiration rate of the product and the gas permeability properties of the packaging film. Depending on product tolerances, the ideal equilibrium O_2 and CO_2 levels would be in the 1-3% and 3-10% range, respectively. [Gorris and Peppelenbos, 1992] However, in many commercial MAP systems extremely low O_2 levels (<1-2%) and high CO_2 levels (>20-30%) have been found. [O'Beirne, 1990a,b] In addition to the potential safety problem of low-oxygen tolerant psychrotrophs, growth and toxin production by *C. botulinum* becomes a serious hazard in these cases where O_2 is almost completely expired [Sugiyama and Yang, 1975]

The objective thus is to assess equilibrium gas atmosphere compositions, to be applied in packagings or coatings, which are optimal for maintaining high product quality (including low levels of spoilage) and safety, using the selected model vegetable products in these studies. Based on this knowledge, suitable combinations of initial gas atmosphere compositions and packaging films or coatings may be identified which provide these equilibrium compositions under practical conditions.

The work comprises studies of effects of different gas atmospheres on the consumer perceived quality of fresh or minimally processed (cut/sliced) vegetables as well as investigation of the effect of the initial product quality (raw materials and raw material

Table 2
Oxygen permeability and water vapour transmission rates f edible and non edible films[*]

Film	Temperature (°C)	Thickness (μm)	Oxygen permeability (g.mm/m^2.day.mmHg)
Polyethylene (low density)	25	25	5.79
Polyethylene (high density)	25	25	1.45
Cellophane	25	-	0.012
Starch	24	790	4.86
Gliadins + glycerol	23	10	1.03
MC/HPMC + beeswax	25	51	0.24
Gluten + glycerol	23	11	0.18
			Water vapour transmission rate
Polyethylene (low density)	37.7	25	0.01
Aluminium foil	37.7	25	0.001
Starch + cellulose acetate	37.7	1190	29.3
Pectin	25	36	8.2
Gluten + glycerol	30	50	1.05
Gluten + monoglyceride	30	110	0.024
Gluten + beeswax	30	90	0.005

[*] Adapted from Gontard *et al.*, 1992b

preparation) on microbial growth (aerobic mesophiles, lactic acid bacteria, yeasts and moulds), sensory quality (odour, appearance, taste) and nutrient (ascorbic acid) retention.

In addition, accurate measurements of low O_2 levels (0-2%) in MAP packaged foods are constructed which are indispensable requisites for testing the suitability of packaging films under practicle conditions. Work focuses on gas sampling techniques (extraction method, volume for analysis) limits of detection, accuracy and precision. It includes effects of interference by other gases (argon, high levels of CO_2), processing and packaging film properties. Also, the feasibility of using electronic probes to measure dissolved O_2 at the surface layers of packaged foods is investigated.

F. DEVELOPMENT OF EDIBLE, MODIFIED ATMOSPHERE COATINGS

The use of edible protective films and coatings, such as so-called Edible Protective Superficial Layers (EPSL), has intrigued packaging and food scientists for a long time because of their relatively simple mode of employ. EPSLs are applied directly to the surface of a food product and act as an additional hurdle for protection of overall food quality and stability against microbial spoilage and loss of intrinsic product quality. The functional characteristics required depend on the primary mode of deterioration of the food product. Since there is an increased awareness of the deleterious effect on the environment of plastic packaging materials used, for instance, in MAP systems, the idea has emerged to use such coatings for this purpose instead. Materials, technologies and applications of EPSLs with moderate moisture foods have been extensively been studied in the recent past. [Guilbert, 1986; Gontard *et al.*, 1992a,b,1993] In this project EPSLs are developed which are fully biodegradable and create a modified atmosphere around vegetable-based model food products (*a.o.* carrot). Active edible coatings in which natural antimicrobial or anti-oxidant additives are included are designed in order to allow the use of strongly reduced amounts of additives because the additives are fixed at the product surface where the prime protection is required.

Hydrophillic EPSLs (0,05 mm thickness) with good resistance to breakage and abrasion appropriate flexibility have been realised on the basis of wheat gluten and pectin. The dependence of the film development on pH and on polymer and ethanol concentration has been established. For instance, a resistant film can be obtained using a film-forming solution (45% ethanol based) with a high gluten content (12.5%) and pH 5. More hydrophobic, composite and multilayered films were obtained using film-forming solutions containing pectin and beeswax.As can be seen in Table 2, the permeabilities of EPSLs for oxygen and water vapour can be manipulated to fall within the ranges of non-degradable packaging materials. [Gontard *et al.*, 1992b]. Additives such as antimicrobial, antioxidants, nutritional factors and flavours can be incorporated in edible films to obtain localized functional effects, usually at the product surface, at rather low additive concentrations relatively to the total weight of the food. Sorbic acid is an antimicrobial additive which may increase the microbial stability of a coated food product. Using pectin films, the diffusion coefficients of K-sorbate and sorbic acid at 18°C were found to be 3.2×10^{-6} and 6.8×10^{-5} cm^2/min, respectively. Presently, experiments are conducted to evaluate the effects of additive concentration, temperature and pH on sorbic acid diffusion through pectin films as well as through films prepared from other polymers. In addition, the microbial and physical stability of successful EPSLs is studied intensively.

G. DATA MODELLING

Given the increased popularity of ready-to-eat vegetable products, an inventory of the health hazards is required. Traditionally, the approach would be to inoculate different foods with various pathogens and to assess their potential to grow to concern levels under conditions that are representative of processing, distribution and retailing. This type of experimentation, though relatively easy to perform, limits the usefulness of the data obtained to the exact conditions of the food in the trial and allows no interpretation of tolerances for controlling factors. In practice, conditions are not fixed and tolerances are decisive tools in managing and controlling the production process. An alternative approach would be to assess the growth responses of relevant pathogens systematically using carefully designed experiments which represent, as isolated parameters, the exact physical or (bio)chemical conditions of a product plus the preservation technique utilised and to integrate the responses obtained in a mathematical computer model. Also with regard to the product quality during the course of processing, distribution and retailing, predictive mathematical models would be very valuable. A great advantage of such models would also be that they do not apply for a single product only. Providing the product data are related well to physical and (bio)chemical factors, predictions may be obtained for other products for which these factors fall within the range of the model.

The research conducted in the various tasks discussed previously, is designed to render data suitable for inclusion in mathematical computer models on three levels: microbiology (safety), product quality and mild preservation technique. On the level of the microbiology, the project may link up with ongoing activities of the predictive modelling of microbial growth in foods supported by the European Community and several national governments. [Roberts, 1991,1992] With regard do product quality, a systematic approach has been proposed as a bases for the construction of a predictive model of vegetable physiology with regard to consumer perceived quality. The model will enable one to derive the permeability characteristics of a suitable packaging material for a product, considering its type and its degree of maturation and processing. It will be developed using model products from several classes of commodities, classified according to key parameters such as respiratory activity and tolerances to chilling or gasses. [Peppelenbos, 1993] This approach and the resulting model have been adopted in the present study as well. To facilitate the choice of packaging conditions, a third model currently is being compiled within a new EC-AIR project that interlinks data on product quality with optimal packaging material characteristics. [Evelo, 1993] The effects of various combinations of physicochemical factors are determined under a number of practical MAP storage conditions and do not rely on, for instance in the case of the packaging materials, specifications provided by manufacturers. A unique situation has arisen during the past several years that several research disciplines have chosen a systematic approach to tackle their specific questions, as a result of which the individual findings may be integrated into an even more informative computer model.

Once a model is constructed, validation experiments using *in situ* tests are used to fine tune the model. Although a model should best be used within the boundaries of the physicochemical parameters (pH, a_w, temperature, gas composition, etc.) investigated empirically, a validated model may be used to predict the growth of a pathogen under a variety and under varying conditions. Validation of the models will be by use of the (semi-)full scale processing and packaging facilities of the participating SMEs. The companies have packaging equipments in place for shrink wrapping, stretch wrapping and pillow packing with gas flushing and can provide bacteriological counts and growth rates.

IV. CONCLUSIONS

With the increasing popularity of ready-to-eat, fresh and processed foods which are preserved only by relatively mild techniques, a new habitat for microbial growth has emerged. In order to control the growth of food poisoning and spoilage microorganisms in these habitats, while keeping loss of product quality to a minimum, sound information on important factors affecting the survival and growth of such microorganisms under the mild preservation conditions is required. This project aims to improve both the safety and quality of such fresh and processed vegetable foods that rely on mild preservation techniques combined with refrigeration for their shelf life. Existing mild preservation techniques are or newly designed to establish adequate barriers to the growth of pathogens which are tolerant to low temperature and low oxygen conditions. Biopreservation and (active) modified atmosphere coatings are among the novel techniques which may render the current techniques more environmentally friendly and effectively inhibit pathogenic bacteria while reducing post-harvest losses due to microbial spoilage and physiological degradation of product quality. The project has only recently been started, but the multidisciplinary approach and the integrated research effort of participants in the project as well as of many other interested researchers warrants considerable expectations.

REFERENCES

Bae, H.Y. and Miller, J., Identification of two proline transport systems in *Staphylococcus aureus* and their possible roles in osmoregulation". *Appl. Environ. Microbiol.*, 58, 471-475, 1992.

Berrang M.E., Brackett, R.E. and Beuchat, L.R., Growth of *Listeria monocytogenes* on fresh vegetables stored under controlled atmosphere. *J. Food Prot.*, 52, 702-705, 1989a.

Berrang M.E., Brackett, R.E. and Beuchat, L.R., Growth of *Aeromonas hydrophila* on fresh vegetables stored under a controlled atmosphere. *Appl. Environ. Microbiol.*, 55, 2167-2171, 1989b.

Beumer, R.R., te Giffel, M.C., Cox, L.J., Rombouts, F.M. and Abee, T., Effect of exogenous proline, betaine and carnitine on the growth of *Listeria monocytogenes* in a minimal medium. *Submitted for publication.*

Brackett, R.E., Microbiological consequences of minimally processed fruits and vegetables. *J. Food Qual.*, 10, 195-206, 1987.

Carlin, F. and Nguyen-the, C., Factors influencing the growth of *Listeria monocytogenes* on fresh green salads. IUMS-ICFMH Symposium "Food Micro '93", 31 August-3 September, Bingen, Germany. Poster P2-02, 1993.

Csonka, L.N., Physiological and genetic responses of bacteria to osmotic stress. *Microbiol. Rev.*, 53, 121-147, 1989.

Evelo, R.G., Personal communication concerning "Modified atmosphere systems in flexible temperature regimes (MASTER) - AIR2-CT92-1326", 1993.

Gontard, N., Guilbert, S. and Cuq, J.L., Edible wheat gluten films: influence of main process variables on films properties using response surface methodology. *J. Food Sci.*, 57, 190-199, 1992a.

Gontard, N., Guilbert, S. and Cuq, J.L., Edible films and coatings from natural polymers. In: New Technologies for the Food and Drink Industries. Campden FDRA, 1992b.

Gontard, N., Guilbert, S. and Cuq, J.L., Water and glycerol as plasticizers affect mechanical and water vapor varrier properties of an edible wheat gluten film. *J. Food Sci.*, 58, 206-211, 1993.

Gorris, L.G.M. and Peppelenbos, H.W., Modified atmosphere and vacuum packaging to extend the shelf-life of respiring food products. *HortTechnology*, 2, 303-309, 1992.

Gorris, L.G.M., Smid, E.J., de Witte, Y. and Jongen, W.M.F., Spoilage and safety of respiring produce packaged under modified atmospheres. In: COST94 workshop *The Post-harvest Treatment of Fruits and Vegetables*, Pala, M, Ed., 1-2 October 1992, Istanbul, Turkey, 1993a. *In press.*

Gorris, L.G.M., Bennik, M.H. and Abee, T., The contribution of biopreservation to the quality and safety of respiring produce kept under moderate vacuum with refrigeration. IUMS-ICFMH Symposium *Food Micro '93*, 31 August-3 September, Bingen, Germany. Poster P5-03, 1993b.

Guilbert, S., Technology and application of edible protective film. In: *Food Packaging and Preservation*, Malthlouthi, M., Ed., Elsevier Applied Science Publishers, London, 371-394, 1986.

Hotchkiss J.H. and Banco M.J., Influence of new packaging technologies on the growth of microorganisms in produce. *J. Food Prot.*, 55, 815-820, 1992.

Jewell, J.B. and Kashket, E.R., Osmotically regulated transport of proline by *Lactobacillus acidophilus* IFO 3532", *Appl. Environ. Microbiol.*, 57, 2829-2833, 1991.

Kallander, K.D., Hitchins, A.D., Lancette, G.A., Schmieg, J.A., Garcia, G.R., Solomon, H.M. and Sofos, J.N., Fate of *Listeria monocytogenes* in shredded cabbage stored at 5°C and 25°C under a modified atmosphere. *J. Food Prot.*, 54, 302-304, 1991.

Killham, K. and Firestone, M.K., Salt stress control in intracellular solutes in *Streptomyces* indigenous to saline soils". *Appl. Environ. Microbiol.*, 47, 301-306, 1984.

Lund, B.M. and Peck, M.W., Heat resistance and recovery of spores of non-proteolytic *Clostridium botulinum* in relation to refrigerated, processed foods with an extended shelf life. *Ssubmitted for publication.*

Nguyen-the, C. and Lund, B.M., The lethal effect of carrot on *Listeria* species. *J. Appl. Bacteriol.*, 70, 479-488, 1991.

Nychas G.J. and Arkoudelos, J.S., Microbiological and physicochemical changes in minced meats under carbon dioxide, nitrogen or air at 3°C. *Int. J. Food Sci. Technol.*, 25, 389-398, 1990.

O'Beirne, D., Chilling combined with modified atmosphere packaging, In: *Chilled Foods: The Revolution in Freshness, Volume 3*, Zeuthen, P. *et al.*, Eds., Elsevier Appl. Sci. Publ. pp 191-203, 1990a.

O'Beirne, D., Modified atmosphere packaging of selected prepared fruit and vegetables, In: *Chilled Foods: The Revolution in Freshness, Volume 3*, Zeuthen, P. *et al.*, Eds., Elsevier Appl. Sci. Publ. pp 230-233, 1990b.

Patchett, R.A., Kelly, A.F. and Kroll,, R.G., Effect of sodium chloride on the intracellular solute pools of *Listeria monocytogenes*. *Appl. Environ. Microbiol.*, 58, 3959-3963, 1992.

Peck, M.W., Fairbairn, D.A. and Lund, B.M., The effect of recovery medium on the estimated heat-inactivation of spores of non-proteolytic *Clostridium botulinum*. *Lett. Appl. Microbiol.*, 15, 146-151, 1992a.

Peck, M.W., Fairbairn, D.A. and Lund, B.M., Factors affecting growth from heat-treated spores of non-proteolytic *Clostridium botulinum*. *Lett. Appl. Microbiol.*, 15, 152-155, 1992b.

Peck, M.W., Fairbairn, D.A. and Lund, B.M., Heat-resistance of spores of non-proteolytic *Clostridium botulinum* estimated on medium containing lysozyme. *Lett. Appl. Microbiol.*, 16, 126-131, 1993.

Peppelenbos, H.W., A systematic approach to research on modified atmosphere packaging of produce. In: COST94 workshop *The Post-harvest Treatment of Fruits and Vegetables*, Pala, M, Ed., 1-2 October 1992, Istanbul, Turkey, 1993. *In press.*

Roberts, T.A., Predicting the growth and survival of bacteria in foods. Flair-Flow document F-FE 37/91, 1991.

Roberts, T.A., Advancing predictive modelling through international cooperation. IFTEC, 15-18 November 1992, The Hague, The Netherlands, Abstract 122, 1992.

Sugiyama, H. and Yang, K.H.A., Growth potential of *Clostridium botulinum* in fresh mushrooms packaged in semipermeable plastic film. *Appl. Microbiol.*, 6, 964-969, 1975.

DISCUSSION

SHEWFELT: I think we need to be careful in considering edible films as a modified atmosphere package. In edible coatings you are modifying the surface of fruit or vegetable and have one gaseous equilibrium between the product and the atmosphere while in MAP there are two interfaces involved.

GORRIS: I agree with you, but we are not only coating respiring products directly with edible materials, we also pursue a second option to design packaging films from the same materials used for coating, which would be in line with a true MAP situation.

KIRBY: 1) With challenge experiment of pathogens, the history of the inoculum is important with regard to growth and survival. 2) It is difficult to draw a line of what level of pathogens is safe or not, objectively.

GORRIS: 1) I agree, but the details of the present study may be discussed more easily on a bilateral basis
2) Toxin level is (maybe) an objective measure to put a level of acceptance to, but holds only for toxigenic and not for infectious pathogens.

ZOTTOLA: Could you specify details on the computer image analysis you use to measure color and texture objectively?

GORRIS: With CJA, using color camera's rendering digital data, we look at color differences of a product in relation to consumer defined quality characteristics. For texture, e.g. with tomatoes we use a color assay which indicates mealiness, quantified by CJA. But we also use sensory data and other, mechanical, measurements.

BUSTA: Has your modeling predications taken into consideration the growth of other microorganisms (e.g. other spore formers. lactics, etc.) against the growth of non-proteolytic *C botulinum*. They may modify pH, increase or decrease, or may influence available nutrients or germinants.

GORRIS: At this stage of our project, the answer is no, but I personally appreciate the organisms you mention, and epiphytes, in general, will play an important role in determining the ability and possibility of pathogen to grow. Looking at on-going predictive modeling projects (e.g. the UK "Food Micromodel project), there is a limitation to the number of different factors you can include in a single model. This number is about 3 to 4, and a priority rating of the various determinative factors that identify the most important factors to be included in a single model to obtain a fair prediction, validated by laboratory experiments and results with foods products in practice.

BANKS: You only studied effects of avian egg-white lysozymes with regard of promoting survival of non-proteolytic *Cl. botulinum*. Lysozymes of (plant) other origin may have quite different effects.

GORRIS: I agree, but as we deal with *sous-vide* products in which hen egg-white may be used, we looked (up to now) only at this type of lysozyme.

BRUHN: How will you address consumer communications when bacteria are used as a control?

GORRIS: The idea of bioconservation is by far a new one and many research groups and industries are involved now in consumer acceptance and legislation issues. The bacteriocin nisin (from *lactococus lactis*) has been acknowledged for use in low acid product canning about 20 years ago. Recently, the FDA has approved the use of this bacteriocin with some fresh products LAB are GRASS (generally recognized as safe) organisms. This fact would enable quick legislative approval, probably as long as the LABs are not genetically modified. As for labeling a yogurt product sold by Biogard is now marketed with a clear indication on the label that two specific types of LABs are used in it. There it is a favorable feature and accepted by the consumer as such.

OSMOTIC PRECONCENTRATION: DEVELOPMENTS AND PROSPECTS

Harris N. Lazarides
Department of Agricultural Industries - Food Science & Technology
School of Agriculture, Aristotelean University of Thessaloniki
Box 255, Thessaloniki 540 06, Greece

ABSTRACT

Osmotic preconcentration is a valuable processing tool with great future in minimal processing of fruits and vegetables. It is also a potential outlet for salvaging agricultural surpluses. It can be applied either as a separate process or as a processing step in alternative processing schemes leading to a variety of end products. Despite its well recognized advantages there is still a lot of room for improvements necessary to justify its wide application in industrial dehydration. This work is an effort to present recent developments regarding proper control of process variables, outline process and modelling limitations, suggest research needs and evaluate future prospects.

I. INTRODUCTION

Osmotic preconcentration (dehydration) is the partial removal of water by direct contact of a product with a hypertonic medium, i.e. a high concentration sugar or salt solution for fruits and vegetables, respectively.

As the cost of energy and other resources increases, there is a progressively greater pressure to minimize the cost of processing, shipping, handling and storing of agricultural products. Osmotic preconcenration is an effective way to reduce overall energy requirements in dehydration processes aiming at product stabilization. It has been found that osmotic dehydration with syrup reconcentration demands 2-3 times less energy (per unit) compared to convection drying. [Lenart and Lewicki, 1988] Mild heat treatment at relatively low process temperatures (up to 50 °C) favors color and flavor retention resulting in products with superior organoleptic characteristics. [Ponting, 1973] Sugars are known to prevent loss of volatile flavor components during vacuum drying. [Wienjes, 1968] Thus, sugar uptake favors aroma retention in osmo-vacuum dehydrated fruits. Due to constant immersion in the osmotic medium there is no need to use sulphur dioxide for protection against oxidative and enzymatic discoloration. [Ponting et al., 1966; Ponting, 1973; Dixon et al., 1976]

The above advantages make osmotic preconcentration an attractive process-step in several alternative processing schemes, as shown in Figure 1.

During osmotic dehydration two major counter-current flows take place simultaneously. Under the water and solute activity gradients across the product-medium interface, water flows from the product into the osmotic medium, while osmotic solute is transferred from the medium into the product (Figure 2). A third transfer process, leaching of product solutes (sugars, acids, minerals, vitamins) into the medium, although recognized as affecting the

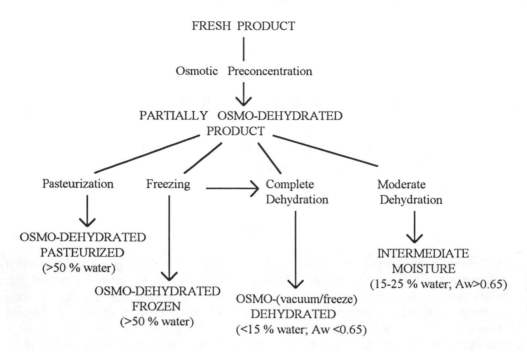

Figure 1. Possible outlets for partially osmo-dehydrated fruits and vegetables.

organoleptic and nutritional characteristics of the product, it is considered quantitatively negligible. [Dixon and Jen, 1977]

Solute uptake during osmotic dehydration modifies the composition and taste of the final product. This so called "candying" (or "salting") effect is some times desirable, as it tends to improve the taste and acceptability of the final product. In many cases, however, extensive solute uptake is undesirable, because of its negative impact on taste and the nutritional profile of the product, which can no longer be marketed as "natural". Leaching of natural acids out of osmo-dehydrated fruits also affects the taste, since it changes the natural sugar to acid ratio.

In every case, solute uptake results in the development of a concentrated solids layer under the surface of the fruit, upsetting the osmotic pressure gradient across the fruit-medium interface and decreasing the driving force for water flow. [Hawkes and Flink, 1978] Early penetration studies showed that the rate of solute penetration is directly related to the solution concentration and inversely related to the size of the sugar molecule. [Hughes et al., 1958]

Besides its negative effect on the rate of water loss during osmotic preconcentration, solute uptake "blocks" the surface layers of the product, posing an additional resistance to mass exchange and lowering the rates of complimentary (vacuum-, convection-, freeze-) dehydration. [Lenart and Grodecka, 1989] On the other hand, partial dehydration and solute uptake protect the product against structural collapse during complimentary dehydration and against structural disorganization and exudation (loss of juices and texture) upon freeze/thawing.

Solute uptake also affects product rehydration (hygroscopicity). Rehydration of osmotically dried fruit is lower than the untreated in both rate and extent. [Ponting et al., 1966; Lerici et al., 1977; Lenart, 1991] This is due to the lower rehydration of sugar in the product, compared to the natural tissue itself. It has been shown that, the longer the osmosis time, the lower the rehydration rate and extent of osmo-convection dried carrots. [Lenart, 1991] Water vapor adsorption followed a similar response.

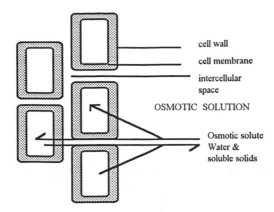

Figure 2. Mass exchange between natural tissue and osmotic medium.

Regarding chemical stability of an osmo-dehydrated product, fruits that have been dehydrated to very low water contents using sugar or syrup solutions tend to become rancid after storage for several weeks at room temperature. [Ponting et al., 1966] This is perhaps due to the greater retention of flavor oils in osmotically dehydrated fruits. Thus, it may be necessary to use a suitable fat antioxidant and/or vacuum packaging with certain osmo-dehydrated fruits for safe storage at room temperature.

The above introductory discussion underlines the importance of certain process parameters with respect to the rate and extent of dehydration, the quality characteristics, the stability and the behavior of the final product. It is also clear that there are definite disadvantages and limitations which have to be faced before we can fully exploit the important advantages and the promising potential of osmotic dehydration in food processing operations.

This work is an effort to present recent developments regarding proper control of process variables, outline process and modelling limitations, suggest research needs and evaluate future prospects of osmotic preconcentration as a tool in low-cost processing of agricultural produce yielding high-quality products. Emphasis is placed on osmotic dehydration of fruits, mostly because of literature availability and immediate industrial interest. With minor deviations, the same information is apparently applicable to vegetable processing as well. Recently there has been a growing interest in using low temperature osmotic dehydration for processing animal products. [Collignan and Raoult-Wack, 1992]

It is therefore obvious that information regarding the osmotic preconcentration process is directly or indirectly applicable to most every food item.

II. ROLE OF PROCESS VARIABLES

A large number of process variables have a significant effect on process efficiency and final product quality. The list of important variables includes: initial (raw) product characteristics (i.e. species, variety, maturity level, size and shape); application of a pretreatment (i.e. blanching, freeze/thawing); osmotic medium composition (i.e. solute species, molecular size, synergistic effect of solute combinations); medium (solution) concentration, process (medium) temperature; phase contacting (i.e. process setup, agitation, solution/product mass ratio, rinsing); process duration.

Process efficiency is evaluated in terms of rate and extent of water removal, aiming at shortening process duration to increase productivity and decrease processing cost. Together with this quantitative aspect of process efficiency, there is a qualitative one referring to final

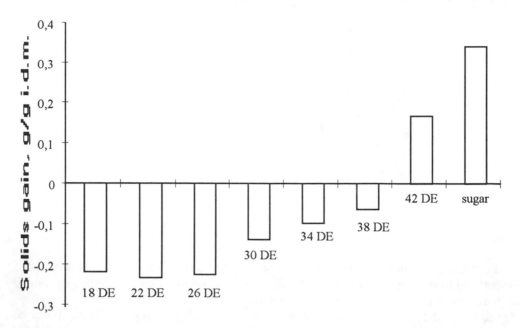

Figure 3. Effect of molecular size of osmotic solute on net solids gain during osmotic preconcentration of apple slices in corn syrup solids of different dextrose equivalents (D.E) and sugar.

product quality. Solute uptake and leaching of valuable product constituents lead to substantial modification of original product composition with a negative impact on the nutritional profile of the final product. With few exceptions (i.e. desirable "candying" or "salting"), extensive solute uptake is undesirable; thus, process efficiency can be judged on the basis of maximal (i.e. fast and extensive) water removal with minimal solids uptake. Organoleptic properties (i.e. taste, flavor, color and texture), rehydration characteristics and final product stability add to the list of quality attributes which are also affected by the process. These attributes are readily assessed by the consumer and play a crucial role in final product acceptance.

In order to improve final product quality and achieve superior process efficiencies it is absolutely necessary to know the role of each process variable and understand the mechanisms underlying mass transport phenomena throughout the process. It is also essential to assess process limitations in order to reach a realistic evaluation of the potential and the future of the process in industrial applications. The following is a condensed description of the impact of important product/process variables on crucial product/process characteristics.

A. PRODUCT CHARACTERISTICS
1. Species, variety, maturity level

Not only different *species*, but different *varieties* of the same species, even different *maturity levels* of the same variety have been found to give substantially different response to osmotic dehydration.

Species, variety and maturity level all have a significant effect on natural tissue structure in terms of cell membrane structure, protopectin to soluble pectin ratio, amount of insoluble solids, intercellular spaces, tissue compactness, entrapped air, etc. In turn, these structural differences substantially affect diffusional mass exchange between product and osmotic medium.

Hartal (1967) showed that under identical process conditions different potato varieties give substantially different (by ca. 25 %) weight reduction (water loss).

2. Size and shape

Size and shape also play a significant role in mass exchange, since they result in different specific surface areas or surface to thickness ratios.

According to Lerici et al. (1985), up to a certain A/L (total surface/half thickness) ratio, higher specific surface samples (i.e. rings) gave higher water loss (WL) and sugar gain (SG) values compared to lower specific surface shapes (i.e. slice and stick). Past this A/L limit, however, higher specific surface samples (i.e. cubes) favored sugar gain at the expense of lower water loss resulting in lower weight reduction.

The lowest water loss associated with the highest A/L ratio was explained as the result of reduced water diffusivity due to high sugar uptake.

B. PRETREATMENT

Product pretreatments and process conditions affecting the integrity of natural tissue have a severe effect on water loss and solids gain. Disruption of structural barriers improves water and solute diffusivities within the product, resulting in faster equilibration in favor of higher solute uptake. [Karel, 1975]

Blanching, freeze/thawing, sulfiding, acidification and high process temperatures all favor solids uptake yielding lower water loss/solids gain (WL/SG) ratios. [Ponting, 1973; Lerici et al., 1988; Biswal and Le Maguer, 1989]

Structural changes seem to result in decreased tortuosity of diffusion paths favoring solids transfer. [Oliveira and Silva, 1992]

C. OSMOTIC MEDIUM COMPOSITION

Several solutes, alone or in combinations, have been used in hypertonic solutions for osmotic dehydration. Sugar and salt have been extensively used for fruits and vegetables, respectively. Other solutes include: sugar/acid and sugar/salt mixtures, corn syrup solids of various degrees of polymerization, glucose, fructose, glucose/fructose and glucose/poly-saccharide mixtures, high fructose syrups, lactose, glycerine, ethanol/salt mixtures,e.t.c.. Besides hypertonic solutions, sugar, salt, sugar/salt and salt/starch mixtures in dry form have also been employed.

Based on effectiveness, convenience and flavor, salt and sugar solutions proved to be the best choises. A comparison of various osmotic solutions at a constant solids concentration showed that mixed sucrose/salt solutions give a greater decrease in product water activity compared to pure sucrose solutions, although water transport rates were similar. [Lenart and Flink, 1984a] This was blamed on extensive salt uptake.

In fact, spatial distribution analysis by the same workers revealed large differences between osmosis distribution curves for dehydration taking place in sucrose or salt solutions. [Lenart and Flink, 1984b] While sucrose accumulated in a thin sub-surface layer resulting in surface tissue compacting (an extra mass transport barrier), salt was found to penetrate the osmosed tissue to a much greater depth. The presence of salt in the osmotic solution can hinder the formation of the compacted surface layer, allowing higher rates of water loss and solids gain. Finally, increasing salt concentration in sugar/salt solutions of constant weight concentration leads to lower water activity solutions with respectively increased driving (osmotic) force.

Besides fruits and vegetables, sugar/salt solutions have also been used successfully for partial dehydration of animal products. Collignan and Raoult-Wack (1992) used concentrated sucrose/salt solutions to partially dewater meat and fish at low temperature (10 °C). Their results indicated that the presence of sugar can enhance water loss and hinder salt uptake. This is of particular importance in meat and fish processing, since it leads to shorter processing times and better control of salt uptake.

Extensive solids uptake is a major drawback against using sucrose, salt or mixed sucrose/salt solutions due to its above mentioned negative impact on both product quality (nutritional and organoleptic) and on the rate of water removal.

Acidification of concentrated sucrose syrups has a favorable effect on water loss. Addition of organic acids (acetic, citric, lactic or hydrochloric) at concentrations up to 3.5 % increased the rate of osmotic dehydration in some tropical fruits (i.e. papaya) but caused no change in others. [Moy et al., 1978] Pectin hydrolysis and depolymerization is probably responsible for the improved dehydration efficiency of acidified solutions, especially at higher process temperatures.

Finally, there is good evidence that the molecular size of the osmotic solute has a significant effect on water loss to solids gain ratio. The smaller the solute, the larger the depth and the extent of solute penetration. [Hawkes and Flink, 1978; Bolin et al., 1983; Lerici et al., 1985; Lenart and Lewicki, 1987 and 1989; Lenart, 1992] In our laboratory we found a strong linear relationship between solute size and solids uptake. Large Dextrose Equivalent (D.E.) corn syrup solids favored sugar uptake resulting in lower water loss to sugar gain ratios and vise versa. In fact, lower DE (large size) corn syrup solids gave negative solid gain values, indicating that solute uptake was inferior to leaching (loss) of natural tissue solids (Figure 3).

D. MEDIUM CONCENTRATION

Increased solute concentrations result in increased WL and SG. The observed increase in WL, however, is much higher than the increase in SG. As a result, increased concentrations lead to increased weight reductions. [Hawkes and Flink, 1978; Conway et al., 1983; Lenart, 1992]

A totally different response can be observed during the early stages of osmosis. Upon osmotic dehydration of apples in a higher concentration sugar solution (65 vs. 45 °Brix) for 3 hours our team observed some benefit in terms of faster water loss (ca. 30 % increase); at the same time, however, there was a severe loss in terms of a much greater uptake of sugar solids (ca. 80 % increase). The net result was that *short term osmosis under increased concentrations favored solute uptake* resulting in lower WL/SG ratios . Figure 4 presents the effect of temperature and concentration on the *"effective mass transfer coefficient"* (K) for solute uptake. As shown in this figure, there is a substantial increase in solute mass tranfer coefficient for increased concentrations.

Ponting et al. (1966) noticed that at high sugar concentrations (above 65 %) additional increase in concentration did not promote further weight loss. Similar response to concentration increases was observed by other workers as well. [Contreras and Smyrl, 1981] There was a difference regarding the concentration cut point; that is, the point above which an increase in concentration was not followed by a significant increase in weight loss. This difference can be explained on the basis of differences in experimental setup among the above workers. Nevertheless the fact is that, under certain experimental conditions *there is a concentration maximum above which water loss can not be improved through increased concentration.*

E. PROCESS TEMPERATURE

It is well recognized that diffusion is a temperature dependent phenomenon. Higher process temperatures seem to promote faster water loss through swelling and plasticizing of cell membranes, faster water diffusion within the product, and better mass (water) transfer characteristics on the product surface due to lower viscosity of the osmotic medium.

At the same time, solids diffusion within the product is also promoted by higher temperatures; only at different rates, mainly dictated by the size and concentration (Figure 4) of the osmotic solute. During osmotic dehydration of apples at process

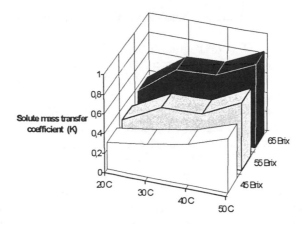

Figure 4. Effect of temperature (20 °C - 50 °C) and concentration (45 °Brix - 65 °Brix) on "effective mass transfer coefficient", K ($h^{-1/2}$) for solute uptake during the early stages (3 hours) of osmotic preconcentration of apple slices in sugar solutions .

temperatures between 30 and 50 °C we found substantially higher sugar gains (up to ca. 55 %) compared to room temperature conditions. The higher uptake values of treatments above 20 °C were most likely due to the membrane swelling/plasticizing effect, which improved the cell membrane permeability to sugar molecules. It is interesting to mention that, increase in process temperature from 30 to 50 °C resulted in substantial increase of water loss (ca. 50 %) without affecting sugar gain. Lenart and Flink (1984b) showed that in the case of sucrose, temperature affects water loss without affecting sucrose penetration; which is not the case with salt. Therefore, *whenever it is desirable to succeed higher water removal and lower solids gain, it pays to use a higher process temperature* within the applicable range.

In the case of fruits and vegetables there is an upper temperature limit beyond which there is a negative impact on final product quality due to softening, browning, flavor loss, etc. This limit is specified by the sensitivity of each particular product and is conventionally placed around 50 °C. [Ponting et al., 1966]

Animal products are very sensitive to microbial decay and growth of pathogens. Therefore it is rather dangerous to use even moderate temperatures (i.e. above 10 °C) for several hours needed to accomplish partial osmotic dehydration. [Collignan and Raoult-Wack, 1992]

In every case, low process temperatures are associated with high viscosities of the osmotic media leading to mixing difficulties and problematic external mass transfer.

F. PHASE CONTACTING

Transport phenomena during osmotic dehydration largely depend on *diffusivity* differences between water and osmotic solutes. Water diffusivity is superior to diffusivities of osmotic solutes (i.e salt, sugar, etc). Therefore, limiting transfer situations can be quite different for the two counter-current fluxes; which means that, under the right process conditions, it is possible to succeed extensive WL with only marginal solute uptake. Slow dehydration conditions favor parallel approach to equilibrium for both water and solute, resulting in a relatively higher solute pickup. [Karel, 1975]

Agitation ensures a continuous contact of the particle surface with concentrated osmotic solution, securing a large gradient at the product/solution interface. Therefore, agitation can have a tremendous impact on WL, whenever water removal is characterized by large external mass transfer resistance. This is the case when water leaving the particle surface hits a *high viscosity*, slow moving or immobile medium and accumulates in a progressively diluted contact zone.

According to Raoult et al. (1989) agitation favors WL, especially at lower temperatures (<30 °C) - where viscosity is high - and during the early stages of osmosis. Agitation during the first hour (1/3 of time needed to reach the WL plateau) gave similar WL with permanent agitation. There was a markedly different response of WL and SG to agitation. The extent of WL increased (with agitation) and reached a different plateau following the same rate. On the other hand it was the rate (slope) of SG which decreased with agitation. For short process periods agitation had no effect on sugar gain. For longer process periods SG decreased drastically with agitation.

It seems possible that agitation has no direct impact on SG throughout the entire osmotic process, since external transfer of the osmotic solute is not limiting. Agitation induced decrease in the rate of SG for longer osmosis periods could be an indirect effect of higher WL (due to agitation) altering the solute concentration gradient inside the food particle.

Since diffusion of solutes into natural tissue is slow, most of the solute accumulates in a thin sub-surface layer. In fact, solute penetration studies showed that, during osmotic preconcentration sucrose penetrates only to a depth of ca. 2-3 mm. [Lenart and Lewicki, 1987] Thus a *quick rinse* with water can remove a substantial amount of picked up solids. Such a rinse would definitely increase the surface moisture content. This is not important, however, in processes where osmotic preconcentration is followed by a complimentary dehydration step (i.e. vacuum or freeze-drying, etc).

The *solution to product mass ratio* is obviously another important element in the contacting process. Most workers use a large ratio (at least 30:1) to avoid significant dilution of the medium and subsequent decrease of the (osmotic) driving force during the process. Some investigators used much lower solution/product ratios (i.e. 4:1) in order to monitor mass transfer by following changes in concentration of the sugar solution. [Conway et al., 1983]

G. PROCESS DURATION

In a study conducted to determine the conditions defining the equilibrium state between product and osmotic solution, it was shown that equilibrium is characterized by an equality of water activity and soluble solids concentration in the product and the solution. [Lenart and Flink, 1984a] While equilibrium was approached within 20 hours, it was found that mass transport data (except for solids gain) were not significantly changed in the period between 4 and 20 hours. For the particular experimental setup this can be considered as the *practical end-point* of the process.

In every case the practical end-point is far before equilibration and is actually specified by the observed relative changes of the mass transport data in combination with the desired final product characteristics.

In most non-equilibrium studies the osmotic process was carried out for a period of 3 to 5 hours. [Hawkes and Flink, 1978; Conway et al., 1983; Biswal et al., 1991] It is rather important to notice that the first period of time is the most significant one, since the transport phenomena are fast and they have a dramatic impact on further evolution of the osmotic process. Within the first hour of osmotic preconcentration of apple slices the rate of water loss dropped to ca. 50 % the initial rate (Figure 5). Within 3 hours the product lost 50 % of its initial moisture while it over-doubled its initial total solids picking-up sugar .

Thus, an efficient way to limit solute uptake and succeed large WL/SG ratios is *early interruption* of osmosis.

As mentioned earlier, a fast rinsing helps remove a good portion of the solute taken up by the product during osmosis.

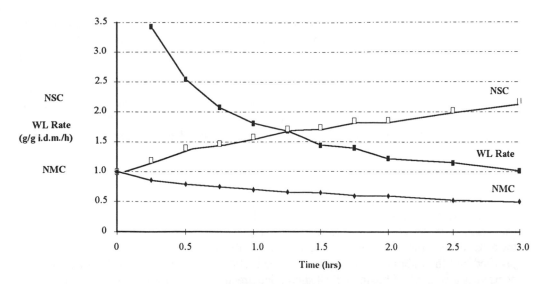

Figure 5. Evolution of moisture loss and solids gain during the early stages of osmotic preconcentration of apple slices in 55 °Brix sugar solution at 50 °C. (WL Rate = Water Loss Rate, g/g initial dry matter/h; NSC = Normalized Solids Content = s/s_o; NMC = Normalized Moisture Content = m/m_o; s, s_o & m, m_o = final and initial solids and moisture contents)

A third way to keep solute uptake at low levels is to create an extra *solute barrier*. Camirand et al. (1968) used a water permeable pectate coating impermeable to solutes to reduce solute "contamination" of the tested products (fruits, olives, meat and fish items). As a result, dehydration times increased drastically (up to 144 hrs), which is a rather heavy cost for the obtained benefit.

III. MODELLING LIMITATIONS

Inadequate control of product characteristics results in major difficulties regarding process modelling and optimization.

Raw material variability is probably the most crucial source of deviations. Variety, maturity level, even cultivation procedures drastically affect the composition and barrier properties of a specific fruit or vegetable.

The complex non-homogeneous structure of natural tissues seriously complicates any effort to study and understand the mass transport mechanisms of several interacting counter-current flows (water, osmotic solute, soluble product solids).

Toupin (1986) made a major systematic effort to consider cell membrane characteristics during osmotically induced mass transfer in biological systems. Toupin et al. (1989) and Toupin and Le Maguer (1989) used a simplified geometrical analogue of the actual cellular matrix to study the influence of various cellular and tissue properties on the dynamics of mass transport phenomena taking place in plant storage tissues. They ended up with a rather complicated model for simulation of water and solute fluxes in cellular tissues. Marcotte et al. (1991) modified the model proposed by Toupin (1986) to give a closer thermodynamical description of the forces involved in the osmotic process. Computer simulation showed good agreement between predicted and experimental values, supporting the validity of the proposed model. [Marcotte and Le Maguer, 1992]

The above studies establish that *mass transfer* in natural tissues *is not simply a diffusion phenomenon* and that *cell membrane represents the major resistance to mass transfer* in such systems.

It is recognized at the same time that the values given to the macroscopic permeability coefficients of the membranes remain the most uncertain links in any analysis. More reliable permeability estimates require experimental determination of these coefficients for plant and animal tissues.

Currently used models are based on the assumption that mass transfer is described by a simplified *unsteady state Fickian diffusion* model. *Effective diffusivities* are calculated by regression analysis of specific mass transport data. [Hawkes and Flink, 1978; Conway et al., 1983] As a result, the use of such models is largely limited to the specific experimental setup.

IV. RESEARCH NEEDS

Organoleptic and nutritional changes caused by extensive solute uptake pose a serious limitation to wider application of the osmotic dehydration process. Every research effort towards minimizing solute uptake without drastically decreasing the rate of WL is largely justified and the information it yields is rather useful. In this context, the WL/SG ratio can be used as a valuable guide to evaluate and compare alternative treatments.

In order to succeed sartisfactory control of process variables we need to better understand the mechanisms involved with simultaneous, interacting, counter - current flows within biological material. The complex structure of natural tissues and the specific problem of liquid/solid contacting found in the osmotic process seem to complicate this task.

We need to delineate probable water-solute flow interactions and calculate separate water and solute transfer coefficients both inside and outside the particle. We also need to study the interaction of cell membrane with solute transfer inside the food particle.

Process economics and final product quality largely depend on the design of continuous process equipment ("osmoreactors") for efficient contacting and thorough control of process variables. Finally, we need to couple the low energy osmotic process with efficient ways for reconcentration (recycling) of the osmotic medium.

V. PROSPECTS

Economically attractive ways of minimal processing of agricultural products can offer a satisfactory solution to seasonal surpluses of fruits and vegetables, through production of existing products at a much lower cost and/or development of new products with improved consumer appeal.

Despite the above mentioned limitations, it seems possible (through better understanding of transfer phenomena within natural tissues) to establish process conditions favoring satisfactory rates of water loss with minimal rates of solute uptake. It is, therefore, possible to minimize the adverse effects of solute uptake on final product quality (organoleptic, nutritional).

With moderate energy demands for both osmotic dehydration and for recycling of the osmotic medium, improved product quality and availability of several alternative complimentary treatments yielding a number of different products, the future of osmotic preconcentration looks exceptionally good.

REFERENCES

Biswal, R. N., Bozorgmehr, K., Tompkins, F. D. and Liu, X., Osmotic concentration of green beans prior to freezing, J. Food Sci., 56, 1008, 1991.

Biswal, R. N. and Le Maguer, M., Mass transfer in plant materials in aqueous solutions of ethanol and sodium chloride: Equilibrium data, J. Food Process Eng., 11(3), 159, 1989.

Bolin, H. R., Huxsol, C. C., Jackson, R. and Ng, K. C., Effect of osmotic agents and concentration on fruit quality, J. Food Sci., 48(1), 202, 1983.

Camirand, W. M., Forrey, R. R., Pepper,K., Boyle,F. P. and Stanley, W. L., Dehydration of membrane-coated foods by osmosis, J. Sci. Food Agric., 19, 472, 1968

Collignan, A. and Raoult-Wack, A. L., Dewatering through immersion in sugar/salt concentrated solutions at low temperature. An interesting alternative for animal foodstuffs stabilization, in *Drying '92, Part B,* Mujumdar, A. S., Ed., Elsevier Science Publ., 1992, p. 1887.

Contreras, J. E. and Smyrl, T. G., An evaluation of osmotic concentration of apple rings using corn syrup solids solutions, Can. Inst. Food Sci. Technol. J., 14(4), 310, 1981.

Conway, J., Castaigne, F., Picard, G. and Vovan, X., Mass transfer considerations in the osmotic dehydration of apples, Can. Inst. Food Sci. Technol. J., 16(1), 25, 1983.

Dixon, G. M. and Jen, J. J., Changes of sugars and acids of osmovac-dried apple slices, J. Food Sci., 42(4), 1136, 1977.

Dixon, G. M., Jen, J. J. and Paynter, V. A., Tasty apple slices result from combined osmotic-dehydration and vacuum-drying process, Food Prod. Devel., 10(7), 60, 1976.

Hartal, D., Osmotic dehydration with sodium chloride and other agents, Ph. D. Thesis, University of Illinois, Urbana, Illinois, 1967.

Hawkes, J. and Flink, J. M., Osmotic concentration of fruit slices prior to freeze dehydration, J. Food Proc. Preserv., 2, 265, 1978.

Hughes, R. E., Chichester, C. O. and Sterling, C., Penetration of maltosaccharides in processed Clingstone peaches, Food Technol., 12, 111, 1958.

Karel, M., Dehydration of Food, in *Principles of Food Science, Part II, Physical Principles of Food Preservation,* Karel, M., Fennema, O. R. and Lund, D. B. , Marcel Dekker Inc., New York, 1975, chap.10.

Lenart, A., Effect of saccharose on water sorption and rehydration of dried carrot, in *Drying '91,* Mujumdar, A. S. and Filkova, I., Eds., Elsevier Science Publ., Amsterdam, 1991, p. 489.

Lenart, A., Mathematical modelling of osmotic dehydration of apple and carrot, Pol. J. Food Nutr. Sci., Vol. 1/42, No. 1, 33, 1992.

Lenart, A. and Flink, J. M., Osmotic concentartion of potato. I. Criteria for the end-point of the osmosis process, J. Food Technol., 19(1), 45, 1984a.

Lenart, A. and Flink, J. M., Osmotic concentartion of potato. II. Spatial distribution of the osmotic effect, J. Food Technol. 19, 65, 1984b .

Lenart, A. and Grodecka, E., Influence of the kind of osmotic substance on the kinetics of convection drying of apples and carrots, Ann. Warsaw Agricult. Univ.-SGGW-AR, Food Technol. and Nutr. 18, 27, 1989.

Lenart, A. and Lewicki, P. P., Kinetics of osmotic dehydration of the plant tissue, in *Drying '87,* Mujumdar, A. S., Ed., Hemisphere Publ. Corp., New York, 1987, p 239.

Lenart, A. and Lewicki, P. P., Energy consumption during osmotic and convective drying of plant tissue, Acta Alimentaria Polonica, Vol. XIV, No. 1, 65, 1988.

Lenart, A. and Lewicki, P. P., Osmotic dehydration of apples at high temperature, in *Drying '89,* Mujumdar, A. S. and Roques, M., Eds., Hemisphere Publ. Corp., New York, 1989, p.501.

Lerici, C. R., Pinnavaia, G., Dalla Rosa, M. and Bartolucci, L., Osmotic dehydration of fruit: Influence of osmotic agents on drying behaviour and product quality, J. Food Sci., 50, 1217, 1985.

Lerici, C. R., Pepe, M. and Pinnavaia, G., La disidratazione della frutta mediante osmosi diretta .I. Risultati di esperienze effettuate in laboratorio, Industria Conserve, 52(2), 125, 1977.

Lerici, C. R., Mastrocola, D., Sensidoni, A. and Dalla Rosa, M., Osmotic concentration in food processing, in *Proc. of the International Symposium on Preconcentration and Drying of Foods*, Eindhoven, The Netherlands , Nov. 5-6, 1987, Bruin, S., Ed., Elsevier, 1988, p.123.

Marcotte, M. and Le Maguer, M., Mass transfer in cellular tissues. Part II: Computer simulations vs experimental data, J. Food Eng., 17, 177, 1992.

Marcotte, M., Toupin, C. J. and Le Maguer, M., Mass transfer in cellular tissues. Part I: The mathematical model, J. Food Eng., 13, 199, 1991.

Moy, J. H., Lau, N. B. H. and Dollar, A. M., Effects of sucrose and acids on osmovac-dehydration of tropical fruits, J. Food Proc. Pres., 2, 131, 1978.

Oliveira, F. A. R. and Silva, L. M., Freezing influences diffusion of reducing sugars in carrot cortex, J. Food Sci., 57 (4), 932, 1992.

Ponting, J. D., Osmotic dehydration of fruits - Recent modifications and applications, Process Biochem., 8, 18, 1973.

Ponting, J. D., Watters, G. G., Forrey, R. R., Jackson, R. and Stanley, W. L., Osmotic dehydration of fruits, Food Technol., 29(10), 125, 1966.

Raoult, A. L., Lafont, F., Rios, G. and Guilbert, S., Osmotic dehydration: Study of mass transfer in terms of engineering properties, in *Drying '89*, Mujumdar, A. S. and Roques, M., Eds., Hemisphere Corp., New York, 1989, p.487.

Toupin, C. J., Osmotically induced mass transfer in biological systems: The single cell and tissue behavior, Ph.D. Thesis, Department of Food Sciense, University of Alberta, Edmonton, Canada, 1986.

Toupin, C. J., Marcotte, M. and Le Maguer, M., Osmoticaly-induced mass tranfer in plant storage tissues: A mathematical model. Part I, J. Food Eng., 10, 13, 1989.

Toupin, C. J. and Le Maguer, M., Osmoticaly-induced mass transfer in plant storage tissues :A mathematical model. Part II, J. Food Eng., 10, 97, 1989

Wienjes, A. G., The influence of sugar concentration on the vapor pressure of food odor volatiles in aqueous solutions, J. Food Sci., 33, 1, 1968.

DISCUSSION

ANDREZEJ: Could you explain influence of sample's shape on mass transfer during osmotic dehydration (it means water loss and solids gain)

LAZARIDES: Above a certain specific surface (A/L ratio) solute uptake is so large that it severally interferes ("blocks") water loss resulting in decreased water reduction rates.

KNORR: Could the use of edible coatings as mass transfer barriers to reduce sugar uptake of samples.

LAZARIDES: Yes, indeed. Camirand et al (1968) used a water permeable pectate coating to reduce solute uptake during osmotic dehydration of several products including meats and fish. The problem was that together with the decrease in solute uptake they had a dramatic increase in dehydration times. This does not mean that edible coatings can not help in the future. On the contrary it is an area that needs further investigation.

BRUHN: Describe how the process takes place. Are there potentials for applications of other technologies with osmotic preconcentration to overcome some of the limitations described.

LAZARIDES: The product is placed in a hypertonic solution of sugar, salt or other solute. Under the strong osmotic gradient it loses a large portion of its moisture (i.e. 50%). If the process calls for complimentary dehydration we can probably use other technologies (i.e. freezing/thawing) to overcome the decreased dehydration rates experienced at the practical end-point of osmosis.

OSMOTIC DEHYDRATION OF FRUITS BEFORE DRYING

Lenart Andrzej

I. INTRODUCTION

The phenomenon of osmosis is utilized in the dehydration of food products. Osmotic substance when in contact with plant tissue causes the diffusion of water from the cell sap owing to the difference of osmotic pressures. In an intensive and longer duration dehydration process, the cell membrane looses the properties of semi-permeability and makes the diffusion of the osmoactive substance into the inside of the cell possible. This process can last as long as the concentration of particular components in the solution and in the cell sap is not different. A too strong penetration of osmotic substance is most frequently not favorable for subsequent treatment of the semi product for instance for drying or freezing it. That is why this process should be interrupted in due time or the raw material looses the properties of semi-permeability of cell membranes. Moreover, technological requirements determine the admissible content of osmoactive substances [1].

Osmotic dehydration can be applied for food raw materials of tissue structure such as fruit, vegetable or meat. The process consists of removing water from the food without a phase change, and is done in concentrated solution of osmoactive substances. Its driving force is the difference in the osmotic pressure of solutions on both sides of semi-permeable cell membranes. Selective properties of cell membranes make it possible for water and low-molecular cell sap components to diffuse into the surrounding solution of higher osmotic pressure. Other cell components are only to a small extent let outside cell membranes. The cell sap concentration lowers its water activity [2].

The diffusion of water and low-molecular weight substances from the food of tissue structure during the osmotic dehydration is accompanied by the counter-current diffusion of osmoactive substances. For this reason, osmotic dehydration, as opposed to convection drying, is characterized by the complex movement of water, substances dissolved in cell sap and osmoactive substances. This significantly influences the process itself and its final effect with respect to the preservation, nutrition and organoleptic properties [3].

The preliminary treatment of fruits and vegetables influences the chemical composition and physical properties of dried products obtained. By both blanching and freezing, the raw material structure is damaged, cell membranes are destroyed causing a greater shrinkage of the dried material. Sulphiting does not cause such a change on the physico-chemical properties of dried products, nevertheless it is considered undesirable due to the toxicity of sulphur compounds [4].

The process of an initial treatment before convection drying is particularly advantageous as far as the quality of the given food product is concerned. Fruits and vegetables dehydrated by osmosis become very attractive for direct use due to their chemical composition and physico-chemical properties. At the same time, the osmo-convection drying narrows the range of other applied methods of inactivating enzymes, such as sulphiting of fruits or blanching or vegetables [5].

Osmotic dehydration in solutions of saccharose and polysaccharides increases their share in the chemical composition of dried product, decreasing the content of monosaccharides and acids. Much higher retention of taste and flavour substances in osmo-convection drying as compared with those dried by convection, has been proven [6].

It has been shown that fruits dried after preliminary dehydration in solutions of sugars showed a much lower rehydration rate and a lowered hygroscopicity. Similar to rehydration and hygroscopicity, sorption properties of the products dehydrated by the osmo-convection method have been subject of a few studies [7].

Fruits and vegetables dried by the osmo-convection method are characterized as materials suitable for storage in the atmosphere of considerable humidity without any threat of caking [8].

Most published results are concerned with vacuum drying and dehydrofreezing and the research has been directed first of all, to the quality of final products. Hence, there is a need to increase the knowledge of the course and the mechanism of osmotic dehydration of fruits in the aspect of its application as an initial processing followed by convection drying [9].

It is necessary to recognize reconstitution properties of fruits preserved by osmo-convection method. There is a lack of a full estimation of influence of osmotic substances and kind of raw material on adsorption properties of dried product. Little information is available with regards to the reconstitution properties of dried fruits obtained by osmo-convection method, especially those concerned with rehydration, hygroscopicity and water binding capacity, i.e., properties related to the state of water in dehydrated food [10].

The aim of this work was to analyze the effect of initial osmotic dehydration on the course of convection drying and on the reconstitution properties of dried fruits. In this work, the desorption kinetics from the raw material during the osmotic dehydration and during the convection drying are analyzed. The research comprises evaluation of selected reconstitution properties of dry material, namely - the rehydration and adsorption of water vapour as related to the degree of initial osmotic dehydration.

II. MATERIALS AND METHODS

The investigations were carried out on apples of the McIntosh variety, on strawberry of Senga Sengana variety and on cherry of Nefris variety. Each experiment was carried out on a standardized portion of the material. The raw material was washed. In the case of apples the raw material was peeled, pitches were removed and cut into cubes 10 mm sides. In the case of cherry, stones were removed and material was cut into half.

Osmotic dehydration was carried out in saccharose solution and starch syrup. Solution of a concentration corresponding to water activity of 0.90 at 30°C were used, i.e., for saccharose - 61.5% and starch syrup - 67.5%. The process was carried out with mixing of the suspension at a ratio of material to solution 1:4 w/w. Dehydration temperature was kept constant and it was within the range of 30-90°C. Dehydration time was from 0 to 24 hours.

Dehydrated samples were blotted with filter paper and analysed. Mass of the sample and its dry matter content were determined. Dry matter content was measured according to Polish Standard: PN-63A-75901.

The measured values were expressed as:
- water content in grams per one gram of dry matter of the dehydrated sample, gH_2O/d.g.m.;
- water loss in grams per one gram of the initial dry matter of the sample, gH_2O/g.i.d.m.;
- solids gain in grams per one gram of the initial dry matter of the sample, g/g.i.d.m.

Drying was carried out in a convection drier in a single layer at a temperature of 70°C and at a constant air velocity 1.6 m/s and air humidity 60-70% at 20°C.

In the samples obtained, the water content and water activity were determined. The water content was determined according to the Polish Standard PN-63/A-75901. The water activity in samples within the range of 0.3-1.0 was determined by the modified method with the standard [11].

In dried fruits, the water content, the rehydration and sorption isotherms were determined. Increase of mass and the change of dry matter content were determined in rehydrated material.

The data were analyzed statistically at 0.05 significance level. Correlation coefficients (r) were obtained.

III. RESULTS AND DISCUSSION

A. OSMOTIC DEHYDRATION

1.   Effect of temperature

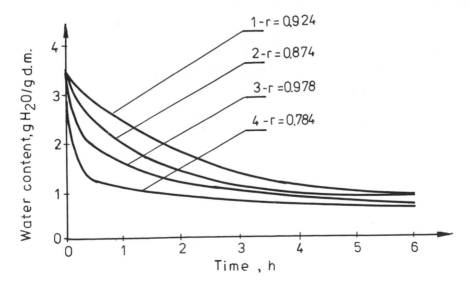

Figure 1. Effect of temperature on the course of osmotic dehydration of apple in saccharose solution. Temperature: 1-30°C, 2-50°C, 3-70°C, 4-90°C.

Changes of water content in dehydrated apple, strawberry and cherry with time duringthe six hours dehydration process were described with the help of exponential function (Fig. 1-3). Most significant changes take place during the first 30 minutes of dehydration (Fig. 1). However, after three hours of dehydration, independently of the temperature of

the process, water content of the material becomes only slightly dependent on the dehydration time. A considerable temperature effect on the process of dehydration is observed in the first stages of water removal. Such effect is also observed in the first stages of water removal from strawberry and cherry (Fig. 2-4).

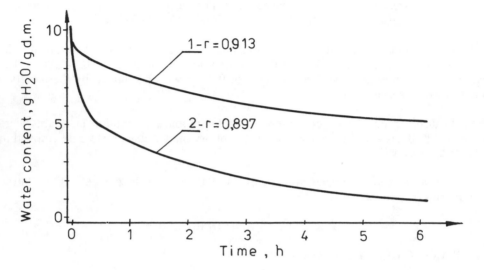

Figure 2.   Effect of temperature on the course of osmotic dehydration of strawberry in saccharose solution. Temperature 1-30°C, 2-90°C.

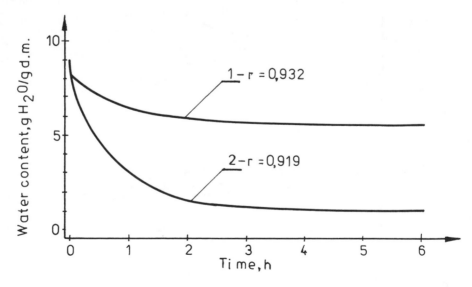

Figure 3.   Effect of temperature on the course of osmotic dehydration of strawberry in starch syrup. Temperature: 1-30°C, 2-90°C.

The analysis of curves illustrating changes of water content in apple, strawberry and cherry (Fig. 1-4), shows that the equilibrium water content of osmodehydrated materials depends on the temperature of the process. Increase of the temperature, on the other hand, significantly shortens the time necessary to obtain that water content.

During osmotic dehydration no period of a constant dewatering rate appears (Fig. 5). Such a considerable decrease of water content is the result of two opposite processes: movement of water outside the sample (water loss) and penetration of osmotic substance into the material (solids gain). Depending on the parameters of the process, the proportions between these two flows of the mass are changing and that fact influences the course of osmotic dehydration.

Changes of solids gain in dehydrated apple, strawberry and cherry with time were described with the help of exponential function and, independently of temperature of the process, high correlation coefficients were obtained (Fig. 6-8). A significant effect of temperature on the parameters of the function were observed for the temperature range of 30-90°C. Analogically, as for water content changes, most significant solids gain occurs at the beginning of the process. There was no significant gain of dry matter after 3-4 hours of dehydration.

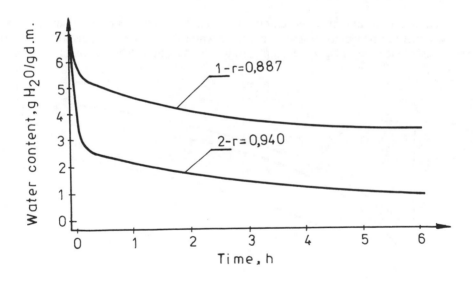

Figure 4.   Effect of temperature on the course of osmotic dehydration of cherry in saccharose solution. Temperature: 1-30°C, 2-90°C.

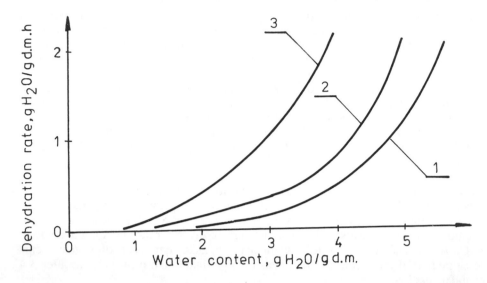

Figure 5.   Influence of temperature on the kinetics of osmotic dehydration of apple. Temperature: 1-20°C, 2-30°C, 3-40°C.

Changes of water loss in dehydrated apple, strawberry and cherry with time were described with the help of exponential function and independently of temperature of the process, high correlation coefficients were obtained (Fig. 9,10). Analogically as for solids gain changes, most significant water loss occurs at the beginning of the process.

Comparing the rate of water removal with that of osmotic substance penetration it is shown, that independently of temperature, the water flow is always greater than that of osmotic substance (Fig. 11). For example, following six hours dehydration of strawberry

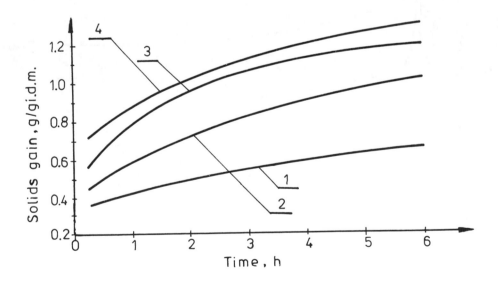

Figure 6.   Effect of temperature on the solids gain of apple. Temperature of osmotic dehydration in saccharose solution: 1-30°C, 2-50°C, 3-70°C, 4-90°C.

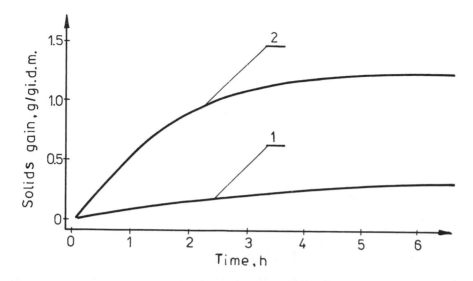

Figure 7.   Effect of temperature on the solids gain of strawberry. Temperature of osmotic dehydration in saccharose solution: 1-30°C, 2-90°C.

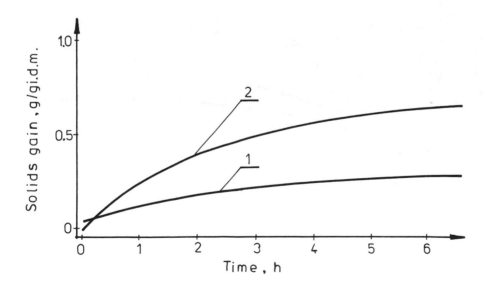

Figure 8.   Effect of temperature on the solids gain of cherry. Temperature of osmotic dehydration in saccharose solution: 1-30°C, 2-90°C.

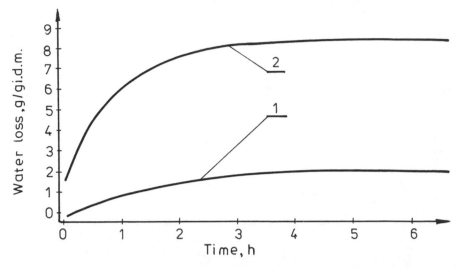

Figure 9.   Effect of temperature on the water loss of strawberry. Temperature of osmotic dehydration in saccharose solution: 1-30°C, 2-90°C.

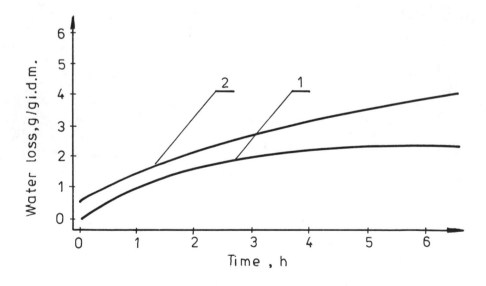

Figure 10.   Effect of temperature on the water loss of strawberry. Temperature of osmotic dehydration in starch syrup: 1-30°C, 2-90°C.

in saccharose solution, the value of the ratio of water loss to solids gain was 10 at temperature 30°C and seven at temperature 90°C.

Analogically as for apple and strawberry, rate of water loss to solids gain was 12 and 6 during osmotic dehydration of cherry, respectively.

The results concerning the effect of temperature on kinetics of osmotic dehydration of apple, strawberry and cherry supported a hypothesis about a two-directional exchange of mass during that process. The increase of temperature intensifies both the process of water removal and the penetration of osmotic substance into the tissue.

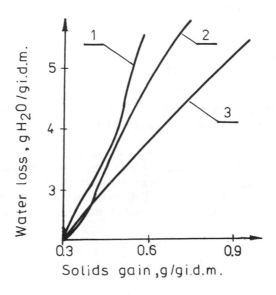

Figure 11. Relationship between water loss and solids gain of apple during osmotic dehydration. Temperature: 1-20°C, 2-30°C, 3-40°C.

However, independently of temperature, the character of exponential function describing the process does not change.

2. Effect of the type of osmotic substance

During osmotic dehydration of apple, strawberry and cherry in the range of temperature 30-90°C, solids gain and water loss of the material were observed. As a result of those changes, a decrease of water content is observed (Fig. 12, 13). From the course of curves it follows, that the greatest water content change in dehydrated material, independently of osmotic substance applied, occurs during the first 2-3 hours of the process. The influence of the kind of osmotic substance on equilibrium water content and that of the time needed to obtain that value was observed.

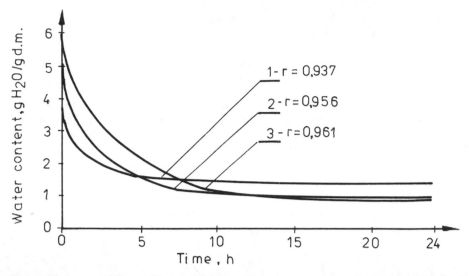

Figure 12.   Effect of osmotic substance on the course of osmotic dehydration of apple. Temperature-30°C. Osmotic substance: 1-glucose, 2-saccharose, 3-starch syrup.

Water content in dehydrated material increases as the molecular mass of the used osmotic substance increases. The results of this work indicate a significant increase of equilibrium water content as a molecular mass of osmotic substance increases. The analysis of water activity changes as a function of time allowed the hypothesis that the

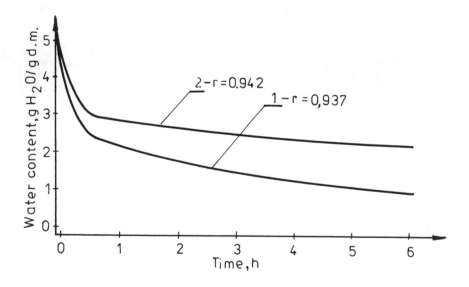

Figure 13.   Effect of osmotic substance on the course of osmotic dehydration of cherry. Temperature-90°C. Osmotic substance: 1-saccharose, 2-starch syrup.

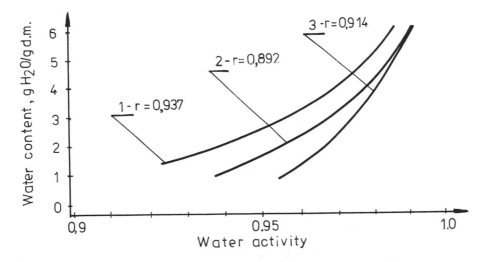

Figure 14.   Effect of osmotic substance on the water desportion isotherms from apple during its osmotic dehydration. Temperature-30°C. Osmotic substance: 1-glucose, 2-saccharose, 3-starch syrup.

water activity in the dehydrated food would depend on the kind of osmotic substance at the same water content. This hypothesis constituted a basis for the analysis of the water content in the water activity function. The statistical analysis proved a significant effect on the osmotic substance on the course of the relationships analyzed. On the other hand, no statistically significant effect of the kind of raw material has been proven. The application of sugar of lower molecular mass enables dehydration of apple, strawberry and cherry to lower water activity in the equilibrium state. At the same time the osmotic substance of lower molecular mass enables one to obtain the same water activity at higher water content. For example, to get the water activity in the osmotically dehydrated apple at the level of 0.96, they should be dehydrated to the water content: in glucose - to 2.91 g/g d.m. in saccharose - 5 to 1.98 g/g d.m. and in the starch syrup - to 0.89 g/g d.m. (Fig. 14).

In order to define the importance of the phenomenon, solids gain and water loss were calculated per unit of initial dry matter of the material (Fig. 15, 16). Analysis of variance

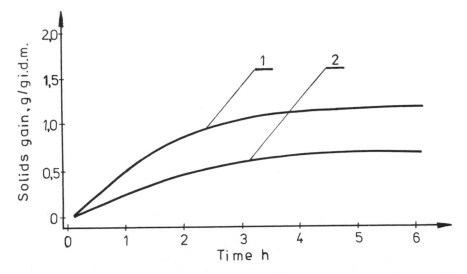

Figure 15.   Effect of osmotic substance on the solids gain of strawberry. Temperature-90°C. Osmotic substance: 1-saccharose, 2-starch syrup.

showed a statistically significant effect of the kind of osmotic substance used on change in solids gain and water loss in osmo-dehydrated apple, strawberry and cherry.

The course of the curves indicates that, independent of osmotic substance used, there are greatest changes in solids gain in the process (Fig. 15). A different quantitative as well as qualitative reason was stated at the beginning of the process and in the state of equilibrium. In strawberry, solids gain in the first hours of the process significantly decreases as the molecular mass of the applied osmotic substance increases. Similarly, as the molecular mass of the osmotic substance decreases, the gain of dry matter in state of equilibrium also increases.

Similarly, like the case of solids gain, the kind of osmotic substance affects the time needed to reach the equilibrium water loss (Fig. 16). A significant increase of the time needed to obtain the state of equilibrium is observed when starch syrup is used as osmoactive agent.

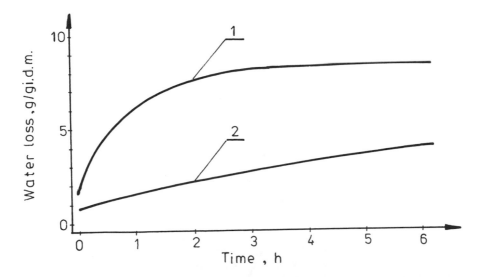

Figure 16.   Effect of osmotic substance on the water loss during osmotic dehydration of strawberry. Temperature-90°C. Osmotic substance: 1-saccharose, 2-starch syrup.

The data presented above illustrate changes of solids gain and those of water loss during osmotic dehydration, and support the hypothesis of a two-direction mass movement during this process. The mechanism of osmotic dehydration at the very beginning of the process is controlled by the penetration of osmotic substance and removal of water from the surface layer of the material in which most of the cells are damaged. In a further stage of osmodehydration, internal diffusion affects the rate of the process. The results of osmotic dehydration of apple, strawberry and cherry in starch syrup indicate a surface character of the process. During dehydration in saccharose solution, its penetration into the tissue is of greater importance.

B.   CONVECTION DRYING

The suitability of applied osmotic substances depends, first of all, on the proposed method of further preservation of fruits. In case of the convection drying saccharose was regarded in most cases as the best osmotic substance with respect to the quality of the product. Comparative investigations of the solution of saccharose and starch syrup as osmotic substances applied for preliminary treatment before drying showed a lack of significant differences in the effect of these substances on the convection drying kinetics.

In the course of convection drying of apple, strawberry and cherry, after the initial osmotic dehydration, no period of a constant drying rate appears. Irrespective of the character of the plant tissue, the osmotic dehydration involves a decrease of drying rate (Fig. 17-19). In the course of convection drying of strawberry, the greatest changes in drying rate were caused by the initial osmotic treatment in the saccharose solution (Fig. 18). In the case of drying cherry, a small difference in influence of saccharose and starch syrup on the kinetics of drying of this material was observed (Fig. 19). Differences in the kinetics of convection drying of apple, strawberry and cherry which were preliminary dehydrated in the saccharose and starch syrup result most probably from differences in the structure of the plant materials under investigation.

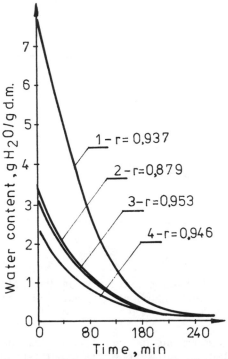

Figure 17.   Effect of osmotic dehydration on the kinetics of convection drying of apple. 1-raw apple. 1-raw apple. Temperature of osmosis: 2-20°C, 3-40°C, 4-40°C.

In the case of initial osmotic dehydration of apple, strawberry and cherry in sugar solutions of various particle mass, no essential influence of osmotic dehydration time on the kinetics of the process was observed. This dependence is particularly evident when starch syrup which contains a considerable share of dextrines is used. The submersion of apples into the starch syrup before convection drying causes a decrease of drying rate (Fig.

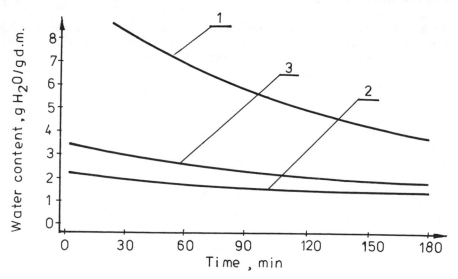

Figure 18.   Effect of osmotic substance on the kinetics of convection drying of strawberry. 1-raw strawberry. Osmotic substance: 2-saccharose, 3-starch syrup.

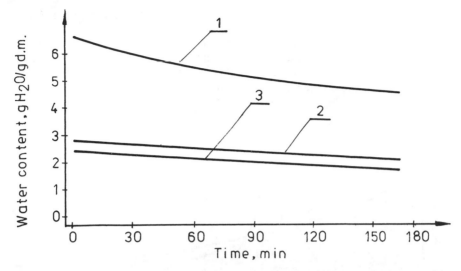

Figure 19.   Effect of osmotic substance on the kinetics of convection drying of cherry. 1-raw cherry. Osmotic substance: 2-saccharose, 3-starch syrup.

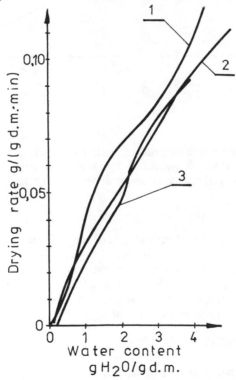

Figure 20.   Influence of the dehydration degree of apple in starch syrup on the kinetics of convection drying. 1-raw apple, 2-submerged in solution, 3-dehydrated.
20).   Drying rate of apple, which are initially osmotically dehydrated is on an approximate level irrespective of the dehydration time.

From the curves on convection drying rate, according to the time of osmotic dehydration, the hypothesis on the essential influence of the surface layer of sugar solution on the exchange of heat and mass is confirmed. This layer is formed at the beginning of the process when the curves additional resistance to the exchange of heat and mass. This hypothesis is proven by the results of experiments covering the analysis of the influence of rinsing apple on the kinetics of drying. Rinsing causes a 20-30%

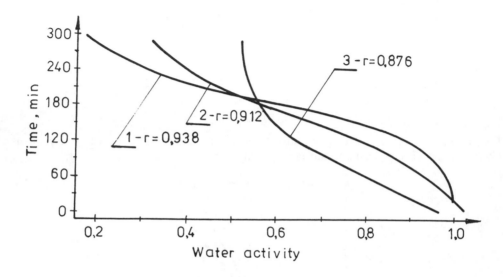

Figure 21.   Water increase of drying rate of apple which was dehydrated both in the saccharose solution and in the starch syrup.activity kinetics in apple during convection drying: 1-raw apple, 2-submerged in saccharose solution, 3-20 h osmotic dehydration.

Curves of changes of convection drying time necessary for obtaining a definite water activity depends on the degree of osmotic dehydration. The preliminary osmotic dehydration changes character of the course of curves. Along with increasing degree of osmotic dehydration, the curves become steeper (Fig. 21). As it follows from the course of curves, the preliminary osmotic dehydration causes a shortening in drying time necessary for ensuring water activity in the material higher than 0.6.

C.   RECONSTITUTION PROPERTIES

One of the most important properties of dried food is its capacity for rapid and total rehydration. Variability of that property can result from properties of dried food itself, from the applied initial processing prior to drying and from drying parameters as well.

The water content in rehydrated product changes with the rehydration time and depends on the initial osmotic dehydration (Fig. 22). The very short time contact of apple with saccharose solution prior to drying (immersion for less than few seconds - time of osmosis, 0 h) results in a two-fold decrease of water content after 0.5 hour rehydration as compared with that of apple dried without osmotic dehydration.

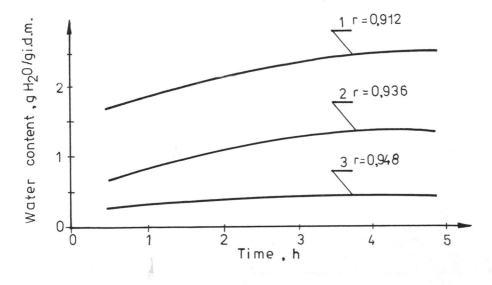

Figure 22.　　　Effect of osmotic dehydration on the rehydration properties of dried apple. 1-raw apple, 2-submerged in saccharose solution, 3-20 h osmotic dehydration.

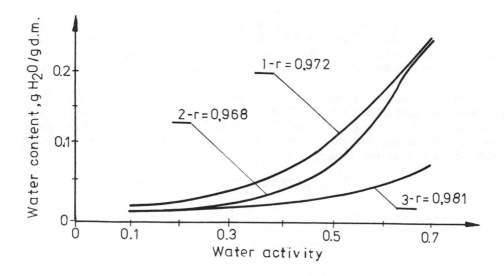

Figure 23.　　　Water sorption isotherms in the dried apple. 1-raw apple, 2-submerged in saccharose solution, 3-20 h osmotic dehydration.

The course of rehydration process of dried material which was immersed into the saccharose solution for few seconds is caused by an increased saccharose concentration on the surface of sample, and not by its structural changes due to osmotic treatment. Since the difference between raw and osmotically treated material is much larger than that between osmotically dehydrated samples, it can be inferred that the rehydration process is affected by saccharose penetrating the surface layer of the material in first place.

The effect of preliminary osmotic dehydration on adsorption properties of dried product is expressed also in the course of sorption isotherms. The course of the isotherms obtained proves (Fig. 23) that the preliminary osmotic dehydration leads to a change of the character of curves within the range of water activity of 0.3-0.7. Along with increasing the degree of osmotic dehydration, isotherms assume a shallower course.

IV. CONCLUSIONS

In the course of the osmotic dehydration of apple, strawberry and cherry in concentrated solutions of osmotic substances, the removal of water from dehydrated material is accompanied by the diffusion of osmoactive substance into dehydrated tissue. Regardless of the osmotic substance type, the quantity of removed water is always larger than quantity of the diffusing osmoactive substance.

The kind of osmotic substance applied in dehydration of apple, strawberry and cherry, affects strongly water content changes in the material. Saccharose causes significant modification of chemical composition of dehydrated fruits as compared with that of starch syrup. Increase of temperature of osmotic dehydration intensifies water removal and osmotic substance penetration into the tissue.

Changes in the chemical composition and structure of the dehydrated plant tissue effects the mechanism of convection drying of apple, strawberry and cherry which are initially osmotically dehydrated. The time of drying fruits after the preliminary osmotic treatment depends on the initial and final water content and on the drying rate.

Initial osmotic dehydration preceding convection drying of fruits results in a decreased rate and degree of rehydration, a decreased water vapour adsorption rate, and equilibrium water content of the material. The extent of changes in rehydration and water vapour adsorption properties of osmo-convection dried fruits depends on the degree of initial osmotic dehydration of the material.

REFERENCES

1. Raoult-Wack, A. L., Lenart, A., and Guilbert, S., Recent advances in dewatering through immersion in concentrated solutions, in Drying of Solids, Mujumdar, A. S., Ed., International Science Publisher, New York, 21, 1992.

2. Lewicki, P. P., and Lenart, A., Energy consumption during osmo-convection drying of fruits and vegetables, in Drying of Soils, Mujumdar, A. S., Ed., International Science Publishers, New York, 354, 1992.

3. Lenart, A., Mathematical modeling of osmotic dehydration of apple and carrot, Pol. J. Food Nutr. Sci., 1/42(1), 33, 1992.

4. Lenart, A., and Lewicki, P. P., Osmotic dehydration of apples at high temperature, in Drying 89, Mujumdar, A. S., and Roques, M., Eds., Hemisphere Publishing Corporation, New York, 501, 1990.

5. Lenart, A., Lewicki, P. P., and Mlynarczyk, G., Effect of osmotic dehydration on the physico-chemical properties of dried apples, Ann. Warsaw Agricult. Univ. -SGGW, Food Technol. and Nutr. 17, 55, 1987.

6. Lenart, A., and Lewicki, P. P., Osmotic dehydration of carrot at high temperature, in Engineering and Food, Spiess, W. E. L., and Schubert, H., Eds., Elsevier Applied Science, London, 731, 1990.

7. Lenart, A., Rehydration properties of osmo-convection dried carrot, in Strategies for Food Quality Control and Analytical Methods in Europe, Vol. 1, Lebensmittelchemische Gesellschaft, Frankfurt a.M., 169, 1991.

8. Lenart, A., Sorption properties of apples and carrot preserved by the osmo-convection method, Ann. Warsaw Agricult. Univ. -SGGW, Food Technol. and Nutr., 19, 27, 1991.

9. Lenart, A., Water desorption from apples and carrot during osmo-convection drying, Ann. Warsaw Agricult. Univ. -SGGW, Food Technol. and Nutr., 19, 35, 1991.

10. Lenart, A., Effect of saccharose on water sorption and rehydration of dried carrot, in Drying 91, Mujumdar, A. S., and Filkova, I., Eds., Elsevier Science Publishers, Amsterdam 489, 1991.

11. Lenart, A., and Flink, J. M., An improved proximity equilibration cell method for measuring water activity of food, Lebensm. -Wiss. -Technol. 16, 84, 1983.

DISCUSSION

GORRIS: If you would further want to reduce the water content after osmotic drying and convection heating does not work, could you use microwave heating instead, because with the shape of strawberries and cherries, microwave heating would be from the inside out and maybe more effective.

ANDREZEJ: At Prof. Schubert's laboratory in Karlsruhe, some experiments have been done. Problem is to control inside of a strawberry during microwave heating.

DEBARDEMAEKER: How does cherry peel affect the rate of osmotic dehydration?

ANDREZEJ: Water loss decreases, solids uptake decreases drastically, during osmotic dehydration of whole cherry (with skin). It is necessary to treat the skin (for example, using chemical means)

VACUUM OSMOTIC DEHYDRATION OF FRUITS

Fito, P.; Andres, A.; Pastor, R.; Chiralt, A.

Department of Food Technology
Universidad Politecnica de Valencia

INTRODUCTION

In the last few years, osmotic dehydration (OD) of foods, as in fruits, vegetables, meat and fish, has received increasing interest (Le Maguer, 1988; Raoult-Wack, 1991; Toupin and Le Maguer, 1989; Toupin et al, 1989; Marcotte et al,1991; Marcotte and Le Maguer, 1992 ; Fito and Pastor, 1993).

The Vacuum Osmotic Dehydration (VOD).

When pressures lower than atmospheric pressure are used, the vacuum osmotic dehydration (VOD) occurs. It has been studied for several fruits and vegetables (Mata, 1991, Fito, 1993, Fito et al, 1993 a, b), using a specially designed pilot plant equipment.

Several advantages are observed for VOD: a faster kinetic for water loss, principally during the first period of drying, and a sugar gain similar than obtained for OD. In fruits, the product obtained by VOD shows better sensorial properties than achieved at the same temperature by OD. Also the former shows more stability in relation to deteriorative reactions (browning and oxidation) (Pastor et al, 1992; Fito, 1993; Mata, 1991).

The Hydrodynamic Mechanism (HDM).

The reasons for the observed differences arise from the consideration of the responsible mechanisms of mass transfer phenomena. It has been established that in many foods with cellular structure there is a volume fraction of air located in different pores, as intercellular spaces (Trakoontivakorn et al, 1988; Puig, 1992 a, b; Mata, 1991). When the porous food is immersed in a liquid under vacuum conditions the hydrodynamic mechanism (HDM) acts (Fito and Pastor, 1993; Fito 1993; Andres and Fito, 1992, 1993). This mechanism leads to the movement of external liquid phase to go in or out of a product as a consequence of driving forces due to pressure differences between external and internal parts of pores. The pressure differences can be due only to capillary, but also to temperature or pressure changes imposed on the system. In such a situation, the occluded gas in the pores was compressed or expanded, while controlling the amount of transfer liquid. Eq. (1) allows one to calculate the volume fraction of external liquid (x) introduced in the food porous structure.

$$x = \varepsilon_e \, x_v = \varepsilon_e \, [\, 1 - (1/r)] \qquad (1)$$

where, x_v is the pore volume fraction occupied by external liquid; ε_e is the solid volume fraction occupied by gas (effective porosity), and r is the actual compression ratio, defined as:

$$r = (p_2 + p_c)/p_1 = (p_2/p_1) + (p_c/p_1) \quad (2)$$

where, p_2: final pressure, p_1: initial pressure, p_c: capillary pressure.

It can be defined:

$$R = p_2/p_1 \qquad (3)$$

as the apparent compression ratio and:

$$p_r = p_c/p_1 \qquad (4)$$

as a reduced capillary pressure. Thus:

$$r = R + p_r \qquad (5)$$

In many cases $\mathbf{p_r}$ is much smaller than $\mathbf{R}$, therefore $\mathbf{r = R}$.

The imposition of pressure gradients to the solid-liquid-gas system in the external pore surface may be a consequence of changes in the external pressure, but also of temperature changes in the system itself. This is the case of traditional thermal treatments used in canning, blanching, etc. Nevertheless, in this case, the eventual positive effects of the pore liquid penetration in the pores (a_w and pH control, food reformulating, kinetic control of food deteriorative reactions, etc.) are usually due to the undesirable thermal treatment.

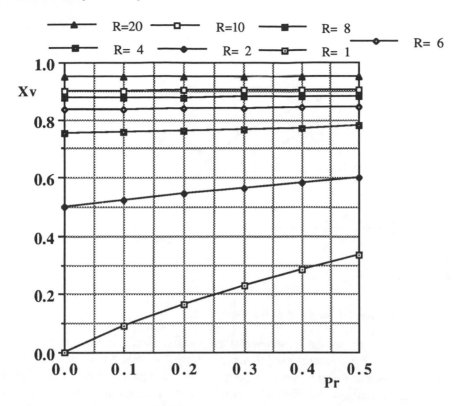

Fig. 1.- Values of x_v as predicted by eq.1.

Fig 1 shows x_v values calculated with Eq. (1) as a function of **R** and p_r. The curve corresponding to **R=1** only describes the capillarity effect, without imposed pressure gradients. It can be noticed that for low p_r values, the x_v values were small. Nevertheless, when increasing p_r (e.g. working at low pressures) the x_v value increases considerably. There is a larger increase when the external pressure gradients (**R>1**) were imposed as can be seen in the corresponding set of curves. The increase in **R** values may by obtained also working at high pressures but this usually results in excessive cost of equipment. As predicted by eq.1, similar values of x_v may be obtained working at 50 mbar as at 20 bars.

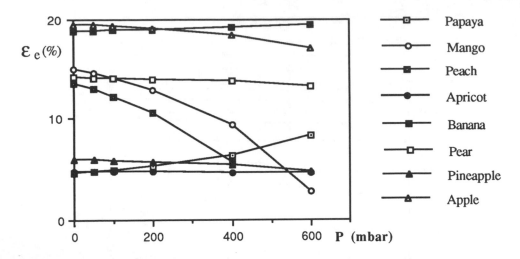

Fig. 2.- Experimental values of ε_e for some fruits.

Fig. 2 shows the experimental ε_e values, obtained for some tested fruits, compared with the pressures used. For pressures below 600 mbar, the ε_e experimental values kept practically constant as expected. Only an increase of effective porosity (ε_e) was noticed for mango and peach when pressure decreases. It was possibly due to a loss of native liquid during the expansion and release stage of the gas occluded in the pores. This loss will cause an increase in the inner volume of pores which was available for the gas phase.

The Pulsed Vacuum Osmotic Dehydration.

The main advantage of the VOD against the OD lies in the mass transfer due to the HDM and to the corresponding increment produced in the solid-liquid interface surface. The obstacle is basically the higher cost of the equipment.

Considering that the most important HDM effect is very quick and it occurs just when the system is placed at atmospheric pressure again, a new procedure was designed to carry out the VOD, which is called Pulsed Vacuum Osmotic Dehydration (P-VOD) (Fito et al., 1993 a, b). Through this procedure, short periods (e.g. 5 min.) of vacuum treatment were applied to the product, while it was immersed in the osmotic solution. After that, they suffer a normal osmotic dehydration at atmospheric pressure. In this way, the filling of the food pores with the same osmotic solution was induced at the beginning of the treatment. This procedure brings about many of the VOD advantages. However, the treatment was carried out most of the time in an atmospheric pressure installation.

The aim of this work is to compare and to analyze the differences of the results obtained in experiments by OD, VOD and PVOD for apple var. Granny smith, working at 40 °C .

MATERIAL AND METHODS

Material.

Apple var. Granny smith, from Lérida (Spain), with approximately the same maturity degree were used for the experiments. Samples were peeled, cored and cut in slices of 8.8 mm thick and 65 mm in diameter.

Methods.

The equipment and procedure used for drying (OD and VOD) was described in previous papers (Mata, 1992; Mata and Fito, 1992). For PVOD, the samples were subjected to 110 mbars of pressure during 5 min., and afterwards atmospheric pressure was reestablished. Experiments constituted by one, two and three pulses respectively were considered. When more than one pulse was used, vacuum conditions were imposed at each 15 min.

A sucrose solution in water at 65% (w/w) was used as Osmotic Solution. In all cases, temperature (T) and pressure (p) in the equipment, and osmotic solution concentration were maintained constant. Also, stirring conditions of this solution was chosen to assure that non external resistance was controlling the mass transfer phenomena.

Five series of experiments, each one in triplicate, were made at 40 ° C:

- **OD** (p = 1030 mbars)
- **VOD** (p = 110 mbars)
- **PVOD-1P** (1 pulse) (p =110 mbars during 5 min. Afterwards p=1030 mbars all the time)
- **PVOD-2P** (2 pulses) (P: 110 mbars, 5 min. / 1030 mbars, 10 min./ 110 mbars, 5 min.. Afterwards p=1030 mbars all the time)
- **PVOD-3P** (3 pulses) (P: 110 mbars, 5 min. / 1030 mbars, 10 min.. 110 mbars, 5 min./ 1030 mbars, 10 min./ 110 mbars, 5 min.. Afterwards p=1030 mbars all the time.)

RESULTS AND DISCUSSION.

Samples were analyzed after (t subscript) and before (0 subscript) each drying experiment for its weight (m), water (Xw) and solid mass fraction (Xs), water activity (aw) and sample thickness (h). From this data, weight reduction (M), water loss (Mw) and solutes gain (Ms) were calculated. These calculations were referred to initial raw material according to equations (6) to (9).

$$M = Mw - Ms \qquad (6)$$

$$M = (m_0 - m_t) / m_0 \qquad (7)$$

$$Mw = (m_0 \, Xw_0 - m_t \, Xw_t) / m_0 \qquad (8)$$

$$Ms = (m_t \, Xs_t - m_0 \, Xs_0) / m_0 \qquad (9)$$

M, Mw and Ms values were calculated from experimental data by eq. (7) to (9), the eq. (6) allows one to verify the confidence of the obtained results and the possible experimental errors not previously detected. In Fig. 3, the values of (M + Ms) are plotted vs. Mw, to assess, according to eq. (3), the confidence of obtained results. A good correlation was obtained.

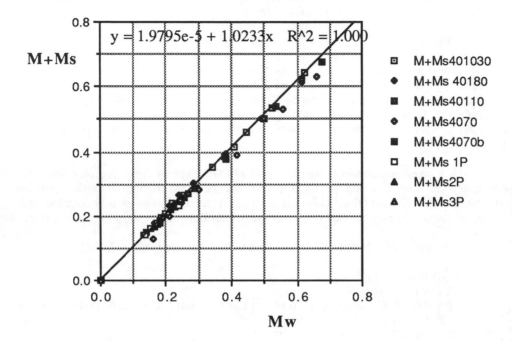

Fig. 3.- Overall mass balance for OD, VOD and PVOD experimental data (eq. 6).

Effects of HDM on raw material

Previous discussion has included the important changes produced in the food composition and structure due to pressure changes occurring during the different treatments. In Fig. 4, the sample pores fraction occupied by liquid phase has been plotted to show solid fraction (cellular structure) for apple samples used as raw material for OD, VOD and PVOD experiments. It can be seen that operating conditions strongly define the evolution of original raw material (Fig. 4A) during the very first moments of treatment.

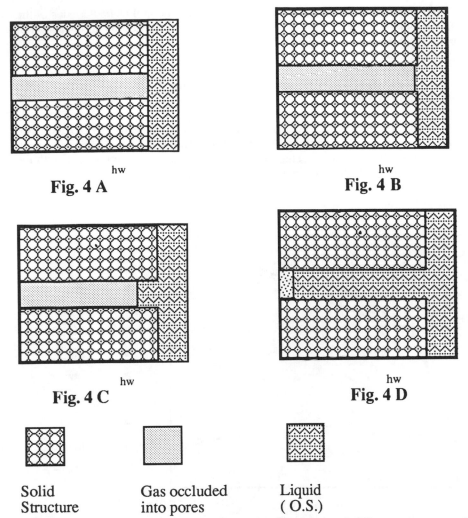

Fig. 4.- Situation of fruit sample structure at the beginning of different treatments. **4 A.-**
Fresh apple; **4 B.-OD:** Pore shows liquid penetration, as predicted by the HDM, when
capillarity acts as the only driving force. P = 1030 mbar; **4 C.-VOD:** More liquid
penetration. The capillarity is more effective because the system is at vacuum condition.(P
= 110 mbar) ; **4D.-PVOD:** Dramatic increase in liquid penetration as combined effect of
capillarity and external pressure driving forces. (P = 110 mbars for 5 min. and after at
atmospheric pressure)

- In OD process the changes in gas fraction are very small (Fig. 4B).
- In VOD process a significant penetration of liquid occurs which assume 15 % of
  the pores fraction, located beside the external surface of the product (Fig. 4C).
- For PVOD process the pores are practically full of liquid (Fig. 4D). Only the 10%
  of the initial gas volume remains in the pore.

Table 1 shows some physico-chemical and microstructural characteristics of the product in the four cases represented in Fig. 4. The important changes due to the action of HDM for the apple during VOD or PVOD can be observed.

Mass Transfer Kinetics and Food Structural Changes during OD, VOD and PVOD.

In Fig. 5, the results M, Mw and Ms are plotted vs. time for PVOD with one (PVOD-1P), two (PVOD-2P) and three (PVOD-3P) pulses respectively. It was observed that no significant differences were achieved in the three cases. So, comparisons with the other treatments (OD and VOD) were made only for PVOD-1P.

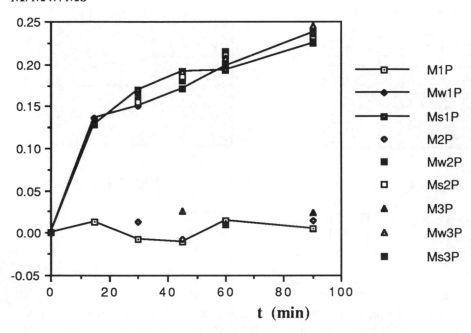

Fig. 5.- M, Mw and Ms value vs: time, for different PVOD treatments.

M/Mw/Ms

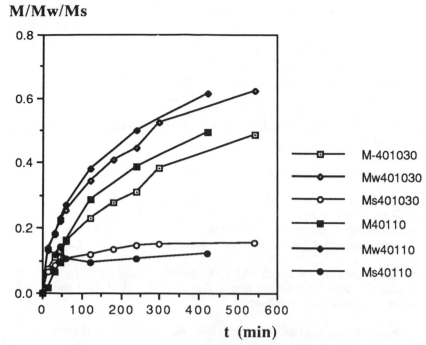

Fig. 6.- Values of M, Mw and Ms for OD and VOD treatments.

M/Mw/Ms

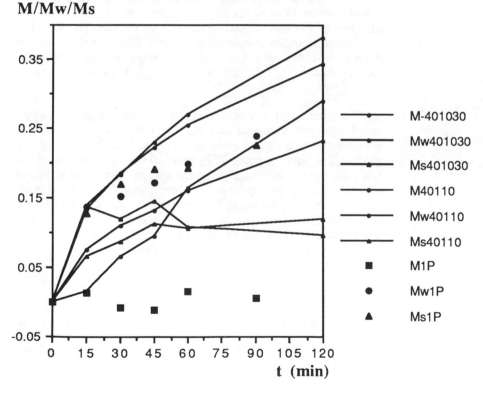

Fig. 7.- Values of M, Mw and Ms for OD, VOD and PVOD during the first 120 min. of treatments.

Fig. 6 and 7 allow one to compare the obtained results (weight reduction, water loss and sugar gain) from three different treatments. It is observed that kinetics of mass transfer phenomena in VOD is faster than in OD at the same temperature (Fig. 6). Nevertheless sugar gain is slightly slower in VOD at long times of treatment, although in the first 60 min. it is quite fast. In Fig. 7, it can be observed that this results for the two first hours of treatment, including those obtained in the PVOD-1P. The behavior of mass transfer phenomenon in the last one is very different to that of OD and VOD, showing Mw and Ms values similar to each other, and for M values close to zero. It is remarkable that behavior of Ms and Mw values of VOD seems to follow the PVOD pattern during the first 15 min. of treatment.

These results can be explained on the basis of the following considerations:

- a) During the first moments of treatment, an input of Osmotic Solution (OS) by HDM occurs into the pores, increasing the specific solid-liquid interface, and favoring the water transfer by osmotic mechanism. The input of OS represents itself an important amount of water and sugar transferred.

- b) The pores fraction occupied by OS offers a "mass transfer path" for solutes diffusion from the external to the internal part of food and also for water loss by the same mechanism.

- c) During drying, shrinking of sample have the following consequences:

  - 1.- A collapse of the cellular structure in the external surface of sample occurs. The pores became more closed, hindering the diffusion mechanism responsible for solutes transfer. The structure changes occurring in sample were confirmed by scanning electron microscopy of sample and quantified by image analyses in previous works (Mata, 1992; Puig, 1992a, b; Domingo and Lluch, 1991).

  - 2.- The fraction of gas occluded from inside the pores, before shrinking, may become partially expelled out of the pore, when the pore volume became too small to contain the gas. In this case the gas expelled from the pore (GEP) could take away some part of the originally transferred liquid. As a consequence, kinetics of diffusion transfer mechanisms, specially solutes transfer, will be slower. Also, as effective solid-liquid interface area decreases, water transfer by osmotic mechanism will also be affected. So, the initial advantage of VOD disappears at long times of treatment. The time when GEP will occur to a significant extent is function of the amount of initially transferred liquid to the pores and therefore of the work pressure.

  - 3.- For the VOD, the new structure achieved in the sample during treatment, with collapsed cells and pores on the external surface will hinder the input of OS (by HDM) when vacuum treatment is completed and atmospheric pressure is reestablished. At least, the kinetics of HDM at the end of VOD will be very slow.

  - 4.- During PVOD a great proportion of the volume of pores was occupied by OS (Fig. 4D), the residual volume of gas being small. When shrinking of pores occurs, the GEP phenomenon can not happen because the large liquid plug at the entry of pore. The path of solutes diffusion remains open to some extent, and therefore, Xs increases during the entire treatment time. Fig. 8 to 11, where evolution of Xw, Xs, a_w and h for the different treatments are plotted, support this hypothesis.

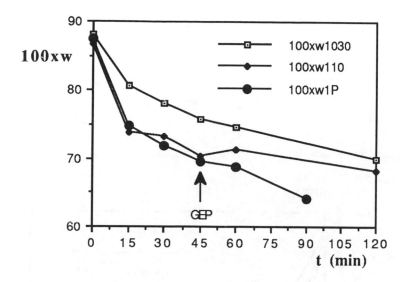

Fig. 8.- Variation of 100Xw with time during OD, VOD and PVOD treatments.

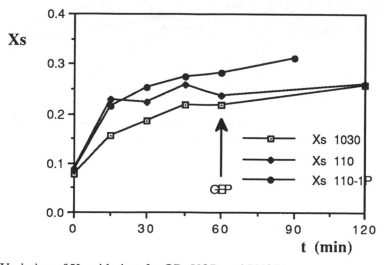

Fig. 9.- Variation of Xs with time for OD, VOD and PVOD.

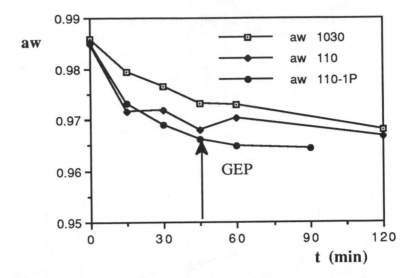

Fig.10.- Values of aw vs: time for the different treatments.

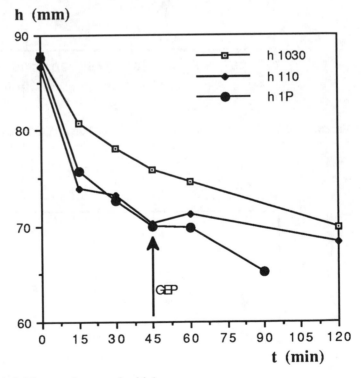

Fig.11.- Effect of shrinking on the sample thickness.

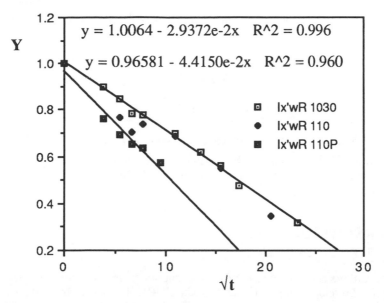

Fig.12.- Values of Y for the different treatments.

For samples treated by PVOD, X_w, X_s, a_w and h change faster than for those treated by OD. The behavior of samples processed by VOD is similar in water loss and h pattern until 45-60 min. of treatment (Fig. 8 and 11) to those processed by PVOD. It can be assumed that osmotic mechanism acts in a similar way in the three cases on the external surface of the sample, but in VOD an PVOD a special contribution of internal surface in the pores increases the kinetic of mass transfer process. In the PVOD case, a larger effect could be expected, because of the large internal surface acting in the operation, but collapse of cells and pores by osmotic effect can diminish progressively the effective transfer surface, lowering the differences among the treatments as the process progresses. From 45-60 min. of treatment, the evolution of samples characteristics in VOD is similar to that of OD.

The values of X_s and a_w are in agreement with the hypothesis of GEP effect. If it acts, it influences directly on the only important mechanism for solutes transfer, the diffusion in the liquid phase within the pores. Then X_s will achieve soon a constant value, corresponding probably to the amount diffused in the liquid penetrated in the pore, after GEP, by HDM due to capillary forces. In this sense the differences of X_s for the different treatments can be explained, moreover these observations allow one to establish the period of time when GEP is occurring (Fig. 9 and 10). In Fig. 12 these observations were reflected, plotting the reduced driving forces vs. square root of time. These are defined as Y and calculated by eq. (10).

$$Y = (z_t - z_{os}) / (z_0 - z_{os}) \qquad (10)$$

where the subscript **OS** concerns the osmotic solution properties and z is the water or solute mass fraction in the liquid phase of the sample. The evolution of reduced driving force confirms that the rate of solute transfer in the VOD differs from PVOD during the first 15 min. to follow a behavior more similar to that of OD, which was attained at approximately 60 min. of treatment. During this period of time (between 15 and 60 min.)

it seems that GEP effect occurs in samples treated by VOD and no more solutes than in OD enter in the product.

From slope of the fitted straight line to the points in Fig. 12, the effective diffusivity of water was calculated according to eq. (11) (Crank, 1985)

$$Y = 1-(4D_et/\pi h^2)^{0.5} \qquad (11)$$

The obtained values for the effective diffusivity (D_e) were $1.7 \ 10^{-10} \ m^2/s$ in OD treatment, and $3.9 \ 10^{-10} \ m^2/s$ in PVOD treatment. These values are of the same order of magnitude as obtained by other authors (Mercado, 1990; Mata, 1992; Pastor et al, 1992) for the same product.

ACKNOWLEDGMENTS

The authors acknowledge the financial support from the Program "Science and Technology for Development - 2" (STD-2) of the EUROPEAN COMMUNITY and from the "Comision Interministerial de Ciencia y Tecnologia" (Ministerio de Educacion y Ciencia. ESPAÑA).This work is a part of the Spanish contribution to the CYTED.

REFERENCES

Andrés, A.; Fito, P.; 1992. "The hydrodynamic penetration mechanism (HDM) in some fruits. In ISOPOW-V. Valencia (Spain)

Andrés, A.; Fito, P.; 1993. "El mecanismo hidrodinamico: Modelización de las operaciones de impregnación de frutas a vacío. In " II Congreso Latino Americano y del Caribe". México D.F. (Mexico).

Crank, J. (1965). The mathematics of diffusion. Pp. 47-49. Clarendon Press. Oxford(U.K.)

Domingo, L.; Lluch, M.A. (1991). Estudio de las modificaciones microestructurales de la manzana Malus Comunis L."Granny Smith" sometida a deshidratacion osmotica. Anales de Investigacion del Master en Ciencia e Ingenieria de Alimentos. Vol.1; 709-732 SPUPV. ISBN84-7721-168-X. Valencia (Spain).

Fito, P.; Pastor, R. (1993). On some non diffusional mechanism occuring during Vacuum Osmotic Dehydration. J. of.Food Eng.(In Press).

Fito, P.(1993). Modelling of Vacuum Osmotic Dehydration of Food. Proceedings of ISOPOW-V Symposium.(1992). J. of Food Eng.. In press.

Fito, P.; Chiralt, A.; Serra, J.; Mata, M.; Pastor, R.; Andrés, A.; Xian, S. (1993a). Procedimiento de Flujo alternado para favorecer los intercambios líquidos en productos alimenticios y equipo para realizarlo. Spanish Pat. n° P 9300805.

Fito, P.; Pastor, R.; Pensaben, M.; Xian, S; Chiralt, A.; Serra, J. (1993b). Pulsed Vacuum Osmotic dehydration of fruits. In CHISA '93 Congress. Prague (Czechoslovakia).

Fito, P.; Shi, X.Q.; Chiralt , A.; Acosta, E.; Andres, A. (1992). Vacuum Osmotic Dehydratation of fruits. In ISOPOW-V .Valencia (Spain).

Le Maguer, M.(1988). Osmotic Dehydration: review and future directions. Proc. Symposium on progress in Food Preservation Processes. (1): 183-309.

Marcotte, M.; Toupin, C.J.; LeMaguer, M. (1991). Mass Transfer in cellular tissues. part 1. The Mathematical model. J.of Food Eng. 13; 199.

Marcotte, M.; LeMaguer, M. (1992). Mass Transfer in cellular tissues; part 2. Computer simulations vs: experimental data. J. of Food Eng. 17; 177

Mata, M. (1991). Aportacion al desarrollo de un proceso de deshidratación osmotica a vacio de alimentos. Ph.D. Thesis. Universidad Politecnica de Valencia. Spain.

Mata, M. Fito, P.(1992). Vacuum Osmotic Dehydration of Foods (VOD) I: Design and evaluation of a pilot plant. In ISOPOW-V.Valencia (Spain).

Mercado, E. (1990). Deshidratación Osmótica de Manzana Granny Smith: Estudio y Modelización del Transporte de Materia y de la evolución de algunas propiedades físicas y químicas. Ph.D.Thesis. Universidad Politécnica de Valencia.

Pastor, R.; Mata, M.; Fito, P. (1992). Deshidratacion Osmotica de manzana.Anales de Investigacion del Master en Ciencia e Ingenieria de Alimentos. Vol.1;857-874 SPUPV. ISBN84-7721-168-X.Valencia (Spain).

Puig, A. (1992a). Effect of atmosferic and vacuum osmotic dehydration on microstructure of pinapple (ananas comosus L; var; Cayena Lisa). In ISOPOW-V.Valencia (Spain).

Puig A. (1992b). Effect of atmosferic and vacuum osmotic dehydration on microstructure of Mango (Mangifera indica L; var.Haden). In ISOPOW-V.Valencia (Spain).

Raoult-Wack, A. (1991). Les procedes de Deshydratation-Impregnation par inmersion dans des solutions concentrees (DII). Etude experimental et modelisation des transferts d´eau et de solute sur gel modele. Ph.D.Thesis. Universite Montpellier II. (France).

Toupin, C.J.; Marcotte, M.; LeMaguer, M. (1989). Osmotically induced mass transfer in plant storage tissues: a mathematical model-part 1. J. of Food Eng., 10; 13-38.

Toupin, C.J.; LeMaguer, M.(1989). Osmotically induced mass transfer in plant storage tissues: a mathematical model- part 2. J.of Food Eng., 10; 97-121.

Trakoontivakorn, G.; Patterson, M.E.; Swansoo, B.G. (1988). Scanning Electron Microscopy of Cellular Structure of Granny Smith and Red Delicius Apples. Food Microestructure 7; 205-212.

OHMIC THAWING OF FROZEN SHRIMP: PRELIMINARY TECHNICAL AND ECONOMIC FEASIBILITY

Henderson, J. T.; Balaban, M. O. and Teixeira, A. A.

INTRODUCTION

The increasing environmental regulation of waste water from industrial processing operations in Florida has pressured the Florida seafood processing industry to search for ways to reduce, recycle, or reuse processing water. Florida's Gulf Coast shrimp processors use nearly 3 billion liters of water annually (Bough and Perkins, 1977). Approximately one-half of all this water is used in the single unit operation of thawing frozen blocks of raw shrimp by immersion in large tanks of warm water. This thawing operation is a necessary first step in the sequence of operations that lead to the processing of various value-added shrimp products for distribution to the restaurant, institutional, and retail trade.

Ohmic heating can be considered as one possible means to accomplish this unit operation without the use of water. Ohmic heating is a form of electrical resistance heating in which heat is internally generated within electrically conductive materials that are sandwiched between electrodes by the passage of electric current in response to a voltage drop (electric potential) between the electrodes (Halden et al., 1990). The use of ohmic heating in food processing is not new. The electrical treatment of foods to control microbes was applied to the pasteurization of milk in the early twenties (Palaniappan et al., 1990). Current food engineering research in ohmic heating deals with the UHT sterilization of particulate food materials in a liquid (Sastry, 1990a, b).

When ohmic heating is applied to the thawing of frozen blocks of shrimp, heating will occur throughout the bulk rather than at the surface (Biss et al., 1989). This reduces the chance of microbial growth in addition to the advantage that little or no water is used in the process. Furthermore, thawing times are reduced, and no leaching of water soluble components will occur. There are some disadvantages, however, that still must be overcome in using ohmic heating to thaw shrimp blocks. Runaway heating may occur within the product as ice changes phase from solid to liquid, thus increasing its electrical conductivity (Biss et al., 1989). This may cause cooking of shrimp and/or burning of tails. One way to solve this problem is to modulate the current as electrical conductivity changes. Temperature distribution patterns are unknown, and there is a lack of ohmic thawing research to establish an engineering basis for equipment systems design. It will also be important to determine that ohmic thawing will not adversely affect the sensory quality of the product or cause yield losses from lack of moisture absorption during thawing that might not be compensated in subsequent water immersion storage and handling operations.

The purpose of this study was to demonstrate preliminary technical feasibility and estimate the economic worth to Florida's shrimp processing industry of adopting ohmic thawing as an alternative to water immersion thawing of frozen shrimp blocks. Such economic worth, if any, would form a basis for economic justification of future research investment in this technology.

METHODS AND PROCEDURES

The scope of work undertaken in this study included the following tasks:

1. Determining technical feasibility through a series of pilot plant experiments on frozen blocks of shrimp in order to:
    a. determine electrical requirements and thawing performance of ohmic heating on frozen shrimp blocks,
    b. determine moisture content and rates of water absorption of shrimp thawed in water and shrimp thawed by ohmic heating, and
    c. determine if differences between water thawed and ohmically thawed shrimp could be detected by consumers through sensory panel evaluations.
2. Visits to Florida shrimp processing plants to obtain data on current processing operations in order to establish base-line costs of traditional water immersion thawing operations.
3. Determining economic feasibility by:
    a. Comparing current operating costs of water immersion thawing with estimated operating costs of thawing by ohmic heating at the same capacity.
    b. Estimating cost savings per annum of ohmic thawing over water immersion thawing, presenting a sensitivity analysis of the cost elements that will effect these savings, and determining the economic worth of ohmic thawing to the Florida shrimp industry.

ELECTRICAL POWER REQUIREMENT

Thawing experiments were designed to determine electrical power requirements by measuring voltage, current and time, and to calculate thawing performance by comparing the heat generated ohmically and the total heat needed theoretically to thaw the block. The apparatus used for the ohmic heating of frozen blocks of shrimp was built in the Food Science and Human Nutrition Department pilot plant at the University of Florida. The device consisted of an outer box made of 1.0 cm thick Lexan (polycarbonate) used to collect water and to separate the person operating the equipment from the high voltage electrode plates. Within this outer box was an inside support stand, also of Lexan, which included a lip, open in the corners, to allow thawed water from the block to run off and keep the frozen block from sliding off the stand or coming in contact with the water previously thawed from the block. The ohmic heating unit used two stainless steel plate electrodes. The support stand also acted as a support for the bottom electrode. The top electrode was moveable and rested on top of the frozen block of shrimp as shown in Figure 1.

The plate electrodes were connected to a voltage regulator for controlling the amount of output current that was monitored by an ammeter placed around the wire leading from the voltage regulator to the top plate. An RMS multimeter was connected to the ammeter to allow digital read-out in amperes of the current flowing through the block. A personal computer equipped with the RS-232 communication link with instruments was used to record both voltage and amperage as shown in Figure 2.

The temperature of the storage freezer was recorded as the initial temperature of the block of shrimp. The frozen block of shrimp was weighed and then placed between the two plate electrodes. The amount of electrical current supplied to the frozen block of shrimp was controlled by manual adjustment of the voltage output from the voltage regulator. Amperage was monitored through the digital display on the ammeter. Periodically the current was turned off in order to check the block for

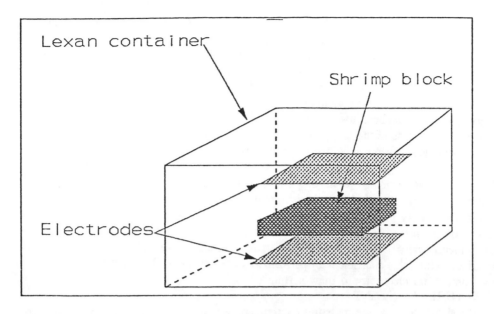

Figure 1. Schematic of experimental ohmic thawing apparatus.

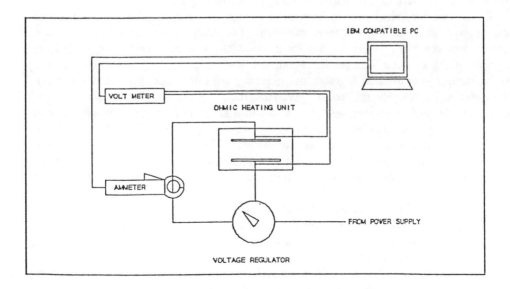

Figure 2. Schematic of experimental ohmic thawing system setup.

hot spots and remove those shrimp that had thawed. After the block was thawed, the shrimp and the drip loss water were weighed.

The voltage, amperage, and time data were used to calculate the total amount of electrical energy used to thaw the frozen blocks of shrimp. This electrical energy input was termed ohmic thawing energy input. Theoretical energy input, (the minimum energy required by thermodynamics) was calculated by summing the latent heat of fusion and sensible heat loss before and after thawing for both shrimp and water making up the mass of the entire frozen block. Heat capacities and freezing points for shrimp and water were taken from the American Society of Heating, Refrigeration, and Air conditioning Engineers, Inc. (1986). Since heat could be absorbed freely from the ambient atmosphere to accomplish thawing without added energy, the actual electrical energy delivered by ohmic heating would be some fraction of the total theoretical energy. This comparison was accomplished by introducing a coefficient of performance defined as the electrical ohmic thawing energy input divided by the energy input theoretically needed to thaw the frozen block of shrimp, multiplied by one hundred. Thus, this coefficient of performance represented the percentage of energy that came from electricity to thaw shrimp.

MOISTURE CONTENT

The shrimp used in the moisture content analysis were thawed using three different thawing methods: still-air thawing (control group), flow-through water immersion thawing (industry group), and ohmic thawing (experimental group). The American Shrimp Canners and Processors Association's (1992) method for determining net weight of block frozen peeled shrimp was used. Four 2.27 kg (5 lbs.) blocks of frozen shrimp (2 blocks of 300-500 count and 2 blocks of 30-40 count) were cut into three pieces. Two of the pieces were approximately the same size. A third piece was cut larger to allow for chill tank experiments after ohmic thawing. After being divided, the pieces were returned to the freezer to equilibrate overnight.

Each piece was measured for dimensional size and weight prior to thawing and then thawed under one of the three methods. For each thawing method, a random sample was removed. Fifty shrimp from the 300-500 count shrimp were blended together using a mortar and pestle. From the 30-40 count shrimp a random sample of five shrimp were blended together using quick pulses of a food processor to remove any variation between the shrimp in a given treatment. From the blended shrimp, four samples were randomly taken and placed in a thin layer on four pre-weighed tins and weighed. These samples were placed in a dryer at approximately 80°C, dried for fourteen to eighteen hours, and then reweighed. The difference between the pre drying weight and the post drying weight was taken to be the available free moisture content taken to be the available free moisture content in accordance with Horwitz (1980).

To simulate industry practices of storing shrimp in a mixture of ice and water in chill tanks until needed on the processing line, the ohmically-thawed shrimp were placed in an ice/water mixture to measure reabsorption rates. Every 15 minutes, a sample, equal to those stated above, was removed, blended together to remove variation from the sample, placed into tins and dried according to the same procedure stated above in order to determine moisture absorption rates.

SENSORY EVALUATION

Sensory panel triangle tests were carried out using 30-40 count shell-on shrimp to determine if ohmically thawed shrimp could be distinguished from traditional water-thawed shrimp. Each panelist was given two sets of identically-thawed shrimp and one set of differently-thawed shrimp to sample, and was then asked to choose the one they felt was not like the others. Each sample set consisted of two shrimp placed together on a section of the sample plate.

The shrimp had been boiled prior to the taste panel, according to the standards set forth on cooking seafood products by Helrich (1990). At the taste panel, the shrimp were placed onto paper plates divided into three sections with two shrimp placed in each section to remove any variation between shrimp that might occur. In order to determine whether a statistically significant number of panelists were able to detect a difference between the thawing methods, the technique described by Meilgaard, et. al. (1992) was used. The number of panelists who claimed to be able to detect a difference between thawing methods was compared against the total number of panelists.

ECONOMIC ANALYSIS

Operating costs for both a typical shrimp processing plant using traditional thawing methods and a hypothetical plant with identical production capacity using the ohmic thawing method were estimated for a side-by side comparison. The production capacity of 25,954 kg of (57,200 lbs.) was selected as representative of medium to large firms in the Florida Gulf Coast region. The amount of product produced, water used, wastewater produced, and the costs of water use and wastewater disposal were all estimated from data gathered during visits to the two cooperating processing plants. For the ohmic thawing system, electrical demand was calculated from the average kilowatt-hours per kilogram measured to thaw the frozen blocks of shrimp in the thirteen laboratory experiments. Electrical cost for ohmic thawing was calculated by taking total kilograms of product, multiplying it by electrical demand, and then multiplying by the industrial cost of electricity per kilowatt-hour. The amount of wastewater that would need to be treated from the ohmic thawing system is only the drip loss upon thawing. From the experiments, thaw drip loss was scaled-up to reflect the amount of drip loss expected for a given amount of production. Total monthly costs were calculated by adding the cost for water use, wastewater disposal, and electricity for each thawing method. Cost savings were thus calculated by simple difference between the total costs from the two methods, and reported as cost saved per kilogram of shrimp thawed.

The cost savings per kilogram of shrimp will vary as economic cost elements change. Four cost elements were considered in the sensitivity analysis: electrical demand, cost of electricity, cost of obtaining water, and cost of discharging wastewater. Sensitivity of cost savings to energy demand and energy cost was determined against constant water and wastewater costs at current levels. The range for the energy demand per kilogram of product from an ohmic thawing system was determined from the range of power requirements found in the technical feasibility section. The range for cost of electricity was based upon information obtained from the local utility, with $0.06 per kilowatt-hour taken as the current industrial rate (Hobart, 1993). Sensitivity of cost savings to water and wastewater costs was determined against constant energy demand and energy cost at current levels. Water costs ranged from zero, for the case of company-owned well water, to $0.38 per thousand liters of water incoming to the plant. The cost for the treatment of wastewater was taken as $0.58 per thousand liters wastewater in accordance with data obtained from plant visits. Two other economic scenarios considered what would happen to the cost savings per kilogram of shrimp if wastewater cost dropped to $0.41 or increased to $0.75 per thousand liters.

The economic worth for a typical plant was assumed to be equal to the present value of the annual cost savings expected over a three year period at a twelve percent (12%) rate of return (typical time and rate considered for most process improvement investments in the United States Food Industry). The cost savings achieved by using ohmic heating rather than water immersion was calculated as dollars per kilogram of shrimp. By multiplying cost savings per kilogram by the kilograms of product produced by the shrimp processors, an annual cost savings was achieved. The present worth was then calculated using the following formula or corresponding interest tables:

$$PW(i) = \sum_{t=0}^{n} A_t (1+i)^{-t}$$

where:

PW(i) = present worth using rate of return of i%
  A_t = net cash flow at the end of period t
  n = planning horizon
  i = minimum attractive rate of return

taken from White, et. al., 1989.

Once the present worth of the process had been determined for the processing plant the fraction devoted to research and development was assumed to be ten percent of the total present worth. The other ninety percent of the economic worth would go into manufacturing, purchasing, and installation of ohmic equipment into the plant. It was also assumed that the Florida shrimp industry is made up of ten typical firms for the purpose of estimating an economic worth to that industry.

RESULTS

ELECTRICAL POWER REQUIREMENT

The experimental results from the ohmic thawing experiments are compiled in Table 1. The table shows the coefficient of performance (i.e. percent of thawing energy attributed to ohmic heating), time needed to thaw product, and electrical energy demand in kilowatt-hours per kilogram of shrimp thawed. These data show the relationship between the coefficient of performance and the time it takes to thaw a frozen block of shrimp. The higher the percentage of ohmic heating used to thaw the frozen blocks of shrimp, the shorter the time needed to thaw. Conversely, the lower the percentage of ohmic heating the longer the thawing times.

Since thawing can be completed within two hours using the traditional method of water immersion, only experiments which were conducted under conditions that achieved thawing in less than two hours were used to determine the appropriate value for coefficient of performance. The very shortest thawing times were disregarded because they had incidents of runaway heating producing cooked shrimp. Under the experimental conditions in which these constraints were satisfied, the coefficient of performance had an average value of 87% achieving thawing times around one hour with an electrical demand of 0.092 KWh/ kg of shrimp thawed. Variation in the measured coefficient of performance is possibly the result of varying amount of ice between shrimp and/or the shells on the shrimp. Thawing performance data clearly show that frozen blocks of shrimp can be thawed using ohmic heating, and in a shorter time than water immersion thawing.

MOISTURE CONTENT DETERMINATION

Table 2 shows the summary of average percent water content for the four samples within each treatment. Analysis of variance table and the F-test statistic were used in accordance with Ott (1988) to determine whether the average percent water content given for each treatment was significantly different from each other. From the data in Table 2, a two percent difference between shrimp thawed in water and those thawed using ohmic heating is apparent. Although this difference has no statistical significance, this loss of weight could alarm the processors, who sell by weight. However, Table 3 shows how water is later reabsorbed in the chill tanks that are used

Table 1. Ohmic Thawing Electrical Demand and Thawing Performance on Frozen Blocks of Shrimp.

Block	Coefficient of Performance (%)	Thaw Time (Minutes)	Energy (KWh/kg)
1	77.57	101.5	0.08273
2	53.08	179.0	0.05728
3	78.35	86.4	0.08352
4	92.98	40.8	0.10005
5	92.28	50.2	0.09373
6	86.85	61.9	0.09447
7	87.92	64.3	0.11254
8	78.69	134.0	0.08525
9	90.16	145.0	0.09666
10	69.82	187.0	0.07524
11	94.04	191.0	0.10468
12	37.45	156.0	0.04012
13	25.72	280.8	0.02751

Table 2. Shrimp Moisture Content After Thawing (averages of 8 samples).

count	air thawed (%)	water thawed (%)	ohmically thawed (%)
300/500	82.8	84.2	81.7
30/40	80.8	82.5	79.6

Table 3. Moisture Content of Ohmically Thawed Shrimp During Time of Holding in Chill Tanks. (in percent)

Time (min)	Block 1 300-500 count	Block 2 300-500 count	Block 3 30-40 count	Block 4 30-40 count
15	83.50	84.50	80.60	80.98
30	84.77	84.86	80.55	82.28
45	84.57	85.53	81.07	80.44
60	85.58	85.37	81.68	80.64
75	85.17	86.18	80.86	80.85
90	85.27	86.15	82.05	81.61

for holding shrimp until needed on the processing line. The data in Table 3 also show that the larger sized shrimp (blocks 3 & 4) do not gain or lose moisture as quickly as the smaller shrimp (blocks 1 & 2) because of the longer time needed for moisture to diffuse through a larger mass of tissue. These rate differences are not important since chill tank moisture content experiments showed that water not gained by shrimp during ohmic thawing, is absorbed later during chill tank storage and reaches the same level of moisture as the shrimp thawed in water within 15 minutes.

SENSORY EVALUATION

Nine of the twenty four panelists claimed that they could distinguish the odd sample. According to Meilgaard's (1992) statistical tables, this result did not show statistical significance in taste differences resulting from the different thawing methods. When asked which sample they preferred, four (4) panelists preferred the ohmically thawed shrimp, four (4) preferred the shrimp thawed in water, and one (1) had no preference at all.

ECONOMIC FEASIBILITY

Key water and energy balances for the two parallel processing operations with different thawing methods are shown on the process flow diagrams in Fig. 3. Based on the average daily plant capacity of 25,954 kg shrimp, monthly production would amount to 571,000 kg (1,260,000 lbs.). In the traditional overflow water thawing operation, this monthly production would consume approximately 13,646,900 kg (3,607,405 gallons) of water. Since no treatment usually occurs before being discharged, the amount of wastewater to be discharged would be the same. In ohmic thawing, no water is required to thaw the frozen shrimp blocks and thus the amount of wastewater is minimal. For the production amount considered above, only 113,514 kg (30,000 gallons) of wastewater would be generated through drip loss in a month. Table 4 summarizes unit costs or requirement for each of the four cost elements used in the sensitivity analysis and economic worth estimates.

SENSITIVITY ANALYSIS ON COSTS SAVINGS

Figure 4 shows the sensitivity of cost savings to energy demand and energy costs while keeping the water and wastewater costs constant at midrange values. Figure 5 shows the sensitivity of cost savings to water and wastewater costs while keeping the energy demand and energy costs constant at midrange values. As Figure 4 shows, cost savings per kilogram of shrimp decrease as electrical demand increases, while Fig. 5 shows how these cost savings increase as water and wastewater costs increase. From both graphs a cost savings of $0.0139 per kilogram of shrimp can be estimated from assuming midrange values for the current economic cost elements shown in Table 4. For the typical plant, this yields an annual cost savings of $95,242, as shown in Table 5.

The present worth of this annual cost savings to a single typical plant is equal to $203,375 using a three-year payback period at a twelve percent rate of return. Assuming total regional industry capacity is equivalent to ten typical shrimp processing plants, the present worth of ohmic thawing is over two million dollars to the Florida Gulf Coast shrimp processing industry. These two million dollars represent the total capital investment that can be economically justified for research, design, development, purchase and installation of commercial ohmic thawing equipment systems for the Florida Gulf Coast shrimp industry. Most of this investment (approximately 90%) will have to go to prospective equipment manufactures for scale-up, tooling, manufacture and sale of equipment systems that will do the job. Thus, the remaining 10% or approximately $200,000 can be justified as a research and development budget to fund development and bench-scale testing of workable equipment system design concepts for ohmic thawing of frozen shrimp blocks at current shrimp processing plant capacities in the Florida Gulf Coast region.

Table 4. Unit cost and Requirement for Cost Elements Used in Economic Alanlyses.

Cost Element	Unit Cost or Demand
water costs	$ 0.237/ 1000 kg water
wastewater costs	$ 0.58/ 1000 kg water
electricity costs	$ 0.06/ KWh
electrical demand	0.092 KW/ kg shrimp

Table 5. Projected Cost Savings from Ohmic Versus Water Thawing for Florida Gulf Coast Shrimp Industry.

Cost Basis	Water	Ohmic	Savings
Thawing Cost per kg shrimp	$0.0195	$0.0056	$0.0139
Annual Cost per plant (6,850,000kg/yr)	$133,575	$38,360	$95,200
Annual Cost Florida Gulf Coast Shrimp Industry (68,500,000kg/yr)	$1,335,750	$383,600	$952,000
Present Worth (3-year cost savings @ 12%)	-	-	$2,000,000

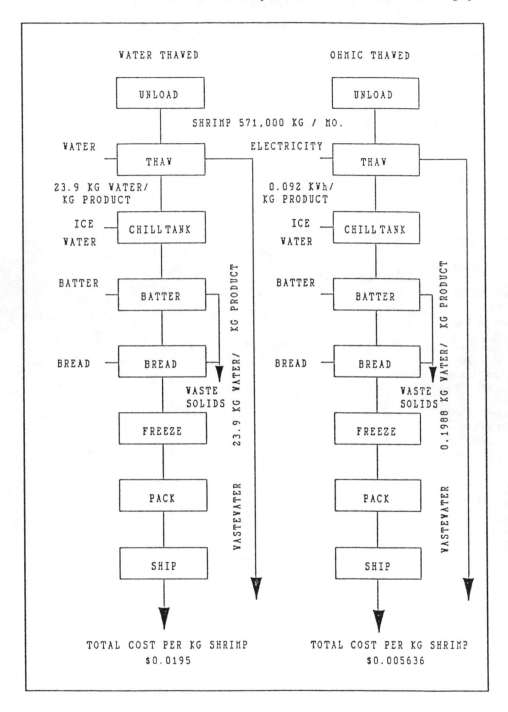

Figure 3. Flow diagrams for typical Florida shrimp processing operations showing key water and energy balances for two different methods of thawing incoming frozen shrimp blocks.

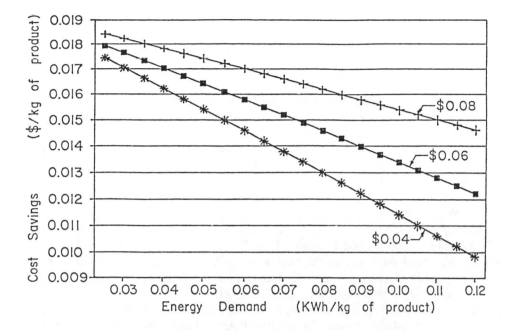

Figure 4. Sensitivity of cost savings to ohmic energy demand for three levels of energy cost ($/kwhr) with water and wastewater costs fixed.

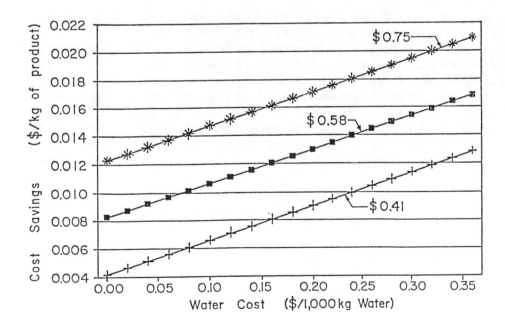

Figure 5. Sensitivity of cost savings to cost of water for three levels of wastewater cost ($/1000 kg) with fixed energy demand and cost.

REFERENCES

American Shrimp Canners and Processors Association, Method for Determining Net Weight of Block Frozen Peeled Shrimp, New Orleans, Louisiana, 1992.

American Society of Heat, Refrigeration, and Air Condition Engineers Inc., *ASHRAE Handbook of Refrigeration Systems and Application;* Inch-Pound edition, Author, Atlanta, Georgia, pg. 26.4, 1986.

Biss, C. H., Coombes, S. A. and P. J. Skudden, P. J., The Development and Application of Ohmic Heating for the Continuous Heating of Particulate Foodstuff, In: R. W. Field and J. A. Howell, eds., *Process Engineering in the Food Industry, Developments and Opportunities,* Elsevier Applied Science, New York, pp. 17-25, 1989.

Bough, W. A. and Perkins, B. E., Recovery of By-Products from Seafood Effluents, In W. R. Hess, Jr., ed., Proceedings of the Interstate Seafood Seminar, VPI Seafood Seminar, VPI Seafood Extension Unit, Hampton, Virginia, pp. 201-273, 1977.

Halden, K., de Alwis, A. A. P., and Fryer, P. J., Changes in the Electrical Conductivity of Foods during Ohmic Heating, International Journal of Food Science and Technology, 25, Cambridge, UK, pp. 9-25, 1990.

Helrich, K., Official Methods of Analysis of the Association of Official Analytical Chemists, fifteenth edition, Association of Official Analytical Chemists, Inc., Arlington, Virginia, pg. 865, 1990.

Hobart, K., Personal communication, Tampa Electric, Inc., Tampa, Florida, 1993.

Horwitz, W., Official Methods of Analysis of the Association of Official Analytical Chemists, thirteenth edition, Association of Official Analytical Chemists, Inc., Arlington, Virginia, 1980.

Meilgaard, M., Civille, G. V. and Carr, B. T., *Sensory Evaluation Techniques, second edition*, CRC Press, Boca Raton, Florida, pp. 60-67, 1992.

Ott, Lyman, *An Introduction to Statistical Methods and Data Analysis*, PWS-Kent Publishing Company, Boston, Massachusetts, pp. 679-695, 1988.

Palaniappan, S., Sastry, S. K. and Richter, E. R., Effects of Electricity on Microorganisms: A Review, Journal of Food Processing and preservation, 14:393-414, 1990.

Sastry, S. K., A Model for Continuous Sterilization of Particulate Suspensions by Ohmic Heating, In W. E. L. Spiess and H. Schubert, eds., *Engineering and Food Preservation Processes and Related Techniques*, Vol. 2, Elsevier Applied Science, New York, 1990a.

Sastry, S. K., Models of Ohmic Heating: Constant Voltage and Constant Current Strategies, Presented at IFT Annual Meeting in Anaheim, California, June 16-20, 1990, 1990b.

COMBINED MICROWAVE/FREEZING METHODS TO IMPROVE PRESERVED FRUIT QUALITY

M. Pilar Cano

I. INTRODUCTION

Microwave processing can offer several advantages when compared to conventional heating methods. These advantages include speed of operation, energy savings, precise process control, and faster start-up and shut-down times [Decareau, 1985]. However, in comparison with microwave household appliances, the microwave processing in food industry has not been as successful. The number of microwave food processing installations is estimated to be no more than 500 units worldwide [Giese, 1992]. The reasons for the slow rate of development of microwave energy for the industrial processes are related to costs and lack of information about the technology [Schiffman, 1992].

Microwave energy is currently being used for several food processing operations, including cooking, drying, tempering, baking, pasteurization, and sterilization (Table 1). These last techniques, sterilization and pasteurization, are the most extended and promising for food industry [Harlfinger, 1992].

Water, proteins, and carbohydrates are among the polar molecules that line up in a microwave field. The alternating action of this field, billions of times each second, causes friction and heating to occur. Heating also results from the movement of the electrically charged ions within the food. The dielectric constant is one measure of microwave heating and penetration, being dependent on the content of moisture and salt of food [Kent, 1987]. Specific heat, density, surface to volume ratio, evaporative cooling, and thermal conductivity are physical factors that determine microwave penetration, overall heating rate, and conventional heat transfer [Decareau, 1991].

Industrial use of microwave energy for food processing applications was emphasized with the development of continuous conveyor ovens in the 1960s. Distribution of microwave energy is non uniform, therefore conveying products through a microwave field exposes them to a greater uniformity of energy than is found in a stationary cavity [Ayoub et al., 1974].

II. MICROWAVE BLANCHING

A. BLANCHING OPERATIONS

The use of microwave energy for enzyme inactivation is not very common. Blanching is a heat treatment applied to fresh produce, fruits and vegetables, to inactivate enzymes directly related to the development of off-flavor and off-color in processed products. This treatment is common to several processing techniques, freezing, canning, dehydration, freeze-drying, etc. Depending on

Table 1
Microwave Food Processing Applications

Process	Products
Tempering	Meat, fish, poultry
Cooking	Bacon, meat patties, sausage, potatoes, sardines, chicken
Drying	Pasta, onions, rice cakes, egg yolk, snack foods, seaweed
Vacuum drying	Orange juice, grains, seeds
Freeze drying	Meat, vegetables, fruits
Pasteurization	Bread, yogurt
Sterilization	Pouch-packed foods
Baking	Bread, donuts
Roasting	Nuts, cocoa beans, coffee beans
Blanching	Corn, potatoes, fruit
Rendering	Lard, tallow

the final method of preserving the products, blanching can fulfill one or several purposes: a) inactivation of enzymes prevents discoloration or development of unpleasant taste during storage; b) proteins are forced to coagulate and shrink under liberation of water; c) air, which is confined to plant tissues, is expelled and oxidation during frozen storage will be reduced; d) microbial status is improved because vegetative cells, yeast and mold are killed; e) cooking time of the finished products is shortened. When water or steam is used for heating, leaching of vitamins, flavors, colors, carbohydrates, and other water-soluble components takes place. If products are going to be frozen after blanching, a chilling step will generally take place before transporting the product into the freezer. If this cooling is done with cold water, additional leaching takes place.

These considerations are of great importance if the products to be frozen are fruits. Frozen fruits usually are processed employing an addition of syrups or crystallized sugars in different proportions [Luh, 1986] in order to preserve quality. Thus most fruits do not need to be blanched, some, such as apples, pears, peaches and apricots, sometimes benefit from such a heat treatment because oxidative enzymes such as polyphenol oxidase would otherwise cause darkening of the color. In these cases, losses of fruit constituents during traditional blanching processes (water or steam followed by cooling in tap water) could be very detrimental to nutritional and sensory quality of frozen/thawed product.

Microwave blanching operations have received some research attention and failed to produce a marked advantage when compared to conventional steam blanching in several crops [Huxloll, 1970; Decareau, 1985]. A successful application may be the enzyme inactivation of whole tomato fruits or whole soybeans [Porretta and Leoni, 1989; Klinger and Decker, 1989].

B. ENZYME INDICATORS FOR BLANCHING

There are four groups of enzymes primarily responsible for the quality deterioration of frozen fruits and vegetables. Lipoxygenases, lipases, and proteases can cause off-flavor development, while pectic enzymes and cellulases can cause textural changes. Polyphenol oxidase, chlorophylase, and peroxidase may cause color changes and ascorbic acid oxidase and thiaminase can cause nutritional changes [Williams et al., 1986]. In secondary reactions,

Table 2
Enzyme indicators for blanching

Enzyme	Food
Ascorbic oxidase	Peach and vegetables
Catalase	Vegetables
Chlorophyllase	Spinach
Lipoxygenase	Peas
Pectin esterase	Citrus juice
Peroxidase	Peas
Phosphatase	Orange juice
	Milk
Polygalacturonase	Citrus juice
	Papaya
Polyphenoloxidase	Several fruits

lipid hydroperoxides and hydroperoxy radicals produced by lipoxygenase cause loss of color due to chlorophyll and carotenoids. Benzoquinones and melanins, produced by polyphenol oxidase, react with the ε-amino group of lysine residues of proteins, thereby affecting the nutritional quality and solubility of proteins.

Process optimization (blanching/freezing) involves measuring the rate of enzyme destruction, such that the blanch time is just long enough to destroy the indicator enzyme. Joslyn (1949) concluded that loss of peroxidase activity paralleled the loss of enzymes responsible for off-flavor more closely than did the loss of catalase activity. Neither catalase nor peroxidase has been shown to be directly involved in quality deterioration of frozen unblanched fruits and vegetables. They were, and are, used because of their presence in most vegetables and fruits and their easy assay. Other enzymes are less frequently used as indicators of adequate heat treatment of vegetables and fruits. These include: polyphenol oxidase (off-colors), polygalacturonase (loss of consistency) and lipoxygenase (off-flavor development). These and some other enzymes used as indicators are shown in Table 2. In general, these enzymes are less stable than peroxidase. Böttcher [1975] concluded that the complete absence of peroxidase activity indicated overblanching and, for best quality products, blanching process that left some percent of residual peroxidase activity was recommended.

Reactivation of peroxidase after blanching was observed during frozen storage. Lu and Whitaker (1974) and others have studied the conditions that lead to this reactivation. suggesting that the presence of isoenzymes having different heat stability could cause this regeneration of activity. Other enzymes also occur in isoenzyme forms with different heat stability, polyphenol oxidase and lipoxygenase. Problems with heat inactivation of different isoenzymes of the same enzyme occur because the inactivation begins at different temperatures and proceeds at different rates.

The advantages of microwave energy for blanching arise from the "deep heating" of material without relying on a temperature gradient; microwave energy can be concentrated to effect a very rapid heating; and, because water is particularly high absorber of microwave energy, water tends to heat more rapidly than most other constituents. Thus, microwave blanching may be exploited to efficiently heat inactivate some plant products that exhibit especially complex located enzymes related to quality deterioration, as banana products.

III. MICROWAVE/FREEZING PROCESS OF BANANA SLICES

A. BANANA PROCESSING
1. State of the Art

Bananas undergo rapid browning as a result of tissue disruption and exposure to oxygen during the peeling and slicing operations prior to further processing. The quantity of bananas used for processing is insignificant by comparison with the international market for fresh fruit, but about 15% of the fruit produced for export is rejected due to size, skins stains or other factors. The successful prevention of banana browning by addition of sodium bisuphite has been widely discussed [Guyer and Erickson, 1954; Tonaki et al, 1973; García et al, 1974], but blanching for enzyme inactivation of whole peeled fruits or slices has been reported by Cano (1990) for *Musa cavendishii* cv. Enana fruits and Giami (1991) for *Musa paradisiaca* fruits. A process for banana puree preservation was described by Garcia et al (1985), which required a mild heat treatment plus the addition of sodium bisulphite, citric acid and potassium sorbate. Other authors described the effect of some heat and chemical treatments on the quality of banana puree stored at 30°C [Mowlah et al., 1982] or on intermediate moisture banana products [Ramirez-Martinez et al., 1977].

Biochemical and physiological changes during prefreezing, processing, freezing and frozen storage and thawing can lower the quality of frozen fruits. Polyphenol oxidase (EC 1.14.18.1; PPO) and peroxidase (EC 1.11.1.7; POD) catalyze the oxidation of phenols which produces changes in raw fruit color, thus the changes in these enzyme activities could be employed as indicator enzymes of blanching effectiveness. These biochemical changes during processing and storage can produce a significant loss of quality in frozen fruit. To optimize this process, it is necessary to study the effects of some mild heat treatments and freezing on the enzymatic activities in banana at different maturity levels. Several authors have reported the characteristics of banana polyphenol oxidase and peroxidase [Haard and Tobin, 1971; Galeazzi et al., 1981], but there are only few published studies on the biochemical aspects of fruit pre-processing. Cano et al (1990) described the effects of some heat pre-treatments and freezing on these enzyme activities including a single experiment by microwave blanching as preliminary study of its possibilities as blanching of banana products.

2. Objectives

The main objective of this study was the evaluation of the use of microwave energy for enzyme inactivation in fruit pieces to be processed by freezing. The effects of this non conventional heating on the fruit tissue constituents at different stage of maturity were envisaged, taking account the possible undesirable changes in fruit quality due to this pre treatment.

3. Experimental Methods

a. *Plant material*

Air-freight shipments of green bananas (*Musa cavendishii* L., var enana), produced in the Canary Islands, were obtained from commercial source in Madrid. Undamaged fruit free from infection were selected and stored at 14°C and 85-90% relative humidity, conditions recommended by Salunke (1984). At each storage interval ten fruits were removed from twelve bunches, and each banana was selected by its situation in the bunch due to the fruit color variability during ripening. Fruits were carefully peeled and sliced (thickness, 1.1 cm), selecting only the slices with homogeneous diameter (3.0-3.2 cm). Analysis were carried out in triplicate from these ten fruits.

b. *Microwave treatments*

The microwave system used in this study was a Toshiba model ER-6860 oven. The power generator for this unit operates at a frequency of 2450 MHz, and the power output and time were adjustable to the following values: powers 200 W, 475 W and 750 W for 30 s time, or power 380 W for 15 s, 30 s and 45 s. After treatments banana slices for biochemical

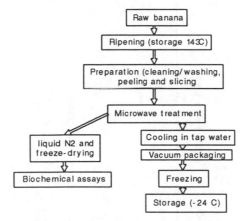

Figure 1. A general flow-chart for banana microwave/freezing process.

assays were immediately frozen in liquid nitrogen and liophylized to stabilize enzymes. Treated banana slices for conventional freezing studies were packed in plastic bags and cooled in tap water during 5 min, vacuum sealed and frozen in an air-blast freezer (Frigoscandia), working at -40°C during 30 min. Frozen banana slices were stored at -24°C for stability studies, Figure 1.

c. *Chemical and physico-chemical analysis*

Fruits for processing were chosen at 0 (green), 7 (green-yellow), 10 (yellow-green), 14 (yellow) and 21 (yellow with spots) days of storage (14°C) and analyzed to stablize their ripening stage. Sugars (total and reducing), titrable acidity, pH, soluble solids, total solids, moisture content, pulp rupture force, pulp/skin ratio and weight were evaluated following the methods described by Cano et al. (1990). Estimation of enzymatic browning (alcohol soluble soluble color, ASC index) was carried out by alcoholic extracts and spectrophotometric readings at 420 nm. Total phenols were analyzed following the method reported by Gaines (1974).

d. *Biochemical analysis*

A crude enzyme extract was prepared by taking 2g of banana lyophilized powder and homogenizing with 0.2M sodium phosphate buffer (pH 7.0)

Table 3

Compositional and physical changes during ripening of bananas, *Musa cavendishii* var enana, at 14°C

Storage time (days)	0	7	10	14	21
Peel color:	Green	Green-Yellow	Yellow-green	Yellow	Yellow with spots
Weight (g)	109.5±10.0	108.6±10.7	107.6±12.7	106.0±6.3	106.2±8.2
Firmness (kg)	5.0±0.5	4.6±0.6	1.2±0.5	0.6±0.1	0.2±0.1
Pulp:skin ratio	1.1±0.1	1.3±0.1	1.3±0.1	1.3±0.1	1.5±0.1
Moisture content (g kg^{-1})	767±3.0	716±5.0	742±5.0	723±1.0	735±3.0
Soluble solids (g kg^{-1})	78±12.0	70±5.0	184±5.0	230±10.0	248±8.0
Total sugars (g kg^{-1})	15.9±9.0	42.3±10.0	95.7±7.0	180.6±8.0	202.4±3.0
pH	5.2±0.4	5.3±0.2	4.6±0.4	4.4±0.4	4.8±0.2
Titratable acidity (mEq kg^{-1})	31±2.4	28±6.0	34±1.2	59±1.1	62±9.0

Values are mean of ten fruits ± standard error

containing 10 g liter^{-1} insoluble polyvinylpyrrolidone (PVP) and 5 g liter^{-1} Trition X-100. The homogenate was centrifuged at 13000g at 4°C for 15 min. The supernatant was decanted and stored at 4°C until assayed. Enzyme activity of PPO was determined by measuring the rate of increase in absorbance at 420 nm and 25°C in a model Lambda 15 double beam Perkin Elmer spectrophotometer. The reaction mixture contained 2.0 mL of distilled water and 25 µL of diluted (1+1) or undiluted enzyme extract depending on the sample to be analyzed. Enzyme activity of POD was determined by measuring the rate of increase in absorbance at 485 nm and 25°C. The reaction mixture contained 2.7 mL of 0.2M citrate/phosphate buffer (pH 6.5) with 200 µL of 10 g liter^{-1} p-phenylenediamine as H-donor, 100 µL of 15 mL liter^{-1} hydrogen peroxide as oxidant and 25 µL of diluted or undiluted enzyme extract. Protein concentration in all preparations was determined employing the Bio-Rad reaction according to Bradford (1976).

4. Results and Discussion

Because metallic temperature indicators cannot be used in the microwave field, the ear temperatures during microwave heating were not recorded. Studies about the use of microwave blanching of corn-on-the-cob showed that the accurate time-temperature data are difficult to obtain with the employed toluene-glass thermometers because of relatively large lag time [Huxsoll et al, 1970]. In the present work, the results are evaluated referring to the power and/or time of microwave treatment versus the different characteristics of the fruit tissue that can explain its effectiveness for processing.

a. *Effect of microwave energy on heating diameter*

Banana slices show a very characteristic location of polyphenol oxidase (PPO) and peroxidase (POD) enzymes. A demonstration of the location of the active enzyme can be done by cutting transversal slices of the fruit and dipping them in cathecol solution (PPO activity) or p-phenylendiamine and oxygen peroxide solution (POD activity). The most reactive tissue is in the central part as shown in Figure 2. Therefore tannins and other phenolic substances of banana fruit are mainly confined to the latex vessels of the pulp and, in green, unripe bananas the content of these cells (the "latex") is a viscous liquid which stained evenly with tannin reagents. Some tannins appeared in cells adjacent to the latex vessels as ripening proceeded.

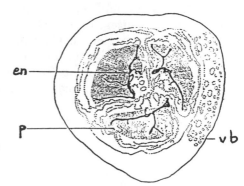

Figure 2. Transversal section of a banana pulp in the yellow-green stage of ripening. Abbreviations used: p, pulp; vb, vascular bundle; en, active enzyme zone.

Microwave energy at different powers (200 W, 475 W and 750 W) during 30 s of exposure and at different times (15 s, 30 s and 45 s) at 385 W produced a high correlated diameter of enzyme denaturation in banana slices. This effect is dependent on the ripening stage of banana fruits and banana variety. Figures 3 and 4, show the relation between these parameters. At green-yellow ripening stage, the most efficient microwave treatment for enzymatic inactivation was achieved in slices of "Gran enana" variety being dependent on the applied power output. However, in time/inactivation study a similar trends for "Gran Enana" and "Enana" were obtained.

Similar results were described in model experiments for predicting time-temperature profiles in one-dimensional slabs and spheres, two-dimensional cylinders, and three dimensional slabs [Mudgett and Keenan, 1983]. These studies, done with agar cylinders of 5-cm radius with different conductivities for evaluating the feasibility of predicting microbial lethalities at different microwave heating periods, showed a pronounced center-

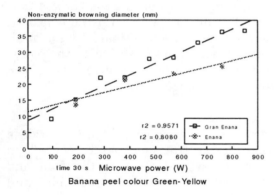

Figure 3. Effect of microwave power output on heating/browning diameter in green-yellow banana slices.

heating effect as the observed one in this present study. First-order inactivation kinetics were seen for all samples assayed, suggesting the possibility of thermal equilibration during the off period of the duty cycle. Thus, banana slices represents a product in which the effects of microwave heating may be exploited. If an efficient central heating can be applied, the most active enzyme systems (PPO and POD) will be inactivated, and no off-colors or off-flavors during frozen storage will be developed. Conventional blanching for processing banana slices by freezing tends to overheat peeled whole banana fruits before slicing and freezing, in order to adequately blanch the central zone of the slices (Cano, 1990).

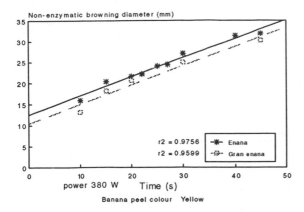

Figure 4. Effect of microwave at 380 W on heating/browning diameter on yellow banana slices after various exposure times

b. *Effect of microwave heating on enzymatic inactivation*

Before evaluating the effects of microwave heating on enzymatic inactivation of banana fruits, a brief description of the compositional changes taking place during ripening may be done (Table 3).

The ripening of banana fruits exhibits a climacteric pattern of respiration, thus the changes in their composition will be the observed ones in these kind of fruits. Transpiration is relatively constant in the mature green fruit. Despite transpiration losses, the moisture content of banana pulp normally increases during ripening, from about 69% (±4%) to about 74% (±3%). The water derived from breakdown of carbohydrates, presumably during respiration, contributes to this increase. The osmotic transfer of moisture from the peel to the pulp is reflected in the changes of the weight ratio of pulp to peel which is about 1.2-1.6 in the green fruit and rises to 2.0-2.7 when the fruit is fully ripe. The pulp to peel ratio has been suggested as a "coefficient of ripeness" (Salunke, 1984). The most striking chemical changes which occur during the post-harvest ripening of the banana are the hydrolysis of starch and the accumulation of sugars. About 20-25% of the pulp of the fresh green fruit is starch, and only 1-2% remaining in the fully ripe fruit. Sugars, normally 1-2% in the pulp of green fruits, increase to 15-20% in the ripe pulp. Total carbohydrate decreases 2-5% during ripening, presumably as sugars are utilized in respiration. The pH of banana pulp falls during ripening, from about 5.3 (±0.4) in the pre-climacteric to about 4.4 (±0.3) in post-climacteric pulp. The titrable acidity of banana oscillates between 30 mEq kg^{-1} to 60 mEq kg^{-1} from unripe to fully ripe fruits, Table 3.

Enzymatic darkening of banana pulp has been attributed to the oxidation of dopamine by the polyphenol oxidase (PPO). Peroxidase (POD) has been implicated in plant senescence and physiological breakdown of fruits and vegetables, being one of the most employed enzyme indicator for blanching treatments. In this work, the changes in the enzymatic activities of polyphenol oxidase and peroxidase of banana fruit during storage at 14°C was carried out, in order to stablize the effectiveness of microwave heating on their inactivation. Figure 5 shows the evolution of these specific enzymatic activities. PPO and POD did not remain constant throughout ripening. Both showed an increase (POD 95%, PPO 98%) at the first stages of fruit ripening (from full green to yellow-green bananas), then PPO slowly diminished (37%) and POD increased to 294 u/min/mg protein following by a larage decrease of PPO in the full ripe bananas (yellow with spots). These changes could be explained taking into account that banana ripening is primarily a differentiation

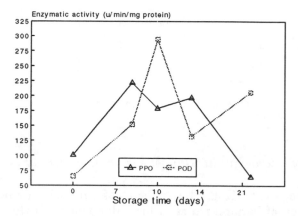

Figure 5. Changes in peroxidase and polyphenol oxidase activities of bananas (*Musa cavendishii*) during storage at 14°C

process involving programmed synthesis of specific enzymes required for ripening. The first stages involved both enzymes following which POD activity increased and PPO slowly diminished until the fruit senescence commenced. At this stage both enzymes began to play a primordial role in fruit metabolism which is related to the darkening of peel and pulp.

Microwave treatment had different effects on PPO and POD activities depending on banana maturity stage at the processing date. Figures 6 and 7 show the effects of microwave prefixed parameters in banana slices with different ripening level. The most effective inactivation of polyphenol oxidase was obtained processing banana fruits with a green-yellow peel color (7 days of storage), attending to the relation between time of treatment, at constant power 380 W, and the residual PPO activity, Figure 6. Inactivation of polyphenol oxidase oscillated between 78% to 98% depending on the time of exposure to microwave energy. Banana slices treated at 380 W during 45 s showed the lower PPO enzymatic activity. PPO activity suffered a similar effect in slices treated at unripe ripening

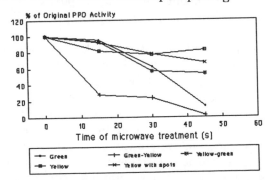

Figure 6. Effect of microwave heating (380 W) on residual PPO activity in banana slices

stage (green peel color). Yellow-green, yellow with spots and yellow banana slices presented small inactivation of this enzyme, obtaining values of 18%-40% depending on microwave heating time.

Figure 7 shows the effectiveness of microwave heating at different power output, assigning a treatment time of 30 s. Again, banana slices at green or green-yellow ripening level showed the greater PPO inactivation. In this case, green banana loss at 78% of the original activity employing a power output of 750 W during 30 s. Ripe (yellow peel color) and over-ripe (yellow with spots) fruits did not show a significant decrease in enzymatic activity.

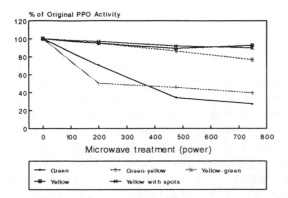

Figure 7. Effect of microwave heating (30 s) on residual PPO activity in banana slices

Microwave heating produced similar effect on peroxidase (POD) activity, Figures 8 and 9. Representation of time of treatment versus original POD activity representation shows that microwave treatment with 380 W in green-yellow banana slices has a great inactivation effect, obtaining values of 20% of original POD activity with only 15 s. Unripe bananas need more prolonged treatment (45 s) to approximately obtain the above mentioned inactivation value for green-yellow slices. Peroxidase of bananas appear to be more resistant to microwave heating than polyphenol oxidase in the same conditions, especially in ripe fruits or in fruits near to senescence.

The effects of changing power output during 30 s on the loss of peroxidase activity were most noticeable in fruits at pre-climacteric stage. Microwave treatment of these banana slices (green and green-yellow) at 200 W during 30 s was enough to produce 40% loss in enzymatic activity. An increase in power output can inactivate until 50%-70% of POD activity. Most ripe banana fruits did not manifest these effects, showing slight decrease in the remaining peroxidase activity after microwave treatments.

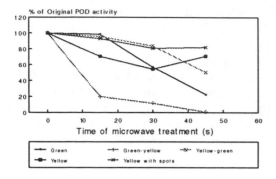

Figure 8. Effects of microwave heating (380 W) on residual POD activity in banana slices

A possible explanation of these microwave effects could be related to the severe changes in moisture content and chemical constituents which are taking place during ripening in banana fruits. Other physical parameters as density or conductivity could have suffered modifications during these physiological transformation of fruit tissue. Water is usually the major influence in how well materials, particularly foods, absorb microwave energy. Usually, if more water is present, the higher the dielectric loss factor and, hence, the better the heating. However, a lower-moisture product may also heat well, since its specific heat decreases. As a product becomes dryer, very often water areas absorb microwave energy preferentially and moisture-leveling effect is seen; this effect can be very useful in drying operations [Schiffman, 1986]. This last aspect of microwave energy on low-moisture foods could be the justification of the greater thermal inactivation of microwave energy in banana fruits at unripe stages when they show the lower-moisture content throughout ripening.

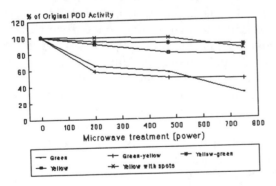

Figure 9. Effects of microwave heating (30 s) on residual POD activity in banana slices

c. *Effect of microwave heating on phenols content and darkening*

Enzymatic darkening in some fruits and vegetables during postharvest and processing is a serious quality defect and is usually undesirable because of the unpleasant appearance and concomitant development of off-flavor and off-color.

The effectiveness of the various conditions of microwave blanching treatments on browning (darkness) in banana fruits is illustrated in Figure 10.

The controls, fruit slices at different maturity level, showed more darkening than the corresponding microwave treated samples. Susceptibility to darkening in banana pulp was continuously increasing until the full ripening level was reached (bananas with yellow peel color). This behavior was

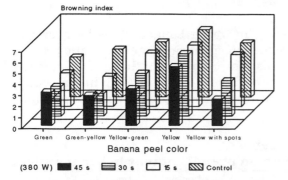

Figure 10. Effect of microwave heating (380 W) on enzymatic browning in banana slices

related to the reactive phenols content and the polyphenol oxidase and peroxidase activities in these fruits. Microwave blanched products were less dark than their respective controls, indicating that properly controlled microwave heating can improve product appearance. Microwave treatment at 380 W, power output, for 30 s or more resulted in products with minimum darkening when green-yellow bananas were employed. Darkening of banana slices was time dependent at all fruit ripening stages except for green banana slices where no significant differences (P≤0.05) beyond 30 s were obtained. Studies of the effects of different microwave power set time showed similar results. The best quality product relating to darkening was obtained employing green-yellow bananas treated for 30 s at 750 W (browning index 0.15). A 45 s blanch at 380 W would produce marginally acceptable products from green, green-yellow and yellow (with spots) bananas. However, banana treated slices from green fruits did not present the necessary banana taste and flavor. In this way, over-ripe banana slices are not suitable for additional processing due to their low firmness.

Microwave blanching did not significantly change the phenol level, Table 4. The only significantly difference in phenols levels was observed between banana fruits at different ripening stages. Green-yellow banana had a high phenol level (387 mg/100 g of sample) than low ripe or senescent fruits. The slight increases in phenol level observed in some microwave treated samples could be explained by a most efficient extraction of these compounds from the tissue due to cellular disruption produced by microwave heating.

To understand how phenols and enzymatic activities, PPO and POD, contribute to darkening and why microwave blanching can prevent darkening, a global study of these parameters could be done. Results show that both phenols and PPO and/or POD are critical to darkening situations. In untreated samples, the darkening is more dependent on the original amount of substrate available than on PPO or POD activities, because enzymes are present in such high amounts relative to substrate levels that at equilibrium most of the substrate has been oxidized to brown pigments [Walter and Purcell, 1980]. Microwave blanched samples, PPO and POD activities decreased dramatically while substrate concentration did not decrease. In these situations, enzymatic activities became the limiting factors; thus, the degree of darkening/browning mainly depends on the residual PPO and POD activities in the samples.

d. *Quality aspects of microwave/frozen banana slices*

Prevention of enzymatic darkening in frozen banana slices by microwave blanching is the result of a large reduction of the original enzymatic activities but not a reduction in phenol levels. In order to obtain a frozen/thawed product of acceptable quality, some precautions could be taken into account in the process. The selection of fruit maturity level for processing is important.

Table 4
Effects of microwave treatments on phenol content in banana slices

Microwave Power, W	treatment time,s	G	G-Y	phenols[a], Y-G	(mg/100g) Y	Y spots
Control[b]		361±7aA	387±9aA	341±5aA	329±8aA	207±6aB
200	30	287±5b	368±8a	299±7a	285±7b	183±5ab
475	30	339±6a	346±5a	280±5ab	278±9b	167±8b
750	30	279±7b	347±8a	395±6a	318±5a	235±4a
380	15	323±8a	341±7a	374±5a	318±9a	235±7a
380	30	314±5a	289±5b	396±4a	306±5a	226±5a
380	45	323±6a	352±5a	379±9a	301±6a	239±7a

[a] Phenols are expressed as milligrams per 100g of wet tissue sample
(ab) Means within column not followed by same letter differ (P≤0.05)
(AB) Means within row not followed by same letter differ (P≤0.05)
All data are the average of three independent experiments.
G (green); G-Y (green-yellow); Y-G (yellow-green); Y (yellow); Y spots (yellow with spots)

In summary, from the results discussed in this study, green-yellow bananas appeared to be the most suitable for microwave treatment in terms of enzyme inactivation.

The effects of freezing, without previous thermal treatments, on PPO and POD activities of banana slices differed depending on fruit maturity and the enzymatic system [Cano, 1990]. Peroxidase is an enzyme linked to the cellular membrane, and its solubility was affected by freezing conditions of fruit tissue. When the moisture content was low, more ice crystals grew and their size was dependent on the freezing method employed. Thus, prefixing freezing conditions, when the amount of ice crystals in the protoplasm increases, more mechanical damage occurs. In consequence, POD extractability could be increased as well as the residual activity in the thawed product. In this way, PPO activity could increase as a result of tissue freezing producing a very fast darkening of the product during thawing process. Unripe banana fruits showed little freezing damage due to their high degree of tissue firmness and low moisture content and, consequently lower PPO and POD extractability.

Microwave/frozen samples showed a different darkening potential during thawing depending on the residual enzymatic activities. If the microwave treatment produced only a slight PPO and POD inactivation, the action of the residual activities during thawing will potentially enhance darkening due to physical changes in the tissue. Microwave heating may increase internal cell pressure leading to rupture and loss of cell contents, oxidation and related deteriorative reactions. Thus, the microwave blanching must completely inactivate polyphenol oxidase and near to complete inactivate peroxidase to avoid these undesirable quality changes. Use of a microwave power (380 W) and exposure time (30 s) for treatment, followed by conventional freezing at -40°C and thawing, the residual

Table 5

Effects of microwave blanching (380 W) on darkening in frozen banana slices

Microwave treatment time,s	G[b]	darkness	rating[a]		
		G-Y	Y-G	Y	Y spots
Control	5.63±0.3a	6.35±0.2a	7.03±0.6a	9.19±0.5a	7.97±0.3a
15	4.45±0.4b	5.08±0.3b	8.64±0.3b	9.32±0.2a	8.32±0.5a
30	5.45±0.7a	6.15±0.5a	8.26±0.6b	8.75±0.6a	6.89±0.7b
45	4.21±0.3b	3.25±0.3c	6.89±0.6a	7.50±0.2b	4.68±0.5c

[a]Darkening was rated on a scale of 1.0-10.0 with 1.0 being totally discolored, 5.0 being moderately discolored, and 1.0 being absent of discoloration
(abc) Means within column not followed by the same letter differ (P≤0.05)
[b]Abbreviations as Table 4.

higher enzymatic activities and darkening were found to be taking place in near and fully ripe banana fruits (Table 5). This fact supports the selection of green-yellow banana for this microwave/freezing process considering the effectiveness of microwave heating at this ripening level.

5. Conclusions

Freezing process could be employed to preserve quality of banana slices. A pre treatment by microwave energy improve the end product due to enzymatic inactivation. A selection of the ripening level of fruit to be processed is critical to perform the conditions of microwave treatment. Other fruits are susceptible to this kind of processing, but more information about the effects of microwave energy on the quality indicators is necessary.

6. Acknowledgments

This work was supported by a Comision Interministerial de Ciencia y Tecnología (Spain) through a project no. ALI91-061.

REFERENCES

Ayoub, J.A., Berkowitz, D., Kenyon, E.M. and Wadswoth, C.K., Continuous microwave sterilization of meat in flexible pouches, *J. Food Sci.*, 39, 309, 1974.

Böttcher, H., Enzyme activity and quality of frozen vegetables. I. Remaining residual activity of peroxidase, *Nahrung*, 19, 173, 1975.

Bradford, H., A rapid and sensitive method for the quantitation of microgram quantities of protein utilizing the principle of protein-dye binding, *Anal. Biochem.*, 72, 248, 1976.

Cano, M.P., Marin, M.A. and Fuster, C., Effects of some thermal treatments on polyphenoloxidase and peroxidase activities of banana (*Musa cavendishii*, var enana), *J. Sci. Food Agric.*, 51, 223, 1990

Decareau, R.V., Microwave food processing throughout the world, *Food Technol.*, 40(6), 99, 1986.

Decareau, R.V., *Microwave foods: New product development*, Food and Nutrition Press, Trumbull, Conn.

Galeazzi, M.A., Sgarbieri, V.C. and Constantinides, S.M., Isolation, purification and physicochemical characterization of polyphenoloxidase (PPO) from a Dwarf variety of banana (*Musa cavendishii*,l), *J. Food   Sci.*, 46, 150, 1981.

Garcia, R., Menchu, J.F.and Rolz, C., Tropical fruit drying, a comparative study, *Proc. IVth Int. Congr. Food Sci. Technol.*, 4, 32, 1974.

Garcia, R., Arriola, M.C., Porres, E. and Rolz, C., Process for banana puree preservation at rural level,   *Lebensm. Wiss. Technol.*, 18, 323, 1985.

Guyer, R.S. and Erickson, F.B., Canning of acidified banana puree, *Food Technology*, 8, 165, 1954.

Haard, N.F. and Tobin, C.L., Patterns of soluble peroxidase in ripening banana fruit, *J. Food Sci*, 36, 854, 1971.

Hardlinger, L., New technology: Microwave sterilization, paper presented at 52nd Ann. Meet. Inst. Food Technol., New Orleans, La., June 24, 1992.

Huxsoll, C.C., Dietrich, W.C. and Morgan, A.I., Comparison of microwave with steam or water blanching of corn-on-the cob, *Food Technology*, 24, 290, 1970.

Joslyn, M.A., Enzyme activity in frozen vegetable tissue, *Adv. Enzymology*, 9, 613, 1949.

Kent, M., *Electric and dielectric properties of foods materials*, Science Technology Publ., London, 1987.

Klinger, R.W. and Decker, D., Microwave heating of soybeans on laboratory and pilot scale, in *Enginneering and Food, Speiss, W.E.L. and Shubert, L., eds., Vol. 2, Elsevier Applied Science, London, 259, 1989.*

Lu, A.T. and Whitaker, J.R.,Some factors affecting rates of heat inactivation and reactivation of horseradish peroxidase, *J. Food Sci*, 39, 1173, 1974.

Luh, B.S., Feinberg, B., Chung, J.I. and Woodroof, J. G., Freezing fruits, in *Commercial Fruit Processing*, Woodroof, J.G. and Luh; B.S., Eds., Avi Publishing Co., Westport, CT, 1986, chap.7.

Mowlah, G., Takano, K., Kamoi, I. and Obara, T., Physicochemical properties and protein behaviours of banana as affected by processing treatments and conditions, *Lebens. Wiss. Technol.*, 15, 211, 1982.

Mudgett, R.E. and Keenan, M.C., Microbial lethality in microwave heating of simulated high moisture foods, *Proc. IMPI Symp.*, 18, 44, 1983.

Porreta, S. and Leoni, C., Preparation of high-quality tomato products using enzyme inactivation by microwave heating, in *Engineering and Food*, Spiess, W.E.L. and Schubert, H., Vol. 2, Elsevier Applied Science, London, 251, 1989.

Ramirez-Martinez, J.R., Levi, A., Padua, H. and Bakil, A., Astringency in an intermediate moisture banana product, *J. Food Sci.*, 5, 1201, 1977.

Salunke, D.K., Banana and Plantain, in *Postharvest Biotechnology of Fruits*, Vol.1, Salunke, D.K. and Desai, B.B., Eds., CRC Press, Boca Raton, Fla, 1984, chap. 2.

Tonaki, K.I., Brekke, J.E., Frank, H.A. and Cavaletto, C.G., Banana puree processing, Rep. 202, Hawaii Agric. Exp. Stn., College of Tropical Agriculture, University of Hawaii, 1973.

Walter, W.M., Jr. and Purcell, A.E., Effect of substrate levels and polyphenol oxidase activity on darkening in sweet potato cultivars, *J. Agric. Food Chem.*, 28, 941, 1980.

Williams, D.C., Lim, M.H., Chen, A.O., Pangborn, R.M., Blanching of vegetables for freezing-Which indicator enzyme to choose, *Food Technology*, 40(6), 130, 1986.

DISCUSSION

GORRIS: Does the microwave treatment affect the texture or sensory aspects (aroma) of the thawed banana-slices?

CANO: Microwave-treated samples showed good texture and flavor after freezing/having process if a correct selection of fruit maturity was achieved. Green-yellow bananas were the most suitable for this process.

A GENERAL ANALYSIS OF THE RESIDENCE TIME DISTRIBUTION OF PARTICLES IN THE ASEPTIC PROCESSING OF PARTICULATE FLUID FOODS IN TUBULAR SYSTEMS

Paulo N. Baptista, Fernanda A. R. Oliveira, Andreas Strauß and Jorge C. Oliveira

I. SUMMARY

The residence time distributions of simulated food particles in a complete aseptic processing system were studied. The parameters considered were the mean and minimum residence time, the variance and the skewness of the distributions. All these parameters were normalized and studied as function of the viscosity of the fluid, particles density, diameter and concentration and the flow rate. Multiple linear regressions were carried out to determine the significant effects on the parameters. The influence of fluid viscosity, particles density and diameter and flow rate were the most significant in relation to the normalized mean and minimum residence time of particles. Interactive effects of fluid viscosity with the three variables above mentioned were also important. The significance level for all those variables or combination of variables, in both parameters, was higher than 95%. With this significance level particle diameter was the only important variable for the distribution variance and no significant effects were observed for the skewness. In order to detect slight effects a graphic approach was used. It was concluded that higher residence times may be obtained with small high density particles in a high viscous fluid at low flow rates.

II. INTRODUCTION

In the recent years aseptic processing has been regarded as a promising technology for the processing of particulate fluid foods. This interest results from the possibility of obtaining a product of higher quality with lower production costs, that can be packaged in many differents ways leading to a larger convenience for the consumer.

In order to achieve the "commercial sterility" of the food product it is necessary to ensure that all the product receives an adequate thermal processing. In canned products that guaranty can be achieved obtaining the temperature history at the coldest point of the product. However, in aseptic processing, when the product is flowing in a system, it is not possible to measure the temperature of the coldest point. Berry [1989] associated this point with the center of the fastest particle in the heating and holding sections. Due to the lack of information related to the residence time distribution - RTD - of particles and heat transfer coefficients between fluid and particles, several models to determine particles temperature use very conservative assumptions. A more consistent knowledge of the residence time distribution of particles is one of the necessary requirements to determine optimum thermal processing conditions and ensure product safety.

RTD studies for particulate fluid foods were carried out both in scraped surface heat exchangers - SSHE - and in holding tubes. Several model foods and real food systems were used in those studies.

Taeymans et al. [1985] determined the RTD of particles in a model food of calcium alginate beads and water. The mean residence time was estimated as function of three non-dimensional numbers: axial Reynolds number, rotational Reynolds number and centrifugal Archimedes number. With this approach variations on the mean residence time due to fluid viscosity, specific gravity and particle diameter were predicted.

Residence time distribution studies using individual spherical particles suspended in a non-Newtonian fluid flowing through a SSHE were conducted by Alcairo and Zuritz [1990]. Different flow rates, fluid viscosity, particles diameter and blade speed were considered. It was conclued that an increase in flow rate, particle diameter and blade speed decreases the standard deviation of the RTD and the mean and the minimum RT were not affected by viscosity, particle diameter and blade speed.

Lee and Singh [1992] used a factorial design at three levels to determine the effect of fluid viscosity, flow rate, particle diameter and concentration and mutator speed in the minimum, maximum and mean normalized particle residence time - NPRT. The SSHE orientation was also considered. With the SSHE in vertical position significant effects were observed in the mean NPRT for all the individual variables as well as the double combinations of flow rate, particles diameter and mutator speed. In the horizontal position the effect of flow rate was neglectable. Minimum and maximum NPRT were not affected by any variable.

Several authors developed mathematical models to predict processing conditions of particulate fluid foods where SSHE were considered as hypothetical heating sections. Sastry [1986] concluded that the effect of RTD in the heat exchanger on achieving commercial sterility was very significative. Manson and Cullen [1974] also verified the same effect. In both works RTD of particles in the holding sections were considered and simulations of thermal processing in those sections were carried out.

Experimental RTD of particles in holding tubes are scarce. Berry [1989] studied the effect of flow rate and particle dimension in the RTD, in a holding tube, using 6% of rubber cubes in a carboxymethylcelullose - CMC - solution. McCoy et al. [1987] studied the effect of the same variables and fluid viscosity in the RTD of single spherical polystyrene particles suspended in CMC solutions. However, as single particles were used, the interaction between particles which exists in real situations was not accounted for. Yang and Swartzel [1992] used individual spherical particles to study the influence of the particles density at different flow rates.

Dutta and Sastry [1990a, 1990b] investigated the velocity distribution in a model food of polystyrene particles in a non-Newtonian viscous CMC solution. Average and fastest particle velocity were studied as function of flow rate, fluid viscosity and particle concentration. Viscosity was found to be the most important variable affecting the velocity distributions. Previously Sastry and Zuritz [1987] developed a mathematical model to predict the single particle trajectories and velocities in a tube flow. Simulations for various particle dimensions and densities and fluid properties were carried out.

Sandeep and Zuritz [1991] studied the influence of flow rate, fluid viscosity and particle diameter and concentration in the RTD of a particulate suspension of polystyrene spheres in CMC solutions. It was observed that an increase of flow rate results in a decrease of the mean residence time -mean RT- and of the standard deviation and an opposite effect was observed with the fluid viscosity. The increase in the particle size resulted in an increase of the mean RT and a decrease in the standard deviation.

Palmieri et al. [1992] investigated the effect of flow rate and particle concentration in the particles RTD of a real particulate food system: an aqueous solution containing 10% w/w of sodium chloride with cubes of Bintje cultivar potatoes. It was observed that the time interval between the slowest and the fastest particles decreases with the increase of flow rate and particle concentration.

Several variables affecting the RT of particles were considered by the different authors. However, in most of those studies the number of variables was limited and the

Table 1
Rheological Characteristics of Solutions [a]

% CMC	Behaviour	Consistency index (Pa.s^n)	Flow behaviour index (n)
0.00	Newtonian	10^{-3}	1.00
0.30	Pseudo-plastic	2.558	0.77

[a] experiments were carried out at room temperature (22 $^\circ$C)

determination of interactive effects between variables was not considered. In the present study five variables were considered: fluid viscosity, particles density, diameter and concentration and flow rate. A factorial design at two levels as described by Box et al. [1978] was used and individual and interactive effects between the variables in the parameters normalized mean and minimum residence time, variance and skewness of the distribution were evaluated.

III. MATERIALS AND METHODS

A. PARTICLES AND FLUID

The simulated food particles used were spheres of polystyrene and spheres of acrylic (Hoover Precision Products, Inc., USA) with a density of 1.05-1.08 g/cm^3 and 1.18-1.19 g/cm^3, respectively. Water and Sodium Carboxymethylcellulose -CMC- solutions (Hoechst AG - Tylose MHB 30000 yp) were used as liquid phases. CMC solutions were prepared adding, slowly, the CMC powder to the water at 60 $^\circ$C with continuous agitation during 40 hours at least. The density of CMC solutions was 1.00 $\pm$ 0.01 g/cm^3. Viscosity values of CMC solutions were determined experimentally with a coaxial cylinder viscometer (Contraves RHEOMAT Model 115, Contraves AG, Zurich, Switzerland). Rheological characteristic of water and CMC solution are presented in table 1. It should be noted that water is a Newtonian fluid and the CMC solution is not. Viscosity could be expressed as function of the fluid consistency index (that is, the apparent viscosity at unit shear rate), instead of CMC concentration, however the change of the rheological behaviour would not be patent in that value. Both the fluid consistency index and the CMC concentration are indicated for the viscosity variable, throughout the work.

B. ASEPTIC PROCESSING SYSTEM

The system used for this study is schematically represented in figure 1 and the basic parts are: a feed tank for the liquid solution (1), a pump (2), a visualization section at the inlet of the aseptic processing unit (3), a magnetic flowmeter (4), a tubular (2.2 cm i.d.) asseptic processing unit with heating, holding and cooling sections (5a, 5b and 5c) and another visualization section at the outlet (6). A vertical tube (7), and the valves (8a and 8b) were used to introduce the particles and remove the air from the system, respectively.

The pump used was a single rotor rotative pump with rubber blades. This type of pump is suitable for particulate fluids and is especially adequate for particles that cannot be crushed (like the model particles used in this work). The rubber blades also avoid mechanical damage in the pump.

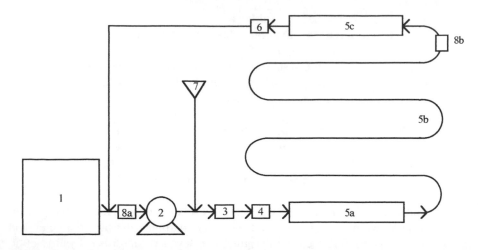

Figure 1 - Diagram of the aseptic processing system

Table 2
Values used for the Operating Variables

Variables	Diameter of particles	Density of particles	Concentration of particles	Viscosity of fluid	Flow rate
Values	(cm)	(g/cm^3)	(% v/v)	(Pa.s^n/%CMC)	(l/h)
lower	0.635	1.065	1.0	10^{-3}/0.00	180
higher	0.952	1.185	3.1	2.588/0.3	360

C. EXPERIMENTS

The feed tank was filled with the adequate liquid solution. This solution was pumped into the aseptic processing system and recirculated, without returning to the feed tank (see figure 1), until all the air had been purged from the system using the valves 8a and 8b. Then, a 3-way valve at the bottom of the vertical tube (7) was opened and the model food particles were introduced at the top of the tube until the desired concentration was reached. The particles were slowly introduced to guarantee an homogeneous particle concentration in the system. Some of those particles were coloured with different colours and also introduced in the system. The residence time of those particles was determined by making a registration of the initial time, at the inlet transparent section, and the final time, at the outlet transparent section. The solid/liquid mixture was recirculated the number of times necessary to record the residence time of one hundred particles, at least. Experiments with 32 different combinations of variables according to a factorial design at two levels were carried out, with a total number of 3292 tracer particles residence time measured. The variables considered were: fluid viscosity (μ), density (ρ_p), diameter (d_p) and concentration (C_p) of particles and the flow rate (Q). For each variable two different values were considered (table 2). Both direct and videotaped visualization were used in this study. Experiments were carried out at room temperature (22 $^\circ$C).

With the residence time of the coloured particle it was possible to draw residence time distribution curves. The statistic parameters: mean ($\bar{t}$), variance (σ^2) and skewness (s^3) of the residence time distribution were calculated. The minimum residence time (t_{min}) was also a parameter considered. The parameters were normalized in order to compare results for different flow rates (Q). The relations between non-normalized and normalized parameters, as described by Levenspiel [1972] are:

Normalized Mean RT:
$$\bar{\theta} = \frac{\bar{t}}{\tau} \tag{1}$$

Normalized Variance:
$$s_\theta^2 = \frac{s^2}{\bar{t}^2} \tag{2}$$

Normalized Skewness:
$$s_\theta^3 = \frac{s^3}{\bar{t}^3} \tag{3}$$

Normalized Minimum RT:
$$\theta_{min} = \frac{t_{min}}{\tau} \tag{4}$$

where:

$$\bar{t} = \sum_{t_1}^{t_n} t \, E(t) \, \Delta t \tag{5}$$

$$s^2 = \sum_{t_1}^{t_n} (t - \bar{t})^2 \, E(t) \, \Delta t \tag{6}$$

$$s^3 = \sum_{t_1}^{t_n} (t - \bar{t})^3 \, E(t) \, \Delta t \tag{7}$$

$$\tau = \frac{V}{Q} \tag{8}$$

V being the volume of the system.

As a result of the different conditions Reynolds numbers were obtained between 2.6 and 5.8×10^3. Thus, experiments were performed in laminar flow when CMC was used and in transient regime when water was used.

IV. RESULTS AND DISCUSSION

A. INDIVIDUAL AND INTERACTIVE EFFECTS

The analysis of the individual or interactive effects can be performed using the factorial design as described by Box et al. [1978]. With this method, drawing the normal plot of the effects for each parameter considered, identifies the significant ones. Drawing a straight line over the points, those that fall outside the line correspond to the significant effects. To avoid the errors associated to an empirical drawing of this line, the variables and combination of variables were adjusted to a multiple linear regression. Selecting a significance level based on a "t" statistical test it was possible to determine which variables, or combination of variables, had a significant coefficient and therefore could not be neglected. The linear multiple regression of the STATVIEW software (Statview, Abacus Concepts, Inc., 1992) was used for these calculations.

1. Analysis of the normalized mean and minimum residence time

Performing the multiple linear regression it was observed that the fluid viscosity, particle density and diameter and flow rate were the individual variables with a significance level higher than 98.5%. The double effect of fluid viscosity/ flow rate also had the same significance level and the double effects of fluid viscosity/particle density and fluid viscosity/particle diameter were not neglectable for a significance level of 97.5%. The flow rate and all the other combinations of variables did not present any significant effect, for those levels of significance. Identical results were obtained for the normalized minimum

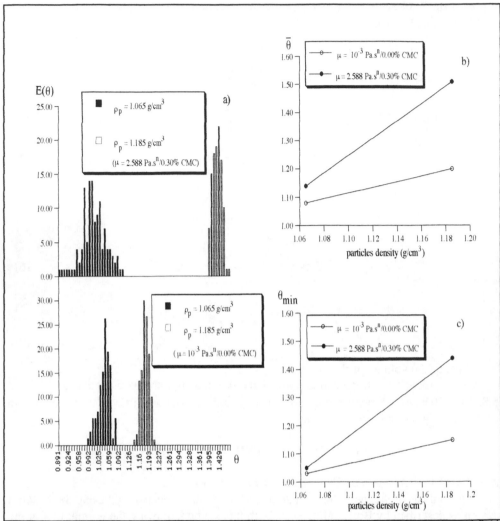

Figure 2 - a) Normalized residence time distribution curve - combined effect of particles density and fluid viscosity (d_p = 0.635 cm, C_p = 3.0% v/v and Q = 360 l/h); b) Average of the combined effect fluid viscosity/particle density in the normalized mean residence time; c) Average of the combined effect fluid viscosity/particle density in the normalized minimum residence time.

RT, although the significance levels were lower: for the individual variables particle density, particle diameter and flow rate and the double interaction fluid viscosity/flow rate the significance level was 97.5% and for the fluid viscosity and double interactions fluid viscosity/particle density and fluid viscosity/particle diameter it was 95%.

Figure 2a) shows the combined effect of fluid viscosity/particle density for a particular set of conditions of the other variables. Figure 2b) and 2c) show respectively the average of these effects in the mean and minimum RT. They were drawn with the data of the 32 experiments and each point represents the average of the 8 experiments where the fluid viscosity and the particle density had the same levels (figures 3 and 4 were built in the same way). It can be seen that particles with a higher density show higher residence times, both mean and minimum. This effect is particulary significant for higher fluid viscosities. It should be stressed that comparison of results with different viscosities implies the analysis of different flow patterns (laminar and transient zone).

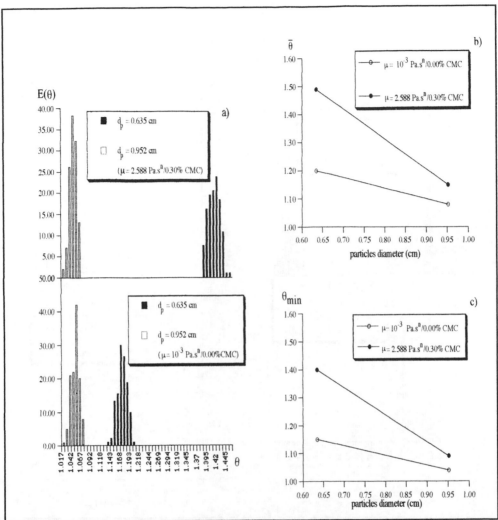

Figure 3 - a) Normalized residence time distribution curve - combined effect of fluid viscosity and particle diameter (ρ_p = 1.185 g/cm^3, C_p = 3.0% v/v and Q = 360 l/h); b) Average of the combined effect fluid viscosity/particle diameter in the normalized mean residence time; c) Average of the combined effect of fluid viscosity/particle diameter in the normalized minimum residence time.

The double effect fluid viscosity/particle diameter is represented in figures 3a), 3b) and 3c). The increase of particle diameter results in a decrease of both mean and minimum residence time. However, this effect is more significant for the higher viscosity. The increase of fluid viscosity is significant for the lower particle diameter and almost neglectable for an higher diameter.

From figures 4a) and 4b) it was possible to conclude that an increase of flow rate also results in a decrease of mean and minimum residence times. This effect is also more significant for higher fluid viscosities. The effect of fluid viscosity is neglectable for high flow rates, being very important for low flow rates.

2. Analysis of the normalized variance and of the normalized skewness

The multiple linear regression was carried out for both these statistical parameters and it was determined that the only variable, or combination of variables, that was not neglectable for a significance level of 95% was the diameter of the particles. However, as

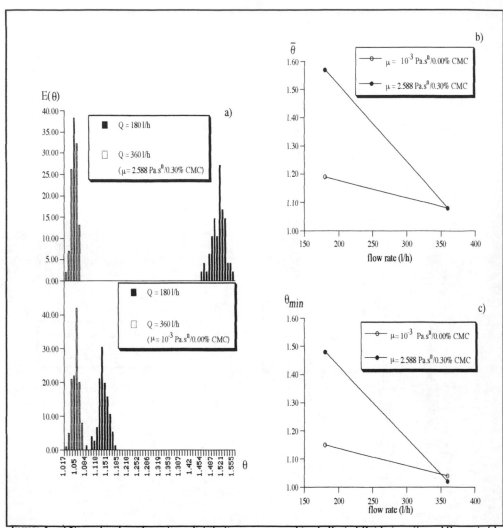

Figure 4 - a) Normalized residence time distribution curve - combined effect of fluid viscosity and flow rate (d_p = 0.952 cm, ρ_p = 1.185 g/cm^3 and C_p = 3.0% v/v); b) Average of the combined effect fluid viscosity/flow rate in the normalized mean residence time; c) Average of the combined effect fluid viscosity/flow rate in the normalized minimum residence time.

others were close to this value, for each individual variable the differences in the two parameters between the higher and the lower level were calculated. For each variable and parameter, a total number of sixteen differences were thus calculated and drawn in a bar chart. The sum of the values of all columns gives the dimension of the individual effect of the variable in that parameter. Graphs where most of the columns are above zero, represent a positive effect in the parameter; when most columns are below zero, the effect is negative; to detect double effects of variables, columns corresponding to the lower and to the higher level of the second variable must be compared. The tables shown under the graphs allow for detection of the referred levels. This type of graphic analysis allows to verify when one effect is masked just by one experiment and is specially useful to detect small effects.

The effect of particle concentration in the variance is not very clear. However, from figure 5a), a slight negative effect can be observed. Comparing columns 1, 2, 3, 4, 9, 10, 11 and 12 respectively with columns 5, 6 , 7, 8, 13, 14, 15 and 16 it is possible to detect a

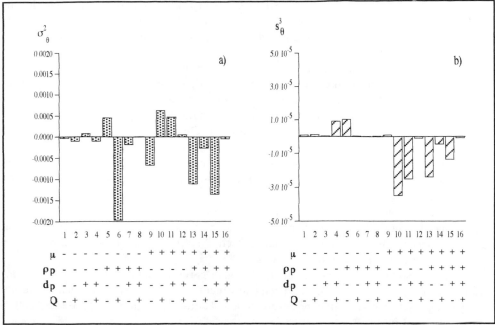

Figure 5 - Effect of particle concentration in the a) variance; b) skewness of the residence time distribution.

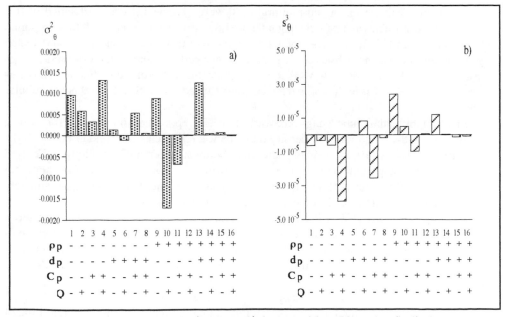

Figure 6 - Effect of fluid viscosity in the a) variance; b) skewness of the residence time distribution.

double effect of particle density/particle concentration: for the same particle concentration an increase of particle density results in a decreased variance. From figure 5b), comparing the first eight columns with the last eight, a decrease of the skewness is patent. That observation corresponds to a double effect fluid viscosity/particle concentration.

This same effect can be detected in figure 6b) confronting columns 1, 2, 5, 6, 9, 10, 13, 14 respectively with columns 3, 4, 7, 8, 11,.12, 15, 16. From the same graph it is possible to visualize an even slighter effect between fluid viscosity and particle density (columns 1, 2, 3, 4, 5, 6, 7, 8 confronted with columns 9, 10, 11, 12, 13, 14, 15, 16). In figure 6a) a majority of positive columns is seen, although two columns significantly negative might be masking a slight positive effect.

Confronting columns 1, 2, 5, 6, 9, 10, 13, 14 respectively with the columns 3, 4, 7, 8, 11, 12, 15, 16 from figure 7a) it is possible to detect an interactive effect between particle density and particle concentration. This effect was already shown in figure 5a). Figure 7b) shows a clear effect of particle density. An increase of particle density results in an increase on the skewness of the normalized RTD. This effect is more significant for the more viscous fluid (columns 9 to 16).

From figures 8a) the flow rate seems to have a neglectable effect in the variance of the normalized RTD. However, from figure 8b) it is possible to detect a small negative effect of flow rate in the skewness.

Figures 9a) and 9b) represent respectively the effect of particle diameter in the variance and in the skewness of the normalized RTD of the particles and it can be seen that particle diameter has a negative effect in the variance. Multiple effects were not found in variance as well as no effect was found in the skewness.

The normalized RTD curve for the two levels of particles diameter (the most significant variable) are represented in figure 10. Each experiment of the RT was normalized and results for particles with the same diameter were added. Therefore, both curves are the result of 3292 individual runs.

B. COMPARISON OF RESULTS WITH RELATED WORKS

In opposition with Alcairo and Zuritz [1990] significant effects of fluid viscosity and particle diameter in the mean and minimum RT were observed. Results obtained in this work for those variables and particle diameter are however in agreement with those reported by Lee and Singh [1992], although the significant effect of particle concentration detected by those authors was not observed. It should be noticed that those authors used a large particle concentration, 40% w/v, that can justify the different conclusion. Another important remark is that both works referred were carried out in SSHEs and not in tubular systems.

In a straight holding tube Yang and Swartzel [1992] reported a similar effect for particle density even with a small range of density ratios between particles and fluid (1.00 to 1.04).

Dutta and Sastry [1990a, 1990b] found that the most important variable was the fluid viscosity and that particle concentration and flow rate had no effect. The apparent contradiction in the effect of flow rate can be the result of higher flow rates (approximately twice the average fluid velocity of the present work) used by those authors.

Sandeep and Zuritz [1991] also reported that an increase of flow rate reduces the mean RT and that viscosity has an opposite effect. The increase of mean RT with the increase of particle diameter was opposite to the effect determined in the present work. The reduction of the variance with the increase of particle diameter was another conclusion reported that was also found in this work.

No analysis of normalized skewness of the RTD was reported.

The differences obtained in the works referred may be related to the different type of systems used and to the different ranges considered for each variable. This consideration is useful to understand some apparent contradictions between different works.

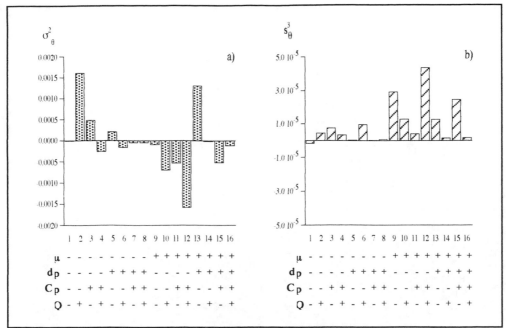

Figure 7 - Effect of particle density in the a) variance; b) skewness of the residence time distribution.

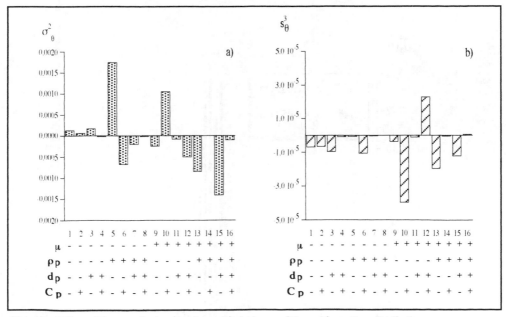

Figure 8 - Effect of flow rate in the a) variance; b) skewness of the residence time distribution.

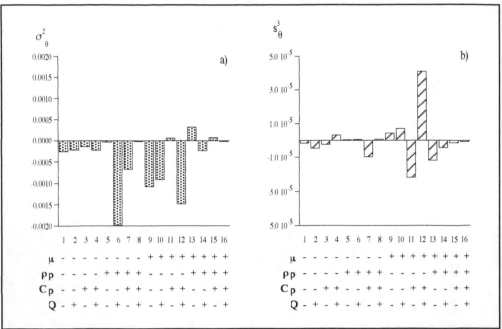

Figure 9 - Effect of particle diameter in the a) variance; b) skewness, of the residence time distribution.

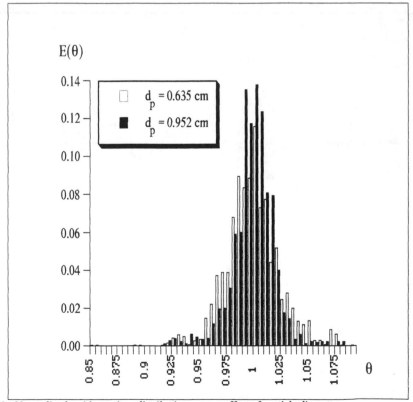

Figure 10 - Normalized residence time distribution curve - effect of particle diameter

V. CONCLUSIONS

For the system used in this work and for the range of values considered, the normalized mean and minimum RT of particles increase with the increase of fluid viscosity and particles density and with the decrease of particle diameter and flow rate. The effect of particles density and diameter and flow rate are specially important, for both parameters, for higher fluid viscosity. As a result, double interactions of particle density and diameter and flow rate with fluid viscosity are significant.

Particle diameter was the most significant variable in the variance of the normalized RTD, decreasing the variance with an increase in particle diameter. However, slight effects exist for the variables particle concentration, fluid viscosity and for the combination of variables particle density/particle concentration.

The effects on the skewness are even less important, although some effects could be detected for particle density and concentration and flow rate. Double effects of particle density and concentration with fluid viscosity also exist.

VI. ACKNOWLEDGMENTS

The authors are thankful to Junta Nacional de Investigação Científica e Tecnológica and to the CEC (FLAIR Programme) for financial support. The authors would also like to acknowledge the invaluable support of ARSOPI, the metallurgical company where the aseptic processing system used in this work was designed and built. A special reference is given to Mr. Armando Pinho and Mr. Ernesto Ferreira for their constant support.

REFERENCES

Alcairo, E. R. and Zuritz, C. A., Residence time distribution of spherical particles suspended in non-Newtonian flow in a scraped-surface heat exchanger, *Transactions of the ASAE*, 33, 1621, 1990.

Berry, M. R., Predicting fastest particle residence time, paper presented at First International Congress in Aseptic Processing Technologies, Indianapolis, IN, 1989.

Box, G. Hunter, W. G. and Hunter, J. S., *Statistics for Experimenters* , John Wiley & Sons, Inc., New York, 1978, chap. 10.

Dutta, B. and Satry, S. K., Velocity distributions of food particles suspensions in holding tube flow: experimental and modeling studies on average particles velocities, *Journal of Food Science*, 55(5), 1448, 1990a.

Dutta, B. and Sastry, S. K., Velocity distributions of food particles suspensions in holding tube flow: Characteristics and fastes-particle velocities, *Journal of Food Science*, 55(6), 1703, 1990b.

Levenspiel, O., Chemical Reaction Engineering, John Wiley & Sons, Inc., New York, 1972, chap. 9.

MacCoy, S., Zuritz, C. A. and Sastry, S. K., Residence time distribution of simulated food particles in a holding tube, paper presented at 1987 International Winter Meeting of the ASAE, Chicago, IL, December 15-18, 1987.

Manson, J. E. and Cullen, J. F., Thermal process simulation for aseptic processing of foods containing discrete particulate matter, *Journal of Food Science*, 39, 1084, 1974.

Palmieri, L., Cacace, D., Dipollina, G. and Dall'Aglio, G., Residence time distribution of food suspensions containing large particles when flowing in tubular systems, *Journal of Food Engineering*, 17, 225, 1992.

Sandeep, K. P. and Zuritz, C. A., Residence time distribution of multiple particles in non-Newtonian tube flow, paper presented at 1991 International Winter Meeting of the ASAE, Chicago, IL, December 17-20, 1991.

Sastry, S. K., Mathematical evaluation of process schedules for aseptic processing of low-acid foods containing discrete particulates, *Journal of Food Science*, 51(5), 1323, 1986.

Sastry, S. K. and Zuritz, C. A., A model for particle suspension flow in a tube, paper presented at 1987 International Winter Meeting of the ASAE, Chicago, IL, December 15-18, 1987.

Singh, R. K. and Lee, J. H., Residence time distribution of foods with/without particulates in aseptic proce ssing systems, in *Advances in Aseptic Processing Technologies*, Singh, R. K. and Nelson, P.E., Eds., Elsevier Applied Science, London, 1992, chap. 1.

Taeymans, D., Roelans, E. and Lenges, J., Influence of residence time distribution on the sterilization effect in a scraped surface heat exchanger used for processing liquids containing solid particles, paper presented at IUFoST Symposium on Aseptic Processing and Packaging of Foods, Tylosand, Sweden, September 9-12, 1985.

Yang, B. B. and Swartzel, K.R., Particle residence time distributions in two-phase flow in straight round conduit, *Journal of Food Science*, 57(2), 497, 1992.

DISCUSSION

DE BAERDEMAEKER: What type of flow (laminar, turbulent) and Reynolds number did you have during your measurements?

BAPTISTA: For both flow rates, the type of flow with the 0.30% CNC solution is laminar: Reynolds number of 2.6 and 5.2 for 180 and 360 l/h respectively were achieved. For water, the type of flow falls into the transient region: Reynolds number of 2.9×10^3 and 5.8×10^3 for 180 and 360 l/h respectively.

ZOTTOLA: Speculate on what foods your chosen particles, styrene & acrylic, would mimic? Was there any destruction of your particles during processing?

BAPTISTA: The simulated food particles of polystyrene and acrylic will be useful to mimic foods with a spherical shape and which dimension and density are close to the values used. This work also gives relevant information for all the type of foods with diameters and densities between those values. An example of a food product that fits in those ranges is peas. As the particles are hard and rubber blades were used in the pump no destruction of particles occurred. Even with real food particles, such as peas, the damages caused by this pump are very limited.

BUSTA: How does a combination of particle sizes influence the residence time of large vs. small particles?

BAPTISTA: The influence of the combination of particle size as well of particle density is one of the subjects of our work at this moment. From the present knowledge, I would expect an increase of the residence time of the large particles as a result of being forced to travel during a certain time with the same (lower) velocity of small particles. It is possible that the residence time of small particles slightly decreases, as a result of being pushed by the large particles behind them. As not all the small particles may be pushed by the larger ones, an increase of the variance distribution will not be surprising.

Part II

Optimization of Food Processes to Minimize Deteriorative Changes in Food Quality and Assure Safety

Chapter

QUALITY CHARACTERISTICS OF FRUITS AND VEGETABLES

Robert L. Shewfelt

I. PRINCIPLES AND CONCEPTS

A. QUALITY

Quality is "the composite of those characteristics that differentiate individual units of a product, and have significance in determining the degree of acceptability by the buyer".[1] If the buyer is the ultimate consumer then quality of fruits and vegetables should be further partitioned into purchase and consumption attributes.[2] Purchase attributes are properties such as appearance, firmness and aroma that are important to the consumer in the buying decision. Consumption attributes such as flavor and mouthfeel influence the degree of liking during eating. In addition, hidden attributes such as nutrient composition and microbiological status contribute to quality but are not readily determined by the consumer.[2,3] Acceptability, or "the level of continued purchase or consumption by a specified population," encompasses quality and other factors important to the consumer such as price, packaging, advertising campaigns, etc.[4]

B. SHELF LIFE

Shelf life is "the time period that a product can be expected to maintain a predetermined level of quality under specific storage conditions."[5] Shelf life of fruits and vegetables can be extended by maintaining optimum temperature, RH and gaseous atmospheric conditions and minimizing mechanical damage. Shelf life of a product is usually estimated by evaluating purchase attributes.

C. PHYSIOLOGY

Fresh fruits and vegetables are living, respiring plant tissue. As such, both quality and shelf life of fresh fruit and vegetable products are dependent on the physiological maturity of the item at harvest, harvesting and handling conditions immediately after harvest and respiration rate during storage.[6,7]

D. CONSEQUENCES OF MINIMAL PROCESSING

Minimal processing is "handling, preparation, packaging and distribution of agricultural commodities in a fresh-like state."[7] Any operation such as slicing, cutting, or dicing that disrupts cellular structure leads to increased respiration and enzyme activity, thus compromising quality and shortening shelf life.[8] Packaging[9] and mild heat treatment can slow respiration and quality degradation, thus extending shelf life. Location of minimal processing operations either near the growing region or near the target market affects the shelf life requirements and the quality of the product when purchased and consumed.[10]

E. SYSTEMS PERSPECTIVE

As a fruit or vegetable moves from the farm as a raw agricultural commodity to the consumer as a minimally processed product, it is transformed through a series of steps that usually results in a loss of weight (trimming and culling), changes in physiology (ripening and senescence) and added value (packaging and convenience).[2]   Quality management of minimally processed products requires an understanding of the entire system and the interaction of the individual steps.[5,7]

II. PURCHASE CHARACTERISTICS

A. APPEARANCE
1. Importance

Appearance of fresh fruits and vegetables is a consideration in the purchase decision of 95% of American purchasers of these items.[11]   Consumers use appearance as a nondestructive means of assessing ripeness and absence of disease or insect damage. Manipulation of harvesting and postharvest handling conditions can produce desired color of fresh fruits and vegetables at the point of sale. Judicious use of preharvest pesticides and postharvest fungicides can decrease the incidence of postharvest defects and decay.[3] Advocates of elimination of pesticides indicate that the reliance on insecticides and fungicides merely results in improved cosmetic quality that is of no consequence to overall fruit and vegetable quality.[12]   In side by side marketing of fresh items grown "organically" and with pesticides, the "organic" produce could not compete effectively.[13] Reasons cited for the disparity in sales were increased price and poorer appearance of the "organic" items.   An implicit assumption associated with decreased use of pesticides is that consumers will learn to accept fresh market produce with lower cosmetic quality when not given a choice.[12]   An alternate scenario is that the dramatic increase in fresh fruit and vegetable consumption observed in the 1980's[14] will be reversed when such items become visually unappealing.

2. Physiological Basis

The desirable colors of fresh fruits and vegetables are a function of the form and concentration of the plant pigments present.[15,16]   Anthocyanins are water-soluble pigments that form the reds, blues and purples of many fruits and vegetables.[17] Betalains are also water-soluble pigments that comprise the intense reds and yellows of beets.[18] Carotenoids, comprised of carotenes and xanthophylls, are lipid-soluble components contributing to the yellow, orange and red coloration of many plant products.[19] Chlorophylls, also lipid-soluble pigments, provide green coloration.[20]

Yellowing and browning are two major appearance quality defects encountered in fresh fruits and vegetables. Yellowing occurs during advanced senescence of many vegetables as chlorophyll, located in the chloroplasts, is oxidized enzymatically, unmasking the yellow xanthophylls which are also present in the chloroplasts.[16] Yellowing is most effectively retarded by inhibiting senescence either by storage at optimal temperatures, atmosphere modification or chemical inhibition.[5] While browning can occur by enzymatic or nonenzymatic processes, most browning of fruits and vegetables is enzymatic.   Disruption of cell structure by mechanical damage brings enzymes (i.e., polyphenoloxidase) and substrates (phenolic compounds) together to form brown pigments.[21]   In addition, chlorophyll can degrade to brown pheophytin[20] and anthocyanins polymerize to brownish forms.[17] Prevention of mechanical damage during handling and transportation is the most effective means of preventing browning in intact fruit and vegetable tissue.

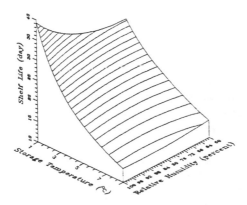

Figure 1. Typical effect of storage temperature and relative humidity of shelf life a vegetable based on visual evaluation of the vegetables. (Figure courtesy of C. N. Thai.)

Other appearance defects occur in fruits and vegetables. Pitting, sunken and necrotic areas in the exocarp of susceptible species, results from chilling injury.[22] Other physiological disorders such as bitter-pit in apples result from nutrient imbalance[23] or russet spotting of lettuce from complex interaction in preharvest and postharvest factors.[24] A typical effect of temperature and relative humidity on shelf life of vegetables based on visual quality evaluation is shown in Figure 1.

Not all postharvest color changes in intact fruit are objectionable. Fruit ripening results in a dramatic change in both purchase and consumption attributes.[25] The degree and magnitude of these changes is affected by cultivar, preharvest cultural practices, harvest maturity, and postharvest handling conditions.[7] Among the most dramatic color changes in fruit ripening are those observed in plastid conversion of oranges, characterized by concurrent degradation of chlorophyll and accumulation of ß-carotene,[26] and tomatoes with lycopene (also a carotene) accumulating as chlorophyll degrades.[27]

3. Consequences of Minimal Processing

Cutting and slicing of fruits and vegetables has adverse consequences on product appearance. Cut surfaces bring enzymes and substrates together which accelerate browning.[8,28] Browning in cut fruits can be inhibited by acid and calcium dips[29] and by modified atmosphere packaging.[30] Most fruits and vegetables have an active wound response mechanism which is triggered by the production of wound ethylene.[31] Ethylene can accelerate color changes associated with ripening and senescence. The wound response leads to suberization in some tissues (e.g. potatoes) which results in objectionable appearance quality. White bloom on cut carrots is a quality defect specifically attributed to lignification on the cut surface.[32]

Minimal processing can decrease the importance of other appearance attributes. If appearance defects are confined primarily to nonedible portions of the fruit or vegetable (melon rind, fruit peel, wrapper leaves), then trimming can remove cosmetic damage without sacrificing edible quality. Packaging can be used to screen out the deleterious effects of light as well as prevent consumer access to appearance factors during purchase. Repeat sales are likely to decrease, however, if packaging is used to hide appearance defects.

B. FIRMNESS AND CRISPNESS

1. Importance

External textural attributes are also important to the consumer in the purchase decision. Firmness-to-the-touch is an indicator of the ripeness of softening fruits such as avocadoes, mangoes, peaches and tomatoes. The perception of crispness of fresh vegetables such as broccoli, carrots, celery and lettuce plays a role in consumer selection.

2. Physiological Basis

Firmness is primarily a function of cell-wall structure and the maintenance of cell wall integrity.[33] Some fruits maintain a desirable crispness during ripening (e.g. apples). Many other fruits undergo a rapid softening during ripening (e.g. bananas, pears, plums). Degradation of cell wall components is important in the softening although the specific components attacked can vary from fruit to fruit (i.e. cellulose is degraded in avocadoes while pectins are attacked in tomatoes and peaches). Although cell wall degradation is enzymatic, the precise mechanism of pectin degradation is not clear.[34] Loss of turgor pressure may also contribute to softening in some fruit.[35]

Turgor pressure is the major factor in crispness of fresh vegetables. Wilting is unsightly and contributes to reduced sales of fresh product. Wilting results from loss of water due to increased transpiration.[36] It is a problem in products with large surface areas that have no protection from water loss. Maintenance of optimal storage temperature and RH is the most effective means of maintaining crispness.

3. Consequences of Minimal Processing

Most minimal processing techniques prevent the purchaser from making firmness-to-the-touch evaluations of whole, intact fruits or vegetables. Cutting and slicing accelerate softening processes in fruits. The appearance of piece integrity as well as maintenance of that integrity during consumer handling and preparation remains an important external quality characteristic.

In addition, cutting and slicing reduce endogenous protection from loss of turgor. Breaking of cells increases loss of cell sap and triggers wounding which in turn increases transpiration. Cutting also increases surface area and exposes surfaces that do not have external protection (e.g. cucumber peel). The sharpness of knives used in cutting has a direct effect on loss of turgor.[37]

Packaging of cut fruits and vegetables protects them from moisture loss. The microenvironment within a package, even one that is not sealed, maintains a moisture equilibrium with the product to decrease evaporation from the surface.

C. AROMA

1. Importance

Aroma provides another indicator of fruit ripeness and vegetable decay. Fruits have a characteristic aroma that serves as a differentiating characteristic to consumers. Aroma development is one of the last events during fruit ripening.[25] In some fruits (e.g. muskmelons, peaches), aroma can be an effective indicator of ripeness, while in others (e.g. tomatoes, watermelons) little aroma is observed in the intact fruit. Off-aromas are formed during microbial attack of fruits and vegetables that provide an indication of the lack of freshness of the item or care taken during postharvest handling and storage.

2. Physiological Basis

Large numbers of volatile aromatic compounds are isolated from fresh fruits. Major categories of compounds include esters, terpenes, aldehydes and alcohols. Some fruits are differentiated by a single character-impact compound. Other fruits are characterized by a mixture of several compounds.[38] The relative contribution of compounds that are not character-impact compounds and the biosynthetic pathways responsible for aroma development are not clear. Recent work is focusing on the role of free amino acids[39] and flavorless glycosides[40] as precursors of fruit aromatic compounds. Rapid accumulation of volatile compounds is associated with the climacteric rise of ripening fruits. In some climacteric fruits like bananas and mangoes, significant accumulation of volatiles is observed during postharvest storage. In other climacteric fruits (apples, peaches, tomatoes), potential accumulation of volatile compounds is greatly reduced after detachment from the plant.[41] The pattern of volatile accumulation in apples on the tree varies significantly from year to year.[42] Fresh vegetable aroma is more subtle and thus more susceptible to detection of off-aromas.

3. Consequences of Minimal Processing

Cutting and slicing accelerates both beneficial and detrimental aroma formation in cut fruits and vegetables. Some compounds such as the potent thiosulfonates that provide onion pungency are not formed until slicing.[43] The profound impact of slicing on aromatic compounds in fruits is emphasized by differences observed in aroma profiles when samples are extracted from whole intact fruit, cut pieces or macerates.[44]

Marked effects on fruit and vegetable aroma are observed during packaged storage. First, packaging usually precludes aroma evaluation of the product at purchase. Aromatic compounds may accumulate in the headspace during storage such that the aroma upon opening of the package may be either unpleasant or uncharacteristic. Atmosphere modification, particularly low oxygen or high carbon dioxide, has been shown to change aroma development pathways (i.e. aerobic to anaerobic) in the tissue.[9] Addition of acetaldehyde[45] and aromatic precursors[46] to the atmosphere can intensify fruit aromas but can also lead to development of off aromas.

III. CONSUMPTION CHARACTERISTICS

A. FLAVOR
1. Importance

Fresh fruits provide a wide range of enticing flavors. Flavor is a driving force behind purchase and consumption of fresh fruits. Flavor is the combination of aroma (described in Section II.C.) and taste as perceived in the mouth. Taste is usually described in terms of bitter, salty, sour and sweet, but it actually encompasses all flavor sensations that can be detected by humans when the nose has been pinched.[47] Sweet, sour, and bitter comprise the primary tastes of fruits and vegetables. Sweet and tangy (desirable sour) are generally considered favorable attributes while bitter and sour are generally considered unfavorable.

2. Physiological Basis

Flavor of sweet fruits is usually result of the combination of aromatic compounds, sugars and organic acids. Sugars, primarily glucose, sucrose and fructose (listed in increasing order of sweetness), contribute directly to sweetness. The presence of acids (e.g. citric, malic, tartaric) tends to decrease the perception of sweetness. Sugars accumulate during ripening either by translocation into the fruit or breakdown of starch reserves. In the former case sugar accumulation cannot occur after detachment from the plant or if defoliation of the plant occurs prior to or during ripening. Postharvest sweetening can occur in fruits that have starch reserves.[48] In many fruits, ripening is also accompanied by a decline in organic acids. In these fruits perception of sweetness is more closely related to the sugar/acid ratio than to total sugar concentration. This ratio is not an effective indicator of flavor in fruits with low levels of acids such as muskmelons or nonsweet fruits such as tomatoes. Total sugar and acid concentration provides a better indicator of tomato flavor.[49]   Bitterness in fruits and vegetables is attributed to specific compounds. Notable bitter compounds include isocoumarin in carrots.[50]

3. Consequences of Minimal Processing
Cutting and slicing are more likely to affect aromatic compounds than taste components although increased respiration can lead to loss of sugars. Off-flavor development in cut carrots can be reduced by acidified storage and modified atmosphere.[51]

B. MOUTHFEEL
1. Importance
In addition to flavor, fruits and vegetables impart textural properties in the mouth. Crispness and crunchiness are desirable attributes in apples, broccoli, carrots, celery, lettuce and some other raw fruits and vegetables. Tenderness is a desirable attribute in other vegetables like asparagus and peas. The slight resistance to chewing prior to a bursting of flavor volatiles is a mouthfeel property of the flesh of some soft fruits including mangoes, muskmelons and peaches. Fuzziness is generally considered an undesirable attribute of unpeeled peaches and okra. Peach and nectarine flesh is also affected by a storage disorder called wooliness resulting in undesirable mouthfeel. Mealiness of apples and astringency in persimmons are additional defects in this category.

2. Physiological Basis
Desirable attributes such as crispness, crunchiness and slight resistance to chewing are a function of cell wall structure and turgor as described in Section II.B.2. The fuzz on okra, peaches and other items is due to trichomes which coat the exterior and are usually removed in packing operations.[2] Wooliness of peaches and nectarines is observed during chilling of the fruit and attributed to a reversible cross-linking of pectin at low temperatures.[52] Astringency in persimmons has been attributed to low molecular weight tannins.[53]

3. Consequences of Minimal Processing
Any changes that adversely affect firmness and crispness as described in Section II.B.3. will also adversely affect mouthfeel. Ameliorative effects of packaging and modified atmospheres are important. Modified atmosphere packaging delays postharvest lignification of asparagus to maintain tenderness.[54]   In addition, use of ethylene absorbants will slow the softening of fruits slices of kiwifruit and bananas in mixed fruit salads.[55]   Addition of calcium to fruit slices has also been shown to be effective in maintaining fruit firmness.[56] Modified atmosphere techniques have been described that decrease astringency in persimmons.[53]

IV. HIDDEN CHARACTERISTICS

A. NUTRIENTS
1. Importance
Fresh fruits and vegetables are good sources of vitamins, minerals and dietary fiber. With a notable exception (avocado), fruits and vegetables are low in lipids, thus providing consumers with a low-fat source of nutrients. Unfortunately, fruits and vegetables also contain antinutrients that bind minerals, making them less available during human digestion. The composition and availability of nutrients in fruits and vegetables can vary widely by genotype, preharvest factors, maturity at harvest, postharvest handling and storage conditions, and processing variables.[57] The nutritional properties of fruits and

vegetables are important in the consumer decision to purchase them,[11] but this decision can only be made on the item's reputation as a good source of specific nutrients as the composition of that item is not readily assessable.

2. Physiological Basis

The wide variation in nutrient composition of fruits and vegetables has been well documented,[58,59] but the physiological mechanisms that cause such variation in composition are not well known. Most nutrients play key roles in intermediary metabolism, and composition is affected by many different cellular processes. Cultural practices including fertilization, irrigation, and application of growth regulators that enhance vitamin or mineral accumulation in some plants or plant organs may decrease the nutrients in others. Effects of cultural practices interact with given environmental factors such as soil fertility, rainfall, relative humidity, growth temperatures and daylength to alter metabolic processes.[60] Once a plant organ is detached from the plant, mineral composition does not change while the chemical form and thus bioavailability can be altered. Net synthesis or degradation of vitamins is affected by postharvest storage and handling conditions.[61] Once again, interaction of genetic and environmental factors affects nutrient stability. For example, loss of ascorbic acid occurs rapidly during storage of asparagus[62] while little loss is observed in broccoli.[63] Misting of broccoli during retail display further decreases the loss of ascorbic acid.[64]

3. Consequences of Minimal Processing

Little information is available on nutrient losses during minimal processing. Cutting and slicing results in loss of cell sap which would be expected to result in the loss of water-soluble nutrients as well as the acceleration of degradative reactions. Washing and soaking provide ideal conditions for leaching of water-soluble vitamins and minerals.[65] The effects of modified atmosphere packaging or storage on nutrient composition are not readily predictable.[66]

B. MICROBIOLOGICAL STATUS
1. Importance

Although safety is usually considered separately from quality, safety issues are more appropriately evaluated in the context of hidden quality attributes. As such, safety becomes the foremost quality attribute. Consumers should reasonably expect the products they buy to be safe and wholesome. Processors and distributors should strive to market products that can withstand moderate levels of abuse without becoming a hazard. Although consumers consider pesticide residues to be the most serious safety concern with fruit and vegetable products, the preponderance of scientific data implicates pathogenic microorganisms.[67] Most consumers associate spoilage and microbial hazard, even though most spoilage microorganisms do not present a health hazard and most human pathogens do not cause spoilage. A guiding principle in the processing and distribution of foods should be that abuse conditions greatly favor spoilage microorganisms over human pathogens[68] such that the product will be rejected due to unacceptable purchase or consumption characteristics prior to sufficient consumption to result in morbidity or mortality.

2. Physiological Basis

Intact fruits and vegetables are resistant to microbial attack. The exocarp provides a physical barrier to microbial invasion. Endogenous compounds that are either normally present in the tissue[69] or are induced by microbial invasion[70] further decrease microbial spoilage. Low temperature storage slows the rate of growth of most decay microorganisms. Prevention of condensation on fresh items also decreases the potential for microbial growth. Mechanical damage resulting in cuts, scratches or abrasion that permit access to the flesh through the exocarp or in bruising and other wound responses that accelerate senescence also tend to accelerate microbial spoilage.

Incidence of microbial decay increases in chilling-susceptible fruits and vegetables when stored at low temperatures. This increased susceptibility to decay, which occurs by a mechanism that is not readily understood, can be decreased by storage above the critical temperature threshold of that commodity or by intermittent warming (cycling storage temperatures above and below the critical temperature threshold).[71]

Human pathogens, although not normally part of the natural flora of fruits and vegetables, can be introduced in the field by contact with the soil (particularly when animal manure is used as a fertilizer), at the packinghouse by fecal contamination in wash water or from handlers, or during transportation in vehicles used for hauling animals or wastes. Factors affecting the growth of decay microorganisms also favor the growth of human pathogens. When conditions favor growth, usually one microorganism or group of microorganisms will predominate. Specific factors such as growth temperature, water activity and pH determine whether the favored growth results in spoilage or a health hazard.[72] Another option is the use of biological control, a microorganism that can outcompete other microorganisms without resulting in spoilage or a safety risk.[73]

3. Consequences of Minimal Processing

Any cutting or slicing operation destroys the protection of the exocarp and provides access of microorganisms to the internal flesh. Knife edges are excellent devices for inoculating and dispersing both spoilage and pathogenic microorganisms. Differences in gaseous atmospheres can alter the competitive balance between competing organisms within the microflora. Studies with intact vegetables suggest that modified-atmosphere packaging (MAP) increases the total counts of spoilage microorganisms while decreasing incidence of spoilage.[74] A major concern of MAP treatments is that extension of shelf life by decreased spoilage may lead to sufficient growth of human pathogens to cause a health hazard.[68,75] Heat and low-dose irradiation can reduce microbial loads. Post-heating contamination and physiological damage to the fruit or vegetable tissue are the primary concerns associated with these techniques.[5]

Increased handling associated with minimal processing leads to increased potential for contamination via human handlers. Compliance with strict sanitation guidelines is critical to minimizing contamination.[75] Recent studies have demonstrated the ability of human pathogens like *Listeria*,[76] *Salmonella*[77] and *E. coli* O157:H7[78] to grow on minimally processed products.

V. QUALITY MANAGEMENT

A. MEASUREMENT

1. Purchase Characteristics

Sophisticated devices are available for both laboratory and on-line measurement of fruit and vegetable quality. Such instruments are useful in products that are uniformly colored and color is a primary indication of color or maturity.[79] Ground color can be used as an effective indicator of maturity for peaches and other fruits that show distinct color differences based on exposure of the fruit to the sun.[80] Delayed light emission can be used in developing a maturity index in fruits that contain chlorophyll.[81] Pattern recognition and video imaging techniques are being studied for potential use to screen out items with visual defects.[82] Visual evaluation is still the predominant means of evaluating quality defects in the laboratory[79] and removing them from the packaging line.[83]

Firmness of fruits and vegetables is usually measured by puncture (Magness-Taylor), deformation or firmness-to-the-touch. Firmness is frequently used as a maturity index in addition to a measure of quality.[79] Recent advances in the development of nondestructive measures of firmness may lead to on-line measurement.[84]

Aroma is the most difficult purchase characteristic to measure by instrumental means because volatile aromatic compounds do not develop until late in the ripening process and clear relationships between specific compounds and aroma quality are not well established. Measurement of respiration gases (carbon dioxide and ethylene accumulation or oxygen depletion) can be used to define ripening fruits physiologically.

2. Consumption Characteristics

The taste component of fresh fruits is usually measured by estimating total soluble solids by refractometry and acidity by titration. Nondestructive measures of soluble solids have been developed.[85] Volatile compounds are extracted by a wide range of methods, separated by gas chromatography and identified by gas chromatography/mass spectrometry.[86] Additional information can be obtained using aroma threshold values which provide information on the odor quality of each compound,[87] aroma dilution techniques[88] or gas chromatography-olfactometry.[89] All of these instrumental techniques are difficult to relate to consumer perception of quality. Thus, sensory analysis of descriptive panel is widely used to characterize flavor. Spider-web diagrams of sensory profiles are a particularly effective means of presenting this data.[90]

Since mouthfeel characteristics are also difficult to measure instrumentally, descriptive panels are frequently used. A texture profile analysis procedure was developed that has an instrumental counterpart.[91] Shear measurements are also used to estimate certain aspects of mouthfeel.[54]

3. Hidden Characteristics

Nutrient composition analysis is performed by standard techniques.[92] The most frequently measured micronutrients in fruits and vegetables are ascorbic acid, which is used as a freshness indicator, and ß-carotene, which is also an indicator of color. Nutritional quality of fruits and vegetables, particularly minerals, is not as much a function of nutrient composition as it is of bioavailability.[61]

Total plate count is a commonly employed but frequently misleading technique for estimating microbiological quality. Such counts provide little information on the potential for spoilage or compromised safety. Yeast and mold counts provide a useful indicator of sanitation practices in minimal processing operations. Coliform tests provide information on fecal contamination.[72] Assays of specific human pathogens are available.[93]

B. SHELF-LIFE EXTENSION
1. Physiological Basis

Shelf-life extension is maintenance of a product at an acceptable level of quality for a longer period of time than currently practiced. Such efforts are frequently for the benefit of a distributor in the handling chain. Shelf-life extension is achieved by slowing ripening and senescence processes. Techniques that decrease respiration and transpiration include temperature reduction, increased relative humidity, modification of gaseous atmospheres and application of chemical additives.[5] Removal of field heat is particularly critical in slowing respiration and transpiration, thus prolonging shelf life. In addition, harvest maturity affects storage stability of fruits and vegetables, particularly climacteric fruits. Generally speaking, the later climacteric fruit is harvested in the ripening process, the shorter its subsequent shelf life during marketing and distribution.

2. Consequences of Minimal Processing

Cutting and slicing operations tend to accelerate physiological degradation and microbial growth. Modified-atmosphere packaging, heat treatment and low-dose irradiation tend to extend shelf life. The net effect of minimal processing depends on many factors including the specific item, type of process and specific process variables.[94]

3. Quality Considerations

Shelf life is usually associated with maintenance of purchase characteristics. Management decisions that are made to extend the life of a product based on appearance or firmness without considering effects on flavor or mouthfeel run the risk of marketing a product that a consumer will be willing to purchase but will find unacceptable to consume.[2] Likewise, shelf-life extension could result in compromised nutritional or microbiological quality if hidden attributes are disregarded. Shelf-life models are needed that can assess the kinetics of degradation of critical attributes that can be followed using time-temperature indicators.[90]

C. INTEGRATED APPROACHES

1. Systems Approach

A systems approach views all unit operations as dynamic components of an integrated whole. Schematic diagrams of handling systems for many types of fresh fruits and vegetables are available.[95] Each operation has the potential of being affected by previous operations and affecting subsequent operations. In addition, feedback mechanisms may exist that affect preceding operations. In the postharvest handling system value is added to a product as defective items are discarded, nonedible material removed and convenience of preparation improved. Money paid through the system serves as a feedback mechanism.[96] Damage incurred during an operation but not evident until a later operation, known as latent damage, can reduce quality and economic return.[2] As we develop a better understanding of the postharvest system, it becomes apparent that preharvest factors are important in postharvest quality and stability. Thus, an understanding of integrated horticultural production that encompasses soil, water, nutrient, crop and pest management should be combined with postharvest management approaches.[60]

Systems thinking alters the perspective of handlers within the postharvest system. It permits optimization of the entire handling system rather than optimization of individual components.[2] For example, methods of shelf-life extension tend to benefit each handler within the system by providing additional time to hold the product with less risk of unacceptable product and the accompanying economic loss. If shelf life is extended, however, by sacrificing consumption quality, the consumer may become dissatisfied. Ultimately, consumer dissatisfaction may lead to reduced sales causing lower economic return. Such cause and effect relationships are difficult to determine if distribution is viewed as a series of fragmented, independent operations. A systems approach can serve as a basis for quality and fiscal management of the collective operations.

The location of processing operations in the handling system is critical for minimally processed products.[10,97] The post-processing shelf life, the sophistication of processing equipment, product value and the distribution system are all important factors. Delivery of the product at a consistent level of acceptable quality and acceptable price can be achieved by a careful matching of consumer requirements and a knowledge of physiological response to processing and storage conditions. Hazard Analysis Critical Control Point (HACCP) is a systematic approach used to identify and eliminate microbiological hazards and is recommended for minimally processed products.[67]

A systems perspective has identified two critical areas of future research: (1) a better understanding of consumer perception of fruit and vegetable quality and (2) identification and quantification of preharvest factors that affect postharvest quality as well as shelf stability.[2] In addition, advances are expected in combining knowledge of basic post-harvest physiology with advanced mathematical modeling techniques to predict postharvest performance of individual fruits under given storage and handling regimes based on non-destructive quality measures made during packing or processing operations.[98] Such models would permit sale of products at acceptable purchase quality and maximal consumption quality rather than maximal purchase quality with unknown consumption quality.

2. Total Quality Management

The quality revolution sweeping the world of manufactured goods is slow coming to the fresh and minimally processed fruit and vegetable industry.[99] Total Quality Management (TQM) has much to offer fresh and minimally processed companies. TQM emphasizes the importance of the consumer in setting quality specifications for products. It is a systematic tool to broaden responsibility of product quality to everyone within the handling system and not a select few who happen to be assigned quality matters in their job description. TQM eliminates inspection as a means to assure quality. Rather, it focuses attention on improving the process to eliminate quality defects.[100,101]

Implementation of TQM in the distribution of fresh fruits and vegetables represents some unique challenges. First, careful consideration must be given to terminology. What represents a "product"? If a carton of fruit leaving a packinghouse is the "product," then does that mean that each fruit is a "component"? If each fruit is a component, then does that mean that if one fruit in the box is unacceptable, then that fruit is a "defect" and thus the whole box is "defective"? Also, is the grading and sorting operation an "inspection" or a "process"? If it is an "inspection" then it should be phased out, but as a "process" it should undergo "continuous improvement" to reduce "defects". The biological variation observed in living, respiring fruits and vegetables makes absence of defects in incoming material seemingly impossible, but the screening out of defects prior to shipment is a difficult but potentially achievable goal. Elimination of defects during growth and development would require viewing crop production as a process with improvement achieved through genetic manipulation and advances in cultural practices. The emphasis of TQM on the measurable defects tends to favor management of purchase characteristics which are most readily measured at the expense of consumption and hidden characteristics which are more variable and more difficult to measure.

In minimally processed products the terminology problems become less difficult, but other concerns must still be faced. Trimming of unacceptable portions of a fruit or vegetable without discarding the whole item introduces a new way to decrease defects not available to distributors of whole, intact commodities. Shelf-life dating and subsequent product removal based on predictive modeling should decrease defective products bought by the consumer.[102] "Best if eaten by" dates with simple but sufficient storage directions place responsibility for quality on the consumer. The emphasis on measurability of defects and the effect that would have on purchase vs consumption characteristics would appear to represent the greatest challenge to implementation of TQM in minimally processed products.

Figure 2. Pictorial diagram of quality enhancement research.

3. Quality Enhancement

An emerging approach to quality improvement of existing products and to provision of a foundation for development of new products is quality enhancement. Quality enhancement is a focused approach to quality management that integrates the concepts of consumer-driven quality, a fundamental understanding of the basis of quality for a specific product, and predictive modeling as a basis of management decisions.

Quality enhancement research of minimally processed products is conducted in four phases: strategic, basic, modeling and product development, as shown in Figure 2. It is based on the premise that each product has a unique set of properties that contributes to quality. Key quality characteristics are determined by consulting the target consumer either through surveys or focus groups. A specific measure is developed in the laboratory using chemical, instrumental or descriptive sensory methods for each key characteristic identified by consumers. The efficacy of each measure is tested to determine if it differentiates accept-ability as evaluated by a consumer panel. When such measures are validated, then basic physiology studies are conducted to identify physiological principles responsible for each key quality characteristic and factors (genetic, cultural, environmental or processing) that affect these physiological principles. Next, a model is proposed that involves the manipu--lation of a factor or factors likely to result in the most significant improvement in quality. A factor analysis experiment is conducted to determine the effects on the measures developed. The final phase then involves consumer testing to determine if prototype products can be produced that represent a significant improvement and a significant decline (to test model validity) in acceptability by modification of a key quality characteristic.

Although no complete application of the quality enhancement approach has been published, specific components of the approach have been studied[103,104] and additional work is being conducted in our laboratories. A bright future in quality of improvement of minimally processed fruits and vegetables is expected through a judicious combination of systems, Total Quality Management, and quality enhancement principles.

REFERENCES

1. **Kramer, A. and Twigg, B. A.** *Quality Control for the Food Industry*, 3rd ed., AVI van Nostrand Reinhold, New York, 1970.
2. **Prussia, S. E. and Shewfelt, R. L.** Systems approach to postharvest handling, in *Postharvest Handling: A Systems Approach*, Shewfelt, R. L. and Prussia, S. E., Eds., Academic Press, San Diego, 1993, chap. 3.
3. **IFT.** Quality of fruits and vegetables - A scientific status summary by the Institute of Food Technologists' Expert Panel on Food Safety and Nutrition. *Food Technol.*, 44(6), 99, 1990.
4. **Land, D. G.** Negative influences on acceptability and their control, in *Food Acceptability*, Thomson, D. M. H., Ed., Elsevier, New York, 1988, chap. 88.
5. **Shewfelt, R. L.** Postharvest treatment for extending the shelf life of fruits and vegetables. *Food Technol.*, 40(5), 70, 1986.
6. **Watada, A. E., Herner, R. C., Kader, A. A., Romani, R. J. and Staby, G. L.** Terminology for the description of developmental stages of horticultural crops. *HortScience*, 19, 20, 1984.
7. **Shewfelt, R. L.** Quality of minimally processed fruits and vegetables. *J. Food Qual.*, 10, 143, 1987.
8. **Rolle, R. S. and Chism, G. W.** Physiological consequences of minimally processed fruits and vegetables. *J. Food Qual.*, 13, 157, 1987.
9. **Kader, A. A., Zagory, D. and Kerbel, E. L.** 1989. Modified atmosphere packaging of fruits and vegetables. *CRC Crit. Rev. Food Sci. Nutr.*, 28, 1, 1989.
10. **Huxsoll, C. C. and Bolin, H. R.** Processing and distribution alternatives for minimally processed fruits and vegetables. *Food Technol.*, 43(2), 124, 1989.
11. **Zind, T.** Fresh trends '90 - A profile of fresh produce consumers. *The Packer Focus*, 96, 54, 37, 1990.
12. **National Research Council.** *Alternative Agriculture.* National Academy Press, Washington, 1989.
13. **Cook, R.** The food safety controversy: Implications for the fresh produce industry. *USDA Outlook '90*, Session 15, 1990.
14. **How, B.** *Marketing Fresh Fruits and Vegetables.* van Nostrand Reinhold, New York, 1990.
15. **Gross, J.** *Pigments in Fruits.* Academic Press, Orlando, 1987.
16. **Gross, J.** *Pigments in Vegetables: Chlorophyll and Carotenoids.* van Nostrand Reinhold, New York, 1991.
17. **Francis, J.** Food colorants: Anthocyanins. *CRC Crit. Rev. Food Sci. Nutr.*, 28, 273, 1989.
18. **von Elbe, J. H.** The betalains, in *Current Aspects of Food Colorants*, Furia, T. E., Ed., CRC Press, Boca Raton, 1978, p. 29.
19. **Karnaukhov, V. N.** Carotenoids: Recent Progress, Problems and Prospects. *Comp. Biochem. Physiol.*, 95B, 1, 1990.
20. **Schwartz, S. J. and Lorenzo, T. V.** Chlorophylls in foods. *CRC Crit. Rev. Food Sci. Nutr.*, 29, 1, 1990.
21. **Vamos-Vigyazo, L.** Polyphenol oxidase and peroxidase in fruits and vegetables. *CRC Crit. Rev. Food Sci. Nutr.*, 15, 49, 1981.
22. **Jackman, R. L., Yada, R. Y., Marangoni, A., Parkin, K. L. and Stanley, D. W.** Chilling injury. A review of quality aspects. *J. Food Qual.*, 11, 253, 1988.
23. **Shear, C. B.** Calcium-related disorders of fruits and vegetables. *HortScience*, 10, 361, 1975.

24. **Ke, D. and Saltviet, M. E.** Regulation of russet spotting, phenolic metabolism, and IAA oxidase by low oxygen in iceberg lettuce. *J. Amer. Hort. Sci.*, 114, 638, 1989.
25. **Tucker, G. A. and Grierson, D.** Fruit ripening, in *Biochemistry of Plants: A Comprehensive Treatise*, Vol. 12, Stumpf, P. K. and Conn, E. E, Eds., Academic Press, New York, 1987, p. 265.
26. **Goldschmidt, E. E.** Pigment changes associated with fruit maturation and their control, in *Senescence in Plants*, Thimann, K. V., Ed., CRC Press, Boca Raton, 1980.
27. **Khudairi, A. K.** The ripening of tomatoes. *Amer. Sci.*, 60, 696, 1972.
28. **Sapers, G. M. and Douglas, F. W.** Measurement of enzymatic browning at cut surfaces in juice of raw apple and pear fruits. *J. Food Sci.*, 27, 465, 1992.
29. **Bolin, H. R. and Huxsoll, C. C.** Storage stability of minimally processed fruit. *J. Food Proc. Pres.*, 13, 281, 1989.
30. **Kim, A. D., Magnuson, J. A., Torok, T. and Goodman, N.** Microbial flora and storage quality of partially processed lettuce. *J. Food Sci.*, 56, 459, 1991.
31. **Yang, S. F. and Pratt, H. K.** The physiology of ethylene in wounded plant tissue, in *Biochemistry of Wounded Plant Tissue*, Kahl, G., Ed., 1978, p. 595.
32. **Bolin, H. R. and Huxsoll, C. C.** Control of minimally processed carrot (Daucus carota) surface discoloration caused by abrasion peeling. *J. Food Sci.*, 56, 416, 1991.
33. **John, M. A. and Dey, P. M.** Postharvest changes in fruit cell wall. *Adv. Food Res.*, 30, 139, 1986.
34. **Fischer, R. L. and Bennett, A. B.** Role of cell wall hydrolases in fruit ripening. *Ann. Rev. Plant Physiol. Plant Mol. Biol.*, 42, 675, 1991.
35. **Shackel, K. A., Greve, C., Labavitch, J. M. and Ahmadi, H.** Cell turgor changes associated with ripening in tomato pericarp tissue. *Plant Physiol.*, 97, 814, 1992.
36. **Yano, T., Kojima, I. and Torikata, Y.** Role of water in withering of leafy vegetables, in *Water Activity: Influences on Food Quality*, Rockland, L. B. and Stewart, G. F., Eds., Academic Press, New York, 1981, p. 765.
37. **Bolin, H. R., Stafford, A. E., King, A. D. and Huxsoll, C. C.** Factors affecting storage stability of shredded lettuce. *J. Food Sci.*, 42, 1657, 1977.
38. **Morton, I. D. and MacLeod, A. J.** *Food Flavours Part C. The Flavour of Fruits*, Elsevier, New York, 1990.
39. **Hansen, K. and Poll, L.** Conversion of L-isoleucine into 2-methylbut-2-enyl esters in apples. *Lebensm.-Wiss. u.-Technol.*, 26, 178, 1993.
40. **Williams, P. J., Sefton, M. A. and Wilson, B.** Nonvolatile conjugates of secondary metabolites as precursors of volatile grape flavor components, in *Flavor Chemistry: Trends and Developments*, Teranishi, R., Buttery, R. G. and Shahidi, F., Eds., American Chemical Society, Washington, 1989, p. 35.
41. **Crouzet, J.** Fruit aromatic components: mechanism and formation after harvest, in *Sixieme colloque sur la researches fruitieres*, 1987, p. 191.
42. **Hansen, K., Poll, L. and Lewis, M. J.** The influence of picking time on the postharvest volatile ester production of 'Jonagold' apples. *Lebensm.-Wiss. u.-Technol.*, 25, 451, 1992.
43. **Thomas, D. J., Parkin, K. L. and Simon, P. W.** Development of a simple pungency indicator for onions. *J. Sci. Food Agric.*, 60, 499, 1992.
44. **Takeoka, G. R., Buttery, R. G. and Flath, R. A.** Volatile constituents of Asian pears (Pyrus serotina), *J. Agric. Food Chem.*, 40, 1925, 1992.
45. **Pesis, E. and Avissar, I.** Effect of postharvest application of acetaldehyde vapour on strawberry decay, taste and certain volatiles. *J. Sci. Food Agric.*, 52, 377, 1990.

46. **Dirinick, D., Depooter, H. and Schamp, N.** Aroma development in ripening fruits, in *Flavour Chemistry: Trends and Developments*, Teranishi, R., Buttery, R. G. and Shahidi, F., Eds., American Chemical Society, Washington, 1989, p. 23.

47. **O'Mahony, M.** Taste perception, food quality, and consumer acceptance. *J. Food Qual.*, 14, 9, 1991.

48. **Kays, S. J.** *Postharvest Physiology of Perishable Plant Products*, van Nostrand Reinhold, New York, 1991.

49. **Stevens, M. A., Kader, A. A., Albright-Holton, M. and Algazi, M.** Genotypic variation for flavor and composition in fresh market tomatoes. *J. Amer. Soc. Hort. Sci.*, 102, 680, 1977.

50. **Chalutz, E., DeVay, J. E. and Maxie, E. C.** Ethylene-induced isocoumarin formation in carrot root tissue. *Plant Physiol.*, 44, 235, 1969.

51. **Juliot, K. N., Lindsay, R. C. and Ridley, S. C.** Directly-acidified carrot slices for ingredients in refrigerated vegetable salads. *J. Food Sci.*, 54, 90, 1989.

52. **Ben-Arie, R., and Lavee, S.** Pectic changes in Elberta peaches suffering from wooly breakdown. *Phytochemistry*, 10, 531, 1971.

53. **Matsuo, T., Ito, S. and Ben-Arie, R.** A model experiment for elucidating the mechanism of astringency in persimmon fruit using respiratory inhibitors. *J. Japan Soc. Hort. Sci.*, 60, 437, 1991.

54. **Everson, H. P., Waldron, K. W., Geeson, J. D. and Browne, K. M.** Effects of modified atmospheres on textural and cell wall changes of asparagus during shelf life. *Int. J. Food Sci. Technol.*, 27, 187, 1992.

55. **Abe, K. and Watada, A. E.** Ethylene absorbent to maintain quality of lightly processed fruits and vegetables. *J. Food Sci.*, 56, 1589, 1991.

56. **Ponappa, T., Scheerens, J. C. and Miller, A. R.** Vacuum infiltration of polyamines increases firmness of strawberry slices under various storage conditions. *J. Food Sci.*, 58, 361, 1993.

57. **Shewfelt, R. L.** Sources of variation in the nutrient control of agricultural commodities from the farm to the consumer. *J. Food Qual.*, 13, 37, 1990.

58. **Karmas, E. and Harris, R. S.** *Nutritional Evaluation of Food Processing*, 3rd ed., van Nostrand Reinhold/AVI, New York, 1988.

59. **Pattee, H. E.** *Evaluation of Quality of Fruits and Vegetables*, van Nostrand Reinhold/AVI, New York, 1985.

60. **Beverly, R. E., Latimer, J. G. and Smittle, D. A.** Preharvest physiological and cultural effects on postharvest quality, in *Postharvest Handling: A Systems Approach*, Shewfelt, R. L. and Prussia, S. E., Eds., Academic Press, San Diego, 1993, chap. 4.

61. **Clydesdale, F. M., Ho, C.-T., Lee, C. Y., Mondy N. I. and Shewfelt, R. L.** The effects of postharvest treatment and chemical interactions on the bioavailability of ascorbic acid, thiamin, vitamin A, carotenoids, and minerals. *Crit. Rev. Food Sci. Nutr.*, 30, 599, 1991.

62. **Hudson, D. E. and LaChance, P. A.** Ascorbic acid and riboflavin content of asparagus curing marketing. *J. Food Qual.*, 9, 217, 1986.

63. **Hudson, D. E., Cappelini, M. and LaChance, P.** Ascorbic acid content of broccoli during marketing. *J. Food Qual.*, 9, 31, 1986.

64. **Barth, M. M., Perry, A. K., Schmidt, S. J. and Klein, B. P.** Misting affects market quality and enzyme activity of broccoli during retail storage. *J. Food Sci.*, 57, 954, 1992.

65. **Klein, B. P.** Nutritional consequences of minimally processed fruits and vegetables. *J. Food Qual.*, 10, 179, 1993.

66. **Weichmann, J.** The effect of controlled-atmosphere storage on the sensory and nutritional quality of fruits and vegetables. *Hort. Rev.*, 8, 101, 1986.

67. **Brackett, R. E., Smallwood, D. M., Fletcher, S. M. and Horton, D. L.** Food safety: Critical points within the production and distribution system, in *Postharvest Handling: A Systems Approach*, Shewfelt, R. L. and Prussia, S. E., Eds., Academic Press, San Diego, 1993, chap. 15.

68. **Hotchkiss, J. H. and Banco, M. J.** Influence of new packaging technologies on the growth of microorganisms in produce. *J. Food Prot.*, 55, 815, 1992.

69. **Nicholson, R. L. and Hammerschmidt, R.** Phenolic compounds and their role in disease resistance. *Ann. Rev. Phytopathol.*, 30, 369, 1992.

70. **Darvill, A. G. and Albersheim, P.** Phytoalexins and their elicitors - a defense against microbial infection in plants. *Ann. Rev. Plant Physiol. Plant Mol. Biol.*, 42, 651, 1991.

71. **Wang, C. Y.** Chilling injury of fruits and vegetables. *Food Rev. Int.*, 5, 209, 1989.

72. **Brackett, R. E.** Microbial quality, in *Postharvest Handling: A Systems Approach*, Shewfelt, R. L. and Prussia, S. E., Eds., Academic Press, San Diego, 1993, chap. 6.

73. **Wilson, C. L. and Wisniewski, M. E.** Biological control of postharvest diseases of fruits and vegetables: An emerging technology. *Ann. Rev. Phytopathol.*, 27, 425, 1989.

74. **Brackett, R. E.** Influence of modififed atmosphere packaging on the microflora and quality of fresh bell pepper. *J. Food Prot.*, 53, 255, 1990.

75. **Hurst, W. C. and Schuler, G. A.** Fresh produce processing - an industry perspective. *J. Food Prot.*, 55, 824, 1992.

76. **Beuchat, L. R. and Brackett, R. E.** Behavior of *Listeria monocytogenes* inoculated into raw tomatoes and processed tomato products. *Appl. Environ. Microbiol.*, 57, 1367, 1991.

77. **Golden, D. A., Rhodehamel, E. J. and Kautter, D. A.** Growth of *Salmonella* spp. in cantaloupe, watermelon and honeydew melons. *J. Food Prot.*, 56, 194, 1993.

78. **Abdul-Raouf, U. M., Beuchat, L. R. and Ammar, M. S.** Survival and growth of *Escherichia coli* O157:H7 on salad vegetables. *Appl. Environ. Microbiol.*, 59, 1999, 1993.

79. **Shewfelt, R. L.** Measuring quality and maturity, in *Postharvest Handling, A Systems Approach*, Shewfelt, R. L. and Prussia, S. E., Eds., Academic Press, San Diego, 1993, chap. 5.

80. **Delewiche, M. J. and Baumgardner, R. A.** Ground color as a peach maturity index. *J. Amer. Soc. Hort. Sci.*, 110, 53, 1985.

81. **Gunasekaran, S.** Delayed light emission as a means of quality evaluation of fruits and vegetables. *CRC Crit. Rev. Food Sci. Nutr.*, 29, 19 1990.

82. **Miller, B. K. and Delewiche, M. J.** Peach defect detection with machine vision. *Trans. ASAE*, 34, 2588, 1991.

83. **Bollen, F., Prussia, S. E. and Lidror, A.** Visual inspection and sorting: Finding poor quality before the consumer does, in *Postharvest Handling: A Systems Approach*, Shewfelt, R. L. and Prussia, S. E., Eds., Academic Press, San Diego, 1993, chap. 9.

84. **Chen, P. and Sun, Z.** A review of nondestructive methods for quality evaluation and sorting of agricultural products. *J. Agric. Res.*, 49, 85, 1991.

85. **Dull, G. G., Birth, G. S., Smittle, D. A. and Leffler, R. G.** Near infrared analysis of soluble solids in intact cantaloupe. *J. Food Sci.*, 54, 393, 1989.

86. **Chitwood, R. L., Pangborn, R. M. and Jennings, W.** CG/MS and sensory analysis of volatiles from three cultivars of capsicum. *Food Chem.*, 11, 201, 1983.

87. **Teranishi, R., Buttery, R. G., Stern, D. G. and Takeoka, G.** Use of odor thresholds in aroma research. *Lebensm. Wiss. u. Technol.*, 24, 1, 1991.
88. **Grosch, W.** Detection of plant colorants in foods by aroma dilution analysis. *Trends Food Sci. Technol.*, 4, 68, 1993.
89. **Acree, T. E., Barnard, J. and Cunningham, D. G.** A procedure for the sensory analysis of gas chromatographic effluents. *Food Chem.*, 14, 273, 1984.
90. **Lyon, G. B., Robertson, J. A. and Meredith, F. I.** Sensory descriptive analysis of cv. Cresthaven peaches - Maturity, ripening and storage effects. *J. Food Sci.*, 58, 1771, 1993.
91. **Szczesniak, A. S., Brandt, M. A. and Friedman, H. H.** Development of standard rating scales for mechanical parameters of texture and correlation between the objective and the sensory methods of texture evaluation. *J. Food Sci.*, 28, 397, 1963.
92. **Eitenmiller, R. R.** Strengths and weaknesses of assessing vitamin content of foods. *J. Food Qual.*, 13, 7, 1990.
93. **Doyle, M. P.** *Foodborne Bacterial Pathogens*, Marcel Dekker, New York, 1990.
94. **Labuza, T. P., Fu, B. and Taoukis, P. S.** Prediction for shelf life and safety of minimally processed CAP/MAP chilled foods: A review. *J. Food Prot.*, 55, 741, 1992.
95. **Kader, A. A.** *Postharvest Technology of Horticultural Crops*, University of California, Division of Agriculture and Natural Resources Publication 3311, 1992.
96. **Schoorl, D. and Holt, J. E.** Quality, reputation and price in horticultural markets. *Agric. Syst.*, 13, 191, 1984.
97. **Cantwell, M.** Postharvest handling systems: Minimally processed fruits and vegetables, in *Postharvest Technology of Horticultural Crops*, Kader, A. A., Ed., University of California Division of Agriculture and Natural Resources Publication 3311, 1992, chap. 32.
98. **Thai, C. N.** Modeling quality characteristics, in *Postharvest Handling: A Systems Approach*, Shewfelt, R. L. and Prussia, S. E., Eds., Academic Press, San Diego, 1993.
99. **Lidror, A. and Prussia, S. E.** Applications of quality assurance techniques to production and handling agricultural crops. *J. Food Qual.*, 13, 171, 1990.
100. **Crosby, P. B.** *Quality is Free*, McGraw-Hill, New York, 1979.
101. **Juran, J. M.** *Quality Control Handbook,* McGraw-Hill, New York, 1988.
102. **Labuza, T. P.** *Open Shelf Life Dating of Foods*, Food and Nutrition Press, Westport, CT, 1982.
103. **Cheng, T.-S. and Shewfelt, R. L.** Effect of chilling exposure of tomatoes during subsequent ripening. *J. Food Sci.*, 53, 1160, 1988.
104. **Galvez, F. C. F. and Resurreccion, A. V. A.** Reliability of the focus group technique in determining quality characteristics of mungbean (*Vigna radiata* (L) wilczec) noodles. *J. Sens. Stud.*, 7, 315, 1992.

DISCUSSION

BUSTA: Discuss the balance between microbial and physiological spoilage of cut or sliced fruits and vegetables. such as cleaned "mini" carrots or cantaloupe (melon) cubes or balls.

SHEWFELT: The balance between microbial and physiological degradation of the plant tissue in these products is the primary factor in shelf life. Many factors including the microbial load, stress on the tissue and storage conditions affect the balance. The most important consideration should be that the product spoil (either by microbial or physiological means) before it becomes a safety hazard.

CAMPBELL-PLATT: Did you suggest that there are increased vitamin losses under refrigeration? What happens under modified atmosphere packaging?

SHEWFELT: Refrigeration slows the rate of loss of vitamins in fruits and vegetables, but large losses can be incurred during refrigeration. Most consumers and many scientists are not aware of the magnitude of these losses. The small amount of data on nutrient losses to modified atmosphere packaging is highly variable, based on treatment and commodity. Much more data are needed to draw meaningful conclusions.

KNORR: In terms of developing concepts for minimal processing, is it useful to induce stress response of plants or to suppress it?

SHEWFELT: In general we wish to minimize wound response as ethylene is released and most changes are deleterious. Use of short term stress can actually help maintain quality in certain cases. There are published instances of stress conditioning to ameliorate chilling injury and preharvest water stress affecting flavor.

LAZARIDES: Regarding consumer appeal it is interesting to notice that the consumers have been conditioned to pay so much attention to the appearance characteristics that they prefer buying a round, well shaped, medium size tomato which lacks satisfactory flavor, over a bad (uneven) shaped, large tomato with excellent flavor (taste/aroma), would you like to comment on this?

SHEWFELT: When quality is defined in terms of the consumer, quality characteristics are targets to be met by the processor. It takes marketing to change human perception.

LAZARIDES: The issue of pesticides is ranked very low in your list of safety concern issues, when we all know that the problem of pesticide residues has become an issue of major concern for today's consumer. Who has suggested this ranking?

SHEWFELT: Regulatory scientists rate microbiological hazard as the most important food borne safety hazard and pesticides as much less threatening. Consumers rate pesticides as a greater concern than microbiological hazards. Dr. Christine Bruhn will have more recent information on this topic in her lecture.

KNORR: Regarding the development of concepts for minimal processing, would it be helpful/useful to postpone size reduction during processing and to try to surface decontaminate.

SHEWFELT: When microbiological stability is the major limitation in shelflife, some type of surface decontamination prior to cutting and slicing is merited. A delay in processing close to distribution is another way to extend post-processing market life. I must emphasize that we need to develop systems that optimize quality at an acceptable shelf life rather than to extend shelf life at an acceptable quality.

PROCESS CONTROL AND QUALITY ASSURANCE THROUGH THE APPLICATION OF HACCP AND PREDICTIVE MICROBIOLOGY

Dr. Jeffrey G. Banks

Within the subject area of 'Process Optimisation and Minimal Processing of Foods', assurance of food safety is of paramount importance. As industry continues the search for new technologies to deliver high quality, inexpensive and novel foods to the increasingly demanding consumer, safeguards must be designed into processes and preservation procedures. This paper will focus on the application of Hazard Analysis Critical Control Point (HACCP) systems and microbiological modelling in safety and quality assurance. Other management and analytical tools are also available to assist the food technologists to achieve consistently good results and these will be discussed in this article - see Table 1.

Western Europeans live longer and healthier lives today than at any time in their history, yet they seem preoccupied with risks of health, safety and the environment. Many advocates, such as industry representatives promoting unpopular technology or government agencies, e.g., United States Environmental Protection Agency defending its regulatory stance, argue that the public has a bad sense of perspective. Evidence suggests, however, that the public is eminently sensible about the risks they face. Recent decades have witnessed significant declines in the rate and social acceptability of smoking, widespread shifts towards low-fat, high-fibre diets and improvements in road safety[1]. All these developments reduce the chance of premature death at little cost. Public trust in risk management has declined, but ironically the discipline of risk analysis has matured as we have seen with the implementation of HACCP. It is now possible to examine potential hazards in a rigorous, quantitative manner.

Table 1
Management and Analytical Tools for Process Control and Assurance of Quality

- Quality management
- Predictive modelling
- HACCP
- Biological process validation
- Laboratory accreditation
- Process monitoring
- Validation of analytical methods
- Hygiene sensing

The traditional approaches to the control of food safety and quality have been largely based on inspection and sampling/testing regimes - 'quality control'. These often fail to discriminate clearly between what is desirable and what is essential and tend to be very unstructured and retrospective mechanisms providing companies with poor assurance of achieving the required product standards. The introduction of quality management systems has provided companies with a structured framework for the definition and implementation of procedures to enable consistent manufacture of products of the required quality. Such as systems allow companies to shift emphasis to a preventative philosophy where product safety and quality standards are achieved by design rather than by inspection and sampling/testing. Food companies across the world (particularly in the U.K. and Europe) are now implementing the International Quality standard ISO 9000[2] which is equivalent to the British standard BS 5750 as well as hazard analysis and critical control points (HACCP) into their work (Table 2). Many are realising measurable benefits in areas such as compliance with legislation, increased confidence in product safety, consistent product quality standards, and third-party acceptance of performance[3].

Table 2
Components of the International Standards Organisation (ISO) System and Corresponding Components of the British Standard (BS) and European Norm (EN) systems

ISO Number	BS Number	EN Number
ISO 9000	BS 5750	EN 29000
ISO 9001	BS 5750 Part 1	EN 29001
ISO 9002	BS 5750 Part 2	EN 29002
ISO 9003	BS 5750 Part 3	EN 29003
ISO 9004	BS 5750 Part 4	EN 29004

In some countries, microbiological specifications for end-products have been built into legislation and have tended to take precedence over factory inspection by representatives of the regulatory agencies. In other countries there has been a tradition of inspection of the processing environment. The U.K. has tended to be in the latter category and only very recently has the U.K. started to consider the incorporation of simple microbiological specifications into the legislation, and so far only in respect to some food products.

It is well known that examination of samples from batches to determine whether they meet an end-product specification cannot ensure that the consumer is protected against receiving product below specification unless 100% inspection can be employed. Such levels of inspection are rarely achievable as all microbiological and most chemical analyses are destructive, so that only a small proportion of the total batch can be examined. Nevertheless, end-product specifications are useful for carrying out acceptance sampling plans, e.g., when a country wishes to control the import of unwholesome foods[3]. Inspection of production facilities should be able to ensure that manufacturing and processing follows appropriate codes of good manufacturing practice. In most countries the activities of the inspectors in their in-factory assessments of hazards and the steps taken to control those hazards have tended to be left to individual interpretation. This is where the application of HACCP and similar systems will increase markedly the efficiency and utility of such inspections. At the present time much effort has been directed towards the communication of the principles and application of HACCP to food businesses. It is of concern that little attention has been given to systems for the food inspector that are based on the same sound principles of monitoring, verification and review.

The increase in the acceptance and adoption rates for quality management systems in the food industry gives some indication of the recent appreciation of the value of their implementation. Predictive and preventative tools such as HACCP[4], Hazard analysis/risk assessment (HARA) and Hazard analysis and operability systems (HAZOP)[5] are perceived as being costly in terms of setting up and maintenance, however, there are significant benefits - see Table 3.

Table 3
Benefits of HACCP Implementation

- Cost-effective control of foodborne hazards
- Preventative approach
- Focusing of resources, e.g., auditing
- Training for team members
- Contribution to due diligence
- Faster introduction of technical change
- Compliance with external pressures

In section 21 of the U.K. Food Safety Act 1990, it is stated that there is the possibility of a defence of due diligence being offered in prosecutions brought under the Act or under Regulations arising therefrom. This legislation has accelerated the introduction of HACCP, total quality management (TQM) systems and certification of premises under ISO 9001 (BS 5750, Part 1)[6] or ISO 9002 (BS 5750, Part 2) as food manufacturers and retailers see these steps as helping to support the plea that they have been exercising due diligence. Some U.K. supermarket chains now require the suppliers/manufacturers of the supermarket's 'own-label products' to have HACCP plans (Table 4).

Table 4
The Seven Basic Principles of HACCP[7]

1) Preparation of a flow diagram of the process.
2) Identification of critical control points (CCP's) using a decision tree.
3) Establishment of target levels and tolerances of each CCP which must be met to ensure that each CCP is under control.
4) Establishment of a monitoring system to ensure control of CCP by scheduled testing or observations.
5) Establishment of corrective action to be taken when monitoring indicates that a CCP is trending to 'out of control'.
6) Establishment of documentation concerning all practices and procedures.
7) Establishment of verification procedures which may include supplementary tests and a review that shows the HACCP system is effective.

Risk analysts start by dividing hazards into two parts: exposure and effect. Exposure studies look at the ways in which a person or product might be subjected to change; effects studies examine what may happen once that exposure has manifested itself. Exposure to a hazard may cause a complex chain of events leading to one of a number of effects, but analysts have found that the overall result can be modelled by a function that assigns a single number to any given level. A simple linear relationship, for instance, accurately describes the average cancer risk incurred by smokers: 10 cigarettes a day generally increases it by a factor of 50. For other risks, however, a simple dose-response function is not appropriate, and more complex models must be used. The study of exposure and effects is fraught with uncertainty. Indeed, uncertainty is at the heart of the definition of risk[1].

For other risks, such as those involving new processing or preservation technologies or those in which bad outcomes occur only rarely, uncertainty enters the calculations at a higher level - overall probabilities as well as individual events are unpredictable. If good actuarial data are not available, analysts must find other methods to estimate the likelihood of exposure and subsequent effects. The development of risk assessment during the past two decades has been in large part the story of finding ways to determine the extent of risks that have little precedent. In one common technique, failure mode and effect analysis (a forerunner to HACCP) workers try to identify all the events that might help cause a system to break down. Then they compile as complete a description as possible of the routes by which those events could lead to a failure. This has been identified as a useful though elaborate technique for the food processing industry and is called HAZOP[5]. When risk specialists must estimate the likelihood that a part will fail or assign a range of uncertainty to an essential value in a model, they can sometimes use data collected from similar systems elsewhere. In other cases, however, historical data are not available. Sometimes workers can build predictive models to estimate probabilities based on what is known about roughly similar systems, but often they must rely on expert subjective judgement. Because of the way people think about uncertainty, this approach may involve serious biases. Even so, quantitative risk analysis retains the advantage that judgements can be incorporated in a way that makes assumptions and biases explicit.

Extensive studies of public risk perception have discovered that the situation is very subtle. When people are asked to order well-known hazards in terms of the number of injuries they cause every year, on average they can give an accurate estimation. If, however, they are asked to rank those hazards in terms of risk, they produce quite a different order. People also rank risks based on how well the process in question is understood, how equitably the danger is distributed, how well individuals can control their exposure and whether risk is assumed voluntarily[1].

These factors can be combined into three major groups. The first is basically an event's degree of dreadfulness (as determined by such features as the scale of its effects and the degree to which it affects "innocent" bystanders). The second is a measure of how well the risk is understood and the third is the number of people exposed. These groups of characteristics can be used to define a "risk space". Where a hazard falls within this space says quite a lot about how people are likely to respond to it. Risks carrying a high level of "dread", for example, provoke more calls for government intervention that do some more workaday risks that actually cause more deaths or injuries.

People tend to underestimate the frequency of very common causes of death-stroke, cancer, accidents - by roughly a factor of 10. They also overestimate the frequency of very uncommon causes of death (botulism poisoning, for example) by as much as several orders of magnitude[1].

Combining management and analytical tools can be particularly effective[8]. For example, predictive microbiological monitoring can be applied to several of the HACCP principles. Recent advances in modelling techniques, data accumulation and database building, particularly for the key pathogens in food, now offer the microbiologist significantly faster and less expensive methods to appraise the likely safety of hypothetical or real situations that may involve microbial growth, survival or death[9]. In the area of shelf life determination, there are several methods that the industry will use (Table 5).

Table 5
Methods Used to Predict Microbial Shelf Life

- "Expert" judgement
- Storage trials
- Accelerated testing
- Challenge testing
- Predictive microbiology

Incorrect prediction of the shelf life may result in increased production and distribution costs if the shelf life were underestimated and poor quality or unsafe material if the shelf life were overestimated. Expert judgement is inexpensive, but is subjective and small changes to the product formulation may not be recognised by an individual or group of experts as having significance. In practice, such changes may have a profound effect on the stability and safety of the product. Storage trials involve the keeping of the product until spoilage is evident or until the numbers of pathogens or production of toxins has reached unacceptable levels. The shelf life that may be eventually set will probably equal the time period established above minus some empirically derived figure that is considered to be sufficient safety margin in terms of time. Accelerated testing may be rapid, however, it may permit the growth of organisms that would not normally proliferate in the product. Therefore the results can be misleading or totally useless and great caution must be used in the interpretation of such data. Similarly, the unusual conditions of storage may accelerate death of organisms and also result in an overestimation of stability. Challenge testing is objective but is relatively expensive and limited by the conditions specified and tested in the challenge. The benefits of predictive microbiological modelling in such a scenario are summarised in Table 6.

Table 6
Benefits of Predictive Microbiological Modelling

- Quantitative results
- Powerful and flexible
- Cost effective
- Rapid results
- Valuable base line information

Predictive modelling that integrates microbial behaviour with other process variables has now begun to find favour with food process engineers. Computational fluid dynamics are now being used to predict flow, temperature and distribution of organisms and consequently provide information on the likely extent of lethality of an aseptic process. Hygiene specialists are considering modelling of microbial survival and growth on food contact surfaces in order to predict the optimal time intervals between necessary cleaning and disinfection. The likely response of a variety of populations of spoilage organisms to different concentrations of disinfectants is also being researched. One result of multifactorial modelling which is often overlooked is the ability to identify the components of most and least significance to the overall fate of the chosen organism. For example, one can test the contribution of several interacting factors such as pH, acid type, water activity and temperature and identify the component that least or most affects growth. This is of immense value in the optimisation of process or preservation procedures since it allows engineering of the process/preservation parameters to effect most benefit at least cost in terms of quality, money and convenience.

In terms of integration of predictive modelling with HACCP principles[10] (Table 4), with principle 1, a database can be interrogated to provide information to assess the likelihood of growth or survival or the organisms of concern. This is of value in hazard assessment/analysis. For principle 2, modelling can provide data to set the lowest preservative level which would inhibit growth, set the scheduled heat process or estimate the consequences of deviations from normal operating procedures. The predictions can be extremely powerful and allow the food technologist to consider reformulation, shelf life limitations and the need for experimental validation work. With principle 3, predictions can estimate if chosen target levels would be exceeded and if so, under what combination of conditions. If deviations to accepted practice may occur, these would be considered under the 5th HACCP principle and predictive data are useful in planning what type and extent of corrective action is necessary. For example, if a monitoring system indicated that a batch of food had not received a full scheduled process, there may be predictive data to estimate the additional lethality required in future (corrective) processing. Principle 6 can be addressed in part by the output from the database. As a strength of the system will be the almost infinite number of "what if?" questions that may be asked, the models will provide substantial and compelling documentation and evidence for a due diligence defence and information for auditing, inspection and verification[11].

Audits and inspections may also form part of the quality management systems and they can provide a monitoring mechanism to gauge the quality of the process. Clearly the quality system will scrutinise the essential operating characteristics of equipment and procedures which are important in quality and safety. For example, attention would be given to the calibration and appropriate servicing of production and monitoring/sensing machinery. Similarly, quality systems will address the quality and reliability of analytical measurements made by supporting services, e.g., laboratories. In the recent past, laboratory accreditation systems have gained wider acceptance as managers begin to appreciate the importance of truly reliable data. Among the many aspects for appraisal, those dealing with standard operating procedures, methodology, proficiency and competence will be critical in terms of provision of reliable results.

Monitoring of processes is intended to supply data that will automatically effect a control or prompt a decision made by 'responsible and competent' personnel. Monitoring is an integral part of a HACCP system; control chart methodology can be a simple yet effective mechanism that is understandable to all levels of personnel and will ensure that corrective action is taken to bring a process that is 'trending out of control', back under control. Decisions made with such control charts can be made with a degree of confidence and where the risk associated with a particular hazard is high, can be constructed to minimise the chance of the process going out of control. Control is active and not passive since the charts will detect impending problems before defective end products are made.

Reliable and timely (therefore useful) monitoring tools are unlikely to be based on conventional microbiological methodologies and recent research has focused on ultrasonic and vibration sensors for detecting the buildup of soil and microbial material on open and closed process equipment. For example, when food products are aseptically processed the pipework of the plant can become covered in a layer of deposit; this fouling has an effect on product flowrate, and in extreme cases, on the integrity of the delivered thermal process. It is clear that a means of measuring the fouling deposits would be beneficial as would a method for measuring the degree of cleanliness achieved by a CIP process in terms of both mineral and microbiological fouling materials. A recent collaborative EC project has evaluated the use of ultrasound for the monitoring and measurement of fouling films as they form on the inside of UHT processing plant[12]. The sensor can be viewed as having three functions: the detection of the presence of a film on a surface, the measurement of buildup of film thickness with time, and the determination of the endpoint of a cleaning operation. For ultrasound sensing, ultrasound is transmitted across a pipe carrying product. The received signal is analysed for time and amplitude changes which are caused by the thickness and type of the fouling film. As the thickness of fouling layer increases the ultrasonic signal is increasingly attenuated. In addition, the time of flight of the ultrasound in the fouling material is seen to change with fouling thickness. Analysis of these two effects leads to an estimate of the thickness of the fouling film deposited on the surface. From information about the velocity of sound as a function of temperature of the likely foulant materials and of the product itself, a measurement of the fouling deposit thickness can be made. The vibration sensor is being developed as a means of monitoring the degree of fouling in food processing equipment such as plate heat exchangers. In principle the sensor detects changes in the vibrational characteristics of the plant as the fouling builds up on the surfaces. Vibrational transducers are attached to the surface of the plate heat exchanger and their response to an excitation signal is used as a means of fouling detection and measurement. The transducers are based on piezoelectric crystals. Laboratory experiments have shown that the sensing system is capable of detecting small amounts of foulant on the plates of the heat exchanger[12]. Monitoring systems that detect microbial ATP are used in some industries to provide immediate feedback information on the cleanliness of a critical control point in a process line. Research efforts are focused on improvement of the sensitivity and reliability of such sensors.

As this paper has a central theme of provision of information to the microbiologist in terms of the quality, speed, reliability and depth of information it is salutary to note that decisions in industry are often based on slow, uncertain and incomplete information which is of little value for monitoring at Critical Control Points (CCP's) or building trend analysis data bases[13]. There are many drawbacks to the current schemes for pathogen testing of primary, intermediate and end products as well as environmental samples. Generally speaking the deficiencies can be related to slow, retrospective, expensive and low quality information. Many currently used methods cannot trace or track contamination and as a result are of limited utility. In the Campden Technical Manual 38 "The Application of HACCP", the seven principles of the HACCP system are described, see Table 4. Whilst much microbiological testing is questionable, appropriate monitoring tools could be used to great effect at CCP's. If one accepts a general scheme for poultry processing as leading from: Breeder flocks to eggs, broilers, killing, scalding, defeathering, automated evisceration, washing, air/water chilling, deboning and butchering, there are at least four recognised CCP's. These are at the plucking, evisceration, washing and chilling points. With particular reference to *Staphylococcus aureus*, it has been known that incoming poultry and equipment may be colonised by various types of the organism. As commonly used and available techniques are only able to give incomplete information about the phenotypic and genetic character of an isolated staphylococcus, poor understanding and process control result. As many different types of *Staphylococcus aureus* may exist in a poultry processing plant, the mere presence of the organism is of limited value to the microbiologist. Indeed, some of the conventional techniques for the further differentiation of the organism can be summarised as follows: At the genus level (*Staphylococcus*) techniques are slow and media based. At the species level (*Staph. aureus*) they are slow, part media and part biochemically based. At the phage type level the methods are slow, expensive and require a higher level of skill. Generally few laboratories have the capability to accommodate the different phage sets and the problem is compounded as such typing is impossible with some strains. At the biotype level there is a lack of agreement on descriptors and at the level of the plasmid profile the information may be variable. Staphylococci isolated from food and farm animals were studied with a general method[14] based on analysis of genomic Eco RI endorestriction nuclease fragments containing portions of the ribosomal Ribonucleic acid (rRNA) operons. Sets of fragments for strains classically identified with *Staphylococcus aureus* and *Staphylococcus warneri* were sorted into two distinguishable and species-specific clusters (Figure 1).

For DNA preparation and processing, bacteria were lysed, DNA was digested to completion with Eco RI. Restriction fragments were separated according to size by gel electrophoresis, transferred to a membrane, dried and probed. Enzyme-triggered chemiluminescence was detected by a customised CCD camera. Size-separated genomic fragments that hybridised with the standard probe are shown in Figure 1. Each strain was represented in a database by a pattern showing the position and intensity of its bands. The patterns were normalised to a universal standard for stretch or compression due to differences in the electrophoresis and for image enhancement. The patterns were ordered by similarity using correlation to determine nearest neighbours[15].

In our experimental design we wished to choose a target in the bacterial genome that had a balance between conserved and variable regions of nucleic acid apparatus. We required sufficiently conserved areas of information to ensure inclusivity of *Staphylococcus* as a genus with sufficiently variable regions to enable differentiation between different species and subtypes of *Staphylococcus*.

From our results we were able to demonstrate high discriminating power between *Staphylococcus* and other bacteria with no interference from non-*Staphylococcus* species as well as excellent discrimination between species and subtypes[16]. On Figure 1, we demonstrate the clear separation between the patterns derived for two different types of *Staphylococcus warneri* and five different types of *Staphylococcus aureus*. We have been able to demonstrate that the five discernible clusters of *Staphylococcus aureus* shown in Figure 1 are derived from different animal hosts - Man, cattle, poultry, rats and monkeys. From factory isolations we have been able to locate the source and track the spread of contamination of particular strains of the enterotoxigenic staphylococci.

In conclusion we have demonstrated a very powerful and generally applicable tool to assist the food microbiologist in the rapid and definitive characterisation of organisms of concern[17].

Figure 1
Computer enhanced membrane image of size separated and probed genomic fragments of rRNA from
Staphylococcus aureus* and *Staphylococcus warneri

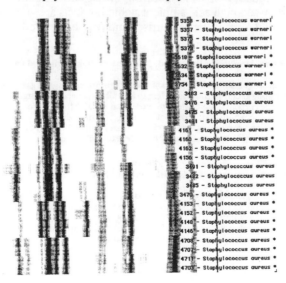

Staphylococcus warneri, cluster A: strains 5358, 5357, 5375 and 5373; *St. warneri*, cluster B: strains 3618, 3632, 3634, and 3754; *St. aureus*, cluster C: strains 3483, 3476, 3475 and 3481; *St. aureus*, cluster D: strains 4161, 4160, 4162 and 4156; *St. aureus*, cluster E: strains 3491, 3482, 3485 and 3470; *St. aureus*, cluster F: strains 4153, 4152, 4148 and 4146; *St. aureus*, cluster G: strains 4708, 4707, 4711 and 4703.

Process design (Table 7) and validation (Table 8) are important aspects of the assurance of both the Safety and Quality of Products. The design of adequate processes for various foods is part of a preventative approach to food safety and quality and as such forms part of any HACCP plan. Indeed, one of the CCP's of many manufacturing processes is the achievement of a set process. The process would be designed to obtain a specific reduction in numbers of pathogenic or spoilage organisms which were considered to be of most significance to the product under the specific formulation and storage conditions.

Table 7
Essential Steps in Setting a Scheduled Process

- Identify product characteristics that control growth in foods, e.g., pH, NaCI, heat
- Determine microorganisms most likely to grow under specified conditions to cause spoilage or safety hazard
- Establish heat resistance of target organisms in food
- Design appropriate heat treatment to achieve defined reduction in numbers of target organisms
- Validate process lethality

Whilst microbiological validation cannot be used as a routine process control device, it forms an essential part of the design and commissioning of new processes or new formulation of existing products.

Table 8
Essential Steps in Validation of a Scheduled Process

- Choose appropriate biological validation tool, e.g., alginate particle technique
- Immobilise target organism in appropriate matrix, e.g., simulated food particulates
- Process biological validation markers with food
- Evaluate efficacy of set process by level of destruction
- Compare microbial data with routine process control device, e.g., calibrated thermocouple
- Monitor process with routine control device

Failure to obtain the set process will result in a loss of quality of the product and it is therefore important that appropriate process monitoring and control measures are in place. The rationale behind the setting of processes is the same regardless of the heat or other factors applied, however, the specific process requirements and controls will change accordingly to each product and manufacturing situation. Examples of processes for aseptically produced low and high acid ambient stable products are given in Tables 9 and 10 as well as an example of a microwave pasteurised and perishable product in Table 11.

Table 9
Process Setting for an Aseptically Processed, Ambient Stable, High Acid (pH <4.6) Product

- Target organisms: Acid tolerant spore forming spoilage organisms
- Controlling factor in process: Heat to kill all target spoilage organisms
- Set process: equivalent to more than 93.3°C for 10 min.
- Validation: Completed during commissioning
- Biological validation used to measure level of kill and compared with thermocouple data
- Routine process control: pH checks and on line temperature check using calibrated thermocouples

Table 10
Process Setting for an Aseptically Processed, Ambient Stable, Low Acid (pH <4.6) Product

- Target organism: Proteolytic *Cl. botulinum* (grows above pH 4.6)
- Controlling factor in process: Heat to kill all target organisms
- Validation: Completed during commissioning
- Biological validation tool used to measure level of kill and compared with thermocouple data
- Routine process control: On line temperature checks using calibrated thermocouples

Table 11
Process Setting for a Microwave Pasteurised, Perishable Product

- Target organism: *Listeria monocytogenes* (psychrotrophic)
- Controlling factors in process: Heat to kill 6 log orders of *L. monocytogenes*
- Chilled storage to prevent growth of mesophilic pathogens
- Set process: Equivalent to more than 70°C for 2 min.
- Validation: Completed during commissioning
- Biological validation used to measure level of kill and compared with thermocouple data
- Routine process control: On line temperature checks using calibrated thermocouples and chill storage temperature checks

CONCLUDING REMARKS

In summary, the future prospects for application of HACCP and predictive microbiological modelling to process control and quality assurance appear good. There is now wide acceptance of the benefits of these management and analytical tools. The challenges will be in their implementation across industry sectors and particularly in the small and medium sized companies. More emphasis should be given to transfer of the information and technology to assist in such implementation.

REFERENCES

1. **Morgan, M.,** Risk Analysis and Management, Scientific American, July, 32-41, 1993.
2. **International Standards Organisation,** Quality Management and Quality Assurance Standards (Guideline for Selection and Use) (ISO 9000), 1987.
3. **Harrigan, W. F.,** The ISO 9000 series and its implications for HACCP Food Control, 4(2), 105-111, 1993.
4. **Mayes, T.,** The application of management systems to food safety and quality, Trends in Food Science & Technology, July (4), 1993.
5. **Kletz, T.,** HAZOP and HAZAN, Identifying and assessing process industry hazards, Institution of Chemical Engineers, 1992.
6. **British Standards Institution,** British Standard Quality Systems BS 5750, 1987.
7. **Campden Food and Drink Research Association,** HACCP: A practical guide, CFDRA Technical Manual 38, Chipping Campden, Glos. U.K., 1992.
8. **Jouve, J. L.,** ed HACCP and Quality Systems, Proceedings of the 3rd World Congress on Foodborne Infections and Intoxications, 16-19 June 1992 Berlin, pp 881-883, Institute of Veterinary Medicine, Robert von Ostertag Institute, Berlin, Germany, 1992.
9. **Walker, S. J., and Jones, J. F.,** Predictive microbiology: data and model bases, Food Technology International Europe, pp 209-212, 1992.
10. **Davis, S. C., and Banks, J. G.,** Predictive microbiology - applications to chilled food microbiology, Proceedings of the 2nd ASEPT Conference, Laval, France, 10-11 June, pp 35-42, 1992.
11. **Adams, C. E., and Banks, J. G.,** Applying HACCP to sous-vide products, Proceedings of the 2nd ASEPT Conference, Laval, France, 10-11 June, pp 151-168, 1992.
12. **Withers, P., and Taylor, J.,** Sanitation of Food Processing Plants, Technical Memorandum, No. 656, Campden Food and Drink Research Association, Chipping Campden, Glos, U.K., 1992.
13. **Banks, J. G., and Doraiswamy, K. C.,** A New Microbiological Monitoring Tool for Critical Control Points, Proceedings of an EC Workshop of Elimination of Pathogenic Microorganisms from Poultry, FLAIR Concerted Action 6, Fribourg, Switzerland, 1993, in press.
14. **Webster, J., Bruce, J., Cole, E., Neubauer, J., Wyer, J., Betts, R., and Banks, J., Bannerman, T., Ballard, D., and Kloos, W.,** Typing of Bacteria Through the Analysis of the Ribosomal RNA Genes, Proceedings of the American Society of Microbiology, Atlanta, USA, 1993.
15. **Webster, J., Hubner, R., Cole, E., Bruce, J., Neubauer, J., Wyer, J., Betts, R., and Banks, J.,** Identification of Members of the Genus *Listeria* to the Level of Species by Analysis of the Organisation of their Ribosomal RNA Genes, Proceedings of the American Society of Microbiology, Atlanta, USA, 1993.
16. **Neubauer, J., Webster, J., Hubner, R., Cole, E., Bruce, J., Iem, C., Bannerman, T., Ballard, D., and Kloos, W.,** Identification of *Staphylococcus warneri* by Observed and Predicted Patterns of EcoR1 Fragments Containing Ribosomal RNA Sequences, Proceedings of the Keystone Molecular Biology Symposium, Keystone, USA, 1993.
17. **Earnshaw, R. G., and Gidley, J. M.,** Molecular methods for typing bacterial pathogens, Trends in Food Science and Technology, 3, 39-43, 1992.

DISCUSSION

GORRIS: 1) How elaborate is the "ribo-typing" (DuPont) data base at the moment? Will it identify any bacterium yet?
2) In your opinion, with regard to predictive mathematical modeling, experts should interface between the databases and the user. The Food Micromodel program assembled in the UK intends to distribute its databases on a floppy disk. You loose the expert interface then. What about liability?

BANKS: 1) The recognition database is extensive and at present is focused on the 4 key pathogens in food: <u>Salmonella</u>, <u>E. Coli</u>, <u>Listeria</u> <u>monocytogenes</u> and <u>Staph. aureus</u>. If an unknown strain is analyzed and belongs to one of the above genera, it will be identified with great certainty. If the unknown strain is not in the computers recognition database, molecular characterization will still provide valuable data.
As the computer will have the capability to build its own data base as unknown strains are analyzed, the value of the characterization will be realized. For example, the food processor could recognize a persistent spoilage organism, locate the source or track the spread of a pathogen or monitor the performance of a beneficial, fermentative organism.
2) To stimulate wider use of the database, a PC-version is essential. Software programmers ensure that the customers interface is appropriate in terms of accuracy, ease of use and also in terms of interpretation of the predictions given. I am certain that liability issues will be completely addressed before the product launch. I do not anticipate that this will be a major problem.

BRUHN: What is the most important factor to increase adoption of HACCP to all industries?
What support is available for the small company?

BANKS: The challenge is to transfer information to these small companies in a simple, jargon-free manner. Until recently, simple user guides were not available. Now the International Life Sciences Institute and the European Communities (FLAIR) group have both published excellent guides. Two other initiatives should also catalyse uptake by food companies:
Firstly, the Campden Food and Drink Research Association will launch a microcomputer-based software application for HACCP that will assist companies in all aspects of HACCP implementation.
Secondly, a multi-media video will be produced in 1994 that will graphically demonstrate how HACCP can be applied to real-life situations.

BRUHN: Do you envision predictive modeling extending to the food production sector of the food chain? If so, how do you believe it will occur?

BANKS: Predictive modeling of microbial behavior in primary ingredients and intermediate materials will indeed be applied. Such predictions should permit appropriate setting of minimal processes to achieve safe products whilst maintaining quality.

BUSTA: What is the time frame for ribotyping of plant samples i.e. time required from sampling to final identification of microorganism?

BANKS: Until recently, molecular biologists would require several days to complete this labor and skill-intensive procedure. With the DuPont automated system, however, all the steps are completed in a user-independent way within 8 hours. Data output is in the form of identification to known organisms in the database or molecular characterization of any strain regardless or whether it is in the recognition database.

MICROBES, HURDLES, FOOD SAFETY AND PROCESS OPTIMIZATION

Edmund A. Zottola

INTRODUCTION

The concept of food safety broadly covers many compounds, chemicals, toxins or organisms that are capable of causing undesirable effects in humans. These include pesticides, herbicides, heavy metals, mycotoxins, poisons, naturally occurring toxins, microbial toxins, and pathogenic microorganisms. Of these, the critical problems of concern in food safety are those associated with microbial contamination of foods. [Zottola and Smith, 1990] The U.S. Food and Drug Administration and the Center for Disease Control have estimated that over the time period 1973 to 1987, 87% of the outbreaks of foodborne illness were caused by bacteria (Table 1). Control of microbial growth of foods has been a major concern of humans since the beginning of recorded history. The fermentation of grape juice to wine or the sun drying of fruits, fish and meat are examples of food preservation methods that have been used by humans for centuries. These methods are still in use today.

When plants or animals are harvested for food, they must be preserved in some manner to control microbial growth or the food will rapidly spoil and become unusable by humans. Therefore, the major concern of food preservation becomes the control of microbial growth to prevent spoilage and the development of microbial toxins making the food undesirable for use. Food preservation processes have to be designed to prevent these undesirable changes in the food. The requirements for food preservation are much different today than when wine making and the sun drying of foods were first developed. Today's consumers want high quality foods that are fresh, safe, and wholesome; thus, the interest in minimizing processing to assure "freshness" and "wholesomeness" and still maintain the safety of the food.

Many of the processes in use today were designed to produce "commercial sterility" foods; that is, food that will not spoil under normal conditions of handling, storage and distribution and are free of pathogenic organisms, their spores and toxins. Such processes have a detrimental effect on the perceived quality of the food, texture, color and flavor. In today's market, these qualities are important to the consumer, and foods processed in this manner may not be attractive. If the safety and quality of these minimally processed foods are to be maintained, the processes used must control those microorganisms of major concern; that is, those that cause spoilage or cause disease.

INTRINSIC AND EXTRINSIC PARAMETERS

Process optimization to obtain safe, high quality, fresh-like foods must be concerned with methods to control spoilage microbes as well as pathogenic microbes. If the processes do not adequately control both of these types of organisms, the food will not meet consumer expectations and will create a number of undesirable problems for the processor. To optimize processes to control microbes, one must be aware of those factors which affect the continued existence of the organisms. In 1955, Mossel and Ingram defined these factors as the intrinsic and extrinsic parameters of foods that influence the response of the organisms to the food. Intrinsic parameters are those characteristics of plant and animal tissues that are an inherent part of the tissue and would include pH, moisture content (water activity), oxidation-reduction potential (Eh), nutrient content, antimicrobial constituents, and biological structures. Extrinsic parameters of foods are those factors external to the food such as the storage environments that affect both the food and the microorganisms. Three major extrinsic parameters are of importance in influencing microbial response in/on the food. These are (1) the relative humidity of the environment, (2) the temperature of storage, and (3) the presence and concentration of gases in the environment. [Mossel and Ingram, 1955]

The intrinsic parameters of a food can be changed and thus influence microbial response to the food. Change in the acid content of a food either by fermentation or by the direct addition of acid will dramatically influence the type and kind of microbes that will grow in the food. The same can be done with the extrinsic factors. Changing the concentration of gases in the environment around the food will alter the microbial response in the food. If these factors are substantially changed, such as the use of very high temperatures, the microbes will be destroyed by the lethal temperatures. Changes of a lesser magnitude, such as the organism being stressed but not killed, will result in injured microbes. Those organisms that receive a sublethal stress are able to recover from the stress and eventually grow once again. In attempting to optimize a process to produce a fresh-like food, it must be recognized that control of microbial spoilage and pathogenic bacteria is achieved by manipulating the intrinsic and extrinsic parameters that influence microbial response in the food. It must also be recognized that less than lethal changes will only injury the organisms and the recovery from the injury will occur in time. Therefore, if one wishes to minimally process food and end up with a product that is safe and acceptable to the consumer, more than one intrinsic or extrinsic parameter must be modified to achieve the goal.

COMBINATION OF PRESERVATION METHODS

The concept of combining food preservation methods to prolong the usable life of a food product is not a new or novel idea. It has been utilized for many years throughout the food industry. The diary industry, for example, has utilized the combination of heat (pasteurization), refrigeration, and packaging to extend the shelf-life of fluid milk from 1 or 2 days, if not processed, to more than 3 weeks. The dairy industry also practices asepsis, or controlling or limiting contamination from farm production to the processing facility. The red meat industry has successfully extended the shelf-life of fresh meats by practicing asepsis during slaughter and processing by using low temperature storage of the raw product and vacuum packaging of the primal cuts. By utilizing these techniques the keeping time of fresh red meat is extended from 3 or 4 days to 3 or 4 weeks. In the thermal processing of foods in a hermetically sealed container, where commercial sterility is achieved, the primary preservation process is the thermal process; but a secondary preservation process, the use of vacuum in the hermetically sealed container, also is involved. Without the vacuum and the hermetically sealed container the food would not remain commercially sterile for very long. It should be apparent that the use of more than one method to preserve food has been utilized successfully for many decades. Unfortunately, these processes do not provide enough shelf-life, such as with fluid dairy products, or are detrimental to the perceived quality of the food product, such as with canned foods.

THE HURDLES CONCEPT

Utilization of control of the intrinsic or extrinsic parameters of the food will markedly affect the response of microbes in/on the food. The use of a single factor, such as reducing the water activity to less than 0.85, will inhibit microbial growth, but such a change in a food product may make it unacceptable to the consumer. Reducing the water activity to 0.90 and utilizing a second limiting factor, such as low temperature storage, may have the desired preservation effect. This concept of utilizing several sublethal preservation methods has been dubbed "the hurdles concept" by Leistner and Rodel (1976). Theoretically, the microbe is stressed by each of these treatments and must overcome each of the "hurdles" before growth can be initiated. If each sublethal treatment is considered as a step in a ladder, then the organism must overcome each step before growth can be initiated. This concept is shown in Figure 1. A ladder with five steps is shown. Each step represents a preservation method; temperature, pH, water activity, Eh, and a preservative. The stressed organism must climb each step (overcome the hurdle) before growth is possible. If the system is properly designed, the organism is unable to reach the top before the end of the shelf-life of the product.

In order to successfully use the "hurdles concept" in the production of minimally processed, fresh-like foods, additional quantitative data are needed. [Scott, 1988, 1989] Extensive data of this type have been developed for *Clostridium botulinum* in meats by Roberts, et al. (1981a, 1981b, 1981c, 1982) and Gibson, et al. (1987); for Salmonellae in meats by Gibson, et al. (1988); and for pasteurized process cheese spreads. [Tanaka, et al., 1986] Data on other pathogenic microorganisms such as *Listeria monocytogenes, Salmonella, Staphylococcus aureus*, and *Aeromonas* have been developed by Petran and Zottola (1989), Palumbo, et al. (1991), Buchanan and Phillips (1990), Buchanan (1991), Buchanan and Klawitter (1992) and Zaika, et al. (1992).

The type of quantitative data that must be generated for these undesirable organisms include the influence of changes in the intrinsic and extrinsic parameters on the growth rate of these microbes. Petran and Zottola (1989) studied the effect of temperature, pH, carbon source, and water activity on the growth of *Listeria monocytogenes*. The data generated in this study are shown in Tables 2, 3 and 4. These investigators used the generation time as the criteria for optimum growth when studying temperature, and pH and maximum population attained when evaluating carbon source and water activity.

If the data in Table 2 are examined, it is noted that the optimum pH for growth of this organism is 7.0, where a generation time of 44.7 minutes was observed. Deviation from the optimum, higher or lower, resulted in increased generation times; 371 minutes at pH 4.7 and 179 minutes at pH 9.2. The optimum growth temperature observed was at 37° C, (Table 3) with a generation time of 39.8 min. If the temperature was increased to 45° C, the generation time increased to 52.3 minutes. A decrease in temperature to 4° C increased the generation time to 33.5 hours.

The data in Table 4 are interesting. Somewhere between a water activity of 0.92 and 0.88 growth is inhibited. But the carbohydrate source data show that little if any growth occurs when lactose or fructose are the carbon source, whereas significant growth was noted with sucrose and galactose. These two appear to be the sugars of choice for growth of this organism.

What happens if all of this information is combined to formulate a system that will control or inhibit the growth of listeria? If the lower limits of each growth factor are

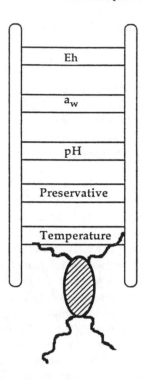

Figure 1. The "hurdles concept." The microorganism must climb each step or overcome each of these sublethal processes: temperature, preservative, pH, a_w and Eh before growth can be initiated.

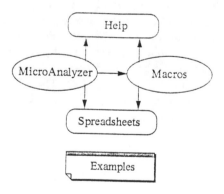

Figure 2. MicroAnalyzer program icons.

Table 1. Number and percentage of foodborne outbreaks and cases by etiologic agent (1973-1987). Totals

Etiologic Agent	Outbreaks		Cases	
	Number	Percent	Number	Percent
Bacterial	1,869	66	108,906	87
Viral	135	5	10,630	9
Parasitic	140	5	1,004	1
Chemical	697	25	4,454	4
TOTAL	2,841	100	124,994	100

Table 2. Growth of Listeria monocytogenes Scott A at different pH values in Tryptic Soy Broth.

Temperature (° C)	pH	Generation Time
30	4.0	no growth
30	4.1	no growth
30	4.3	no growth
30	4.5	no growth
30	4.7	371 min
30	5.0	182 min
30	6.0	52.0 min
30	7.0	44.7 min
30	8.0	50.1 min
30	9.0	146 min
30	9.2	179 min
30	9.4	no growth
30	9.6	no growth
30	9.8	no growth
30	10.0	no growth

Table 3. Growth of Listeria monocytogenes Scott A at different temperatures in Trypic Soy Broth.

Temperature (°C)	pH	Generation time
4.0	7.0	33.5 hr
7.0	7.0	13.1 hr
10.0	7.0	9.6 hr
13.0	7.0	287 min
20.0	7.0	113 min
30.0	7.0	45.4 min
35.0	7.0	42.9 min
37.0	7.0	39.8 min
45.0	7.0	52.3 min

Table 4. Growth of Listeria monocytogenes Scott A in different carbohydrates for 24 hr at 30°.

Carbohydrate (in TSB base)	A_w	pH Initial	pH Final	Log max. population[a]
0.5% lactose	0.99	7.0	7.0	3.11
0.5% fructose	0.99	7.0	7.0	3.11
0.25% galactose	0.99	7.1	7.0	8.11
0.5% sucrose	0.99	7.0	6.1	8.11
9.1% sucrose	0.97	7.0	4.4	8.77
17.0% sucrose	0.96	7.0	4.5	8.75
23.0% sucrose	0.96	7.0	4.6	8.16
28.6% sucrose	0.94	7.0	5.1	8.47
33.3% sucrose	0.94	7.0	6.7	8.38
35.%% sucrose	0.93	6.9	5.9	7.57
37.5% sucrose	0.93	6.8	6.2	7.60
39.4% sucrose	0.92	6.8	6.5	6.49
50.0% sucrose	0.88	6.8	6.6	n.g.[b]
60.0% sucrose	0.87	6.8	6.7	n.g
65.0% sucrose	0.82	6.7	6.7	n.g

used, i.e., pH between 4.7 and 5.0, temperature of 4° C or less, water activity between 0.92 and 0.88, and fructose instead of sucrose, the growth of listeria may be retarded or prevented. Thus, these factors should be considered when formulating a food product that will not support the growth of listeria.

Quantitative data such as those described above have been developed and used by several investigators to develop predictive models for the response of several microorganisms to changes in the intrinsic and extrinsic parameters of food products.

PREDICTIVE MICROBIOLOGY

The use of predictive models to determine microbial responses in food systems is a relatively new concept. These predictive models utilize intrinsic and extrinsic parameters of microbial growth in defined food systems and describe microbial growth kinetics under the defined conditions. Several predictive microbiology modeling systems have been developed, including those developed by Roberts and his group at the AFRC Institute of Food Research, Reading, England, [Roberts, et al., 1981a, 1981b, 1981c, 1982; Gibson, et al., 1987, 1988] Buchannan and his group at USDA, ARS, Philadelphia, [Buchanan, et al., 1989; Buchanan, 1991; Buchanan, 1993; Buchanan and Phillips, 1990; Buchanan and Klawitter, 1992] and, more recently, one developed by Fu and Labuza at the University of Minnesota. [Fu and Labuza, 1993; Labuza, et al., 1992] The "MicroAnalyzer Program" developed in the laboratories of my colleague, T. P. Labuza, will be briefly described in this communication. The program has five icons which are shown in Figure 2. The starting macro is the Microanalyer.

The core of the program involves the analysis of available quantitative microbial growth data which includes growth curves, temperature dependence models and temperature/a_w models. The detailed structure and design of the program are shown in Figure 3. The program uses as it basic format, Microsoft Excel 4.0. It is relatively easy to use and allows input of specific growth parameters by the user. The user selects the organism(s) of concern, enters the selected growth parameters, and the program does the rest. If desired, a printout of the results can be obtained.

For example, if predicted growth of a pathogenic bacteria is desired, the particular pathogen is selected from the menu. The growth parameters are selected and predicted growth curve prepared. In the examples shown here, *Listeria monocytogenes* and *Salmonella* were selected. For *L. monocytogenes*, the following conditions were selected: 200 hours, aerobic, NaCl, 3.0%, pH, 5.5, 12° C with an initial population of 10 cfu/gm and a final population of 10^6/gm. The predicted growth curve and data unpitied are shown in Figure 4. The predicted time to reach 10^6/gm is 308.1 hr; the lag phase duration is 233.88 hr. If only the atmospheric conditions are changed from aerobic to anaerobic, the predicted growth curve would be that shown in Figure 5. Under anaerobic conditions, the time to reach 10^6 cells/gm is 159.2 hr. The change in the Eh brought about by the change in the atmosphere, shortened the generation time of the organism.

The growth parameters selected for *Salmonella* were pH 5.2, NaCl, 2.0%, 10°C, 220 hr, and an initial population of 1,000 cfu/gm and final population of 10^6 cfu/gm. Under these conditions, time to reach 10^6 cfu/gm would be 338 hr and the lag phase would be over 188 hr. The predicted growth curve is shown in Figure 6. If only the pH is changed to 6.5, the time required to reach 10^6 cfu/gm would be reduced to 158 hr and the lag phase 69 hr.

These two examples illustrate the utility of these predictive microbiology modeling programs. Use of this type of information should be very useful in the optimization of food processes to maintain maximum safety, wholesomeness and quality. But, if these

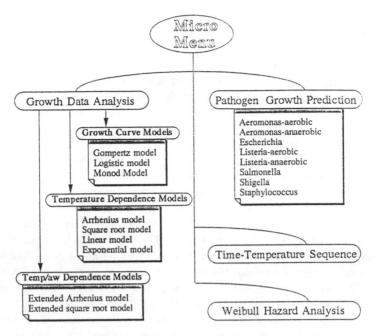

Figure 3. Structures and design of the MicroAnalyzer program.

Predicted Growth of   *Listeria monocytogenes*
(Aerobic)

Food Composition			Storage Condition		
Sodium Chloride (NaCl): (0.5—4.5 %)	3.0	%	Temperature (5—37 °C)	12	°C
Initial pH: (4.5—7.5)	5.5		Time (0—600 hr)	200	hr
Sodium Nitrite (NaNO2): (0-1000 μg/g)	0	μg/g			

	Assumed initial contamination level (No):	**10**	*cfu/g*

Gompertz Parameters		Derived Growth Kinetics Parameters		
A:	1.000	Exponential growth rate:	0.069	Log(cfu/g)/hr
C:	8.500	Generation time:	4.37	hr
M:	279.30	Lag phase duration:	233.88	hr
B:	0.022	Maximum population density:	9.50	log(cfu/g)

If the final level of concern (Nf) is:	**1.00E+06**	*cfu/g,*	*then*
the predicted time to reach this level is:	**308.1**	hr.	

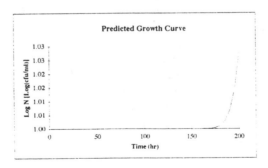

Predicted Growth Curve

Figure 4. MicroAnalyzer program predicted growth of *Listeria monocytogenes* under aerobic conditions and other parameters listed.

Predicted Growth of *Listeria monocytogenes*
(Anaerobic)

Food Composition			Storage Condition		
Sodium Chloride (NaCl): (0.5—4.5 %)	3.0	%	Temperature (5—37 °C)	12	°C
Initial pH: (4.5—7.5)	5.5		Time (0—600 hr)	200	hr
Sodium Nitrite (NaNO2): (0-1000 µg/g)	0	µg/g			

	Assumed initial contamination level (No):	10	cfu/g

Gompertz Parameters		Derived Growth Kinetics Parameters		
A:	1.000	Exponential growth rate:	0.303	Log(cfu/g)/hr
C:	8.340	Generation time:	0.99	hr
M:	152.37	Lag phase duration:	142.23	hr
B:	0.099	Maximum population density:	9.34	log(cfu/g)

If the final level of concern (Nf) is: **1.00E+06** *cfu/g,* *then*
the predicted time to reach this level is: **159.2** *hr.*

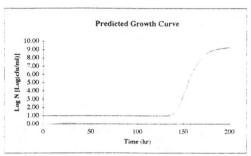

Figure 5. MicroAnalyzer program predicted growth of *Listeria monocytogenes* under anaerobic conditions and other parameters listed.

Predicted Growth of *Salmonella*

Food Composition			Storage Condition		
Sodium Chloride (NaCl): (0.5—4.5 %)	2.0	%	Temperature (10—30 °C)	10	°C
Initial pH: (5.6—6.8)	6.5		Time (0—600 hr)	220	hr
Sodium Nitrite (NaNO2): (0 µg/g)	0	µg/g			

	Assumed initial contamination level (No):	1000	cfu/g

Gompertz Parameters		Derived Growth Kinetics Parameters		
A:	3.000	Exponential growth rate:	0.034	Log(cfu/g)/hr
C:	6.230	Generation time:	8.93	hr
M:	137.08	Lag phase duration:	69.07	hr
B:	0.015	Maximum population density:	9.23	log(cfu/g)

If the final level of concern (Nf) is: **1.00E+06** *cfu/g,* *then*
the predicted time to reach this level is: **158.4** *hr.*

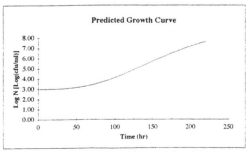

Figure 6. MicroAnalyzer program predicted growth of Salmonella at pH 5.2 and other parameters listed.

Predicted Growth of *Salmonella*

Food Composition			Storage Condition		
Sodium Chloride (NaCl): (0.5—4.5 %)	2.0	%	Temperature (10—30 °C)	10	°C
Initial pH: (5.6—6.8)	5.2		Time (0—600 hr)	220	hr
Sodium Nitrite (NaNO2): (0 µg/g)	0	µg/g			

Assumed initial contamination level (No):	**1000**	*cfu/g*

Gompertz Parameters		Derived Growth Kinetics Parameters		
A:	3.000	Exponential growth rate:	0.020	Log(cfu/g)/hr
C:	6.230	Generation time:	14.95	hr
M:	302.35	Lag phase duration:	188.57	hr
B:	0.009	Maximum population density:	9.23	log(cfu/g)

If the final level of concern (Nf) is:	*1.00E+06*	*cfu/g,*	*then*
the predicted time to reach this level is:	*338.0*	*hr.*	

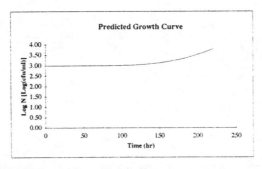

Figure 7. MicroAnalyzer program predicted growth of Salmonella at pH 6.5 and other parameters listed.

modeling programs are used without any actual inoculation and incubation studies to determine if what is said is what you get, then the user may have some foreseen problems with microorganisms. Unfortunately, microbes do not follow the same rules as humans and they do not always react in the expected manner. Therefore, it is imperative that actual shelf-life or pathogen inhibition studies be carried to make sure the predictions hold true.

I would like to leave you with one last thought, and that is the Harvard Law: "Under the most rigorously controlled conditions of pressure, temperature, humidity and other variables, the organism will do as it damn well pleases!" [Bloch, 1979].

REFERENCES

Bloch, A., *Murphy's Law and Other Reasons Why Things Go Wrong*, Price/Stern/Sloan Publishers, Inc., Los Angeles, CA, 1979.

Buchanan, R. L., Stahl, H. G., and Whiting, R. C., Effects and interactions of temperature, pH, atmosphere, sodium chloride and sodium nitrite on the growth of *Listeria monocytogenes, J. Food Protection*, 52, 844, 1989.

Buchanan, R. L., Using spreadsheet software for predictive microbiology applications, *J. Food Safety*, 11, 123, 1991.

Buchanan, R. L. , Predictive food microbiology, *Trends Food Science and Technology*, 4, 6, 1993.

Buchanan, R. L. and Phillips, J. G., Response surface model for predicting the effects of temperature, pH, sodium chloride content, sodium nitrite concentration and atmosphere on the growth of *Listeria monocytogenes, J. Food Protection*, 53, 370, 1990.

Buchanan, R. L. and Klawitter, L. A., The effect of incubation temperature, initial pH, and sodium chloride on the growth kinetics of *Escherichia coli* 0157:H7, *Food Microbiology*, 9, 185, 1992.

Fu, B. and Labuza, T. P., Shelf-life prediction: theory and application, *Food Control* (In press), 1993.

Gibson, A. M., Bratchell, N., and Roberts, T. A., The effect of sodium chloride and temperature on the rate and extent of growth of *Clostridium botulinum* type A in pasteurized pork slurry, *J. Appl. Bacteriol.*, 62, 479, 1987.

Gibson, A. M., Bratchell, N., and Roberts, T. A., Predicting microbial growth: growth responses of salmonellae in a laboratory medium as affected by pH, sodium chloride and storage temperature, *Intl. J. Food Microbiol*, 6, 155, 1988.

Labuza, T. P., Fu, B., and Toukis, P. S., Prediction for shelf life and safety of minimally processed CAP/MAP chilled foods, *J. Food Protection*, 55, 741, 1992.

Leistner, L. and Rodel, W., *The Stability of Intermediate Moisture Foods with Respect to Microorganisms*, Applied Science, London, 1976.

Mossel, D. A. A., and Ingram, M., The physiology of the microbial spoilage of foods, *J. Applied Bacteriology*, 18, 232, 1955.

Palumbo, S. A., Williams, A. C., Buchanan, R. L., and Phillips, J. G., Model for the aerobic growth of *Aeromonas hydrophila* K144, *J. Food Protection*, 54, 429, 1991.

Petran, R. L. and Zottola, E. A., A study of factors affecting growth and recovery of *Listeria monocytogenes* Scott A, *J. Food Science*, 54(2), 458, 1989.

Roberts, T. A., Gibson, M., and Robinson, A., Prediction of toxin production by *Clostridium botulinum* in pasteurized pork slurry, *J. Food Technology*, 16, 337, 1981a.

Roberts, T. A., Gibson, M., and Robinson, A., Factors controlling the growth of *Clostridium botulinum* types A and B in pasteurized, cured meats, I., Growth in pork slurries prepared from "low" pH meat (pH range 5.5-6.3), *J. Food Technology*, 16, 239, 1981b.

Roberts, T. A., Gibson, M., and Robinson, A., Factors controlling the growth of *Clostridium botulinum* types A and B in pasteurized, cured meats, II., Growth in pork slurries prepared from "high" pH meat (range 6.3-6.7), *J. Food Technology*, 16, 267, 1981c.

Roberts, T. A., Gibson, M., and Robinson, A., Factors controlling the growth of *Clostridium botulinum* types A and B in pasteurized, cured meats, III., The effect of potassium sorbate, *J. Food Technology*, 17, 307, 1982.

Scott, V. N., Safety considerations for new generation refrigerated foods, *Dairy and Food Sanitation*, 8(1), 5, 1988.

Scott, V. N., Interaction of factors to control microbial spoilage of refrigerated foods, *J. Food Protection*, 52(6), 431, 1989.

Tanaka, N., Traisman, E., Plantinga, P., Finn, L., Flom, W., Meske, L., and Guggisberg, J., Evaluation of factors involved in antibotulinal properties of pasteurized process cheese spreads, *J. Food Protection*, 49, 526, 1986.

Zaila, L. L., Phillips, J. G., and Buchanan, R. L., Model for aerobic growth of *Shigella flexneria* under various conditions of temperature, pH, sodium chloride and sodium nitrite concentrations, *J. Food Protection*, 55, 509, 1992.

Zottola, E. A. and Smith, L. B., The microbiology of foodborne disease outbreaks: an update, *J. Food Safety*, 11, 13, 1990.

THE APPLICATION OF MOLECULAR BIOLOGY TO UNDERSTANDING THE GROWTH AND SURVIVAL OF FOOD POISONING AND SPOILAGE BACTERIA

Annette Ellison, Martin B Cole & Gordon S A B Stewart

INTRODUCTION

In the food industry today new product development is a major growth area and process conditions are frequently modified to enhance product quality and improve production efficiency. To maintain food safety and quality, further understanding of the growth and physiology of food poisoning and spoilage bacteria is required. This is especially true with regard to interactions with the environment and other bacteria. Molecular biology based methods are being used increasingly to aid understanding in this area. For instance, using recombinant DNA techniques bacteria of specific relevance to the food industry can be engineered to emit light. Such light emission allows for the real time non destructive observation of cell metabolism as dead cells produce no light and injured cells produce less than optimal light. The use of bioluminescence in food microbiology is increasing (Baker *et al.*, 1992; Stewart and Williams, 1992) and one very important area is the development of rapid methods to detect and enumerate micro-organisms so that microbial assays can be brought into real-time rather than their current retrospective position. It is the aim of this chapter to briefly overview bioluminescence, to present results from a research area where its application has been beneficial and to discuss the potential of future applications.

Bacterial Bioluminescence

The bioluminescent reaction (described in detail by Meighen, 1988 and 1991) is catalysed by the enzyme luciferase and involves the oxidation of $FMNH_2$ and a long chain fatty aldehyde. This results in the production of FMN, the conversion of aldehyde to the corresponding fatty acid and the emission of blue-green light. Most of the energy for the production of light is obtained from the oxidation of aldehyde (60%), oxidation of $FMNH_2$ provides the remainder of the energy required (40%). The presence of aldehyde therefore plays an important role in the bioluminescent reaction.

The organisation and regulation of *lux* genes which are essential for expression of a bioluminescent phenotype has been determined and has recently been comprehensively reviewed (Meighen, 1991). The *luxA* and *B* genes code for the luciferase subunits α and β respectively, *luxC, D* and *E* code for reductase, acyl-transferase and synthetase respectively which constitute the fatty acid reductase complex required for the synthesis of the aldehyde substrate. There are two regulatory genes *luxI* and *R*, a number of additional *lux* genes (*luxF, luxG* and *luxH*) coding for proteins of uncertain function and two structural genes *luxY* and *luxL* which encode proteins that modify the wavelength of photon emission (O'Kane and Prasher, 1992).

Many marine micro-organisms are capable of emitting light and are classified into four major genera: *Vibrio, Photobacterium, Altermonas* and *Xenorhabdus*. Most micro-organisms do not possess the genes for luciferase and fatty acid reductase but are able to supply $FMNH_2$ therefore for a dark bacterium to become bioluminescent all that is needed is the genetic transfer of the *lux* genes (Stewart and Williams, 1992; Hill *et al.*, 1991; Hill *et al.*, 1993)). Aldehyde can be added exogenously in chemical form to bacterial cultures as it is freely diffusible across the prokaryotic membrane and can be supplied in sufficient excess so it is not limiting (Blisset and Stewart, 1989). Therefore, given the essential *luxA* and *luxB* genes the potential exists to confer a bioluminescent phenotype on any prokaryotic organism.

The data presented in this paper focuses on the value of bioluminescence as a reporter of metabolic activity during microbial stress and recovery. This has significant implications for studying industrial protocols designed to recover injured micro-organisms; the overall objective being process optimisation.

MATERIALS AND METHODS

Bacterial Strain

All of the experiments carried out during this study used *Salmonella typhimurium* LT2 obtained from the laboratory collection and transformed to the bioluminescent phenotype with either the constitutive *lux* expression plasmid pSB100 (Blisset and Stewart, 1989) or pSB230 (Hill *et al.*, 1991). The strains were maintained in Luria broth (LB) (Maniatis *et al.*, 1982). The corresponding solid medium was made by the addition of 1.5% (w/v) agar to LB. Ampicillin was added to the medium at a concentration of 30µg/ml to maintain selective pressure for the bioluminescent phenotype.

Freeze Injury Method

An overnight culture of *S. typhimurium* LT2[pSB100] was diluted into LB (1:50). The culture was then grown with shaking to an A_{600} of 0.7 - 0.75. After diluting in 0.1% peptone water (final volume 50ml) to a concentration of 10^5 cells/ml, incubation was continued in a static water bath at 30°C. After 1 hour an aliquot of the culture (8ml) was dispensed in 1ml amounts and transfered for freezing either in a domestic -20°C freezer, a Planer Kryo 10/16 Programmable freezer (Planer Biomed, Windmill Road, Sunbury-on-Thames, Middx., U.K.) or by immersion in liquid nitrogen. Incubation continued for the remaining control culture at 30°C while the experimental sample was frozen.

When the freezing cycle was complete (1 hour incubation with -20°C freezer) the samples were thawed in a 30°C water bath or at a controlled rate in the Kryo 10/16. The samples were then pooled and diluted in parallel with the control culture 1:10 into LB and finally incubated at 30°C to allow recovery. Bioluminescence and viable count readings were determined from the experimental and control cultures both prior to freeze treatment and during and after recovery. To record bioluminescence a sample (1ml) was mixed with 40µl of a 1% (v/v) solution of dodecyl aldehyde (Aldrich) in ethanol and transferred to a Turner Designs 20E Luminometer (Steptech Ins., Herts, U.K.). Viable counts were determined by spread plating 100µl of an appropriate dilution onto either non selective agar (Tryptone soya agar containing 0.3% yeast extract [TSYA]) or selective agar (Xylose Lysine Desoxycholate agar [XLD] or brilliant green agar). Only counts representing between 30 - 300 colonies were used to estimate viable numbers.

Heat Injury Method

An overnight culture of *S. typhimurium* LT2[pSB230] was inoculated into LB

(1:50) and grown with shaking at 37°C to an A_{600} of 0.55 - 0.6. A 40ml portion of the culture was then centrifuged at 2.5g for 10 min and the pellet resuspended in 20ml LB.

Cell cultures were injected into a preheated, submerged heating coil (Cole and Jones, 1990), facilitating rapid and uniform sample heating in the range 20-90°C. Samples (200μl) were displaced from the coil into 5ml LB at preselected time intervals. A time zero sample was taken prior to injection by pipetting 200μl of the cell suspension into 5ml LB. Bioluminescence and viable count readings were taken immediately after heating and the sample bottles were incubated in a static water bath at 37°C and bioluminescence was monitored throughout the recovery period.

Bioluminescence was recorded by pipetting each sample (250μl) into a micro-titre tray which was then transferred to a Luminoscan Luminometer (Labsystems (UK) Ltd.). The luminometer was set to read preprogrammed wells and when a reading was to be taken 10μl of aldehyde mix was automatically added and the bioluminescence integrated over a 10 second counting period.

RESULTS

Overview

Most food production processes involve the application of physical treatments or chemical agents with the aim of reducing the microbial load and extending the shelf life of food products. Although most of the cells may be non viable a fraction of the population is often sub-lethally injured and are difficult to detect using conventional methods as they are unable to grow without a recovery period (Busta, 1976). Sub-lethally injured bacteria may retain their pathogenic traits (Ray, 1979) so it is very important for food safety to enumerate such cells in microbial analysis. Additionally, if they remain undetected it may be wrongly assumed that particular food production processes produce uncontaminated products (Hurst, 1977; Ray, 1979). The extent of damage depends on many factors including the type and length of treatment, the physiological state of the cell and the processing medium. This makes it difficult to know the length of time required for recovery. In response to this problem tests for bacteria typically utilise an overnight incubation in an appropriate pre-enrichment medium before attempting detection. Although there are current efforts to minimise the length of time required for bacterial resuscitation, in general this is the limiting step which will compromise efforts to reduce bacterial detection times based on new rapid biotechnology methods (reviewed by Dodd *et al.*, 1990). There is, therefore, still a need to improve understanding of the mechanisms of injury and recovery so that resuscitation times can be reduced.

Current methods to assess injury and recovery are based on the ability of cells to form colonies as effectively on selective as on non selective agar. This is purely an endpoint determination and gives no information regarding the biochemical and physiological process of recovery. Using *lux* recombinant bacteria it is possible to explore in real-time the capability of cells to provide high energy intermediates (FMNH$_2$) following stress. To establish bioluminescence as a tool to monitor bacterial injury and recovery it is important to initially establish the relationship between bioluminescence and classical viability measures such as plate counts.

Freeze injury

Figure 1 represents bioluminescence and viable plate count data for freeze injury of *S. typhimurium* LT2[pSB100] in a -20°C domestic freezer (A) and a Planer controlled rate freezer (freezing rate -2°C/min) (B). Plots of the control cultures have been adjusted (solid line) to subtract the extent of growth which occurred during the freezing period of the experimental culture. The bioluminescence (obtained in real time) and viable count data (obtained after 18-24 hour incubation) shows an excellent correlation. The results from both data sets in Figure 1A are indistinguishable with both showing a 0.79% survival rate. The results for the Planer controlled rate freezer indicate a 2.2% and 2.4% survival respectively for bioluminescence and viable count data.

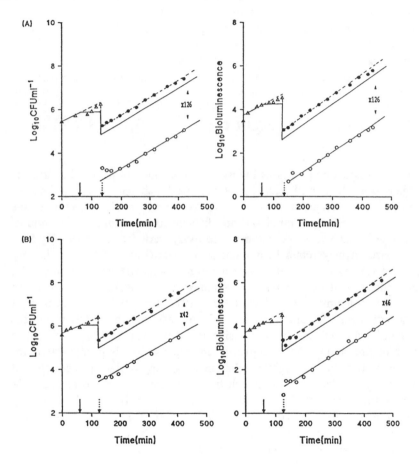

Figure 1. A comparison between viable count and *in vivo* bioluminescence for *S. typhimurium* LT2[pSB100] subjected to a domestic -20°C freezer (A) and a programmable rate freezer (B).
KEY: ▲ Growth in 0.1% peptone water; ● control culture incubated in L. broth; o experimental culture (after freezing) incubated in L. broth; ↓ samples frozen; ⋮ samples thawed, followed by 1:10 dilution into L.B.

The extent of survival following freeze damage can therefore be monitored by viable counts or bioluminescence with both producing equivalent results. To determine the extent of injury a series of experiments was performed using the same protocol but plating on selective and non selective agar. Figure 2 represents data where brilliant green agar (A) and XLD (B) plates were used in addition to TSYA. Non selective agar allows for the growth of both uninjured and injured cells whereas only uninjured cells are capable of growth on selective agar. During the first 50 min of recovery of the experimental culture the differential between the two types of agar is clearly evident. This indicates the presence of injured cells and directly after thawing the proportion of injured cells is very high, comprising 98.5% (A) and 97.1% (B) of the total population. The rapid increase in count on selective agar after this point indicates recovery of damaged cells. During this period the non selective count is constant confirming that the population consists of mainly injured cells. After the 50 min recovery period the cells start to multiply and the count on both agars increases at a parallel rate. It is apparent that the bioluminescence data produces a similar trend to that of the selective plates. This substantiates the fact that sub-lethally injured bacteria produce significantly less light than obtainable from an optimally growing culture. The population returns to optimum bioluminescent potential at the end of the recovery period. Sensitivity to selective agents and extended lag on non-selective media has been observed by many workers (Clark and Ordal, 1969; Ray and Speck, 1972; Mackey and Derrick, 1982) however using bioluminescence recovery can be observed in real time providing additional information about the state of cells during the recovery process.

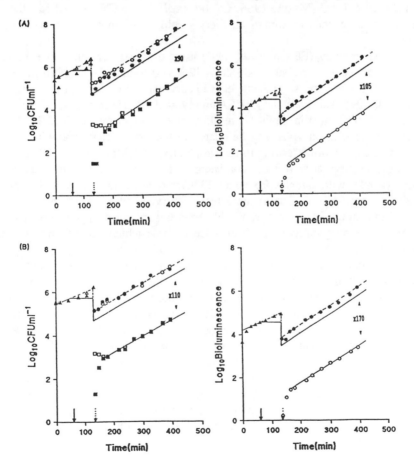

Figure 2. Determining the extent of injury within the freeze-thaw treated population by plating on selective and non selective plates. KEY: Δ growth in 0.1% peptone

water (TYSA); ▲ growth in 0.1% peptone water (sel. agar); o control culture (TSYA); ● control culture (sel. agar); □ experimental culture (TYSA); ■ experimental culture (sel. agar) (Key for bioluminescence data see Figure 1.).

Many factors influence the extent of bacterial death and injury during freeze-thaw damage (Macleod and Calcott, 1976; Mackey, 1984). The freezing rate determines whether water freezes intracellularly and this has an important influence on cell survival. At slow cooling rates the cell loses water as it attempts to maintain equilibrium with the extracellular solution as water is unavailable in the form of ice crystals (Mazur, 1977). The cells are extremely shrunken and dehydrated (Ray and Speck, 1973) and the high concentration of salts produced may denature proteins and destabilise the membrane. At rapid freezing rates water freezes internally as little water is lost by the time the freezing point is reached. This results in injury due to mechanical damage by internal ice crystals. The optimum freezing rate for a bacterial species is one which will be slow enough to prevent the formation of intracellular ice and yet fast enough to minimise exposure of cells to concentrated salts (Ray and Speck, 1973; Ray, 1984). For many bacterial species there is a peak of survival in the freezing rate range between 6-11°C (Mazur, 1966; Calcott *et al*, 1976; Calcott and Macleod, 1974) and then a second maximum at very rapid rates between 6,000 and 10,000°C/min.

In a previous study (Ellison *et al.*, 1991) using a Planer controlled rate freezer at a freezing rate of -30°C/min a survival rate of 12% was found. During this present study, however, there were no significant peaks in survival at freezing rates between -2°C/min (av = 0.909% survival) and -30°C/min (av = 3.1215% survival). Freezing at a higher rate (170°C/min) in liquid N_2 resulted in a survival rate of 2.72%. The small differences found in survival may be related to the extent of supercooling with lower amounts of supercooling leading to increased survival (Ellison *et al.*, 1991).

Changes in survival were more consistently observed if the thawing rate was varied. For example, with samples frozen at -30°C/min and thawed in a 30°C water bath (av thawing rate = 16.7°C/min) the extent of survival was 3.8%, however, if thawed at 6°C/min in the Planer programmable freezer survival dropped to 0.219%. Slow thawing allows recrystallisation of small ice crystals which subsequently cause damage to the cells.

To determine where the greatest extent of injury occurs on freezing, samples were removed at multiple points during the freezing profile. Figure 3 represents the freezing rate profile and survival data recorded by bioluminescence for *S. typhimurium* LT2[pSB100] frozen at -2°C/min. No cellular death occurs during cooling from ambient to 0°C. As the temperature falls below 0°C there is a reduction in viability and the greatest reduction occurs after the supercooling point and during the latent heat extraction period. During the latent heat extraction period the cell suspension freezes and consequently this is when most cellular damage will occur. Such experiments provide additional information about freeze damage and can be used to assess at which point maximum damage occurs in different conditions. For this type of work bioluminescence is an invaluable tool as it is less labour intensive and faster than conventional plate counts.

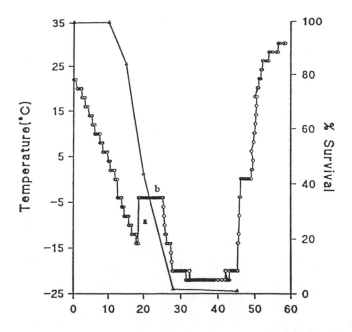

Figure 3. Freezing rate and survival profile for *S. typhimurium* LT2[pSB100] subjected to freezing at -20°C. KEY: o temperature; Δ % survival; (a) temperature increase due to supercooling beyond freezing point; (b) latent heat extraction period.

Heat Injury

Critical margins for the safe thermal processing of foods are traditionally derived from models of thermal death kinetics. For example, the well established model for the destruction of *Clostridium botulinum* spores (Gould, 1989). Predictions of thermal death kinetics in a range of environmental conditions is becoming important as the nature of the heating menstrum is known to have a profound effect on the heat resistance of micro-organisms (Olson and Nottingham, 1980). For instance, many research papers have demonstrated that reduction of A_w by the presence of NaCl, sucrose, glucose, sorbitol, fructose or glycerol enhances heat resistance of Salmonellae (Baird-Parker *et al.*, 1970; Goepfert *et al.*, 1970; Corry, 1974; Lee and Goepfert, 1975). Models are now being developed which are able to quantify the interaction between two or more factors (Cole *et al.*, 1993). This will be of benefit to the food industry as processing conditions for changed product formulations can be modified easily.

Temperatures above 45°C are too high to maintain the *in vivo* stability of the bacterial luciferase. As expected, therefore, initial experiments comparing the reduction in bioluminescence with viable counts at 56°C demonstrated that bioluminescence is inactivated at a rate in advance of viability (Figure 4). To use bioluminescence as a monitor of thermal inactivation it was necessary to develop an alternative method. Figure 5 shows recovery with time following varying lengths of heating at 56°C. The curves consist of an initial steep recovery period followed by exponential growth. The recovery period represents re-synthesis of luciferase enzyme rather than cellular recovery as no injury was detected by plating on selective agar. By extrapolating the exponential portion of the curve back to the y axis a bioluminescence value is obtained which represents the level if there had been no inactivation of luciferase. These points correlate very well with the reduction in viable count immediately after heating (Figure 4).

Figure 4   **Figure 5**

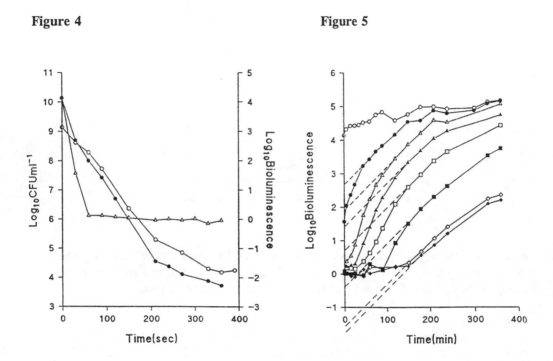

Figure 4. Initial and adjusted bioluminescence data plotted with viable counts showing reduction in survival of *S. typhimurium* LT2[pSB230] following treatment at 56°C. KEY: o viable count data; ● adjusted bioluminescence data (see Figure 5); Δ initial bioluminescence data.

Figure 5. Recovery of *S. typhimurium* LT2[pSB230] after injury at 56°C. KEY: (time at 56°C) o 0s; ● 30s; Δ 90s; ▲ 90s; □ 120s; ■ 150s; ◇ 210s; ◆ 240s.

Using this method the thermal inactivation of *S. typhimurium* LT2[pSB230] was analysed at 52°C, 54°C, 58°C and 59°C. Figure 6 shows the bioluminescence and viable count data from these experiments and the close correlation in clearly evident (correlation coefficient of 0.96).

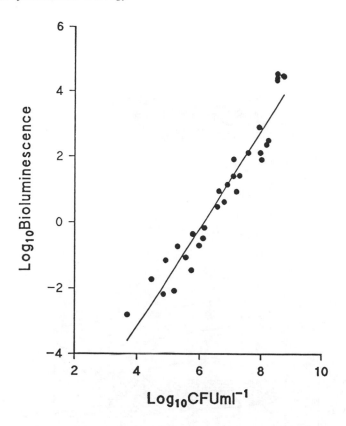

Figure 6. Calibration curve of viable count data and adjusted bioluminescence data for heat treatment of *S. typhimurium* LT2[pSB230], data points from heat inactivation experiments at a range of temperatures.

The thermal inactivation data has been modelled (Ellison *et al.*, 1993) following a method developed by Cole *et al.* (1993). This describes the data using a logistic function fitted to the logarithm of survivors using the SAS statistical package (SAS Software Ltd.). A predictive equation can now be used to obtain bioluminescence and viable count data for heating at a specified time and temperature within the limits of the experimental conditions.

The bioluminescence and viable count data were also analysed for a fit to a quadratic model using the SAS statistical package (Figure 7). The program statistically analyses raw data so that points in between the actual data points can be predicted to complete the surface. The similarity between the two plots confirms the good relationship between bioluminescence and viable count data.

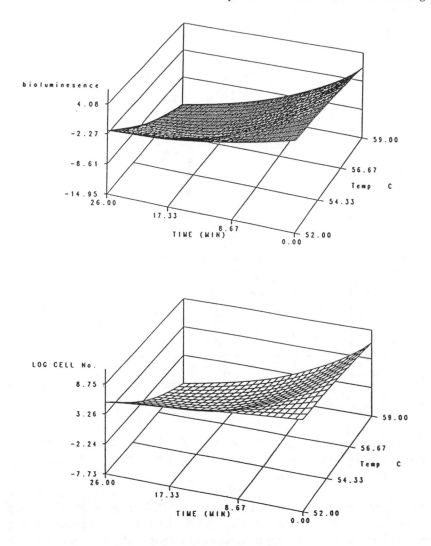

Figure 7. Quadratic response surface fit of bioluminescence and viable count data for heat treatment of *S. typhimurium* LT2[pSB230].

DISCUSSION

The data presented in this study has demonstrated an equivalence between *in vivo* bioluminescence and classical plate count data for monitoring freeze and heat injury of *Salmonella typhimurium* LT2.

Bioluminescence has the advantage of monitoring recovery in real-time and also reduces the laboratories media preparation and sampling procedures which encumber plate counts. This aspect was demonstrated when sampling during freezing where bioluminescence proved to be an invaluable tool as samples needed to be taken rapidly as well as recording recovery profiles.

Different recovery broths can be rapidly compared by measuring the recovery of bioluminescence following freeze-thaw treatment. Use of this technique will have a number of advantages in a food microbiological testing laboratory as recovery systems can be compared using the bacteria of concern transformed to a bioluminescent phenotype in a food containing a mixed microbiological flora. This would be an accurate and fast system to optimise recovery regimes which is an important issue as regards continued improvements in food safety.

The data for survival of *S. typhimurium* from heat treatment were good enough to enable computer generated models to be fitted (by logistic function or quadratic response surface) and these also demonstrated the equivalence of bioluminescence and viable count data. Such methods allow the interaction between two or more factors to be quantified and can be used for prediction of thermal inactivation under conditions which have not been tested, using interpolation within the range of the experimental matrix. For these reasons predictive models are a great asset to the food industry as thermal processing requirements can be predicted with ease and confidence.

As the equivalence of bioluminescence and viable count data has been proven it is now possible to use bioluminescence in conditions where plate counts are impractical. At the present time survival of micro-organisms in response to exposure to changes in temperature, pH and A_w is monitored in pure culture whereas in the environment they are part of a heterogeneous microflora and usually associated with surfaces. Current work is therefore examining the survival from heat treatment of a target micro-organism in a complex mixture of bacteria. This technique would be very difficult using conventional plate counts. Also specific bacterial pathogens can be monitored in real food systems where the food matrix and micro-flora content of the food will influence survival. This should lead to the development of more relevant models and would be a considerable aid to the optimisation of food preservation methods and bacterial recovery techniques and, in consequence, will lead to safer food.

ACKNOWLEDGEMENTS

This work was supported by grants from the Agricultural and Food Research Council and the Ministry of Agriculture Fisheries and Food. The authors wish to thank Dr Martin Jones and colleagues at Unilever Research and Stephen Perry and Dr Tim Harrison at the Central Public Health Laboratory, Colindale. Thanks also to Sheila Godber for typing the manuscipt.

REFERENCES

Baird-Parker, J.M., Boothroyd, M. and Jones, E., The effect of water activity on the heat resistance of heat sensitive and heat resistant strains of *Salmonellae, Journal of Applied Bacteriology*, 33, 515-522, 1970.

Baker, J.M., Griffiths, M.W. and Collins-Thompson, D.L., Bacterial bioluminescence: Applications in food microbiology, *Journal of Food Protection*, 55(1), 62-70, 1992.

Blisset, S.J. and Stewart, G.S.A.B., *In vivo* bioluminescent determination of apparent Km's for aldehyde in recombinant bacteria expressing *luxA/B, Letters in Applied Microbiology*, 9, 149-152, 1989.

Busta, F.F., Practical implications of injured micro-organisms in food, *Journal of Milk and Food Technology*, 39(2), 138-145, 1976.

Busta, F.F., Practical implications of injured micro-organisms in food, *Journal of Milk and Food Technology*, 39(2), 138-145, 1976.

Calcott, P.H., Lee, S.K. and Macleod, R.A., The effect of cooling and warming rates on the survival of a variety of bacteria, *Canadian Journal of Microbiology*, 22, 106, 1976.

Calcott, P.H. and Macloed, R.A., Survival of *Escherichia coli* from freeze-thaw damage: a theoretical and practical study, *Canadian Journal of Microbiology* 20, 671-681, 1974.

Clark, C.W and Ordal, Z.J., Thermal injury and recovery of *Salmonella typhimurium* and its effect on enumeration procedures, *Applied Microbiology*, 18, 332-336, 1969.

Cole. M.B., Davies, K.W., Munro, G., Holyoak, C.D. and Kilsby, D.C., A vitalistic model to describe the thermal inactivation of *Listeria monocytogenes*, *Journal of Industrial Microbiology*, in press, 1993.

Cole, M.B. and Jones, M.V., A submerged-coil heating apparatus for investigating thermal inactivation of micro-organisms, *Letters in Applied Microbiology*, 11, 233-235. 1990.

Corry, J.E.L., The effect of sugars and polyols on the heat resistance of Salmonellae, *Journal of Applied Bacteriology*, 37, 31-43, 1974.

Dodd, C.E.R., Stewart, G.S.A.B., and Waites, W.M., Biotechnology based methods for the detection, enumeration and epidemiology of food poisoning and spoilage organisms, in *Biotechnology and Genetic Engineering Reviews*, Vol. 8, Tombs, M.P., Ed., Intercept Ltd., 1990, ch 1.

Ellison, A., Perry, S.F. and Stewart, G.S.A.B., Bioluminescence as a real-time monitor of injury and recovery in *Salmonella typhimurium*, *International Journal of Food Microbiology*, 12, 323-332, 1991.

Ellison, A., Anderson, W., Cole, M.B. and Stewart, G.S.A.B., The use of bioluminescence data to develop a model to describe the thermal inactivation of *Salmonella typhimurium*, *International Journal of Food Microbiology*, in press, 1993.

Goepfert, J.M., Iskander, I.K. and Amundson, L.H., Relation of the heat resistance of Salmonellae to water activity of the environment, *Applied Microbiology*, 19(3), 429-433, 1970.

Gould, G.W., Predictive mathematical modelling of microbial growth and survival in foods, *Food Science and Technology Today*, 3(2), 89-92, 1989.

Hill, P.J., Rees, C.E.D., Winson, M.K. and Stewart, G.S.A.B., The application of *lux* genes, *Biotechnology and Applied Biochemistry*, 17, 3-14, 1993.

Hill, P.J., Swift, S. and Stewart, G.S.A.B., PCR based gene engineering of the *Vibrio harveyi lux* operon and the *Escherichia coli trp* operon provides for biochemically functional native and fused gene products, *Molecular General Genetics*, 226, 41-48, 1991.

Hurst, A., Bacterial injury: A review, *Canadian Journal of Microbiology*, 23(8), 936-942, 1977.

Lee, A.C. and Goepfert, J.M., Influence of selected solutes on thermally induced death and injury of *Salmonella typhimurium*, *Journal of Milk Food Technology*, 38, 195-200, 1975.

Mackey, B.M., Lethal and sub-lethal effects of refrigeration, freezing and freeze drying on micro-organisms, in *The Revival of Injured Microbes*, Andrews, M.H.E. and Russell, A.D., Eds., Academic Press, 1984, 45-75.

Mackey, B.M. and Derrick, C.M., A comparison of solid and liquid repair media for measuring the sensitivity of heat injured *Salmonella typhimurium* to selenite and tetrathionate media, and the time needed to recover resistance, *Journal of Applied Bacteriology*, 53, 233-242, 1982.

Macleod, R.A. and Calcott, P.H., Cold shock and freezing damage to microbes, in *The Survival of Vegetative Microbes*, Twenty-sixth symposium of the society for General Microbiology, Gray, T.R.G. and Postgate, J.R., Eds., Cambridge University Press, 1976, p81-109.

Maniatis, T., Fritsch. E.F. and Sambrook, J., Molecular Cloning: A Laboratory Manual, Cold Spring Harbour Laboratory, Cold Spring Harbour, New York, p68, 1982.

Mazur, P., Physical and chemical basis of injury in single celled micro-organisms subjected to freezing and thawing, in *Cryobiology*, Meryman, H.T., Ed., London Academic Press, 1966, p213-315.

Mazur, P., The role of intracellular freezing in the death of cells cooled at superoptimal rates, *Cryobiology*, 14, 251-272, 1977.

Meighen, E.A., Enzymes and genes from the *lux* operons of bioluminescent bacteria, *Annual Reviews of Microbiology*, 42, 151-176, 1988.

Meighen, E.A., Molecular biology of bacterial bioluminescence, *Microbiological Reviews*, 55(1), 123-142, 1991.

O'Kane , D.J. and Prasher, D.C., Micro-review:- Evolutionary origins of bacterial bioluminescence, *Molecular Microbiology*, 6(4), 443-449, 1992.

Olson, J.C. and Nottingham, P.M., Temperature in *Microbial Ecology of Foods. Factors Affecting Life and Death of Micro-organisms*, Vol. 1, ICMSF, Academic Press, 1980, ch 1.

Ray, B., Methods to detect stressed micro-organisms, *Journal of Food Protection*, 42, 346-355, 1979.

Ray, B., Reversible freeze injury, in *Repairable Lesions in Micro-organisms*, Academic Press, 1984, ch 8.

Ray, B. and Speck, M.L., Repair of injury induced by freezing *E. coli* as influenced by recovery medium, *Applied Microbiology*, 24(3), 258-263, 1972.

Ray, B. and Speck, M.L., Freeze injury in bacteria, *CRC Critical Reviews in Clinical Laboratory Sciences*, 4, 161-213, 1973.

Stewart, G.S.A.B. and Williams, P., Review article: *Lux* genes and the application of bacterial bioluminescence, *Journal of General Microbiology*, 138, 1289-1300, 1992.

IMPORTANCE OF INJURED FOOD-BORNE MICROORGANISMS
IN MINIMAL PROCESSING

F. F. Busta

When processes are modified for optimization or adjusted to minimal levels, reversible damage may occur in microbial cells exposed to sublethal stress from physical or chemical insults. Much study of injury has focused on the inability to measure viability based on characteristics altered by exposure to sublethal stress. Examples of situations influenced by injury to microorganisms include microbial stability of new minimal processes; survival rates of genetically engineered microorganisms; rapid detection methods for pathogens or indicators; and synergistic processes used as optimal treatments to control bacterial spores. Use of damage to deter spoilage or health hazards by preventing resuscitation and subsequent growth is also a consideration. Injury may be insignificant when detection or measurement of cellular constituents such as toxins, ATP, or genetic material instead of replication of cells is used. The major question remains -- is microbial viability or capability to grow and multiply important in newly developed or minimal processes? This must be answered before effective minimal processes can be defined.

Anytime that processes are adjusted for optimization to a level that minimizes deterioration of quality from process damage, the microbiological ramifications of such modifications must be addressed. Inactivation of microorganisms to ensure safety and stable quality is a primary consideration in developing these processes. As processes are minimized, they may inadvertently cause reversible damage in the microbial cells exposed to sublethal stresses of the minimal process but may not actually inactivate the microorganism.

What is the practical importance of reversible sublethal damage to microorganisms in food products? This paper will review the influence of injury, inhibition, and inactivation of microorganisms associated with minimally processed foods; consider the many food environments and how they influence the response of these microorganisms that have been exposed to sublethal damage; and describe situations influenced by injury and how this damage might be used to control the microorganisms. The influence of injury on new methodology will also be addressed in light of whether or not viability is an important factor in assessing the effectiveness of minimal processes.

Viability in microorganisms is traditionally measured by determining their ability to grow and multiply into sufficient numbers to be detected chemically, physically, or visually. Traditionally, if an organism cannot reproduce and grow into higher populations, it is considered inactivated or non-viable. The inability to measure viability or activity of a microorganism is based on the characteristics altered by the stress applied to the organism.

What are some of these altered characteristics? As a result of sublethal stress, various microorganisms demonstrate increased sensitivity to compounds which they tolerated prior to the sublethal stress. Increased sensitivity to surface active agents such compounds such as brilliant green, crystal violet, or neutral red increase in their inhibitory activity when cells are sublethally damaged.

Injured microorganisms also develop increased sensitivity to various food constituents such as benzoate or sorbate and other fatty acids such as octanoic and decanoic acids. The more traditional preservative acids such as acetic, lactic, malic, and citric have greater effects on certain microorganisms that have been sublethally damaged. This may relate to low pH or to the anion. Other organisms demonstrate increased sensitivity to hydrolytic enzymes such as lysozyme and RNase or to oxygen atmospheres that would not influence them prior to a damaging treatment.

The altered characteristics are not only increased sensitivity. Sublethal damage frequently causes loss of cellular materials including proteins, peptides, amino acids, RNA, enzymes, and various ions. In some cases, sublethal treatment will result in increased hydrophobicity on the surface of the cell or loss in the ability of bacteriophage to absorb.

The growth response frequently reflects a damaged population by delayed germination of spores, extended lag phases of growth, and the inability to multiply until repair of the damage has taken place. This period of permitting cells to repair has been termed resuscitation. Procedures are used regularly to permit damaged cells to resuscitate prior to analysis.

The sublethal injury of cells may be caused by more than one physical or chemical treatment. Damage has been observed after a low temperature storage, refrigeration, or freezing. When heat is applied to a system, many combinations of times and temperatures below the lethal treatment can result in damaged survivors. Air drying or freeze drying or exposure to high concentrations of sugars or salts can damage vegetative cells of microorganisms. Finally, selected radiation, either UV or X-ray, can result in some sublethal injury.

Although we are concentrating on minimal processes, an awareness of effects of chemicals that are known to cause injury is useful. Organic and inorganic acids, sanitizers, chlorine or quaternary ammonia compounds, preservatives such as sorbate or benzoate and toxic chemicals such as mercuric chloride all have been reported to cause sublethal damage to cells.

It is tempting to declare this a laboratory phenomenon and suspect that it is limited to a few experimental cultures. However, most organisms that have been critically evaluated have exhibited injury after exposure to some lethal stress. These include the entire cadre of emerging pathogens, bacterial spores, yeasts, molds, and lactic acid cultures. So as minimal processes are evaluated, potential for injury must be monitored constantly.

What specific food environments produce reversible damage to microbial cells? Preservation treatments, especially those that are minimal are frequently involved. Cleaning and sanitizing or disinfection are also situations that will produce sublethally injured cells. Basically, whenever microorganisms are exposed to sublethal stress, a potential exists for a population of injured microorganisms.

As minimal processes are developed and evaluated, some food factors associated with the assessment of injured organisms should be considered. The conditions of treatment, of storage, and of measurement all influence our observations. Time, under any set of conditions, is influential. Consequently, reduced time may increase the number of damaged cells. If multiple treatments are the key to the successful new process, the impact of those multiple stresses also must be considered on the inactivation and injury of microorganisms. The composition of the food influences survival and injury, as well as the possibilities for growth. In many cases, the method of detecting injured microorganisms or recovering microorganisms depends on the medium. Time and temperature of the method may identify potential problems, or may miss them. As indicated earlier, an opportunity for resuscitation in the appropriate medium for the appropriate time, at the appropriate temperature, is always recommended if one wants to know the true survival rate of any spoilage or hazardous microorganism under consideration.

Resuscitation determines the impact of injury. If the injured cells are not permitted to repair, they act as though they were inactivated and are of little consequence to quality or safety of the food product. However, if any conditions during storage, distribution, or consumer handling give an opportunity for repair, the importance of that cell injury is amplified.

How does the presence of injured organisms influence the topic of minimal processing of foods? Apparent safety is frequently reflected in the ability to determine the presence of a hazard. Consequently, methodology that utilizes pre-enrichments, membrane filters, layers of medium, most probable number techniques, or special media that have been developed to address the presence of sublethally damaged cells may detect the presence of pathogens in food that would otherwise not be observed. Frequently, this utilizes a resuscitation technique with an appropriate time and temperature and non-detrimental conditions. The methodology to detect damaged cells is time consuming and expensive. However, undoubtedly of greater impact, is the possibility of missing the presence of hazardous microorganisms that could subsequently resuscitate and grow into a potential public health problem.

Many of the new processes that were discussed at the Process Optimization and Minimal Processed Foods Workshop have the potential to damage and not totally inactivate microorganisms. This would include mild preservation techniques to extend shelf life of refrigerated products, new methods for pre-concentrating food systems or utilizing high pressure to preserve products. Whenever a reduced time or temperature is proposed or some alternative heating method that may not be as comprehensive or broad spectrum as the traditional methods is suggested, the possibility of surviving damaged cells exist. Even the responses to controlled or modified atmosphere may be influenced by prior treatments which include radiation, preservatives, or reduced water activity. If processes are to involve genetically engineered microorganisms that are released into the environment, a population of organisms may be introduced which previously has not been characterized for appropriate processing. This also influences disinfection effectiveness. Many of the processes that are considered for minimal processing are effective against vegetative cells but permit spores to survive. In some cases, these treatments are sufficiently severe to damage the spores but not totally inactivate them. Also, modelling of inactivation processes must consider injury and heat activation as well as inactivation of spores. [Rodriguez, et al., 1988; Rodriguez, et al., 1992] The consequence of such a treatment or added new food constituents may result in dramatically unusual survival demonstrated in the inactivation kinetics. [Ababouch and Busta, 1987; Ababouch, et al., 1987; Ababouch, et al., 1992]

Some techniques for minimally processed foods utilize starter cultures and fermentations to enhance the preservation or to serve as a fail-safe factor if the product were to be abused. Cryopreservation or dehydration not only will potentially sublethally damage indigenous microorganisms but also damage starter cultures added to the food product. Frequently, it is extremely difficult to assess the relative influence of injury on spoilage microorganisms as compared to pathogenic microorganisms if one intentionally relies upon the spoilage organism to serve as a fail-safe control of abused products. It would not be totally obscure to imagine that an imbalance in the relative populations of surviving microorganisms would be quite different if one took into consideration the damage.

Many authors have reviewed and researched the topic of sublethally injured microorganisms. [Gould, 1984; Hurst, 1984; Ray, 1989; Russell, 1982] Generally, these authors have cautioned the food processing industry to be aware of and be concerned over the potential hazard arising over injured organisms. [Sofos, 1989] The major question remains: is microbial viability or the capability to grow and multiply important in newly developed or minimally processed foods or is damage and delay sufficient? If we are to detect survival, is the measurement of cellular constituents such as ATP, RNA, DNA through such sophisticated methods such as probes or PCR valid in finding organisms that cannot multiply but are present and not totally inactivated? The major question that remains and must be answered is -- should the capability to grow and multiply under the normal storage conditions of a food processed under newly described minimal conditions be the key determinant, or does the survival detected in anyway or with any method whatsoever determine the process?

REFERENCES

Ababouch, L. and Busta, F. F., Effect of thermal treatments in oils on bacterial spore survival, *J. Appl. Bacteriol*, 62, 491, 1987.

Ababouch, L. and Busta, F. F., Tailing of survivor curves of clostridial spores heated in edible oils, *J. Appl. Bacteriol*, 62, 503, 1987.

Ababouch, L., Chaibi, A., and Busta, F. F., Inhibition of growth and germination of bacterial spores by fatty acids and their sodium salts, *J. Food Protection*, 55, 980, 1992.

Gould, G. W., Injury and repair mechanisms in bacterial spores, in *The Revival of Injured Microbes*, Andrew, M. H. E. and Russell, A. D., Eds., Academic Press, London, 1984.

Hurst, A., Injury, in *The Bacterial Spore*, Vol. 2, Hurst A. and Gould, G. W. Eds., Academic Press, London, 1984.

Ray, B., *Injured Index and Pathogenic Bacteria:Occurrence and Detection in Foods, Water and Feeds*, CRC Press, Boca Raton, FL, 1989.

Rodriguez, A. C., Smerage, G. H., Teixeira, A. A., and Busta, F. F., Kinetic effects of lethal temperatures on population dynamics of bacterial spores, *Transactions of the ASAE*, 31, 1594, 1988.

Rodriguez, A. C., Teixeira, A. A., Smerage, G. H., Lindsay, J. A., and Busta, F. F., Population model of bacterial spores for validation of dynamic thermal processes, *J. Food Process Engineering*, 15, 1, 1992.

Russell, A. D., *The Destruction of Bacterial Spores*, Academic Press, London, 1982.

Sofos, J. N., Detection of injured spore-forming bacteria from foods, in *Injured Index and Pathogenic Bacteria*, Ray, B. Ed., CRC Press, Boca Raton, FL, 1989.

COMBINED TREATMENTS TO EXTEND THE
SHELF LIFE OF FRESH FISH

R.M. Kirby, C.J. Capell and P. Vaz-Pires

I. SUMMARY

Initially the application of heat alone on bacterial isolates from fresh fish was studied in mixed culture and on whole fish. Secondly the effect of heat combined with acetic acid was studied on whole fish. Bacterial isolates obtained from fresh horse mackerel (*Trachurus trachurus*) were heat treated at 60°C for 20 s and stored in nutrient broth on ice. The flora were shown to be heat sensitive and the initial numbers reduced by over 2.0 $\log_{10}$ cycles. The lag phase was also increased. Subsequent growth rates were shown to be similar to those seen on the skin of whole fish treated in the same way. Results on the sensory effects of heat treatment on whole fish show no perceiveable change in external characteristics of the fish. A shelf life increase of one day was also demonstrated. Combination of heat and acid showed visible effects in external characteristics of fish, but no decrease in bacterial load. Shelf life was not extended using this combination.

II. INTRODUCTION

Microbial activity is one of the principle causes of the spoilage of meat and fish. Different methods to reduce the psychrophilic contamination of meat animals have been proposed, including a naked flame, hot air-steam [Smith & Graham, 1978], and dilute acetic acid [Eustace *et al.*, 1979]. Animals used to supply meat are homeothermic and their predominant natural microflora is not psychrophilic. Contamination with psychrophiles comes mainly from environmental post-slaughter sources [Tompkin, 1973]. Nevertheless treatments to reduce the psychrophilic contamination have resulted in extended shelf life. Graham *et al.* [1978] reported the successful use of hot water spray washes to decontaminate lamb carcasses. In this work, a temperature of 80°C held for 10 s reduced the bacterial load of the meat surface by 2.5 $\log_{10}$ units. Decontamination by immersion or spraying with hot water has also been successfully applied to meat carcasses to reduce the microbial load [Graham *et al.*, 1978; Kelly *et al.*, 1982; Everton, 1985; Crouse *et al.*, 1988].

Fish, being poikilothermic, possess a microflora influenced by water temperature; in cold European waters, they have a large psychrophilic component. The presence of large numbers of psychrophilic bacteria means that refrigeration is less efficient at delaying spoilage [Liston, 1982]. A reduction of the microbial load resulting in an extension of shelf life of fish and fish products has previously been achieved with chemical and physical methods [Reddy *et al.*, 1992].

Psychrophilic bacteria are generally heat sensitive [Anon., 1980], mild heat treatments may therefore have a beneficial effect upon the keeping qualities of fish. Moderate thermal processing could therefore be used to extend the refrigerated shelf life of some pre-packaged seafoods by reducing initial numbers and increasing the time before the onset of exponential growth. The relatively mild heating conditions result in colour, texture, andflavour characteristics which are similar to "fresh" products [Hackney, 1990]. The use of heat combined with acetic acid is an attempt to potentiate the effect of the mild heat treatments and maximize the reduction in the microbial load. Successful use of acetic acid treatments have been reported on pure cultue isolates [Young & Foegeding, 1993].

III. MATERIALS AND METHODS

A. RAW MATERIALS AND SAMPLE PREPARATION

Horse-mackerel, also known as scad (*Trachurus trachurus*), were caught by bottom trawling, immediatly transfered to storage in freshwater ice, and stored in the ship's bulk room for 2 days. After landing, a sample of fish was taken from a single catch, and transported to the laboratory in a clean insulated container. The fish were weighed and kept in a refrigerated room (3°C ±1°C) boxed and iced.

The average weight of each fish was 223 g (minimum 186 g /maximum 258 g; male to female ratio approximately 1:1).

B. MICROBIOLOGICAL ANALYSIS

1. Media and reagents

Serial dilutions were made in sterile 1/4 strength Ringers solution (Lab M). Nutrient agar and nutrient broth (Lab M) were used as solid and liquid growth media. Media and reagents were prepared according to manufacturers' instructions.

2. Isolation of bacteria

Pieces of skin (about 2 cm^2), gill (1 whole gill arch) and abdominal cavity (all organs) aseptically taken from the fish were separately homogenized in 20 ml Ringers solution for 2 min and bacteria isolated at 10°C on nutrient agar by the spread plate technique. From 80 initial purified isolates, gram reaction, cell shape and colony morphology were noted and 30 isolates chosen for their diversity of reactions and appearance. All 80 isolates were found to be gram negative, catalase and oxidase positive.

3. Heat stress of bacterial isolates

The 30 bacteria isolated above were separately grown in nutrient broth for 48 h at 30°C, these cultures were then mixed in equal volume to form a cocktail. 0.5 ml of the mixture was added to 49.5 ml of stirred Ringers solution at 60°C, after 20 seconds, 0.5 ml were taken from the Ringers solution and put into 49.5 ml of nutrient broth pre-stored on ice. A control was simultaneously performed using Ringers at 30°C. Counts were made on nutrient agar plates (incubated 30°C, 48 h) over the following 3 weeks.

C. ORGANOLEPTIC EVALUATION

1. Following fish spoilage

Three trained panelists graded 3 fish of each group every 2 days, according to the European Community (E.C.) fish sensory scheme.

2. Total counts on fish

a. Heat treatment

A sample of twelve fish was equally divided and treated by total immersion in water, six at 30°C and six at 60°C for 20 s. Afterwards the fish were immediately returned to ice which was regularly replenished and drip water removed. Fish boxes were stored at 3°C (±1°C) in a refrigerated room. At regular intervals a fish from each sample was randomly selected and circle punch samples of skin with approximately 10 cm^2 were taken aseptically from the region behind the operculum and above the pectoral fin and homogenized in 20 ml of sterile Ringers solution. Counts were made on nutrient agar plates (22°C, 48 h) as previously described.

b. Heat combined with acetic acid

The above procedure was repeated but the water was acidified with acetic acid 100% (2% w/v final concentration) prior to treatment of the fish.

3. Raw fish

Twenty fish were randomly assigned time/temperature combination treatments (in the range of 55°C to 80°C and 10 to 60 s), then processed by total immersion. The fish were distributed on a white surface in a 4x5 matrix, with one identified control as a visual reference. One unidentified control was also included. Eleven panelists were asked to classify the fish with reference to the control, looking only at external aspects, and using the words "same", "similar" or "different". Comments about the differences were also requested.

IV. RESULTS

A. HEAT RESISTANCE OF BACTERIAL ISOLATES

When subjected to heat stress using a variety of time temperature combinations, cocktails of bacterial isolates from fresh sea fish consistently demonstrated a sensitivity to heat (Figure 1-A); the collective D_{60} was 0.17 min. A treatment of 60°C for 20 s resulted in a reduction of 2 log cycles and also resulted in an increase in the lag phase of 4 to 5 days when compared with the control.

B. EFFECT OF HEAT TREATMENT ON FISH SURFACE FLORA

Dipping fish in water held a 60°C did not reduce the total viable count (TVC) of samples of skin taken immediately after treatment, as compared to the controls. Growth curves derived from skin samples on treated and untreated fish are shown in Figure 1-B. The onset of growth was delayed for approximately 4 days following heat treatment. The subsequent growth rate of bacteria on the treated fish was greater than that observed with the control and the two counts converged after 4 days.

Results shown in Figure 1-A and B show that following heat treatment the subsequent growth curves, of bacterial isolates stored in nutrient broth on ice and the natural flora of fish skin, are similar. The reduction in initial numbers is however, much higher in liquid broth cultures.

Figure 1

Graph showing the total viable counts during 10 days of storage in ice.
O: 30°C; ▫: 60°C

A - FISH ISOLATES ON BROTH   B - FISH SKIN BACTERIA ON FISH

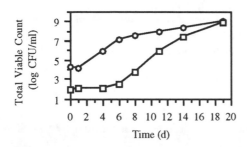
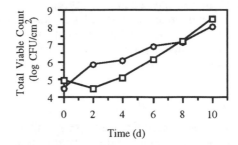

C. SENSORY ANALYSIS OF THE EFFECTS OF TREATMENTS

1. Heat

Results in Figure 2 show that fish treated at temperatures of between 55 to 70°C for times of between 10 to 30 s were indistinguishable from untreated fish. A score of less than 10 indicates that less than 3 panelists scored the fish as different. Fish where analysed by a panel of 30 people composed of professional and non-professional buyers.

Figure 2

The effect of time/temperature combinations of heat treatments on the external organoleptic characteristics of horse mackerel.
Scoring system: 0 points-same, 1 point-similar, 3 points- different.

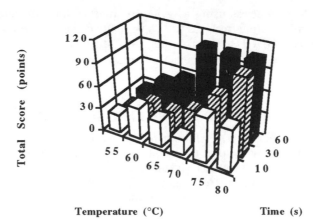

2. Heat and acetic acid

Total viable counts of fish skin after heat and acid treatments are shown in Figure 3. Individual treatments do not show significant differences in the total number of microorganisms, when compared to the control. Acid and heat combined however show a significant reduction in the TVC during storage.

Figure 3

Graph showing the total viable counts per cm² of fish skin

during 10 days of storage in ice.

O: 30°C; ▫ : 60°C; ●:30°C + 2% acetic acid; ■ :60°C + 2% acetic acid

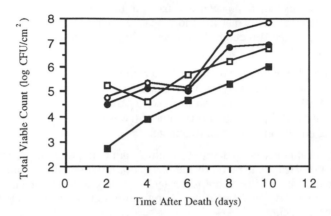

D. SPOILAGE FOLLOWING TREATMENT

Spoilage of treated and untreated fish was followed organoleptically and microbiologically. The results are shown in Fig. 4. The organoleptic analysis used quality bands E (extra quality), A (good), B (acceptable) and C (unacceptable), therefore when the fish leaves band B it is no longer acceptable for consumption as human food. Linear regression analysis shows that organoleptic rejection of the fish was reached after 10 days in the case of untreated fish, 11 days in the case of fish treated with heat alone and 8 days in the case of fish subjected to a combined heat and acid treatment.

Figure 4

The progression of organoleptic panel scores of *Trachurus trachurus*

over time scored in ice after heat and acid treatment.

O : 30°C; ▫ : 60°C; ●:30°C + 2% acetic acid; ■ :60°C + 2% acetic acid

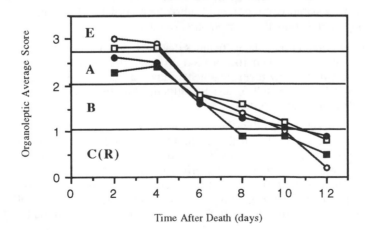

V. DISCUSSION

The results presented here relating to the heat resistance of bacteria isolated from fresh marine fish are in agreement with those previously published for psychrophilic and psychrotrophic bacteria [Liston, 1982]. The overall reduction in TVC was perhaps not as great as could have been expected; however the fact that the bacterial resistance's were not determined in pure culture means that the more heat resistant organisms will distort the curve pushing it more in the direction of an over estimate of the heat resistance of the individual organisms.

The growth curves following treatment of bacteria in nutrient broth and fish skin are similar. Reproducible microbial analysis of fish is a formidable problem [Huss, 1988]. Results of the microbial analysis of the fish in these experiments showed large variations, however, the trends support the second observation of the isolate experiments which noted an increase in the lag phase following heat treatment.

Following heat treatments the reduction in initial numbers is much greater in liquid culture than on the fish. Heat penetration of the fish is not instantaneous or uniform due to protection from direct contact with the hot water by body structures and mucus. The actual stress suffered by the bacteria on the fish may therefore reasonably be expected to be less than in the isolate experiments.

The growth curves of bacteria following heat treatment both in culture experiments and on the fish are different to the controls. Following initial reductions in the viable count and an extended lag phase, bacteria after heat stress grow at a rate similar to the controls. Following exponential growth however the curves appear to converge.

The heat treatment regime chosen for shelf life studies with whole fish was not at the limits which visual analysis of the effects of treatment indicated was possible. The reason for this was that the authors recognized the difficulties in practice of controlling rigorously the time/temperature of exposure in industrial conditions. A treatment regime was chosen therefore which would give a reasonable safety margin allowing for over processing.

The spoilage after treatment was followed organoleptically and microbiologically: the results for the two types of analysis do not appear to agree. In the early stages following treatment bacterial activity is reduced as demonstrated by the increased lag phase, however no apparent benefit in quality is demonstrated in the organoleptic quality analysis. This would seem to indicate that initial losses in quality are not related to bacterial activity but to enzymatic autolysis. These findings are in agreement with the findings of Huss [1988].

It has previously been reported [Hobbs, 1991] that extending the lag phase of bacteria will extend shelf life. The results presented here demonstrate that even with an extended lag phase there remains the possibility of more rapid subsequent growth, which may reduce or eliminate the advantage gained from the delayed onset of growth.

In general the results demonstrate the difficulties that have previously been encountered with other attempts to extend the shelf life of food products [Rippen & Hackney, 1992; Ellerbroek *et al.*, 1993]; that is low level treatments aimed at reducing the initial microflora may not be successful. The result of these treatments may be to alter the organisms spoiling the product and not delay rejection.

Microbiological analysis showed that the use of heated water at 60°C with 2% acetic acid gives a reduction of 2 log cycles in the initial TVC, when compared with a simple wash, heated wash and cold acid wash. No increase in quality or shelf life is however noted when organoleptic analysis is used to assess the effectiveness of the treatment. The major factor involved in organoleptic rejection of the fish were obvious changes in eye opacity and skin mucus transparency. Other organoleptic indicators of spoilage such as off-odour production were however less pronounced in fish subjected to a combined acid and heat treatment.

VI. ACKNOWLEDGEMENTS

This work was supported by JNICT - Junta Nacional de Investigação Científica e Tecnológica (Portugal). We wish to thank José Carlos Soares for donations of samples, Vítor Sousa Pereira for technical information and support and Luciano Ferreira for his excellent technical assistance.

VII. REFERENCES

Anon., Factors affecting life and death of microorganisms. In *Microbial Ecology of Foods 1*. International Commitee on the Microbiological Specifications of Food, Academic Press, London, 1980.

Crouse, J.D., Anderson, M.E. & Naumann, H.D., Microbial decontamination and weight of carcass beef as affected by an automated washing pressure and lenght of time of spray. *Journal of Food Protection* vol.51, 6, 471-474, 1988.

Ellerbroek, L.I., Wegener, J.F. & Arndt, G., Does spray washing of lamb carcasses alter bacterial surface contamination? *Journal of Food Protection*, vol.56, 5, 432-436, 1993.

Eustace, I.J., Powell, V.H. & Bill, B.A., Vacuum packaging of lamb carcasses: use of acetic acid to extend chill storage life. A preliminary investigation. *Meat Research Report* 3/79. Division of Food Research, Meat Research Laboratory, Australia, 1979.

Everton, A.R., Attempts to reduce the microbiological contamination of raw meat. *The Meat Hygienist* Sept.1985, 3-10, 1985.

Graham, A., Eustace, I.J. & Powell, V.H., Surface decontamination - a new processing unit for improved hygiene on carcass meat. *Proceedings of the 24th European Meeting of Meat Research Workers*, Kulmbach, paper B8, 1978.

Hackney, C.R., Processing moluscs. In *Seafood Industry*. Van-Nostrand-Reinhold, USA, pp.165-173, 1990.

Hobbs, G., Fish: microbiological spoilage and safety. *Food Science and Technology Today* 5, (3) 166-173, 1991.

Huss, H.H., In *Le poisson frais: qualité et alterations de la qualité*. Organization des Nations Unies pour l'Alimentation et l'Agriculture. Collection FAO: Pêches, n° 29, 1988.

Kelly, C.C., Lynch, B. & McLoughlin, A.J., The effect of spray washing on the development of bacterial numbers and storage life of lamb carcasses. *Journal of Applied Bacteriology* 53, 335-341, 1982.

Liston, J., Recent advances in the chemistry of iced fish spoilage. In *Chemistry and Biochemistry of Marine Food Products*, ed. R. Martin, G. Flick, C.E. Hebard and D.R. Ward, pp. 27-36. AVI Publishing Co. Westport, CT, 1982.

Reddy, N.R., Armstrong, D.J., Rhodehamel, E.J. & Kautter, D.A., Shelf-life extension and safety concerns about fresh fishery products packaged under modified atmospheres: a review. *Journal of Food Safety* 12, 87-118, 1992.

Rippen, T.E. & Hackney, C. R., Pasteurization of Seafood: Potential for shelf-life extension and pathogen control. *Food Technology* Dec. 1992, 88-94, 1992.

Smith, M.G. & Graham, A., Destruction of *Escherichia coli* and salmonellae on mutton carcases by treatment with hot water. *Meat Science* 2, 119-128, 1978.

Tompkin, R.B., Refrigeration temperature as an environmental factor influencing the microbial quality of food - a review. *Food Technology* Dec.1973, 54-58, 1973.

Young, K.M. & Foegeding, P.M., Acetic, lactic and citric acids and pH inhibition of *Listeria monocytogenes* Scott A and the effect on intracellular pH. *Journal of Applied Bacteriology* 74, 515-520, 1993.

DISCUSSION

GORRIS: You isolated bacteria from the fish in order to prepare an inoculum-cocktail. Did you compose this cocktail according to the relative abundance of the isolates.

KIRBY: Only in so much as they were all selected from the highest dilutions, but the objective was to select as many different organisms as possible.

BUSTA: What population level was observed at the time of organoleptic unacceptability?

KIRBY: $10^7 \log_{10}$ cfu/cm^2

QUALITY DESIGN AND PLANT OPERATION IN FOOD PROCESSING

Ryohei Taniguchi and Masazumi Nihmura

I. QUALITY DESIGN

Food processing technology has greatly contributed to the growth of food industry and the improvement of eating habits. Advanced preservation technology of food has enabled the expanded distribution of products and the creation of new foods, contributing to improved quality of life. At present, a consumer wants not only quality products but products that meet his/her choice. This tendency of consumer behavior also is seen in common with food products which have been diversified, gourmet and main-stream foods. Complex and high technology, e.g., bio-technology, and newly developed food materials, require the food industry to be responsible in assuring food safety.

The basic procedure for designing food quality and for preventing hazards caused by foods are discussed in the following.

A. PRODUCT CHARACTERISTICS FOR SPECIFICATION

Quality concept is most important for product development, which should define who is the target consumer and what characteristics of product best fit with consumer needs. Each person has his/her preference of product characteristics, but the preference is not completely different. Consumers are segmented in several groups which have similar preference of product characteristics. Quality concept is defined to describe target characteristics with sensory terms obtained from consumer research. These sensory terms used by consumers, however, are not so explicit for use directly for product development. These terms should be translated to technical words. Target quality is developed with the translated technical words and characteristics which should be possessed by a product with following points:

> Quality Concept
> Consumer usage with no problems
> Distribution and sales conditions
> Manufacturing conditions

Examples for soluble coffee characteristics are as follows:

Characteristics	Reasons to be specified
Flavor	Consumer preference
Moisture contents	To prevent caking of powder
Bulk density	A serving quantity per spoonful
Oxygen content in a jar	To keep taste characteristics in shelf life
Flowability	Fitness to a filler machine

These characteristics are all meaningful for a product, and should be fixed in order to avoid problems in manufacturing through final consumption.

B. PRODUCT SPECIFICATIONS

1. FACTORS FOR SPECIFICATIONS

After identifying each required product characteristic, product specification is developed to fix a target value with upper and lower limits. A target value should be fixed as an optimum point by taking the following factors into consideration:

> Consumer acceptance
> Raw material cost/availability
> Process constraints
> Policy/regulation

Target values must meet consumer acceptance with an aim towards high quality during manufacturing. The acceptance is measured as a consumer response to product quality by various methods, for example, Home Use Test, Central Location Test, Group Interview, Questionnaire Survey, etc. These methods can be applied to be appropriate for objective, importance, cost , etc. The target must meet with cost and availability of raw materials and process constraints, and satisfy company policy and regulation. Target is best fit between consumer needs and business objectives. Next step is to develop acceptable range of quality, namely upper and lower limits. The limits should satisfy :

> Points at which product is different enough to impact the consumer behavior

> Points based on regulations or company policy

2. SYSTEM CAPABILITY AND SPECIFICATIONS

All product quality characteristics that denote any variation or total variation, namely process performance, should be in conformity with consumer acceptance and regulation. System performance (σ^2) which reflects total variation is given as the following equation :

$$\sigma^2 = \sigma_p^2 + \sigma_c^2$$

σ_p^2 : Process Capability =	Variation over short time period of time using same raw material lots and process conditions
σ_c^2 : Control Variation =	Additional variation traceable to some assignable causes. Some can be eliminated by operators or management. Others may be inherent in the system (between raw material lots and machines, etc.)

System capability is system performance when design controls are in place and used properly. System capability will depend on:

> Assignable causes impacting system
> Controls in place (Manual/computer)
> Process capability

System capability cannot be precisely predicted, but must be evaluated over extended period of time to allow all assignable causes to be experienced, to have their impact, and thus to test effectiveness of control system. From experience it is known that eventual system capabilities range from :

$$\pm 4\sigma_p \quad \text{to} \quad \pm 6\sigma_p$$

Thus if the specification limits are tighter than $\pm 4\sigma_p$, the process will be probably not capable of meeting them. A warning flag should be raised. System performance should be evaluated over the long term with all design controls in place and used properly. The audit data σ will be an estimate of system capability. If the specification limits are tighter than $\pm 3\sigma_p$ from the audit data, the system capability is not adequate to meet the limits. The issue must be resolved through:

Improved control system, and/or
Improved process, and/or
Redefined specification limits

The specification limits should meet consumer acceptance, and system performance, at first, must be examined for improvement. System performance (σ^2) is total of process capability (σ_p^2) and control capability (σ_c^2), and σ_c^2 can be reduced by :

Eliminating assignable causes
Improving control system

Also, σ_p^2 can be reduced by :

Redesigning the process
Finding operating conditions which cause less variation

II. HACCP

A. BACKGROUND

The concept of HACCP (Hazard Analysis Critical Control System) was initiated to assure safety of foods in the 1960s by NASA. In the first half of 1980s, there was a mass out-break of Listeria among individuals who consumed dairy products infected with Listeria pathogens. There were several fatalities as a consequence. This incident accelerated the development of HACCP by WHO and other authorities. NACMCF (National Advisory Committee on Microbiological Criteria of Foods) has strongly recommended the food industry to apply HACCP to their process.

B. OUTLINE

HACCP is a systematic approach to be used in food production as a mean to assure food safety. Seven basic principles underlie the concept. These principles include an assessment of the inherent risks that may be present from harvest through ultimate consumption. Six hazards characteristics and a ranking schematic are used to identify those points through out the food production and distribution system whereby control must be exercised in order to reduce or eliminate potential risks. In the application of HACCP, the use of microbiological testing is seldom an effective means of monitoring of critical control points (CCP) because of the time required to obtain results. In most instances, monitoring of CCP can best be accomplished through the use of physical and chemical tests, and through visual observations. Microbiological criteria do, however, play a role in verifying that overall HACCP system is working.

C. HACCP PLAN IMPLEMENTATION

HACCP plan is implemented for each product/process line/plant individually in accordance with the following steps.

1. Obtain Specifications : Raw Materials, Packaging Materials, Formula, Manufacturing Procedure, and Finished Product.
2. Describe the product and its intended use.
3. Develop a production flow diagram.
4. Perform a Hazard/Risk Assessment (Principle 1).
   - Ingredients prior to processing
   - Processing and handling steps
   - Storage and distribution - End product
5. Determine Critical Control Points (Principle 2).
   - Enter on flow diagram in numerical order.
   - List CCP number and description.
6. Establish and list critical limits (Principle 3).
7. Establish and list monitoring requirements (Principle 4).
8. Establish and list corrective action to be taken when there is a deviation identified by monitoring of the CCP (Principle 5).
9. Establish effective record-keeping systems (Principle 6).
   - That document the HACCP Plan
   - That document CCP monitoring results
10. Establish procedures for verification that the HACCP system is working properly. Verification measures may include physical, chemical and sensory methods, and if needed, the establishment of microbiological criteria (Principle 7).

D. EXPLANATION OF PRINCIPLES

PRINCIPLE NO.1

The objective is to assess hazards associated with growing, harvesting, raw materials and ingredients, processing, manufacturing, distribution, marketing, preparation and consumption of the food. The principle provides for a systematic evaluation of a specific food and its ingredients or components to determine the risk from hazardous microorganisms or their toxins. Hazard analysis is most useful for guiding the safe design of a food product and defining the CCP that eliminate or control hazardous microorganism or their toxins at any point during the entire production sequence. The hazard assessment is a two-part process consisting of ranking a food according to six hazard characteristics, followed by the assignment of risk category which is based upon the ranking.

Hazard Characteristics

Hazard A: A special category that applies non-sterile products designated and intended for consumption by at risk population, e.g., infants, the aged, the infirmed, or immunocompromised individuals.

Hazard B: The product contains sensitive ingredients in terms of microbiological hazards.

Hazard C: The process does not contain a controlled processing step that effectively destroys or excludes harmful microorganisms.

Hazard D: The product is subject to recontamination after processing before packaging.

Hazard E: There is substantial potential for abusive handling in distribution or in consumer handling that could render the product harmful when consumed.

Hazard F: There is no terminal heat process after packaging or when cooked in the home.

Note :   Hazards can also be stated for chemical or physical hazards, particularly if a food is subject to them.

Assignment of Risk Category

Category ·IV: A special category that applies to non-sterile products designated and intended for consumption by at risk populations, e.g., infants, the aged, the infirmed, or immunocompromised individuals. All six hazard characteristics must be considered.

Category ·V: Food products subject to all five general hazard characteristics.

Category ·VI: Food products subject to four general hazard characteristics.

Category ·VII: Food products subject to three of the general hazard characteristics.

Category ·VIII: Food products subject to two of the general hazard characteristics.

Category ·IX: Food products subject to one of the general hazard characteristics.

Category X :   Hazard Class -- No hazard.

PRINCIPLE NO.2

The objective is to determine CCP required to control the identified hazards. A CCP is defined as any point or procedure in a specific food system where loss of control may result in an unacceptable health risk. CCP must be established where control can be exercised.

PRINCIPLE NO.3

The objective is to establish the critical limits which must be met at each identified CCP. A critical limit is defined as one or more prescribed tolerances that must be met to insure that a CCP effectively controls a microbiological health hazard. There may be more than one critical limit for a CCP. If any one of those critical limits is out of control, the CCP will be out of control and potential hazard can exist. The criteria most frequently utilized for critical limits are temperature, time, humidity, moisture level (Aw), pH, titratable acidity, preservatives, salt concentration, available chlorine, viscosity and in some cases, sensorial information such as texture, aroma and visual appearance. Many different types of limit information may be needed for safe control of a CCP.

PRINCIPLE NO.4

The objective is to establish procedures to monitor CCP. Monitoring is the scheduled testing or observation of CCP and its limits. Monitoring results must be documented. From the monitoring standpoint, failure to control a CCP is a critical defect. A critical defect is defined as a defect that judgement and experience indicate may result in hazardous or unsafe conditions for individuals using and depending upon the product. Because of the potentially serious consequences of a critical defect, monitoring procedures must be extremely effective.   Ideally, monitoring should be at the 100 % level. Continuous monitoring is possible with many types of physical and chemical methods. When it is not possible to monitor a critical limit on a full-time base, it is necessary to establish that the monitoring interval will be reliable enough to indicate that the hazard is under control.

PRINCIPLE NO.5

The objective is to establish corrective action to be taken when there is a deviation identified by monitoring of a CCP. Action taken must eliminate the actual or potential hazard which was created by deviation from the HACCP plan, and assure safe disposition of the product involved. Because of the variations in CCP for different food and diversity of possible deviations, specific corrective actions must be developed for each CCP in the HACCP plan. The action must demonstrate that the CCP has been brought under control.

PRINCIPLE NO.6
       The objective is to establish effective recordkeeping systems that document the HACCP plan. The HACCP plan must be on file at the food establishment. Additionally, it is to include documentation relating to CCP and any action on critical deviations and disposition of product. These materials are to be made available to government inspectors upon request.

PRINCIPLE NO.7
       The objective is to establish procedures for verification that the HACCP system is working correctly. Verification consists of methods, procedures and tests used to determine that the HACCP system is in compliance with the HACCP plan. Both the producer and the regulatory agency have a role in verifying HACCP plan compliance. Verification confirms that all hazards were identified in the HACCP plan when it was developed. Verification measures may include physical, chemical and sensory methods and testing for conformance with microbiological criteria when established.

III. CONCLUSION

       Process design that incorporates quality and HACCP for food processing is discussed in this paper. Excellent design and manuals of quality control/operation do not always assure good quality if they do not work well. A key issue deserving attention is the people engaged in jobs. The intention/attitude to do a job well assures better operation and higher quality of product. Plant people from manufacturing, quality control, engineering and production planning sections contribute largely to the project based on their experience and knowledge. Involvement of plant personnel in keeping quality and productivity at high level is essential for manufacturing quality products.

DISCUSSION

IRWE: 1) What type of package do you have for the ready to drink coffee products?
2) Are all these packages for ambient storage and distribution.

TANIGUCHI: 1) We have four types of packages for the ready-to-drink coffee products; can, bottle, glass bottle and Tetra-brick pack.
2) Yes, these packages are all for ambient storage and distribution.

BUSTA: Do you expect to encounter many microbiological hazards in coffee products?

TANIGUCHI: Coffee products are contaminated with some microorganisms, but I do not expect to encounter any microbiological hazards.

SINGH: What criteria is used to relate the bulk density with the tablespoon of coffee? Do you use any quantitative measures other than sensory analysis?

TANIGUCHI: We use "grams per ml of coffee" for bulk density and use concentration of coffee in a cup.

BANKS: 1) What specific microbial hazards are present in ambient stable coffee products?
2) What are the control measures?

TANIGUCHI: 1) No experience of specific microbial hazard.
2) Temperatures of coffee extracts during sterilizing and storage.

BANKS: What new technologies are being developed in Japan to address the needs of high hygienic operation necessary in production of minimally processed food.

TANIGUCHI: I am not aware of newly developed technologies addressed to high hygienic operations.

Part III

Advances in Process Modeling and Assessment

THE ROLE OF TRANSPORT PHENOMENA

Jorge C. Oliveira, Pedro M. Pereira and Fernanda A. R. Oliveira

I. INTRODUCTION

Transport phenomena is a general term used for physical processes where a given variable changes with space and eventually with time in such a way that the change can be related to a given flux. In processing engineering, this term refers mainly to heat, mass and momentum transfer. In heat transfer, a heat flux takes place through one or more materials, causing a change in temperature; in mass transfer molecules move causing a change in concentration; in momentum transfer shear stress is transmitted causing a change in velocity. The similarity of these definitions is an oversimplification in the case of momentum transfer, applied to the most common situation in food processing engineering; the flow of liquids, gases and solids. Given the limited nature of this text, we will focus on heat and mass transfer, that have a more similar approach. The heat or mass fluxes are induced by the existence of a gradient in the physical property. In the case of heat transfer, it is correct to say that a temperature gradient corresponds to the existence of a heat flux with the purpose of equilibrating temperatures. In relation to mass transfer, however, there are some slight differences, that will be detailed later; mass transfer is related to a difference in chemical potential, not in concentration, meaning that it is possible to have a mass flux without a concentration gradient and that a concentration gradient does not induce necessarily a mass flow. Mass transfer will occur with the purpose of equilibrating the chemical potential.

We will be concerned with the description of these processes, that is, in relating the observed fluxes with the changes in the physical variables, not with what causes these fluxes. The application of mathematical tools to describe the physical reality has been very successful, but in most cases the issue of what causes the observed phenomena is an open question that is not solved with mathematics. We will not discuss what is the cause and what is the effect; we simply note that we can observe relations and try to describe them mathematically.

Although heat and mass transfer are physical phenomena they have a very important role in most processes occuring in foods beside physical ones, including microbial metabolism and biochemical reactions. Virtually all processes occuring in food processing are strongly temperature dependent and therefore the heat transfer rate usually has a controlling role. All biochemical and microbial reactions involve the movement of reacting species and of products; in the case of microbial metabolism, this transfer may be through cell membranes; in immobilized enzyme applications, the diffusion through the immobilizing agent is involved. Mass transfer is a slow process and reactional mechanisms are many times faster. Therefore, mass transfer can play a controlling role in biochemical and microbial processes. Mass transfer is also important in many food processing operations, being a controlling step in terms of product quality; loss of water soluble components in water immersion processes, diffusion of additives for improving quality characteristics (e.g. calcium chloride for improving firmness).

Figure 1 lists food processing operations following the synoptic table suggested by Peri (1990). For each, it is indicated whether heat and mass transfer play a significant role in relation to the processing operation as such and in relation to the quality of the final product. The different mechanisms of heat and mass transfer, that will be later discussed, are also specified in this table. Since temperature affects most processes, temperature is always important. However, if the process occurs at nearly isothermal conditions, heat transfer itself is not relevant: it matters what is the temperature, but it has a fixed value and not an evolution in time or space.

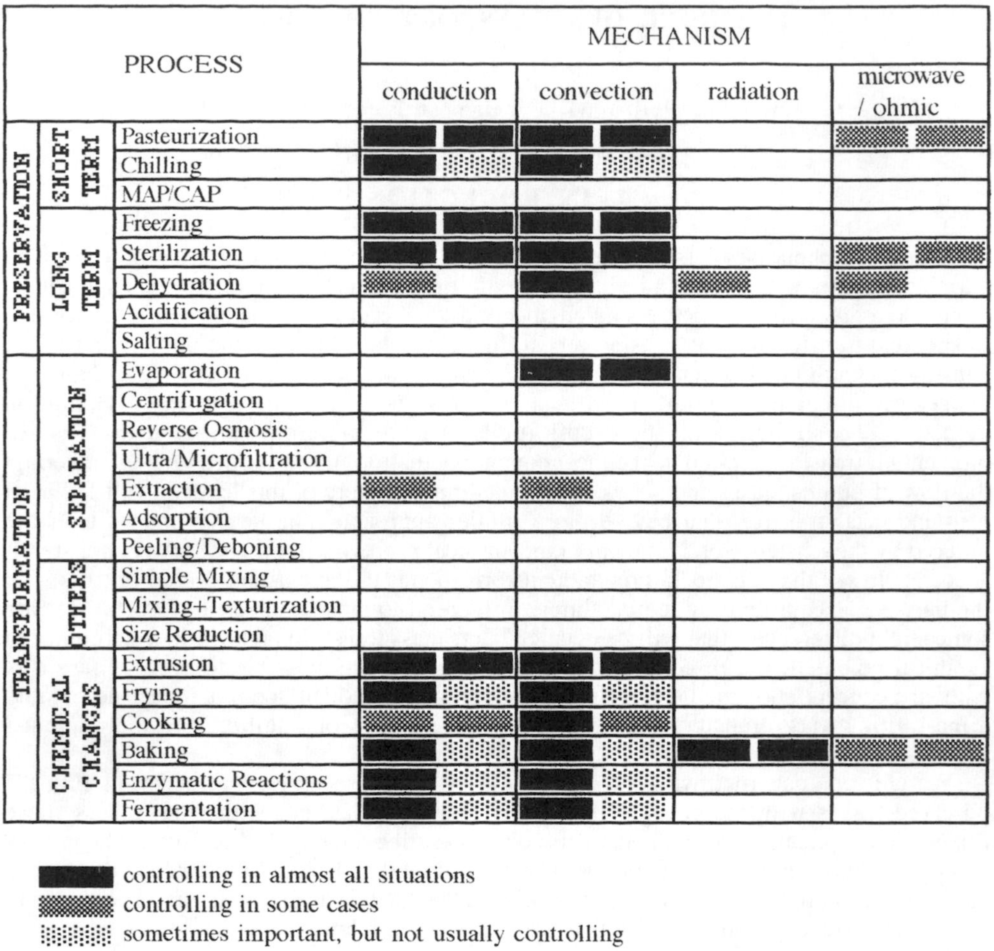

PROCESS			MECHANISM			
			conduction	convection	radiation	microwave / ohmic
PRESERVATION — SHORT TERM		Pasteurization	■ ■	■ ■		░ ░
		Chilling	■ ░	■ ░		
		MAP/CAP				
PRESERVATION — LONG TERM		Freezing	■ ■	■ ■		
		Sterilization	■ ■	■ ■		░ ░
		Dehydration	▨	■	▨	▨
		Acidification				
		Salting				
TRANSFORMATION — SEPARATION		Evaporation		■ ■		
		Centrifugation				
		Reverse Osmosis				
		Ultra/Microfiltration				
		Extraction	▨	▨		
		Adsorption				
		Peeling/Deboning				
TRANSFORMATION — OTHERS		Simple Mixing				
		Mixing+Texturization				
		Size Reduction				
TRANSFORMATION — CHEMICAL CHANGES		Extrusion	■ ■	■ ■		
		Frying	■ ░	■ ░		
		Cooking	▨ ▨	■ ▨		
		Baking	■ ░	■ ░	■ ■	░ ░
		Enzymatic Reactions	■ ░	■ ░		
		Fermentation	■ ░	■ ░		

■ controlling in almost all situations
▨ controlling in some cases
░ sometimes important, but not usually controlling

LEFT SIDE OF THE COLUMN: In relation to process design and control

RIGHT SIDE OF THE COLUMN: In relation to product quality

Figure 1 - Importance of heat and mass transfer in food processing
       a) Heat transfer

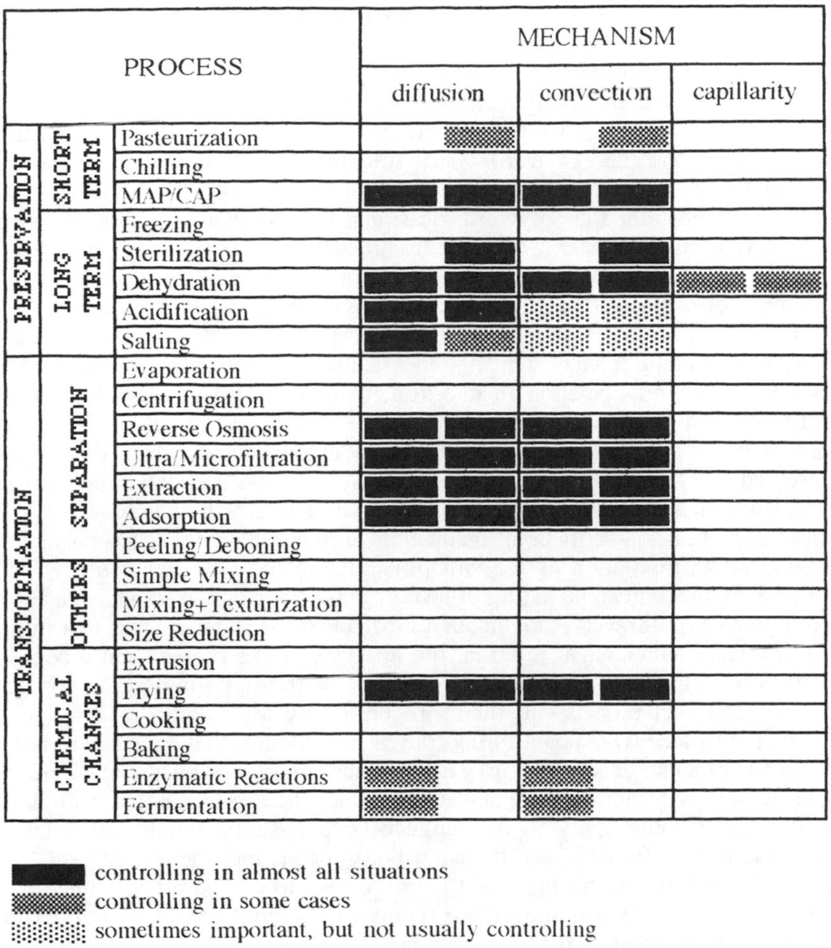

PROCESS			MECHANISM		
			diffusion	convection	capillarity
PRESERVATION	SHORT TERM	Pasteurization	▓ (some cases)	▓ (some cases)	
		Chilling			
		MAP/CAP	█ █	█ █	
	LONG TERM	Freezing			
		Sterilization	█	█	
		Dehydration	█ █	█ █	▓ ▓
		Acidification	█ █	░ ░	
		Salting	█ ▓	░ ░	
TRANSFORMATION	SEPARATION	Evaporation			
		Centrifugation			
		Reverse Osmosis	█ █	█ █	
		Ultra/Microfiltration	█ █	█ █	
		Extraction	█ █	█ █	
		Adsorption	█ █	█ █	
		Peeling/Deboning			
	OTHERS	Simple Mixing			
		Mixing+Texturization			
		Size Reduction			
	CHEMICAL CHANGES	Extrusion			
		Frying	█ █	█ █	
		Cooking			
		Baking			
		Enzymatic Reactions	▓	▓	
		Fermentation	▓	▓	

█ controlling in almost all situations
▓ controlling in some cases
░ sometimes important, but not usually controlling

LEFT SIDE OF THE COLUMN: In relation to process design and control

RIGHT SIDE OF THE COLUMN: In relation to product quality

Figure 1 - Importance of heat and mass transfer in food processing
        b) Mass transfer

II. OVERVIEW OF PUBLISHED RESEARCH

Figure 2 indicates the number of references in the Food Science and Technology Abstracts from 1969 to 1992 involving heat and/or mass transfer studies applied to food processing operations. References were searched by keywords mentioned in the title or abstract of the publications. The keywords "heat transfer", "mass transfer", "diffusion", "conduction" and "model" were searched in combination with a food processing operation. Publications that only mentioned the keywords ocasionally, without an actual consideration of the subject of heat or mass transfer, were then discarded. This does not provide with a full review, but allows for a comparison in relative terms, since all have the same basis. A total of 1325 references were found, 1074 with considerations on heat transfer issues and 490 on mass transfer (239 therefore involved both). The purpose of most of these studies was to describe the phenomena and/or understand the mechanisms with the aim to improve and eventually optimize the unit operation in question, in relation to process efficiency or product quality. Since most physical phenomena can be described by more or less simple mathematical relations, researchers usually try to identify mathematical models that describe the results observed. This would provide the basic tool for process optimization. The exact nature of the mathematical language implies that if a given model, which is based on given assumptions and hypothesis, is verified, all mathematical manipulations of it are also valid, provided there is no extrapolation of the situation outside the range of validity of the assumptions made. Mathematical modelling of heat or mass transfer can also have a second application; if a given hypothesis is made about a mechanism or phenomenum that implies the validity of a given mathematical relation, the analysis of the fit between model and experimental results yields information on the validity of the hypothesis. Experimental validation of mathematical models can therefore be a way to look for whether given mechanisms are taking place with a controlling role or not. Models that have been tried are either based on a fundamental understanding of heat or mass transfer or on a purely empirical fit, that should be obtained with a strong statistical basis. The use of empirical models is limiting because results only apply to the range of experimental conditions used and important factors may be left unnoticed. Its use usually means that there is not sufficient fundamental understanding on the process to provide a more theoretical route to the mathematical model. Figure 2 also shows how many of the published studies provide a theoretical or empirical mathematical model and how many do not involve any modelling work (including reviews). It can be seen from this figure that heat transfer has been more studied than mass transfer. We will discuss the reason for this later.

Figure 3 indicates the distribution of mass transfer studies in the different processing operations considered. Figure 4 shows for each processing operation the number of studies involving heat or mass transfer. From these two figures some interesting points can be taken: a little amount of work has been done on mass transfer in thermal processing operations, in spite of its importance in terms of product quality; there are fewer studies on mass transfer in frying than on heat transfer and it could be found that they relate to the movement of water and very little is found on the movement of oil; most studies involving mass transfer in modified/controlled atmosphere packaging do not attempt to model the experimental observations.

It is apparent from this short overview that there is a lack of information particularly on mass transfer, in processes where this phenomenum is important.

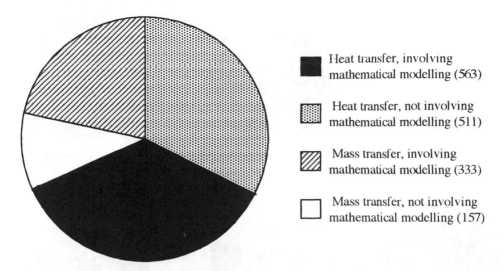

Figure 2 - Number of studies involving heat and/or mass transfer in food processing (referenced in the Food Science and Technology Abstracts 1969-92)

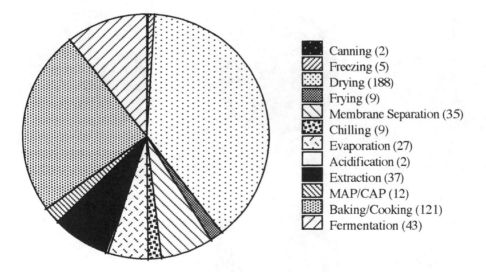

Figure 4 - Number of mass transfer studies in food processing operations (referenced in the Food Science and Technology Abstracts 1969-92)

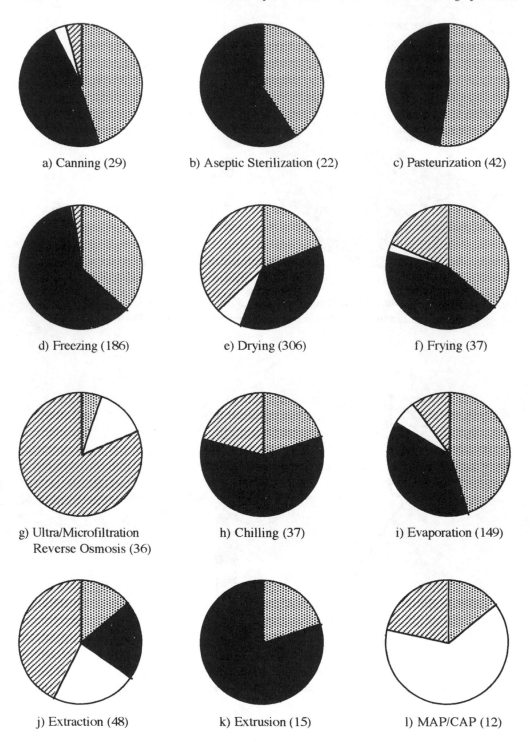

a) Canning (29)   b) Aseptic Sterilization (22)   c) Pasteurization (42)

d) Freezing (186)   e) Drying (306)   f) Frying (37)

g) Ultra/Microfiltration
Reverse Osmosis (36)   h) Chilling (37)   i) Evaporation (149)

j) Extraction (48)   k) Extrusion (15)   l) MAP/CAP (12)

Figure 4 - Number of heat and/or mass transfer studies for each unit operation
          (FSTA 1969-92)

heat transfer: ■ involving modelling   ▦ not involving modelling
mass transfer: ▨ involving modelling   ☐ not involving modelling

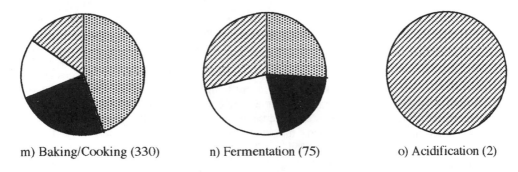

m) Baking/Cooking (330) n) Fermentation (75) o) Acidification (2)

Figure 4 - Number of heat and/or mass transfer studies for each unit operation (continued) (FSTA 1969-92)

heat transfer: ■ involving modelling ▦ not involving modelling
mass transfer: ▨ involving modelling □ not involving modelling

If we can describe well heat and/or mass transfer, we know what is the temperature in time and space and what is the food composition at any given time by simply solving mathematical equations. Furthermore, we may have obtained an understanding of the process sufficiently good to say what variables, parameters and conditions affect our processes, how and to what extent. This allows us to identify changes that are required to produce a given result. Accurate mathematical modeling of heat transfer has opened the possibility to determine optimum processing conditions, regarding process efficiency or product quality, by application of statistical methods. This would give us the assurance that the optimum points determined are the best possible ones. Without this theoretical tool, we would have to experiment with the real equipment for a long time at the cost of a lot of material. When optimizing in terms of process efficiency, we look for operating conditions that lead to minimum processing time (higher productivity) and minimum energy costs. The effect of temperature on food components is different: micro-organisms in particular have a significantly different response to temperature than nutrients and quality factors, such as texture and colour. Therefore, different temperature histories and profiles can be obtained having the same effect on microbial destruction but different effects on quality. When optimizing a process in terms of product quality, we look for the operating conditions that minimize detrimental effects while ensuring the process objectives are met (adequate sterility, sufficiently low water activity, etc.). For example, in thermal processing, we can minimize energy costs, maximize productivity, ensure sufficient microbial destruction, minimize loss of sensory or nutritional attributes (minimizing the result of heat on thermally liable components and on texture and the loss of water soluble components by leaching); in acidification we can minimize acid requirements, maximize productivity, minimize acid flavour, minimize loss of water soluble components, optimize textural characteristics and minimize detrimental effects of heating for processing at high temperatures.

II. FUNDAMENTAL PRINCIPLES

A. Heat Transfer

In heat transfer, thermal energy is transmitted in space. There are different mechanisms by which this process can occur: conduction, convection, radiation, microwaves/radio frequency and ohmic heating.

In conduction, thermal energy is transmitted without any relative movement of the molecules of the material. The increasing vibrational movements are transmitted between adjacent molecules. This process therefore occurs in solids and in stagnant fluids. In convection, there is a mixing of the molecules that contributes to a significant increase in heat transfer rates. The convective movements can be caused by external sources (forced convection) or simply due to the fact that different temperatures result in different densities that induce a hydrodynamic movement (natural convection). This process therefore requires freedom of movements of the molecules and occurs only in fluids. It is possible to visualize convection as the result of a perfectly mixed region bounded by a stagnant layer through which heat moves by conduction. Following this theory, heat transfer by convection would be equivalent to heat transfer by conduction through a boundary layer. The thickness of this layer would depend on the particular hydrodynamic conditions.

These two processes are the most important in most food processing operations. The temperature at any given point in space and in time can be computed by manipulations of a very simple basic hypothesis: that the heat flux per unit cross section area in one direction, in steady state, is proportional to the temperature gradient (the variation of temperature with space). The inverse of the proportionality constant is the resistance to heat transfer (note the similarity of this hypothesis with the basic law of electricity) and is a physical property of the material (the thermal conductivity). This statement forms the basis of Fourier's 1st law, predicting a linear evolution of temperature with space, at steady state, if the conductivity is truly constant. In some situations, the conductivity is actually temperature-dependent, which would complicate a little bit the mathematical description, but without any major problem if the dependence is known. It has been verified that for solids and liquids in fairly small temperature ranges (such as the ones occuring in food processing), thermal conductivity can be assumed constant with temperature.

Mathematical handling leads from this simple relation to apparently very complex equations, when applied to practical situations. In the first place, many important food operations occur in unsteady state. A fundamental law relating temperature with time and space can be deduced from Fourier's 1st law (appropriately called the 2nd law). The most important point to note is that this equation depends on the geometry of the material. Two other physical properties of the material become important as well; the density and the heat capacity.

The fundamental equations of heat transfer (Fourier's 1st and 2nd laws) provide a very good description of heat transfer by conduction, but they are differential equations. This means that its mathematical form is a relation between temperature variations in time and space and the heat flux. In order to obtain the actual temperature values, Fourier's laws must be integrated. An integration requires the specification of the real situation, in terms of what are called the limit conditions: these specify temperature values at some limit point, such as the initial time, an infinite time, the centre of the material, the boundary between two materials. The final result depends as much on the limit conditions as it does on the equation itself, an important point that is sometimes overlooked. The correct selection of the limit conditions is usually the most crucial point in the establishment of a given mathematical model for describing heat transfer. The limit conditions require the definition of the geometry of the problem; since the final result is an integration, it is obviously fundamental to describe the space that is to be integrated. This space may be constituted by more than one material. The point in space where two different materials touch each other is the interface. Since two different materials have different properties, the temperature evolution in space (named the temperature profile) and in time (the temperature history) will be different and therefore there will be a discontinuity at the interface. However, the temperature in the interface of one material must be equal to that of the other, if they are in perfect contact. Otherwise, if a thermal contact resistance exists, a temperature drop will be noted at the interface.This situation may occur with two solid bodies with a rough surface.

Depending on the particular situation, the mathematical model may be a more or less simple equation relating temperature with time and space. Unfortunately, there are some cases for which no analytical integration of Fourier's laws is possible. If we cannot integrate, we can no longer have an analytical equation that, by replacing the physical properties, time and space variables, tells us immediately the temperature and/or the heat flux. This can be due to the simple fact that the material(s) does not have a simple geometry (sphere, infinite or finite cylinder, infinite slab, cube or brick). Another example are non-homogeneous materials, where the physical properties vary with space (imagine a fruit cake, for instance) and anisotropic materials, that is, those where the physical properties vary according to the direction (take the example of an asparagus). Another situation where such problem occurs is when the process itself changes the physical properties of the material (consider baking, for instance).

In cases where analytical integration of the fundamental equations is not possible, the only way forward is to use numerical methods. Finite differences, finite elements and transmission line matrix are the three possibilities that can be used. The basic idea is to imagine that instead of a continuous material we have a mesh. We divide the space in the finite combination of very small elements. Each element is like a block, with the same temperature and physical properties everywhere. According to the particular numerical method being used, a system of equations relating the temperature in all the blocks or nodes of the mesh can be written, from Fourier's laws. When solved, the system of equations leads us to the temperature values in all points. This alternative consumes much more computer power and therefore requires good hardware. It provides us with a complete mapping of temperatures in time and space, therefore lacking the compressed form of a mathematical equation. The most important point to note, however, is that we no longer have an exact solution. So far, it was true that if the basic hypothesis hold (Fourier's 1st law, adequate description of the limit conditions), the final equation is exact, since all mathematical manipulations are exact. Numerical modeling is not, it is an approximation of the real situation and therefore it can be more or less close to the real solution. It is necessary to check results carefully.

Since convective heat transfer can be regarded as conduction through a stagnant boundary layer adjacent to the fluid interface (with the bulk of the fluid being a perfect mix), convective heat transfer would only have one extra issue to solve: the thickness of the boundary layer. In many applications of interest, convection and conduction occur in sequence: e.g. for the transfer of heat between two fluids in a plate heat exchanger we have convection in one fluid, conduction through the metal wall and convection in the other fluid; in heat transfer from a fluid medium to a solid particle, we have convection in the fluid followed by conduction in the solid food. In the latter case, we can solve the equation for the solid, with the convection in the surrounding fluid being expressed as a limit condition (the boundary condition). The problem is dealt with in one step only.

Experimental work has proven that mathematical models for heat transfer by conduction and/or convection that are developed with the required care regarding limit conditions have a very good relation with reality. There are no major concerns regarding the models. This means that we can generate theoretical predictions of temperature profiles and histories with good accuracy. The most important problem on which there is insufficient knowledge is on the values of the convection heat transfer coefficient (or thickness of the boundary layer, if we accept this theory), since it depends on the hydrodynamic conditions. In forced convection resulting from flow, there is some work of importance while on natural convection very few information exists. Without the knowledge of the heat transfer coefficient, the models cannot be used. Since these values depend on the actual situation, it is also not possible to determine them in some laboratory piece of equipment; they would have to be determined *in situ* and many times this is very difficult or impossible. More work is needed on this subject, particularly in relation to the processing of particulate fluid products.

Radiation is also a very common process; radiation exchanges between two different bodies exist always. Heat transfer by radiation is a consequence of the radiation that is emitted naturally by any body. Materials will absorb part of the radiation they receive from others, reflect another part and transmit a third part. Transmition is virtually zero in solids. Since waves travel even in vacuum, radiation does not require any medium to take place. In practice, on earth, it takes place across the air. Heat flux by radiation therefore takes place in parallel with any other that may be occuring (conduction, convection). However, its magnitude, which is proportional to the difference of the 4th power of the temperatures of the two bodies exchanging radiation, is very small and therefore if another heat transfer process is occuring, we can neglect radiation. An exception is made to cases where the temperature difference is high; in practice, this will occur mainly on baking. Drying under the sun is also a case where energy is transfered by radiation. However, note the very long processing times involved, which are not the case with any other food process. Due to its limited application, radiation has been less studied.

Microwave and radio frequency heating are similar to radiation in the sense that an emiting body emits radiation to the food. However, not only this is not natural radiation but a result of a special emission source, but the heating mechanism is different. The principle of microwave heating is the internal generation of heat. Microwaves are electromagnetic waves whose polarity is reverted continuously. This results in a force exerted in polar molecules, such as water, that would cause its rotation if they were free. The result is a generation of heat. Since the material itself absorbs some part of the incident waves, their action is lost along the material, making it impossible to heat up thick pieces of food in this way. Up to such point, however, the material would heat up very evenly and very fast. Evenly heating depends however on the existence of a perfectly homogeneous microwave field and material, which is not the case, and this results in the existence of cold spots and/or hot spots. Heat generated in this way will obviously spread by conduction as well. The most important physical properties of foods affecting microwave heating are the dielectric properties. The mathematical handling of this heating mechanism is very similar to conduction, with the inclusion of a term for the internal generation of heat due to the molecular friction (Mudget, 1986). Good models have been proposed in literature (Ohlsson and Bengtsson, 1971, Ohlsson and Risman, 1978, Ohlsson, 1983), making it possible to use mathematical models in microwave heating with a good accuracy as well. Experimental validation is more limited, particularly on account of difficulties in accurately measuring temperatures in a microwave field, but was achieved in the works mentioned.

Ohmic heating is comparable to microwave heating in the sense that an internal generation of heat is caused in the food. The basic principle is to pass an electrical current through the food: the heating is the result of the action of the food material itself as a resistance to electricity.

Microwave and ohmic heating are fairly new heating technologies in food processing, with potential applications in thermal processing, drying and baking. Most are however still at the prototype stage, due to the high capital costs involved. It is not clear to most investors that the quality benefits overweigh overall costs. Therefore, work carried out in this area is at a less developed stage than that on conduction and convection.

B. Mass transfer

In mass transfer, molecules move in space. Mass transfer is very similar to heat transfer in the mathematical description of the most common processes; diffusion (comparable to heat conduction) and convection. We would have diffusion when the species is moving in an otherwise static environment and convection when a mixing would be promoted by the movement of the medium itself. However, the difference is not so clear cut as in heat transfer. Diffusion involves movement in itself and therefore, it can be considered a convective transport as well. If we consider that there is a movement of molecules (brownian molecular motion, also called self-diffusion in this context), the mass flux has to consider this movement added to diffusion. The basic laws of diffusion are very similar to Fourier's laws and are called Fick's laws. The 1st law expresses the proportional relation between mass flux and the concentration gradient in steady state for a diffusional mechanism. From the moment Fick's 1st law could be established as being the basic description of a mass transfer process, the following mathematical handling would be equal to heat transfer and equally good mathematical models would be obtained. In the same way, convection could be interpreted as the result of a diffusion through a stagnant layer, with a perfectly mixed bulk. For a solid surrounded by a fluid, we can consider the convection in the fluid simply in the boundary condition of the equation for diffusion in the solid, similarly to heat transfer. Unfortunately, there are many reasons why mass transfer is a much more complicated problem than heat transfer:

- Mass transfer is not just diffusion and convection. In any situation, other processes may be occuring, including at cell level, that significantly influence the overall mass transfer rate. If processes are not totally controlled by diffusion, than the basic assumption of Fick's 1st law does not apply. A phenomena that can be observed in some drying processes, for instance, is capillarity. When dealing with mass transfer in a solid piece of food, the cells play a fundamental role. Active transport and facilitated diffusion are processes that can be occuring at low temperatures. In both, there is a given bond between the diffusing species and another component that can increase or decrease mass transfer in relation to simple diffusion.

- the variability of foods affects mass transfer parameters much more than it affects heat transfer. In solid foods, we can have diffusion simultaneously in the solid matrix (which would be a slow process) and in the occluded solution. The latter is usually considered the controlling step. However, most processes involve a change in the food structure (due to heating, or to water loss in the case of drying and frying). These changes will greatly affect the medium in which mass transfer takes place and therefore the parameters can be expected to vary as a function of the evolution of the process. Diffusivity is many times dependent on concentration, time, and/or the temperature history. The variability between food pieces and within one piece itself should be considered. However, given the added complexity, it is assumed that a constant, homogeneous, initial concentration in the diffusing species exists and that material properties are the same everywhere. There are very few exceptions and a lot of work is needed in this topic.

- Diffusion is in most situations a multicomponent problem. There are virtually no situations of interest in food processing where there are not many components diffusing at the same time. When a food piece is immersed in a solution, all components soluble in it will leach out and all components in the solution not present in the food will diffuse into it. Water also moves, it can enter in the food, come out of it, or both, depending on temperature, time and the nature of the surrounding solution. We can assume the diffusion of only one species, as if unchanged by the existence of all other movements, but it is no surprise if we find synergistic effects.

- Equilibrium is theoretically considered in terms of an equality of chemical potential. However, if we have two different physical states, the relationship between chemical potential and concentration is different on each and therefore the concentrations are not equal in equilibrium. Since chemical potential cannot be measured as such, we prefer to keep working with concentration and establish a given relationship between the equilibrium concentrations. We further consider that at the interface equilibrium is immediately established and then the boundary condition would state that at the interface the concentrations will be always equal to those that would be established everywhere at equilibrium. However, there is no experimental evidence of this and some authors have suggested interpreting unexpected results by assuming that the equilibrium concentrations are not established immediately at the interface, but over some period of time. If this were the case, the concentration history would exhibit an inflection (i.e. sigmoid type). There can be several models for the relationship between concentrations in different phases at equilibrium. The most usual one considers a linear relationship between both, with the possibility of a given amount of solute being bound on one of the phases (in which case there would be a non-zero concentration in that phase for a zero concentration in the other - this would be an amount of solute not available for diffusion). However, we may have to consider non-linear relationships. This is a very important subject that has been largely overlooked and not enough understanding exists in most cases on the issue of interface concentrations and equilibrium in mass transfer processes.

- Heat transfer can be measured at the temperature level. We can use simple thermocouples or more sophisticated techniques, such as infrared, to indicate accurate values of temperature in a defined position. Unfortunately, we cannot do that for measuring concentration. Usually, the only option is to measure the total amount of a given component in a defined volume and obtain the average concentration. The equation for average concentration or total amount of solute that has diffused is obtained by integration of the equation for concentration as function of time and space. Therefore, we can only check our models with experimental results at an integrated level in comparison to heat transfer. Given the variability between different food pieces and within each piece in relation to the relevant properties, this results in a larger error. Dispersion of experimental points is therefore larger than what would be admissible in heat transfer. It has been found that the experimental procedure becomes very important in mass transfer experiments to minimize this problem, otherwise it is even possible that results are affected more by the natural variability of food composition than by the mass transfer process itself.

- convection parameters in mass transfer are more difficult to determine than in heat transfer on account of the problem of measuring concentrations. As a result, very little information on film resistance to mass transfer exists and external resistance to mass transfer is many times neglected. Unfortunately, reality does not necessarily avoid our modeling difficulties and external resistance around solid foods is many times very important. A particular case which is insufficiently studied is hindered mass transfer as a result of proximity of food particles. If food pieces are not sufficiently far apart, external resistance will be significantly different from what can be obtained when studying one particle on its own. If particles touch, than the mass transfer area is smaller than the overall surface area and transfer between solid particles themselves could have to be considered.

- published data on diffusivities are less reliable than conductivity ones. There are fairly standard procedures for determining the heat transfer properties and good equipment to do so and good results exist. Unfortunately, it is common to find errors in the determination of diffusion coefficients. Many reasons for these errors have been previously stated: variability of food products not correctly dealt with by the experimental procedure or too large, diffusion of other components (including water) affecting results, external resistance neglected in the model but important in the reality, incorrect limit conditions (particularly the boundary condition), other mechanisms of mass transfer affecting the results. A common problem is the use of an incorrect way of determining the diffusion coefficient (also called diffusivity, like the heat conductivity in heat transfer). For simple Fickian diffusion, the logarithm of the non-dimensional average concentration or the non-dimensional amount of solute that has diffused tends to a linear variation with time, after an initial period. The mathematical model expresses this function in terms of the diffusivity that establishes the slope of this linear variation. However, a straight line is defined not only by the slope but also by the intercept. According to the mathematical solutions, the intercept has a fixed value, that depends on the geometry of the solid and on the external resistance. Therefore, the fact that a plot of the log of the average concentration as a function of time tends to a straight line does not mean that the diffusivity can be obtained directly from the slope of that line: only if the intercept would be statistically similar to the value indicated by the proper mathematical solution would we be able to state this. This detail is many times unnoticed. The best way to determine the diffusivity may be to use a non-linear regression of the experimental data.

The points raised above explain why mass transfer has been much less studied than heat transfer and why there are so many empirical models. Many of these, however, can be reduced to adequate theoretical models using Fick's laws and some extra considerations, such as the variation of diffusivity with concentration, variation of the diffusional path (e.g. drying), dependence on other mass transfer processes (including diffusion of other components) or chemical reactions (for the case of adsorption). Another fundamental approach has been suggested to describe mass transfer beside Fick's diffusion laws, named irreversible thermodynamics (De Groot, 1959 and Prigogine, 1955). This approach is particularly useful if multicomponent diffusion or the existence of several controlling processes has to be considered, otherwise, it will not bring sufficient simplification of the mathematical problem to be an improvement. Irreversible thermodynamics and empirical models share in common the fact that an understanding of the mechanisms is not known nor established by the experimental validation of the models.

Without more understanding of the phenomena that are actually taking place at molecular and cell level, it will be difficult to develop sufficiently in-depth studies on mass transfer occuring in many food processes and therefore describe them properly in some situations. It is true, however, that such understanding would likely yield the most important information regarding the ability to influence and control mass transfer. The use of electronic microscopical techniques has opened the possibility to develop studies of mass transfer at cell level. Only some pioneering work exists and it is considered that such studies are a top priority in food processing engineering.

IV. APPLICATIONS IN FOOD PROCESSING OPERATIONS

The application of heat or mass transfer studies to food processing operations was already summarized in terms of published research. Some case studies can be selected for providing an idea of the type of work that was carried out, noting the potential of studying heat or mass transfer in relation to process or product development and helping to identify areas where more studies would be invaluable. We will focus on a small number of unit operations, enough to cover the most relevant issues: blanching, canning, drying and frying. Water blanching provides an interesting situation, where heat treatment is the main objective of the operation, but the quality issues depend on mass transfer. Canning is perhaps the most studied process in terms of quality optimization, where the role of heat transfer modeling can be well seen. However, it can also be said that there are still substantial issues to be better understood and solved. Drying provides for a complex situation, where the understanding of mass transfer is a fundamental aspect. Frying will provide for an example of how valuable information obtained at microstructural and cell level can be for the development of an adequate mathematical description. Overall, it will be found that the development of process engineering studies, applying heat and mass transfer theory for process or product development, relies on multidisciplinary studies and particularly on cell physiology and microscopy.

A. BLANCHING

Blanching is a pre-processing operation, used mainly for fruits and vegetables that will be subsequently sterilized, frozen or dehydrated. The purpose of blanching is to cause enough heating to inactivate enzymes that would otherwise cause detrimental changes during processing, such as browning. Most problems of concern would occur at the surface, where cells may have been damaged by peeling and the contact with oxygen may lead to such detrimental enzymatic activity. Blanching will also provide additional benefits: expel air from the product, decrease somewhat microbial contamination (particularly at the surface, which is a main concern in vegetable processing). A firming of the tissues can also be promoted by blanching, but would require a two-step process, with a first treatment at 70 °C to activate the enzyme PME to which this effect is attributed. Since enzymes of concern (mostly oxidases) are not very thermally resistant and usually it is necessary to inactivate at the surface only, the thermally process required is very mild (immersion in water at 90 °C for only 3 minutes is an example for turnips). Due to this fact, blanching can be carried out by direct immersion in a hot environment; either steam or liquid water. Each of these options has its advantages and drawbacks, with immersion in liquid water being the most common in Industry. The use of microwaves is being developed as a new alternative for heating. It has been found to have advantages in terms of processing time, but causes some loss of water by drying, which can be a drawback. Furthermore, capital costs are high. Blanching in hot water implies that water soluble components present in the food can leach to the blanching water and therefore be lost. In some cases, this is a benefit: loss of toxic residues from pesticides and loss of sugars, that are the reagents in browning reactions and could cause browning in products such as potatoes and apples (not all browning reactions are enzymatic and therefore blanching may not be enough to avoid browning totally). In other cases, this is a detrimental effect: loss of nutritionally valuable components (e.g. water soluble vitamins), loss of water (that is, of weight and hence of monetary value). Blanching temperatures commonly used (90 - 100 °C) may lead to undesirable softening (Bourne, 1987), but it is possible to develop conditions that lead to absence of significant detrimental effect on texture (Moreira et al, 1993).

It follows that blanching in hot water will be defined in terms of the heat transfer process: considering the inactivation kinetics of the most heat resistant enzyme (usually, peroxidase is considered), the time-temperature relationship must be enough to cause its inactivation. However, the effect on product quality will be affected by mass transfer (leaching of water soluble components). Warming up to processing temperature is usually very fast, making it possible to assume isothermal mass transfer. Califano and Calvelo (1983) verified this for potato spheres of 2.25 cm diameter. Blanching with solutions of additives (e.g. calcium chloride) can be used to infuse the additive, taking advantage of the blanching step (diffusion rates increase exponentially with temperature). However, if we have too concentrated solutions, then the water in the product will diffuse out into the surrounding solution and we would be talking about osmotic drying.

In general terms, it is ensured that adequate thermal processing was achieved by carrying out a peroxidase test rather than by heat transfer calculations and in this process it is mass transfer that becomes a more interesting issue to study.

The first idea that would come to mind is that the loss of valuable components would be minimized if we would use the same water; its concentration in the leachable components would increase up to equilibrium with the solids concentration and material immersed thereafter would not loose components by leaching, since the driving force would be zero. Unfortunately, using the same water for a long time would have serious repercussion in microbial contamination. In continuous processing, it would be possible to recycle and try to decontaminate the water regularly. There are few studies on water recycling in blanching. Considering that the effluent water from a processing with no recycling is a pollutant and that water is increasingly considered a valuable resource that must be preserved, a good future is predicted for more studies on water recycling and in particular the development of adequate processing to eliminate potential microbial hazards.

The evolution of the food composition with time has been well described in many cases by assuming that a simple Fickian diffusion is occuring, with one component leaching out or diffusing into (Schwartzberg and Chao, 1982, Garrote et al, 1984, Rice and Selman, 1984, Garrote et al, 1988, Oliveira and Lamb, 1989). This is a very fortunate fact, given all the potential complexities of mass transfer that were listed earlier. Since in blanching the thermal processing is very mild, we are not too surprised that the diffusivity does not vary with time in many situations, but we are more surprised with the ability to describe with a single diffusion coefficient what is inevitably a multicomponent diffusion problem with many species diffusing at the same time. If we obtain a diffusion coefficient, we can analyse the mass transfer process with this property: diffusivity is inversely proportional to the resistance to mass transfer and the higher its value, the higher the mass flux that is induced by the same concentration gradient. In any case, the movement of water should always be analysed. Mostly, concentrations in solid samples are determined by determining the total amount of the solute and dividing by the total volume. If there is a water intake or loss, the calculation will not be correct, being the result of both solute and water changes. This does not require any added difficulty, only the determination of initial weight and water content, as shown by Oliveira (1989).

Molecular size has been shown to affect inversely the diffusivity. Rice et al (1990) found lower diffusivities for sucrose (molecular weight 342) than for reducing sugars (molecular weight 180). Tomasula and Kozempel (1989) verified that magnesium and potassium had higher diffusivities than glucose. Filipe et al (1993) found glucose and fructose leaching from turnips to have similar diffusivities. This relation is already found for free diffusion in water and in foods the difference is magnified by the importance of obstruction effects being higher the larger the molecule.

Rice et al (1990) did not find any influence of the dry matter and alcohol insoluble solids contents on diffusivities.

By determining diffusion coefficients, researchers were therefore able to identify what affected mass transfer rates.

Garrote et al (1986 and 1988) and Oliveira and Silva (1992) studied the influence of freezing and thawing on diffusion rates, in the first case for ascorbic acid in potatoes and in the second for reducing sugars from carrot cortex tissue. Both found a higher diffusivity but a lower dependence of it on temperature for pre-frozen samples. This means that the changes in tissues caused by the freezing/thawing process increased the leaching rate, but made the process less sensitive to temperature.

Few authors studied the mass transfer process at low temperatures. Although this is not relevant to blanching in itself, it gives some insight into the underlying phenomena. Garrote et al (1984) and Oliveira and Lamb (1989) found that at temperatures lower than cell denaturation temperature losses by leaching became much lower than would be expected by extrapolating the data for higher temperatures and moreover that results are not well fitted by a Fickian model. This led to the suggestion that at these temperatures the solute studied (reducing sugars in both cases) was transfered by osmosis. Moreira et al (1992) did not find a discontinuity at lower temperature for acetic acid intake and proved later that this was due to the permeability of the cell walls to the acetic acid (Moreira et al, 1993). Although these results are for acidification, not blanching, they help indicating that at low temperatures immersion in water solutions leads to mass transfer processes that are controlled by cell walls.

External resistance has not been studied; authors prefer to work with only one piece or with few pieces sufficiently far apart and with sufficient agitation, to avoid external resistance. Lathrop and Leung (1980) compared the results obtained with different agitation speeds, but did not obtain any correlation between agitation rate and ascorbic acid retention in green peas. Azevedo et al (1993) studied external resistance effects in acidification of turnips and of a model food (agarose gel) and found that when external resistance was significant and affected mass transfer rates very noticeably, the experimental data could also be fitted with a diffusion equation, obviously with a lower diffusivity. Although this is a result on acidification, not on blanching, it shows that external resistance can be important, but may go unnoticed, because diffusion models without external resistance can provide good fits as well. However, the diffusivity values thus obtained are evidently incorrect. External resistance would be very important, particularly for analysing continuous processing and it is hoped that more work will be developed on this issue.

Some more complex mass transfer processes have however been found in blanching as well. These were described in terms of a diffusional model with a varying diffusivity. Rice and Selman (1989), Rice et al (1990) and Abdel-Kader (1991) attributed this to changes in the material structure resulting from the thermal processing. They found that diffusivity increased with along the process for temperatures lower than 60 °C and decreased with time for temperatures above 70 °C (those of interest in blanching), tending to a constant value in both cases. Rice and Selman (1989) attributed this decrease to the heat denaturation of cell membrane which results in an initial loss of cell turgor. The tissue may contract and "express" some cell solution rapidly enough so that some solution escapes, particularly via the micropyle. As this effect terminates, losses occur by diffusion means alone and the diffusivity tends to a constant value. Other effects could be added to this explanation: thermal and enzymatic degradation (e.g. production of reducing sugars in potato caused by hydrolysis of starch, Califano and Calvelo, 1983). It should be noted that the diffusivities calculated by the authors mentioned were average diffusivities, not instant values of diffusivity in time. Therefore, it is natural that as the experiment time proceeds the average diffusion coefficient tends to a constant value, since each new point in the calculation causes a smaller change, as the time increase becomes less and less important in comparison with processing time. In any case, these studies provide an indication on added complexities that can arise in a situation where mass transfer is involved.

B. CANNING

Canning is a unit operation that provides a very good case study on the application of fundamental transport phenomena theory to process and product development. The purpose of canning is to provide for long shelf life foods by heat destruction of microbial and enzymatic activity in air tight containers. A more adequate name would be in-container sterilization, since cans are not the only packaging material that can be used. Food is sealed inside the container and then subjected to heat for a sufficiently long time to achieve the desired sterility.

This process is far from isothermal and heat transfer will therefore be an essential tool for defining processing conditions. It is evident from the simple fact that the food exerts a resistance to heat transfer that if all food inside the can had the same physical properties than the slowest heating would occur in the axis of the can. This would be called the cold spot. Since we must ensure adequate sterility everywhere in the product, we will design the process in such a way that the temperature history of the cold spot would be enough for our target sterility. Inevitably, all the remaining part of the food is subjected to a greater heat treatment than this. Heating affects foods very significantly. Quality factors such as colour, flavour, texture properties and nutrients are also heat labile. For the severity required to kill microbial activity, all such changes are detrimental. Therefore, in order to sterilize the cold spot, we overprocess most of the food and cause a significant loss of quality factors.

If we had a liquid product (e.g. milk), it would therefore be much better to process the product in thin layers. This can be done in continuous, with the added advantage of simplicity and efficiency, in heat exchangers. The material would be sealed in air-tight containers after processing, in a sterile environment, and therefore we call this aseptic processing. The products of interest in canning are all solids that will be immersed in some solution (a sauce or syrup). If we have different solids, this may imply that the cold spot is no longer in the axis of the can. It will be located in the centre of a solid piece, but which piece depends on their positions inside the can and on the physical properties of the different pieces.

For low acid foods, the most heat resistant micro-organism considered is *Clostridium botulinum* and therefore we can design the process focusing on the inactivation of this spore forming bacteria. The sensitivity of microbial destruction rates to temperature changes is much higher than that of quality factors degradation. This gives much more room to process optimization (Teixeira et al, 1969a, Lund, 1977, Holdsworth, 1985). When processing at high temperatures, sterility is achieved on a short time and this combination leads to a much smaller loss of quality. However, the high temperature-short time combination (HTST) is defined for the cold spot, as previously mentioned. A sufficiently fast heat transfer rate may not be possible if resistance to heat transfer of the material itself is high. True HTST conditions can usually be found only in aseptic processing, where the use of thin layers of fluid implies a much smaller path for heat transfer and therefore a much faster heating rate of the cold spot. In general, it will be true that the faster the heat transfer rate we can produce, the better the final product quality that can be obtained for the same sterility. The combination of microbial death kinetics with quality degradation kinetics is the key to ensure sterility while maximizing quality.

If we improve heat transfer and energy efficiency, we can optimize processing costs. Barreiro et al (1984) provide an example of such optimization work. However, currently we seem to be much more interested in maximizing product quality. Teixeira et al (1969b) was a pioneering study with that objective. Actually, it would be very useful to combine both aspects; we should be able to say what does the quality improvement cost. It would also be possible to find a balance between the two aspects, if production costs would be very sensitive to quality improvement. Provided we can model mathematically, we can easily find an answer to all relevant questions.

The basic heat transfer equations defined by Fourier's laws provide the description of temperature with time and space. We can use them applied to the cold spot and for the boundary conditions defined by the retort operation we obtain the temperature history. We also need to quantify in equally adequate mathematical terms the change in the microbial population and of the quality property value as a function of temperature, for a non-isothermal process. If we have that as well, than we can calculate both the final sterility and the quality retention. A statistical procedure will identify the best operating conditions that give the target sterility with maximum quality retention.

It follows from the above discussion that the choice of the quality attribute is fundamental (e.g. vitamin C content, textural properties). Since their kinetics can also be different, the best conditions for one factor may not be the best for another. Virtually no research has been made on optimizing a balance between a number of quality factors and therefore we will now assume that one given property or component will be a representative overall quality indicator and it is its value that must be optimized.

We stated earlier that heat transfer equations are very good descriptions of reality, with the main problem in cases where external resistance to heat transfer is important, given the difficulty in many cases in knowing the parameters for this mechanism. Kinetics of microbial death or quality factors degradation must be studied in an equally criterious way. Little work has been done without assuming a first order rate for these processes. This implies that the destruction rate at any temperature and time is proportional to the value of the concentration or microbial population at that point. Furthermore, it is assumed that the proportionality (or rate) constant of this simple law varies exponentially with temperature. The degradation or death kinetics are therefore expressed by two parameters: one indicates the rate constant at some reference temperature and the other the sensitivity of the rate constant to temperature changes. Unfortunately, these two parameters are in many situations highly colinear and an adequate statistical interpretation of the results may be required.

Overall, we cannot rely on theoretical predictions that assume given values for heat transfer coefficients and kinetic parameters wholeheartedly. Experimental validation of models and procedures is very important.

Silva et al (1993a) present a review of published research on heat transfer and sterilization of in-container conduction heating foods (that is, products that heat by conduction only; this requires the package content to be perfectly stagnant, without natural convection affecting the heat transfer rate). Two tables were presented summarizing modeling of heat transfer in conduction heating foods and optimization of calculations for sterilization processes. While some experimental validation of heat transfer models was found, none existed for optimization procedures. Modeling of non-homogeneous and/or anisotropic materials and of non-regular container shapes was considered only by few authors (Naveh et al, 1983, Perez-Martin et al, 1988, Bhowmik and Shin, 1991).

Rao and Anantheswaran (1988) have reviewed work on mathematical simulation for food products with natural convection as the heating mechanism. Experimental results obtained by Zechman and Pflug (1989) and mathematical predictions by Kumar et al (1990) have shown that in this case the cold spot moves towards the bottom of the can.

No modeling work has been carried out for forced convection (this would be the situation in rotating retorts).

The case of particulate products would be even more complex. In this case, both the convection in the liquid and the conduction in the solid must be considered. This requires the numerical solution of the whole can, identifying in each node the velocity and temperature values for a liquid node and the temperature for a solid node. Furthermore, the convection parameters for heat transfer between liquid and solid should be known. If we could assume a simplified situation with uniform initial temperatures for both liquid and solid, constant heat transfer coefficients, uniform bulk fluid temperature, particles of equal size and evenly distributed, than the fit of a mathematical model with experimentally determined temperatures at the centre of some food particles would give us the external resistance values. Lenz and Lund (1979), Deniston et al, 1987 and Stoforos and Merson (1990b) have done this. However, Rumsey et al (1984) have compared model results with experimental determinations and found inaccurate predictions of the temperature at the surface of the solid particles.

Silva et al (1993b) have provided with the first experimental validation of optimum quality calculations, using the acid hydrolysis of sucrose as a quality indicator. They found good agreement between model results and experimental values for quality retention at the surface (near the can walls), for different processing temperatures. Developing more experimental work on validation of theoretical calculations of optimum processing conditions would be very important.

For more details on the mathematical issues and in particular of the definition of the boundary conditions used in literature, the review by Silva et al (1993a) is suggested.

C. DRYING

In drying, solid pieces or liquid layers or droplets are exposed to dry air. As a consequence, water evaporates from the food. From the moment we have a solid matrix, it is easy to visualize that free water must move up to the saturation layer, from where it evaporates. If this movement is slow, it is theoretically understandable that the saturation layer will recede into the matrix, leaving a solid crust over it. In that case, there are two movements of water: in the liquid state inside the still moist material and in the gas state through the solid crust, to the surrounding air. Since both processes occur in parallel, the slowest one will have a controlling role. Therefore, the movement of water could be interpreted as diffusion, but it would be possible to find two diffusional regimes if the diffusion of gas in the solid crust becomes controlling. The existence of an external resistance to mass transfer can be coupled to a diffusional model in terms of the boundary condition and then Fick's laws could be applied. There are some obvious complications in this case, however: the material changes a lot as it dries and it is likely that the diffusivity changes significantly during the process; the size and geometry of the material change, causing a variation in the geometric parameters used in the model; even if the change in the size of the material is small, if the saturation layer is receding like in the mechanism above described, the diffusional path becomes smaller. Beside the complexity these facts can bring into drying, making it difficult to assume a simple Fickian diffusion as a good model, the movement of water can also be significantly affected by the interaction of water with the solid matrix itself. Around 10% of the water contained in solid foods is bound to the solid matrix. Water can be physically adsorbed, chemically adsorbed, trapped in small pores and/or acting as a solvent. In all cases, the loss of water from the food will require additional energy (to break physical or chemical bonds or to compensate for vapour pressure depression). Further to the added energy requirements, the release of this bound water to free state (in which it can diffuse) may be a slow process, in which case it will play a controlling role. In this situation we would have to consider the desorption rate. Further to this, if the material is not in an isothermal condition, the heat transfer rate must also be considered, since temperature affects significantly the diffusional rates, the changes in the material structure and the rate of desorption processes.

The phenomenological description of drying therefore becomes a fairly complex problem to be solved mathematically and subjected to hypothesis that are difficult to validate by experimental observations. For this reason, many authors have tried to apply empirical models. Empirical models usually assume some type of exponential relation between the total amount of water lost and time. Three such models were compared by Diamante and Munro (1991) between themselves and with a simple Fickian diffusion (constant diffusional path and coefficient). According to this study, the equation suggested by Page (1949) would give a best fit between model and experimental results. This conclusion was also taken by Chandra and Singh (1984). However, it can be said that if one uses a model with 3 parameters that can be fitted by a non-linear regression (which is the case of Page, 1949, model), it is no wonder that results are better than Fick's simple diffusion model, that has only one parameter. However, there is no knowledge of what would happen outside the range of the experimental results. Nevertheless, empirical modeling was already a useful tool for studying the effect of drying conditions on the process rate. Saravacos and Charm (1962) and Bains et al (1989) analysed the effect of air flow rate, relative humidity and temperature on solid fruit products. In the first case, the effect of the sample thickness was also studied.

The most important drawback in models based on Fickian simple diffusion is the consideration of a constant diffusion coefficient. Saravacos and Charm (1962), Peraza et al (1986) and Yusheng and Poulsen (1988) have considered this model and tried to prove that the assumption of diffusion of liquid water in the moist pores is the controlling step and that therefore diffusion theory would fit the results. Unfortunately, anomalies were found: Saravacos and Charm (1962) postulated that diffusion of gaseous water in the dry crust might play an important role and Peraza et al (1986) suggested the existence of two different periods with different behaviours while Yusheng and Poulsen (1988) were happy with the fitting of the model with the experimental results. Similarly to the empirical models, these were used by the researchers to study the influence of drying operating conditions on drying rate.

Considering the variation of the diffusion coefficient along the process would therefore lead to better results. Perry et al (1984) suggest a method for calculating the diffusivity as a function of the average water content at a given time. The average diffusivity from the initial time to that point is calculated and then the instant value of diffusivity is obtained. Saravacos and Raouzeos (1984) used this method, but obtained good results only at 50 °C for the drying of starch/glucose gels and not for lower temperatures. However, the authors used the average diffusivity values instead of instant diffusivity values in the calculation of model results, which is incorrect. Karathanos et al (1990) developed a similar study, but determined diffusivity values carefully. The fit with experimental results was good. The authors however point out quite clearly the inherent drawback: in a way, we no longer have a truly phenomenological model, because in the result all relevant phenomena are fitted by the variable diffusion coefficient (e.g. simple Fickian diffusion with variable diffusional path or with desorption rates becoming important after a given period being fitted by a variable diffusivity model). We stated earlier that mathematical models are also useful in giving information on the processes actually occuring because if a given relationship between variables is observed experimentally that follows a given model, we can infer that the assumptions made are valid and that the underlying mechanism explains the reality. In a case such as this, the fact that we can fit experimental results with a model does not imply that we know what are the mechanisms and assumptions that are valid; there are many options, most of which can be well fitted by the model.

Instant values of diffusivity can also be obtained numerically by a procedure well described by Crank (1975). It can be proven that if diffusivity varies with time directly (not as a result of the variation of another variable with time), we can use the solution of Fick's equation with constant diffusivity, with the diffusion coefficient being the average value in the time period considered. If diffusivity depends on another variable, such as concentration, this is not correct. However, we can use the average diffusivity thus obtained as a first estimate. Instant diffusivities are obtained by simple derivation of the average diffusivity. An iterative procedure follows, reproducing a theoretical curve by numerical solution of Fick's equation with the variable diffusivity values found, until the model curve fits well the experimental data. Note the numerical nature of the solution; we obtain a set of diffusivity values for different processing times, that is, different concentrations.

Karathanos et al (1990) obtained instant diffusivity values also in a numerical way and compared the procedure to the one suggested by Perry (1984), previously mentioned. The procedure was comparable to the one established by Crank (1975). It is interesting to note that the authors found very good agreement in systems where clearly only a diffusion process takes place (such as gels) and less good agreement with porous materials, thereby suggesting that the procedure might be good in cases where internal diffusion is the only controlling step. In cases where other phenomena are relevant, Perry (1984) method provided better results.

The work of Patil (1988) with corn grains is interesting to mention because the author did not consider negligible external resistance; something which is almost always assumed, without verification. On the other hand, the author did not estimate instant diffusivities, but considered an empirical relation between diffusivity and moisture content suggested by Sabbah (1972). Nevertheless, the model gave good results.

More complex models have been proposed, with particular interest in cases where the process does not occur at constant temperature. There are two temperature variations possible: in the initial period, if the food does not heat instantaneously and after some time, if the food leaves the initial wet bulb temperature and evolves to the dry bulb temperature. In the first case, temperature variation can be considered simply by applying Fourier's laws. In the second case, the thermodynamic relation between air humidity and wet bulb temperature is also required. Moyle (1981) used Dalton's equation (proposed by Eckert and Drake, 1959) for describing the thermodynamic relation with temperature, but this added consideration is actually not sufficiently relevant to provide for much better results.

Sereno and Medeiros (1990) used Fourier's laws to describe heat transfer and considered external resistance both to heat and to mass transfer. Energy and mass balances were used to provide for the thermodynamic relations for wet bulb temperature and the apparent density was considered to vary as a result of the loss of water. The model fitted very well the experimental results published by Crapiste et al (1988) for apple slices and less well those of Bimbenet at el (1985) for carrot cubes. Zuritz and Singh (1980) have suggested a complex model based on mass and heat transfer equations for a continuous drying operation, applied to a batch process in a spouted bed, for rice grains. The authors have found a significant variation of the diffusion coefficient. The fit for moisture content data was very good and the one for temperature was less good, but still acceptable. This model required the determination of a number of physical properties and parameters of the material, some of which are not very easy to obtain experimentally. This requirement is also necessary for the model developed by Hayakawa and Furuta (1989), that considered the material anisotropic, a chemical reaction having a controlling role and shrinkage of the material. Such model is very thorough and can be expected to fit well experimental results, since it accounts for a variety of phenomena that can be taking place. Comparison with experimental results was good, but it should be said that a model system was used.

Models using a different approach deserve a special mention. Rothstein and co-workers, cited by Gekas (1992), developed the so-called Bahia-Blanca model. The interest of this model is that the material is assumed as a multiphase cellular material, comprising the vacuole, the cytoplasm, the cell wall and intercellular space. The medium is therefore described in terms of its reality and not compared to porous inorganic materials, such as in all other cases. In order for the model to be valid, however, the cellular structure of the material must be preserved during the process. Tupin (1991) also tried to model mass transfer at cellular level, but applied to osmotic processes. Husain et al (1972) considered two periods in the drying process with a decreasing drying rate and applied a model based on irreversible thermodynamics for drying potato slices, with good results.

The basic problem in trying to describe drying mathematically is that it is not known what are the phenomena that actually take place. As models become more complex, they present a number of parameters, which gives some room for the regression to accommodate the experimental results. The point is that there is no certainty that the values found for the parameters used are correct measurements of the importance of the phenomena assumed. A more complex model may fit results better not because it considered phenomena that were important, but simply because it provided more parameters for the fit. A good fit between a complex model and experimental results is therefore no indication that the model is valid in its phenomenological interpretation. It would be most useful to try to carry out experiments to find out the importance of different phenomena and how to identify them in experimental results, but there are obvious problems in being able to do so experimentally, when all that can be easily measured is the total mass of a product, from which the water content is calculated. Work that allows to understand the process at cellular level would be a top priority.

D. FRYING

Frying consists basically in immersion of food pieces in hot vegetable oil. The high temperature causes an evaporation of the water, that moves from the food and through the surrounding oil. Oil is absorbed by the food, replacing some of the water lost. As a result, this process in similar to drying, with the difference that an oil absorption occurs. In vegetables, water content is decreased from over 80% to about 1-2% while final fat content reaches 30-40 % as a result of the oil intake. Vegetables therefore become more porous and crunchy, as a result of the decrease in water content. Due to oil absorption, shrinkage is much smaller than with drying; Mottur (1989) indicated an average of 65% decrease in thickness for potato chips. We immediately identify mass transfer as a fundamental phenomena in this process and we note that we should consider its multicomponent nature, with water and oil moving in opposite directions. Since temperatures are fairly high (180 °C has been indicated as optimum temperature in many processes), for larger slices of food heat transfer may have to be considered. For thin layers, such as potato chips, the food might be able to be considered isothermal during the process. Fat (type and content) and texture are the most important characteristics of the product, both being controlled by the mass transfer processes in question. The fat content of a deep fried food influences cost, palatability, energy density and cholesterol content. Texture affects organoleptic quality; together with the type of oil and the fat content, it creates the particular mouth feeling of the product. Potato composition, oil quality and type and existence of pre-processing treatments (e.g. blanching, calcium addition) will also affect product quality.

An important factor that might have to be considered for a given industrial process is the visible chemical changes the oil suffers as a result of the continuous heat exposure (darkening, increasing viscosity, decreasing smoke point and increasing foaming). These are caused by hydrolyzation to free fatty acids and diglycerides, oxidation and polymerization. As a result, the medium in which the food is immersed will change in time, over a long period of time, which can affect the mass transfer rates and the product quality. In relation to the changes the food itself suffers, beside the changing structure resulting from the loss of water and oil absorption, starch containing vegetables such as potatoes will suffer significant changes due to starch gelatinization. This will affect, for instance, the thermal properties.

In a first phase of the frying process, evaporation occurs at the surface and surface tissues will shrink more, starting to build a crust that constitutes a barrier to mass transfer. This causes a build-up of pressure inside the crust layer, observable in the form of swollen pockets (Ashkenazi et al, 1984), which is released on bursting. In this stage, the rate of evaporation has been observed constant (Ashkenazi et al, 1984). Further shrinkage is significantly affected by solids content, which may be the reason for the familiar saddle shape of potato chips; the low starch containing centre shrinks more than the higher starch containing edge, producing a warping of the chip (Mottur, 1989). This is evidently very relevant for the transport phenomena involved: the material geometry changes during the process.

A very important observation was made by Keller et al (1986): using an oil soluble dye, it was found that oil does not go further than the crust. This result was recently supported by Farkas et al (1992) in SEM microphotographs. Both works were carried out in potato slices with the thickness of normal french fries. Mottur (1989) reported an earlier work of Reeve and Neel (1960) where by using an oil-specific stain and microscopical observation, oil was found mainly in intercellular spaces and in the cell wall material. The few cells with oil inside that were observed were at the surface. The author also refered earlier observations that indicated that nearly all cell walls were intact. These facts are very important to the description of the oil intake: we can postulate that oil will diffuse in the intercellular space and only up to the crumb (the core, inside the crust). We therefore know much more about what happens to oil during frying and therefore we can describe this movement with greater certainty than what we could do with water movement in blanching or drying. This is a good example of cell physiology and microscopic techniques being fundamental tools to provide an insight into the reality we want to describe mathematically for all our engineering tasks.

However, we did not complete our understanding of oil uptake, because all we could observe was the final result. Evaporated water forms a steam current that can make a big pressure against oil movement. It is possible that most of the oil does not enter the food during the immersion, but on removal from the fryer, when the condensation of steam produces a local vacuum that pulls the surrounding oil into the food. This suggestion was presented by Gamble et al (1987).

The water movement may be more difficult to describe. Liquid water can diffuse in the crumb, but it is possible that vapour also diffuses in this core section. In the crust, we should have only vapour diffusing, since temperature there is well over 100 °C. However, we now have the vapour moving in oil. According to the above discussion, we may or may not have oil moving in the crust at the same time. From the surface of the chip, vapour will then rise through the surrounding oil in bubbles. These bubbles cause a significant turbulence around the chips, leading to a minimization or eventually elimination of the external resistance, a fact that will simplify our models. In heat transfer calculations, Ashkenazi et al (1984) reported a high convection heat transfer coefficient (low resistance). Other components of the food may be lost in the water flow. Gamble and Rice (1988) reported differences in yield (product weight per raw material weight) when using potatoes of different initial solids content. The loss of components such as reducing sugars is a curious point, because the water leaves the food in the gaseous phase and sugars do not vaporize at these temperatures. How are they lost?

Since the crust is an important barrier to water and oil flow, its formation will be a crucial aspect of any description of frying.

Ashkenazi et al (1984) suggested two models for potato chips, one of which assumed that the rate of evaporation was controlled by internal diffusion. This model was considered to fit well the experimental results. Rice and Gamble (1989) also used a diffusional model, with Fick's simple diffusion being used to describe water loss. The authors considered the model adequate for an intermediate period, but not for short and for long frying times.

The similarity of frying with drying in relation to water loss makes drying models a starting point for describing frying. We note that since cells remain mostly intact, we can use models such as the Bahia-Blanca. Gamble et al (1987) reported a linear relationship between oil content and water content during frying, thereby establishing the need to consider multicomponent diffusion.

Phenomena such as the change in size and geometry and in the physical properties (particularly if starch gelatinization occurs) must be combined with a multicomponent diffusional model for trying to obtain more comprehensive models, that are yet to be applied to frying. These issues will make it necessary to use numerical modeling techniques. If the process cannot be assumed isothermal, heat transfer must also be considered.

V. RESEARCH NEEDS

We can summarize the research needs identified in the previous discussion.

In heat transfer, the more obvious aspect to study is the external resistance coefficient, that is, the heat transfer process between a solid body and the surrounding liquid. This involves studies in forced convection cases, applicable to rotating retorts in canning and to aseptic processing and also in natural convection cases, which is the case for canned particulate products. Some work has been carried out and it is expected that much more data will be accumulated in a near future.

A limited amount of work has been carried out considering food products that are not homogeneous, isotropic and exhibit constant properties. In particular, no work has been carried out in situations where there is an evolution of the thermal properties with the process. This would be the case, for instance, with starch containing products in processes where starch gelatinization occurs. This means studying the dynamic behaviour of the physical properties themselves.

Another important issue that has been insufficiently looked at is the experimental validation of mathematical models. More work should be carried out to clearly demonstrate the accuracy of theoretical optimization procedures in real situations.

In relation to mass transfer, it can be said that all aspects mentioned above for heat transfer should be looked into. In particular, there is a significant lack of information on continuous processes. External resistance to mass transfer should be studied in more detail.

However, the most important issue is the nature of the mass transfer processes, that should be better understood. It is hoped that more studies involving advanced microscopical techniques and linking the cell physiology aspects observed with mass transfer will clarify the phenomena actually taking place.

REFERENCES

Abdel-Kader, Z.M., A study of the apparent diffusion coefficients for ascorbic acid losses from peas during blanching in water, *Food Chemistry*, 40, 137, 1991

Ashkenazi, N.M., Mizrahi, S. and Berk, Z., Heat and mass transfer during frying. In *Engineering and Food. VI. Engineering Sciences in the Food Industry*, McKenna, B.M. (Ed.), Elsevier Applied Science Pub., pp. 109, 1984

Azevedo, I.C.A. and Oliveira, F.A.R., The use of an agarose gel as a model food system for mass transfer studies in the acidification of vegetables, Poster presentation. IFT Annual Meeting, Chicago, U.S.A., July, 1993

Bains, M.S., Tray drying of apple puree, *Journal of Food Engineering*, 9, 195, 1989

Barreiro, J.A., Perez, C.R. and Guariguata, C., Optimization of energy consumption during the heat processing of canned foods, *Journal of Food Engineering*, 3, 27, 1984

Bhowmik, S.R. and Shin, S., Thermal sterilization of conduction-heated foods in plastic cylindrical cans using convective boundary condition, *Journal of Food Science*, 56 (3), 827, 1991

Bimbenet, J.J., Daudin, J.D. and Wolff, E., Air drying kinetics of biological particles. In *Drying '85*, Toei R. and Mujumdar, A.S. (Eds.), Hemisphere Pub. Co., Washington, DC, U.S.A., 1985

Bourne, M.C., Effect of blanch temperature on kinetics of thermal softening of carrots and green beans, *Journal of Food Science*, 52, 667, 1987

Califano, A.N. and Calvelo A., Heat and mass transfer during warm water blanching of potatoes, *Journal of Food Science*, 48, 220, 1983

Chandra, P.K. and Singh, R.P., Thin layer drying of parboiled rice at elevated temperatures, *Journal of Food Science*, 49, 905, 1984

Crank, J., The Mathematics of Diffusion, Oxford University Press, Oxford, U.K., 1975

Crapiste, G.H., Whitaker, S. and Rotstein, E., Drying of cellular material - II: Experimental and numerical results, *Chemical Engineering Science*, 41, 2929, 1988

Deniston, M.F., Hassan, B.H. and Merson, L.R., Heat transfer coefficients to liquids with food particles in axially rotating cans, *Journal of Food Process Engineering*, 52 (4), 962, 1987

Diamante, L.M. and Munro, P.A., Mathematical modelling of hot air drying of sweet potato slices, *International Journal of Food Science and Technology*, 26, 99, 1991

Eckert, E.R:G. and Drake, R.M., Heat and Mass Transfer, McGraw-Hill, New York, U.S.A., 1959

Farkas, B.E., Singh, R.P. and McCarthy, M.J., Measurement of oil interface in foods during frying. In *Advances in Food Engineering*, Singh, R.P. and Wirakartakusumah, M.A. (Eds.), CRC Press, Boca Raton, FL, U.S.A., pp. 237, 1992

Filipe, C.M., Moreira, L.A., Oliveira, F.A.R. and Singh, R.P., A comparative study of solute uptake of either naturally present or absent solutes in foods. Poster presentation. IFT Annual Meeting, Chicago, U.S.A., July, 1993

Gamble, M.H. and Rice, P., Effect of pre-fry drying on oil uptake and distribution in potato crisp manufacture, *International Journal of Food Science and Technology*, 22, 535, 1987

Gamble, M.H., Rice, P. and Selman, J.D., Relationship between oil uptake and moisture loss during frying of potato slices from c.v. Record U.K. tubers, *International Journal of Food Science and Technology*, 22, 233, 1987

Garrote, R.L., Bertone, S.A. and Silva, E.R., Effect of soaking blanching conditions on glucose losses in potato slices, *Canadian International Food Science and Technology Journal*, 17 (2), 111, 1984

Garrote, R.L., Silva, E.R. and Bertone, S.A., Effect of freezing on diffusion of ascorbic acid during water heating of potato tissue, *Journal of Food Science*, 53 (2), 473, 1988

Garrote, R.L., Silva, E.R. and Bertone, S.A., Effect of surface freezing on ascorbic acid retention in water blanched potato strips, *Journal of Food Science*, 54 (4), 1090, 1989

Gekas, V., Transport Phenomena of Foods and Biological Materials, CRC Press, Boca Raton, FL, U.S.A.., 1992

Giner, S.A. and Calvelo, A., Modelling of wheat drying in fluidized beds, *Journal of Food Science*, 52 (5), 1358, 1987

Hayakawa, K. and Furuta, T., Thermodynamically interactive heat and mass transfer coupled with shrinkage and chemical reaction. In *Food Properties and Computer Aided Engineering of Food Processing Systems*, Singh, R.P. and Medina, A.G. (Eds.), Kluwer Academic Pub., Boston, U.S.A., 1989

Holdsworth, S.D., Optimization of thermal processing - a review, *Journal of Food Engineering*, 4, 89, 1985

Karathanos, V.T., Villalobos, G. and Saravacos, G.D., Comparison of two methods of estimation of the effective moisture diffusivity from drying data, *Journal of Food Science*, 55 (1), 218, 1990

Kell, C., Escher, F. and Solms, J., A method for localizing fat distribution in deep-fat fried products, Lebensmittel Wissenchaft und Technologie, 19 (4), 346, 1986

Kumar, A. Battacharya, M. and Blaylock, J., Numerical simulation of natural convection heating of canned thick viscous liquid food products, *Journal of Food Science*, 55 (5), 1403, 1990

Lathrope, P.J. and Leung, H.K., Thermal degradation and leaching of vitamin C from green peas during processing, *Journal of Food Science*, 45, 995, 1980

Lenz, M.K. and Lund D.B., The lethality Fourier number method. Heating rate variations and lethality confidence intervals for forced convection heating foods in containers. *Journal of Food Process Engineering*, 2, 227, 1979

Lund, D.B., Design of thermal processes for maximizing nutrient retention, *Food Technology*, 2, 71, 1977

Mottur, G.P., A scientific look at potato chips - the original savory snack, *Cereal Chemistry*, 34 (8), 620, 1989

Moyls, A.L., Drying of apple purees, *Journal of Food Science*, 46, 939, 1981

Mudget, R.E., Microwave properties and heating characteristics of foods, *Food Technology*, 6, 84, 1986

Naveh, D., Kopelman, I.J. and Pflug, I.J., The finite element method in thermal processing of foods, *Journal of Food Science*, 48, 1086, 1983

Ohlsson, T., Low power microwave thawing of animal foods, Paper presented at the CoST '91 Seminar, Athens, Greece, 1983

Ohlsson, T. and Bengtsson, N., Microwave heating profiles in foods: comparison between heating experiments and computer simulation, *Microwave Energy Applications Newsletter*, 4, 3, 1971

Ohlsson, T. and Risman, P.O., Temperature distribution of microwave heating of spheres and cylinders, *Journal of Microwave Power*, 13, 303, 1978

Oliveira, F.A.R., Mass transfer analysis for the leaching of water soluble components from food. PhD thesis, University of Leeds, U.K., 1989

Oliveira, F.A.R. and Lamb, J., The mass transfer process of water soluble solids and reducing sugars in carrot cortex tissue. In *Food Properties and Computer Aided Engineering of Food Processing Systems*, Singh, R.P. and Medina, A.G. (Eds.), Kluwer Academic Pub., Boston, U.S.A., 1989

Oliveira, F.A.R. and Silva, C.L.M., Freezing influences diffusion of reducing sugars in carrot cortex tissue, *Journal of Food Science*, 57 (4), 932, 1992

Page, G.E., Factors influencing the maximum rate of air drying shelled corn in thin layers, MSc thesis, University of Purdue, IN, U.S.A., 1949

Patil, N.D., Evaluation of diffusion equation for simulating moisture movement within an individual grain kernel, *Drying Technology*, 6 (1), 21, 1988

Peraza, A.L., Peña, J.G., Segurajauregui, J.S. and Vizcarra, M., Dehydration and separation of grape pomace in a fluidized bed system, *Journal of Food Science*, 51 (1), 206, 1986

Perez-Martin, R.I., Banga, J.R. and Galhardo, J.M., Simulation of thermal processes in tuna can manufacture. In *Proceedings of the International Symposium on Progress in Food Preservation Processes*, Vol. 1, CERIA, Brussels. Belgium, 1988

Peri, C., A synoptic table of food processes and products. Inserto redazionale, *Italian Journal of Food Science*, 2, 1990

Perry, R.H., Green, D.W. and Maloney, J.O., Perry's Chemical Engineering Handbook, 6th edition, McGraw-Hill, New York, NY, U.S.A., 1984

Rao, M.A. and Anantheswaran, R.C., Convective heat transfer to fluid foods in cans, *Advances in Food Research*, 32, 39, 1988

Rice, P. and Gamble, M.H., Modelling moisture loss during potato slice frying, *International Journal of Food Science and Technology*, 24, 183, 1989

Rice, P. and Selman, J.D., Technical note: Apparent diffusivities of ascorbic acid in peas during water blanching, *Journal of Food Technology*, 19, 121, 1984

Rice, P. and Selman, J.D., Apparent diffusivities of ascorbic acid in peas during blanching, *Journal of Food Technology*, 19, 121, 1989

Rice, P., Selman, J.D. and Abdul-Rezzak, R.K., Nutrient loss in the hot water blanching of potatoes, *International Journal of Food Science and Technology*, 25, 61, 1990

Rumsey, T.R., Modelling heat transfer in cans containing liquid and particulates. *1984 Winter Meeting of the American Society of Agricultural Engineers*, ASAE, Michigan, paper No. 84-6515, 1984

Sabbah, M.A., Foster, G.H., Haugh, G.C. and Peart, R.M., Effect of tempering after drying on cooling shelled corn, *Transactions of the American Society of Agricultural Engineers*, 15 (4), 763, 1972

Saravacos, G.D. and Charm, S.E., A study of the mechanism of fruit and vegetable dehydration, *Food Technology*, 1, 78, 1962

Saravacos, G.D. and Raouzeos, G.S., Diffusivity of moisture during air drying of starch gels. In *Engineering and Foods. Vol. I.*, McKenna, B. (Ed.), Elsevier Applied Science Pub., London, U.K., pp. 499, 1984

Schwartzberg, H.G. and Chao, R.Y., Solute diffusivities in leaching processes, *Food Technology*, 73, 1982

Sereno, A.M. and Medeiros, G.L., A simplified model for the prediction of drying rates for foods, *Journal of Food Engineering*, 12, 1, 1990

Silva, C.L.M., Oliveira, F.A.R. and Hendrickx, M., Modelling optimum processing conditions for the sterilization of prepackaged foods, *Food Control*, 4 (2), 67, 1993a

Silva, C.L.M., Oliveira, F.A.R., Lamb, J., Pinheiro-Torres, A. and Hendrickx, M., Experimental validation of models for predicting optimum surface quality sterilization temperatures, submitted to the *International Journal of Food Science and Technology*, 1993b

Stoforos, N.G. and Merson, R.L., Estimating heat transfer coefficients in liquid/particulate canned foods using only liquid temperature data, *Journal of Food Science*, 55 (2), 478, 1990

Teixeira, A.A., Dixon, J.R., Zahradnik, J.W. and Zinsmeister G.E., Computer determination of spore survival distributions in thermally processed conduction heated foods, *Food Technology*, 23 (3), 78, 1969a

Teixeira, A.A., Dixon, J.R., Zahradnik, J.W. and Zinsmeister G.E., Computer optimization of nutrient retention in the thermal processing of conduction heated foods, *Food Technology*, 23 (6), 137, 1969a

Tomasula, P. and Kozempel, M.F., Diffusion coefficients of glucose, potassium and magnesium in Maine Russet Burbank and Maine Katahdin potatoes from 45 to 90 °C, *Journal of Food Science*, 54 (4), 985, 1989

Yusheng, Z. and Poulsen, K.P., Diffusion in potato drying, *Journal of Food Engineering*, 7, 249, 1988

Zechman, S. and Pflug, P., Location of the slowest heating zone for natural convection heating fluids in metal containers, *Journal of Food Science*, 54 (1), 205, 1989

Zuritz, C.A. and Singh, R.P., Mathematical modelling of rough rice drying in a spouted bed, *Journal of Food Process Engineering*, 4, 19, 1980

DISCUSSION

KNORR: In many foods processes heat and mass transfer are dynamic processes that go in opposite directions . How can this problem be addressed in modeling?

J. OLIVEIRA: There is no significant problem in adding heat and mass transfer equations. We will have a set of equations to solve, but mathematically it puts the problem more complex yet perfectly possible to handle. However, I did like the mention of the word "dynamic", because our models usually assume a dynamic behavior as a sum of infinite static points, situations where the temperature history affects the result just as much as the actual value of temperature. One has to consider the dynamic behavior of the food properties; an interesting point, insufficiently dealt with at this time.

SHEWFELT: Let me respond to these excellent presentations from two perspectives - cell physiology and quality. Cell membranes provide the primary barrier properties of both plant and animal cells. In physiological active tissue (fresh fruits and vegetables) ion transfer is as important if not more important factor than molecular transfer. Also in physiologically active tissue, relative humidity, carbon-dioxide, and oxygen concentration, and the presence of ethylene can be as important as temperature on final product quality. In chilling sensitive tissue, shelf life is best extended by cycling the temperature above and below the optimum temperature (intermittent warming) which makes time/temperature indicators meaningless. Having said that, visual purchase characteristics (color, wilting, etc.) are intrinsic indicators. Thus by focusing on purchase characteristics (flavor in fruits or texture in vegetables) a time temperature indicator in a defined product, such as a MAP kept within a reasonable temperature range, we have an achievable objective. I also think that we need to abandon the concept of quality index and attempt to model consumer acceptability.

J. OLIVEIRA: Ion transport is one of those examples I mentioned broadly as "active transport" and therefore we agree on the fact that an understanding of processes occurring at cell level is very important for describing mass transfer in foods. I also take the point of Prof. Shewfelt in relation to cell walls: these only exist in vegetables tissue and the mentions I made to cell walls in my communication are meant for fruits and vegetables not animal products. Since I do not work with animal products, I forgot to clarify that point.
I do not share the view, however, that TTI's will not be applicable in situations when there are temperature fluctuations. In fact, that is the whole purpose of TTI's. All that is necessary is that the TTI kinetics has been determined in the range of values of those temperature variations.
I would also like to mention that I also consider that research on linking quality indicators to consumer acceptability or preference would be most important. However, I would prefer work with a quality indicator and its correlation to consumer preference rather than directly with consumer preference, since the latter may not be very sensitive.

FOOD PROCESS MODELING WITH UNCERTAIN DATA

Vassilis   Gekas

I. INTRODUCTION

Minimal processing of foods is a big challenge for the food industry, as being a response to the needs for more "natural" food combined with even higher food safety, less energy usage and less waste output to the environment. In order to achieve these goals an enhanced modeling and simulation capability is required, not only for the sake of the predictability but also for the possibility of on-line control development to guarantee the quality standards within accepted confidence intervals.

It is a common place that the materials, with which the food infustry is dealing with, exhibit variability in chemical and physical properties which in their turn may cause a variability, if not taken into account, may cause a variability in the quality properties of the product coming out from the process. This property variability presents a first source of uncertainty in food processing. It may arise from seasonal variability, from batch to batch random property variations and - when concerning plant origin foods (fruits, vegetables) - even from variations in the same plant. The uncertainty arisen from this kind of variability is an uncertainty on the macroscopic, statistical level.

A second source of uncertainty in food processing may be found on the microscopic or mechanistic level due to random events taken place during food processing, as it may occur during processing with any other material in general. This paper will distinguish between the two types of uncertainty and will provide examples of ways to handle with both cases. The first one will be referred to as uncertainty in food property data, the second one as uncertainty manifestated on the level of mechanism or mechanistic uncertainty. The second case, the mechanistic uncertainty, has been traditionally approached using discrete dynamic methods, known as Monte Carlo methods or techniques. Monte Carlo techniques, in this sense, have been extensively used in a lot of application fields in the natural sciences, including food processing.

It is possible to use the Monte Carlo principle for the first category of uncertainty, i.e.to handle with uncertain property data. In food processing the classical application is in establishing confidence intervals of heat process lethality. Normal distributed properties have been considered[1-5]. Recently, Andersson[6] and Persson[7] have used a Monte Carlo simulation algorithm to predict variation in ultrafiltration membrane permeability, in which, besides normal, also bimodal distributions of the properties involved have been investigated. Johns[8] has compared the Monte Carlo methodology with his own "quadrature integration" procedure in a thermal processing application with normally distributed heat diffusivity.

II THE MONTE CARLO APPROACH

The principle of the Monte Carlo method is very simple. It is the approach of making a random choice using a roulette. With the emergence of powerful computational capacities in the few last decades the roulette has been replaced by the computer. In every computer language, a specific algorithm (in the form of a function or procedure or subprogram) exists which enables randomizing in a given integer interval. On the purely computational level, the Monte Carlo algorithm, although with the risk of a slight imperfection, has dominated among the alternative randomizing algorithms (Las Vegas and Atlantic City), most probably, because it always stops.[9]

In the "mechanistic" uncertainty case the application of the Monte Carlo approach is straightforward. If one system (for example a particle, a drop of a fluid, a macro-molecule, a microorganism, etc) can follow one out of a small number, n, of equi-probable routes or events in a process, each simulation will make use of the randomizing algorithm to choose an integer in the interval 1..n. The randomization is uniform. More subtle methodology is required in the case of a property which follows a certain distribution pattern and we are interested to draw a random value from that distribution.

Let us consider a property, expressed by the independent variable x, and following a given distribution, known or assumed, $\phi(x)$. Our aim is at studying a second property, y, depending on x, through a direct or indirect relationship of the type

$$y = f(x) \text{ or}$$

$$f(x, y) = 0 \tag{1}$$

Here, the Monte Carlo simulation, each time, gives a value of the stochastic variable x, by randomly choosing u_i, i. e. u_i gives the number of standard deviations (σ) by which the chosen value differs from the mean value of x (see Figure 1). This is the first step of the simulation and it is usually very fast. Using Eqs (1) a value of the dependent variable or property, y, is obtained for each x value. The set of the simulated y values is then studied for possible distribution pattern with particular attention to the estimation of percentages of failed cases or the prediction of confidence intervals, when certain "failing" percentages are accepted.

Assuming for example that y is a quality property, the value of which must exceed (or inversely, must not exceed) a critical value, y_{crit}. After a number of simulations a percentage of the failed y values can be calculated. Alternatively if a percentage of failure is accepted the critical value can be estimated from the distribution $\phi(y)$ of the y property.

The confidence limits may also concern an interval centered to a desired critical value. If more than one parameters have uncertain data the confidence regions centered to a critical value are usually ellipsoids. Traditonal percentages for confidence levels are: 68.3% (the lowest confidence worthy of quoting), 90%, 95.4%, 99%, 99.73% and 99.9..9%. Use of the 68.3, 95.4 and 99.73 is made because these numbers have a connection with the properties of the normal distribution.[10]

III APPLICATIONS IN FOOD PROCESSING: A BRIEF REVIEW

Whereas the Monte Carlo applications in other fields show a rapid increase the last years, relatively a few of them concern the field of food processing. In a literature survey in the whole area of natural sciences and technology the number of articles containing in their title "Monte Carlo" has showed the following tendency the last four years:

1989	473
1990	559
1991	620
1992	728
(1993)	(802)

Only 20 articles were found in the FSTA 1969-1991 containing in their abstracts the name Monte Carlo. (one of them was outside the scope of this paper, since it was about the Monte Carlo tomato cultivar).

The Monte Carlo approach has been applied in different areas of Food Science and Food Technology, including: genetics, food microbiology, food irradiation, enzyme technology and heat process lethality (Table 1). Below, I shall focus on the two types of application mentioned in the Introduction i.e. those relevant to food processing: the first one concerning uncertainty arisen from food-property variability ("macroscopic" uncertainty).and the second one which concerns "mechanistic" (or "microscopic" uncertainty, where the decision or choice, is made on a microscopic level).

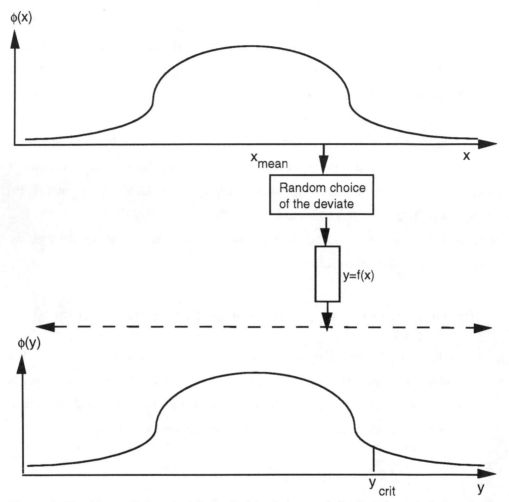

Figure 1. The Monte Carlo principle applied in the case of distributed properties; x is the independent variable distributed with a distribution function $\phi(x)$, y is the dependent variable either given directly as a function of x: $y=f(x)$ or by the solution of a differential equation of the form $f(x,y) = 0$. Simulated values y_i are obtained from the equations above, where x_i-values have been derived by choosing u_i. Then the distribution $f(y)$ of the dependent property, y, can be studied particularly with respect to the establishment of confidence intervals.

A.   HANDLING MACROSCOPIC UNCERTAINTY

1   Heat Process Lethality

Very important food processes are those asociated with the thermal destruction of microorganisms. The calculations, thereby, should take into account uncertain food property data, biological variability, as well as, uncertainties arisen from process parameter fluctuations. Lenz and Lund,[1] have estimated confidence intervals of calculated lethality and mass-average retention for heat-conducted canned foods. The influencing parameters studied were microorganism biological variability and heat diffusivity. As the result of these influences, the F-value was found to show standard deviations of 10-15 % for small Fourier numbers and 5-8% for larger (>0.7) Fourier numbers. The mass-average retention showed a maximum standard deviation of 5%. Lund has presented a review concerning the certainty in lethality calculation.[2]

Table 1
Applications of the Monte Carlo Method in Food Science and Technology

Field of application	Parameters or qualities studied	Reference
Heating process	Lethality	1-5
Heating process	Centre temperature and heating time	7
Particles in suspensions	Particle stability	11-13
Phase inversion	Cell membrane behaviour	11
Food microbiology	*Esc.coli,* and salmonella level	14-17
Food genetics	Weak gene expression	18
Food enzymology	Amylolusis product concentrations	19
Food irradiation	Irradiation dose level	20
Food toxicity	Aflatoxin level in peanuts and in corn	21-23
Food storage	Moisture content and temperature	24
Food serving	Serving size	25
Flow dynamics	Path in a cavity transfer mixer	26

Lenz and Lund have also studied heating rate variations and lethality confidence intervals for foods heated by forced convection[3]. The 95% confidence intervals for the lethality ranged from 20-60% of the median lethality value, depending on the conditions of the process. Particularly, the fluid/particle interface heat transfer coefficient was found of a significant effect on the lethality distributions.

Massaguer has presented in his PhD Thesis[4], the effect of the variability of 9 parameters on the thermal process lethality. Conduction heated spagetti in tomato sauce was the product investigated. Hayakawa et al.,[5] have extended the method to food packages heated by conduction. Normal but also gamma distributions have been used to describe the variations in thermal properties and in reaction kinetics properties.

2   Other Property Variation

Johns has presented results of the effect of the variability in heat diffusivity on the obtained centre (coolest point) temperature in canned foods[8]. Lo *et al.* [24] simulated moisture and temperature of storaged wheat influenced by weather variability.

B.   HANDLING MECHANISTIC UNCERTAINTY

1   Problems Involving Particles And Their Interactions

It is known that the Monte Carlo method has been applied to collections of molecules or small particles, in various areas of Physics. Middlehurst and Parker[12] have reviewed the application of the idea in Food Science. Both "hard" particles (not interacting with each other, unless they come into contact) and interacting, at distance, particles have been considered. The Monte Carlo method was used to generate snapshots of collections of particles and then estimate the probability of any particular arrangement, taking into account particle overlapping in the case of "hard" particles and Boltzmann-distributed interaction energies in the case of "soft" particles.

Middlehurst and Parker, however, restrict the Monte Carlo method, arbitrarily, to a method used for checking theories concerning the movement of particles *(sic)*, and, furthermore, they claim that the Monte Carlo method has been used in Food Science, for the first time, in 1976, when Snook and van Megen[13] simulated colloids stabilized by electrostatic forces, only. The idea expressed in the present paper is that Monte Carlo is a more universal mehod, than the description given to it by Middlehurst and Parker, and useful to handle with uncertain data in a much broader spectrum of applications.

Another important area in Food Science, mentioned by the same authors,[12] into which Monte Carlo application is possible (but the experimental data are yet not fully reliable), is studies of phase transitions. A particular application is presented concerning cell membranes in photosynthetic cells. The membranes are considered as protein or protein aggregate "particles" floating in a sea of lipids.

2   Enzymatic Conversion

In enzymatic hydrolysis, more than one enzymes may act on the same kind of substrate, for example the two enzymes α-amylase and β-amylase, hydrolyze amylose and amylopectin, acting either separately or conjointly. Toner and Potter,[19] used a Monte Carlo technique in order to simulate this action in the commercial mashing of malted cereals. Product concentration of the amylolysis was the output parameter and the modelled results were found in agreement with the experiment, especially at low product concentrations.

3   Flow dynamics

In a recent paper, van der Meer et al.,[26] have used the Monte Carlo technique to simulate the flow and mixing of food fluids in a cavity transfer mixing

IV. EXAMPLE OF A SIMULATION

A   THE PROBLEM

For the sake of an illustration of the Monte Carlo application, a simple food processing example is considered, of a fish chunk undergoing cooling from 20 to 0 $^{\circ}$C. The dimensions of the product are such (thickness=27 mm) so that to justify the assumption of a semi-infinite slab . The heat diffusivity of the fish, is assumed to be constant in the said temperature interval, following a normal distribution with a mean $1.47 \ 10^{-7} \ m^2/s$ and a standard deviation of $0.16 \ 10^{-7} \ m^2/s$. If the temperature of the cooling medium is 0 $^{\circ}$C, let us assume that we accept a centre (warmest product point) temperature if this is < 1 $^{\circ}$C, at the end of the cooling process. A last assumption concerns the heat transfer coefficient or the Biot number which is taken practically infinity for the sake of simplicity. (With other values of the Biot number, however, or even with different initial and boundary conditions, or in cases of two or three dimesions, it is also possible to proceed; it is the solution of the transport equation.which becomes somewhat more difficult).

The questions to be answered are i) how to account for the uncertainty in the values of the heat diffusivity of the product and ii) how to estimate the confidence limits of the process. The Monte Carlo method can be used to give the answers here, since the independent variable, x, is the heat diffusivity, a, whereas the dependent y variable, y, can be either the cooling time, t, required to obtain a given center temperature (T<1 $^{\circ}$C) or the center temperature after a fixed cooling time, for example 30 min or 45 min.

B.   THE METHOD

A program has been written in Turbo Pascal for the simulation of random, yet distribution-weighed, values of the heat diffusivity. One thousands values have been received applying the program. The heat transfer equation

$$\frac{\partial T}{\partial t} = a \frac{\partial^2 T}{\partial z}$$

with $T(0)=0$, for all t and $T(z\neq0)=20$ at $t=0$ (2)

has been solved with an explicit discretization scheme, implying that

$$\Delta t = (\Delta z)^2 / 2a \tag{3}$$

where Δt and Δz are the time and space intervals of the discretization. If a number n of time-steps is required to obtain a pre-set centre temperature, then the corresponding total cooling time, t_C, can be obtained as a function of a, using Eq.3, as follows:

$$t_C = n (\Delta z)^2 / 2a \tag{4}$$

For the centre temperature, T_f, at a fixed cooling time, t_f, it is possible to obtain, also, an algebraic expression, from the analytical solution of Eq.2 [27].

$$\frac{T_f T_o}{T_m - T_o} = \frac{8}{\pi} \exp \left\{ -\frac{\pi}{4} a \frac{t_C}{b^2} \right\}$$

valid for large values of the Fo number   (5)

where T_m, is the temperature of the cooling medium ($0\ ^\circ C$), T_o, the initial product temperature $20\ ^\circ C$, a the heat diffusivity, b the half thickness of product and t_C the cooling time.

Equations (4) and (5) are, then, useful for the study of the distributions of the dependent parameters t_C and T_f. It is useful to study if the obtained distributions are normal, since the normal distribution is well studied and its properties are known and tabulated in every book of statistics. One means of testing the hypothesis of the normal distribution is the application of the so-called χ^2-test, which is well described in statistics textbooks.

When the distribution, experimental or simulated, of a parameter, is not a normal one, the recipe, then, is try to find a function of the parameter which is a normal. For example, there is a tendency for a number of properties to exhibit log-normal distri-butions (molecular weights, pore sizes etc.). In case of failing to find such a function, it is possible for an irregular distribution $w(x)$, to apply the rejection method. According to this method one finds the closer normal distribution $\phi(x)$ so that $w(x)$ is everywhere less than $\phi(x)$. Then each time a random, x_o, of the normal distribution $\phi(x)$, is generated with a simulated probability $p(x_o)$, and in a second step this random is accepted if $p(x_o) \leq w(x_o)$ otherwise it is rejected. In case of rejection another random from $f(x)$ is generated and the procedure is repeated a number of times before the final random belonging to $w(x)$ is obtained.[10]

C   RESULTS

1.   Simulated Heat Diffusivity Data

The results of simulations is shown in Fig. 2. The heat diffusivity was assumed as normally distributed. Fig.2 and application of the χ^2 test verified the assumption.

2.   Confidence Limits

a   Pre-set Centre Temperature

The simulated distribution of the cooling time for the final centre temperature to be lower than 1^oC, is shown in Fig.3. The distribution is asymmetric (not a normal one) and this is expected, since it is the inverse of the cooling time to be normally distributed, as being proportional to heat diffusivity, by virtue of Eq. 3.

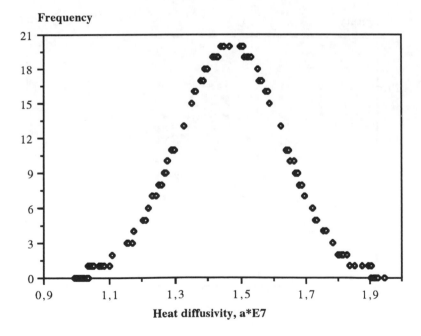

Figure 2 Simulated distribution of the heat diffusivity of the fish chunks

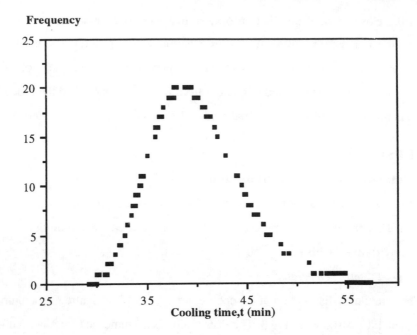

Figure 3. Obtained distribution for the cooling time required due to heat diffusivity variability, applied into the cooling processing of fish chunks.

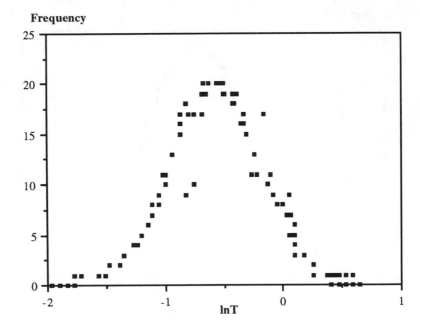

Figure 4. Simulated distribution of the final centre temperature during cooling for 45 min, of fish chunks.

From Fig.3, it can be concluded that a cooling time of 60 min is required for a 99.99.% confidence level, while a cooling time of 45 min gives a confidence interval of 94.7 % which is close to the customary con-fidence level of 95.4%. Assuming that, the above level is acceptable, taking into account the objectives of the cooling process and in agreement, always, with a food microbiologist, the requirement for minimal processing, then, could be satisfied by choosing a cooling time of 45 min and not 60 min.

b　*Pre-set Cooling Time*

For a pre-set cooling time 45 min, the logarithm of the final centre temperature of the product is distributed as in Fig.4. By virtue of Eq.4 this temperature is expected to follow a log-normal distribution as it is shown in Fig.4.

ACKNOWLEDGMENTS

The thanks of the author go to NORDRECO, AB, subsidiary of the NESTLE Company in Sweden, and the Swedish Board for Technical Development (NUTEK) for financial assistance. Lars-Göran Skanz, Malmö, is acknowledged for discussions.

REFERENCES

1. **Lenz, M. K. and Lund, D. B.**, The lethality-Fourier number method: confidence intervals for calculated lethality and mass-average retention of conduction-heating, canned foods, *J.Food Sci.*, 42(4), 1002-7, 1977

2. **Lund, D. B.**, Statistical analysis of thermal process calculations, *Food Tecnology*, 32(3), 76-78 , 1978

3 **Lenz, M. K. and Lund, D. B.**, The lethality-Fourier number method. Heating rate variations and lethality confidence intervals for forced-convection heated foods in containers, *J. Food Process Eng.*, 2(3)　227-271, 1978

4 **Massaguer, P. R.de,** Influence of random variation of process parameters on the variation of thermal process lethality for conduction-heating food, *Dissertation Abstracts Intl,* B 45 (2) 503, 1984

5 **Hayakawa, K. I., Massaguer, P. R.de, and Trout, R.J.,** Statistical variability of thermal process lethality in conduction heating food computerized simulation, *J.Food Sci.*, 53(6), 1887-93,1988

6 **Andersson, Ch** , A study of flux distribution in ultrafiltration membranes using Monte Carlo simulation, Internal report, Food Engineering div., Lund University, 1992, contact person V.Gekas.

7 **Persson, K., Gekas, V.and Trägårdh, G.,** On the presence of double log-normal permeability distributions in UF membranes, *J.Memb.Sci.*, submitted, 1993

8 **Johns, W. R.**, Simulation of food processes with uncertain data, in *Food Engineering in a Computer Climate*, Inst. of Chem. Engineers, Rugby, UK, (Eds), Hemisphere Publishing Co., NY, Philadelphia, London, 1992, 1-24.

9 **Raywarth-Smith, V. J.**, *A First Course in Computability*, Blackwell Scientific Publications, Oxford, 1986, chapter 6

10 **Press, W. H., Flannery, B. P., Teukolsky, S. A. and Vetterling, W. T.**, *Numerical Recipes in Pascal*, Cambridge University Press, 2nd edition, 1990, chapters 7 and 14.

11 **Middlehurst, J. and Parker, N. S.**, The Monte Carlo method and its use in food science, *CSIRO Food Research Quarterly*, 45(1), 12-17, 1985

12 **Snook, I. K., and van Megen, W. J.**, Prediction of ordered and disordered states in colloidal suspensions, *J. Chem. Soc. Faraday Trans*, 72(2),216-223, 1976

13 **Barker. G. C., and Grimson, M. J.**, Food colloid science and the art of computer simulation, *Food Hydrocolloids*, 3(5), 345-363, 1989

14 **Arndt, G., Hildebrandt, G., Weiss, H. and Siems, H.**, Use of the most-probable-number technique for quantitative detection of Salmonella. I. Monte Carlo study (computer simulations) to examine the statistical properties of the most-probable-number method, *Fleisch-wirtschaft*, 61(9), 1373-1380,1981

15 **Hildebrandt, G., and Weiss, H.**, Mittelwertbildung bei der Keimzahl-bestimmung an einem Probenhomonisat, *Fleischwirtschaft*, 64(1), 83-86, 1984

16 **Hildebrandt, G., Weiss, H. and Hirst, L.**, A propos Forrmiloe's formula. Calculating mean values when determining bacterial counts in a sample homogenate, *Fleischwirtschaft*, 66(7), 1128-30,1986

17 **Haas, C. N., and Heller, B.**, Test of the validity of the Poisson assumption for analysis of most-probable-number results, Applied and Environ.Microb., 54 (12), 2996-3002, 1988

18 **Yonezawa, K., Sato, Y. I., Nomura, T., and Morishima, H.**, Computer simulated evalua- tion of the hybrid weakness gene system as a means of preventing contamination of rice cultivars, *Plant Breeding*, 104(3), 241-247,1990

19 **Toner, M. C., and Potter, O. E.**, Computer simulation of beta amylolysis of starch compounds,*J. Inst. Brewing*, 83(2), 78-81, 1977

20 **Humphreys, J. C., Chappel, S. E., McLaughlin, W. L., and Jarret, R. D.**, Measurements of dose distributions in various materials irradiated by 10 MeV electrons, in *Radiation Processing*, (FSTA (1979) 11 5G346), 1977, 749-761

21 **Whitaker, T. B., and Wiser, E. H.**, Theoretical investigations into the accuracy of sampling shelled peanuts for aflatoxin, *J. American Oil Chem Soc.*, 46(7), 377-379, 1969

22 **Whitaker, T. B., Dickens, J. W., and Wiser, E. H.**, Monte Carlo technique to simulate aflatoxin testing programs for peanuts, *J. American Oil Chem Soc.*, 53(8), 545-547, 1976

23 **Whitaker, T. B., and Dickens, J. W.**, Evaluation of a testing program for aflatoxin in corn, *J. Assoc. Offic. Anal. Chem.*, 66(5), 1055-1058, 1983

24 **Lo, K. M., Chen, C. S., Clayton, J. T., and Adrian D. D.**, Simulation of temperature and moisture changes in wheat storage due to weather variability, *J.Agric. Eng. Res.*, 20(1), 47-53, 1975

25 **Ward, R. C., Harper, J. M., and Jansen, N. B.**, Monte Carlo simulation of nutrient based serving sizes of foods, *J. Food Process. and Preser.*, 2(3), 155-174, 1978

26 **van der Meer, J. J., Fryer, P. J., and Rielly, C. D.**, A Monte Carlo simulation of the flow and mixing of food fluids in a cavity transfer mixer, in *Food Engineering in a Computer Climate*, Inst. of Chem. Engineers, Rugby, UK (Eds), Hemisphere Publishing Co., NY, London, 1992, 37-46

27 **Gekas, V.,** *Transport Phenomena of Foods and Biological Materials*, R. P. Singh and D. R. Heldman, Eds, CRC Press, Boca Raton, FL., 1992, p 15

Computer–aided design of cook–chill foods under uncertain conditions

Bart M. Nicolaï*, **Mia Schellekens**[†], **Toon Martens**[†] and
Josse De Baerdemaeker*

* Agricultural Engineering Department, K.U. Leuven, Kardinaal Mercierlaan 92, B–3001 Heverlee, Belgium
[†] Universiteitsrestaurants ALMA v.z.w., Van Evenstraat 2C, B–3000 Leuven, Belgium

1 ABSTRACT

The sources of variability which affect the accuracy of cook–chill food process design calculations are reviewed, based on a literature survey and measurements in a university restaurant. A first order perturbation algorithm, which is well suited for incorporation in computer–aided process design software, is presented. A simulation study of lasagna undergoing cold storage and reheating under random variable temperature conditions reveals large variabilities of the microbial load at the end of the reheating phase. The need for more statistically sound process design techniques is emphasized.

2 INTRODUCTION

There appears to be an increasing consumer demand for high quality meals which require minimal preparation time and maintain a fresh–like taste. These foods are, after preparation, stored at low temperatures (but above freezing temperatures) and are reheated immediately before consumption. In order to keep the sensorial and textural quality as high as possible, relatively mild thermal treatments are applied. During these processes temperatures occur in the center of the food that are relatively low and only sufficient to obtain some degree of pasteurization.

Recently there has been some concern on the microbial safety of minimally processed foods (see e.g. Betts and Gaze, 1993). According to the Guidelines of the Chilled Food Association (Anonymous, 1993), a 6D reduction of *Listeria monocytogenes* (under the assumption of $D_{70°C}= 2$ min and $z = 7.5°C$) is required in the U.K. in the cooking processes which are involved in the preparation of chilled foods (excluding sous–vide products) and, if applicable, during domestic reheating. The equivalent thermal treatment is by no means sufficient to destroy psychotropic *Clostridium botulinum* spores (typically $D_{70°C}= 1675$ min; $z = 9°C$). The advised storage temperature ($<5°C$) is not low enough to inhibit germination and growth of *C. botulinum* (typical generation times at 5°C: 42 hours for type E and 30.3 hours for type B (Betts and Gaze, 1993). For products with an extended shelf–life (more than 10 days) and which are not heated prior to consumption, the heating process must be sufficient to achieve a 6D reduction of psychotropic strains of *Clostridium botulinum*. This is equivalent to a 10 min treatment at 90°C and leads to a considerable reduction of the sensorial and textural quality of the food which thereby looses much of its advantages in comparison with commercially sterile foods. Similar guidelines exist for sous–vide products. The regulations in other European countries differ to some extent but lead to the same conclusion that under the currently advised processing conditions the risk for survival and subsequent growth of pathogenous microorganisms cannot be excluded. The *due diligence* concept is included in the Food Safety Act of 1990 in the U.K. and in the proposal for the EC directive on hygiene of foodstuffs, and forces the manufacturer to adequately design and control the food preparation and handling processes.

Processing and food core temperature measurements as well as microbial counts at different stages of the preparation process are necessary and also required according to the Guidelines of the Chilled Food Association. However, there is a need for an easy to use but reliable process design methodology which allows to quickly evaluate the effect of a change in the processing conditions on the core temperatures, the microbial loads and the food texture. Schellekens et al. (1993a) presented a prototype of a software package for computer–aided design of cook–chill foods. This package combines expert systems technology with advanced numerical software and state–of–the–art models for heat transfer, microbial growth/inactivation and texture kinetics. The expert system kernel is described in Schellekens et al. (1993a; 1993b) and an overview of the models and numerical algorithms is given in Nicolaï et al. (1994).

Since there is now an increasing amount of evidence that the equipment characteristics as well as the food and microbiological properties are very variable, deterministic calculations (in which all the parameters are assumed to be known and valid for all parts of all batches) are limited to the analysis of worst–case scenario's. However, because of the large number of parameters, assuming the worst possible values for all the parameters will lead to a very conservative process design which will result in considerable overprocessing. The alternative is a stochastic process design, in which a certain number (e.g. one out of 10^6) of defective units (e.g. which have a *C. botulinum* level above a certain threshold) is tolerated. The aforementioned package for computer–aided process design offers the operator the ability to compute approximate confidence intervals of the course of the temperature, the microbial load and the texture kinetics at an arbitrary location inside the food.

In this contribution first the mathematical models for heat transfer and microbial inactivation are summarized. Only conductively heated foods will be considered. The solution of the heat conduction equation using the finite element method is briefly discussed. A survey is given of the sources of uncertainty which arise in thermal process computations of foods in general, based on literature data and measurements in a university restaurant. Some methods for the computation of the propagation of uncertainties through the heat transfer processes are summarized. A first order perturbation method is described, which is attractive in terms of computer time in comparison to the Monte Carlo method. Some numerical examples are given.

It must be emphasized that here only *process design* in uncertain conditions is addressed. Once the process is actually run, some of the process deviations such as fluctuations of the oven or refrigerator temperature can be measured on–line. In this case, these parameters are known and thus deterministic, no matter how irregular their behavior, and appropriate *process control* actions can be taken (see e.g. Datta et al., 1986).

3　MODELING OF HEAT TRANSFER AND MICROBIAL GROWTH AND INACTIVATION

Linear heat conduction in foods is governed by the Fourier equation, a partial differential of the parabolic type, which is given by

$$k\nabla^2 T = \rho c \frac{\partial T}{\partial t} \tag{1}$$

where T is the temperature [°C], k the thermal conductivity [W/m K], ρ the density [kg/m^3] and c the heat capacity [J/kg°C]. The initial condition is defined as

$$T = T_0 \qquad \text{at} \quad t = t_0 \tag{2}$$

On the boundary Γ of the heated object specific temperature and convection boundary conditions may apply

$$T = T_\infty \qquad \text{on} \quad \Gamma_D \tag{3}$$

$$k\frac{\partial}{\partial n_\perp} T = h\,(T_\infty - T) \qquad \text{on} \quad \Gamma_C \tag{4}$$

with $n_\perp$ the outward normal to the surface, h the convection coefficient [W/m^2°C], T_∞ the (known) ambient temperature and $\Gamma = \Gamma_D \cup \Gamma_C$. Nonlinear boundary conditions such as radiation can be important during oven heating by infrared radiation or when the oven walls are at relatively high temperatures (e.g. 200°C).

The thermal inactivation of microorganisms at a given location in the food is often assumed to follow first order reaction kinetics with temperature dependent rate constant according to a D–z type model (Stumbo, 1973). (minutes) required to reduce the bacterial The growth of microorganisms in cook–chill foods is often predicted by means of analytical expressions which describe, for given ambient conditions, the evolution of the logarithm of the microbial load y as a function of the time. A polynomial dependency of the model parameters on the temperature, water activity and pH is then assumed (see e.g. Zwietering 1990 and 1991 for a comparison of some widely used expressions). These models are now being commercially used in the U.K (Food MicroModel) for hygienic consultancy purposes but are static in the sense that they only apply for conditions which do not change during the time interval considered. A dynamic non-autonomous model for microbial growth and inactivation is described recently (Baranyi et al., 1993). Van Impe et al. (1992) proposed a combined dynamic model for microbial growth and inactivation which can be summarized by

$$\frac{d\mathbf{y}}{dt} = \mathbf{g}(\mathbf{y}, T) \tag{5}$$
$$\mathbf{y} = \mathbf{y}_0 \qquad \text{at} \quad t = t_0$$

with $\mathbf{y} = [\, y \; y_r \,]^T$ and $\mathbf{g}$ a 2×1 vector function. y_r is a reference level from which growth restarts after a thermal inactivation phase. For further details the reader is referred to Van Impe et al. (1992).

4   COMPUTATION OF HEAT TRANSFER AND MICROBIAL KINETICS

The finite element method is widely used for the simulation of conductive heat transfer. In this method the continuum is subdivided in (finite) elements of variable size and shape which are interconnected in a finite number n_{nod} of nodal points. In every element the unknown temperature can be approximated by a low order interpolating polynomial in such a way that the temperature is uniquely defined in terms of the (approximate) temperatures $u_i(t)$ at the nodes. It is possible to show (Segerlind,

1984) that the unknown nodal temperature vector $\mathbf{u} = [\, u_1 \ldots u_{n_{nod}} \,]^T$ may be found by solving the following linear differential system

$$\mathbf{C\dot{u}} + \mathbf{Ku} \;=\; \mathbf{f} \tag{6}$$

$$\mathbf{u} \;=\; \mathbf{u}_0 \qquad \text{at} \quad t = t_0 \tag{7}$$

with $\mathbf{C}$ the capacity matrix and $\mathbf{K}$ the stiffness matrix, both $n_{nod} \times n_{nod}$ - matrices and $\mathbf{f}$ a $n_{nod} \times 1$ vector. $\mathbf{C}$ involves ρ and c, $\mathbf{K}$ involves k and h, $\mathbf{f}$ involves T_∞ and h. If the continuum is initially at a uniform temperature T_0 then equation (7) reduces to

$$\mathbf{u} = T_0[\, 1\ 1\ \cdots\ 1\,]^T \qquad \text{at} \quad t = t_0 \tag{8}$$

This system is usually discretized with a finite difference scheme (forward and backward Euler, Crank–Nicholson,...). This results in an algorithm in which at each time step an algebraic system has to be solved using well–known techniques (Gauss and variants, iterative methods such as Gauss–Seidel, ...).

Currently, there are several commercial finite element packages (ANSYS, NASTRAN, ABACUS, SAMCEF, ...) available on most hardware platforms (from PC to super computer). These packages are very user friendly and provide extensive graphical support for the assembly of the finite element grid and for the visualization of the computational results.

The differential equations which describe the microbial kinetics can be numerically solved using subroutine libraries (NAG, IMSL, Harwell,...) or numerical analysis packages (Matlab, Mathematica, CSMP,...). During the thermal treatment the temperature distribution is in general non–uniform. Due to the consecutive heating and cooling steps and also to the often complicated geometry of the food it is not always obvious to determine the lowest process lethality point. Therefore, at each node i of the finite element grid a differential equation of the form (5) has to be solved

$$\frac{d\mathbf{y}_i}{dt} \;=\; \mathbf{g}_i(\mathbf{y}_i, \mathbf{u}) \tag{9}$$

$$\mathbf{y}_i \;=\; \mathbf{y}_{i,0} \qquad \text{at} \quad t = t_0, \qquad i = 1, \ldots, n_{nod}$$

and the lowest process lethality point can determined by inspection of the computational results. Note that equation (9) suggests a dependence of $\mathbf{g}_i$ on the complete nodal temperature vector, although it is actually only dependent on the temperature u_i at node i

$$\frac{\partial \mathbf{g}_i}{\partial u_j} \equiv \mathbf{0}_{n_{nod} \times 1}, \qquad i \neq j \tag{10}$$

where $\mathbf{0}_{n_{nod} \times 1}$ is a $n_{nod} \times 1$ zero vector; this notation is convenient for further use. In equation (9) it is also assumed that the microbial population is not subject to spatial dispersion (e.g. as a consequence of mixing or stirring) during the heat transfer process so that the microbial load in node j adjacent to node i does not affect the load in node i.

5　SOURCES OF UNCERTAINTY IN THERMAL FOOD PROCESSING CALCULATIONS

The errors which arise in thermal food processing calculations can broadly be divided into random errors and systematic errors. *Random errors* include measurement errors caused by the unability to reproduce exactly the same experimental conditions in which a quantity was measured, errors due to unpredictable fluctuations between samples and within a sample. *Systematic errors* comprise errors caused by limitations of the relevant mathematical models used in the design calculations, application of a model outside the range of conditions it was intended for and inappropriate calibration of the measurement devices. It is convenient to identify four domains of random fluctuation: 1. Thermophysical properties of the food, 2. Microbiological properties, 3. Process conditions encompassing initial and boundary conditions and 4. Shape and dimensions of the food.

Until now, little attention has been paid in the literature to the uncertainty involved in thermal food process design. Also, there is not much consistency in the format to describe the uncertainty. Some authors only mention the lower and upper values of a measured quantity, others report the mean value and the standard deviation or the coefficient of variation C_v which is defined as the ratio of the standard deviation and the mean value. Sometimes a 95% confidence interval is given. This can approximately be converted to the sample mean and standard deviation since under the assumption of Gaussian independent measurement errors the confidence margins of the parameter X are given by $\overline{X} \pm t_{0.05}^{n-1} s_X$, with n the number of observations, $\overline{X}$ and s_X the sample mean and standard deviation, respectively, and $t_{0.05}^{n-1}$ the Student-t statistic with $n-1$ degrees of freedom and at the 0.05 probability level. If n is sufficiently large ($\succ 15$) then $t \rightarrow 1.96$.

5.1. Thermophysical properties of the food

In sterilization technology it is common to plot the logarithm of the product center temperature versus the time in minutes. This curve is then empirically described in terms of the slope factor f, which is the time in minutes required for the temperature in the straight–line portion to traverse one log cycle; and the lag factor j, which is defined as

$$j = (T_R - T_0^*)/(T_R - T_0) \tag{11}$$

with T_R the retort temperature, T_0 the initial temperature and T_0^* the temperature at which an extension of the straight–line portion of the heating curve intersects the ordinate axis. The f–value is related to the thermal diffusivity $\alpha \overset{\Delta}{=} k/\rho c$ [m^2/s]. For a cylindrical can with radius r_0 [m] and half–height L [m] the following relationships are obtained for constant fixed temperature boundary conditions and a uniform initial temperature (Teixeira and Shoemaker, 1989)

$$\alpha = 0.398((1/r_0^2 + 0.427/L^2)f)^{-1} \tag{12}$$

$$j = 2.03970 \tag{13}$$

A first order Taylor expansion of equation (12) around its mean value $\overline{\alpha}$ yields

$$\Delta\alpha \cong (\partial\alpha/\partial f)\Delta f = -\overline{\alpha}\Delta f/\overline{f} \tag{14}$$

so that

$$\sigma_{alpha}^2/\overline{\alpha}^2 = \mathcal{E}(\Delta\alpha)^2/\overline{\alpha}^2 \cong \mathcal{E}(\Delta f)^2/\overline{f}^2 = \sigma_f^2/\overline{f}^2 \tag{15}$$

with $\mathcal{E}$ the mean value operator. Hence, the C_v of α is approximately equal to the C_v of f. For convection boundary conditions the equality (13) does not apply and there does not exist an explicit relationship between j and the heat transfer parameters. The reported confidence intervals or other measures of uncertainty of j therefore cannot be converted to an equivalent uncertainty measure of some product or process parameter and are therefore not mentioned here.

Based on a literature survey, Hicks (1961) reported a C_v of 5–15% on the f–value. Lenz and Lund (1977) investigated uncertainties in the estimation of heating rates in conduction heating foods. They reported a C_v of the thermal diffusivity of about 3.33% for pureed peas, pureed lima beans, baked beans and applesauce. The thermal diffusivity was assumed to be Gaussian distributed. Patino and Heil (1985) carried out a series of heat penetration experiments with potatoes in brine and established a C_v of the f–value from 15.90% up to 25.56%, giving an overall C_v of 20.01%. Hayakawa et al. (1988) estimated coefficients of variation of f–values from temperature data collected by processing 211×300 and 307×409 cans of spaghetti in tomato sauce. The coefficients of variation were 1.549% and 4.684% for the cooling phase in the large and small cans, respectively, and 4.684% and 13.40% for the heating phase. For both sizes, application of the Kolmogorov–Smirnov test for goodness–of–fit showed a slightly better representation of the f–value in the heating phase by a gamma distribution than by a Gaussian distribution. In the cooling phase there was nearly no difference in goodness–of–fit. The authors also report a frequently observed C_v–value of 7% and 12% for the f–value in the heating and the cooling phase, respectively. Wang et al. (1991) conducted a series of heat penetration experiments using bentonite suspension in 211×300 and 307×409 cans. The coefficients of variation of f were 3.39% and 3.70% for the cooling phase in the large and small cans, respectively, and 3.15% and 4.35% for the heating phase.

Several authors (Hicks, 1961; Thompson et al., 1979) emphasize that the natural variability of foods seems to be the major cause of random fluctuation, and that the measurement errors are relatively small if a proper experimental setup is applied. This is somewhat contrary to the results of the COST 90 and COST 90bis projects, summarized by Meffert (1983). In these European research programs a series of collaborative measurements of the thermal properties of foods and model foods were carried out (Kent et al., 1984). The between replicates C_v of the thermal conductivity of glass beads, 70% and 98% carrageenan gels was found to be from 0.1% up to 6.2%, depending on the used method, while the between labs C_v extended from 3.0% up to 15.7%. For the thermal diffusivity they found values between 0.0% and 8.1% for the between replicates C_v, and between 1.9% and 13.8% for the between labs C_v. The maximum overall C_v of the thermal conductivity of yoghurt, milk powder, apple pulp, meat paste and fish paste was 1.3%, 19.3%, 4.3%, 14.5% and 11.5%, respectively. The corresponding maximum overall C_v of the thermal diffusivity was 8.6%, 16.7%, 12.4%, 14.6% and 15.8%, respectively. These maximum values varied with the temperature. Meffert (1983) estimated the between-laboratory deviations, in the absence of recommended methods for measurement and calculations, to be as large as 40% at 95% confidence level. After calibration, this value could be reduced to 20%. The author concluded that the possible maximum error in the experimental determination of the thermal conductivity can be as high as 30–50% at 95% confidence level. In this respect Sweat (1974) found a 4 to 24% between–samples variation of the thermal conductivity at the 95% confidence level.

During the past two decades several equations have emerged, which correlate the thermophysical parameters of foods with their chemical composition. Miles et al. (1983) compared the accuracy of a large set of published equations. Their results are summarized by Meffert (1983) who estimates that the accuracy of the calculated values of the thermophysical properties using these equations is of the same order as the accuracy of experimental data. This is very important since these equations can easily be implemented in computer–aided design software.

5.2. Microbiological properties

Hicks (1961) distinguished four main factors causing variations in the microbiological properties in the context of canning:

- The nature of the organism

- The history of the spores, in particular that immediately prior to use and including the conditions of sporulation

- The substrate in which the spores are heated

- The substrate used for detecting survivors

The author discussed these factors extensively and assumed values for the C_v of z from 2% up to 9% . The experimental errors in D were considered at least as substantial as those in z and coefficients of variation of up to 25% were reported. Because both parameters are usually derived from the same experimental data set, they are not statistically independent according to this author. Lenz and Lund (1977b) used C_v–values up to 17.5% for the destruction rate constant in their Monte Carlo experiments. They also assumed (Lenz and Lund, 1977b) a C_v of 7.7% and 11.1% for the activation energy and z value, respectively. The activation energy was presumed to be Gaussian distributed. Pflug and Odlaug (1978) reviewed D and z–values for *Clostridium botulinum* in phosphate buffer and in various foods and found mostly very large 95% confidence intervals. For a 10°C z–value they obtained a C_v of 11%. Patino and Heil (1985) reported a C_v of 3% up to 6% for D–values. Hayakawa et al. (1988) assumed a gamma–distributed z and an associated C_v of 3.6% for their Monte Carlo calculations. Based on laboratory and literature data, these authors report a typical value of 11% for the C_v of z.

It must also be emphasized that the assumption of a linear relationship between log survival numbers and time is seldom valid (Casolari, 1989) and can lead to serious systematic process design errors.

Only very recently some dynamic mathematical models for growth of microorganisms in given pH and a_W conditions have appeared in the literature (Van Impe et al., 1992; Baranyi et al., 1992). Although no information on the confidence intervals of the model parameters is known yet, it is well–known to food microbiologists that the history of the microorganisms prior to the inoculum has important consequences on some features of the growth curves such as the lag time.

5.3. Process conditions

The process conditions can be subdivided in ambient temperature, surface heat transfer coefficient, initial temperature and initial microbiological load.

Ambient temperature

Sheard and Rodger (1993) compared the time required for vacuum–packed and non–vacuum–packed potato slabs to rise from 20°C to 75°C with an oven setpoint of 80°C at different locations in commercial steamers and for different oven types. They found substantial differences in heating time between packs of the same shelf. According to these authors, the observed heating times variations were due to the intermittent inputs of the steam used to maintain temperatures below 100°C.

Several reports on fluctuation of temperatures in the distribution chain were compiled by Zeuthen et al. (1990) . Gormley (1990a) recorded in–flight air temperatures from 2.5°C up to 21.1°C in the cargo holds of scheduled passenger flights were chilled foods usually are kept during the flight. Standard deviations varied from 0.2 to 4.3°C. The author also performed spot tests on temperatures of air freighted smoked salmon and of chilled foods being transported by road. In another study (Gormley, 1990b) the same author measured product temperatures in chilled display cabinets and found mean temperatures from -1.6 up to 5.2°C and standard deviations from 0.8 up to 2.8°C. Mean temperatures of yoghurt varied from 2.0 up to 9.6°C with differences of up to 2.4°C between different locations in the cabinet. For milk mean temperatures of 5.2 up to 11.2°C with location dependent differences of up to 4°C where recorded. Mean air temperatures in domestic refrigerators varied from 1.6 up to 8.8°C with standard deviations from 0.4 up to 1.6°C and location dependent temperature differences of up to 7.2°C.

James and Evans (1990) recorded air temperature fluctuations at a given point in chilled display cabinets from 5 to 20°C. They found location dependent temperature fluctuations of up to 11°C. They also concluded that the spatial temperature and air speed distribution is often caused by poor refrigerator coil and fan design. Temperatures at different positions within chilled food compartments of a number of domestic refrigerators revealed overall mean temperatures ranging from 3 to 10.4°C. Even higher temperatures were recorded during transport from retail shops to the home.

Olsson (1990) concluded from literature data and own investigations on retail shop open chill cabinets that the temperatures vary widely, from freezing temperatures up to about 20°C. Usually the temperature at the front is higher than at the back of the shelves (difference about 5°C), while the temperature at the top shelves is higher than at the bottom shelves.

Similar results were obtained by Willocx et al. (1993), who investigated the temperature fluctuations in two vertical multi–deck open display cabinets for vacuum products. They observed that the mean temperature at different locations in the cabinet varied over a time span of one week from 0.5 up to 2.9°C and from 6.3 up to 8.7°C in the first and the second cabinet, respectively. The standard deviation at each location was between 0.9 and 2.6°C. These authors also investigated the temperature variations in 14 domestic refrigerators. They concluded that the average temperature was in between 2.8°C and 14.7°C, with standard deviations of 0.2 up to 1.5°C. There was also a significant difference between the temperatures at different decks.

Numerous other reports with similar measurements and conclusions have appeared during the last years and the reader is referred to the references in the articles discussed above.

In Figure 1 the time course of the oven temperature is shown at 11 locations in a combined convection/microwave oven using a hot air/steam mixture as heating medium. Clearly, the temperature fluctuates in an unpredictable way as a function of time but depends also on de measurement location. A similar noisy temperature course is observed in the refrigerator rooms (Figure 2).

Surface heat transfer coefficient

Almost no literature data are available on the variability of surface heat transfer coefficients during the processing of cook–chill meals.

A substantial amount of information is now being gathered in the ALMA–university restaurants on the variability of the temperatures and surface heat transfer coefficients in ovens and chill rooms. In Figure 3 the evolution of the surface heat transfer coefficient, is shown at 4 locations in a combined convection/microwave oven using hot air as a heating medium.

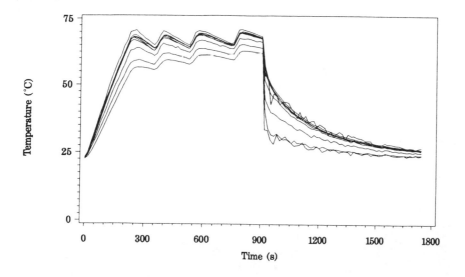

Figure 1. Temperature at nine different locations in oven using a hot air/steam mixture as heating medium

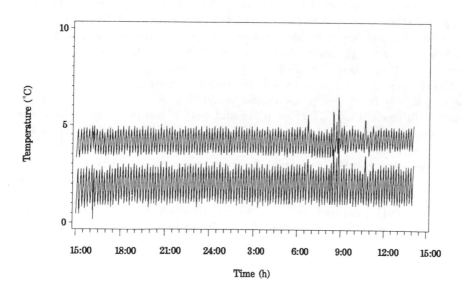

Figure 2. Temperature at two different locations in a refrigerated room

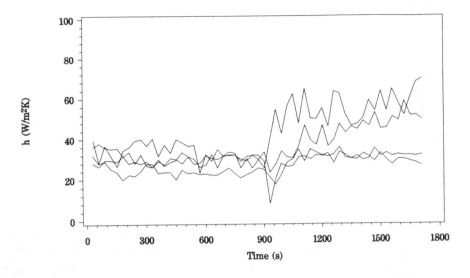

Figure 3. Surface heat transfer coefficient at four different locations in oven using a hot air/steam mixture as heating medium

Initial temperature

The variability of the initial temperature is difficult to assess since usually the start time of the simulations is determined based on engineering insight. E.g. in the preparation process of lasagna which is served at the ALMA university restaurant, after (separate) preparation the sauce and pasta are stored over night in a refrigerator room; the next day the lasagna's are assembled, followed by cold storage and transportation to the ALMA university restaurant where they are stored for another few days in a refrigerator room. Most likely, if appropriate (low) temperatures are applied during the cold storage of the sauce and pasta immediately after preparation, no significant growth will occur due to the lag phase in the bacterial growth curve. Therefore, if the storage conditions at the production plant and during transport are known, the simulations could be started at the moment the lasagna's are assembled. If, on the other hand, these conditions are not known, the simulations can be started at the point the lasagna's arrive at the restaurant. In the latter case, the initial temperature may vary considerably, depending on the season (hot summer days versus cold winter days) and the refrigeration capability of the refrigerator vans.

Wang et al. (1991) estimated an initial temperature Cv of 5.85% from 108 experimental temperature history curves for a bentonite suspension in two different can sizes.

Initial microbial load

The initial microbial load of the food may vary considerably, due to natural variability as well as errors involved in the count procedures. Jarviss (1989) gives an excellent account of both sources of uncertainty.

Jarviss distinguishes three types of spatial dispersion of micro–organisms: random, regular and contagious. In a *random* distribution any one individual cell occupies any one specific position in a sample with equal chance. In this case a Poisson distribution is usually adequate to describe the probability that a sample is contaminated with

Table 1. Initial contamination level of several microorganisms in fresh lasagna samples

Sample	1	2	3	4	5	6	7
Aerobic mesophilic count	$2.9\ 10^3$	$8.1\ 10^3$	$3.0\ 10^4$	$1.3\ 10^4$	$1.9\ 10^4$	$6.0\ 10^3$	$4.8\ 10^3$
H_2s producing anaerobic count	10–100	–	–	< 10	< 10	30	< 10
Molds	<10	<10	< 10	< 10	< 10	<10	< 10
Yeasts	50	30	30	50	$1.0\ 10^2$	< 10	$9.0\ 10^2$
Lactobacilli	$2.5\ 10^4$	$1.6\ 10^4$	$1.1\ 10^5$	$7.8\ 10^4$	$9.0\ 10^4$	$1.8\ 10^4$	$6.2\ 10^4$
Enterobacteriaceae	< 10	< 10	< 10	< 10	<10	<10	10
Fecal Streptococci	< 10	< 10	30	40	20	<10	80
Staphylococcus aureus	< 10	–	$6.0\ 10^2$	$1.0\ 10^2$	$2.0\ 10^2$	<10	<10
Clostridium perfringens		–	1–10	< 10	< 10	< 10	<10

a certain number of cells. Consequently, the variance is equal to the mean. When the cells are crowded together, the population is *regular* and can be described by the binomial distribution. The variance is then less than the mean. Sometimes clumps and aggregates of cells appear, reflecting the colonial growth of micro–organisms which is a consequence of the spatial dispersion of environmental factors on growth. In this case the population is is *contagious* and a negative binomial distribution is frequently assumed, implying that the variance is greater than the mean. Although in all three cases the distribution is non–normal, the logarithm of the number of microbial cells usually approximates a normal distribution.

Jarviss also discusses the errors involved in the estimation of micro–organisms in foods which can be subdivided into those associated with the preparation of the laboratory samples for analysis, and those related to the count procedures. The preparation error is due to laboratory sampling errors, diluent volume errors, pipette volume errors, maceration errors and errors caused by inadequate mixing. Pipette volume errors seem to be the most important ones, with a between–pipettes C_v as large as 13.75% for some types of capillary pipettes. The error sources due to the count procedures vary according to the type of procedure. In case of colony count procedures the following error sources are distinguished: specific technical errors, such as hiding of colonies inside the agar by colonies at the surface, pipetting errors and distribution errors, caused by sublethal damage during the sample preparation process. Jarviss estimates the overall C_v of the colony count procedure of the order of 0.25, although values up to 0.75 have been reported (Hall, 1977).

In table 1 the initial contamination level of several microorganisms in seven fresh lasagna samples is shown.

5.4. Shape and dimensions of the food

In sterilization processes the shape and the dimensions of the containers are known and constant. There does not appear to exist much information on the variability of dimensions of cook–chill foods, although these dimensions are known to have a large effect on the core temperature (De Baerdemaeker and Nicolaï, 1993. As an example, the spatial C_v of the thickness of the lasagna samples which are served in the ALMA university restaurant at Leuven, measured at 9 distinct locations, varies from 5 to 11%.

6 UNCERTAINTY PROPAGATION ANALYSIS

The random features of the foods, microorganisms and process conditions will propagate through the heat transfer processes and will eventually affect the obtained pasteurization effect. Several numerical methods exist now to compute the uncertainty propagation through the heat transfer process, depending on the mathematical uncertainty models used to describe the stochastic parameters. In this contribution only random variable parameters (parameters which may fluctuate from one food to another, but which do not vary as a function of the spatial and temporal coordinates) are considered. More complicated stochastic models such as random fields and random processes have been considered elsewhere (Nicolaï and De Baerdemaeker, 1992a; 1992b

A straightforward statistical approach is a Monte Carlo experiment, in which a sample vector of the stochastic parameters is generated by the computer and the corresponding thermobacteriological problem is numerically solved. This is repeated several times and finally the mean values and (co)variances but also higher order moments can be estimated using the usual statistical techniques, and tests of hypothesis and significance can be performed. While the Monte Carlo technique yields a rather complete picture of the stochastic properties of the process variables, the large number of runs (typically >1000) necessary to get results with an acceptable accuracy and the considerable amount of computer time required limits its applicability. The Monte Carlo method has been used before for thermobacteriological problems (Lenz and Lund, 1977b; Hayakawa et al., 1988; Wang et al., 1991). Alternatively, (semi-)analytical methods may be applied. Although there exist a formalism which relates the probability density function of the process variables to the stochastic parameters when their mathematical relation is known, in general this is a complicated mathematical problem. (Melsa and Sage, 1973). However, often it is possible to compute at least some probabilistic characteristics such as the mean values and (co)variances of the process variables. This is commonly known as *second order moment analysis*. In some cases knowledge of these probabilistic characteristics implies a complete probabilistic knowledge. Depending on the probabilistic nature of the stochastic quantities, even second order moment techniques may become extremely complicated. This is especially true for nonlinear processes such as microbial kinetics and distributed processes such as heat transfer. For problems in which the parameters are stochastic but do not change during the process, a *probabilistic perturbation method* may be used. This method is based on the computation of the propagation of an infinitesimal perturbation of the (stochastic) parameters through the process. Mean values and (co)variances of the process variables can then easily be evaluated. This method is widely used for stochastic elliptic and hyperbolic problems in structural engineering but applications in parabolic problems are limited.

Problems with stochastic parameters which change in an unpredictable way during the process can be considered as (nonlinear) systems with stochastic inputs. The latter are the subject of stochastic systems theory, which provides a general *variance propagation algorithm* for the computation of the second order statistics of the process variables (Melsa and Sage, 1973). This algorithm is applicable to processes which are described by systems of ordinary differential equations, and are as such not immediately applicable to distributed systems.

In the following subsection a first order perturbation algorithm for the computation of mean values and (co–)variances of the temperature and microbial load field in conduction heated foods with random variable parameters is derived.

7 A FIRST ORDER PERTURBATION ALGORITHM

A first order perturbation algorithm for the computation of mean values and variances of the temperature and microbial load at an arbitrary time and location inside a conduction–heated food undergoing thermal treatment under random variable conditions was proposed by Nicolaï and De Baerdemaeker (1993a; 1993b) .

A general overview of the algorithm is presented below. It is assumed that the parameter vectors $\mathbf{p}_H$ and $\mathbf{p}_M$ of respectively the heat transfer and the microbial model are subjected to random variable disturbances with means $\overline{\mathbf{p}}_H$ and $\overline{\mathbf{p}}_M$ and covariance matrices $\mathbf{V}_{\mathbf{p}_H,\mathbf{p}_H}$ and $\mathbf{V}_{\mathbf{p}_M,\mathbf{p}_M}$. The food is initially at a uniform random variable temperature T_0 with mean $\overline{T}_0$ and variance $\sigma_{T_0}^2$. It is also assumed that initially the microbial load at each node i is a random variable with mean $\overline{\mathbf{y}}_0$ and covariance matrix $\mathbf{V}_{\mathbf{y}_0,\mathbf{y}_0}$. Finally, $\mathbf{p}_H$, $\mathbf{p}_M$, $\mathbf{y}_0$ and T_0 are considered to be statistically independent.

The nodal temperature vector $\mathbf{u}$ at an arbitrary time is dependent on its initial condition T_0 and the random parameter vector $\mathbf{p}_H$ and can be expanded into a first order Taylor series around its (so far unknown) mean solution

$$\mathbf{u} \cong \overline{\mathbf{u}} + \frac{\partial \overline{\mathbf{u}}}{\partial \overline{\mathbf{p}}_H}\Delta \mathbf{p}_H + \frac{\partial \overline{\mathbf{u}}}{\partial \overline{T}_0}\Delta T_0 \tag{16}$$

in which the partial derivatives must be evaluated using $\overline{\mathbf{p}}_H$, and $\overline{T}_0$. Similarly, $\mathbf{C}$, $\mathbf{K}$ and $\mathbf{f}$ are function of $\mathbf{p}_H$ and are expanded around their mean values $\overline{\mathbf{C}}$, $\overline{\mathbf{K}}$ and $\overline{\mathbf{f}}$ which can be assembled as usual but using the mean parameter vector $\overline{\mathbf{p}}_H$

$$\mathbf{C} \cong \overline{\mathbf{C}} + \frac{\partial \overline{\mathbf{C}}}{\partial \overline{\mathbf{p}}_H}\Delta \mathbf{p}_H \tag{17}$$

$$\mathbf{K} \cong \overline{\mathbf{K}} + \frac{\partial \overline{\mathbf{K}}}{\partial \overline{\mathbf{p}}_H}\Delta \mathbf{p}_H \tag{18}$$

$$\mathbf{f} \cong \overline{\mathbf{f}} + \frac{\partial \overline{\mathbf{f}}}{\partial \overline{\mathbf{p}}_H}\Delta \mathbf{p}_H \tag{19}$$

Substitution of equations (16)–(19) into equation (6) and combining appropriate terms in $\Delta \mathbf{p}_M$ and ΔT_0 yields the following system

$$\overline{\mathbf{C}}\frac{d}{dt}\overline{\mathbf{u}} + \overline{\mathbf{K}}\,\overline{\mathbf{u}} = \overline{\mathbf{f}} \tag{20}$$

$$\overline{\mathbf{C}}\frac{d}{dt}\frac{\partial \overline{\mathbf{u}}}{\partial \overline{T}_0} + \overline{\mathbf{K}}\frac{\partial \overline{\mathbf{u}}}{\partial \overline{T}_0} = 0 \tag{21}$$

$$\overline{\mathbf{C}}\frac{d}{dt}\frac{\partial \overline{\mathbf{u}}}{\partial \overline{\mathbf{p}}_H} + \overline{\mathbf{K}}\frac{\partial \overline{\mathbf{u}}}{\partial \overline{\mathbf{p}}_H} = \frac{\partial \overline{\mathbf{f}}}{\partial \overline{\mathbf{p}}_H} - \frac{\partial \overline{\mathbf{C}}}{\partial \overline{\mathbf{p}}_H}\frac{d}{dt}\overline{\mathbf{u}} - \frac{\partial \overline{\mathbf{K}}}{\partial \overline{\mathbf{p}}_H}\overline{\mathbf{u}} \tag{22}$$

Equation (20) expresses that the mean value of the temperature $\overline{\mathbf{u}}$ may be found by solving the original heat transfer problem using mean values of the parameters. From equation (16) it can be derived that the temperature covariance matrix $\mathbf{V}_{\mathbf{u},\mathbf{u}}$ can be computed from

$$\mathbf{V}_{\mathbf{u},\mathbf{u}} = \frac{\partial \overline{\mathbf{u}}}{\partial \overline{T}_0}\sigma_{T_0}^2\frac{\partial \overline{\mathbf{u}}}{\partial \overline{T}_0}^T + \frac{\partial \overline{\mathbf{u}}}{\partial \overline{\mathbf{p}}_H}\mathbf{V}_{\mathbf{p}_H,\mathbf{p}_H}\frac{\partial \overline{\mathbf{u}}}{\partial \overline{\mathbf{p}}_H}^T . \tag{23}$$

Further, since

$$\mathbf{V}_{\mathbf{u},\mathbf{u}}(t=0) \triangleq \begin{bmatrix} 1 & 1 & \cdots & 1 \\ 1 & 1 & \cdots & 1 \\ \vdots & \vdots & \ddots & \vdots \\ 1 & 1 & \cdots & 1 \end{bmatrix} \sigma_{T_0}^2 \tag{24}$$

and because the T_0 is independent of $\mathbf{p}_H$ it follows that initial conditions corresponding to the system (20)–(22) are given by

$$\overline{\mathbf{u}}(t=0) \quad = \quad \overline{T}_0[\,1\,1\,\ldots\,1\,]^T \tag{25}$$

$$\frac{\partial \overline{\mathbf{u}}}{\partial \overline{T}_0}(t=0) \quad = \quad [\,1\,1\,\ldots\,1\,]^T \tag{26}$$

$$\frac{\partial \overline{\mathbf{u}}}{\partial \overline{\mathbf{p}}_H}(t=0) \quad = \quad [\,0\,0\,\ldots\,0\,]^T. \tag{27}$$

The same procedure can be applied to the microbial model (9). $\mathbf{y}_i(\mathbf{y}_0, \mathbf{u}, \mathbf{p}_M)$ and $\mathbf{g}_i(\mathbf{y}_i, \mathbf{u}, \mathbf{p}_M)$ can be expanded in a first order Taylor series a around their mean values $\overline{\mathbf{y}}(\overline{\mathbf{y}}_0, \overline{\mathbf{u}}, \overline{\mathbf{p}}_M)$ and $\overline{\mathbf{g}_i}(\overline{\mathbf{y}}_i, \overline{\mathbf{u}}, \overline{\mathbf{p}}_M)$, respectively.

$$\mathbf{y}_i \quad \cong \quad \overline{\mathbf{y}}_i + \frac{\partial \overline{\mathbf{y}}_i}{\partial \overline{\mathbf{u}}}\Delta\mathbf{u} + \frac{\partial \overline{\mathbf{y}}_i}{\partial \overline{\mathbf{p}}_M}\Delta\mathbf{p}_M + \frac{\partial \overline{\mathbf{y}}_i}{\partial \overline{\mathbf{y}}_0}\Delta\mathbf{y}_0 \tag{28}$$

$$\mathbf{g}_i \quad \cong \quad \overline{\mathbf{g}}_i + \frac{\partial \overline{\mathbf{g}_i}}{\partial \overline{\mathbf{y}}_i}\left(\frac{\partial \overline{\mathbf{y}}_i}{\partial \overline{\mathbf{u}}}\Delta\mathbf{u} + \frac{\partial \overline{\mathbf{y}}_i}{\partial \overline{\mathbf{p}}_M}\Delta\mathbf{p}_M + \frac{\partial \overline{\mathbf{y}}_i}{\partial \overline{\mathbf{y}}_0}\Delta\mathbf{y}_0\right)$$

$$+ \frac{\partial \overline{\mathbf{g}_i}}{\partial \overline{\mathbf{u}}}\Delta\mathbf{u} + \frac{\partial \overline{\mathbf{g}_i}}{\partial \overline{\mathbf{p}}_M}\Delta\mathbf{p}_M. \tag{29}$$

Substitution of equation (28) and (29) in equation (9), combining appropriate terms in $\Delta\mathbf{u}$, $\Delta\mathbf{p}_M$, and $\Delta\mathbf{y}_0$, and taking into account equation (10) the following differential system is obtained

$$\dot{\overline{\mathbf{y}}}_i \quad = \quad \overline{\mathbf{g}_i}(\overline{\mathbf{y}}_i, \overline{\mathbf{u}}) \tag{30}$$

$$\frac{d}{dt}\left(\frac{\partial \overline{\mathbf{y}}_i}{\partial \overline{u}_i}\right) \quad = \quad \frac{\partial \overline{\mathbf{g}_i}}{\partial \overline{\mathbf{y}}_i}\frac{\partial \overline{\mathbf{y}}_i}{\partial \overline{u}_i} + \frac{\partial \overline{\mathbf{g}_i}}{\partial \overline{u}_i} \tag{31}$$

$$\frac{d}{dt}\left(\frac{\partial \overline{\mathbf{y}}_i}{\partial \overline{\mathbf{p}}_M}\right) \quad = \quad \frac{\partial \overline{\mathbf{g}_i}}{\partial \overline{\mathbf{y}}_i}\frac{\partial \overline{\mathbf{y}}_i}{\partial \overline{\mathbf{p}}_M} + \frac{\partial \overline{\mathbf{g}_i}}{\partial \overline{\mathbf{p}}_M} \tag{32}$$

$$\frac{d}{dt}\left(\frac{\partial \overline{\mathbf{y}}_i}{\partial \overline{\mathbf{y}}_0}\right) \quad = \quad \frac{\partial \overline{\mathbf{g}_i}}{\partial \overline{\mathbf{y}}_i}\frac{\partial \overline{\mathbf{y}}_i}{\partial \overline{\mathbf{y}}_0} \tag{33}$$

An expression for the covariance matrix $\mathbf{V}_{\mathbf{y},\mathbf{y}}$ can then be derived from equation (28)

$$\mathbf{V}_{\mathbf{y}_i,\mathbf{y}_i} = \frac{\partial \overline{\mathbf{y}}_i}{\partial \overline{u}_i}\sigma_{u_i}^2 \frac{\partial \overline{\mathbf{y}}_i}{\partial \overline{u}_i}^T + \frac{\partial \overline{\mathbf{y}}_i}{\partial \overline{\mathbf{y}}_0}\mathbf{V}_{\mathbf{y}_0,\mathbf{y}_0}\frac{\partial \overline{\mathbf{y}}_i}{\partial \overline{\mathbf{y}}_0}^T + \frac{\partial \overline{\mathbf{y}}_i}{\partial \overline{\mathbf{p}}_M}\mathbf{V}_{\mathbf{p}_M,\mathbf{p}_M}\frac{\partial \overline{\mathbf{y}}_i}{\partial \overline{\mathbf{p}}_M}^T. \tag{34}$$

where $\sigma_{u_i}^2$ is the i–th diagonal entry of the nodal temperature covariance matrix $\mathbf{V}_{\mathbf{u},\mathbf{u}}$.
It can be derived easily that

$$\frac{\partial \overline{\mathbf{y}}_i}{\partial \overline{u}_i}(t=0) \quad = \quad [\,0\,0\,\cdots\,0\,]^T \tag{35}$$

$$\frac{\partial \overline{\mathbf{y}}_i}{\partial \overline{\mathbf{p}}_M}(t=0) \quad = \quad [\,0\,0\,\cdots\,0\,]^T \tag{36}$$

$$\frac{\partial \overline{\mathbf{y}}_i}{\partial \overline{\mathbf{y}}_0}(t=0) \quad = \quad \mathbf{I}_{2\times 2} \tag{37}$$

with $\mathbf{I}_{2\times 2}$ a 2×2 unity matrix.

The overall algorithm can be summarized as follows: first the mean temperature vector and the temperature sensitivity functions $\partial \overline{\mathbf{u}}/\partial \overline{T}_0$ and $\partial \overline{\mathbf{u}}/\partial \overline{\mathbf{p}}_H$ are computed by solving the differential system (20)–(22). Then the nodal temperature covariance $\mathbf{V}_{\mathbf{u},\mathbf{u}}$ is computed using expression (23). Subsequently, the microbial load sensitivity functions $\partial \overline{\mathbf{y}}_i/\partial \overline{u}_i$, $\partial \overline{\mathbf{y}}_i/\partial \overline{\mathbf{p}}_M$ and $\partial \overline{\mathbf{y}}_i/\partial \overline{\mathbf{y}}_0$ are computed from equation (30)–(33). Finally, the microbial load covariance is computed from equation (34).

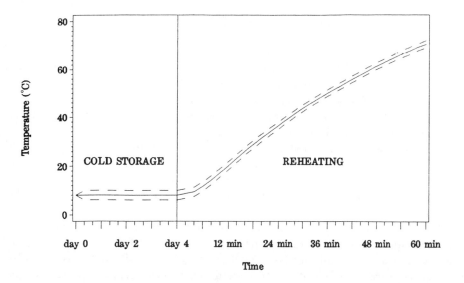

Figure 4. Mean and approximate 95% confidence interval of the temperature in a nodal point near the center of the lasagna during cold storage and reheating

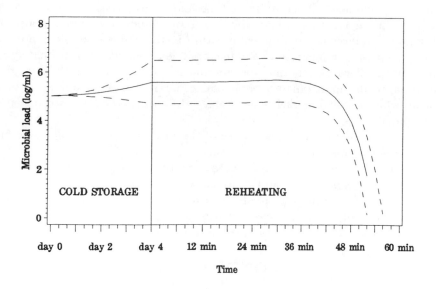

Figure 5. Mean and approximate 95% confidence interval of the microbial load in a nodal point near the center of the lasagna during cold storage and reheating

8   NUMERICAL EXAMPLE

The algorithm which was outlined in the previous subsection was implemented by modification of the finite element code DOT (Polivka and Wilson, 1976). The system (20)–(22) was solved using an implicit Euler finite difference algorithm; the system (30)–(33) was solved using a 4th order Runge-Kutta-Gill solver.

The test problem consisted of a lasagna in a porcelain recipient which was stored for 4 days in a refrigerator and subsequently heated for 60 min in an oven (Nicolaï and De Baerdemaeker, 1993b). The spoilage of the lasagna is mainly caused by lactic acid bacteria which most probably originate from the ground cheese on top of the lasagna and which cause a considerable decrease of the lasagna pH.

All parameters except the refrigerator and oven temperature were assumed to be deterministic. The mean and standard deviation of the refrigerator temperature were respectively $\overline{T}_{fridge} = 8°C$ and $\sigma_{fridge} = 1°C$; the corresponding surface heat transfer coefficient was equal to 10 W/m^2K. For the oven, $\overline{T}_{oven} = 100°C$, $\sigma_{oven} = 1°C$ and the surface heat transfer coefficient was equal to 25 W/m^2K. Both the refrigerator and oven temperature were considered to have a Gaussian distribution. The shape, dimensions, thermophysical parameters and finite element grid are described elsewhere (Nicolaï et al. 1993c). The parameters of the microbial growth and inactivation model for *Lactobacillus plantarum* were extracted from Van Impe et al. (1992). The initial microbial load was equal to 10^5 organisms/ml. In a previous study (Nicolaï and De Baerdemaeker, 1993b), a good agreement was found between the results of a Monte Carlo simulation and the perturbation method.

In the first order perturbation method an approximately linear correspondence is assumed between the random variable conditions on the one hand, and the temperatures and microbial loads at an arbitrary time and location inside the food on the other hand. A linear function of normally distributed random variables is also normally distributed. Since the ambient temperature in the above example is normally distributed, the product temperatures and microbial loads are also approximately normally distributed. Because a normal distribution is fully characterized by its mean and (co–)variance, the approximate 95% confidence interval of the temperature or microbial load, which is equal to their mean value ±1.96 times the standard deviation, can be computed.

In Figure 4 and Figure 5 the time course of the mean temperature and microbial load and their approximate confidence intervals are shown. Clearly, the width of the confidence band of the microbial load increases sharply during the inactivation. In view of the relatively small uncertainties which were assumed for this example, it is clear that the use of deterministic predictions of the microbial inactivation during heating in commercially available ovens, which usually have poor temperature and surface heat transfer control capabilities, is questionable.

9   CONCLUSION

Jarviss (1977) studied the validity of absolute microbiological criteria and concluded that 'Since both the intrinsic variability in the food and the variability associated with methodology is large in magnitude, the use of absolute critical values which take no account of the variation cannot be considered to be scientifically sound'. Although the author focused on the uncertainty involved in the quantitative analysis of micro–organisms in foods, his conclusions are also applicable to process design of cook–chill meals based on predictive models for heat transfer and microbiological growth and inactivation.

A survey of the parameters involved in the design of heat transfer processes of cook–chill foods reveals large variabilities which propagate through the thermal treatment and result in large variabilities of the achieved pasteurization effect. As a consequence, a deterministic statement such as 'The process value is equivalent to 6D units' is meaningless and should be replaced by a stochastic expression such as 'The process value is in between 4 and 8 with 95% probability'. Note that the in sterilization technology widely used notion 'Probability of a Non-Sterile Unit' is misleading as it refers to a highly degenerate case of a conditional probability with all random parameters fixed at constant values (usually these values which lead to the most conservative process design). This concept is meaningless for cook–chill foods considering the large number of random parameters which are involved.

A first order perturbation algorithm was presented for the computation of the mean value and variance of the temperature and microbial load at an arbitrary time and location inside the food. Numerical computations indicate that even small uncertainties on some parameters such as the refrigerator but in particular the oven temperatures result in a large variability of the microbial load at the end of the thermal treatment. Given the considerable fluctuations of the temperature and air velocity in both time and space inside commercially available ovens, the intensity of the thermal treatment experienced by individual food samples may vary so extensively that hardly any reasonable process design is possible. As long as the temperature and air velocity uniformity in space and time of the commercially available ovens do not improve, any process design will result in an end–product with a very variable quality.

10 ACKNOWLEDGEMENTS

The authors wish to thank the European Communities (FLAIR project AGRF-CT91-0047(DTEE)) and the F.K.F.O (grant 2.0095.92) for financial support.

11 NOMENCLATURE

α	: thermal diffusivity [m^2/s]
C_v	: coefficient of variation
c	: capacity matrix
c	: heat capacity
D_{ref}	: time to reduce the bacterial concentration by 10 [min]
$\mathcal{E}$	: mean value operator
f	: slope of the heat penetration curve [1/min]
f	: thermal load vector
Γ_C	: convection boundary condition surface
Γ_D	: fixed temperature boundary condition surface
Γ	: boundary surface
h	: surface heat transfer coefficient [W/m^2°C]
j	: lag factor of the heat penetration curve
k	: thermal conductivity [W/m°C]
K	: stiffness matrix
n_{nod}	: number of nodes
$n_\perp$	: outward normal
$\mathbf{P}_H$	: parameter vector of the heat transfer model
$\mathbf{P}_M$	: parameter vector of the microbial kinetics
r_0	: radius of a can [m]
ρ	: density [kg/m^3]
σ^2	: variance
t	: time [s]

T	: temperature [°C]
T_D	: fixed temperature boundary condition function
T_∞	: ambient temperature [°C]
$\mathbf{V}$	: variance–covariance matrix
$\mathbf{u}$	: nodal temperature vector
y	: log microbial load [1/ml]
y_r	: reference level of log microbial load [1/ml]
z	: increase in temperature to reduce D_{ref} by 10 min [°C]

12   REFERENCES

Anonymous (1993), *Guidelines for Good Hygienic Practice in the Manufacture, Distribution and Retail Sale of Chilled Foods*, 2nd edition, Chilled Food Association, London.

Baranyi , J., T.A. Roberts , and McClure P. (1993), " A non-autonomous differential equation to model bacterial growth ", *Food Microbiology*, 10, 43–59.

Betts , G.D. and J.E. Gaze (1993), " Growth and heat resistance of microorganisms in "sous vide" products heated at low temperature (54-65°C) ", in T. Martens and M. Schellekens , editors, *Proceedings of the First European "Sous Vide" Cooking Symposium*, Leuven, Belgium, 27–39.

Casolari , A. (1988), " Microbial death ", in M.J. Bazin and J.I. Prosser , editors, *Physiological Models in Microbiology*, volume 2, Boca Raton, CRC press, Inc., 1–44.

Datta , A.K., A.A. Teixeira , and J.E. Manson (1986), " Computer-based control logic of on-line correction of process deviations ", *Journal of Food Science*, 51, 480–483.

De Baerdemaeker , J. and B.M. Nicolaï (1993), " Equipment considerations for sous vide cooking ", in T. Martens and M. Schellekens , editors, *Proceedings of the First European "Sous Vide" Cooking Symposium*, Leuven, Belgium, ALMA Universiteitsrestaurants v.z.w., 101–116.

Gormley , T.R. (1990), " Temperature measurements in chilled retail cabinets and household refridgerators ", in P. Zeuthen , J.C. Cheftel , C. Eriksson , T.R. Gormley , P. Linko , and K. Paulus , editors, *Processing and Quality of Foods Vol. III* , London, Elsevier Applied Science, 295–304.

Gormley , T.R. (1990), " Temperature monitoring of chilled foods during air and road freighting ", in P. Zeuthen , J.C. Cheftel , C. Eriksson , T.R. Gormley , P. Linko , and K. Paulus , editors, *Processing and Quality of Foods Vol. III* , London, Elsevier Applied Science, 273–278.

Hall , L.P. (1977), " A study of the degree of variation occurring in results of microbiological analyses of frozen vegetables ", Technical Report 182, Chipping Campden, U.K.

Hayakawa , K., P. De Massaguer , and R. Trout (1988), " Statistical variability of thermal process lethality in conduction heating food - computerized simulation ", *Journal of Food Science*, 53(6), 1887–1893.

Hicks , E.W. (1961), " Uncertainties in canning process calculations ", *Journal of Food Science*, 26, 218–226.

James , S. and J. Evans (1990), " Temperatures in the retail and domestic chilled chain ", in P. Zeuthen , J.C. Cheftel , C. Eriksson , T.R. Gormley , P. Linko , and K. Paulus , editors, *Processing and Quality of Foods Vol. III* , London, Elsevier Applied Science, 273–278.

Jarvis , B. (1989), *Statistical Aspects of the Microbial Analysis of Foods*, volume 21 of *Progress in Industrial Microbiology*, Elsevier, Amsterdam, Oxford, New York, Tokyo.

Kent , M., K. Christiansen , I.A. van Haneghem , E. Holtz , M.J. Morley , P. Nesvadba , and K.P. Poulsen (1984), " Cost 90 collaborative measurements of thermal properties of foods ", *Journal of Food Engineering*, 3, 117–150.

Lenz , M.K. and D.B. Lund (1977a), " The lethality-Fourier number method: experimental verification of a model for calculating temperature profiles and lethality in conduction-heating canned foods ", *Journal of Food Science*, 42(4), 989–996.

Lenz , M.K. and D.B. Lund (1977b), " The lethality-Fourier number method: confidence intervals for calculated lethality and mass-average retention of conduction-heating, canned foods ", *Journal of Food Science*, 42(4), 1002–1007.

Meffert , H.F. (1983), " Story, aims, results and future of thermophysical properties work within COST-90 ", in R. Jowitt , F. Escher , B. Hallstrom , H. Meffert , W. Spiess , and G. Vos , editors, *Physical Properties of Foods*, London, Applied Science Publishers, 229–267.

Melsa , J.L. and A.P. Sage (1973), *An Introduction to Probability and Stochastic Processes*, Prentice-Hall, Englewood Cliffs, New Jersey.

Miles , C.A., G. Van Beek , and C.H. Veerkamp (1983), " Calculation of thermophysical properties of foods ", in R. Jowitt , F. Escher , B. Hallstrom , H. Meffert , W. Spiess , and G. Vos , editors, *Physical Properties of Foods*, London, Applied Science Publishers, 269–312.

Nicolaï, B.M. and J. De Baerdemaeker (1992a), " Simulation of heat transfer in foods with stochastic initial and boundary conditions. ", *Transactions of the IChemE part C*, 70, 78–82.

Nicolaï, B.M. and J. De Baerdemaeker (1992b), " Stochastic finite element analysis of conduction heat transfer in foods with random field thermophysical properties ", in F. Maceri and G. Iazeolla , editors, *Eurosim '92 preprints*, 539–544.

Nicolaï, B.M. and J. De Baerdemaeker (1993a), " Computation of heat conduction in materials with random variable thermophysical properties ", *International Journal for Numerical Methods in Engineering*, 36, 523–536.

Nicolaï, B.M. and J. De Baerdemaeker (1993b), " A first order probabilistic perturbation analysis of the growth and inactivation of Lactobacillus cells during cold storage and reheating of lasagna ", in *Proceedings of ICEF-6*, Chiba, Japan.

Nicolaï, B.M., P. Van den Broeck , M. Schellekens , G. De Roeck , T. Martens , and J. De Baerdemaeker (1993c), " Finite element analysis of heat conduction in lasagna during thermal processing ", *Submitted.*

Nicolaï, B.M., W. Obbels , M. Schellekens , B. Verlinden , T. Martens , and J. De Baerdemaeker (1994), " Computational aspects of a computer aided design package for the preparation of cook-chill foods ", in *To be presented at the Food Processing and Automation Conference III* , Florida.

Olsson , P. (1990), " Chill cabinet surveys ", in P. Zeuthen , J.C. Cheftel , C. Eriksson , T.R. Gormley , P. Linko , and K. Paulus , editors, *Processing and Quality of Foods Vol. III* , London, Elsevier Applied Science, 273–278.

Patino , H. and J.R. Heil (1985), " A statistical approach to error analysis in thermal process calculations ", *Journal of Food Science*, 50, 1110–1114.

Pflug , I.J. and T.E. Odlaug (1978), " A review of z and F values used to ensure the safety of low-acid canned foods ", *Food Technology*, 63–70.

Schellekens , M., T. Martens , T.A. Roberts , B.M. Mackey , B.M. Nicolaï, J.F Van Impe , and J. De Baerdemaeker (1993), " Computer aided microbial safety design of food processes ", in *Proceedings of Food Micro '93: Novel Approaches towards Food Safety Assurance*, Bingen, Germany.

Schellekens , M., W. Obbels , T. Martens , B.M. Nicolaï, B.E. Verlinden , J.F. Van Impe , and J. De Baerdemaeker (1993), " Computer aided process design procedures to improve the quality and safety of products with a limited shelf life ", in *Proceedings of the AIFA Conference on Artificial Intelligence for Agriculture and Food*, Nimes, France.

Segerlind , L. (1984), *Applied Finite Element Analysis*, 2nd edition, John Wiley and sons, New York.

Sheard , M. and C. Rodger (1993), " Optimum heat treatments for "sous vide" cook-chill products ", in T. Martens and M. Schellekens , editors, *Proceedings of the First European "Sous Vide" Cooking Symposium*, Leuven, Belgium, ALMA Universiteitsrestaurants v.z.w., 118–126.

Stumbo , C.R. (1973), *Thermobacteriology in Food Processing*, 2nd edition, Academic Press, Inc., New York.

Sweat , V.E. (1974), " Experimental values of thermal conductivity of selected fruits and vegetables ", *Journal of Food Science*, 39, 1080–1083.

Teixeira , A.A. and C.F. Shoemaker (1989), *Computerized Food Processing Operations*, Van Nostrand Reinhold, New York.

Thompson , D.R., I.D. Wolf , K. Larson Nordsiden , and E.A. Zottola (1979), " Home canning of food: risks resulting from errors in processing ", *Journal of Food Science*, 44(1), 226–233.

Van Impe , J.F., B.M. Nicolaï, T. Martens , J. De Baerdemaeker , and J. Vandewalle (1992), " Dynamic mathematical model to predict microbial growth and inactivation during food processing ", *Applied and Environmental Microbiology*, 58(9), 2901–2909.

Wang , J., R.R. Wolfe , and K. Hayakawa (1991), " Thermal process lethality variability in conduction-heated foods ", *Journal of Food Science*, 56, 1424–1428.

Willocx , F., M. Hendrickx , and P. Tobback (1993), " Temperatures in the distribution chain ", in T. Martens and M. Schellekens , editors, *Proceedings of the First European "Sous Vide" Cooking Symposium*, Leuven, Belgium, ALMA Universiteitsrestaurants v.z.w., 81–99.

Zeuthen , P., J.C. Cheftel , C. Eriksson , T.R. Gormley , P. Linko , and K. Paulus , editors (1990), *Processing and Quality of Foods*, volume 3, London, Elsevier Applied Science.

Zwietering , M.H., J.T. de Koos , B.E. Hasenack , J.C. de Wit , and K. van 't Riet (1991), " Modelling bacterial growth as a function of temperature ", *Applied and Environmental Microbiology*, 57, 1094–1101.

Zwietering , M.H., I. Jongenburger , F.M. Rombouts , and K. van 't Riet (1990), " Modeling of the bacterial growth curve ", *Applied and Environmental Microbiology*, 57, 1875–1881.

DISCUSSION

KNORR: It is possible to use stochastic modeling techniques to determine the equipment which leads to a process with least uncertainty, e.g. conventional heating versus microwave heating?

DE BAERDEMAEKER: So far, no microwave heating models have been incorporated in the computer-aided design software. However, with the existing software it is already possible to evaluate different types of conventional ovens or process conditions (steam, hot air, air/steam mixtures) and their effect on the uncertainty of the microbial quality of the end product. For example, the temperature in steam ovens is usually much better controlled than that in hot air ovens, while the surface heat transfer coefficient variance is much smaller in the latter case.

CHAU: You presume that it is feasible to model the uncertainty in case of microwave heating?

DE BAERDEMAEKER: Although microwave heating has been modeled successfully using finite element and finite difference techniques, the geometry is essentially 3D which results in huge grids with many nodes (often several ten thousands). Moreover, in general this requires also the solution of the Maxwell equations so that the necessary computer time may be prohibitive, especially for use in an interactive computer-aided design environment. Extension of the microwave models to include uncertainties is feasible but will require even more computer time.

VAN DIJK: Are the models validated?

DE BAERDEMAEKER: The finite element model for conduction heat transfer in lasagna is indeed validated. The simulated temperatures inside the lasagna agreed well with the measured ones under different process condition (steam, air/steam mixture). The model for microbial growth and inactivation of *Lactobacillus planetarium* is validated for growth in artificial media. However, the purpose of this contribution is to show that, even if the models are valid, the uncertainties on all the parameters which are involved are amplified through the process calculations so that the final microbial quality might be very variable. The applicability of the perturbation method has been validated for relatively small uncertainties using the Monte Carlo method. For large uncertainties such as the parameter uncertainties of the microbial growth kinetics, the perturbation method underestimates the variance of the microbial load.

BRUHN: It would be nice to extend the models to incorporate quality deterioration due to overheating and to address the uncertainty when left-overs are reheated to serve another day?

DE BAERDEMAEKER: The computer-aided design package for the design of cook-chill preparation processes which is now under development at the K.U. Leuven does include these features.

VAN DIJK: Apparently uncertainty increases with time. On the other hand, if you heat a product, the microbial load decreases. What happens then?

DE BAERDEMAEKER: In this case the width of the confidence interval increases. In other words, although the upper 95% confidence margin decreases, it may decrease not as fast as the mean value. Similarly, the lower 95% confidence margin decreases faster than the mean value.

ADVANCES IN PROCESS MODELLING AND ASSESSMENT : THE PHYSICAL MATHEMATICAL APPROACH AND PRODUCT HISTORY INTEGRATORS

Hendrickx, M., Maesmans, G., De Cordt, S., Noronha, J., Van Loey, A., Willocx, F. and Tobback, P.

I. INTRODUCTION

Preservation and distribution of foods is one of the cornerstones in the organization of our modern society. Food processors are aggressively seeking preservation techniques which deliver convenience products that are fresh-like, chef-like and have the image of invisible manufacturing. [Malkki, 1987, Lund, 1989] This has resulted in new heating techniques, such as continuous processing in rotary retorts and aseptic units, ohmic and microwave heating, of foods comprising liquids and solids, minimally processed foods (Refrigerated Processed Foods with Extended Durability - combined processing) and thermal treatments in combination with high hydrostatic pressure. [Hermans, 1988, Skudder, 1988, Richardson, 1991, Mertens and Knorr, 1992]

For these technologies, adequate evaluation of the impact on both safety (microbiological safety, risk analysis) and quality (consumer acceptability, nutritional value) aspects of foods subjected to a specific history (time-temperature) profile is essential. Here, we will discuss the current approach and status of the methodologies used to evaluate the integrated history (time-temperature) effect[1]. Both the current status and expected evolutions in the use of physical-mathematical approaches as well as the use of product history integrators are being discussed.

II. A UNIFYING THEORY ON FOOD PRESERVATION

A. THE GENERAL PRINCIPLE OF FOOD PRESERVATION

The classical approach to food preservation is to state that any preservation technology relies on both inhibition and destruction of pathogenic and non-pathogenic spoilage organisms. [Pivick and Petrasovicks, 1973] It is the food technologist/scientist's responsibility to satisfy the safety *conditio sine qua non* for any product appearing on the market within the restraints set forth by producers, consumers and the authorities.

$$\text{PROTECTION} = \text{DESTRUCTION} + \text{INHIBITION} \tag{1}$$

This principle can, according to us, be expressed more quantitatively as

$$\text{PROTECTION} = \text{INTEGRAL(rate, time)} \tag{2}$$

---

1 The term integrated time-temperature effect refers to the integral over time with respect to the conditions on a single point. The terms 'mass average' or 'volume average' impact will be used to indicate volume and time integral

indicating that the microbial load of a food product at the moment of consumption will be a net effect of the initial microbial load and the microbial growth and/or inactivation rate as a function of time. In these terms, inhibition stands for a very low or ideally even a zero growth rate. The impact of a food preservation treatment on other than microbial quality attributes can be expressed in a similar way as

$$\text{LEVEL OF QUALITY ATTRIBUTE} = \text{INTEGRAL(rate, time)} \tag{3}$$

Food preservation is the control of residence time (distribution) of the product and the rates of desired and un-desired responses (physical, chemical and biological) of the product through a proper choice and control of intrinsic and extrinsic factors (processing conditions) during the total (shelf)life of the product. Intrinsic factors refer to characteristics of the product itself while extrinsic factors refer to the environmental conditions of the product. In some cases, there might be an interaction between extrinsic and intrinsic factors. This reasoning is as applicable to thermal processing (pasteurization and sterilization) as it is to cooling, freezing, drying and high pressure treatments and combinations of these processes.

B. CONCEPTS TO EXPRESS THE INTEGRATED TIME-TEMPERATURE EFFECT

In the context of the theory outlined above, we will further limit to systems where temperature is the only extrinsic factor changing with time. This implies thermal processes such as sterilization, pasteurization, blanching, cooling, frozen storage and combinations of these. Concepts have been formulated to express the impact of the thermal process. These concepts are needed in design, evaluation and optimization of such preservation processes.

Concepts that have been formulated in the past are (i) an equivalent time at a chosen reference temperature, (ii) an equivalent temperature at a chosen process time and (iii) the equivalent point method (an equivalent time at an equivalent temperature). These concepts have been developed based on an n-th order model to describe the safety and/or quality attribute as a function of time under constant extrinsic and intrinsic factor conditions in combination with an Arrhenius type model to express the temperature dependency of the rate constant.

$$-dX/dt = k_x X^n \tag{4}$$

$$k_T = k_{ref}.\exp\left[(Ea/Rg)(1/T_{ref}-1/T)\right] \tag{5}$$

with X = actual level of a quality attribute, X_0 = initial level of the quality attribute, k = rate constant, Ea = activation energy, Rg = universal gas constant, T = temperature and t = processing time

This model in general and the first order model $(n=1)$ in particular is applied to thermal destruction (wet-heat) of microorganism (spore), enzyme inactivation and loss of quality during heating, cooling and frozen storage. [Bigelow, 1920, 1921, Stumbo, 1973, Tobback *et al.*, 1987, Teixeira, 1992]

Scientists and technologists in the food and pharmaceutical area have used and still use field specific descriptors, e.g. the decimal reduction time (D_T), more frequently than the rate constant (k_T). The TDT-model and the Arrhenius equation are the common models to quantify the temperature sensitivity of the rate constants, respectively leading to the z-value (C°) and activation energy (J/mol). [Lund, 1975] Both models can be reduced to the equation

$$\ln(k) = C_1 + C_2.T^m \tag{6}$$

where C_1 and C_2 are constants and $m=1$ in the TDT-model or $m=-1$ according to the Arrhenius model.

For reviews on thermostability of microorganisms and their spores and of chemical and physical quality attributes, the reader is referred to open literature. [Lund, 1975, 1986, Holdsworth, 1985, Pflug and Odlaug, 1988, Adams, 1991, Villota and Hawkes, 1986, 1992, Thompson and Norwig, 1986, Wilson, 1986, Hällström et al., 1988]

Data are also available on growth of several pathogenic and spoilage microorganisms and enzyme activity at refrigeration temperatures. Predictive modelling of food microbiology (bacterial growth) including the temperature dependency of lag-phase and growth rate currently receive a high interest. [Zwietering et al., 1990, 1991, Ratkowsky et al., 1991, Baranyi et al., 1993, Willocx et al, 1993] Theoretical and application aspects on predictive microbiology were recently thoroughly reviewed. [McMeekin et al., 1993] There is a complete lack of kinetic data in the transition domain between refrigeration temperatures (growth) and pasteurization temperatures (vegetative cell inactivation). In order to develop and use full-history models, e.g. [Van Impe et al., 1992] that allow to describe microbial growth and inactivation in prepared foods which are first pasteurized and subsequently distributed at refrigeration temperatures (e.g. sous-vide products), this information will have to be established.

III. EVALUATION OF THERMAL PROCESSES : COMMONLY USED APPROACHES

A. INTRODUCTION

The evaluation of a thermal process can be performed in three ways : by (i) an *in-situ* method, (ii) a physical-mathematical approach or (iii) by using Product History Integrators. In case temperature is the only variable changing with time, product history integrators become Time-Temperature-Integrators (TTI's). The first two methods constitute the common evaluation techniques whereas the latter embodies a promising more recent research direction.

B. THE *IN SITU* APPROACH

In the '*in situ*' method, the level of the food quality attribute of interest is evaluated before and after the process. This technique is applied e.g. in the judgment of the nutritional quality of canned foods after sterilization (vitamin loss) [Mulley et al., 1975a, 1975b], in sensorial appreciation of taste, colour, texture,... [Hayakawa et al., 1977], in the physical evaluation of colour [Hayakawa and Timbers, 1977]. A typical example is the evaluation of the adequacy of blanching processes prior to frozen storage where residual enzyme activity of the enzyme responsible for quality degradation during storage is monitored (e.g. peroxydase, catalase, lipoxygenase).

The main advantage of the *in situ* method is that the impact of the process on the parameter of interest is directly and accurately known. Disadvantages associated with this methodology are that (i) in some cases, the reduction of a quality attribute due to the heat process may be beyond the detection limit of current analytical methods and/or the sample size required may be unacceptable large (e.g. the microbiological bioburden for the safety of sterilized low-acid foods: a Probability of Non Sterile Unit of 10^{-9} can not be monitored, [Pflug, 1987]); (ii) the analysis of the parameter under investigation can be quite laborious, time consuming and expensive which renders the method unfeasible for routine checks. Other, more convenient methods have been and have to be developed.

C. THE PHYSICAL-MATHEMATICAL APPROACH : AN ESTABLISHED APPROACH FOR EVALUATION OF STERILIZATION PROCESSES

One of the oldest and most generally applicable alternatives for the *in situ* method of thermal process evaluation is based on determining the temperature history imposed on the food. Together with knowledge on the kinetic parameters of the safety (quality) attribute, the impact of the treatment can be calculated. The temperature profile can be obtained either from direct physical measurement or from constructive computation.

For in-pack heat treatment, the time-temperature profile can be measured directly (as in a classical heat-penetration study, [CFDRA, 1977, Myers, 1978, FPI, 1989]) but for reasons of practical versatility, it is quite often reconstructed theoretically or empirically from knowledge of the physical heating characteristics of the food (thermal conductivity, heat capacity, heating rate factor (f_h), lag time (j), *etc.*) and the measured heating medium temperature. The lethal rates associated with the physically measured temperatures at the slowest heating point in the packaged food can be integrated directly in the general method with an accurate integration procedure, but this methodology has little or no predictive power for processing conditions different from those under which experiments were run and it is unsuited for reverse calculations (estimating the process time needed to reach a certain lethality). To overcome this inconvenience, predictive methods have been developed, based on empirical formulae and theoretical models for heat transfer.

Empirical formulae have been widely used to describe the temperature response characteristics of foods to be sterilized. Their success arises from the fact that these methods allow one to handle complex heating systems which can be encountered in real food processing systems (conduction, convection and quite often mixed conduction-convection heat transfer modes). These cases are difficult to solve theoretically because of the large computing times needed and the detailed knowledge on the complex boundary conditions and thermophysical properties required. Where 'formula methods' are used, design and evaluation of the heat treatment follows the strategy: (i) a heat penetration study is performed under well defined processing conditions (ii) the time-temperature data are analysed in order to obtain the heating characteristics of the food (f- and j-values); (iii) the process at hand is evaluated by generating the time-temperature profile in a specific point in the food from the heating medium temperature and integrated in the impact-value (F). These methods, pioneered by Ball [Ball, 1923, Ball and Olson, 1957] and revised [Steele and Board, 1979, Steele *et al.*, 1979], refined [Stumbo, 1973, Hayakawa *et al.* 1977, 1978] and made more easily accessible [Vinters *et al.*, 1975, Kao *et al*, 1981], do allow for both calculation of the processing value of a heat treatment (forward calculations) and prediction of the process time needed to obtain a desired F-value (reverse calculations). Datta [1990] proved the theoretical soundness of the 'f' and 'j'- heating parameters used in these empirical formula methods for conduction and forced convection heating of an arbitrary shaped object but indicated the lack of a fundamental foundation to describe natural convection heating of unagitated liquid foods with these parameters. In this case, the method can yield only an approximate description of the heating/cooling behaviour of the product over small ranges of processing time. Extended reviews on this topic are available. [Hayakawa, 1972, Hayakawa, 1977, Clark, 1978]

Theoretically, conductive heating/cooling can be adequately predicted from the Fourier equation. [Bird *et al.*, 1960] Not only the heating behaviour of the coldest point in a food product can be evaluated but also predictions on the mass average lethality of the packaged products are possible. Elaborated and well documented (i) analytical [Hayakawa, 1969] and (ii) numerical [Teixeira *et al.*, 1969, Teixeira and Manson 1983] calculation procedures are reported in literature and have been used extensively in food research. The complexity of the mathematical treatment has limited the development of theoretical models for convection and mixed convection/conduction heating, such as for food comprising a liquid and solids. [Stoforos, 1988] Empirical dimensionless correlation equations have been used to describe heat transfer in these cases. Natural and forced convection of liquid foods in cans has been modelled. [Hiddink, 1975, Datta and Teixeira, 1988, Kumar *et al.*, 1990] A number of authors [Bimbenet and Duquenoy, 1974, Stoforos and Merson, 1992] have proposed mathematical solutions to the heating of canned solid/liquid systems. Especially for heating conditions which are not readily accessible to physical experimentation (e.g. batch and continuous rotary retorts), these models may contribute substantially to our understanding and the evaluation of thermal processes. For both conductive and convective modelling, the lack of accurate input parameters (thermophysical properties of foods, flow characteristics, motion and temperature of particles inside the package) limits the application of existing models.

By analogy, models and calculation procedures have been proposed to evaluate aseptic processing of liquid foods and foods comprising solid particulates in liquids, based on (i) manual formula methods [Dail, 1985, Larkin, 1989] or (ii) numerical calculations. [de Ruyter and Brunet, 1973, Teixeira and Manson, 1983, Sastry, 1986] These models have been very useful in estimating the relative importance of critical factors for these processing technologies such as (but not limited to) particle size, shape, integrity, temperature distribution, fluid-to-particle heat transfer coefficient, residence time distribution,... but await accurate input data to be implemented in reality. [Defrise and Taeymans, 1988, Pflug *et al.*, 1990, Maesmans *et al.*, 1992]

When the temperature response of a food to an externally enforced heating profile can be regenerated *ad libidem* for a given product from empirical or theoretical formulae, evaluation of process deviations, optimization and even on-line control of the thermal process are within reach. The availability of software packages for process evaluation, design and optimization is no longer restricted to research institutions since low-cost personal computers have made commercial dissemination of these techniques possible. Examples of such commercially available packages, based on empirical and theoretical approaches, are Calsoft, TPRO, Thermal Profiler, ProCalc, CTemp. (This list is not exhaustive).

D. LIMITATIONS AND NEW TRENDS FOR CURRENT EVALUATION METHODS

The *in situ* evaluation of a thermal process is often restricted by the detection limit of the analytical techniques at hand. Also the time and effort needed to assess the impact of the heating process on any quality attribute (e.g. quantitative microbiology) with this method constrains its use in practical processing conditions. More sensitive and fast analytical methods have to be developed to increase the applicability of the *in situ* approach (this includes on-line sensors for monitoring food quality).

The use of the aforementioned physical-mathematical approach may seem quite appealing and promising, but it has to be stressed that the accuracy for evaluation of a thermal process depends entirely on the accuracy of the physical (empirical) input parameters required. These input parameters are not always available (e.g. data on viscosity or conductivity at high temperatures, order of magnitude of fluid-to-particle heat transfer coefficient, residence time distribution of particles in a continuous process). In addition, in the step of model validation direct measurement of the time variable food temperature is needed and can be very difficult or is impossible under some processing conditions without disturbing the heating behaviour of the food. The use of existing wireless measuring systems (e.g. Datatrace from Ball Electronic Systems Division, West Ridge, Colorado, USA, Tracker from Datapaq, Cambridge,UK, Testostor from Testotherm, Lenzkirch, FRG, Thermophil store from Ultrakust, Ruhmannsfelden, FRG) is limited by their size [May and Cossey, 1991], although improved miniaturization is being strived for.

For the more recently introduced heating techniques such as aseptic processing, ohmic heating and microwave heating models are under development. With regard to these processes the further development of the physical-mathematical approach is focussed on the solution of all equations describing the (coupled) transport phenomena involved. The same remark holds for the more theoretical approach of in-pack thermal processing of agitated containers.

In addition the physical-mathematical approach is extended to the evaluation of food safety and quality for pasteurised and/or refrigerated products. In this area the current research interest is mainly focussed on predictive microbiology and needs further input on kinetic models and data (see above). The integration of modern hardware (sensors, timers and information storage) and software has lead to commercially available temperature loggers/integrators (e.g. Smartlog, Control One, Autolog, Delphi, Don Whitley TTFI etc..).

The limitations on the *in situ* and the physical-mathematical approach have lead to the development of, and current interest in, Time-Temperature-Integrators as an alternative means for process evaluation.

IV. THE EVALUATION OF PRESERVATION PROCESSES : THE USE OF TIME-TEMPERATURE INTEGRTORS

Next to the *in situ* and physical-mathematical evaluation of a thermal process, Time-Temperature-Integrators (TTI's) have been and are being developed to overcome the inherent disadvantages associated with these methods. A TTI can be defined as '*a small device that shows a time-temperature dependent, easily, accurately and precisely measurable irreversible change that mimics the changes of a target quality parameter undergoing the same variable temperature exposure*'. [Taoukis and Labuza, 1989a, Weng et al., 1991b, De Cordt et al., 1992] Table 1 presents a classification of TTI's in terms of working principle, type of response, origin, application in the food material and location in the food.

A. GENERAL ASPECTS ON TTI'S

From the definition of a Time-Temperature Integrator given above, general criteria such a measuring device should meet can be formulated : (i) the TTI has to meet convenience criteria such as inexpensive, easy and fast in preparation, easy to recover, exhibit an accurate and user-friendly read-out, etc.. (ii) it must be possible to incorporate the TTI in the food product without disturbing the heat transfer, the presence of the TTI can not change the time-temperature profile of the food and the TTI should experience the same time-temperature profile. This conditions can lead to specific requirements in thermophysical properties (thermal diffusivity, dielectric properties etc..) for the TTI. (iii) the TTI shall quantify the impact of the process on a target quality and/or safety

Table 1
General Classification of Time Temperature Integrators

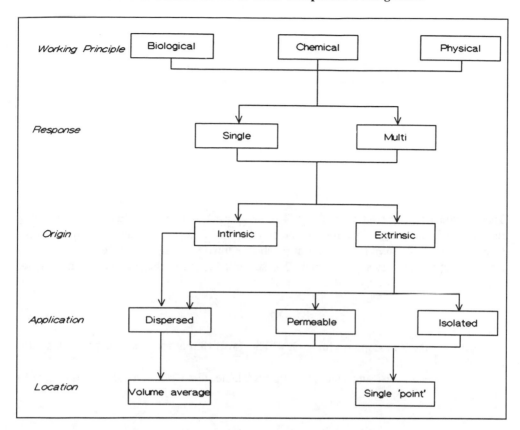

attribute. This results in specific kinetic requirements for the TTI.

1. Working principles of TTI's.

Depending on the working principle, TTI's can be subdivided into (i) biological, (ii) chemical and (iii) physical systems. For biological TTI's the change in biological activity, such as microorganisms, their spores (viability) or enzymes (activity) upon heating is the basic working principle. Chemical and physical systems are based on a purely chemical or physical response, respectively towards time and temperature.

2. Response type and kinetic requirements.

The kinetic requirement which a TTI should fulfil can be derived theoretically. [Taoukis and Labuza, 1989b, Tobback *et al.*, 1992, Maesmans, 1993] When the processing value is chosen as the concept to express the integrated impact of time and temperature, the equation that should be satisfied by a time-temperature integrator (TTI) that undergoes the same thermal treatment (time-temperature profile) as a target index can be written as :

$$(^{Z}F_{Tref})_{target} = (F_{Tref})_{TTI} \tag{7}$$

Table 2
TTI-response Functions for Different Reaction Orders

Reaction order	TTI function ($f(X)$)
1	$Ln(X_o/X)$
n ($n \neq 1$)	$[1/(n-1)](X^{1-n}-X_o^{1-n})$

That is, the lethality read from the TTI $((F_{Tref})_{TTI})$ shall be identical to the lethality response of the target quality attribute $((^Z F_{Tref})_{target})$ as integrated from time-temperature data or from the concentration change introduced by the process.
Considering an n-th order TTI-system (X), subjected to an isothermal heat treatment, we obtain :

$$f(X) = k_X t \qquad (8)$$

where $f(X)$ is the TTI response function which, for different reaction orders as illustrated in Table 2.

If the TTI is subjected to a time-variable temperature profile, Eq.4 can be integrated as :

$$f(X) = \int_0^t k_X dt \qquad (9)$$

With k expressed according to the Arrhenius equation.

It can be easily derived that for a system to function as a TTI, its activation energy has to be identical to that of the target index.

$$Ea_{TTI} = Ea_{target} \qquad (10)$$

Although it was recognized early [Kelsey, 1958, Horne *et al.*, 1976, Hayakawa, 1978] that for first order reaction kinetics any agent used for monitoring the lethal effect of heat processes should have the same z-value as the target microorganisms, but could have a different D-value, this basic requirement for proper functioning of a TTI has frequently been neglected.

It has been demonstrated [Wells and Singh, 1988, Taoukis and Labuza, 1989a, 1989b, Hendrickx *et al.*, 1992] that exactly the same prerequisite for proper functioning of a TTI (equal activation energies of TTI and Target) is valid when the equivalent temperature at a chosen reference time concept is used to evaluate the quality retention in chilled foods and the impact of a heat treatment, provided of course that temperature is the only extrinsic factor responsible for changing the rate of safety or quality determining reactions.

Whenever the integrated impact of time and temperature on a quality or safety attribute can be described with a single activation energy, a TTI can be looked for or developed which has an identical activation energy. When such a single-component TTI is used, a direct relation between the change in status of TTI and target quality attribute it monitors can be proposed without any further restriction. Problems arise when the change in quality attribute can not be described by a single activation energy or no TTI can be found with a matching Ea.

In case the activation energy of TTI and target attribute differ, information on the time-temperature history is needed to predict the impact on the target quality attribute from the response of a single-component TTI. This approach may be feasible for well-controlled processing conditions e.g. heating of liquids at constant known holding temperature in an aseptic heating system where (Berry *et al.*, 1989) or by approximating the temperature profile at the centre of a moving particle in an aseptic system by a general power law expression with a known maximum temperature (Kim and Taub, 1993).

Multi-component TTI's have been suggested to quantify the change in status of a quality attribute in case the activation energy of TTI and target attribute differ, and information on the time-temperature history is not available. The reading of the impact of a process on a 'set' of individual components, each responding according to their own activation energy, and predicting from this multiple response the change in status of a quality attribute with a different activation energy can not be done without an appropriate transformation function. An approximative transformation function has been proposed which summarizes a variable temperature history by an equivalent time at an equivalent temperature, independent of the activation energy [Swartzel, 1982, Swartzel, 1986, Nunes and Swartzel, 1990] but since a theoretical basis for this 'equivalent point method' is not at hand [Maesmans *et al.*, 1993a], this transformation function has to be validated for each temperature profile and for each set of temperature sensitive components (Ea's). [Maesmans *et al.*, 1993b, Maesmans, 1993]

3. Origin, application and location in the food.

With respect to the origin of the TTI, extrinsic and food intrinsic TTI's can be distinguished. An extrinsic TTI is a system added to the food while intrinsic TTI's are intrinsically present in the food. In the latter case, another intrinsic marker than the actual quality attribute is monitored to represent the behaviour of the target index. Analysis of the impact of a process on these components is not considered as an *in situ* evaluation because they have to meet the same requirements as extrinsic TTI's.

With regard to the application of the TTI in the food product three approaches can be distinguished (i) dispersed, (ii) permeable or (iii) isolated. Dispersed systems allow to evaluate the volume average impact while all three approaches can be the basis for single point (specific location in the food) evaluations.

When using intrinsic components as TTI, the TTI is more or less homogeneously distributed in the food or drug. The close contact between food and TTI eliminates heat transfer limitations. Since kinetic characteristics of any reaction are influenced by the micro-environment of the TTI, a kinetic calibration study is needed for each of these TTI's in direct contact with each food product. [Bowman, 1969, Skaug and Berube, 1983, McCormick, 1988, Philipp and Sucker, 1988, 1990] An inoculated pack is an example of an extrinsic TTI, dispersed in the food system. [Yawger, 1978, Pflug, 1987] Also in aseptic processing studies, microorganisms or their spores have been dispersed as such in the (liquid) food product. [Burton *et al.*, 1977, Dodeja *et al.*, 1990]

To avoid the influence of the food environment on the kinetic behaviour of the TTI, encapsulated TTI-systems have been proposed : the extrinsic TTI is then completely isolated by embedding it in an inert carrier material such as glass [Hershom and Shore, 1981], plastic [Hunter, 1972, Pflug, 1976], metal [Rodriguez and Teixeira, 1988] or paper (e.g. spore strips, [Smith *et al.* 1976]). To avoid recontamination during recovery of the TTI, carrier systems have been proposed with a separate compartment for a nutrient solution and a means of bringing nutrient and TTI in contact after the thermal treatment. [Orelski, 1981, Karle and Kralovic, 1983, McCormick and Scoville, 1987, Brown and Buglino, 1988, Welsh and Dyke, 1989]   In case the TTI is completely isolated from contact with the food, its kinetics can be determined independent of the type of food and nature of the surrounding medium.

As an intermediate between these two modes of applying a TTI, systems have been proposed where the TTI is enclosed in a carrier material but the carrier material itself is permeable for the environment. This allows for fast identification and recovery of the TTI on the one hand and in addition, the TTI is exposed to the same (variable) intrinsic conditions as the actual quality attribute. It is clear that the influence of these intrinsic properties on the kinetic behaviour of the TTI have to be known. Examples of such permeable TTI-carriers are model food particles of alginate [Brown *et al.*, 1984] of polyacrylamid. [Rönner, 1990a, 1990b]   Care has to be taken to avoid leakage of the temperature sensitive component out of the carrier. [Fink and Cerny, 1988, Matthiasson and Gudjonsson, 1991]   It should be verified upon embedding the system in a food product or particle that heat transfer still is limited by the food product and not by the carrier material of the TTI.

In order to make heat transfer rates to a TTI and a real target quality attribute in a solid food particle comparable, thermal diffusivity of the chosen casing surrounding/incorporating the TTI shall be in the range of food-stuff thermal diffusivity. Density of the TTI-vehicle should be identical to that of the real food particle when the influence of free rotation and translation of the monitored particle is of importance as in agitated or aseptic thermal processes.

B. EXISTING TTI'S
1. Biological TTI's

The use of *microbiological systems* to monitor efficacy of a sterilization process is the most endeavoured application of TTI's in the food and pharmaceutical industry. The z-values of spores utilized to monitor safety closely resemble the z-value reported for *Clostridium botulinum*. Any divergence from $z=10C°$ will make the processing value determined from these spores deviate from F_0. For low acid foods, spores of *Bacillus stearothermophilus, Bacillus subtilis* 5230, *Bacillus coagulans* and *Clostridium sporogenes* are used as biological indicators for wet heat sterilization. [Jones and Pflug, 1981, Pflug and Odlaug, 1986]   Only *calibrated* spores can be used to determine the killing power of a heat treatment.

An inoculated pack is a practical example of the use of such Biological Indicator Unit (BIU). It is a biological, single-component extrinsic dispersed TTI to measure the volume average impact of the thermal treatment. An inoculated pack study is classified as a TTI approach and not as an '*in situ*' approach because calibrated spores with known kinetic properties have to be used. They are not identical to the actual bioburden of the food; they are used to mimic the inactivation of the bioburden. Where the inoculated pack was used at first to verify process calculations and heat penetration measurements for still retorts, for almost forty years now it has been applied to design and monitor heat sterilization processes in machines where it is almost impossible to determine the F_0-value by physical-mathematical methods. [Pflug, 1987]

Different tools have been designed to control the location of microorganisms or their spores at a specified site in the food product instead of dispersing it over the entire can contents. Spore solutions have been enclosed in carrier units which largely simplify the recovery procedure after processing. Biological, single-component extrinsic isolated TTI's are thus constructed to measure the impact of the heating process on a single-point location in the food (normally the coldest spot). A plastic rod, filled with a calibrated *Bacillus stearothermophilus* spore solution was suggested by Pflug [Pflug, 1976] and has been used successfully to validate sterilization processes for green beans in a Sterilmatic [Jones *et al.*, 1980], whole kernel corn in a Steritort [Pflug *et al.*, 1980a] and peas in brine in a Steritort [Pflug *et al.*, 1980b]. Sterilizing values determined from the physical-mathematical approach were compared with their BIU-counterpart. The authors concluded that plastic rod biological indicator units can be used effectively to determine the sterilizing value delivered to cans processed in agitating retorts. According to Pflug [Pflug *et al.*, 1980b], this method allows routine determination of F_0 with a 15% accuracy. An aluminium biological indicator carrier has been proposed because of the improved heat transfer characteristics and mechanical strength of this material, which is of particular importance in the thermal processing of rapidly heating low viscosity foods. [Rodriguez and Teixeira, 1988] Paper strips have been inoculated with a known amount of spores to produce a spore strip. [Pflug and Smith, 1977] The performance of such strips has been discussed. [Smith *et al.*, 1976, 1982] To avoid contamination upon recovery of the spores (after processing), spore strips have been placed in a vial, sometimes together with a nutrient solution and a means to bring spores and recovery medium into contact after processing (=contamination-free test package). [Orelski, 1981, Wels and Dyke, 1989] Beside complete recovery of all the heat-treated microorganisms at the end of a process, a measurement with these 'biological thermocouples' can not be affected by the contact of the TTI with the food environment and hence be influenced by pH, oxidation-reduction potential, nutrient condition, *etc.* [Pflug *et al.*, 1990]

With the advent of aseptic processing technologies for viscous liquid foods containing particulates, carrier systems for microbiological TTI's have been miniaturized in order to measure sterilizing values in these conditions. Biological, single-component, extrinsic TTI's have been described where the TTI is (i) isolated from direct contact with the food material [Hunter, 1972, Hersom and Shore, 1981] (ii) dispersed in a food or food model particle [Bean *et al.*, 1979, Brown *et al.*, 1984, Heppel, 1985, Holdsworth 1987, Sastry *et al.*, 1988, Gaze *et al.*, 1990] or (iii) applied in a permeable food model particle [Rönner, 1990a, 1990b].

The major disadvantage of any microbiological heat treatment efficacy test is the length of the assay. The large incubation time between process and read-out of the system does not allow for fast intervention upon any kind of systematic failure or deviation. Quantification of a microbiological TTI requires skill [Pflug and Smith, 1977] and the analytical precision of the techniques are rather low [Jason, 1983]. The likelihood of contamination during microbiological sterility testing of foods has been recognized and documented. The inherent limitations of these methods in determining the efficacy of thermal processing and the time, tedium and expense associated with these methods have prompted the investigation of alternatives. [Mulley *et al.*, 1975c] Beside chemical and physical alternatives, the potentials of enzyme systems currently receive considerable interest.

It has been shown how by encapsulation of covalently immobilized horseradish peroxidase together with an organic solvent, the z-value of the thermal inactivation kinetics of horseradish peroxidase was shifted towards a z-value of 10C°. [Weng *et al.*, 1991a, Weng *et al.* 1991b, Tobback *et al.*, 1992, Hendrickx *et al.*, 1992, Weng, 1992] By embedding this unit in a food system, pasteurization efficiency, aimed at destroying D-streptococci (z=10C°), could be monitored in the temperature domain between 65 and 90°C. The same group of researchers [De Cordt *et al.*, 1992] revealed potentials of *Bacillus licheniformis* α-amylase as a TTI (90-115°C). Here, a z-value of $\pm$ 10C° was obtained through immobilization and changes in pH and Ca^{2+}-concentration of the enzyme micro-environment. These systems are examples of potential biological (enzymic) single-component extrinsic isolated systems to measure the impact at specific locations in the food.

The time-temperature dependent action of an enzyme has been proposed to construct a biological, extrinsic isolated TTI to monitor shelf life of chilled foods. [Blixt and Tiru, 1976, Agerhem and Nilsson, 1982] This I-point Time Temperature Monitor (I-point biotechnologies, Malmö, Sweden) consists of lipase and a lipid substrate, immobilized by adsorption to a PVC-carrier in an aqueous solution with a pH-indicator, separately encapsulated. When substrate and enzyme are brought into contact, the temperature-controlled hydrolysis of the lipid substrate will decrease the pH of the aqueous solution, inducing a gradual colour change. The temperature dependence of the reaction rate can be controlled by chosing an appropriate enzyme/substrate combination. This allows construction of TTI's with activation energies between 62 and 146 kJ/mol. The performance of this TTI is discussed by McMeekin *et al.* [1993].

2. Chemical TTI's

Detection of the concentration change of a chemical compound, added to the food product, as a measure for the impact of a thermal process was advocated more than thirty years ago [Struppe, 1968, Mulley *et al.*, 1975c] to overcome inherent disadvantages associated with quantitative microbiology. [Lee *et al.*, 1979] Thiamin (z=26C°, $D_{250°F}$=243.5 min), [Mulley *et al.*, 1975b] was mixed in pea puree, beef puree (conduction heating) and added to the liquid fraction of peas-in-brine (convection heating) in 307x409 cans. Predicted and measured thiamin reduction were proven to match one another in all circumstances. After carefully determining and calibrating the thermal hydrolysis-kinetics of disaccharides (z=18C°) added to a food system, Wen Chin [Wen Chin, 1977] attempted to predict the safety of the heating process (z=10C°) from this chemical reaction. For a convection heating product (no temperature gradients) with a negligible come-up-time and processed at the reference temperature (121.1°C) -which makes the F_o-value independent of z-agreement between the $^{10}F_{121.1}$ and $^{18}F_{121.1}$ was reasonable (as could be expected). When a conduction heating product (5% bentonite solution), characterised by a considerable come-up-time and thermal gradients throughout the system was tested, agreement quickly vanished since the basic requirement of a TTI was not met. Acid hydrolysis of sucrose (pH=2.5, Ea= 94.6 kJ/mol) and the alkaline destruction of Blue #2 at different pH's have been suggested as chemical TTI's. By altering pH, the activation energy of the Blue#2 solution could be modified (Ea= 58.2 kJ/mol at pH=11.3 and Ea=74.5 kJ/mol at pH=9.5), but the temperature dependence of *Clostridium botulinum* spores could not be matched. [Sadeghi and Swartzel, 1990] By using the equivalent point method, the authors are trying to elude this disadvantage. The response of this chemical TTI underestimates the actual impact of the thermal treatment (F_o) in most cases reported. [Sadeghi and Swartzel, 1990] Favetto *et al.* [Favetto *et al.*, 1988, 1989] monitored the darkening of a paper disc impregnated with a reducing sugar and an amino acid (Maillard's reaction) upon heating between 85 and 100°C and tried to correlate this with the inactivation of the foot and mouth disease virus. Since no z-value was reported, the system could neither be compared with the inactivation of its target nor be evaluated in terms of its processing or pasteurization value. Another disadvantage is the practical need to attach the TTI to the packaging surface, which makes evaluation of the coldest spot of the product impossible. Thermal degradation of methylmethionine sulfonium between 121.1 and 132.2°C (20.0C° $\leq$ z $\leq$ 22.8C°) in a citrate buffer (4 $\leq$ pH $\leq$ 6) was correlated [Berry *et al.*, 1989] with the reduction of micro-organisms. However, the presented relation is only valid for a food system heated at constant heating medium temperature and where no temperature gradients, come-up-time nor cool-down-time exist (rectangular barema in the product). The safety of rather small volumes of perfectly mixed low-viscosity liquids (which heat by convection) can be monitored in this way.

Recently, attempts have been reported to predict the impact on food safety (or any other quality attribute) from the response of innate chemical constituents (chemical intrinsic TTI) of a food product. The gradual formation under influence of heat of 2,3-dihydro-3,5-dihydroxy-6-methyl-(4H)-pyran-4-one from fructose, an intrinsic component of many foodstuffs was monitored. [Kim and Taub, 1993] Assuming the shape of the thermal history to be known, these authors could correlate the increase of this intrinsic chemical marker (Ea= 96 kJ/mol) with the destruction of *Clostridium botulinum* spores (or any other target). However, use of such chemical constituents is limited to well-known and controlled processing conditions for which a transformation function is established that allows to relate the impact on the TTI to the response of a target quality attribute with a different z- or Ea-value.

High hopes are put on chemical TTI's as promising tools for the evaluation of a thermal process. [Danielson, 1982] The only, yet crucial, deficiency at this moment is that no reactions thus far have been identified in open literature on heat treatments of foods that feature the temperature dependency (z-value or activation energy) required to monitor food safety in the sterilization temperature range and only a few are available that can be used to follow other quality attributes deterioration.

The temperature dependent latice-controlled polymerization of di-substituded acetylene crystals is used as working principle of a chemical TTI for chilled foods. [Patel and Yee, 1980, Fields, 1985] (Lifelines Technology Inc., Morris Plains, NJ, USA) The resulting polymer is coloured and the colour intensity can be read with a hand-held laser optic wand. The side groups R on the monomer can be changed in order to change reaction properties. These tags have activation energies between 83 and 100 kJ/mole and response (reflectance) as a function of time shows first order behaviour. Lifelines 'Fresh-Check' TTI has the same working principle but allows visual evaluation of shelf-life by consumers. Good correlation between the response of these TTI's and the sensorial appreciation of processed milk, orange juice and irradiated cod fillets freshness were reported for storage at constant temperatures with UHT. [Fields, 1985] Zall *et al.* [1986] illustrated the feasibility of this TTI for the evaluation of freshness of UHT milk under variable temperature conditions (21 - 34°C). Wells and Singh [1988] mention that the response of this TTI does not always correspond to the changes in sensorial quality attributes of tomatoes, naked and wrapped lettuce, canned fruitcake or UHT-milk.

Bhattacharjee [1985, 1986, 1987, 1988] described and patented a photo-activatable chemical TTI, consisting of a leucobase-photoacid mix impregnated on a paper disc covered with a transparent polyester film and attached to the perishable product to monitor freshness at sub-ambient temperatures. In the controlled oxygen atmosphere under the polyester film, the transformation of the colourless leucobase to a diphenylmethane or triarylmethane dye (e.g. malachite green, brilliant green, crystal violet,...) is a chemical reaction controlled by the presence of acid. Bhattacharjee [1988] therefore mixed the leucobase with o-nitrobenzaldehyde which could be converted to o-nitrobenzoic acid by ultraviolet light ($\pm$ 350 nm). Reflectance of the TTI, measured with an optical wand and a He-Ne laser (632 nm) as light source, was used to monitor the extent of the reaction (colour formation). The o-nitrobenzaldehyde was selected as photoacid because it was not activated by ambient light, nor by the light of the light source used to measure reflectivity, nor by temperature. This TTI showed an activation energy of $\pm$ 60 kJ/mole, but it was claimed that this could be changed by varying the concentration of nature and concentration of the photoacid, the intensity or duration of the photoactivation or the concentration of the leuco dye.

3. Physical TTI's

The temperature sensitive mechanism of physical TTI's for the evaluation of thermal processes is based on diffusion. Witonsky [1977] proposed a TTI functioning for monitoring steam sterilization processes. A dry chemical in an embossed well of an aluminium plate was placed at the end of a paper wick. The system was subsequently covered by a transparent plastic film of known steam permeability. The water vapour that permeates the film depresses the melting point of the coloured chemical. The molten chemical is wicked up in the paper film and the distance of the colour front from the base is a function of the increased saturation steam pressure which increases the permeation of steam, increasing temperature which increases permeability of the film and the decreasing difference between melting point and processing temperature upon increasing temperature which results in a lower water requirement before melting of the coloured chemical. Activation energies could be controlled by the choice of the plastic film and melting point of the chemical. Since a TTI could be constructed which temperature dependence between 115 and 135°C was characterized by a z-value of 10°C (70 kcal/mole), the device could be calibrated in terms of F_O-value to monitor product safety. The mean precision was reported to be $0.75*F_O$. Bunn and Sykes [1981] later reconsidered proper function of this TTI, showing it could be used to monitor F_O-values ranging from 4 to 23 in a steam environment between 115 and 123°C with readings within 0.5 units of the conventionally calculated F_O. The second physical TTI described in literature is the 'Thermal memory cell' [Swartzel *et al.* 1991], based on the diffusion of ions in the insulator layer of a metal-insulator-semiconductor capacitor. The rearrangement of the ions is easily and accurately read out by measuring the capacitance change of the cell before and after thermal treatment. This change is capacitance can be described as an n-th order reaction with an activation energy depending on the type of mobile charged carriers. The authors try to correlate the respective readouts of at least two of these systems (with different Ea's) via the equivalent point method with the effect of the process at hand on any food property, irrespective of its matching the kinetic characteristics (Ea-value) of the ions used in the TTI.

Both physical TTI-systems cited are described as easy and accurate to prepare and calibrate, user-friendly in read-out and easy to recover. The disadvantage associated with the system described by Witonsky [1977] is that it is activated by steam and can hence not be used to monitor other types of heating media nor be embedded in a solid particle. (Also its size limits this). If the equivalent point method can truly be used to interpolate processing values for one food characteristic (e.g. microbial safety with $z=10$°C) from the status of several other ($z \neq 10$°C) temperature sensitive reactions, the thermal memory cell constitutes a very promising and powerful tool for thermal process design, optimization and evaluation. However, more proof and thorough testing will have to be provided, both on the patented TTI-system and the general validity of the equivalent point method.

C. CONCLUSIONS ON USE AND DEVELOPMENT OF PRODUCT HISTORY INTEGRATORS

The excellent agreement between predicted and measured thiamin retention after heat processing at 121.1°C tempted Mulley, Stumbo and Hunting [Mulley *et al.*, 1975c] to conclude : '*While the whole area of chemical indicators promises to be a fertile ground for research and patent hunters, its practical application may be just beginning. The authors believe that the use of a chemical index in sterilization processing has the potential of effecting a revolutionary change in the food and pharmaceutical industries.*' The recognition that such a monitoring system could be encapsulated so that its kinetic characteristics can be determined independently from the application (composition of food, heating mode, heating technology) made these authors almost two decades ago, the founders of what we consider today to be a TTI. Although several researchers pursuited their legacy, the ultimate Time-Temperature Integrator for evaluating thermal processing of foods has not yet been revealed. Whereas until now calibrated microbiological TTI's have received most interest, the endorsement that from calibrated enzymatic, chemical or physical monitoring systems the same information on the impact of the heat treatment can be obtained in a more feasible way, renewed attention is given to possibilities and restrictions of these systems as Time-Temperature Integrator in thermal processing of foods.

Since all TTI's are by definition *post-factum* indicators of the efficacy of a thermal process, this technique is not suited for on-line monitoring and control of the heat treatment to which a particulate in a rotating retort or continuous process is subjected. Because the present level of technology does not seem to allow the construction of a device suited for this purpose, the exploration of possibilities and restrictions of a wireless TTI-system seems justified. With the basic requirements for the development of such a monitoring system in mind, adequate accomplishment of this task should be possible.

V. GENERAL CONCLUSIONS AND RESEARCH NEEDS

Three approaches can be considered in the evaluation of the impact of preservation processes on foods (in terms of safety and quality).

The further development of the in situ method needs more sensitive and rapid detection methods for evaluation of microbial safety and non-destructive rapid detection methods to assess quality.

The physical-mathematical method is well elaborated in thermal processing but awaits further input on kinetic modelling and kinetic data, thermophysical properties and modelling of transport phenomena to handle the more complex heating behaviour of foods and to extend its applicability to more recently developed preservation techniques.

Product history integrators in general and time temperature integrators in particular are receiving an increased interest and further developments in this area will help to overcome some of the severe restrictions of *in situ* and physical-mathematical methods.

REFERENCES

Adams, J.B., Review : enzyme inactivation during heat processing of food-stuffs, *Int. J. Food Sci. Technol.*, 26(1), 1-20, 1991.

Agerhem, H., Nilson, H., Substrate composition and the use thereof, European patent, 0 019 601, 1982

Ball, C.O., Olson, F.C.W., *Sterilization in Food Technology*, McGraw-Hill Book Company, Inc., New York, USA, 1957.

Ball, C.O., Thermal process time for canned food, Bull. 7-1(37), Natl. Res. Council, Washington D.C., USA, 1923.

Baranyi, J., Roberts, T.A., Mc. Clure, P., A non-autonomous differential equation to model bacterial growth, *Food Microbiol.*, 10, 43-59, 1993.

Bean, P., Dallyn, H., Ranjith, H., The use of alginate spore beads in the investigation of ultra-high temperature processing, *J. Food Technol.*, 281-294, 1979.

Berry, M.F., Singh, R.K., Nelson, P.E., Kinetics of Methylmethionine sulfonium in buffer solutions for estimating thermal treatment of liquid foods, *J. Food Proc. Preserv.*, 13, 475-488, 1989.

Bhattacharjee, H., Activatable time temperature indicator, European patent, 0 231 499, 1985

Bhattacharjee, H., Photoactivatable leuco base time temperature indicator, International patent WO 87 03367, 1987

Bhattacharjee, H.R., Photoactivatable time-temperature indicators for low-temperature applications, *J. Agric. Food Chem.*, 36(3), pp 525-529, 1988.

Bigelow, W.D., Esty, J.R., Thermal death point in relation to time of typical thermophilic organisms, *J. Infect. Dis.*, 27(6), 602-624, 1920.

Bigelow, W.D., The logarithmic nature of thermal death time curves, *J. Infect. Dis.*, 29(5), 528-536, 1921.

Bimbenet, J.J., Duquenoy, A., Simulation mathématique de phénomènes intéressant les industries alimentaires, *Industries alimentaires et agricoles*, 4, 359-365, 1974.

Bird, R.B., Stewart, W.E., Lightfoot, E.N., *Transport Phenomena*, John Wiley & Sons, New York, USA, 1960.

Blixt, K.G., Tiru, M., An enzymatic time/temperature device for monitoring the handling of perishable commodities, *Develop. Biol. Stand.* 36, 237-241, 1976.

Bowman, F.W., The sterility testing of pharmaceuticals, *J. Pharm. Sci.*, 58(11), 1301-1308, 1969.

Brown, J., Buglino, S., Fragile container with rupturing device, US patent, 4 732 850, 1988.

Brown, K.L., Ayers, C.A., Gaze, J.E., Newman, M.E., Thermal destruction of bacterial spores immobilized in food/alginate particles, *Food Microbiol.*, 1, 187-198, 1984.

Bunn, J.L., Sykes, I.K., A chemical indicator for the rapid measurement of F_0-values, *J. Appl. Bacteriol.*, 51, 143-147, 1981.

Burton, H., Perkin, A.G., Davies, F.L., Underwood, H.M., Thermal death kinetics of *Bacillus stearothermophilus* spores at ultra high temperatures, III, Relationship between data from capillary tube experiments and from UHT sterilizers, *J. Food Technol.*, 12, 149-161, 1977.

CFDRA (Campden Food and Drink Research Association), Guidelines for the establishment of scheduled heat transfer processes for low-acid foods, Technical Memorandum No. 3, Campden Food and Drink Research Association, Chipping Campden, Gloucestershire, UK, 1-80, 1977.

Clark, P.J., Mathematical modelling in sterilization processes, *Food Technol.*, 32(3), 73-75, 1978.

Dail, R., Calculation of required hold time of aseptically processed low acid foods containing particulates using the Ball method, *J. Food Sci.*, 50(6), 1703-1706, 1985.

Danielson, N.E., Sterilization process indicators: biological vs chemical, *Med. Instr.*, 16(1), 52, 1982.

Datta, A.K., On the theoretical basis of the asymptotic semilogarithmic heat penetration curves used in food processing, *J. Food Eng.*, 12, 177-190, 1990.

Datta, A.K., Teixeira, A.A., Numerically predicted transient temperature and velocity profiles during natural convection heating of canned liquid foods, *J. Food Sci.*, 53(1), 191-195, 1988.

De Cordt S., Hendrickx M., Maesmans G., Tobback P., Immobilized α-amylase from Bacillus licheniformis : a potential enzymic time-temperature-integrator for thermal processing, *International J. Food Sci. Technol.*, 27, 661-673, 1992

de Ruyter, P.W., Brunet, R., Estimation of process conditions for continuous sterilization of foods containing particulates, *Food Technol.*, 27(7), 44-51, 1973.

Defrise, D., Taeymans, D., Stressing the influence of residence time distribution on continuous sterilization efficiency, Proceedings of the international symposium on 'Progress in Food Preservation Processes', Ceria, Brussels, Belgium, April 12-14, 171-184, 1988.

Dodeja, A.K., Sarma, S.C., Abichandani, H., Thermal death kinetics of *B. stearothermophilus* in thin film scraped surface heat exchanger, *J. Food Process. Preserv.*, 14, 221-230, 1990.

Favetto, G.J., Chirife, J., Scorza, O.C., Hermida, C., Color-changing indicator to monitor the time-temperature history during cooking of meats, *J. Food Protect.*, 51(7), 542-546, 1988.

Favetto, G.J., Chiriffe, J., Scorza, O.C., Hermida, C.A., Time-temperature integrating indicator for monitoring the cooking process of packaged meats in the temperature range of 85-100 degrees Celsius, United States Patent, 4 834 017, 1989.

Fields, S.C., Computerized freshness monitoring system. IIF-IIR Commissions D1, D2, and D3, Orlando, Florida, USA, 319-327, 1985.

Fink, A., Cerny, G., Microbiological principles of short-time sterilization of particulate foods, Proceedings of International Symposium on Progress in Food Preservation Processes, Ceria, Brussels, Belgium, April 12-14, 1988, Vol. 2, 185-190, 1988.

FPI, *Canned foods. Principles of thermal process control, acidification and container closure evaluation*, 5th Ed., The Food Processors Institute, Washington DC, USA, 1989.

Gaze, J.E., Spence, L.E., Brown, G.D., Holdsworth, S.D., Microbiological assessment of process lethality using food/alginate particles, Technical Memorandum No. 580, Campden Food and Drink Research Assoc., 1-47, 1990.

Hallström, B., Skjöldebrand, C., Trägårdh, C., *Heat transfer and food products*, Elsevier Applied Science, Ltd., Barking, Essex, U.K., 1988

Hayakawa, K.-I., A critical review of mathematical procedures for determining proper heat sterilization processes, *Food Technol.*, 32(3), 59-65, 1978.

Hayakawa, K.-I., Estimating temperatures of foods during various heating or cooling treatments. *ASHRAE J.*, Sept., 65-69, 1972.

Hayakawa, K.-I., Mathematical methods for estimating proper thermal processes and their computer implementation, *Adv. Food Res.*, 23, 75-141, 1977.

Hayakawa, K.-I., New parameters for calculating mass average sterilizing value to estimate nutrients in thermally conductive food, *Can. Inst. Food Technol. J.*, 2(4), 165-172, 1969.

Hayakawa, K.-I., Timbers, G., Stier, E., Influence of heat treatment on the quality of vegetables : organoleptic quality, *J. Food Sci.*, 42(5), 1286-1289, 1977.

Hayakawa, K.-I., Timbers, G.E., Influence of heat treatment on the quality of vegetables : changes in visual green colour, *J. Food Sci.*, 42(2), 778-781, 1977.

Hendrickx, M., Zhijun Weng, Maesmans, G., Tobback, P., Validation of a time-temperature-integrator for thermal processing of foods under pasteurization conditions, *Int. J. Food Sci. Technol.*, 27(1), 21-31, 1992.

Heppell, N.J., Measurement of the liquid-solid heat transfer coefficient during continuous sterilization of foodstuffs containing particles, Proceedings of a IUFOST symposium on aseptic processing and packaging of foods, September 9-12, Tylosand, Sweden, 108-114, 1985.

Hermans, W.F., In-flow fraction specific thermal processing (FSTP) of liquid foods containing particulates, Proceedings of the international symposium on 'Progress in Food Preservation Processes', Ceria, Brussels, Belgium, April 12-14, 1207-212, 1988.

Hersom, A.C., Shore, D.T., Aseptic processing of foods comprising sauce and solids, *Food Technol.*, 35(4) 53-62, 1981.

Hiddink, J., Natural convection heating of liquids with reference to sterilization of canned food, Ph. D. Thesis, Landbouwhogeschool Wageningen, the Netherlands, 1975.

Holdsworth, S., Optimization of thermal processing - a review, *J. Food Eng.*, 4, 89-116, 1985.

Holdsworth, S., Use of an alginate particle technique to study continuous flow particle sterilization,. in *'HTST treatment: severity of processing and influencing parameters'*, 4th workshop Cost 91bis, Subgroup 1, Helsinki, November 5, 1987.

Horn, H., Machmerth, R., Witthauer, J., Biologische und Chemische Indikatoren der thermischen Keimtötung und deren möglicher einflüss auf ihrer generelles Konzept, *J. Hygiene. Epidemiol. Micro. Immun.*, 20(2), 164-170, 1976

Hunter, G.M., Continuous sterilization of liquid media containing suspended particles, *Food Technol. Aust.*, April, 158-165, 1972.

Jason, A.C., A deterministic model for monophasic growth of batch cultures of bacteria, *Antonie van Leeuwenhoek J. Microbiol. Serol.*, 49, 513-536, 1983.

Jones, A., Pflug, I., *Bacillus coagulans*, FRR B666, as a potentional biological indicator organism, *J. Parenter. Sci. Technol.*, 35(3), 82-87, 1981.

Jones, A., Pflug, I., Blanchett, R., Effect of fill weight on the F-value delivered to two styles of green beans processed in a Sterilmatic retort, *J. Food Sci.*, 45(2), 217-220, 1980.

Kao, J., Naveh, D., Kopelman, I.J., Pflug, I.J., Thermal process calculations for different z and j_c-values using a hand-held calculator, *J. Food Sci.*, 46(1), 193-197, 1981.

Karle, D., Karlovic, E., Contamination free sterilization indicating system, European Patent, 0 078 112 A2, 1983.

Kelsey, J.C., The testing of sterilizers. *Lancet*, i, 306-309, 1958.

Kim, H.-J., Taub, I.A., Intrinsic chemical markers for aseptic processing of particulate foods, *Food Technol.*, 47(1), 91-99, 1993.

Kumar, A., Bhattacharya, M., Blaylock, J., Numerical simulation of natural convection heating of canned thick viscous liquid food products, *J. Food Sci.*, 55(5), 1403-1411, 1990.

Larkin, J.W., Use of a modified Ball's formula method to evaluate aseptic processing of foods containing particulates, *Food Technol.*, 43(3), 124-131, 1989.

Lee, J.H., Singh, R.K., Larkin, J.W., Determination of lethality and processing time in a continuous sterilization system containing particulates, *J. Food Eng.*, 11(1), 67-92, 1990.

Lund, D. Kinetics of physical changes in foods, in *'Physical and Chemical properties of Food'*, Ed. Okos, M.R., American Society of Agricultural Engineers, St. Joseph, Michigan, USA, 367-381, 1986.

Lund, D.B., Food processing : from art to engineering, *Food Technol.*, 43(9), 242-247, 1989.

Lund, D.B., Heat processing, in *'Principles of Food Science, Part II, Physical principles of Food Preservation.'* Eds. Karel, M., Fennema, O.R., Lund, D.B.. Marcel Dekker Inc., New York, NY, USA, 1975.

Maesmans, G., Hendrickx, M., De Cordt, S., Fransis, A., Tobback, P., Fluid-to-particle heat transfer coefficient determination of heterogeneous foods: a review, *J. Food Process. Preserv.*, 16, 29-69, 1992.

Maesmans, G., Hendrickx, M., De Cordt, S., Tobback, P., Theoretical considerations on the general validity of the equivalent point method, Accepted for publication in *J. Food Eng.*, 1993a.

Maesmans, G., Hendrickx, M., De Cordt, S., Tobback, P., Theoretical considerations on the design of multi-component time temperature integrators in the evaluation of thermal processes, Accepted for publication in *J. Food Proc. Preserv.*, 1993b.

Maesmans, G., Possibilities and limitations of thermal process evaluation techniques based on time temperature integrators, Ph.D.-thesis, Centre for Food Science and Technology, Unit Food Preservation, KULeuven, Leuven, Belgium, 1993.

Malkki, Y., Research priorities in food processing. Proc. 7th international congress on food science and technology, Singapore, September 27-October 2, 1987.

Matthiasson, E., Gudjonsson, T., Indicators for estimation of the time-temperature distribution in heat treatment of particulate food, ITI, Reykjavik, 1991.

May, N., Cossey, R., Review of data acquisition units for in-container temperature measurement and initial evaluation of the Ball datatrace system- part 1, Technical Memorandum No. 567, 1-50, Campden Food and Drink Research Association, Chipping Campden, Gloucestershire, UK., 1991.

Mc Cormick, P., Scoville, J., Biological indicator for sterilization processes, UK patent, 2 186 974A, 1987.

McCormick, P.J., Brief Report : Biological Indicators, *Infect. Cont. Hosp. Epidemiol.*, 9(11), 504-507, 1988.

McMeekin, T.A., Olley, J.N., Ross, T., Ratkowsky, D.A., *Predictive Microbiology : theory and application*, John Wiley & Sons Inc., New York, 1993.

Mertens, B., Knorr, D., Developments of nonthermal processes for food preservation, *Food Technol.*, 46(5), 124-133, 1992.

Mulley, A., Stumbo, C., Hunting, W., Kinetics of thiamine degradation by heat : A new method for studying reaction rates in model systems and food products at high temperatures, *J. Food Sci.*, 40(5), 985-988, 1975a.

Mulley, A., Stumbo, C., Hunting, W., Kinetics of thiamine degradation by heat : Effect of pH and form of the vitamin on its rate of destruction, *J. Food Sci.*, 40(5), 989-992, 1975b.

Mulley, A., Stumbo, C., Hunting, W., Thiamine : a chemical index of the sterilization efficacy of thermal processing, *J. Food Sci.*, 40(5), 993-996, 1975c.

Myers, R.B., Practical system for validating heat sterilization processes, *J. Parent. Drug Assoc.*, 32(5), 216-225, 1978.

Nunes, R.V., Swartzel, K.R., Modelling thermal processes using the equivalent point method, *J. Food Eng.*, 11, 103-117, 1990.

Orelski, P., Biological indicator unit for thermal process evaluation, US Patent, 4 291 122, 1981.

Patel, G.N., Yee, K.C., Diacetylene time-temperature indicators, United States Patent, 4 228 126, 1980.

Pflug, I.J., Berry, M.R., Dignan, D.M., Establishing the heat preservation process for aseptically packaged low acid foods containing large particulates, sterilized in a continuous heat-hold-cool system, *J. Food Protect.*, 53(4), 312-321, 1990.

Pflug, I.J., Jones, A., Blanchett, R., Performance of bacterial spores in a carrier system in measuring the F_0-value delivered to cans of food heated in a Steritort, *J. Food Sci.*, 45(4), 940-945, 1980a.

Pflug, I.J., Method and apparatus for sterility monitoring, US patent, 3 960 670, 1976.

Pflug, I.J., Odlaug, T.E., A review of z- and F-values used to ensure the safety of low-acid canned food, *Food Technol.*, 32(6), 63-70, 1978.

Pflug, I.J., Odlaug, T.E., Biological indicators in the pharmaceutical and medical device industry, *J. Parenteral Sci.Technol.*, 40(5), 242-248, 1986.

Pflug, I.J., Smith, G., Holcomb, R., Blanchett, R., Measuring sterilizing values in containers of food using thermocouples and biological indicator units, *J. Food Prot.*, 43(2), 119-123, 1980b.

Pflug, I.J., Smith, G.M., The use of biological indicators for monitoring wet heat sterilization processes, in *'Sterilization of Medical Products'*, Eds. Gaughran, E.R.L. and Kereluk, K., Johnson and Johnson, New Brunswick, N.J., 193-230, 1977.

Pflug, I.J., *Textbook for an introductory course in the microbiology and engineering of sterilization processes*, Sixth edition, Environmental Sterilization Laboratory, 100 Union St., Minneapolis, MN 55455, USA, 1987.

Philipp, B. Sucker, H., Heat sterilization of bioindicators in propylene glycol and propylene glycol-water mixtures : Arrhenius equation, thermodynamic data and z-values, *Pharm. Res.*, 7(12), 1273-1277, 1990.

Philipp, B., Sucker, H., Heat sterilization of bioindicators in propylene glycol and propylene glycol water-mixtures, *Pharm. Ind.*, 50(3), 360-363, 1988.

Pivick, K., Petrasoviks, A., A rationale for the safety of canned shelf-stable cured meat: protection=destruction+inhibition, in *19th reunion Europeenne des Rechercheurs de Viande*, Paris, France, 1973.

Ratkowsky, D.A., Ross, T., McMeekin, T.A. and Olley, J., Comparison of Ahrrenius-type and Belehradek-type models of bacterial growth in foods, *J. of Appl. Bacteriol.*, 71, 452-459, 1991.

Richardson, P.S., Microwave technology - the opportunity for food processors, *Food Sci. Technol. Today*, 5(3), 146-148, 1991.

Rodriguez, A.C., Teixeira, A.A., Heat transfer in hollow cylindrical rods used as bioindicator units for thermal process validation, *Trans. ASAE*, 31(4), 1233-1236, 1988.

Rönner, U., A new biological indicator for aseptic sterilization, *Food Technology International Europe*, 90, 43, 45-46, 1990a.

Rönner, U., Bioindicator for control of sterility, *Food Laboratory News*, 22,(6:4), 51-54, 1990b.

Sadeghi F. and Swartzel K.R, Time temperature equivalence of discrete particles during thermal processing, *J. Food Sci.*, 55(6), 1696-1698,1739, 1990.

Sastry, S.K., Li, S.F., Patel, P., Konanayakam, M., Bafna, P., Doores, S., Beelman, R.B., A bioindicator for verification of thermal processes for particulate foods, *J. Food Sci.*, 53(5), 1528-1531, 1988.

Sastry, S.K., Mathematical evaluation of process schedules for aseptic processing of low-acid foods containing discrete particulate matter, *J. Food Sci.*, 51(5), 1323-1328, 1986.

Skaug, N. and Berube, R., Comparative thermoresistance of two biological indicators for monitoring steam autoclaves, I. Comparison performed in a gravity BIER/Steam vessel, *Acta path. microbiol. immunol. scand. Sect. B 91*, 435-441, 1983.

Skudder, P.J., Development of the ohmic heating process for continuous sterilisation of particulate food products. Proceedings of the international symposium on 'Progress in Food Preservation Processes', Ceria, Brussels, Belgium, April 12-14, 267-270, 1988.

Smith, G., Kopelman, M., Jones, A., Pflug, I., Effect of environmental conditions during heating on commercial spore strip performance, *Appl. Env. Microbiol.*, 44, 12-18, 1982.

Smith, G., Pflug, I.J., Chapman, P., Effect of storage time and temperature on ther variation among replicate tests (on different days) on the performance of spore disks and strips, *Appl. Env. Microbiol.*, 32, 257-263, 1976.

Steele, R., Board, P., Ammendments to Ball's formula method for calculating the lethal value of thermal processes, *J. Food Sci.*, 44(1), 292-293, 1979.

Steele, R., Board, P., Best, D., Willcox, M., Revision of the formula method tables for thermal process evaluation, *J. Food Sci.*, 44(2), 954-957, 1979.

Stoforos, N.G., Heat transfer in axially rotating canned liquid/particulate food systems, Ph.D. Thesis, University of California, Davis, CA, USA, 1988.

Stoforos, N.G., Merson, R.L., Physical property and rotational speed effects on heat transfer in axially rotating liquid/particulate canned foods, *J. Food Sci.*, 57(3), 749-754, 1992.

Struppe, H.F., Untersuchungen zur Eignung von Sterilisationsindikatoren für Schnellautoklaven, *Archiv. Hyg. Bacteriol.*, 152(4), 360-365, 1968.

Stumbo, C.R., *Thermobacteriology in food processing*, Academic Press, New York, NY, USA, 1973.

Swartzel, K.R., Arrhenius kinetics as applied to product constituent losses in ultra high temperature processing, *J. Food Sci.*, 47(6), 1886-1891, 1982.

Swartzel, K.R., Equivalent-point method for thermal evaluation of continuous-flow systems, *J. Agric. Food Chem.*, 34, 396-401, 1986.

Swartzel, K.R., Ganesan, S.G., Kuehn, R.T., Hamaker, R.W., Sadeghi, F., Thermal memory cell and thermal system evaluation, US Patent 5,021,981, 1991.

Taoukis, P.S., Labuza, T.P., Applicability of time-temperature indicators as shelf life monitors of food products, *J. Food Sci.*, 54(4), 783-788, 1989a.

Taoukis, P.S., Labuza, T.P., Reliability of time-temperature indicators as food quality monitors under nonisothermal conditions, *J. Food Sci.*, 54(4), 789-792, 1989b.

Teixeira, A., Dixon, J., Zahradnik, J., Zinsmeister, G., Computer determination of spore survival distributions in thermally-processed conduction-heated foods, *Food Technol.*, 23(3), 78-80, 1969.

Teixeira, A., Manson, J., Thermal process control for aseptic processing systems, *Food Technol.*, 37(4), 128-133, 1983.

Teixeira, A.A., Thermal process calculations. Chapter 11 in '*Handbook of Food Engineering*', Eds. Heldman, D.R and Lund, D.B., Marcel Dekker Inc, New York, USA, 563-620, 1992.

Thompson, D.R, Norwig, J., Microbial populations, enzyme and protein changes during processing, in '*Physical and Chemical properties of Food*', Ed. Okos, M.R., American Society of Agricultural Engineers, St. Joseph, Michigan, USA, 202-265, 1986.

Tobback, P., Hendrickx, M., Van Tongelen, J., Van Pottelbergh, E., Heat inactivation of peroxidase in Brussels sprouts (*Brassica oleracea*): a comparison of different cultivars. Paper presented at the 7th World Congress on Food Science and Technology, Singapore, September 28- October 2., 1987.

Tobback, P., Hendrickx, M.E., Weng, Z., Maesmans, G.J., De Cordt, S.V., The use of immobilized enzymes as Time-Temperature Indicator system in thermal processing, in *'Advances in Food Engineering'*, Eds. Singh, R.P. and Wirakartakusumah, M.A., CRC Press, Boca Raton, Florida, USA, 561-574, 1992.

Van Impe, J.F., Nicolaï, B.M., Martens, T., de Baerdemaeker, J., Vandewalle, J., Dynamic mathematical model to predict microbiological growth and inactivation during food processing, *Appl. Env. Microbiol.*, 58, 2901-2909, 1992.

Villota, R., Hawkes, J.G., Kinetics of nutrients and organoleptic changes in foods during processing, in *'Physical and Chemical properties of Food'*, Ed. Okos, M.R., American Society of Agricultural Engineers, St. Joseph, Michigan, USA, 266-366, 1986.

Villota, R., Hawkes, J.G., Reaction kinetics in food systems, Chapter 2 in *'Handbook of Food Engineering'*, Eds. Heldman, D.R and Lund, D.B., Marcel Dekker Inc, New York, USA, 39-144, 1992.

Vinters, J.E., Patel, R.H., Halaby, G.A., Thermal process evaluation by programmable computer calculator, *Food Technol.*, 29(3), 42-48, 1975.

Wells, J.H. and Singh, R.P., Application of time-temperature indicators in monitoring changes in quality attributes of perishable and semiperishable foods, *J. Food Sci.*, 148-156, 1988

Wells, J.H., Singh, R.P., Response characteristics of full-history time-temperature indicators suitable for perishable food handling, *J. Food Process. Pres.*, 12, 207-218, 1988.

Welsh, J.D., Dyke, D.G., Disposable biological indicator test pack for monitoring steam and ethylene oxide sterilization cycles, United States Patent, 4 839 291, 1989.

Wen Chin, L., Disaccharade hydrolysis as a predictive measurement for the efficacy of heat sterilization in canned foods, Ph. D. thesis, department of Food Science and Nutrition, University of Massachusetts, Amhers, Massachusetts, USA, 1977.

Weng Zhijun, A time-temperature integrator for thermal processing of foods: a case study on immobilized peroxidase, Ph.D.-thesis, Centre for Food Science and Technology, Unit Food Preservation, KULeuven, Leuven, Belgium, 1991.

Weng Zhijun, Hendrickx, M., Maesmans, G., Gebruers, K., Tobback, P., Thermostability of soluble and immobilized horseradish peroxidase, *J. Food Sci.*, 56(2), 574-578, 1991a.

Weng Zhijun, Hendrickx, M., Maesmans, G., Tobback, P., Immobilized peroxidase: a potential bioindicator for evaluation of thermal processes, *J. Food Sci.*, 56(2), 567-570, 1991b.

Willocx, F., Mercier, M., Hendrickx, M. and Tobback P., Modelling the influence of temperature and carbon dioxide upon the growth of *Pseudomonas fluorescens*, *Food Microbiol.* 10, 159-173, 1993.

Wilson, L.A., Kinetics of flavor changes in foods, in *'Physical and Chemical properties of Food'*, Ed. Okos, M.R., American Society of Agricultural Engineers, St. Joseph, Michigan, USA, 382-407, 1986.

Wittonsky, R.J., A new tool for the validation of the sterilization of parenterals, *Bull. Parenter. Drug Assoc.*, 11(6), 274-281, 1977.

Yawger, E.S., Bacteriological evaluation for thermal process design, *Food Technol.*, 32(6), 59-62, 1978.

Zall, R., Chen, J., Fields, S.C., Evaluation of automated time temperature monitoring system in measuring freshness of UHT milk, *Dairy and Food Sanitation*, 6(7), 285-290, 1986.

Zwietering, M.H., De Koos J.T., Hasenack, B.E., De Wit J.C. and Van 't Riet K., Modeling of bacterial rowth as a function of temperature, *Appl. Env. Microbiol.*, 57(4), 1094-1101, 1991.

Zwietering, M.H., Jongenburger, I., Rombouts, F.M., and Van 't Riet K., Modeling of bacterial growth curve, *Applied and Environmental Microbiology*, 56(6), 1875-1881, 1990.

DISCUSSION

CAMPBELL-PLATT: How widely used are TTI's on retail packs?

HENDRICKX: The use of there TTI's (I-POINT, Fresh Check, Fresh-Scan) is rather limited. The major technical barrier towards the practical implementation is a lack on identification of quality targets and their kinetic properties.

BUSTA: Are you proposing these TTI enzymatic systems be used as substitutes for microbial inoculated packs in process biological verification.

HENDRICKX: The TTI's for thermal processing can be looked at as a kind of biological thermocouple (see Pflug 1987). Anyone accepting process verification both on temperature measurement (and subsequent interaction in F_0-value) should accept a TTI-approach. After all, the inoculate pack is also carried out with well calibrated spores with known kinetic properties. A well-defined TTI with the correct kinetic properties will exactly integrate the lethal effect of temperature.

OBTAINING A WELL BALANCED PRODUCT QUALITY IN THERMALLY PROCESSED CONDUCTION HEATING FOODS BY ANALYZING SURFACE AND VOLUME AVERAGE QUALITY OPTIMUM PROCESSING CONDITIONS

Cristina L.M. Silva, Fernanda A.R. Oliveira, and Mark Hendrickx

I. ABSTRACT

Sterilization temperatures that maximize each of the two objective functions - surface or volume average quality - were calculated as a function of the relevant influential variables using computer modeling. The resulting volume average and surface quality retentions were both determined for each case study. The effect of using sterilization conditions defined for optimal surface quality on volume average quality and the effect of using sterilization temperatures defined for optimum volume average quality on surface quality were qualitatively and systematically investigated using the statistical 2^n factorial design method. If the degradation kinetics for the quality factors considered at the surface and in volume average terms are similar, the use of sterilization temperatures for maximizing surface quality is suggested, when the aim is to obtain a product with a well balanced maximum final quality. On the other hand, if the quality factors have different degradation kinetics, the use of optimum sterilization conditions maximizing the quality at the surface or in volume average terms has to be analyzed on a case by case basis.

II. INTRODUCTION

The criteria to optimize the sterilization conditions for maximizing the quality of a given food product should not be understood only in terms of surface or volume average quality retention. Quality at the surface is more critical for primary sensorial evaluation, while volume average quality is also very important for quality factors such as total nutritional value. Therefore, it would be useful to consider both criteria. Although several research work has been carried out to calculate optimal sterilization conditions for minimizing quality degradation [Silva *et al.*, 1993], an objective function taking into consideration the maximization of both volume average and surface quality is yet to be developed.

With the exception of Ohlsson [1980a, 1980b, 1980c] and Banga *et al.* [1991], few authors compared average and surface quality optimum conditions. Ohlsson [1980a, 1980b] stated that for infinite slab and finite cylinder geometries optimum temperatures for minimum average cook value were approximately 7.5 to 10°C and 2.5°C above the ones for minimum surface quality degradation, respectively. However, the range of cases studies carried out was very limited and the objective function used to maximize the average quality retention was not the most suitable [Silva *et al.*, 1992a]. Furthermore, the case studies assumed the same degradation kinetics both for the quality factor at the surface and in volume average terms. Banga *et al.* [1991] studied the influence of optimum variable retort temperature profile on average quality and concluded that this policy does not offer significant advantages over constant retort temperature. On the contrary, when an optimized variable retort temperature is applied the surface quality can be significantly improved.

Studies on the effect of using optimal conditions for maximizing surface quality on the final volume average quality, or vice-versa, have not been reported. Furthermore, there is a need to quantify thoroughly how these effects are influenced by the thermal diffusivity, dimensions and geometry of the product, z and $D_{121.15°C}$ values for the quality factor degradation kinetics, surface heat transfer coefficient and target sterility value (F_t).

The main objective of this work was a systematic study of the influence of the relevant variables on: 1) the difference between optimal sterilization temperatures that maximize each objective function, 2) the effect of using optimum surface quality sterilization conditions on the final volume average quality and 3) the effect of using optimum volume average quality sterilization conditions on the final surface quality.

III. METHODOLOGY

The influence of several variables on three different responses was investigated. The variables considered (listed in Table 1) are: the RF number (i.e. the ratio between the thermal diffusivity and the square of the characteristic dimension), z and D values for the quality factor degradation kinetics (z_q and $D_{121.15°C}$), Biot number, target sterility value at the 'coldest' or least-lethality point (F_t) and geometry. The difference between optimal sterilization temperatures for maximizing surface quality $[(T_{op})^{surf}]$ and volume average quality $[(T_{op})^{ave}]$ [Silva *et al.*, 1992a], the decrease (in relation to the optimum value) in volume average quality when sterilization conditions for maximizing surface quality are used, and finally the decrease (in relation to the optimum value) in surface quality when sterilization conditions for maximizing volume average quality are used, were the parameters used to quantify the responses:

$$P_1 = (T_{op})^{surf} - (T_{op})^{ave} \qquad (1)$$

$$P_2 = [(N/N_o)_{ave}^{optimum\ surface} - (N/N_o)_{ave}^{optimum\ average}] \times 100 \qquad (2)$$

$$P_3 = [(N/N_o)_{surf}^{optimum\ average} - (N/N_o)_{surf}^{optimum\ surface}] \times 100 \qquad (3)$$

with

$(N/N_o)_{ave}^{optimum\ surface}$ — volume average quality retention for sterilization conditions that maximize surface quality

$(N/N_o)_{surf}^{optimum\ average}$ — surface quality retention for sterilization conditions that maximize volume average quality

$(N/N_o)_{ave}^{optimum\ average}$ — volume average quality retention for optimum sterilization conditions

$(N/N_o)_{surf}^{optimum\ surface}$ — surface quality retention for optimum sterilization conditions

The parameter P_1 is always negative, since sterilization temperatures for maximizing volume average quality are always higher than the ones for maximizing quality at the surface. To maximize quality in terms of volume average quality the sterilization conditions must be more rigorous in order to accelerate the heat transfer into the food and decrease the lag times through the whole product. Therefore, although the use of an higher temperature implies a decrease in surface quality, the resulting volume average value is maximized. The parameters P_2 and P_3 may vary from 0 to -100%, meaning a quality retention not sensitive or with maximum sensitivity to sterilization conditions,

Table 1
Levels of the Variables Used to Calculate
Optimal Sterilization Temperatures

Variable	Minimum	Maximum
	(-)	(+)
RF (s^{-1})	2.844×10^{-5}	6.400×10^{-3}
z_q (°C)	15	45
$D_{121.15°C}$ (min)	5	900
Bi	5	∞
F_t (min)	3	15
Geometry	sphere	infinite slab

respectively. The quality retentions were evaluated both at the surface and in terms of volume average assuming quality factors with the same degradation kinetics.

A factorial design at two levels [Box *et al.*, 1978; Walpole and Myers, 1993] was used to evaluate the influence of the variables (in Table 1) on the parameters P_1, P_2 and P_3. This statistical method has the advantages of: i) evaluating the influence of varying all variables simultaneously over a wide range of conditions, ii) estimating the influence of all possible interactions among the variables and iii) optimizing the number of case studies required [Gacula and Singh, 1984; Diamante and Munro, 1991]. This method was used in several works. Bhowmik and Hayakawa [1983], for example, applied it to evaluate the influence of can size, food consistencies, retort temperatures, initial food temperature and target lethalities on steam consumption and mass average quality degradation.

Optimal sterilization temperatures for maximizing surface

$$(N/N_o)_{surf} = 10^{-\left(\int_0^t 10^{(T_{surf}-121.15°C)/z_q} \, dt\right) / D_{121.15°C}} \tag{4}$$

or volume average quality retention

$$(N/N_o)_{ave} = \frac{1}{V_T} \int_0^{V_T} 10^{\left[-\frac{1}{D_{121.15°C}} \int_0^t 10^{(T-T_{121.15°C})/z_q} \, dt\right]} dV \tag{5}$$

were calculated as a function of the variables listed in Table 1 using the computer program described by Hendrickx *et al.* [1993] and Silva *et al.* [1992b]. For each objective function (surface or volume average quality) 64 case studies ($2^6 = 64$) were executed, therefore combining the six variables at two levels. For each case study the surface and volume average quality retentions (Equations 4 and 5) were both calculated.

IV. RESULTS AND DISCUSSION

The estimated influences of the six variables and their interactions in the parameters P_1, P_2 and P_3 were calculated according to the procedures specified in Box *et al.* [1978] and Walpole and Myers [1993] for non replicated experiments. A total of 63 influences had to be considered: 6 simple, 15 double, 20 triple, 15 quadruple, 6 quintuple and 1 sextuple effects. These influences were plotted on normal probability paper (Figures 1, 2 and 3) to analyze their importance. Each point in a normal probability paper represents the individual or joint effects. Points that lie on a straight line are normally distributed and these influences are attributed to random variations or chance. Points that are not fitted by a straight line represent a significant effect [Box *et al.*, 1978; Diamante and Munro, 1991].

To analyze the 2^6 factorial study, a regression approach to the analysis of variance (as specified in Walpole and Myers [1993]) was also carried out using the software STATVIEW [Anonymous, 1992] (Table 2). The analysis of variance model can be treated as a special case of a multiple linear regression model. The dependent variable is the parameter to be analyzed (i.e. P_1, P_2 or P_3) and the independent variables are the influential variables and their interactions. A t-test is then applied to each coefficient to evaluate the hypothesis that the regression is insignificant for each coefficient. If the probability of a coefficient to be equal to zero is smaller than 5% that interaction is considered to affect significantly the dependent variable [Walpole and Myers, 1993].

Figures 4, 5 and 6 present the extension of the most important effects (with 0.05% level of significance) in the parameters P_1, P_2 and P_3, respectively. Individual effects are represented as straight lines, double effects as square edges and triple effects as cube edges. Bold lines represent the magnitude of the variation of the parameter. The more significant the effect, the longer the line. A line drawn from left to right or from bottom to top means a positive effect, i.e. an increase in the parameter, while a line drawn from right to left or from top to bottom means a negative effect, i.e. a decrease in the parameter. The - sign represents the low level of the variable and the + sign the high level. Small numbers indicate the value of the parameter on that corner and the bold numbers the variation in the parameter from one corner to another.

A. COMPARATIVE STUDY BETWEEN OPTIMAL TEMPERATURES

Figure 1 shows the normal probability plot for parameter P_1. The points that fall outside the normal distribution represent the most important influences. The more relevant effects (with 0.05% level of significance) were the following: i) single - $D_{121.15°C}$, RF, Geometry, ii) double - $D_{121.15°C}$/RF, $D_{121.15°C}$/z_q and iii) triple - $D_{121.15°C}$/z_q/Geometry interactions (see Figure 1 and Table 2).

Figure 4 represents the extension of the most important effects in the difference between optimal temperatures for maximizing surface and volume average quality (parameter P_1). The modulus of P_1 decreases with $D_{121.15°C}$ and RF and increases with Geometry (i.e. from a sphere to an infinite slab). Products that have an high heat transfer rate (i.e. large RF number - corresponding to large thermal diffusivity or/and small characteristic length, or spherical shape), where the temperature distribution over the volume approaches the temperature at the surface a bit more, have smaller differences between optimal temperatures for maximizing surface or volume average quality. Products for which the quality factor to optimize is very resistant to thermal degradation (i.e. with high $D_{121.15°C}$) also show a smaller difference between optimal temperatures for maximizing surface or volume average quality.

The decrease of the difference between $(T_{op})^{ave}$ and $(T_{op})^{surf}$ with $D_{121.15°C}$ depends on RF, being particularly significant if the product heats slowly (i.e. small RF) and almost negligible if the heating is fast. Similarly, if the quality factor is not resistant to thermal degradation (i.e. low $D_{121.15°C}$) the heating rate has a very important effect, while for high $D_{121.15°C}$ this effect is also negligible. The effect is minimized for products with large heating rates (i.e. large RF) and very resistant to thermal degradation (i.e. large $D_{121.15°C}$).

Table 2
Regression Approach to Analysis of Variance on Parameters P_1, P_2 and P_3.

Source of Variation	t - Value for P_1[a]	Significance for P_1[b]	t - Value for P_2[a]	Significance for P_2[b]	t - Value for P_3[a]	Significance for P_3[b]
$D_{121.15°C}$	13.45	***	3.02	**	-6.27	***
RF	10.86	***	-3.43	**	4.59	***
F_t	-1.45	—	0.43	—	1.17	—
z_q	-3.80	**	-5.71	***	-7.31	***
Bi	-0.04	—	-2.64	*	-1.78	—
Geometry	-8.36	***	-7.07	***	-8.42	***
$D_{121.15°C}$/RF	-8.92	***	12.30	***	16.99	***
$D_{121.15°C}$/F_t	0.77	—	-2.51	*	-4.90	***
$D_{121.15°C}$/z_q	-4.42	***	0.03	—	-4.79	***
$D_{121.15°C}$/Bi	2.00	—	0.97	—	-1.33	—
$D_{121.15°C}$/Geometry	2.13	*	0.70	—	-8.36	***
RF/F_t	0.10	—	2.20	*	4.28	***
RF/z_q	-0.22	—	0.02	—	5.24	***
RF/Bi	0.29	—	-1.08	—	0.81	—
RF/Geometry	0.60	—	-0.98	—	7.14	***
F_t/z_q	2.41	*	-1.18	—	-0.49	—
F_t/Bi	-0.05	—	0.16	—	-0.04	—
F_t/Geometry	1.07	—	0.46	—	0.71	—
z_q/Bi	0.46	—	-1.05	—	-0.29	—
z_q/Geometry	0.35	—	-3.82	**	-2.90	**
Bi/Geometry	-0.10	—	-1.34	—	0.30	—
$D_{121.15°C}$/RF/F_t	0.40	—	-0.53	—	-1.79	—
$D_{121.15°C}$/RF/z_q	2.01	—	5.97	***	7.77	***
$D_{121.15°C}$/RF/Bi	-0.72	—	2.50	*	1.26	—
$D_{121.15°C}$/RF/Geometry	0.91	—	6.74	***	7.20	***
$D_{121.15°C}$/F_t/z_q	-2.98	**	1.16	—	2.26	*
$D_{121.15°C}$/F_t/Bi	0.32	—	-0.48	—	0.35	—
$D_{121.15°C}$/F_t/Geometry	-1.62	—	-1.59	—	-2.81	*
$D_{121.15°C}$/z_q/Bi	1.57	—	0.45	—	0.97	—
$D_{121.15°C}$/z_q/Geometry	-4.13	***	-0.35	—	-5.55	***
$D_{121.15°C}$/Bi/Geometry	0.99	—	0.40	—	-1.25	—
RF/F_t/z_q	-0.26	—	-0.92	—	-1.89	—
RF/F_t/Bi	-0.20	—	0.41	—	-0.56	—
RF/F_t/Geometry	-1.61	—	1.46	—	2.38	*
RF/z_q/Bi	0.28	—	-0.40	—	-0.72	—
RF/z_q/Geometry	-0.20	—	0.38	—	5.79	***
RF/Bi/Geometry	-0.52	—	-0.45	—	0.92	—
F_t/z_q/Bi	0.25	—	-0.29	—	0.92	—
F_t/z_q/Geometry	-1.00	—	-0.29	—	0.69	—
F_t/Bi/Geometry	0.23	—	0.03	—	0.03	—
z_q/Bi/Geometry	-0.03	—	-1.00	—	0.39	—

[a] All quadruple and higher interactions were pooled together to estimate a residual for computing t-value. The value of the estimate of variance for this residual was 35.73, 1.78 and 1.32 respectively for P_1, P_2 and P_3. These residuals have 22 degrees of freedom.

[b] Single, double and triple asterisks respectively signify 5%, 1% and 0.05% levels of significance. Sources without any asterisk did not significantly influence the parameters P_1, P_2 and P_3.

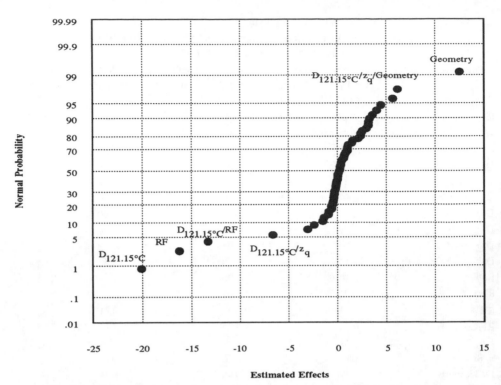

Figure 1. Normal probability plot for the estimated effects in the parameter P_1 (Equation 1).

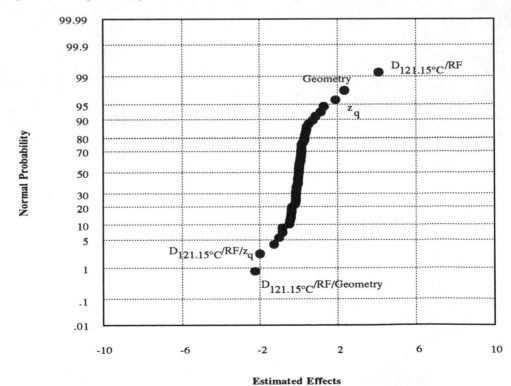

Figure 2. Normal probability plot for the estimated effects in the parameter P_2 (Equation 2).

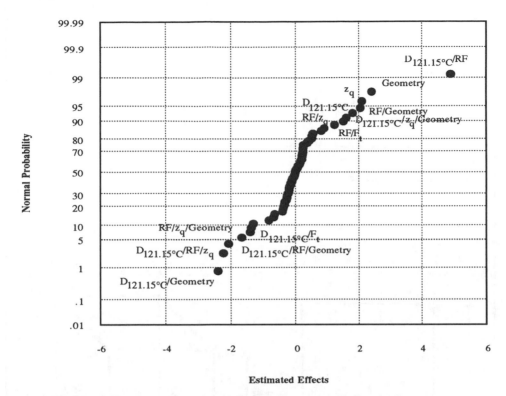

Figure 3. Normal probability plot for the estimated effects in the parameter P_3 (Equation 3).

There is an interaction between the variables: $D_{121.15°C}$ and z_q, but it may be included in the triple interaction between Geometry, z_q and $D_{121.15°C}$ that is also observed. The difference between optimal temperatures is always higher for an infinite slab geometry. This means that products with geometries resulting in a decrease of the heat transfer rate to the slowest heating point show larger differences between optimal sterilization conditions. However, this effect is more important when the z_q and $D_{121.15°C}$ values are both low or high. For a spherical shape $[(T_{op})^{ave} - (T_{op})^{surf}]$ increases with z_q and decreases with $D_{121.15°C}$ and the magnitude of these variations does not depend on $D_{121.15°C}$ or z_q, respectively. However, this type of behaviour is not observed for infinite slab geometry products. The modulus of P_1 continues to decrease with $D_{121.15°C}$, but is more significant when the quality factor is more sensitive to temperature variations (i.e. for low z_q). Furthermore, the difference between optimal temperatures increases with z_q for quality factors with low thermal resistance to thermal degradation (i.e. large $D_{121.15°C}$) and decreases with z_q if the quality factor is very thermal resistant.

The Biot number and F_t value do not affect significantly the difference between optimal sterilization temperatures. This is an important result, since it implies that the considerations made above apply equally independent of using a still or a rotational retort and of the target microorganism lethality.

B. COMPARATIVE STUDY BETWEEN OPTIMAL QUALITY RETENTIONS

The parameters P_2 and P_3, defined in Equations 2 and 3, allow for the evaluation of the decrease in volume average and surface quality, when temperatures for maximizing surface and average quality are used, respectively. Figures 2 and 3 represent the normal probability plots for each parameter.

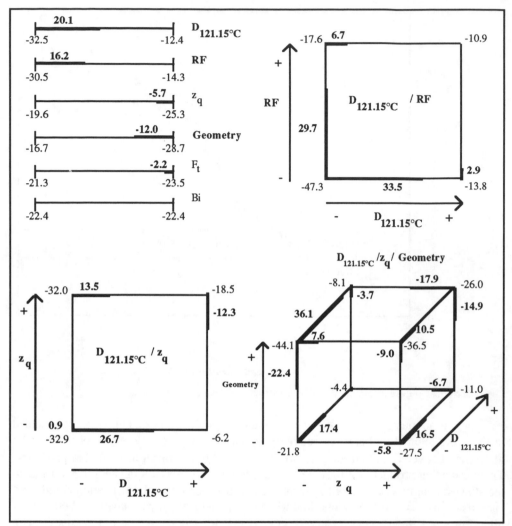

Figure 4. Extension of the most important effects in the parameter P_1 (Equation 1).

The parameters P_2 and P_3 are both affected by the following: i) single - z_q, Geometry, ii) double - $D_{121.15°C}$/RF and iii) triple - $D_{121.15°C}$/RF/z_q, $D_{121.15°C}$/RF/Geometry interactions with 0.05% level of significance (see Figures 2 and 3 and Table 2). The parameter P_3 is also affected by other: i) single - $D_{121.15°C}$, RF, ii) double - $D_{121.15°C}$/F_t, $D_{121.15°C}$/z_q, $D_{121.15°C}$/Geometry, RF/F_t, RF/z_q, RF/Geometry and iii) triple - $D_{121.15°C}$/z_q/Geometry, RF/z_q/Geometry interactions.

Figures 5 and 6 show the extension of the most important effects (with 0.05% level of significance) in the parameters P_2 and P_3, respectively. The decrease in quality (modulus of P_2 and P_3) is larger for products with quality factor kinetics not very sensitive to temperature variations (i.e. for higher z_q values) and with geometries with slower heat transfer rate (i.e. infinite slab geometries). These conditions correspond also to case studies where the difference between optimal sterilization temperatures (i.e. modulus of P_1) is larger. However, the magnitude of the surface quality decrease (modulus of P_3) is more significant than for volume average quality (modulus of P_2). The decrease in surface quality (parameter P_3) is also affected by the $D_{121.15°C}$ and RF variables. Products with slow

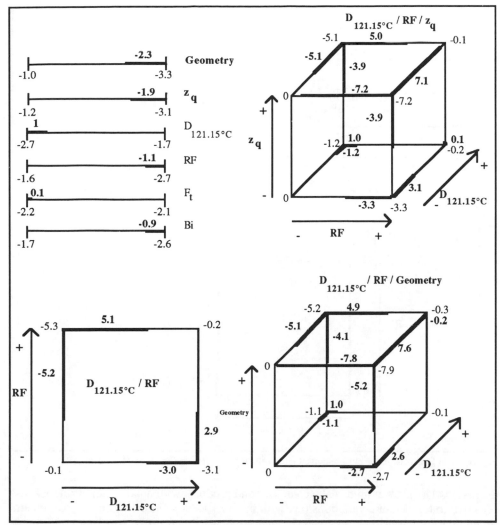

Figure 5. Extension of the most important effects in the parameter P_2 (Equation 2).

heating rate (i.e. small RF) and more resistant to thermal degradation (i.e. large $D_{121.15°C}$) show, in average terms, larger decreases in surface quality when conditions for maximizing overall quality are used.

The decrease in quality (parameters P_2 and P_3) is affected by the interaction $D_{121.15°C}/RF$. However, there are two triple effects: $D_{121.15°C}/RF/z_q$, $D_{121.15°C}/RF/Geometry$. The variables RF and $D_{121.15°C}$ are common to both triple effects and show the same interaction behaviour, independently of z_q or Geometry. When RF and $D_{121.15°C}$ are both at the lower or higher level, the decrease in quality is insignificant. These situations correspond to case studies with very low or very large final quality retention, respectively, and where final quality is not sensitive to sterilization conditions. However, when RF is large (i.e. products with high heating rate) and $D_{121.15°C}$ is low (i.e. quality factor with rapid degradation kinetics) or vice-versa, the decrease in quality is maximized. Furthermore, the modulus of P_2 and P_3 increase or decrease with $D_{121.15°C}$ if RF is low or high, respectively, and increase or decrease with RF if $D_{121.15°C}$ is low or high, respectively. The magnitude of the quality

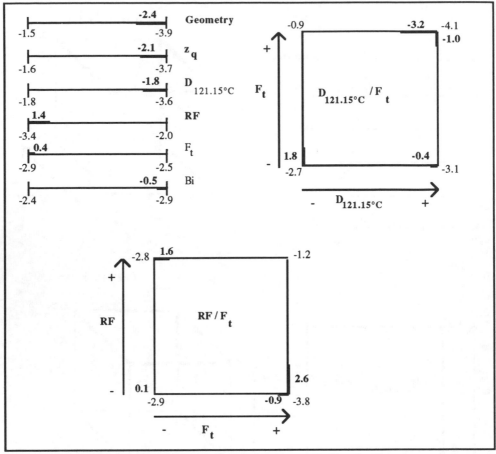

Figure 6. - a) Extension of the most important effects in the parameter P$_3$ (Equation 3).

decrease is larger for infinite slab geometries and products with quality factor kinetics less sensitive to temperature variations (i.e. with large z_q value). Furthermore, the parameters P$_2$ and P$_3$ are insensitive to z_q and Geometry when RF and D$_{121.15°C}$ are both low or large, i.e. conditions resulting in a very low or very large final quality retention, respectively. The decrease in quality increases with z_q and from a spherical to an infinite slab geometry. However, the extension of these effects are more important when RF is large and D$_{121.15°C}$ is low for average quality (i.e. parameter P$_2$), and when RF is low and D$_{121.15°C}$ is large for surface quality (i.e. parameter P$_3$).

The parameter P$_3$ has the double effects: D$_{121.15°C}$/F$_t$ and RF/F$_t$ (see Figure 6 - b)). The decrease in surface quality (i.e. modulus of P$_3$) increases for products more resistant to thermal degradation (i.e. with large D$_{121.15°C}$) and small heat transfer rate (i.e. with small RF). However, this effect is not significant for small target F$_t$ values. Furthermore, the decrease in quality increases or decreases with F$_t$ value if D$_{121.15°C}$ is large or small, respectively. The modulus of P$_3$ also decreases with F$_t$ for products with large heating rate (i.e. large RF), while for small RF the F$_t$ effect is not significant.

The parameter P$_3$ is also affected by the triple interactions: D$_{121.15°C}$/z_q/Geometry and RF/z_q/Geometry. The double effect z_q/Geometry is common to both triple effects. The decrease in surface quality (i.e. modulus of P$_3$) increases with z_q and from a spherical to an infinite slab geometry. However, this effect is only relevant when the quality factor is resistant to thermal degradation (i.e. large D$_{121.15°C}$) or when the product has a low heat

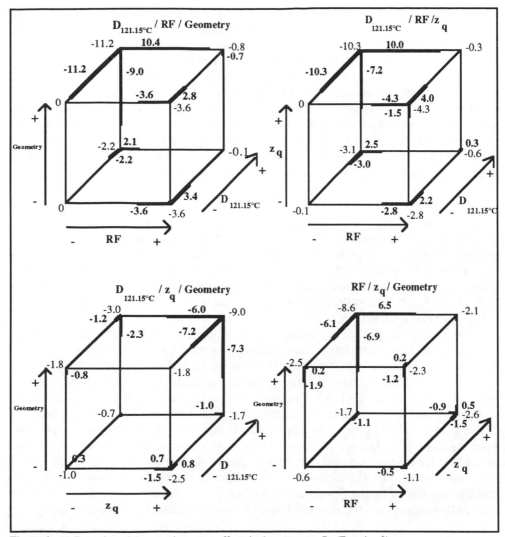

Figure 6. - b) Extension of the most important effects in the parameter P3 (Equation 3).

transfer rate (i.e. small RF). Furthermore, the modulus of P3 increases with $D_{121.15°C}$ and decreases with RF, but these effects are significant only for infinite slab geometries and for quality factors with degradation kinetics not sensitive to temperature changes (i.e. large z_q).

The surface resistance to heat transfer or Biot number does not show a relevant effect on the parameters P2 and P3. The F_t value only affects significantly the parameter P3.

The decrease in surface quality when volume average quality optimal conditions are used, and vice-versa, is minimized when the z_q value is low and the Geometry conducts to an higher heat transfer rate (i.e. spherical shape). If the z_q and the Geometry are fixed these decreases are minimized when RF and $D_{121.15°C}$ are both low or large. Furthermore, the decrease in surface quality when sterilization conditions for optimum volume average quality are used is maximized for infinite slab geometries and when the quality factor kinetics is very resistant to thermal degradation and not very sensitive to temperature variations (i.e. large $D_{121.15°C}$ and z_q values) or when the z_q value is large and the product has a low heat transfer rate (i.e. small RF).

The use of optimal sterilization conditions for surface quality does not imply a decrease in volume average retention greater than 8% and in some cases no reduction at all is verified. On the other hand, when temperatures for maximizing volume average quality are used decreases of 11% in surface quality can be observed, and in some cases no reduction at all is verified. Therefore, it can be suggested the use of optimal conditions to maximize surface quality when the aim is to obtain a final product with a well balanced maximum quality (considering quality factors at the surface or in volume average terms with similar degradation kinetics). This benefit is, however, not very important.

The conclusions that have been drawn are based on the assumption that the volume average or surface quality retentions to be optimized are for quality factors having the same degradation kinetics. However, this assumption may not be applicable in every situations. To investigate the validity of the conclusions when the quality factors have different degradation kinetics, a 2^4 factorial design study was carried out. The z_q and $D_{121.15°C}$ values were fixed and parameters P_1, P_2 and P_3 were calculated as a function of the variables in Table 1. The quality factors to optimize at the surface or in volume average terms were the Chlorophyll and the Thiamin, respectively (Lund, 1975) (Table 3).

A regression approach to the analysis of variance on parameters P_1, P_2 and P_3 is presented in Table 4. It can be observed that only the variables RF and Geometry and the interaction RF/Geometry affect significantly the parameters P_1, P_2 and P_3. This conclusion agrees with previous results (when the quality factors considered at the surface and in volume average terms had the same degradation kinetics), since the $D_{121.15°C}$ and z_q values were now assumed constant.

However, the magnitude of the decrease in volume average quality in relation to the optimum value when sterilization conditions for maximizing surface quality are used (i.e. modulus of P_2) is now greater than the decrease in surface quality when optimum sterilization conditions for volume average quality are used. This is particularly important when the RF value is low and for infinite slab geometries (i.e. for slow heating rate products). Under these conditions the parameter P_2 can show values of approximately 8%. The low $D_{121.15°C}$ value for the Chlorophyll (i.e. rapid degradation rate) may explain the small decrease in surface quality (in relation to the optimum value) when sterilization conditions for maximizing volume average quality are used. Therefore, when a well balanced maximum quality product is to be obtained, the use of optimum sterilization conditions maximizing the quality at the surface or in volume average terms has to be analyzed on a case by case basis, depending on the quality factors degradation kinetics.

V. CONCLUSIONS

The most important variables affecting the: i) difference between sterilization temperatures for maximizing surface and volume average quality or ii) decrease in surface quality retention when optimal conditions for volume average quality are used, and vice-versa are the z_q, $D_{121.15°C}$, RF and Geometry and some of their interactions. The Biot number does not affect significantly the referred parameters. The F_t value only affects significantly the decrease in surface quality when conditions for optimum volume average quality are used.

If the degradation kinetics for the quality factors considered at the surface and in volume average terms are similar, at this state of the art the use of sterilization temperatures for maximizing surface quality is suggested when the aim is to obtain a product with a well balanced maximum final quality. This approach will be more adequate for case studies with low z_q value and spherical geometry, when RF and $D_{121.15°C}$ are both low or large and for low F_t value. On the other hand, if the quality factors have different degradation kinetics, the use of optimum sterilization conditions maximizing the quality at the surface or in volume average terms has to be analyzed on a case by case basis.

Table 3
Kinetics Parameters Used to Investigate the Validity of the Conclusions for Quality Factors with Different Degradation Kinetics.

Quality Attribute	
Chlorophyll	
z_q (°C)	44.2
D_{refq} (min)	14.0
T_{refq} (°C)	121.0
Thiamin	
z_q (°C)	25.9
D_{refq} (min)	246.9
T_{refq} (°C)	121.0

Table 4
Regression Approach to Analysis of Variance on Parameters P_1, P_2 and P_3, Considering Different Degradation Kinetics for the Quality Factor at the Surface or in Volume Average Terms.

Source of Variation	t - Value for P_1[a]	Significance for P_1[b]	t - Value for P_2[a]	Significance for P_2[b]	t - Value for P_3[a]	Significance for P_3[b]
RF	19.32	*	17.10	*	-34.56	*
F_t	-12.00	—	-5.99	—	-0.33	—
Bi	-2.85	—	-5.08	—	-12.44	—
Geometry	-31.00	*	-18.00	*	-33.33	*
RF/F_t	11.59	—	5.58	—	-0.33	—
RF/Bi	0.54	—	4.63	—	-12.44	—
RF/Geometry	17.51	*	16.59	*	-33.33	*
F_t/Bi	-0.98	—	-1.60	—	-1.56	—
F_t/Geometry	-10.00	—	-6.07	—	-0.44	—
Bi/Geometry	1.45	—	-4.61	—	-12.33	—
RF/F_t/Bi	1.03	—	1.51	—	-1.56	—
RF/F_t/Geometry	10.58	—	5.66	—	-0.44	—
RF/Bi/Geometry	0.72	—	4.17	—	-12.33	—
F_t/Bi/Geometry	-1.48	—	-1.10	—	-1.00	—

a) The quadruple effect was used to estimate a residual for computing t-value. The value of the estimate of variance for this residual was 0.975, 0.066 and 0.002 respectively for P_1, P_2 and P_3. The residual had 1 degree of freedom.

b) Single, double and triple asterisks respectively signify 5%, 1% and 0.05% levels of significance. Sources without any asterisk did not significantly influence the parameters P_1, P_2 and P_3.

REFERENCES

Anonymous, *Statview - The Ultimate Integrated Data Analysis & Presentation System*, 1st ed. Abacus Concepts, Inc., Berkeley, CA, 1992.

Banga, J.R., Perez-Martin, R.I., Gallardo, J.M. and Casares, J.J., Optimization of the thermal processing of conduction-heated canned foods: study of several objective functions, *Journal of Food Engineering*, 14, 25, 1991.

Bhowmik, S.R. and Hayakawa, K., Influence of selected thermal processing conditions on steam consumption and on mass average sterilizing value, *Journal of Food Science*, 212, 1983.

Box, G.E.P., Hunter, W.G. and Hunter, J.S., *Statistics for Experimenters*, John Willey & Sons, New York, 1978, chap. 10.

Diamante, L.M. and Munro, P.A., Mathematical modelling of hot air drying of sweet potato slices, *International Journal of Food Science and Technology*, 26, 99, 1991.

Gacula, M.C. and Singh, J., *Statistical Methods in Food and Consumer Research*, Academic Press, Inc., New York, 1984, chap. 6.

Hendrickx, M., Silva, C., Oliveira, F. and Tobback, P., Generalized (semi)-empirical formulas for optimal sterilization temperatures of conduction heated foods with infinite surface heat transfer coefficients, *Journal of Food Engineering*, 19, 141 , 1993.

Ohlsson, T., Optimal sterilization temperatures for flat containers, *Journal of Food Science*, 45, 848, 1980a.

Ohlsson, T., Optimal sterilization temperatures for sensory quality in cylindrical containers, *Journal of Food Science*, 45, 1517, 1980b.

Ohlsson, T., Optimization of heat sterilization using C-values, in *Food Process Engineering*, P.137, Applied Science Publishers, U.K, 1980c.

Silva, C., Hendrickx, M., Oliveira, F. and Tobback, P., Critical evaluation of commonly used objective functions to optimize overall quality and nutrient retention of heat preserved foods. *Journal of Food Engineering*, 17, 241, 1992a.

Silva, C., Hendrickx, M., Oliveira, F. and Tobback, P., Optimal sterilization temperatures for conduction heating foods considering finite surface heat transfer coefficients, *Journal of Food Science*, 57(3), 743, 1992b.

Silva, C.L.M., Oliveira, F.A.R. and Hendrickx, M., Modeling optimum processing conditions for the sterilization of prepackaged foods, *Food Control*, 4(2), 67, 1993.

Walpole, R.E. and Myers, R.H., *Probability and Statistics for Engineers and Scientists*, MacMillan Publishing Company, Toronto, 1993, chap. 13.

DISCUSSION

LAZARIDES: It strikes me that Biot number has no effect on the quality responses. How do you explain this?

SILVA: The Biot number strongly affects optimum temperatures and final quality retention. However, it affects both temperatures (to maximize surface or volume average quality retention) in the same way. Since our responses are always a difference between optimum temperatures (P1) or final retention (P2 and P3) under both optimum temperatures, they are not affected by Biot. The Biot number influences the temperature at the surface in the same way as the temperatures within the product. Finite Biot number introduces a lag time in the temperature in each point of the product.

Assessing Overprocessing in the Aseptic Processing of Fluid Foods: An Analysis of the Contribution of Diferent Portions of the Fluid to Average Quality

Andreia Pinheiro Torres and Fernanda A. R. Oliveira

I. Abstract

An analysis of the contribution of different portions of fluid to the average quality of the final product can be applied for the aseptic processing of foods, similarly to what is usually done in in-container processing with the concept of surface quality. The contribution of different portions is extended to all fractions with different residence times in the system, which include the fastest and slowest portions of fluid. It is assumed that the process is designed so that microbiological safety is attained in the fastest part of fluid for the worst case of residence time distribution of the fluid. A case study simulates the aseptic processing of liquid foods, where the product in a tubular holding tube is considered to behave as a dispersed plug flow. Different processing conditions (holding times and temperatures) and quality factors kinetics are represented jointly by the Damköhler number. Results from the simulations are compared and related to the average quality retention attained in the final product.

The effect of a broader residence time distribution would apparently lead to a larger overprocessing, as the part of the product with greater residence time is subjected to larger processing times and temperatures. Though in this case there is a significant fraction of fluid flowing much faster than in cases with less dispersion and the final average quality retention does not decrease as much as in low dispersion of the fluid. Fractional quality retention is higher for the first fractions of fluid, being highest in cases of higher dispersion, corresponding to lower Peclet numbers. On the other hand the quality retention of the last fractions decrease most, being this effect stronger for the same low Peclet numbers. The logarithm of fractional and average quality retention decrease significantly with the Damköhler number. Therefore, to obtain a minimally processed product, the optimization and control of aseptic processing should target specially on the minimization of the severity of the process and this can be achieved by guaranteeing a tight control in the processing temperature. Safety margins obtained by increased processing temperature are particularly detrimental for quality retention.

II. Introduction

Thermal processing of foods has been extensively used in conjunction with other preservation techniques for extending the shelf-life of foods. The main function of the thermal process is to reduce significantly the microorganisms and/or enzymes that would deteriorate the product during storage or that could endanger consumers health [Lund, 1977]. This is achieved by raising the product temperature above ambient for a certain time, sufficient to guarantee the above mentioned safety requirements. Depending on the product, the main process objective and the desired shelf-life, the process can be more or less severe, and thus be classified as sterilization , pasteurization or blanching [Fellows, 1988].

However, together with the destruction of microorganisms, its spores, or enzymes, the thermal treatment has also an effect on the nutrient content and the overall quality of the product. Therefore, another objective of the process will be the maximization of the final product quality [Lund, 1975; Lund 1977]. Due to the differences in the temperature dependence of quality and microbiological kinetics [Lund, 1977], it is possible to find different time/temperature combinations that result in the same lethal effect but different quality losses. This is the basis for the principle of high temperature - short time (HTST) and ultra-high temperature (UHT) in aseptic processing.

Aseptic processing consists in a rapid heating of the product to the lethal temperature, holding for the processing time and cooling of the product prior to filling in aseptic conditions. The use of higher temperatures and shorter processing times results in more uniform product characteristics and higher quality retention.

Nevertheless, to take full advantage of aseptic processing, a tight control of temperature and the distribution of residence times of the product in the process (specially in the holding section) is needed, so that overprocessing is minimized while safety is guaranteed. As holding tubes are sized according to the minimum residence time achievable in the holding section [Jordan and Holland, 1953], all other parts of the product are overprocessed. In addition, when the minimum residence time is not experimentally obtained, Newtonian laminar flow is considered as basis for conservatively calculating this parameter [Dickerson et al, 1968; Kessler, 1988]. The holding temperature, is usually maintained at least some degrees above required. Both these common procedures guarantee a large margin of safety, but at the expense of quality. The effects on quality have not been quantified.

The processing effects on microbial destruction have been largely studied. The effect of residence time is considered of great importance [Rao and Loncin, 1974], and lethality of microorganisms can be calculated using both single-point and integrated effects [Teixeira and Manson, 1983]. Most references to the optimization of nutrient and/or average quality retention are general and consider only overall effects [Brown, 1990; Fellows, 1988; Lund, 1975; Lund 1977]. Surface quality determination and optimization refers only to in-container processing [Ohlsson, 1980; Silva et al. 1992], where its importance on product quality is enhanced.

The evaluation and optimization of aseptic processing requires therefore the definition of pertinent parameters for calculating the overprocessing effects on reactions, such as microbial destruction, enzyme inactivation and nutrient, sensory attributes and/or overall quality retention. These parameters will be defined to include the residence times of the product in the system, specially in the holding section, where the higher temperatures are maintained for longer times. The processing effects on these objective functions will be evaluated in order to determine targets in the optimization and control of aseptic processing for obtaining a minimally processing, that is, minimal detrimental effects while ensuring that process objectives are met.

III. THEORETICAL CONSIDERATIONS

A. KINETICS OF QUALITY ATTRIBUTES

The mathematical treatment of kinetic data for the reactions occurring in food during thermal treatment is well described in literature [Datta, 1993; Karel et al., 1975 and Lund,

1975]. Two different nomenclatures are commonly used, namely the classical kinetic parameters used by reaction engineers, and an empirical nomenclature mostly used by microbiologists and food processors. The first one was adopted and will be continuously used in this text, but the most useful equations and the relations between parameters of both nomenclatures are also presented.

Most chemical and biochemical changes in foods during thermal processing are considered to follow apparent first-order reaction kinetics [Datta, 1993], where the rate of change of the reactant, c, at constant temperature, T, is:

$$\left(\frac{c}{c_0}\right) = e^{-k_T t} \tag{1}$$

The kinetic constant k_T is temperature dependent, and this dependence is most commonly described by an Arrhenius type equation:

$$k_T = k_0 e^{-Ea/RT} \tag{2}$$

where k_0 is the pre-exponential factor or Arrhenius constant, Ea the activation energy, R the universal gas law constant and T the absolute temperature.

The empirical Bigelow model is more commonly used by microbiologists and food processors to describe the inactivation of microorganisms during a thermal treatment. Its parameters are called D- and z-values, and they are equivalent to the kinetic parameters k_T and Ea.

The decimal reduction time D is the time (generally expressed in minutes) at a defined temperature T needed to destroy 90% of the spores or vegetative cells of a given microorganism and is related to the kinetic constant k_T by the following equation:

$$D_T = \frac{\ln(10)}{k_T} \tag{3}$$

The thermal death time constant - z value - is defined as the number of degrees for the thermal resistance curve ($\log D$ versus T) to transverse one log-cycle and is equal to the reciprocal of the slope of the curve. The z-value is constant when this thermal resistance curve follows a straight line, given by the following equation:

$$\log\left(\frac{D_T}{D_{Tref}}\right) = \frac{-(T - T_{ref})}{z} \tag{4}$$

The z-value can be related to the activation energy Ea by the following equation [Karel et al. 1975]:

$$Ea \cong \frac{\ln(10) R T_{ref}^2}{z} \tag{5}$$

This relation is only valid for temperatures around the chosen reference temperature T_{ref} [Datta, 1993].

1 . Kinetics of Quality Changes in Food Processing

To describe the effect of time-temperature combinations on the food quality changes, two parameters are needed [Lund, 1975]: i) the rate of change at a reference temperature (k_T D_T or k_0 at infinite reference temperature), and ii) the dependence of the rate of change on temperature (Ea or z value). Typical values describing the kinetics of some quality attributes in liquid foods are shown in table 1.

2 . Combining Quality Kinetics with Processing Parameters

The combination of kinetic parameters of relevant reactions occurring in the food during processing and the thermal process itself is useful for comparing processes and defining process evaluation parameters.

Minimum processing parameters for aseptic processes can be found in table 2. These values are used for the design of holding tubes, where the minimum time corresponds to the fastest particle of fluid flowing in the system. If this same particle satisfies the minimum processing requirements, then the rest of the food is guaranteed to be safe. Contributions from heating and cooling zones are also conservatively neglected.

Several empirical parameters have been used to characterize the severity of a thermal process with respect to a particular reaction. The F value [Ball and Olson, 1957] for microbial lethality in sterilization, C or cook value [Mansfield, 1962] for sensory degradation and PU or pasteurization value [Shapton, 1966] for lethality in pasteurization represent the same concept of an equivalent processing time at a reference temperature for a particular reaction; by reducing the processing parameters to a reference, they allow for comparing the severity of thermal processes with respect to reactions with identical temperature dependence (equal activation energy).

For the case where isothermal conditions are assumed, The F, C and PU values can be obtained from the following expression:

$$value = 10^{((T-T_{ref})/z)} \times \tau \quad \text{(in time units)} \tag{6}$$

which gives an equivalent processing time at the reference temperature. Consistent with the classical nomenclature, the dimensionless Damköhler number Da also combines the reaction rate at constant processing temperature k_T with a processing time. Da for first order kinetics is defined as [Bailey and Ollis, 1986]:

$$Da = k_T \tau \tag{7}$$

where τ is the mean residence time in the reactive system (here the holding tube), defined as:

$$\tau = \frac{V}{Q} \tag{8}$$

with Q standing for product flowrate and V for reactor (holding tube) volume.

Table 1
Typical Kinetic Parameters For The Thermal Degradation Of Quality Attributes In Liquid Foods

Component	Medium	Range (°C)	Z (°C)	Ea (kcal/mole)	D_{121} (min)
thiamin	vegetable purees[‡][*]	109-149	45	27	134-163
	meat purees[‡][*]	109-149	45	27	115-157
	vegetable purees[‡][*]	121-138	48	27.5	247-254
	milk[#]	-	-	24	103.5
lysine	milk[#]	-	-	26	556
chlorophyll I	vegetable purees[‡][*]	115-138	66-81	16.1-12.6	13.9
	vegetable purees[‡][*]	79-138	52-59	22-19	113-166
Maillard reaction	apple juice[‡][*]	38-130	45-55	27-20.7	271-285
anthocianins	fruit juices[‡][*]	20-121	42-60	28-19	17.8-110
browning	goat's milk[‡][*]	93-121	45	27	0.98-1.08
vitamins	general[§][†]	-	25-45	20-30	120-1200
color	general[*][†]	-	20-60	-	10-120
overall quality	general[§][†]	-	25-44	10-30	5-500
sensory quality	general[‡]	-	25-33	-	188
enzymes	general[§]	-	6-60	12-100	1-15
vegetative cells	general[§]	-	4-6.6	100-120	0.002-0.003
spores	general[§]	-	6-12	53-83	0.1-5.0
(heat resistant)		-	6-12	-	0.2-50

Data reported by [†]Jelen [1983], [‡]Karel et al. [1975], [#]Kessler [1988], [*]Lund [1975] and [§]Lund [1977]

B. RESIDENCE TIME DISTRIBUTIONS

Real flow systems deviate many times from the ideal flow patterns of plug flow or flow patterns predicted from mechanistic models. The investigation and quantification of the behavior of real flow systems was first introduced by Danckwerts [1953] with the residence time distribution (RTD) analysis. This concept and its application to reaction engineering is extensively described in literature [Levenspiel 1984; Levenspiel and Bischoff, 1963]. Various models have been developed since then to describe the flow behavior of vessels, tubes and other reactors. In food engineering, RTD has its widest application in aseptic processing [Bailey and Ollis, 1986; Burton, 1958; Rao and Loncin, 1974].

The residence time of an element of fluid is defined as the time elapsed since it entered the reactor until it reaches the exit. The distribution of these times is called the residence time distribution function of the fluid E, and represents the fraction of fluid for each time, having units of time^{-1}. Therefore we have:

$$\int_0^\infty E(t)dt = 1 \tag{9}$$

Table 2

Minimum Holding Time / Temperatures for Continuous Thermal Processing of Liquid Foods

Product	Temperature (°C)	Processing time	Reference
acid products	93.3	5' - 10'	Brown [1990]
concentrated milk	75	15"	Brown [1990]
cream	80	25"	Brown [1990]
dairy products	72	15"	Brown [1990]
beer	72-75	1' - 4'	Fellows [1988]
fruit juices, holding	77	1'	Fellows [1988]
fruit juices, HTST	88	15"	Fellows [1988]
ice cream, holding	71	10'	Fellows [1988]
ice cream, HTST	80	15"	Fellows [1988]
liquid egg	65	2.5'	Fellows [1988]
dairy, holding	71 - 74	15" - 40"	Kessler [1988]
dairy, HTST	85 - 90	1" - 4"	Kessler [1988]
dairy, UHT	135 - 150	2" - 20"	Kessler [1988]
tomato paste	108 - 114	1' - 3'	Mallidis [1990]

The mean residence time $\bar{t}$ is the first moment of the distribution, defined as:

$$\bar{t} \equiv \int_0^\infty t \times E(t)dt = \tau \tag{10}$$

and is equal to the mean holding time τ (equation 8) for vessels with ideal flow.

It is often convenient to use a dimensionless time θ and the corresponding residence time $E(\theta)$, so that (Levenspiel, 1973):

$$\theta = \frac{t}{\tau} \quad \text{and} \quad E(\theta) = \tau \times E(t) \tag{11}$$

One application of RTD is to predict performances of real reactors directly or in conjunction with flow models. For first order reactions (which will be the only ones discussed here), RTD information is sufficient to predict conversion in the reactor. In this case, the final average concentration in the exit stream of the reactor is [Levenspiel and Bischoff, 1963]:

$$c_{final} = \int_0^\infty c(t) \times E(t)dt \tag{12}$$

C. CALCULATION OF QUALITY RETENTION

Thermal processing effects on product safety and quality may be calculated using both single-point or integrated effects. The determination of minimum required thermal processing is for safety reasons based on single-point evaluation of the least-lethality point in the product (container center in batch, maximum velocity particle in aseptic processes). Process evaluation or optimization on the other hand uses integrated (overall) effects on safety [Teixeira and Manson, 1983] or quality parameters (Table 3), respectively. Only for in-container processing, surface retention for quality and safety attributes has been applied, [Ohllson, 1980; Hendricks et al. 1990; Silva et al. 1992], and these are again single-point evaluated. A similar approach in aseptic processing would have little sense, as the slowest, "surface" equivalent point of the product may represent less than an infinitesimal part of the product, irrelevant in the final container.

Average quality may be obtained directly from equation (1), which combined with equation (7) gives:

$$\left(\frac{c}{c_0}\right)_{ave} = e^{-Da} \tag{13}$$

where the subscript *ave* stands for average quality retention.

Equation (13) does not consider residence time distributions in the product, but can be a good approximation and is used in most calculations for average quality retentions under isothermal conditions.

A more exact calculation requires the knowledge of the E-curve and the average quality retention is the integrated quality obtained considering all product parts with different residence times. By combining equations (1) and (12), an expression for calculating the overall quality retention at the exit of the holding tube can be deduced [Bailey and Ollis, 1986; Levenspiel, 1973]:

$$\left(\frac{c}{c_0}\right)_{ave} = \int_0^\infty e^{-kt} \times E(t)dt \tag{14}$$

The same principle used in equation (14) may be applied to define the quality of a fraction of the fluid: a given fraction of the product, f_i, with residence time in the aseptic processing system between t_{i-1} and t_i is defined as:

$$f_i = \int_{t_{i-1}}^{t_i} E(t)dt \tag{15}$$

The value of the residence times t_{i-1} and t_i may be chosen to analyze the effects of processing on different fractions of the fluid and the quality retention observed in that fraction may therefore be calculated as:

$$\left(\frac{c}{c_0}\right)_i = \frac{\int_{t_{i-1}}^{t_i} e^{-kt} \times E(t)dt}{\int_{t_{i-1}}^{t_i} E(t)dt} \tag{16}$$

Table 3

Reported Average Quality Retentions (%) in Aseptic Processes

process	thiamin	vitamin C	vitamin B12
milk, hold	90	80	90
milk, HTST	93.2	90	100
milk, UHT	90	90	80
tomato juice, HTST	97.2	--	--

Data collected from Fellows [1986] and Lund [1975]

This retention will be referred to as fractional quality retention. Different fractions of fluid may be considered and in the limit, if t_{i-1} equals zero and t_i equals infinity, this quality retention will be equivalent to the average quality retention.

If n sequential fractions of fluid covering the whole range of residence times are considered, then the average quality retention can also be obtained from the following equation:

$$\left(\frac{c}{c_0}\right)_{ave} = \sum_{i=1}^{n} f_i \times \left(\frac{c}{c_0}\right)_i \tag{17}$$

Fractional quality retention for fraction with long residence times may be a good measure of the overprocessing in relation to a particular quality factor, when significant tailing in the residence time distribution curve occurs. For liquid foods this is not such an important concept, since the perceivable quality is the average one, but it may be very important for a particulate fluid food. Also, the average retention can be a result of a similar effects of all portions of the fluid or the result of a drastic effect of some parts of the fluid while others have little effect. Identifying the fraction responsible for the greater quality losses may therefore be very important in process optimization.

IV. METHODOLOGY

Simulations to calculate average and fractional quality retentions were made to assess effects of the quality kinetics, the processing time/temperature combination and the residence time distribution. These parameters were condensed in the two dimensionless numbers, Damköhler Da and Peclet Pe, as will be presented next. Isothermal conditions as required for the holding section were assumed.

A. KINETICS OF QUALITY CHANGES

Combining tables 1 and 2 in equations (1) and (7), minimum Damköhler numbers were calculated (Table 4). The reported minimum holding times (Table 2) were considered as the time required for the fastest part of fluid in the case of the highest dispersion of residence times. This ensures that in all the situations considered all parts of the fluid are processed at least for the minimum time required. Da values from 0 (no reaction) to 2.5 were considered in the simulations.

Table 4
Calculated Minimum Da Values

attribute	minimum Da
thiamin, UHT	up to 0.3
thiamin, tomato paste	0.075 to 1.25
vitamins, acid products	up to 2.5
browning, UHT	up to 1.25
overall quality, HTST	up to 1.15

B. RTD: THE DISPERSED PLUG FLOW MODEL

The dispersed plug flow has been widely applied to describe the flow in tubes, and was therefore chosen to simulate the case of product flow in a holding tube. This model considers that axial diffusion is superimposed to plug flow, and the resulting dispersion D is expressed in a dimensionless Peclet number:

$$Pe = \frac{vL}{D} \tag{18}$$

where L is the length of the tube and v is the average velocity in the tube. For small extents of dispersion, or large Pe numbers (Pe≥500), this model is mostly valid and solutions do not depend on the boundary conditions. The resulting E curve and its mean $\bar{\theta}$ and variance σ^2 are (Levenspiel, 1973):

$$E(\theta) = \sqrt{\frac{Pe}{4\pi}} \exp\left\{-\frac{Pe(1-\theta)^2}{4}\right\}$$

$$\bar{\theta} = 1 \quad \text{and} \quad \sigma^2 = \frac{2}{Pe} \tag{19}$$

Solutions for small values of Pe depend on the type of boundary conditions imposed. For open vessels, the solution in equation (20) is obtained, and the corresponding RTD curves are represented in figure 1.

$$E(\theta) = \sqrt{\frac{Pe}{4\pi\theta}} \exp\left\{-\frac{Pe(1-\theta)^2}{4\theta}\right\}$$

$$\bar{\theta} = 1 + \frac{2}{Pe} \quad \text{and} \quad \sigma^2 = \frac{2}{Pe}\left(1 + \frac{4}{Pe}\right) \tag{20}$$

The dispersed plug flow model from equation (20) was used to simulate the RTD in a holding tube, in the range of Peclet numbers from 5 to 500. These values cover the range of reported Peclet values for holding sections in continuous aseptic units [Aiba and

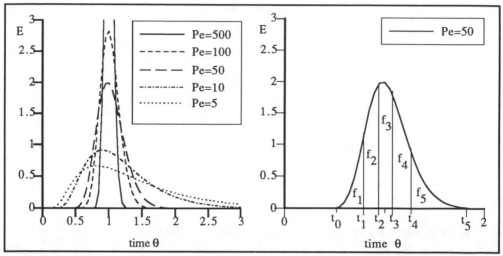

Figure 1. a) E-curves for different Péclet numbers, using the dispersion model; b) E-curve illustrating the areas considered for calculating the fractional quality retentions

Sonoyama, 1965; Sancho and Rao, 1992]. The minimum residence time for the worst case of dispersion considered is lower than the values suggested when assuming Newtonian laminar flow [Dickerson et al, 1968; Kessler, 1988] but guarantees a conservative approach.

C. CALCULATION OF AVERAGE AND FRACTIONAL QUALITY RETENTIONS

Quality retention was calculated using the expressions in equations (14) and (16). Five sequential fractional quality retentions were considered in the calculations, corresponding to a constant f_i value of 0.20. The first fractional quality retention corresponds to the 20% fastest fraction of fluid flowing in the system, while the fifth corresponds to the slowest 20% of product. An example of the fractions considered in the simulations are illustrated in figure 1b.

V. RESULTS AND DISCUSSION

Average and fractional quality retentions were plotted in semi-log scale versus the Damköhler number (figure 2), and showed in all cases approximately linear decreases with *Da*. This decrease in quality was more significant for fractions corresponding to larger residence times.

A. DEPENDENCE OF AVERAGE QUALITY RETENTION ON THE THERMAL PROCESS

Average quality retention as function of Damköhler number is shown in figure 2a in a semi-log scale. A similar plot [Aiba et al, 1973] for *Da* ranges from 20 to 140, which are the ranges applicable to microorganisms, showed also an approximately linear relationship between the logarithm of retention values and *Da*. A 25% loss in quality may be achieved for a Da of 0.3 (e.g. thiamin in table 4), which is higher than reported values for thiamin retention in the UHT processing of milk (table 3). Also for tomato processing, the reported

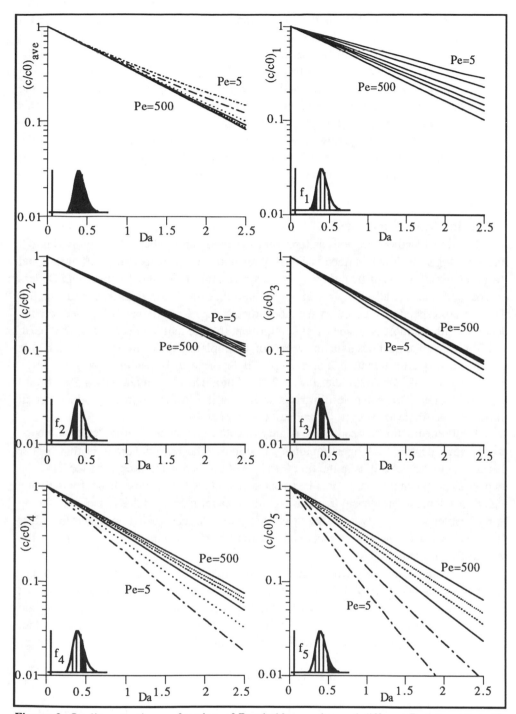

Figure 2. Quality retention as function of Damköhler number and different Peclet numbers (5, 10, 25, 50, 100, 500) for 6 cases (f=1 -average- and f_i=0.20: i=1-5)

thiamin retention of 97.2% (table 3) is above the minimum value achieved for the calculated Da in table 4, and can go down to 70%, according to figure 2.

For low Da numbers, the Peclet number has little or no effect in the decrease in average quality retention, and this is due to the residence time curves having all their mean approximately equal to the holding time in this range (equation 10). This is not true for Da above 1, as open boundary conditions were applied to the dispersed plug flow model. At larger processing times or temperatures, the average quality retention decreases more the higher the Peclet numbers. This implies that small dispersions lead to a lower quality product, which is not very straightforward. The reason hereof is due to the process being defined based on a fastest particle that flows much faster than the mean residence time. This fact is far from reality for high Peclet numbers, as can be seen in figure 1a, and leads therefore to higher average quality losses. If the process schedule would be based on the effective minimum residence time, this effect would be inverted, but such analysis would not allow the comparison of effects for equal processing conditions.

B. DEPENDENCE OF FRACTIONAL QUALITY RETENTION ON THE THERMAL PROCESS

Fractional quality retentions as function of Damköhler number are shown in figure 2 on a semi-log scale. The first fractional quality retention corresponds to the 20% fastest and least processed fraction of fluid flowing in the system, thus being the decrease with Da less pronounced than for all other fractions. On the other hand, the fifth fractional quality retention corresponds to the slowest and most processed 20% of product, being therefore the decrease with Da most pronounced. It is important to note that the decrease in the fourth and fifth fractional qualities assumes values of a completely different order of magnitude than average quality retention. For example, if reported to browning during the UHT processing of milk (Da=1.25), the slowest 20% of product will contain only 4.5% of non-browned matter. This example illustrates well how the last fraction of product has an important contribute to the overall quality of the product.

While for fractions f_1 and f_2 the quality retention is higher for low Peclet numbers, this is not true for the longer remaining fractions. The reason hereof is that higher dispersion (lower Pe) results in the first fractions of product being processed for less time than the corresponding fractions in a less dispersed product. On the other hand, for the same high dispersion, the subsequent fractions are much longer processed due to the large tailing in the residence time curves, resulting therefore in a higher quality loss than in a less dispersed product flow. These facts become more clear by analysing the shape of the residence time curves in figure 1a.

The result of the different effects of individual fractions in the average quality retention is equivalent for Da below 1, but above this value low Peclet numbers result in a better average quality product. This dependence on Peclet numbers is however just relevant for values lower than 100. Figure 3 clearly shows that for Peclet number above this value, both fractional and average quality retentions become almost independent of Peclet number.

VI. CONCLUSIONS

Parameters for calculating the overprocessing effects on reactions, such as microbial destruction, or nutrient, sensory and/or overall quality retention were defined. These parameters included the residence times of the product in the holding section, where the higher temperatures are maintained for longer times. The processing effects on these objective functions were evaluated and it was concluded that for processes defined on the minimum residence time of the worst case of dispersion of product flow, a better quality is attained for low Peclet numbers if the process is severe enough to give Damköhler numbers above 1.

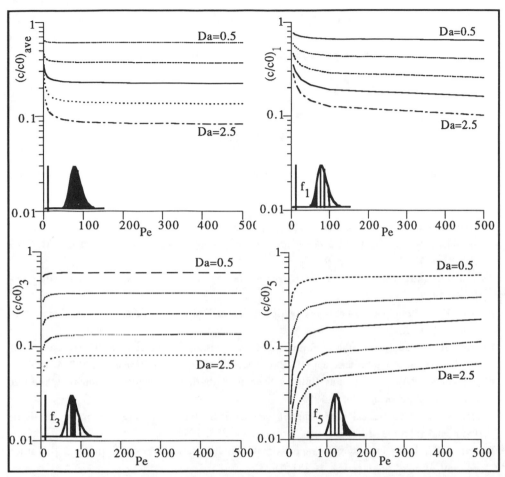

Figure 3. Quality retention as function of Peclet numbers for different fractions of fluid (average and $f_i=0.20$: $i=1$, 3 and 5) and different Damköhler numbers (0.5, 1.0, 1.5, 2.0 and 2.5)

Therefore, to obtain a minimally processed product, the optimization and control of aseptic processing should target specially on the minimization of the severity of the process. This can be achieved by guaranteeing a tight control in the processing temperature. Safety margins obtained by increasing processing times or temperatures above design values increase significantly the Damköhler number, and are therefore particularly detrimental for the quality retention in the processed product.

On the other hand , scheduling the process based on the real minimum residence time of the product in the system, and not on a hypothetical and in most cases unreal "worst case" of minimum residence time, will also reduce significantly the Damköhler number with the evident benefits in the overall quality of the processed product.

The obtained results on the analysis to fractional quality can be extended to cases where the dispersed plug flow model is not valid, or to the processing of heterogeneous products. When the product has other flow characteristics than the dispersed plug flow model, it is common to find tailing of residence time distribution curves, as characteristic in dispersed plug flow at lower Peclet numbers. In this cases, the conclusions taken from the analysis to fractional quality can also be applied.

In the processing of heterogeneous products, the fractional quality concept becomes most important, as the longer processed parts of the product are not homogenized in the final mix, and are therefore perceived separately as overprocessed by the consumer.

REFERENCES

Aiba, S. and Sonoyama, T., Residence-time distribution of a microbial suspension in a straight pipe, *Journal of Fermentation Technology*, 43, 534, 1965

Aiba, S., Humphrey, A.E. and Millis, N.F., Biochemical Engineering, 2nd. ed., University of Tokio Press, Tokio, 1973, pp. 263

Bailey, J.E. and Ollis, D.F., *Biochemical Engineering Fundamentals*, 2nd ed. McGraw Hill, New York, 1986, Chap. 9

Ball, C.O. and Olson, F.C.W., *Sterilization in Food Technology*, McGraw Hill, New York, 1957

Brown, K.L., Principles of heat preservation, in *Heat Preserved Foods*, Permon Press, New York, 1990, Chap. 2

Burton, H., An analysis of the performance of an ultra-high-temperature milk sterilizing plant II. Calculation of the bactericidal effectiveness, *Journal of Dairy Research*, 25, 324, 1958

Danckwerts, P.W., Continuous flow systems (distribution of residence times), *Chemical Engineering Science*, 2(1), 1, 1953

Datta, A.K., Mathematical modelling of biochemical changes during processing of liquid foods and solutions, *Biotechnology Progress*, 7, 397, 1991

Datta, A.K., Error estimates for approximate kinetic parameters used in food literature, *Journal of Food Engineering*, 18,181, 1993

Dickerson, R.W., Scalzo, .M., Read, R.B. and Parker, R., Residence times of milk products in holding tubes of HTST pasteurizers, *Journal of Dairy Science*, 51(4), 1731, 1968

Hendricks, M., Van Genechten, K. and Tobback, P., Optimizing quality attributes of conduction heated foods: a simmulation approach, In Engineering and Food: Preservation Processes and Related Techniques, Vol. 2, W.E.L. Spiess and H Schubert, Eds., Elsevier, London, U.K., 1990

Jelen, P., Review of basic technical principles and current research in UHT processing of foods, *Canadian Institute of Food Science and Technology Journal*, 16(3), 159, 1983

Jordan, W.K. and Holland, R.F., Studies on thermal methods for measuring the holding time in HTST pasteurizers, *Journal of Milk and Food Technology*, 16(1), 15, 1953

Karel, M., Fennema, O.R. and Lund, D.B., *Principles of Food Science, Part II: Physical principles of food preservation*, Marcel Dekker, New York, 1975

Kessler, H.K., *Lebensmittel- und Bioverfahrenstechnik - Molkereitechnologie*, Verlag A. Kessler, Freising, Germany, 1988, Chap. 6

Levenspiel, O., *Chemical Reaction Engineering*, 2nd. edition, John Willey, New York, 1973

Levenspiel, O. and Bischoff, K.B., Patterns of flow in chemical process vessels, in *Advances in Chemical Engineering*, Vol. 4, Drew, T.B., Hoopes, Jr., J.W. and Vermeulen, T., Eds., Academic Press, New York, 1963

Lund, D.B., Effects of blanching, pasteurization and sterilization on nutrients, in *Nutritional Evaluation of Food Processing*, 2nd ed., Harris, R.S. and Karmas, E., Eds., Avi, Connecticut, 1975, Chap.1 Part I

Lund, D.B., Design of thermal processes for maximizing nutrient retention, *Food Technology*, 32(2), 71, 1977

Mansfield, T., High temperature, short time sterilization, In *Proc. First International Congress of Food Science and Technology*, Vol. 4, Gordon and Breach, London, 1962

Ohlsson, T., Optimal sterilization temperatures for sensory quality in cylindrical containers, *Journal of Food Science*, 45, 1517, 1980

Rao, M.A. and Loncin, M., Residence time distribution and its role in continuous pasteurization, *Lebensmittel-Wissenshaft und Technol.ogie* 7(1), 5, 1974

Sancho, M.F. and Rao, M.A., Residence time distribution in a holding tube, *Journal of Food Engineering*, 15(1), 1, 1992

Shapton, D.A., Evaluating pasteurization processes, *Process Biochemistry*, 1(5), 121, 1966

Silva, C.L.M., Hendricks, M., Oliveira, F. and Tobback, P., Critical evaluation of commonly used objective functions to optimize overall quality and nutrient retention of heat-preserved foods, *Journal of Food Engineering*, 17(3), 241, 1992

Teixeira, A.A. and Manson, J.E., Thermal process control for aseptic processing systems *Food Technology*, 37(4), 128, 1977

THE OPTIMISATION OF PRODUCT QUALITY
DURING THERMAL PROCESSING
A CASE STUDY ON WHITE BEANS IN BRINE

**Van Loey, A., Fransis, A., Hendrickx, M., Noronha, J.,
Maesmans, G., Tobback, P.**

I. ABSTRACT

Based on thermal degradation kinetics, on a model for heat transfer and a simplified optimisation approach optimal sterilisation processes for thermal softening of white beans (*Phaseolus vulgaris*, subsp. *nanus* Metz., variety *Manteca de Leon*) were calculated considering constant retort profiles in a still and end-over-end rotary water cascading retort (Barriquand Steriflow). The temperature dependence of the rate index of thermal softening of white beans, described using the thermal death time model (z- and D-value) and the Arrhenius model (E_a and k-value), was determined within the temperature range of 90°C to 122°C using an industrial tenderometer and by use of a trained taste panel. Because of the difference in reaction order and in kinetic parameter estimates between an objective and a subjective hardness evaluation, the tenderometer does not describe thermal softening measured by a taste panel. Consequently kinetic data of taste panel evaluation were selected for the optimisation study. Optimal processes were calculated and evaluated by a trained taste panel. In general, agreement was found between calculated and experimental results. The panelmembers could distinguish between optimal rotary and optimal still processes on the 1% level. Deviations from the optimal temperature of 4 and 8°C were distinguished on the 1% level.

II. INTRODUCTION

Thermal processing of food results in an extended shelf-life but affects both the nutritional and the sensorial quality of the product. One of the challenges to the food canning industry is to minimize these quality losses meanwhile providing an adequate process to achieve the desired degree of sterility. Optimization is possible because of the higher temperature dependence of the bacterial spore inactivation as compared to the rate of quality destruction, sensorial as well as nutritional (Lund, 1977). Several optimisation studies can be found in literature (Teixeira *et al.*, 1969, 1975; Saguy and Karel, 1979; Ohlsson, 1980b, c, d; Nadkarni and Hatton, 1985; Tucker and Holdsworth, 1990; Hendrickx *et al.*, 1992; Banga *et al.*, 1991; Silva *et al.*, 1992). All studies however focus on conduction heating foods only and are limited to theoretical considerations. Experimental validation of optimum conditions defined using mathematical procedures has not yet been attempted. It is however essential to assess the usefulness of any mathematical procedure and should therefore be considered as a priority in this field (Silva *et al.*, 1993). The objective of the present study is the design and evaluation of an optimal process for a particulated in-pack food, namely optimizing thermal softening of white beans in brine considering still and end-over-end rotary processes in a water cascading retort.

Quantification of a sensory parameter (thermal softening) is not as straightforward as it is for nutrients. Physical and chemical measuring techniques for sensory quality indicators are available (e.g. use of a colour difference meter to measure changes in green colour of peas, Hayakawa and Timbers , 1977), but there are some restrictions since the usefulness of an instrumental method as a substitute for subjective testing is determined by how well it correlates with sensory data, the assumption being that the latter is relevant. Physical and chemical measurements are usually linearly correlated to the intensity whereas panel response could be nonlinear (Lund, 1982). Moreover, many quality attributes cannot be measured with reliability by use of objective instrumentation because of their inherent complexity. The perceptive response of a consumer to a food product is based on several interacting factors, while objective measurements only measure one factor (Sawyer, 1971). As a consequence the temperature dependence of sensorial quality attributes should be determined using sensory analysis since objective measurements which highly correlate with an evaluation by a taste panel still aren't available. However, in quality control and in research and development there is a pressing need for instrumentation which will enable rapid and easy measurement of flavour and quality attributes to bypass wherever possible the more expensive and time-consuming panel work (Sawyer, 1971). Hereto the instrumental method should highly correlate with a subjective evaluation of the quality attribute. In this study an objective (FMC tenderometer) and a subjective method (trained taste panel) for the evaluation of thermal softening of beans (*Phaseolus vulgaris,* subsp. *nanus* Metz., variety *Manteca de leon*) are compared within the temperature range of 90°C to 122°C.

III. MATERIALS AND METHODS

A. Strategy

The optimisation approach is shown in figure 1. Optimisation of sensory quality deterioration must be based on the heat transfer in the food and also on the knowledge of the kinetics of microbiological inactivation and of quality degradation (Ohlsson, 1980d). For sterilisation of low-acid foods, processed in the temperature range of 110 to 130°C, the heat resistance of *Clostridium botulinum* spores is chosen as a reference, characterised by a z-value of 10°C and a D-value at reference temperature (121.15°C) of 0.21min. Kinetic parameters for thermal softening of white beans were determined using sensory analysis. Since the analytical and numerical solution of Fourier's equation is restricted to conduction heating foods and the general method do not provide a means of predicting time-temperature relationships of food, heat transfer in a mixed type heating food (i.e. white beans in brine) should be modelled using an empirical formula method. The great advantage of the empirical formula method of Ball is the simplicity and the wide applicability independently of the mode of heat transfer involved. The use of the empirical formula method of Ball implies a heat penetration study to determine the heat penetration characteristics f_h and j_h. Optimisation can be considered in terms of maximizing surface quality, when e.g. appearance or aroma are considered or volume average quality as an indicator for consistency, taste or nutrient retention (Ohlsson, 1980b). Although hardness degradation should be optimised in terms of overall quality retention, surface cook-value is chosen as optimisation parameter since for the optimisation of volume average quality retention the time-temperature profile as a function of position is needed, which cannot be predicted for a mixed type heating food.

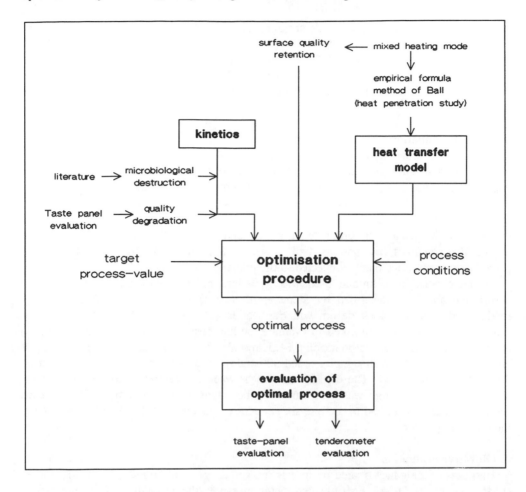

Figure 1. Schematic representation of the optimisation approach to minimize surface quality destruction for a mixed type heating food

B. Kinetic study
1. Description of the product

White beans (*Phaseolus vulgaris,* subsp. *nanus* Metz., variety *Manteca de leon*) were harvested in the summer of 1991 in Spain. They were stored dry (0.19 g moisture/g dry weight) at 15°C. Before heat treatment, they were soaked in distiled and demineralized water at 15°C for at least 16 hours. Preliminary experiments showed that beans reached maximal moisture content after 16 hours (1.22 g moisture/g dry weight). The mean sizes of the dry beans were 15.14mm±1.30mm length, 10.58mm±0.80mm width and 8.71±0.80mm height.

2. Heat treatment

Soaked beans were heated in small cans of 73mm diameter and 27mm height, each can containing ± 50g of beans. The use of flat cans reduced the existence of a temperature gradient in the can during heating so that the difference in heat treatment at the surface and in the centre of the cans was minimal. Before sealing cans were filled with distiled and demineralized water, headspace was minimal. Cans were placed in a calibrated oil bath of 30l (Grant instruments, Cambridge, Limited HB30). Processing temperatures were chosen to cover the temperature range used in the food industry : namely 90°C, 100°C, 110°C, 116°C and 122°C. Samples were heated for pre-determined heating times. Immediately after heating, cans were cooled in icewater to minimize quality destruction during cooling. Prior to evaluation cans were stored at 4°C during ± 48 hours.

3. Sensory evaluation

Prior to evaluation, cans that had the same heat treatment were mixed to obtain homogeneous samples. Individual samples (± 20g of white beans) were served in plastic cups (King disposables, 72mm diameter and 22mm height), coded with three random digits. Hardness of white beans was examined under red light to mask differences in appearance. For hardness testing a dinner-plate, a fork and a spoon were placed at the disposal of the judges. In order to eliminate distraction and prevent communication among the panellists, evaluations took place in individual booths. Each booth was provided with a service-hatch connected to the preparation room.

The evaluation technique used was a variant of the QDA-method (quantitative descriptive analysis), described by Stone *et al.* (1974). Beans were analysed using a randomized balanced block design with replication (Cochran and Cox, 1957). Twelve trained panelmembers were asked to evaluate the hardness of at random coded (3 digits) samples by placing a vertical mark on a 14.5cm scale. Two references were indicated at 1 cm of both ends : the fresh sample, coded 000 and the product processed at 100°C during 300 minutes, coded 300. The distance in mm between the reference coded 000 and the mark placed by the judge was a measure for the intensity of a treatment. During each session, panelmembers had to analyse six samples of different quality levels processed at the same temperature.

4. Objective evaluation

Hardness of 200g heat treated beans was measured by use of an industrial tenderometer (FMC, model 4011). An FMC tenderometer measures the resistance of the product to compression and shear between two sets of blades. The equipment was calibrated properly. Each heat treatment was evaluated in duplicate with intermediate cleaning in place of the blades. Results were expressed in tenderometer units, i.e. pounds per square inch.

5. Data-analysis

First, residues of regression analysis were checked on normal distribution and homoscedasticity. Meeting both requirements, the following statistical techniques could be applied to control whether differences in quality level were detectable : the page-test (O'Mahony, 1986), the calculation of R-indices (O'Mahony *et al.*, 1980; O'Mahony, 1986; Vie *et al.*, 1991) and analysis of variance, combined with multiple comparison tests (Wonnacott & Wonnacott, 1985). The page-test and the calculation of R-indices were programmed in Turbo Pascal. Anova was performed on the Statistical Analysis System package (SAS, 1982). Performance of the panel as a whole was checked using factoranalysis and by calculating the Pearson's correlationmatrix. Both methods were programmed using SAS (1982).

The reaction order of the heat induced hardness degradation was determined using nonlinear regression analysis on the sensorial and the tenderometer data and by examining the tendency of residues.

The thermal destruction kinetics of hardness of beans was described using two models: the thermal death time concept (Bigelow, 1921) (eq.1) and the Arrhenius model (Arrhenius, 1889) (eq.2).

$$D = D_{ref} \; 10^{(T_{ref}-T)/z} \tag{1}$$

D	=	decimal reduction time (min)
D_{ref}	=	decimal reduction time at T_{ref} (min)
T	=	temperature (°C)
T_{ref}	=	reference temperature (°C)
z	=	temperature dependence of D-value (°C)

$$k = k_0 \exp (-E_a/RT) \tag{2}$$

k	=	reaction rate constant
k_0	=	reaction rate constant at $T = \infty$
E_a	=	activation energy (J/mole)
T	=	temperature (K)
R	=	8.3065 J/K mole

The kinetic parameters (k, E_a) and (D, z) were estimated by use of three different least squares methods (Ratkowsky, 1983; Haralampu *et al.*, 1985) : two-steps linear regression, multiple linear regression and nonlinear regression. All calculations were performed using SAS (1982).

The correctness of the z-values determined by the taste panel and by the tenderometer was validated using the Kramer test (Wonnacott & Wonnacott, 1985).

C. Heat penetration study to model heat transfer

Beans were processed in glass jars (Carnaud-Giralt Laporta S.A., Spain) of 172mm height, 40.5mm diameter and a glass thickness of 2.6mm, each jar containing 440g of soaked beans and filled with distiled and demineralized water until the desired gross headspace was reached. Still and end-over-end rotary processes were simulated in a Steriflow simulator (single-door unit with one cage), microflow type 911R n°877 of Barriquand (France). The datalogger (MDP 8250 analog input-relay output system from Mess + System Technik GmbH) was installed with a thermocouple box containing the reference junction and connected to a personal computer. The μV signal was converted to °C with a maximum error of this conversion of 0,01°C in a temperature range of 0 to 170°C. The datalogger was calibrated using the PVG77 thermocouple voltage calibrator from Ellab (Denmark). The output was recorded with an accuracy of 0.1°C. For rotational processes a slipring contact (DCS85-12, Ellab, Denmark) was needed. All temperature sensors were copper-constantan thermocouples (type T) of Ellab (Denmark). For the heat penetration measurements needle type thermocouples with rounded tip (160x1.2mm, SSA-12xxx-G700-SF, where xxx stands for the length of the needle) were used while for the coldest spot determination temperatures were registrated using four point probes (ST4-11120-G700-SL). Thermocouples were calibrated in icewater (0°C) and at the temperature used during the heat penetration tests (121°C). The entire length of the lead wire was in the retort during the test. All thermocouples measured temperatures within 0.2°C and could be used for the heat penetration study.

For the coldest spot determination four-point probes measured temperatures at different depths along the central axis of three glass jars. The coldest spot in a container was defined according to Zechmann and Pflug (1989) as the location in the container which received the lowest sterilisation value (F_0-value) from the process. The F_0-value was calculated using the general method of Bigelow (1921). The heating characteristics (f_h and j) were determined by running experiments on five containers using needle type thermocouples with a rounded tip (160x1.2mm, SSA-12xxx-G700-SF) placed in a bean at the previous determined coldest spot. The containers of interest were placed along the central axis of the cage to avoid variations due to different rotation angles. Simultaneously, two thermocouples (SSR-60020-G700-SF) registrated the heating medium temperature every 15 seconds. An equilibration phase to an initial product temperature of 40°C preceded all processes. Processes consisted of a linear increase in temperature upto 121°C during 16min followed by a holding phase (13min) at 121°C. As the degree of agitation depends mainly on the rotational speed, the size of the headspace bubble and the consistency of the product (Berry and Bradshaw, 1980), the influence of headspace (10 and 20mm) and of rotational speed (0, 5, 10, 15 and 20 rpm) on the heat penetration characteristics was studied.

D. Optimisation procedure

The optimal processing barema, defined as the temperature-time combination which results in a food product with minimum surface cook-value after achieving the desired degree of sterility, was estimated using a computer program developed at the Unit of Food Preservation of the KUL. In the development of the program some simplifying assumptions were made : there is no heat transfer resistance at the product surface (infinite Biot number), the time-temperature profile consists of a come-up-time with a linear increase of temperature from the initial temperature to the constant holding temperature, the food product is homogeneous and isotropic, the initial temperature of the food is uniform and kinetic parameters are described by the empirical D-z-model. The flowdiagram of this optimisation program is shown in figure 2.

The process parameters were $T_0 = 40°C$, $T_w = 15°C$, CUT=16min, HT=13min and $t_c = 17$min. The kinetic parameters for *Cl. botulinum* spore inactivation ($z_m = 10°C$, $D_{ref} = 0.21$min, $T_{ref} = 121.15°C$) were selected while for thermal softening of white beans kinetic parameters were determined as described earlier (*cfr.* B. Kinetic study). Based on literature values for low acid canned foods in general (Pflug, 1987a, b) 6 min was chosen as target process-value. The heating characteristics of white beans were determined in a heat penetration study. Berry and Bush (1989) found the extrapolation of data to other retort temperatures a safe practice for conduction or induced convection heated products.

As starting value for the iterative estimation of the optimal temperature the generalized (semi)-empirical formula for conduction heated foods with infinite surface heat transfer coefficient (eq. 3) developed by Hendrickx *et al.* (1993) is applied.

$$T_{opt} = 86.68 + 9.74 \log(F_t/f_h) + 10.45 \ln z_q + 0.025 T_0 \qquad (3)$$

Secondly, the holding time, initialised at zero, is incremented iteratively by a Δt_h (eq.5) until the desired degree of sterility (target process-value calculated using eq.4) is achieved at the coldest spot of the product.

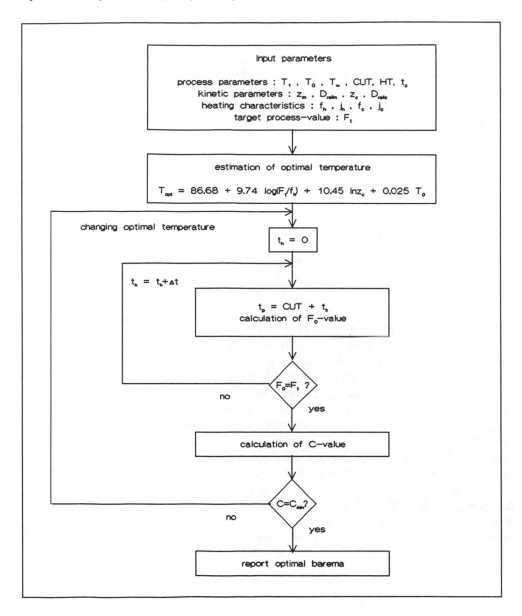

Figure 2 : flowsheet of the computer program to calculate optimal processes

$$F_t = \int_0^t 10^{(T-T_{ref})/z_m} \, dt \qquad (4)$$

$$\Delta t_h = \frac{F_t - F}{10^{(T_1-T_{ref})/z_m}} \qquad (5)$$

Knowing the time-temperature profile to reach the target process-value the resulting surface cook-value can be calculated using eq. 6, 7 (contribution of respectively the come-up-time and the holding time to the surface cook-value) and 8. Since an infinite surface heat transfer coefficient was assumed the temperature profile of the product surface equals the retort temperature profile. To optimise for the surface cook-value the computer program is integrated in an optimisation routine, using the Davis, Swann and Campey method (Saguy, 1983).

$$C_1 = \frac{z_q \cdot CUT}{2.303 \cdot (T_1 - T_0)} \left(\exp\left(2.303 \, \frac{T_1 - T_{ref}}{z_q} \right) - \exp\left(2.303 \, \frac{T_0 - T_{ref}}{z_q} \right) \right) \quad (6)$$

$$C_2 = \exp\left(2.303 \, \frac{T_1 - T_{ref}}{z_q} \right) HT \quad (7)$$

$$C_{tot} = C_1 + C_2 \quad (8)$$

E. Evaluation procedure of optimal processes

Since in the optimisation procedure a few simplifying assumptions were made a trained taste panel and an industrial tenderometer checked the theoretically computed optimal processes. Test conditions were the same as in the kinetic study. The sensory evaluation technique was a pairwise ranking test with Friedman analysis (Meilgaard, 1987). The samples were presented using a randomized balanced block design with replication (Cochran and Cox, 1957). Four or five pairs of white beans were presented simultaneously and twelve trained judges were asked to indicate the hardest sample of each pair. The first aim of these evaluations was to check whether the taste panel and the tenderometer were able to distinguish between the optimal process and equivalent processes at temperatures deviating 4°C and 8°C from the optimal temperature. Hereto, the optimal and deviating still processes with a headspace of 10mm and since a rotational speed above 15 rpm did not improve the heat penetration neither the quality retention (more broken beans) the optimal and deviating processes at 15 rpm and a headspace of 20mm were executed in the pilot retort. A second purpose was to indicate an in-batch variation in the optimal processes. If still and rotary processes resulted in distinguishable quality levels, the third aim of the evaluation, was checked comparing the optimal still process with the optimal rotating process.

IV. RESULTS AND DISCUSSION

A. Kinetic study
1. Statistical evaluation of the sensory analysis

Statistical analysis, including examination of the panel performance was applied to the sensorial data. To evaluate whether differences in quality levels were detectable, two non parametric tests, the page-test and the calculation of R-indices and one parametric test, an analysis of variance, were used.

The page-test (O'Mahony, 1986) is a related-samples non parametric test to evaluate a ranked tendency. If the calculated L-value exceeds the tabled value, the null hypothesis (no ranked tendency) is rejected in favour of the alternative hypothesis. The tendency to be expected for a given temperature was the shortest process to be ranked first and further ranking according to the process time. Each test rejected the null hypothesis at a level of significance of 0,1%, what a ranking of the quality intensity according to the processing time justified. The R-index (O'Mahony *et al.*, 1980; O'Mahony, 1986; Vie *et al.*, 1991) is the probability of a given judge distinguishing correctly between two items. The greater the degree of difference, the higher the probability of discrimination. $R = 1$ indicates samples which are perfectly distinguishable, $R = 0.5$ represents correct discrimination by accident. Intermediate values indicate the magnitude of discrimination by accident. At each temperature R-indices were calculated for six processing times. Hardness of samples treated during short times was clearly different ($R = 1$). Confusion (decrease of R-indices) appeared as processing time increased. As a consequence hardness decay faster in the beginning of the process, indicating a non-linear degradation with time (n^{th} order with $n \neq 0$). Analysis of variance on the sensorial data indicated differences between samples, processed at a given temperature during increasing processing time, at the 0,1% level. Tukey HSD tests (Wonnacott & Wonnacott, 1985) pointed out, as the R-indices, that differences between samples treated during short times were obvious while samples undergoing longer heat treatments differed slightly.

Performance of the individual judges (validity and reliability) was evaluated using two-out-of-five tests, by presenting duplicate test samples and replications. None of the panelmembers showed serious deviations in their judgements during the sessions. Factor analysis and the calculation of Pearson's correlation matrix examined the performance of the panel as a whole. The length of the vectors in the biplot presentation indicates a very good reconstruction in the two dimensional plane. The high correlation of these vectors (representing different panelmembers) with factor 1 and the small angles between the vectors prove that all panellists judged the quality parameter hardness in the same way. The same conclusion can be made by inspection of the correlation matrix. All the judgements of the panelmembers were highly correlated with each other (correlation coefficient ranging from 0.754 to 0.916).

2. Estimation of reaction order

The reaction order of thermal softening of white beans was estimated by nonlinear regression analysis on the sensorial and the tenderometer data, respectively 0.36 ± 0.17 and 1.18 ± 0.10. To verify these results, for each temperature a linear regression analysis was performed with the score values on one hand and with the logarithmic values on the other hand as a function of processing time. The tendency of residues was a measure to determine if the reaction was zero or first order just as the coefficient of determination (R^2) of the linear regressions. The residues of the regression on the sensorial data showed no systematic tendency and higher correlation coefficients were observed for a first order reaction. Based on the results of nonlinear regression, the tendency of residues and the coefficient of determination, a first order reaction (eq. 9) for the heat induced hardness degradation evaluated by a trained taste panel was assumed to calculate kinetic parameters.

$$\log A = \log A_0 - kt / 2.303 = \log A_0 - t / D \qquad (9)$$

A	=	quality at time t
A_0	=	initial quality
k	=	reaction rate constant (min^{-1})
t	=	time (min)
D	=	decimal reduction time (min)

On the contrary, the residues of a linear regression analysis on the logarithms of the tenderometer data of beans treated for longer processing times showed a systematic tendency (fig. 3). The rate of thermal softening of beans could not be described by a first order reaction. Based on a study of Huang and Bourne (1983) a two phase model was used to describe the tenderometer hardness evaluation. The heat induced hardness degradation can be seen as a sum of two independent first order processes. These processes take place simultaneously but with different reaction rate constants.

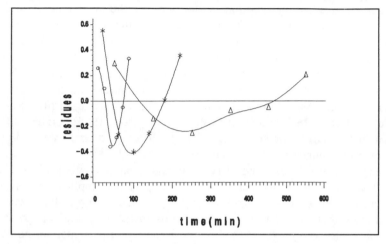

Figure 3 : plot of the residues of a linear regression analysis on the logarithms of hardness of beans evaluated using a tenderometer, as a function of processing time and temperature
Δ = 90°C, * = 100°C, o = 110°C

3. Determination of kinetic parameters for hardness degradation

Based on a sensorial hardness evaluation of beans kinetic parameters were estimated using three least squares regression methods considering both the thermal death time (TDT) method and the Arrhenius model.

The most common method to estimate TDT or Arrhenius parameters is a successive two-steps linear least squares fit. First, common logarithms of average score values for hardness degradation are plotted versus heating times to obtain the reaction rate constant k and consequently the decimal reduction time D, since $k = \ln 10/D$ at each temperature (fig. 4). The second step in the Arrhenius model is to regress the natural logarithm of the reaction rate constant, lnk, versus the reciprocal of the absolute temperature, $1/T$, to obtain the estimates of $\ln k_0$, the intercept and E_a/R, the slope index of the regression line (eq. 2) (fig. 5). In the TDT method, common logarithms of decimal reduction times are plotted against heating temperatures. From the linear regression equation the z-value can be derived (eq. 1), representing the temperature change necessary to change the rate of the sensory quality deterioration one order of magnitude. The results of a two-steps linear regression analysis are given in table 1. Two-steps linear regression has the disadvantage of applying regression on regression coefficients. Errors on the first regression coefficients were not transformed to the second regression coefficients. Also according to Haralampu *et al.* (1985) and Cohen and Saguy (1985) two-steps linear regression gives the least accurate estimates for Arrhenius parameters, probably because it estimates too many intermediate values and by not considering the data set as a whole.

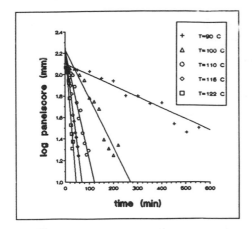

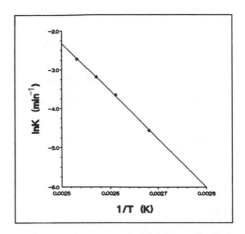

Figure 4 : linear regression analysis on the logarithms of the panelscores for the evaluation of hardness of beans treated at different temperatures to obtain the reaction rate k and the decimal reduction time D at each temperature.

Figure 5 : Arrhenius plot to determine the activation energy for the heat induced hardness degradation of white beans, measured by a trained taste panel.

To overcome this inaccuracy a multiple linear regression analysis was suggested. This regression technique gains strength by estimating fewer parameters and by analyzing the data set as a whole (Haralampu *et al.*, 1985). By substituting eq.(1) in (9) and eq.(2) in (9) and subsequently taking the common logarithms, formula (10) and (11) can easily be derived :

$$\log \left(\log \frac{A}{A_0} \right) = \log D_{ref} - \frac{1}{z} (T - T_{ref}) - \log (t) \tag{10}$$

$$\ln \left(\ln \frac{A}{A_0} \right) = \ln k_{ref} - \frac{E_a}{R} \left(\frac{1}{T} - \frac{1}{T_{ref}} \right) - \ln (t) \tag{11}$$

Equations (10) and (11) can be modelled as :

$$Y = a + bX_1 + cX_2$$

Estimated values of z and E_a by multiple linear regression are reported in table 1. Deviation from one of the estimated c-parameter can be taken as a measure of accuracy of the regression. In both cases, TDT and Arrhenius, c was estimated as 1,05 which indicated a rather accurate regression analysis. Haralampu *et al.* (1985) indicated biased but very precise regression coefficients using multilinear regression analysis to estimate Arrhenius parameters. Bias could appear because the reaction rate estimates were not independent (i.e. they are correlated through the multiple linear regression) thus violating an assumption of least squares regression. Although, Haralampu *et al.* (1985) preferred multilinear regression to two-steps linear regression.

Table 1
Activation energies (kJ/mole) and z-values (°C) for thermal softening of white beans processed in a temperature range of 90°C to 122°C evaluated by a trained taste panel and using a tenderometer

	TDT-model z (°C)	Arrhenius-model E_a (kJ/mole)
Taste panel		
two-steps	27.6 ± 0.9	102.0 ± 1.3
multilinear	21.3 ± 0.4	130.8 ± 1.3
nonlinear	29.0 ± 1.8	97.0 ± 6.1
Tenderometer		
phase 1	25.4 ± 2.8	105.1 ± 20.1
phase 2	25.2 ± 2.8	105.2 ± 12.4

A third method, nonlinear regression, estimated z- and E_a-values using the Marquardt procedure. As for multilinear regression, it performed a single regression on all data without calculating rates at each temperature. Formula 10 and 11 together with the derivatives of $\ln(A/A_0)$ to k_{ref} and E_a for Arrhenius parameters and to D_{ref} and z for TDT parameters are required as input equations. Results of the previous two methods were used as first estimates for z and E_a. The results of a nonlinear regression analysis are reported in table 1. Nevertheless, one must be aware that also the estimates of a nonlinear regression can be biased. Least squares estimators in linear models are unbiased, normally distributed and achieve the minimum variance bound given the assumption of an independent homoscedastic normally distributed error. For nonlinear models, the least squares estimators have these properties only assymptotically, i.e. for infinite sample size (Ratkowsky, 1983). Secondly, good starting values are vital according to Myers (1990). Poor values may converge in a local minimum of the error mean squares function.

As E_a and z-values differ according to the least squares method applied, the question arises which values can be used if all restrictions of the methods are considered. Least accurate estimates, largest confidence intervals, estimation of unnecessary parameters and regression on regression coefficients made the two-steps linear regression inferior to the other methods. While Myers (1990) favoured nonlinear regression, the arguments of Ratkowsky (1983) are a plea to linearize data. The possibility of biased estimates injures the multilinear regression technique while problems with convergence and dependence of the estimated parameters on the initial values makes the nonlinear regression less attractive. Since calculations in multiple regression are less complex than in the nonlinear case and indications of accurate estimations were present, results of multiple linear regression analysis were selected for the optimisation study.

The objective evaluation of thermal softening of beans could be described using a two-phase model. For both phases activation energies were estimated using two-steps linear regression analysis. First, for each temperature natural logarithms of relative tenderometer values were plotted versus heating times to obtain the reaction rate constants k_1 and k_2 of respectively phase 1 and phase 2 of thermal softening (fig. 6). Processing times at 116°C and 122°C were too short to distinguish two first order processes. These temperatures were eliminated for the two phase approximation. At each temperature the rate constants of phase 1 and phase 2 are obviously different as can be seen from the different slopes of the regression lines in fig. 6. The second step in the Arrhenius model is to regress the natural logarithm of the rate constants, $\ln k_1$ and $\ln k_2$, versus the reciprocal of the absolute temperature, $1/T$, to obtain the estimates of the activation energies E_{a1} and E_{a2} from the slope index of the regression lines (*cfr.* eq. 2). Regression lines to determine activation energies of thermal softening of beans are more or less parallel so that activation energies for both phases of hardness degradation of beans measured by a tenderometer are equal (105 kJ/mole). This contradicts the results of Huang & Bourne (1983). They indicated a higher activation energy for the first phase of thermal softening of white beans (*Phaseolus vulgaris*) soaked during 16hours in tap water ($E_a = 104$ kJ/mole) than for the second ($E_a = 54$ kJ/mole).

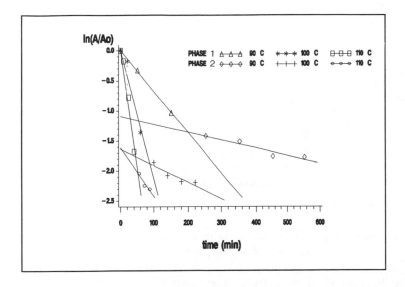

Figure 6 : Two phase approximation of thermal softening of beans. The natural logarithms of relative hardness are plotted versus heating times at 90°C, 100°C and 110°C to obtain rate constants at different temperatures.

4. Evaluation of the kinetic parameters

To evaluate the correctness of the determined parameters, the judges had to evaluate a series of samples of the same quality level processed at different temperatures (T = 100°C, 110°C, 116°C and 122°C). Processing times were calculated according to the method described by Ohlsson (1980 a, b) using the z-value for thermal softening previously determined by the taste panel. The panelmembers were asked to rank the samples on the same scale as used for the determination of the kinetic parameters. The Kramer test (Wonnacott & Wonnacott, 1985) was used to check if the panel as a whole was able to pin-point differences between the samples. No differences were found which meant that either the ranked data were insufficient to indicate differences or - and this is only hypothetical because statistical tests are not designed to indicate similarities - that quality levels were really the same. There was concluded that the panelmembers could not distinguish between samples of the same quality level processed at different temperatures.

5. Comparison subjective versus objective measurement

Factoranalysis was applied to examine the correlation between the taste panel and the tenderometer evaluation of thermal treated beans. The length of the vectors in the biplot presentation indicates a very good reconstruction in the two dimensional plane. The vector representing the tenderometer evaluation is highly correlated with those representing the panelmembers. The correlationcoefficient for an objective versus a subjective evaluation of thermal softening of beans was 0.88. However when relative objective data are plotted versus relative subjective data (fig. 7) a difference in hardness evaluation according to the measuring technique can be noticed.

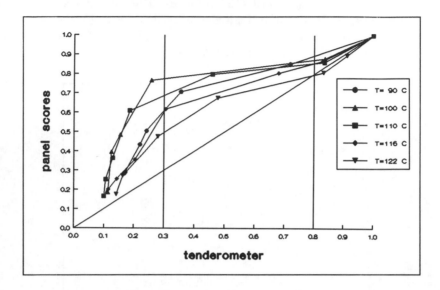

Figure 7 : plot of relative panel scores versus relative tenderometer values for hardness evaluation of beans treated at different temperatures.

Figure 7 can be divided in three parts. For short processing times a tenderometer evaluation and a taste panel evaluation give analogous results. In the second part the tenderometer scores decrease faster than those of the taste panel which indicates a higher sensitivity of the tenderometer hardness evaluation. For longer processing times the taste panel seems to be more sensitive to a further hardness decay. Based on the difference in reaction order, in kinetic parameter estimates and in sensitivity to hardness degradation according to a subjective or an objective measurement the tenderometer cannot be used to describe thermal softening of beans measured by a trained taste panel. Consequently kinetic data of taste panel evaluation were selected for the optimisation study.

B. Heat penetration study

For different processing conditions the coldest spot was determined as the location in the container which received the lowest sterilisation value from the process. An increase in rotational speed or headspace moved the coldest spot from the bottom towards the centre of the jar.

The obtained heat penetration characteristics at the coldest spot (table 2) are specific for the specified product, fill weight, headspace, type of container, dimensions of the container, rotational speed, come-up-time of the retort, heating medium and initial temperature. The influence of headspace on the heat penetration rate for different rotational speed was studied using analysis of variance (SAS, 1982). For still processes no differences in f_h-values were found on the 5% significance level due to headspace whereas for rotary processes (rpm=5, 10, 15 or 20) f_h-values were distinguished on the 5% level.

Table 2
influence of rotational speed and headspace on the heating characteristics measured at the coldest spot of a container and on the calculated optimal temperature

Rpm	Headspace (mm)	f_h (min)	j_h	T_{opt} (°C)	HT
0	10	13.4±0.6	6.3±0.6	118	25.5
	20	12.9±0.5	6.6±0.5	118	23.9
5	10	9.5±0.4	12 ± 2	121	13.1
	20	8.5±0.3	19 ± 3	122	10.5
10	10	8.1±0.3	21 ± 3	122	9.6
	20	7.4±0.3	31 ± 3	123	8.1
15	10	7.5±0.3	29 ± 5	123	7.9
	20	6.7±0.3	41 ± 6	123	7.6
20	10	7.0±0.3	40 ± 8	123	7.9
	20	6.5±0.4	56 ± 24	123	7.6

Comparison of the mean f_h-values for different rotational speed at a certain headspace indicated differences at the 5% level except for 20mm headspace the difference in heat penetration characteristics between 10 and 15 rpm and between 15 and 20 rpm was not significant. Berry and Dickerson (1981) studied whole kernel corn in brine processed in a Steritort and found that the effect of headspace on still processes was not significant. For agitating retorts however an increase in headspace within normal limits tended to increase the sterilisation value because of the increased ability of the brine to move effectively through the agitated can. As can be seen in table 2 increasing rotational speed results in a faster heat penetration (lower f_h-value). Several authors studying axial or end-over-end rotation came to the same conclusion (Conley et al., 1951; Berry et al., 1979; Berry and Bradshaw, 1980; Naveh and Kopelman, 1980; Berry and Dickerson, 1981; Berry and Bradshaw, 1982; Berry and Kohnhorst, 1985). However the influence of rotational speed on the heat penetration rate is limited. Besides as rotational speed was increased to 20 rpm, more broken beans and consequently leakage of starch appeared. Therefore an optimal heat penetration with minimal quality destruction is obtained at a rotational speed of 15 rpm.

C. Optimisation procedure

Table 2 shows the computed optimal processes for different processing conditions. Both for still and rotary processes an increase of headspace had little effect on the optimal temperature, which was expected since headspace had a main effect on the j-value and little effect on the f_h-value (table 2). The optimal temperature is loglinearly correlated to the f_h-value while the heat penetration parameter j has a minor impact on the optimal temperature (Hendrickx et al., 1993). Agitation resulted in higher optimal temperatures and better quality retention as compared to still processes. However, the advantage of processing at higher rotational speed is limited. Above 15 rpm, increasing rotational speed has a limited influence on the quality retention. Ohlsson (1980 b, c) and Hendrickx et al. (1993) plotted optimal temperatures as a function of f-slope values for different z_q and F_0-values. A change of f-slope value from 5 min to 10 min, which was the range of agitated white beans, corresponded to a change of ± 3°C in optimal temperature.

Table 3
Optimal and deviating processes executed in the pilot retort to check whether the optimisation
results are valid

Rpm/HS	Process	Temp. (°C)	HT (calculated) (min)	HT (requested) (min)
0 /10	opt.	118	25.5	25.2
	dev.1	122	16.9	15.1
	dev.2	110	90.3	85.4
15/20	opt.	123	7.6	7.0
	dev.1	119	13.8	12.9
	dev.2	115	28.6	25.6

D. Taste panel and tenderometer evaluation of optimal processes

To validate the computed optimal processes optimal as well as deviating processes, reported in table 3, were executed in the pilot retort and evaluated by a trained taste panel and a tenderometer. Deviations of 4 and 8°C from the optimal temperature were distinguished on the 1%level except for still processed beans at 122°C which showed better quality than the computed optimal process at 118°C. The beans processed at 122°C were found harder ($P<0.05$) than the beans processed at the calculated optimal temperature. The process deviating 8°C was softer ($P<0.01$) than the process at the calculated optimal temperature. In this case the taste panel rejected the calculated optimal process and accepted the still process at 122°C as optimal. Results of the tenderometer evaluation confirmed the ranking of these processes. A still process at 122°C was also judged harder than the calculated optimal process at 118°C. This can be explained by the assumptions made in the optimisation procedure. Surface resistance to heat transfer was considered negligible which is only valid for condensing steam sterilisation and processing of metal containers (Tucker and Holdsworth, 1990) while white beans were processed in glass jars. The influence of a finite surface heat transfer coefficient on optimal temperature has not yet been systematically studied but it is expected that for low Biot numbers, when thermal gradients occur only during a very small time period compared to the total holding time, the optimal temperature increases (Silva *et al.*, 1993). A second reason is the optimisation of the surface cook value whereas for hardness degradation the volume average quality should be considered. Optimal temperature to maximize overall quality is higher than the one to maximize surface quality because with higher temperatures, higher overprocessing at the surface is compensated by higher overall quality (Silva *et al.*, 1993). Ohlsson (1980d) stated that optimal temperature for maximizing overall quality of an infinite slab are about 7.5 to 10°C higher than the correspondent optimum to maximize surface quality. For rotary processes the taste panel distinguished between deviations of 4°C and deviations of 8°C on the 5%level. Data were not sufficient to indicate in-batch variation. The question if quality retention is improved by agitation was positively answered by the taste panel. Agitated beans were significantly harder ($P<0.01$) than still processed beans.

For an initial temperature of 40°C, a come-up-time of 16min and a process-value of 6min, optimal hardness of beans was reached in a still process at 122°C during 15min. Even better quality was reached by agitation : for a headspace of 20mm and a rotational speed of 15rpm, the optimal processing temperature was 123°C during 7min holding time.

V. CONCLUSION

Kinetic parameters for thermal softening of white beans were estimated using a multilinear regression analysis on the sensorial data whereas for hardness evaluation using a tenderometer a successive two-steps linear regression analysis estimated activation energies of both phases of the two phase approximation. Because of the difference in reaction behaviour, in kinetic parameters estimates and in sensitivity to hardness degradation the tenderometer does not predict thermal softening measured by a taste panel. Consequently kinetic parameters measured by a taste panel are recommended in the optimisation study

Based on heat penetration data, kinetics of microorganisms and quality indices, process conditions and target process-value, optimal still and rotational processes were calculated for white beans in brine. The computed optimal processes were evaluated by a taste panel and a tenderometer. Rotary versus still processes could be distinguished on a 1% level. Deviations from the optimal temperature were also distinguished on the 1% level except for still processed beans where the first deviation ($T = T_{opt}$-4°C) showed better quality than the calculated optimal process. Although these optimal processes are only valid for the considered container, product, fill weight, retort, location in the retort, processing medium and processing conditions this work confirms the usefulness of the surface cook-value as optimisation parameter and the efficiency of the simulation approach to design optimal processes.

VI. ACKNOWLEDGEMENTS

This research has been performed as a part of the Food-Linked Agro-industrial Research programme (project AGRF-CT90-0018), supported by the European Commission. This research was also funded by the "Vlaamse Executieve" (project EG/FLAIR/20).

VII. NOTATION

a, b, c	parameters to be estimated
A	quality parameter
CUT	time to reach sterilisation temperature (min)
D	decimal reduction time (min)
dev	deviation
E_a	activation energy (J/mole)
F_0	sterilising value at reference temperature 121.15°C and $z_m = 10$°C
F_t	target process-value (min)
f_h	slope factor of a heating curve (min) (Ball terminology)
HS	headspace (mm)
HT	holding time (min)
j	heating lag factor (dimensionless)
k	reaction rate constant (min^{-1})
opt	optimal
P	probability level
R	universal gas constant = 1.9872 cal/(K mol)
R^2	coefficient of determination
rpm	rotations per minute
T	temperature (°C or K)

T_1 processing temperature (°C)
T_w cool water temperature
T_{opt} optimal processing temperature
TDT thermal death time
t time (min)
t_c cooling time
z_m temperature dependence of the decimal reduction time of an index
microorganism (°C)
z_q temperature dependence of the decimal reduction time of a quality index
(°C)

Subscripts

0 initial value
ref reference temperature

VIII. REFERENCES

Arrhenius, S., Über die reaktionsgeschwindigkeit bei der inversion von rohrzucker durch Säuren, Zeitschrift für physikalische chemie, 4(2), 226-248, 1889.

Ball, C.O., Olson, F.C.W., Sterilisation in food technology, McGraw-Hill Book Company, Inc., New York, 1957.

Banga, J.R., Perez-Martin, R.I., Gallardo, J.M., Casares, J.J., Optimization of the thermal processing of conduction-heated foods : study of several objective functions, *J. Food Eng.*, 14, 25-31, 1991.

Berry, M.R. jr., Bradshaw, J.G., Heating characteristics of condensed cream of celery soup in a steritort: heat penetration and spore count reduction, *J. Food Sci.*, 45, 869-874, 879, 1980.

Berry, M.R. jr., Bush, R.C., Establishing thermal processes for products with straight-line heating curves from data taken at other retort and initial temperatures, *J. Food Sci.*, 54(4), 1040-1042, 1046, 1989.

Berry, M.R. jr., Dickerson, R.W. jr., Heating characteristics of whole kernel corn processed in a steritort, *J. Food Sci.*, 46, 889-895, 1981.

Berry, M.R. jr., Savage, R.A., Pflug, I.J., Heating characteristics of cream-style corn processed in a steritort: effects of headspace, reel speed and consistency, *J. Food Sci.*, 44, 831-835, 1979.

Berry, M.R.Jr., Bradshaw, J.G., Heat penetration for sliced mushrooms in brine processed in still and agitating retorts with comparison to spore count reduction, *J. Food Sci.*, 47, 1698-1704, 1982.

Berry, M.R.Jr., Kohnhorst, A.L., Heating characteristics of homogeneous milk-based formulas in cans processed in an agitating retort, *J. Food Sci.*, 50, 209-214, 253, 1985.

Bigelow, W.D., The logarithmic nature of thermal death time curves, *The journal of infectious diseases*, 29, 528-536, 1921.

Cochran, W.G., Cox, G.M., Experimental design. Second edition John Wiley & Sons, New York, 1957.

Cohen, E., Saguy, I., Statistical evaluation of Arrhenius model and its applicability in prediction of food quality losses, *J. Food Proc. Preserv.*, 9, 273-290, 1985.

Conley, W., Kaap, L., Schumann, L., The application of "End-over-end" agitation to the heating and cooling of canned food products, *Food Technol.*, 5(11), 457-460, 1951.

Haralampu, S.G., Saguy, I., Karel, M., Estimation of Arrhenius model parameters using three least squares methods, *J. Food Proc. Preserv.*, 9, 129-143, 1985.

Hayakawa, K.-I., Timbers, G.E., Influence of heat treatment on the quality of vegetables: changes in visual green color, *J. Food Sci.*, 42(3), 778-781, 1977.

Hendrickx, M., Silva, C.,Oliveira, F., Tobback, P., Generalized (semi)-empirical formulae for optimal sterilization temperatures of conduction-heated foods with infinite surface heat transfer coefficients, *J. Food Eng.*, 19, 141-158, 1993.

Hendrickx, M., Silva, C, Oliveira, F., Tobback, P., Optimization of heat transfer in hermal processing of conduction heated foods, In *Advances in food engineering*, Singh, R.P., Wirakartakusumah, A., eds., CRC Press, Boca Raton, FL., 1992.

Huang, Y.T., Bourne, M.C., Kinetics of thermal softening of vegetables, *J. Texture Stud.*, 14, 1-9, 1983.

Lund, D.B., Design of thermal processes for maximizing nutrient retention, *Food Technol.*, 2, 71-78, 1977.

Lund, D.B., Quantifying reactions influencing quality of foods : texture, flavour and appearance, *J. Food Proc. Preserv.*, 6, 133-153, 1982.

Meilgaard, M., Civille, G.V., Carr, B.T., Sensory evaluation techniques, CRC press, Inc., Boca Raton, Florida, 281p, 1987.

Myers, R.H., Classical and modern regression with applications, Second edition. PWS - KENT Publishing Company, 488p, 1990.

Nadkarni, M.M., Hatton, T.A., Optimal nutrient retention during the thermal processing of conduction-heated canned foods : Application of the distributed minimum principle, *J. Food Sci.*, 50, 1312-1321, 1985.

Naveh, D., Kopelman, I.J., Effects of some processing parameters on the heat transfer coefficients in a rotating autoclave, *J. Food Proc. Preserv.*, 4, 67-77, 1980.

O'Mahony, M., Garske, S., Klapman, K., Rating and ranking procedures for short-cut signal detection multiple difference tests, *J. Food Sci.*, 45, 392-393, 1980.

O'Mahony, M., Sensory evaluation of food : statistical methods and procedures, Marcel Dekker, Inc., New York and Basel, 487p, 1986.

Ohlsson, T., Temperature dependence of sensory quality changes during thermal processing, *J. Food Sci.*, 45, 836-839, 847, 1980a.

Ohlsson, T., Optimal sterilisation temperatures for flat containers, *J. Food Sci.*, 45, 848-852, 859, 1980b.

Ohlsson, T., Optimal sterilization temperature for sensory quality in cylindrical containers, *J. Food Sci.*, 45, 1517-1521, 1980c.

Ohlsson, T., Optimization of heat sterilization using C-values, In: Food Process Engineering. Applied Science Publishers, U.K., 137-145, 1980d.

Pflug, I.J., Calculating F_T-values for heat preservation of shelf-stable, low-acid canned foods using the straight-line semilogarithmic model, *J. Food Protection*, 50(7), 608-615, 1987a.

Pflug, I.J., Factors important in determining the heat process value, F_T, for low-acid canned foods, *J. Food Protection*, 50(6), 528-533, 1987b.

Ratkowsky, D.A., Nonlinear regression modeling. A unified practical approach, Marcel Dekker, Inc., New York and Basel, 276p, 1983.

Saguy, I., Karel, M., Optimal retort temperature profile in optimizing thiamin retention in conduction-type heating of canned foods, *J. Food Sci.*, 44, 1485-1490, 1979.

Saguy, I., Optimization methods and applications, In: Saguy, I. (ed.) Computer-aided techniques in food technology, Marcel Dekker, New York, 1983.

SAS, SAS Institute Inc. SAS User's guide: statistics, 1982 edition Cary, NC: SAS Institute Inc., North Carolina, 1982.

Sawyer, F.M., Interaction of sensory panel and instrumental measurement, *Food Technol.*, 25(3), 51-52, 1971.

Silva, C., Hendrickx, M., Oliveira, F., Tobback, P., Optimal sterilization temperatures for conduction heating foods considering finite surface heat transfer coefficients, *J. Food Sci.*, 57(3), 743-748, 1992.

Silva, C., Oliveira, F. Hendrickx, M., Modelling optimum processing conditions for the sterilization of prepackaged foods, *Food Control*, 4 (2), 67-78, 1993.

Stone, H., Sidel, J., Oliver, S., Woolsey, A., Singleton, R.C., Sensory evaluation by quantitative descriptive analysis, *Food Technol.*, 28(11), 24-33, 1974.

Teixeira, A.A., Dixon, J.R., Zahradnik, J.W., Zinsmeister, G.E., Computer determination of spore survival distributions in thermally-processed conduction heated foods, *Food Technol.*, 23(3), 352-354, 1969.

Teixeira, A.A., Zinsmeister, G.E., Zahradnik, J.W., Computer simulation of variable retort control and container geometry as a possible means of improving thiamine retention in thermally processed foods, *J. Food Sci.*, 40, 656-659, 1975.

Tucker, G.S., Holdsworth, S.D., Optimization of quality factors for foods thermally processed in rectangular containers, In: Field, R.W., Howell, J.A. (eds.), Process Engineering in the Food Industry. 2, Convenience foods and quality assurance, Elsevier Applied Science, 59-74, 1990.

Vie, A., Gulli, D., O'Mahony, M., Alternative hedonic measures, *J. Food Sci.*, 56(1), 1-5, 46, 1991.

Wonnacott, R.J., Wonnacott, T.H., Introductory Statistics, fourth edition, John Wiley and sons, Inc., 649p., 1985.

Zechmann, L.G., Pflug, I.J., Location of the slowest heating zone for natural-convection-heating fluids in metal containers, *J. Food Sci.*, 54(1), 205-209, 226, 1989.

DISCUSSION

SASTRY: Where were thermocouples placed for determination of heat penetration curves?

VAN LOEY: In the bean at the coldest spot of the container, During the heat penetration study, 2×5 containers were placed at the central axis of the retort.

BUSTA: An Fo of 6 indicates no safety concerns for safety in the process optimization, is that correct?

VAN LOEY: The aim of the work was the validation of this optimization approach for particulated foods. In the project, there was an agreement between partners to $F_O = 6$ min. as it was not the objective to optimize safety, but the optimization routine used (Davis, Swanna and Campey method) is an optimization routine for one parameter. The study could easily be repeated for other target process-values.

FEASIBILITY OF PROTEIN-BASED TTI DEVELOPMENT

De Cordt, S., Maesmans, G., Hendrickx, M. and Tobback, P.

I. INTRODUCTION

The need for time-temperature-integrators (TTI's) for efficient and correct evaluation of heat sterilization processes on food, pharmaceutical or medical products has been repeatedly proclaimed in the pertinent literature. [e.g. Mulley *et al.*, 1975; Pflug and Smith, 1977; Witonsky, 1977; Rönner, 1990; Weng, 1991; Weng *et al.*, 1991b] It follows from the inadequacies of both the *in situ*- and physical-mathematical methods. [Maesmans, 1993] A tentative definition says that a TTI is a small, inexpensive system having an easily and correctly measurable property that responds irreversibly to time-temperature combinations in such a way that it mimics the changes in a well defined target quality or safety parameter of a product undergoing the same variable-temperature exposure. An application scheme showing how a TTI works is outlined by Taoukis and Labuza [1989] and De Cordt *et al.* [1992b].

The measurement of changes in a quality or safety attribute, or in general, of the process impact F_{par} on such parameter, has several applications. [Maesmans, 1993]

Much attention has been paid to development of microbiological TTI's, but inherent limitations and problems of these systems are numerous. [Mulley *et al.*, 1975; Pflug and Smith, 1977; Witonsky, 1977; Pflug and Odlaug, 1986; Rönner, 1990; Weng, 1991; Weng *et al.*, 1991b; Hendrickx *et al.*, 1992a]

II. CONDITIONS

For a system to be able to function as a TTI, a number of conditions must be fulfilled. Some of the practical and/or economic requirements are explicitly stated in the definition above, but four more conditions, all bearing upon the kinetics of the change in TTI-response, should be read between the lines.

The impact of a heating process (F) on a "system" (TTI or product safety or quality attribute) is defined as

$$F=\int_0^t \frac{k}{k_{ref}}dt \qquad (1)$$

where $k_{(ref)}$ is the rate constant (at reference temperature) and t the time.

A very basic requirement following from (1) is that k_{TTI} and k_{par} must obey the same law as to their temperature dependence. Most often the Arrhenius relation applies, so that

$$F_{TTI}=\int_0^t \exp[\frac{E_{ATTI}}{R}(\frac{1}{T_{ref}}-\frac{1}{T})]dt \qquad (2)$$

$$F_{par}=\int_0^t \exp[\frac{E_{Apar}}{R}(\frac{1}{T_{ref}}-\frac{1}{T})]dt \qquad (3)$$

where E_A is the activation energy, R the universal gas constant and $T_{(ref)}$ the (reference) temperature. Equations 2 and 3 clearly show that, except for the case of isothermal heating at T_{ref}, equality of F_{TTI} and F_{par} can be guaranteed only if $E_{ATTI}=E_{Apar}$. There is a way to convert ^{EA1}F to ^{EA2}F ($E_{A1} \neq E_{A2}$), but then one needs to know either the actual time-temperature profile or good estimates of f_h, f_c and j_c [Pflug and Christensen, 1980], whereas the very omission of such necessity is the main purpose of using TTI's. [De Cordt *et al.*, 1992b] Attempts to derive F_{par} from a set of ^{EAi}F-values where none of the E_{Ai} is equal to E_{Apar}, e.g. with a multi-component TTI, can bring about errors of more than 50%. [Maesmans, 1993]

However useful Eq.2 and 3 are to make clear the condition of equality of E_{ATTI} and E_{Apar}, they are not of practical use in relation with TTI's because there F-values are calculated relying solely on the TTI-response. Therefore, it is also necessary that the TTI response kinetics obey a rate equation that allows separation of the variables, so that a response function (R) can be defined as

$$R(X)=\int_0^t kdt \qquad (4)$$

where X is the TTI-response.

F can then be calculated as

$$F=\frac{R(X)}{k_{ref}} \qquad (5)$$

The explicit form of R depends on the prevalent rate law. In most cases one deals with decay processes and very often the "nth order model" satisfactorily fits the data, so that we can write

$$\frac{dX}{dt}=-kX^n \qquad (6)$$

and

$$R(X)=\ln\frac{X_0}{X_t} \quad \text{if } n=1 \qquad (7)$$

and

$$R(X)=\frac{1}{n-1}[X_t^{1-n}-X_0^{1-n}] \quad \text{if } n \neq 1 \qquad (8)$$

Moreover, Equations 5 and 7-8 show that the latter condition evenso applies to the monitored quality or safety factor, if one wants to derive its actual endpoint level Y_t or its reduction level Y_t/Y_0 from F_{par} (=F_{TTI} if all necessary conditions are fulfilled).
The last kinetic requirement concerns the value of k_{TTI}. In the considered temperature range, this should be of a magnitude that allows measurable TTI-response changes. That is, if there is a response decrease in time (which is the case most often), k_{TTI} should be sufficiently low so as to prevent X_t from falling below the detection limit on the one hand. On the other hand it should be high enough to allow a measurable decrease in X to take place.
It is important to note that there is however no condition of equality of $k_{ref\,TTI}$ and $k_{ref\,par}$.

III. DEVELOPMENT OF PROTEIN-BASED TTI'S

In principle, a TTI could be based on a chemical or a physical reaction, as well as on microbiological (spore) reduction. In neither reaction group, however, it is straightforward to find a system meeting all the conditions to be able to function as a TTI for e.g. the safety of heat treated foods. A possible approach then is to look for systems that have some of the required properties, and to make them fulfill the remaining conditions.
From this point of view, thermostable proteins offer an appropriate basic material because they allow the development of small systems, many of them are relatively low-priced, and they deteriorate with a practically measurable velocity in the pasteurization or sterilization temperature range. Either the enzymic activity or the deterioration heat (ΔH) measured by Differential Scanning Calorimetry (DSC) may serve as response property. Their thermostability characteristics, i.e. the rate constant k and activation energy E_A, can be manipulated in various ways. Thus k can be decreased, bringing into reach a higher range of measurable F-values, and E_A can be shifted towards the value of the E_A characterizing a quality decay or safety progress reaction of interest. The latter objective seems manageable in view of the fact that the range covered by E_A-values of protein denaturation includes both the typical values for safety (thermal spore destruction) as well as quality (e.g. nutrient destruction). [Okos, 1986]

IV. STABILIZATION OF PROTEINS AGAINST HEAT DETERIORATION

A. INTRODUCTION
It is appropriate to distinguish clearly between reversible and irreversible protein deterioration. Whereas reversible changes are widely discussed in the more fundamental field of "conformational" stability of proteins in general, irreversible decay studies mostly deal with the "operational" stability of enzymes in particular.

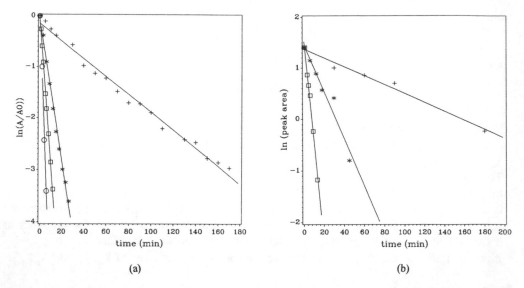

Figure 1. Irreversible thermal decay of proteins, studied by (a) following the inactivation, and (b) measuring the DSC peak area after various heating times. The samples were (a) dissolved *Bacillus licheniformis* α-amylase, heated at 90°C (+), 95°C (*), 97.5°C (□) and 100°C (○) [De Cordt *et al.*, 1992a], and (b) dissolved *Bacillus amyloliquefaciens* α-amylase, heated at 75°C (+), 80°C (*), and 85°C (□)°C [De Cordt *et al.*, 1993d].

Some authors [Mozhaev and Martinek, 1982; Ahern and Klibanov, 1985] define "reversibility" as the spontaneous recovery of the native state, within a reasonable experimental time, upon removal of the denaturing factors. A general scheme of irreversible thermal enzyme inactivation was proposed by Klibanov [1983]:

$$N \overset{K}{\leftrightarrow} U \overset{k_{ir}}{\to} I \qquad (9)$$

where N is the native, U the partially unfolded (reversibly denatured) and I the irreversibly inactivated enzyme, K the unfolding equilibrium constant (=[U]/[N]) and k_{ir} the rate constant of the irreversible step. Whereas reversible denaturation practically implies that only non-covalent events are taking place, irreversible decay may result from covalent and/or conformational changes. For the purpose of using proteins as TTI, the undergone thermal decay must be irreversible. At pasteurization or sterilization temperatures, this is generally the case.

In accordance with the duality of the concept protein "stability", various interpretations of "stabilization" exist. In the present work, protein stability and stabilization are evaluated in terms of changes in the rate constant k and the activation energy E_A that are observed if one follows the irreversible heat decay process, either by measuring the course of the enzymic activity (A) (Fig. 1a) or of the deterioration heat (ΔH) (Fig. 1b) as a function of heating time. The thus obtained rate constant k can be formulated in terms of the constants used in scheme 9 as follows:

$$k=k_{ir}\frac{K}{1+K} \qquad (10)$$

As to the elucidation of the mechanisms that rule changes in protein stability, most studies are concerned with enhanced resistance against reversible denaturation rather than with influences on the rate of irreversible decay. [Gray, 1988] Probably this is because the complete mechanisms of irreversible decay are usually not known. [Ahern and Klibanov, 1985] Anyhow, since it appears that reversible unfolding is a universal initial step in the irreversible deterioration process, it is indeed sensible to try to suppress reversible unfolding with a view to reducing the overall (observed) reaction rate (Eq.10). [Klibanov, 1979;1983; Mozhaev *et al.*, 1988] Stabilization methods are sometimes inspired on studies of the forces contributing to the conformational stability such as the work of Khechinashvili [1990] or Pace *et al.* [1991], or on comparisons of structurally related proteins with different thermostability. [Perutz, 1978; Klibanov, 1979;1983; Monsan and Combes, 1984a; Mozhaev and Martinek, 1984; Mozhaev *et al.*, 1988; Gianfreda and Scarfi, 1991] However, most methods have a mere empiric basis.

The thermostability of a protein can be manipulated by genetic or protein engineering, by chemical modification, or by changes in its "environment" ("solvent engineering") or in its "state". [Klibanov, 1983; Mozhaev *et al.*, 1988; Gianfreda and Scarfi, 1991] The latter two methods and their probable mechanisms are briefly discussed.

B. SOLVENT ENGINEERING

Modification of the environment or the surrounding medium of a protein in view of a specific goal can be called solvent engineering. Generally this comprises variation in the pH, addition of protein stabilizing substances of widely varying nature, and full substitution for the aqueous environment by an organic solvent.

1. pH

Opinions on the importance of electrostatic interactions in maintaining the native state of a protein are differing. In some instances their influence is large [Perutz, 1978; Mozhaev *et al.*, 1988; Gianfreda and Scarfi, 1991] but in other cases it seems to be very small compared to other effects. [Pace *et al.*, 1991]

2. Specifically binding ligands

Allosteric effectors, e.g. metal ions like Ca^{2+}, have specific binding sites on the protein. Stabilization due to binding of specific ligands can be explained by considering a shift in the equilibrium between the native and denatured protein. [Moriyama *et al.*, 1977; Pace and McGrath, 1980] If the native state has m binding sites for ligand L, scheme 9 can be extended as follows:

$$NL_m \leftrightarrow ... \leftrightarrow NL \leftrightarrow N \overset{K}{\leftrightarrow} U \overset{k_{ir}}{\to} I \qquad (11)$$

Gray (1988) argues that many ligands bind by multipoint attachment, the binding site usually being composed of several amino acid residues, and thereby rigidify the protein molecule by intramolecular cross-linking. Rigidification improves the resistance against unfolding and hence decreases k (Eq.10). Moreover, ligands can shield the functional, reactive groups of their binding site from chemical as well as non-covalent deteriorative interactions.

3. Polyalcoholic substances

Polyols of 3 or more carbon atoms generally stabilize proteins. [Gray, 1988] Timasheff and co-workers [Arakawa and Timasheff, 1982; Timasheff and Arakawa, 1989] state that addition of a stabilizing solute increases the protein's chemical potential and that this rise is proportional to the surface of the protein, so that the native, folded state is favoured over the denatured one. It seems to be a general rule that stabilizing substances are preferentially excluded from the protein's vicinity, or in other words, that the protein is preferentially hydrated in their presence. Timasheff and Arakawa [1989] attribute the exclusion of sugars (as well as of amino acids and some salts) to their water surface tension increasing effect. Glycerol, on the other hand, should be excluded from the protein environment as a consequence of a predominating "solvophobic" effect, i.e. contact of the nonpolar protein surface regions with a glycerol-water mixture is even more entropically unfavourable than contact with water. Other authors try to link the preferential hydration of protein to several mutually related features. Monsan and Combes [Monsan and Combes, 1984a,b; Combes *et al.*, 1987] notice that glycerol, sorbitol and sucrose depress the water activity (a_w), but they add that the protective effect of these additives on proteins cannot be ascribed to the a_w decrease alone. They suggest that stabilization stems from the increased degree of water organization induced by the solutes. Gerlsma [1968] attributes the improved protein stability to the loss of hydrogen-bond rupturing capacity of water in the presence of sugars and polyols, and also to the strengthening of the hydrophobic bonds in the protein. [Gerlsma, 1970; Mozhaev and Martinek, 1984] Moriyama *et al.* [1977] try to explain the stabilizing effect of glycerol by its lowering effect on the solvent dielectric constant, so that Coulomb forces in the protein are reinforced. Back *et al.* [1979] argue that stabilization by sugars and polyols is due to the enhancement of pairwise hydrophobic interaction in the presence of "structure-making" solutes, i.e. solutes that reinforce the hydrogen-bonded organization of water.

Stabilization with some polymers and by the lyotropic effect of some salts in high concentrations are evenso explained in terms of preferential exclusion.

4. Organic solvents

It seems to be quite general that enzymes are stabilized in dry organic media, and that this stabilizing effect increases with the solvent hydrophobicity. Klibanov and co-workers [Zaks and Klibanov, 1984; 1988; Klibanov, 1989; Volkin *et al.*, 1991] propose the following explanation. To maintain its native conformation, a protein needs an "essential water layer", but substitution for the rest of the water by an organic solvent can rigidify the molecule and thereby increase its resistance to denaturation (scheme 9). This results in a decrease in the overall irreversible inactivation rate (Eq.10). Water is considered as a lubricant, i.e. it is believed that it confers higher conformational flexibility or mobility to the protein. Whether an organic solvent is stabilizing or destabilizing should depend on its interaction with the protein essential water layer. According to these authors, the quite hydrophobic media interact less with the essential water layer and therefore stabilize the enzyme, whereas the more polar solvents "strip" it "off" and thus denature the protein. Laane *et al.* [1987] stress that the use of the term "distortion" is preferable to "stripping off" because enzymes can be denatured in anhydrous as well as in water saturated polar solvents. From their results, it can be concluded that solvents having a logP<2 (P is the partition coefficient in the octanol-water two-liquid phase system) strongly distort the essential water-enzyme interaction, thereby denaturing the biocatalyst. Solvents with a logP between 2 and 4 are weak distorters and their effect on the activity is unpredictable, whereas those having a logP>4 do not distort the essential water coat and leave the enzyme in the native state. Increasing the water content of an organic medium depresses the stabilizing effect. Probably the protein molecules are less rigidified then. [Zaks and Klibanov, 1985; Reslow *et al.*, 1988; Volkin *et al.*, 1991]

C. MODIFICATION OF THE "STATE"

1. Immobilization

A wide variety of immobilization procedures has been developed [Goldstein and Manecke, 1976; Royer, 1978] and many different definitions and classifications exist. A recent general definition is: the physical localization in a certain region of space, or the conversion from a water-soluble mobile state to a water-insoluble immobile one. [Gianfreda and Scarfi, 1991]

Apart from some specific cases where immobilization is applied for prevention of autolysis of proteases or coagulation, the mechanisms of its stabilizing influence can be summarized as follows. Both covalent links and non-covalent interactions between protein and carrier entail rigidification of the enzyme molecule, which suppresses unfolding (scheme 9). Indeed, various physicochemical analyses point out that the conformational mobility or flexibility of immobilized proteins is reduced. [Pye and Chance, 1976; Martinek and Berezin, 1977;1980; Klibanov, 1979;1983; Ulbrich *et al.*, 1986] The increase in rigidity is generally positively correlated with the number of linkages. [Martinek and Berezin, 1977;1980; Klibanov, 1979] Rigidification is probably the most important effect. Its stabilizing influence may be due to the decrease of the positive conformational denaturation entropy ($\Delta S = S_U - S_N$) [Engelborghs, 1992] which is normally a major driving force of unfolding. [Pace *et al.*, 1991]

Moreover, immobilization can create a micro-environment whose pH and solute concentration may differ largely from those prevailing in the bulk solution. This is attributed to partition effects arising from noncovalent interactions between the support material and solutes and/or to diffusion, especially with porous carriers. [Pye and Chance, 1976; Martinek and Berezin, 1977; Royer, 1978; Klibanov, 1979] Monsan and Combes [Monsan and Combes, 1984a; Combes *et al.*, 1987; Monsan and Combes, 1988] state that stabilization against unfolding of immobilized proteins is also due to the improved structural organization of water in the vicinity of the interface between a solid carrier and the liquid.

Increasing the loading density of enzyme on the support can further improve the protein stability [Weng, 1991; De Cordt *et al.*, 1992a] or have no effect. [Ulbrich *et al.*, 1986]

Stabilization by immobilization can be only an apparent effect [Klibanov, 1979] and there are numerous reports of cases where no stabilization was achieved. According to Klibanov [1979;1983] and Martinek and Berezin [1977;1980], incongruency of the enzyme molecule and the support surface may cause conformational stress, and immobilization does not necessarily lead to rigidification of the whole protein molecule. As to the latter feature, Ulbrich *et al.* [Ulbrich *et al.*, 1986; Schellenberger and Ulbrich, 1989; Ulbrich, 1993] argue that the unfolding process is nucleated in a well determined site, and only if this "unfolding nucleus" is rigidified by the immobilization, there can be a stabilizing influence.

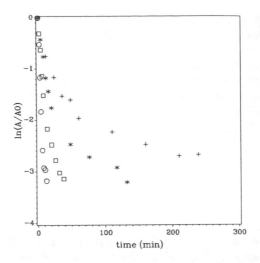

Figure 2. Biphasic inactivation of immobilized *Bacillus licheniformis* α-amylase, heated at 96°C (+), 100°C (*), 104°C (□) and 108°C (O) (De Cordt *et al.*, 1992a).

From the point of view of kinetics and potential use of a protein as a TTI, it is important to note that immobilization may give rise to multiphasic heat decay kinetics (Fig.2). [Fischer *et al.*, 1980; Monsan and Combes, 1984b; Ulbrich *et al.*, 1986; Germain *et al.*, 1989; Schellenberger and Ulbrich, 1989; De Cordt *et al.*, 1992a,b; 1993a,b,c] Usually such behaviour can be adequately described by the nth order model, with n>1 (Eq.6). A reaction order in excess of unit can reflect various mechanisms. [Laidler and Bunting, 1973] In some instances, e.g. when the decay of both dissolved and immobilized enzyme proceeds in a multiphasic way and if there are indications that this is caused by the presence of different fractions in the molecule population each inactivating in a first-order way but with different rate constants, the next model is adopted:

$$A=\sum_i A_i exp[-k_i t] \qquad (12)$$

where A is the total, experimentally measured activity, A_i the activity of the ith fraction at time 0 and k_i its inactivation rate constant. Often two terms suffice to obtain a good data fit [Ulbrich *et al.*, 1986; Schellenberger and Ulbrich, 1989] and usually model 12 is indeed used in the simplified form

$$A=A_l exp[-k_l t]+A_s exp[-k_s t] \qquad (13)$$

where A_s and A_l are respectively the activities of the thermostable and thermolabile fractions at time 0, and k_s and k_l their rate constants. [Weng, 1991; Weng *et al.*, 1991a,b; De Cordt *et al.*, 1992b;1993a,b,c; Hendrickx *et al.*, 1992a,b] Model 13 does not allow for separation of the variables, which is one of the requirements for possible functioning of a system as a TTI (Eq.4). Nevertheless, it can be applied for characterization of a TTI if some additional conditions are met and taking into account an important restriction as to the range of measurable F-values. [De Cordt *et al.*, 1992b]

2. Drying

When an enzyme is thoroughly dried, e.g. lyophilized, some water remains thightly bound to the molecules and this is vital for the maintainance of the native structure. Moreover, an enzyme lyophilized from a buffer solution having a pH optimal for activity and then suspended in an apolar solvent, gets fixed in its active, native conformation and is thus more resistant to heat denaturation. [Zaks and Klibanov, 1985] The enhanced stability of dried enzymes can be explained by considering the deleterious effects of water, both as a lubricant (B.4.) and as a participant in most chemical reactions leading to irreversible protein decay at high temperatures. [Klibanov, 1983]

V. CASE-STUDIES ON *BACILLUS SP.* α-AMYLASES AND HORSE RADISH PEROXIDASE

Horse radish peroxidase (HRP) and the α-amylases of *B. licheniformis* (BLA) and *B. amyloliquefaciens* (BAA) are thermostable enzymes. Their activity loss under influence of heat can be described by the nth order model (Eq.6), with n either equal to or higher than 1. The inactivation rate constants follow the Arrhenius law (Eq.2-3). For both HRP and α-amylase, there are well established procedures for easy and correct measurement of the enzymic activity, which serves as TTI-response property. Recently it was shown with BAA that the heat of deterioration (ΔH) (Fig.1b), measured as the area of a DSC peak, can evenso be used as TTI-response. [De Cordt *et al.*, 1993d] The above outlined techniques for modification of the thermostability characteristics of proteins were applied with a view to further decreasing the decay rate constant and shifting the activation energy. Some examples of the results are schematically shown in Figure 3. The kinetic characteristics of irreversible heat decay that are relevant with relation to potential use of the system as a TTI, are stated. The value of the rate constant is approached by indicating a typical temperature at which the system could be heated for a few minutes without being completely destroyed. That is, after such heat treatment the response X_t was still detectable.

Varying the pH or the Ca^{2+} concentration altered only k [De Cordt *et al.*, 1992a], but immobilization, addition of polyhydric substances or drying brought about changes in both k and E_A. Upon lyophilization of HRP, its thermostability increased dramatically. The activation energy of thermal decay depended strongly on the a_w (<1). [Hendrickx *et al.*, 1992b] Similar results were obtained with BAA. There, the E_A was varying between 109 and 176 kJ/mol, depending on a_w, and the rate constant was of a magnitude allowing a heat treatment of a few minutes at about 155°C with retention of a detectable response (activity). [Saraiva *et al.*, 1993] HRP, immobilized on glass beads and suspended in dodecane, has the right E_A to make it a possible TTI for monitoring e.g. the thermal death of D-*Streptococci* in pasteurization processes. [Weng, 1991] BLA, immobilized and suspended in Tris buffer or dissolved in Tris buffer with 70 ppm and 50 vol% glycerol, has an E_A that makes it a potential TTI for measuring the thermal destruction of *Clostridium botulinum* spores. [De Cordt *et al.*, 1992a,b; 1993a,b] Only, the rate constant of this system should be further decreased in order to enable its use at sterilization temperatures.

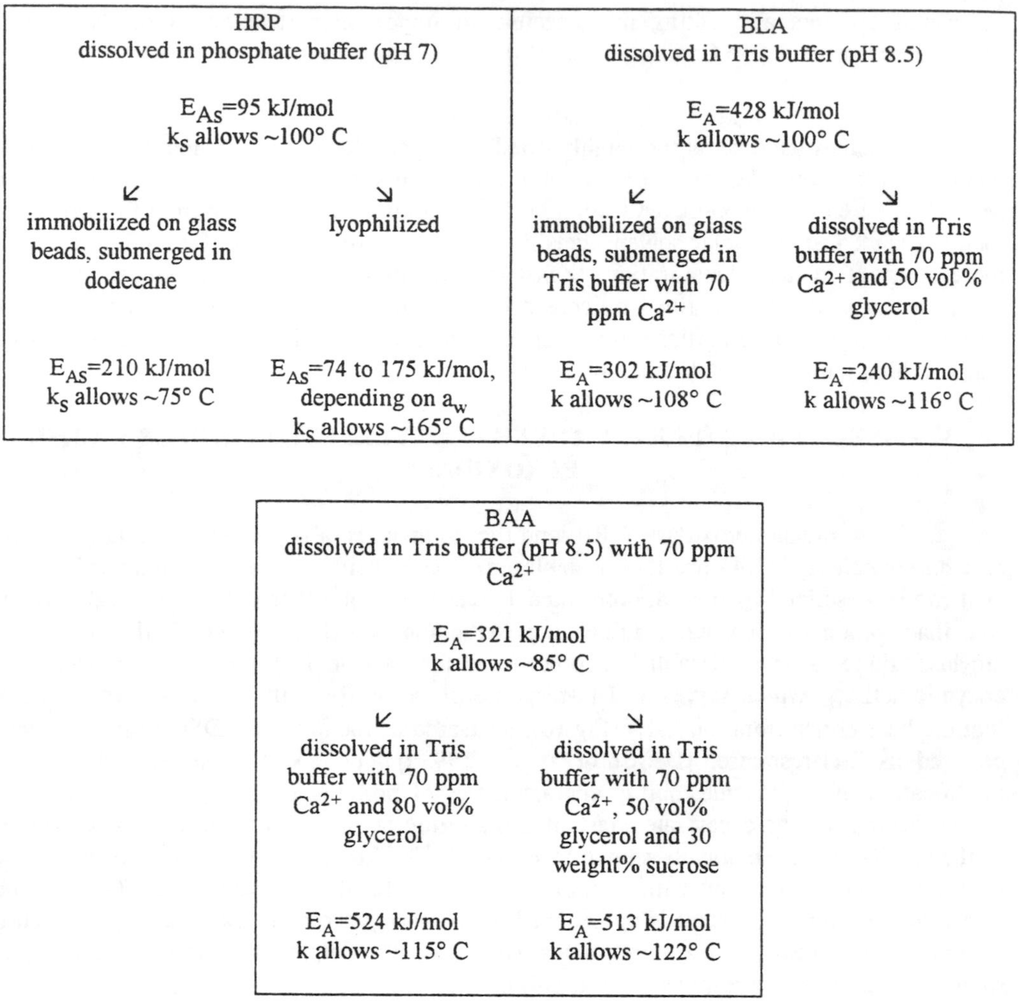

HRP dissolved in phosphate buffer (pH 7) E_{As}=95 kJ/mol k_s allows ~100° C		BLA dissolved in Tris buffer (pH 8.5) E_A=428 kJ/mol k allows ~100° C	
↙	↘	↙	↘
immobilized on glass beads, submerged in dodecane	lyophilized	immobilized on glass beads, submerged in Tris buffer with 70 ppm Ca^{2+}	dissolved in Tris buffer with 70 ppm Ca^{2+} and 50 vol % glycerol
E_{As}=210 kJ/mol k_s allows ~75° C	E_{As}=74 to 175 kJ/mol, depending on a_w k_s allows ~165° C	E_A=302 kJ/mol k allows ~108° C	E_A=240 kJ/mol k allows ~116° C

BAA dissolved in Tris buffer (pH 8.5) with 70 ppm Ca^{2+} E_A=321 kJ/mol k allows ~85° C	
↙	↘
dissolved in Tris buffer with 70 ppm Ca^{2+} and 80 vol% glycerol	dissolved in Tris buffer with 70 ppm Ca^{2+}, 50 vol% glycerol and 30 weight% sucrose
E_A=524 kJ/mol k allows ~115° C	E_A=513 kJ/mol k allows ~122° C

Figure 3. Application of different methods to change protein stability on horse radish peroxidase and the α-amylases of *Bacillus licheniformis* and *B.amyloliquefaciens*.

The work with HRP and BLA was carried out using the enzymic activity as response. Conversely, the results with BAA were obtained by means of DSC analysis. Using the DSC peak area as response property, BAA in Tris buffer with 70 ppm Ca^{2+}, 50 vol% glycerol and 30 weight% sucrose, could serve as a TTI for evaluating the impact of a sterilization process on another protein, e.g. one of interest to the quality of a food product. [De Cordt *et al.*, 1993d]

The effective use as TTI of immobilized HRP, immobilized BLA and dissolved BAA, was validated by experimental demonstration of their time-temperature integrating performance under variable-temperature conditions. [Weng, 1991; De Cordt *et al.*, 1992b;1993d; Hendrickx *et al.*, 1992a] Moreover, immobilized HRP and dissolved BLA found application in studies of the spatial distribution of F-values in in-pack heat treated food model systems. [Maesmans, 1993]

REFERENCES

Ahern, T.J. and Klibanov, A.M., The mechanism of irreversible enzyme inactivation at 100°C, Science, 228, 1280, 1985.

Arakawa, T. and Timasheff, S.N., Stabilization of protein structure by sugars, Biochemistry 21, 6536, 1982.

Back, J.F., Oakenfull, D. and Smith, M.B., Increased thermal stability of proteins in the presence of sugars and polyols, Biochemistry, 18(23), 5191, 1979.

Combes, D., Yoovidhya, T., Girbal, E., Willemot, R.-M. and Monsan, P., Mechanism of enzyme stabilization, in *Enzyme Engineering 8*, Laskin, A.I., Mosbach, K., Thomas, D., and Wingard, L.B.Jr., Eds., Annals N.Y. Ac.Sci. New York 501, 1987.

De Cordt, S., Vanhoof, K., Hu, J., Maesmans, G., Hendrickx, M. and Tobback, P., Thermostability of soluble and immobilized α-amylase from *Bacillus licheniformis*, Biotechnology and Bioengineering, 40, 396, 1992.

De Cordt, S., Hendrickx, M., Maesmans, G. and Tobback, P., Immobilized α-amylase from *Bacillus licheniformis*: a potential enzymic time-temperature-integrator for thermal processing, International Journal of Food Science and Technology, 27, 661, 1992.

De Cordt, S., Saraiva, J., Hendrickx, M., Maesmans, G. and Tobback, P., Changing the thermostability of *Bacillus licheniformis* α-amylase, in *Stability and stabilization of enzymes*, van den Tweel, W.J.J., Harder, A. and Buitelaar R.M., Eds., Elsevier Science Publishers, Amsterdam, 1993.

De Cordt, S., Hendrickx, M., Maesmans, G. and Tobback, P., The influence of polyalcohols and carbohydrates on the thermostability of α-amylase, Biotechnology and Bioengineering, in press, 1993.

De Cordt, S., Hendrickx, M., Maesmans, G. and Tobback, P., Convenience of immobilized *Bacillus licheniformis* α-amylase as time-temperature-integrator (TTI), Accepted for publication in Journal of Chemical Technology and Biotechnology, 1993.

De Cordt, S., Avila, I., Hendrickx, M., Maesmans, G. and Tobback, P., DSC and protein-based time-temperature-integrators: case study on α-amylase stabilized by polyols and/or sugar, Submitted for publication, 1993.

Engelborghs, Y., personal communication, 1992.

Fischer, J., Ulbrich, R., Ziemann, R., Flatau, S., Wolna, P., Schleiff, M., Pluschke, V. and Schellenberger, A., Thermal inactivation of immobilized enzymes: a kinetic study, Journal of Solid-phase biochemistry, 5(2), 79, 1980.

Gerlsma, S.Y., Reversible denaturation of ribonuclease in aqueous solutions as influenced by polyhydric alcohols and some other additives, The Journal of Biological Chemistry, 243(5), 957, 1968.

Gerlsma, S.Y., The effects of polyhydric and monohydric alcohols on the heat induced reversible denaturation of chymotrypsinogen A, European Journal of Biochemistry, 14, 150, 1970.

Germain, P., Slagmolen, T. and Crichton, R.R., Relation between stabilization and rigidification of the three-dimensional structure of an enzyme, Biotechnology and Bioengineering, 33, 563, 1989.

Gianfreda, L. and Scarfi, M.R., Enzyme stabilization: state of the art, Molecular and cellular biochemistry, 100, 97, 1991.

Goldstein, L. and Manecke, G., The chemistry of enzyme immobilization, in *Applied biochemistry and bioengineering 1: Immobilized enzyme principles*, Wingard, L.B.Jr., Katchalski-Katzir, E., Goldstein, L., Eds., Academic Press, New York, 1976.

Gray, C.J. Additives and enzyme stability, Biocatalysis, 1, 187, 1988.

Hendrickx, M., Weng, Z., Maesmans, G. and Tobback, P., Validation of a time-temperature-integrator for thermal processing of foods under pasteurization conditions, International Journal of Food Science and Technology, 27, 21, 1992.

Hendrickx, M., Saraiva, J., Lyssens, J., Oliveira, J. and Tobback, P., The influence of water activity on thermal stability of horseradish peroxidase, International Journal of Food Science and Technology, 27, 33, 1992.

Khechinashvili, N.N., Thermodynamic properties of globular proteins and the principles of stabilization of their native structure, Biochimica et Biophysica Acta, 1040, 346, 1990.

Klibanov, A.M., Enzyme stabilization by immobilization, Analytical biochemistry, 93, 1, 1979.

Klibanov, A.M., Review on enzyme stabilization, Advances in applied microbiology, 29, 1, 1983.

Laane, C., Boeren, S., Vos, K. and Veeger, C., Rules for optimization of biocatalysis in organic solvents, Biotechnology and Bioengineering, XXX, 81, 1987.

Laidler, K.J. and Bunting, P.S., General kinetic principles, in *The chemical kinetics of enzyme action*, Clarendon Press, Oxford, 1973.

Maesmans, G., Possibilities and limitations of thermal process evaluation techniques based on time temperature integrators, Doctoral dissertation no.240, Faculty of Agricultural and Applied Biological Sciences, K.U.Leuven, 1993.

Martinek, K. and Berezin, I.V., General principles of enzyme stabilization, Journal of Solid-phase biochemistry, 2(4), 343, 1977.

Martinek, K. and Berezin, I.V., The stabilization of enzymes - a key factor in the practical application of biocatalysis, Russian chemical reviews, 49, 737, 1980.

Monsan, P. and Combes, D., Stabilization of enzyme activity, in *The world biotech report 1984 1*, Online, Pinner, England, 1984a.

Monsan, P. and Combes, D., Effect of water activity on enzyme action and stability, in *Enzyme Engineering 7*, Annals N.Y.Sci. New York 343, Laskin, A.I., Tsao, G.T., Wingard, L.B.Jr., Eds., 1984b.

Monsan, P. and Combes, D., Enzyme stabilization by immobilization, Methods in Enzymology, 137, 584, 1988.

Moriyama, S., Matsuno, R. and Kamikubo, T., Influence of dielectric constants and ligand binding on thermostability of glucoamylase, Agricultural and Biological Chemistry, 41(10), 1985, 1977.

Mozhaev, V.V., Berezin, I.V. and Martinek, K., Structure stability relationship in proteins: fundamental tasks and strategy for the development of stabilized enzyme catalysts for biotechnology, Critical reviews in biochemistry, 23(3), 235, 1988.

Mozhaev, V.V. and Martinek, K., Inactivation and reactivation of proteins (enzymes), Enzyme Microbial Technology, 4(9), 299, 1982.

Mozhaev, V.V. and Martinek, K., Review: Structure-stability relationships in proteins: new approaches to stabilizing enzymes, Enzyme Microbial Technology, 6(2), 50, 1984.

Mulley, E.A., Stumbo, C.R. and Hunting, W.M, Thiamine: a chemical index of the sterilization efficacy of thermal processing, Journal of Food Science, 40, 993, 1975.

Okos, M.R., Physical and chemical properties of foods, American Society of Agricultural Engineers, St. Joseph, Michigan, USA, 1986.

Pace, C.N., Heinemann, U., Hahn, U. and Saenger, W., Ribonuclease T1: Structure, function and stability, Angew. Chem. Int. Ed. Engl., 30, 343, 1991.

Pace, C.N. and McGrath, T., Substrate stabilization of lysozyme to thermal and guanidine hydrochloride denaturation, The Journal of Biological Chemistry, 255(9), 3862, 1980.

Perutz, M.F., Electrostatic effects in proteins, Science, 201, 1187, 1978.

Pflug, I.J. and Christensen, R., Converting an F-value determined on the basis of one z-value to an F-value determined on the basis of a second z-value, Journal of Food Science, 45, 35, 1980.

Pflug, I.J. and Odlaug, T.E., Biological indicators in the pharmaceutical and the medical device industry, Journal of parenteral Science and Technology, 40, 242, 1986.

Pflug, I.J. and Smith, G.M., The use of biological indicators for monitoring wet heat sterilization processes, in *Sterilization of medical products*, Gaughran E.R.L. and Kereluk, K., Eds., New Brunswick, NJ, Johnson and Johnson, 1977.

Pye, E.K. and Chance, B., Investigation of the physical properties of immobilized enzymes, Methods in Enzymology, 44, 357, 1976.

Reslow, M., Adlercreutz, P. and Mattiasson, B., On the importance of the support material for bioorganic synthesis, European Journal of Biochemistry, 172, 573, 1988.

Rönner, U., A new biological indicator for aseptic sterilization, Food Technology International Europe, 43, 1990.

Royer, G.P., Immobilization methods, Immobilized enzymes catalysis reviews 1978, 30, 1978.

Saraiva, J., De Cordt, S., Hendrickx, M., Oliveira, J. and Tobback, P., Inactivation of α-amylase from *Bacillus amyloliquefaciens* at low moisture contents, in *Stability and Stabilization of enzymes*, van den Tweel, W.J.J., Harder, A. and Buitelaar, R.M., Eds., Elsevier Science Publishers, Amsterdam, 1993.

Schellenberger, A. and Ulbrich, R., Protein stabilization by blocking the native unfolding nucleus, Biomedica et Biochimica Acta, 48(1), 63, 1989.

Taoukis, P.S. and Labuza, T.P., Applicability of time-temperature indicators as shelf life monitors of food products, Journal of Food Science, 54, 783, 1989.

Timasheff, S.N. and Arakawa, T., Stabilization of protein structure by solvents, in *Protein structure: a practical approach*, Creighton, T.E., Ed., IRL Press at Oxford University Press, 1989.

Ulbrich-Hofman, R., Golbik, R. and Damerau, W., Fixation of the unfolding region - a hypothesis of enzyme stabilization, in *Stability and Stabilization of enzymes*, van den Tweel, W.J.J., Harder, A. and Buitelaar, R.M., Eds., Elsevier Science Publishers, Amsterdam, 1993.

Ulbrich, R., Schellenberger, A., and Damerau, W., Studies on the thermal inactivation of immobilized enzymes, Biotechnology and Bioengineering, XXVIII, 511, 1986.

Volkin, D.B., Staubli, A., Langer, R. and Klibanov, A.M., Enzyme thermoinactivation in anhydrous organic solvents, Biotechnology and Bioengineering, 37, 843, 1991.

Weng, Z., A time-temperature-integrator for thermal processing of foods: a case study on immobilized peroxidase, Doctoral dissertation nr.201, Faculty of Agricultural Sciences, K.U.Leuven, 1991.

Weng, Z., Hendrickx, M., Maesmans, G. and Tobback, P., Thermostability of soluble and immobilized horseradish peroxidase, J.Food Sci., 56(2), 574, 1991a.

Weng, Z., Hendrickx, M., Maesmans, G. and Tobback, P., Immobilized peroxidase: a potential bioindicator for evaluation of thermal processing, Journal of Food Science, 56(2), 567, 1991b.

Witonsky, R.J. A new tool for the validation of the sterilization of parenterals, Bulletin of the parenteral drug association, 31, 274, 1977.

Zaks, A. and Klibanov, A.M., Enzyme catalysis in organic media at 100°C, Science, 224, 1249, 1984.

Zaks, A. and Klibanov, A.M., Enzyme-catalyzed processes in organic solvents, Proc. Natl. Acad. Sci .USA, 82, 3192, 1985.

Zaks, A. and Klibanov, A.M., Enzymatic catalysis in nonaqueous solvents, The Journal of Biological Chem.istry, 263(7), 3194, 1988.

Part IV

Packaging Systems for Minimally Processed Foods and Consumer Acceptance of Minimally Processed Foods

DESIGN OF MODIFIED ATMOSPHERE PACKAGES
FOR FRESH FRUITS AND VEGETABLES

Khe V. Chau and P.C. Talasila

IMPORTANCE

Fresh fruits and vegetables are a very important part of our diet. Unfortunately, they are very seasonal and perishable. Great care has to be taken in the harvesting, handling, storage, packaging, and transportion of the products to reduce damage, maintain quality, and extend shelf life.

Reduction of postharvest losses is probably more important than increasing yields through improved production practices. This is apparent when one considers that all of the cost inputs and energy inputs for production, harvesting, packing, and transportation are wasted when the product is lost near the point of consumption. Proper handling, storage, and shipment of fruits and vegetables will reduce losses, maintain higher product quality, and provide producers with a competitive edge in their competition for domestic and foreign markets.

Fruits and vegetables after harvest are still living things. The process of deterioration begins at harvest. It cannot be stopped but it can be slowed down if the products are handled properly. The deterioration process is affected by biological, environmental, and mechanical factors. Biological factors include the natural processes of respiration, ethylene production, and transpiration. Environmental factors include temperature, humidity, oxygen, carbon dioxide, nitrogen, and ethylene. Mechanical factors include bruises and damages caused during harvesting, handling, storage, and transportation. Of all the possible steps taken to reduce losses and maintain quality, lowering the product temperature immediately after harvest is probably the most important single operation that will provide the greatest benefit in prerserving quality. Temperature affects the rate of respiration, ethylene production, and transpiration. All these factors contribute heavily to the deterioration of the product. Most fresh products are now stored and shipped under refrigeration to reduce losses and maintain quality.

Modification of atmosphere surrounding the product is another method used to maintain quality. It is used in many cases to supplement refrigeration. This technique is referred to as controlled atmosphere (CA) or modified atmosphere (MA) storage. CA implies a greater degree of precision than MA in maintaining specific levels of O_2, CO_2 and other gases. Modified atmosphere is created by packing fresh products in polymeric film bags or in containers with specially designed gas diffusion windows. It can be established via active or passive modifications or a combination of the two. In active modification, an atmosphere is established by replacing the atmosphere of the package with the desired gas mixture. Additionally, absorbers or adsorbers may be included in the package to scavenge oxygen, carbon dioxide and/or ethylene. In the case of passive modification, modification of the atmosphere is attained through controlling the gas exchange across the package. The process relies on the natural process of respiration and diffusion of O_2 and CO_2 across the package to provide the desired environment.

Use of modified atmosphere packaging (MAP) may allow handling of products at temperatures higher than their optimum storage temperatures. It is especially beneficial for chilling-sensitive commodities to avoid their exposure to chilling temperatures. Another important benefit from MAP is reduction in transpiration rate to a great extent. The package will act as a barrier to the movement of water vapor and aid in the maintenance of high relative humidity inside, which in turn reduces transpiration rate.

MAP is a dynamic system in which four main processes namely product respiration, product transpiration, permeation of gases through the packaging material, and heat transfer occur simultaneously. Respiration which is a function of temperature, stage of maturity of the product, and concentrations of CO_2, O_2, and ethylene inside the package varies continuously because of the changes in the variables. The respiration process produces heat energy which in turn alters the temperature of the product. Transpiration is a function of surface temperature of the product and the temperature and relative humidity of the surroundings. Permeability of the package is a function of the characteristics and temperature of the packaging material, material thickness, type of permeating gas, and difference in gas concentration across it. After a period of adjustment, steady state conditions may be established inside the package. Thus passive atmosphere modification is a complex process with many interactions among different components. It is important to understand the interrelationships of the different parameters involved to design a suitable modified atmosphere packaging system for a chosen commodity.

PERTINENT EQUATIONS

Diffusion of a gas through a film follows Fick's law of molecular diffusion:

$$\frac{dn_A}{dt} = - D A \frac{dc_A}{dz} \tag{1}$$

The concentration of a gas on the surface of a film may be expressed in terms of the solubility S of the film to the gas and the partial pressure p of the gas outside the surface of the film as (Geankoplis, 1978):

$$c_A = \frac{S_A p_A}{22.414} \tag{2}$$

Equation (1) may be rewritten as:

$$\frac{dn_A}{dt} = \frac{A D S_A}{E (22.414)} (p_1 - p_2) \tag{3}$$

where E is the thickness of the film, and p_1 and p_2 are the partial pressures of the gas on each side of the film.

In the case of a modified atmosphere package of fruit or vegetables, the process of respiration of the packaged product releases CO_2 and consumes O_2. Taking respiration into consideration, the equations for the rate of change in concentration in O_2, CO_2 and N_2 may be expressed as:

$$\frac{dn_{A_1}}{dt} = \frac{A\ P_{M_A}}{22.414E}(p_{A_2} - p_{A_1}) - R_A W \qquad (4)$$

$$\frac{dn_{B_1}}{dt} = \frac{A\ P_{M_B}}{22.414E}(p_{B_2} - p_{B_1}) + R_B W \qquad (5)$$

$$\frac{dn_{C_1}}{dt} = \frac{A\ P_{M_C}}{22.414\ E}(p_{C_2} - p_{C_1}) \qquad (6)$$

The partial pressures at steady state are readily available by setting the right hand side of equations (4), (5), and (6) equal to zero:

$$\frac{A\ P_{M_A}}{22.414\ E}(p_{A_2} - p_{A_1}) = R_A\ W \qquad (7)$$

$$\frac{A\ P_{M_B}}{22.414\ E}(p_{B_2} - p_{B_1}) = R_B\ W \qquad (8)$$

$$p_{C_1} = p_{C_2} \qquad (9)$$

In the majority of cases, the rate of respiration varies with time due to variations of temperatures and gas concentration inside the package. In that case, there are no analytical solutions to equations (4), (5), and (6). A numerical technique approach is required.

If the respiration rate does not change with time, exact solutions are possible. Assuming that the perfect gas law is applicable, the mass n of the gas is expressed as a function of the gas partial pressure and equations (4) and (5) can be integrated readily:

$$p_{A_1} = p_{A_{1,ss}} - [p_{A_{1,ss}} - p_{A_{1,0}}].\ \exp\left(\frac{-A\ P_{M_A}\ G\ T_1}{22.414\ E\ V_1}\ t\right) \qquad (10)$$

$$p_{B_1} = p_{B_{1,ss}} - [p_{B_{1,ss}} - p_{B_{1,0}}]\ \exp\left(\frac{-A\ P_{M_B}\ G\ T_1}{22.414\ E\ V_1}\ t\right) \qquad (11)$$

From equations (10) and (11), the time taken by the package to reach a certain partial pressure p_{A1} and p_{B1} (and therefore a certain concentration) for O_2 and CO_2 is given by:

$$t_A = \ln[\frac{p_{A_{1,ss}} - p_{A_{1,0}}}{p_{A_{1,ss}} - p_{A_1}}] \frac{22.414 \ V_1 \ E}{G \ T_1 \ A \ P_{M_A}} \tag{12}$$

$$t_B = \ln[\frac{p_{B_{1,ss}} - p_{B_{1,0}}}{p_{B_{1,ss}} - p_{B_1}}] \frac{22.414 \ V_1 \ E}{G \ T_1 \ A \ P_{M_B}} \tag{13}$$

Equations (12) and (13) may be used to get a preliminary estimate of the time required to reach steady state by setting the final pressure to be within 10% of the steady state pressure.

The design of modified atmosphere packages when they are maintained at constant temperature throughout the handling and storage period is actually fairly simple, based on equations (7) and (8). Combining equations (7) and (8) will give the required permeability ratio:

$$\frac{P_{M_B}}{P_{MA}} = \frac{R_B}{R_A} \cdot \frac{(p_{A_2} - p_{A_1})}{(p_{B_1} - p_{B_1})} \tag{14}$$

The partial pressures p_{A1} and p_{B2} are given by:

$$p_{A_1} = P_1 \cdot (v_{A_1}/V_1) \tag{15}$$

$$p_{B_1} = P_1 \cdot (v_{B_1}/V_1) \tag{16}$$

where the total pressure P_1 is given by:

$$P_1 = 0.79 \ (\frac{1}{1 - v_A/V_1 - v_{B_1}/V_1}) \tag{17}$$

p_{A2} and p_{B2} are equal to 0.21 and 0 respectively, when the package is stored in ambient air.

In practice, it is very rare that a film that has the exact permeability ratio given by equation (14) is available. Usually, a film that has a permeability ratio closest to the required ratio is selected. Depending on the permeabilities of the film selected, the required area of the film is given by equations (7) and (8):

$$\left(\frac{A}{W}\right)_A = \frac{22.414 \ R_A \ E}{P_{M_A} \cdot (0.21 - p_{A_1})} \tag{18}$$

$$\left(\frac{A}{W}\right)_B = \frac{22.414 \ R_B \ E}{P_{M_B} \cdot p_{B_2}} \tag{19}$$

If the permeability ratio of the film selected is not exactly equal to that required by equation (14), the film areas A in equations (18) and (19) will not be the same. Usually, the larger of the two areas is selected as the area for the film to be on the safe side. If the area given by equation (18) or (19) is too large for the package under consideration, a film with higher permeabilities will be selected and the required area is recalculated. Once the final film area has been selected, the final concentrations of O_2 and CO_2 are recalculated. They will be slightly different from the original design values because of the permeability ratio of the film selected is usually not exactly the same as the required ratio.

The time taken to reach gas concentrations close to the final steady-state conditions may be calculated from equations (12) and (13). If this time is excessively long for the film selected, the package must be flushed with a gas mixture close to the steady-state composition; otherwise, the product may spoil before the beneficial gas concentrations are reached.

The above presentation shows that, conceptually, the design of a modified atmosphere package is a fairly simple process. The simplicity of the process is due to some simplifying assumptions such as the respiration rate and the permeability coefficients are constant. In real situations where the modified atmosphere packages are exposed to varying temperatures during handling and storage, these parameters are not constant. In fact, in many cases, the exact values of these parameters are not known.

Variations in Respiration Rate.

The change in respiration rate due to temperature is well documented. Hardenburg et al., for example, summarizes the changes in respiration rate as a function of temperature for most common fruits and vegetables in USDA Handbook 66. The increase in respiration rate with temperature is more than linear in most cases. Even though the effect of temperature on respiration rate is well documented, this does not mean that the respiration rate for the particular batch in the package under consideration is known accurately. Respiration rate varies quite significantly from sample to sample, even within each cultivar.

In addition to temperature, respiration rate is affected by the gas concentration surrounding the product. For example, Talasila et al. (1992) have found that respiration rate for strawberries decreases with a decrease in O_2 concentration and the change is more rapid at lower O_2 concentrations and at higher temperatures. The CO_2 concentration has relatively less of an effect on the respiration of strawberries. Other products may have a different response. For example, Emond (1992) has found that CO_2 has a significant effect on the respiration rate of blueberries. Information on the effect of gas concentration on respiration is much less readily available than the effect of temperature.

Variation in Permeability Coefficient.

The permeability coefficients of most commercially available film may be obtained from company data. However, these values were determined at some fixed temperature, usually in the ambient temperature range, 20-25 C. Most packaging films used for fresh fruits and vegetables are exposed to much lower temperatures. Doyon et al. (1991) and Doyon et al. (1992) have provided a limited number of information on permeabilities of films at different temperatures. The permeability P_M of most films tends to follow a relationship of the type:

$$P_M = a \ e^{(-b/T)} \tag{20}$$

Permeability increases with temperature. If this increase is in the same magnitude as the increase in respiration due to temperature, then the net effect of temperature would be negligible, as far as the gas composition inside the package is concerned. In most cases however, the rates of increase of permeability and respiration due to temperature are not the same. Usually, the respiration rate tends to increase faster than the permeability. As the respiration rate is increased, less oxygen is available, and that, in turn, decreases the respiration rate of many products. Thus, there is a dampening effect due to a decrease in O_2 concentration surrounding the product.

Mechanical damages such as punctures and stretching will greatly affect the permeability of a film.

Varying Surrounding Temperatures.

If a package is designed for an application where it is subjected to varying teperatures such as in the case where the temperature during shipment is quite different from the storage temperature, the simplified analysis presented earlier will not provide the proper answer. A numerical solution of equations (4), (5), and (6) has to be developed. This solution has to be coupled with the solution of the heat transfer between the product inside the package and the surrounding space because the rate of heat transfer will affect the respiration rate and the permeability of the packaging film. Talasila (1992) developed such a model. In his analysis, the temperature and gas concentration gradients inside the package were assumed uniform; water vapor transfer and the evaporative cooling effect of product transpiration were neglected, based on the assumption that water vapor saturation inside the package occurred quickly. The product was discretized in elemental volumes and finite difference heat conduction, convection, and radiation equations were written. Simulation results agreed well with experimental results. This type of simulation model needs to be used in the design of modified atmosphere packages that are subjected to varying temperature conditions because this is the only way to determine the gas composition and the temperature inside the package. This type of analysis requires detailed information on the changes in respiration rate due to temperature and gas concentration and changes in film permeability due to temperature as discussed earlier.

As a general rule, when a modified atmosphere package is expected to be subjected to varying surrounding temperatures, the package has to be designed for the higher temperature because at the higher temperature, the respiration rate is at its highest value and rapid changes in gas concentration will occur. This is when gas concentration levels may reach unacceptable limits, especially in the case of oxygen; anaerobic conditions are usually not tolerated by products very long. Of course, if the package is exposed to the high temperature for only a relatively short time, it is possible that the effect is not significant enough to cause concern. A transient analysis discussed earlier will provide to necessary data to make an informed decision.

Multiple Levels of Atmosphere Modification.

One possible approach worth exploring when dealing with packages exposed to two different temperatures is the use of two levels of atmosphere modification, such as a modified atmosphere package inside another modidified atmosphere package or a modified atmosphere package inside a controlled atmosphere storage. Consider the case of a product that is shipped at a different temperature than the final storage temperature. We can design a package that will provide the proper atmosphere when it is exposed to the final storage temperature. During shipment, especially long distance shipment, several of these packages may be stored inside another larger modified atmosphere package. The purpose of the outer package is to provide a surrounding different than ambient (21% O_2 and 0% CO_2) to the inner packages. By changing the surrounding atmosphere, it may be possible to overcome the change in respiration rate and film permeability due to a change in temperature. Instead of being inside another modified package, the inner packages may be kept in a large controlled atmosphere container. This concept has not been investigated but it appears to merit further study. It may be important to note that this approach will not work in all cases. For example, to overcome the increase in CO_2 production, the CO_2 concentration outside the inner package has be negative in order to have enough gradient to difuse out CO_2. This is obviously not possible.

Mixed Loads.

Shippers would like to be able to ship several commodities inside one shipping container. One can visualize a system in which the commodities are held in modified atmosphere packages which are then put in controlled atmosphere shipping containers. With proper design of the modified atmosphere packages and proper control of the atmosphere inside the shipping container, the various commodities in their different packages would be exposed to conditions that are close to optimal for quality maintenance. There is a need for research on possible new applications of the atmosphere modification process.

Use of Perforations to Modify Atmosphere.

Perforations may be used to control the flow of gases in and out of a package. The rate of gas diffusion across a perforation depends on the size and shape of a perforation. One disadvantage of perforations is the lack of selective permeabilities for CO_2 and O_2. The permeability ratio for perforations is typically one to one. If perforations are used for modified atmosphere packages that are kept in ambient conditions (21% O_2 and 0% CO_2), these packages will provide only modified atmosphere within a certain range of values only. By combining equations (7) and (8):

$$\frac{P_{M_B}}{P_{M_A}} = \frac{R_B}{R_A} \left(\frac{P_{a_2} - P_{a_1}}{P_{B_2} - P_{B_1}} \right) \tag{21}$$

Under ambient conditions, p_{A2} is equal to 0.21 and p_{B2} to 0. The permeability ratio of perforations is approximately one and the respiratory quotients R_B/R_A for most fruits and vegetables are in the range of 0.8 - 1.2. Assuming a value of one for both the permeability ratio and the respiratory quotient, equation (21) gives:

$$p_{A1} + p_{B1} = 0.21$$

The sum of the final volume concentrations of O_2 and CO_2 inside the package will be in the neighborhood of 21%, with the exact value depending on the respiratory quotient of the product being stored. Products requiring optimum atmospheres of 5% O_2 and 15% CO_2, for example, would be a good candidate for a modified atmosphere package controlled by perforations. Actually there are several products whose optimal conditions requirements are such that perforations can potentially provide the proper atmosphere. They include blackberries, blueberries, raspberries, strawberries, sweet cherries, parsley, spinach, lima beans, asparagus, avocados, fig, grapefruit, lemons, limes, and many others.

Even though perforations do not provide the selective characteristic of permeability to O_2 and CO_2 that polymeric films do, they offers some advantages. Polymeric films may stretch or get punctured and their permeability characteristics are altered drastically. Perforations can be made to be very strong and they can be incorporated into larger containers whereas films may not have the strength necessary for the larger weight of products inside the larger packages. Perforations have very high permeability. The permeability per unit area is 8.5 million times larger than low density polyethylene for O_2 and 1.5 million times for CO_2 (Mannaperruma, 1989). The size and number of perforations needed can be very small. The total pressure inside a perforated package is the same as the atmosphere surrounding the package. Most modified atmosphere packages fitted with polymeric films have either a negative or a positive gage pressure inside the package.

Perforations appear to have good potential for applications in modified atmosphere packaging in many situations. More work is needed to explore fully the the capability of this system. Basic data on permeabilities of perforations as a function of size and shape are needed. The effect of an induced draft when more than one perforation are present has never been investigated.

Uniformity of Temperature and Gas Concentration.

Most modified atmosphere packages are small, consumer pack size. In these cases, it may be assumed that the temperature and gas concentration within the package are uniform. This obviously will make the analysis of the system much easier. However, when the size of the modified atmosphere package gets larger, the question of uniformity of temperature and gas concentration inside the package has to be asked. Is it possible that at some location in the package there is an anaerobic condition even though the rest of the package is at some tolerable limit of O_2? To obtain an answer to such a question, an analysis of the gas transfer at the film window or perforations and the transport of gases within the bulk of the of the product coupled with the respiration process has to be made. Conceptually, this type of analysis is quite straightforward when a finite difference method is used. In practice, there are some difficulties due to the tortuous path that the gases follow when they move through the product bulk and the small time steps taken to simulate the slow process of gas diffusion. Different types of nodes are required to describe the void spaces in the product bulk, the free head space, the interface between the product and the head space, and the various boundaries created by the walls and corners. Emond (1992) analyzed such a system when he considered the case of a container (3-dimensional) with a perforation at one end of the head space. He used blueberries to validate the results. To simplify the problem, he considered only the case where the temperature is uniform. Experimental data and simulation results show that for large packages, up to 1 meter long, the gas concentration gradient from one end of the head space to the other could be significant, up to 5 or 6 percentage points, when the head space is only a few millimeters thick. The gradient decreased with an increase in head space thickness. However, gas concentration gradient within the product bulk was minimal, under the conditions tested. There was a linear relationship between the decrease in the length of the package and the decrease in gas concentration gradient.

SUMMARY

The design of modified atmosphere packages for fresh fruits and vegetables under constant temperature is a straighforward process. However, the results are not always reliable because the values of the design parameters such as respiration rate and permcability are not always known exactly. The design process becomes much more difficult when the packaged products are exposed to varying surrounding temperatures during handling and storage.

There is a great need for data on product respiration under different temperatures, O_2 and CO_2 levels. More data are needed for the permeabilities of polymeric films at different temperaturs. Simulaton models can provide the necessary information for manufacturers to formulate the proper films for the various products.

NOMENCLATURE

A	: area of the packaging film, m^2
c_A	: concentration of gas A, $kmol.m^{-3}$
E	: thickness of the packaging film, m
G	: universal gas constant, $m^3.atm/kmol/K$
n_{A_1}	: mass of O_2 inside the package, kmol
n_{B_1}	: mass of CO_2 inside the package, kmol
n_{C_1}	: mass of N_2 inside the package, kmol
P_1	: total pressure inside the package, atm
p_{A_1}	: partial pressure of O_2 inside the package, atm
$p_{A}1,ss$	: $p_{A}1$ at steady state, atm
$p_{A}1,0$	: $p_{A}1$ at time equal zero, atm
p_{A_2}	: partial pressure of O_2 outside the package, atm
p_{B_1}	: partial pressure of CO_2 inside the package, atm
p_{B_2}	: partial pressure of CO_2 outside the package, atm
p_{C_1}	: partial pressure of N_2 inside the package, atm
p_{C_2}	: partial pressure of N_2 outside the package, atm
P_{M_A}	: permeability of O_2 through the packaging film, m^3 at $STP.m/m^2/s/atm$
P_{M_B}	: permeability of CO_2 through the packaging film, m^3 at $STP.m/m^2/s/atm$
R_A	: O_2 consumption rate of the product, kmol/kg/s
R_B	: CO_2 production rate of the product, kmol/kg/s
t	: time, s
t_A	: time required to reach a certain O_2 partial pressure of P_{A1} inside the package, s

t_B : time required to reach a ceratain CO_2 partial pressure of P_{B1} inside the package, s
T_1 : temperature inside the package, K
V_1 : volume of the free space in the package, m^3
v_{A_1}/V_1 : volume concentration of O_2 in the package, m^3/m^3
v_{B_1}/V_1 : volume concentration of CO_2 in the package, m^3/m^3
W^1 : mass of the product in the package, kg

REFERENCES

Doyon, G., Gagnon, J., and Brunet, F., Permeability of packaging films. Project report no. Flo-1-E. Food Research and Development Center, Saint-Hyacinthe, Canada, 1992.

Doyon, G., Gagnon, J., Toupin, C., and Castaigne, F., Gas transmission properties of polyvinyl chloride (PVC) films studied under subambient and ambient conditions for modified atmosphere packaging applications. Packaging Technology and Science. (4), 157, 1991.

Emond, J.P., Mathematical Modeling of Gas Concentration Profiles in Perforation Generated Modified Atmosphere Bulk Packaging, Ph. D. dissertation, University of Florida, 1992.

Geankoplis, C.J., Transport processes and unit operations, Allyn and Bacon, Boston, 1978, chap.5.

Hardenburg, R.E., Watada, A.E., and Wang, C.Y., The Commercial Storage of Fruits, Vegetables, and Florist and Nursery Stocks. Agriculture Handbook No. 66, ARS, USDA, Washington, DC., 1986.

Mannapperuma, J.D., Zagory, D., Singh, R.P. Kader. A.A., Design of Polymeric Packages for Modified Atmosphere Storage of Fresh Produce, paper presented at 5th International Controlled Atmosphere Research Conference , Wenatchee, WA, 1989.

Talasila, P.C., Chau, K.V. and Brecht, J.K., Effects of gas concentrations and temperature on O2 consumption of strawberries, ASAE Trans.,35(1), 221, 1992.

Talasila, P.C., Modeling of Heat and Mass Transfer in a Modified Atmosphere Package, Ph. D. dissertation, University of Florida, 1992.

MODIFIED ATMOSPHERE PACKAGING OF MEATS[1]

George - John E. Nychas

**National Agricultural Research Foundation,
Institute of Technology of Agricultural Products,
Sof. Venizelou 1, Lycovrisi 14123, Greece**

I. INTRODUCTION

It is well established that Modified- Atmosphere Packaging (MAP) can extend the shelflife of refrigerated meats (Luiten *et al* 1982; Gray *et al* 1983). Almost any combination of carbon dioxide, nitrogen and oxygen may be used in Modified Atmosphere Packs (MAP) (Clark *et al* 1976; Gill and Molin 1991) in order to sustain visual appearance and/or to extend shelf - life of meat and meat products (Erichsen and Molin 1981; Blickstad and Molin 1983; Nychas and Arkoudelos 1990). Much work has been focused on the prevention of spoilage of meat and meat products by organisms such as pseudomonads (Newton *et al* 1977; Erichsen and Molin 1981). However, concerns have been expressed by regulatory authorities (Gill 1988), food Industry groups (Anon 1988) and others that the practice may represent an undue safety hazard. Indeed despite the fact that the increasing commercial interest in the use of MAP to extend the self-life of many perishable products such as poultry, the concern about the potential growth of those pathogenic bacteria, which could survive and grow even at refrigeration temperatures (Silliker and Wolfe 1980; Palumbo 1987), remains the limiting factor to further expansion of the method. Indeed only few studies have examined the effect of MAP on the growth/survival of the foodborne pathogens-particularly the psychrotrophic strains.

Since there is increasing interest in storing food in vaccum or defined atmospheres, a system for testing growth of eg *Salmonella enteritidis*, *Listeria monocytogenes* or *Yersinia enterocolitica* under MAP and VP would appear useful. Although Hintlian and Hotchkiss (1986) reported some inhibitory effect on growth of *Salmonella* with MAP/VP in a wide range of beef and chicken products in contrast Elliot and Gray (1981) reported that levels up to 60% carbon dioxide stimulated growth of *Salmonella enteritidis* in artificial medium at 10°C.

While the microbiological changes on MAP meat are well established, the chemical changes accompanying the growth of bacteria on meat during storage, either aerobically or anaerobically, are not equally investigated. It was suggested that in both cases the metabolism of glucose, lactic acid, certain amino acids, nucleotides, urea and sarcoplasmic proteins can all occur during storage (Gill 1986; Jay 1986; Nychas *et al* 1988; Nychas and Arkoudelos 1990; Gill and Molin 1991). The off-odours formed on spoilage of meat have been attributed to various catabolites produced by all the various types of bacteria growing under different storage conditions (Dainty and Hibbard 1983; Dainty *et al* 1985; Nychas and Arkoudelos 1990; Ordonez *et al.* 1991; Kakouri and Nychas 1994). The identity of the end products of bacterial metabolism on meat or poultry stored under vacuum or in air has been well established (Jay 1986; Nychas and Arkoudelos 1990; Kakouri and Nychas 1994). There is little information however on the products produced during carbon dioxide, carbon dioxide/oxygen, and nitrogen storage.

---

[1]This research is conducted under EC contract agrf-ct90-0024 (tsts). The results of the work described are coming from the project leaders and their teams: G.J.E Nychas (NAgReF-ITAP, Athens), Dr.P.A Gibbs (Leatherhead Food R.A, Leatherhead) Prof. R.G Board (Univ. of Bath, Bath), Dr. J.J. Sheridan (Teagasc-NFC, Dublin)

Lactic acid bacteria are the dominant organisms on meats stored under carbon dioxide (Nychas and Arkoudelos 1990). It is well known that the end products of their metabolism depend upon environmental conditions. Thus oxygen and glucose limitation as well as pH affect the pattern of fermentation products, of these organisms during growth in laboratory media (Tseng and Montville 1990; Borch *et al* 1991). No information is available, however, about the chemical changes that occur during the refrigerated storage of meat of normal and high pH, stored in 100% carbon dioxide and 100% nitrogen.

As the pressure on the small food manufacturers in many countries increases to adopt the EC specifications for quality and safety standards (Anon 1988), the European Community decided, in the Framework of Food Linked Agro-Industry Research (FLAIR) (Anon 1992), to fund four Institutes/University from three different countries (Institute of Technology of Agricultural Products, Greece; Leatherhead Food R.A and University of Bath from UK; National Food Center IR) in order to examine the influence of MAP/VP on the growth of the food - poisoning organisms and to identify chemical changes that could be used to monitor safety and freshness. The objectives of the present study were the following:

a: To determine the effect of MAP/VP on the selection of a spoilage association from the initial microflora. As the spoilage flora develops,the changes with physico-chemical attributes of meat and meat products will be quantified.

b: To determine the influence of MAP/VP and the members of the association, particularly lactic acid bacteria, as well as the MAP effect on the following food poisoning organism will be determined: *Listeria monocytogenes, Staphylococcus aureus, Salmonella* spp, *Yersinia enterocolitica, Aeromonas hydrophila* etc.

c: To analyse the physico-chemical characteristics of MAP/VP meats during storage to identify changes that could be used to monitor safety and freshness The physico-chemical characteristics will include volatile compounds such as sulphur compounds, acetoin, diacetyl, and changes in protein patterns as shown in electrophoresis. Changes in these components will be correlated with microbial ecology. Moreover, suitable sensing or detection systems for changes will be studied (hydrogen sensors for hydrogen and volatiles such as organic sulphur compounds etc)

d: To examine changes in the organoleptic properties of the products. Organoleptic acceptability of meat and meat products will correlated with the microbial ecology and with physico-chemical changes.

e: To correlate time-temperature integrals assessed by various electronic, enzymic or chemical devices, with changes in organoleptic properties and organoleptic acceptability.

II. THE EFFECT OF VP/MAP ON GROWTH/SURVIVAL OF MEAT MICROBIAL FLORA.

Since the publication of the classic review in which Mossel and Ingram (1955) defined food microbiology as a branch of microbial ecology, the selection of spoilage flora has been considered in the contex of four interacting systems 1. the nutrient composition of a food (intrinsic factors), 2. the physiological attributes of spoilage organisms 3. extrinsic factors (eg gaseous composition of a storage environment, temperature etc) and 4. processing factors.

As far as concerns the last mentioned factor, methods and devices have been developed to clean carcasses surfaces completely. However sterilization of raw meat is difficult to achieve. It is well known that microbial spoilage of meat is influenced not only by their initial attachment to the surface, but also by subsequent proliferation after attachment (Chung *et al* 1989). Indeed Chung *et al* (1989) observed that bacteria attached more readily to meat did not necessarily proliferate more readily. For example, although *Pseudomonas aeruginosa* and *Listeria monocytogenes* showed no significant competive attachment to the meat samples, the numbers of the former were signifficantly greater after incubation of meat at abuse temperature (Chung *et al* 1989). The storage temperature and the packaging method could affect the proliferation (growth/survival) of these organisms.

al Processing of Foods

different microorganisms
on dioxide (Eklund and
monocytogenes and the
growth conditions is also
he inoculum reduces the
art *et al.* 1991). During
al for *L. monocytogenes*
ass and *vice versa* from
led carcases. In poultry
ogenes is mainly *via*
n and Mead 1989) and
ation.

s stored at 0 and 5° C
ed in 100% CO_2

5°C

4
6
0

ike most other enteric pathogens it can
blinger (1980) this bacterium is one of
lop easily in refrigerated foods.

y heterogenous group of bacteria which
ns which are ubiquitous in terrestrial and
with consumption of contaminated foods,
ied incidences of yersiniosis have increased
litica from food. This is probably due to
; detection (Schofield 1993). The growth/
meat the temperature and the environmental
(Zee *et al.* 1984) with *Y.enterocolitica* have
dioxide (>40%) inhibits the growth of this
have found that, compared to the percentage
under 100% carbon dioxide atmosphere, was
lthough Kleinlein and Untermann (1990) have
. beef at 4° C in 20% CO_2 / 80% O_2, while at
ge at 10° C was only slightly delayed the growth
ica on lamb pieces (Sheridan *et al* 1992), it was
nosphere containing 100% CO_2 or 80% O_2/ 20%
ll other atmopshere (50% N_2/ 50% CO_2; vacuum
and at 5° C there was no inhibition of growth,

$_{10}$ cfu/g) on lamb pieces and minced at 42 days in
tial inoculum *ca* $\log_{10}$ 1.8 cfu)

	minced		
A	B	C	D
4.0	NG	NG*	NG
6.0	6.0	4.0	NG

; C: 80%/20% O_2/CO_2 D: 100% CO_2

found also that
ogen (Table 3)

here containing 80% O_2/ 20% CO_2 is difficult to explain, in
re the CO_2 level was much higher (50% CO_2/ 50% N_2). The
periment suggests that this is real effect of some significance
hermore the inhibited effect of this particular gas atmosphere
ed from Church *et al.* 1992.
the map have no advantage on the quality of poultry inoculated
Grant (1989) suggested that carbon dioxide may offer a solution
ica at 5° C

rne transmission of *Listeria monocytogenes* can play a role in the
osis has led to the need for information on the characteristics of this
in particular meat, poultry and seafood products.
n implicated in epidemic foodborne outbreaks of human listeriosis,
oncern for several reasons, including substantial incidence rates of
evel samples and ability of the microorganism to survive or grow
normal processing, distribution or marketing (Buchanan *et al* 1989).

v on chilled
beef of pH
the higher
) conclude
ng factors;

Few studies have been made to obtain a quantitative comparison between grown under defined atmospheric conditions, except in the case of carb Jarmund 1993). This is particularly so for the "emerging pathogens" *L* verotoxigenic *Escherichia coli*. The effect on laboratory studies of inoculum a cause of increasing concern. It is reported that low temperature growth of lag phase of *L. monocytogenes* under chill conditions (Walker *et al*. 1990; H slaughter and subsequent chilling of carcases at the abattoir, there is the potent (if present) to be spread from an incoming warm carcase on to a chilled car contaminated chilled working surfaces and equipment on to either warm or chi processing it is reported that contamination of carcases with *L. monocy* contaminated working surfaces and equipment (Genigeorgis *et al* 1989; Huds hence low temperature growth of the inoculum may be the more relevant sit

Church *et al*. (1992) found that *L.monocytogenes* did not grow on beef steak (Table 2). *Listeria monocytogenes* found to be more sensitive in samples pac in both temperatures used in their study.

Table 2
Growth\Survival of *Listeria monocytogenes* on beef steaks stored at 0° and

days of storage	vacuum pack		50%/50% N_2/CO_2		80%/20% O_2/CO_2		100% CO_2	
	0	5	0	5	0	5	0	5
0	4.94	4.94	4.94	4.94	4.94	4.94	4.94	4.9
2	–	4.80	–	4.68	–	4.61	–	4.7
7	–	4.58	–	4.71	–	4.84	–	5.2
9	4.69	–	4.57	–	4.65	–	4.68	–
10	–	4.86	–	5.14	–	4.89	–	4.6
14	–	4.52	–	4.47	–	4.72	–	4.34
20	–	4.51	–	4.49	–	4.69	–	4.39
22	4.56	–	4.60	–	4.75	–	4.72	–
24	–	–	–	4.63	–	4.74	–	
28	–	4.35	–	–	–	–	–	4.07
29	–	–	–	4.45	–	–	–	–
34	–	4.12	–	–	–	–	–	3.59
36	4.29	–	4.67	–	4.50	–	4.38	–
49	4.02	–	4.41	–	4.43	–	3.77	–
63	3.98	–	3.67	–	3.43	–	3.83	–
77	3.93	–	3.37	–	–	–	2.82	–

Sheridan *et al*. (1992) in their experiments with *L. monocytogenes* in lamb pieces 100% CO_2 was the most effective storage condition to inhibit the growth of this path

Table 3
Growth\Survival of *Listeria monocytogenes* on lamb pieces stored at 5° C

days of storage	vacuum pack	50%/50% N_2/CO_2	80%/20% O_2/CO_2	100% CO_2
0	2.2	2.2	2.2	2.2
7	1.8	2.3	2.6	2.0
14	3.3	2.4	3.7	1.8
21	5.9	4.1	5.8	2.0
28	6.1	4.8	6.8	1.7
35	6.3	5.3	6.9	2.9
42	5.7	6.9	7.3	2.4

In the literature there is conflicting evidence about the ability of this organims to gro beef. Grau and Vantderlinde (1988) report growth of *L. monocytogenes* on vacuum-packe 5.6 and 6.0 at both 5.3° C and 0° C. *Listeria monocytogenes* growth was faster at temperature and pH and was initiated more rapidly on fatty tissue. Kaya and Schmidt (199 that growth of *L. monocytogenes* on vacuum-packed beef is chiefly dependent on the follow temperature, pH and competing flora. They observed no increase in *L. monocytogenes* numbers at 2, 4, and 7° C on beef of pH 5.6 but did observe an increase at 7° C on beef of pH 5.8 and 5.9. Shelef (1989) found that *L. monocytogenes* numbers in ground beef remained unchanged during storage at 4° C or 25° C and Jonhson *et al.* (1988) report survival but no growth of two strains of *Listeria* in ground beef held at 4° C for two weeks in either oxygen -permeable or oxygen-impermeable bags. Likewise, Gouyet *et al* (1978) found that in sterile ground beef inoculated with *L. monocytogenes* and stored at 8° C *Listeria* numbers decreased initially and then remained constant.

3. *Salmonella* spp.

It is well known that the probability of growth of one *Salmonella* surviving/ growing is dependent on numerous factors such as atmosphere, pH, competing microflora, temperature, -time of storage. In order to examine the effect of atmosphere on the growth /survival of this organism, Church *et al.* (1992) and Nychas (1993) inoculated beef steaks, minced beef and poultry (thigh and breast) with *S. tympimurium* (Tables 4, 5) and *S. enteritidis* (Table 6) respectively.

At 5^OC, *S. typhimurium* numbers remained static in all packs over the 36 day storage period (Table 4) in beef steaks. At 0^OC, in the vacuum, 50% CO_2/50% N_2 and 100% CO_2 atmospheres, *S. typhimurium* numbers declined slightly (*ca* 0.5 $\log_{10}$) over the 91 day storage period (Table 4) whilst in the 20% CO_2/ 80% O_2 pack there was a somewhat greater decline towards the end of the storage period.

Hintlian and Hotchkiss (1987) also found *S. tymphimurium* inhibited by MA environments. Moreover Luiten *et al.* (1982) found that, at 10ºC, numbers of *S. typhimurium* increased significantly during storage on beef steaks wrapped with oxygen permeable film, but remained low and fairly constant for

Table 4
Growth\Survival of *Salmonella typhimurium* on beef steaks stored at 0° C

days of storage	vacuum pack	50%/50% N_2/CO_2	80%/20% O_2/CO_2	100% CO_2
0	5.11	5.11	5.11	5.11
2	-	-	-	-
6	-	-	-	-
7	5.26	5.13	5.25	5.18
10	-	-	-	-
14	-	-	-	-
20	-	-	-	-
21	5.12	5.00	4.99	5.13
28	-	-	-	-
35	5.06	4.85	4.96	4.80
36	-	-	-	-
49	4.41	4.23	4.82	4.57
63	4.39	4.22	4.10	4.29
77	4.23	4.09	3.46	4.38
91	3.73	4.03	2.45	3.74

Table 5
Growth\Survival of *Salmonella typhimurium* on beef steaks stored at 5° C

days of storage	vacuum pack	50%/50% N_2/CO_2	80%/20% O_2/CO_2	100% CO_2
2	5.17	5.20	5.18	5.15
6	5.19	4.88	5.22	4.80
7	-	-	-	-
10	4.99	4.70	4.92	5.14
14	5.26	4.88	4.97	5.15
20	5.04	4.98	4.92	5.31
21	-	-	-	-
28	4.79	4.90	5.14	4.84
35	-	-	-	-
36	4.43	4.46	4.63	4.47

vacuum or gas-packed (60% CO_2/40% O_2) steaks. A CO_2 dependent bacteriostasis of *Salmonella* at chill temperatures, as observed in our study and by Luiten *et al.* (1982), has also been reported by Baker *et al.* (1986) and Eklund and Jarmund (1983).

The growth/ survival of *S. enteritidis* on breast and thigh stored at 10° C in vacuum pack, 100% carbon dioxide, 100% nitrogen and in 20% carbon dioxide/ 80% air are shown in Table 6. *Salmonella enteritidis* survived but did not grow in all gaseous atmospheres at 3° C (Nychas 1993). At 10°C the numbers of *S.enteritidis* increased rapidly in samples flushed with 100% nitrogen or with 20% carbon dioxide/80 % air and to a lesser extent in vacuum pack samples. It is worth to note that in samples stored under 100% carbon dioxide *S. enteritidis* numbers were decreased *ca* 1 log unit after 12 days of storage in breast, while in thigh (lower leg) remained at the initial level. Similar results were obtained from Gray *et al* (1984). The type of muscle (lower leg or breast) used in this study did not affect significantly the survival at 3°C or the rate of growth at 10° C of *S. enteritidis* (Table 6).

Table 6
Growth\Survival of *Salmonella enteritidis* on chicken thigh or breast stored at 10° C

days of storage	thigh				breast			
	A*	B	C	D	A	B	C	D
0	4.45	4.45	4.45	4.45	5.45	5.45	-	5.45
1	4.00	5.45	4.70	5.15	-	-	-	-
2	-	-	-	-	5.23	6.05	-	4.85
3	5.60	5.55	5.15	4.85	-	-	-	-
6	5.70	5.90	5.85	5.20	5.81	6.30	-	4.50
8	6.05	6.65	-	5.15	-	-	-	-
11	5.45	-	-	5.00	-	-	-	-
12	-	-	-	-	5.90	6.25	-	4.60

*: A: vacuum pack; B: 50%/50% N_2/CO_2; C: 80%/20% O_2/CO_2 D: 100% CO_2

Other studies have also demonstrated partial inhibition of either *S. tymphimurium* in TSB (Eklund and Jarmunt 1983) or *S. enteritidis* on chicken by 100% CO_2 (Gray *et al.* 1984)

4. verotoxigenic *Escherichia coli*

Not much work related to growth of this organism under vp/map has been published. Buchanan and Klawitter (1992) studied the effect of initial pH, sodium chloride, incubation temperature on the aerobic and anaerobic growth of *Escherichia coli* V157:H7 strain. Church *et al* (1992) studied the effect of different gaseous conditionsy on the growth survival of verotoxigenic *Escherichia coli*, the results are shown in Tables 7 and 8.

Table 7
Growth\Survival of verotoxigenic *Escherichia coli* on beef steaks stored at 0, 5 and 12° C

days	0°C				5°C				12°C			
	A*	B	C	D	A	B	C	D	A	B	C	D
0	4.52	4.52	4.52	4.52	4.52	4.52	4.52	4.52	4.52	4.52	4.52	4.52
1	-	-	-	-	-	-	-	-	4.60	4.52	4.52	4.52
3	-	-	-	-	4.42	4.55	4.94	4.59	4.74	5.07	5.21	4.80
8	4.55	4.75	4.56	4.76	4.62	4.81	4.71	4.49	4.60	4.88	5.24	4.50
13	-	-	-	-	-	-	-	-	4.70	4.95	6.33	4.69
14	-	-	-	-	4.68	4.58	4.69	4.71	-	-	-	-
16	-	-	-	-	-	-	-	-	4.69	4.98	5.87	4.82
21	4.69	4.66	5.28	5.22	4.63	4.74	4.79	4.70	4.31	4.48	4.19	4.43
29	-	-	-	-	-	-	-	-	4.78	4.36	4.87	4.82
35	4.81	4.84	4.65	4.78	4.79	4.97	4.91	4.87	-	-	-	-
49	4.59	4.17	4.00	4.71	-	-	-	-	-	-	-	-
70	4.72	4.53	4.43	4.77	-	-	-	-	-	-	-	-
91	4.72	4.45	4.11	4.71	-	-	-	-	-	-	-	-

*: A: vacuum pack; B: 50%/50% N_2/CO_2; C: 80%/20% O_2/CO_2 D: 100% CO_2

Table 8

Growth\Survival of verotoxigenic *Escherichia coli* on minced beef stored at 0, 5 and 12° C

days	0°C A*	B	C	D	5°C A	B	C	D	12°C A	B	C	D
0	4.86	4.86	4.86	4.86	4.86	4.86	4.86	4.86	4.86	4.86	4.86	4.86
1	-	-	-	-	-	-	-	-	4.72	4.84	4.84	4.81
3	4.66	4.79	4.92	4.83	4.68	4.65	4.87	4.69	4.58	4.84	5.27	4.60
6	4.61	4.67	4.74	4.33	4.62	4.61	4.75	4.74	4.80	4.71	6.49	4.68
8	-	-	-	-	-	-	-	-	4.66	4.72	6.09	4.79
10	4.60	4.69	4.90	4.76	4.74	4.78	4.85	4.72	4.70	4.63	6.30	4.74
13	4.88	4.82	4.92	4.86	4.80	4.77	4.90	4.74	-	-	-	-
14	-	-	-	-	-	-	-	-	4.71	4.64	5.53	4.64
16	4.67	4.73	4.88	4.72	4.66	4.76	4.86	4.74	-	-	-	-
21	-	-	-	-	-	-	-	-	4.31	4.48	4.19	4.43
22	4.56	4.71	4.65	4.78	4.58	4.54	4.65	4.70	-	-	-	-

*: A: vacuum pack; B: 50%/50% N_2/CO_2; C: 80%/20% O_2/CO_2 D: 100% CO_2

In all cases the verotoxigenic *Esch. coli* did not grow and remains in the same initial inoculation level. It needs to be noted that this organism survive and its numbers remain constant even during the storage at 0° C in all gaseous atmopsheres used in this study

5. *Staphylococcus aureus*

It is well established that *St.aureus* is a facultative anaerobe but grows best in the presence of oxygen. *St.aureus* can grow and produce enterotoxin under anaerobic conditions; however neither the extent nor the amount of toxin produced is as great as it is under aerobic conditions (Genigeorgis 1989). Moreover other environmental factors such as pH, water activity, storage temperature and chemical composition of a food (eg. type of oil, different types of muscles - red or white muscles) (Bergdoll 1989; Nychas and Board 1991) could also affect its growth/survival as well as enterotoxin B production. When poultry meat (thigh and breast) were inoculated with *St. aureus* no enterotoxin production was found after 7 days storage with vp/map (Table 9).

Table 9

Growth\Survival and toxin* production by *Staphylococcus aureus* S-6 on chicken thigh stored at 3 and 22° C

days of storage	3° C vacuum pack	100% CO_2 breast	22° C vacuum pack	100% CO_2	3° C vacuum pack	100% CO_2 thigh	22° C vacuum pack	100% CO_2
0	5.7	5.7	5.7	5.7	5.4	5.4	5.4	5.4
1	5.7	6.2	-	-	4.4	4.8	-	-
2	-	-	7.8	7.6	-	-	-	-
3	5.6	6.2	-	-	-	-	-	-
5	5.9	5.9	-	-	6.5	5.9	-	-
7	5.9	5.8	-	-	7.0	6.5	-	-

* : no enterotoxin B was found at both temperatures after 7 days of storage

Staphylococcus aureus survived but did not grow in either (vp or 100% CO_2) atmosphere at 3° C (Nychas 1992). At 22° C, *St.aureus* numbers increased in both atmosheres after 24h storage but again no toxin was detected. The numbers of *St. aureus* remained reasonably static at chill temperature and did not decline as was observed by Baker *et al.* (1986) in meat culture at 2°C. It is worth to note that the microorganim's minimum growth temperature is *ca* 7-10°C. Recently Buchanan *et al* (1992) reported that small increases in the levels of *St. aureus* was observed in the adequately (5° C) refrigerated turkey breast. Silliker and Wolfe (1980) found no growth of *St. aureus* inoculated into ground beef which was next stored under 60% carbon dioxide/25 % oxygen/ 15% nitrogen atmopshere for 10 days at 10°C. In general it has been reported that carbon dioxide affected the growth of *St.aureus* in minced meat, thigh, minced chicken, sausages and turkey. No toxin production was observed in the storage temperature of 8 and 12° C after a period of 31 days. However in sausages samples toxin production was occured at 26°C (Genigeorgis 1989).

It needs to be stressed that the inhibition of pathogens (*S. enteritidis, S. tymphimurium, L. monocytogenes, S. aureus, Y. enterocolitica* etc) described above could be due to synergetic effects of carbon dioxide and temperature (its inhibition action is greater at low temperature, Baker *et al.* 1986), low oxygen tension or due to the domination of lactic acid bacteria (see below) and to their antimicrobial agents (bacteriocins) and other metabolic products (eg acetic acid; Kakouri and Nychas 1994).

B. SPOILAGE FLORA

The surfaces of raw meat stored in an environment of high humidity are very susceptible to the growth of aerobic flora. The level of contamination of fresh meat is affected by many processes during the slaughtering of the animal, as highlighted by the studies of Newton *et al.* 1978, Gill and Newton 1982. Coliforms, *Brochothrix thesmosphacta*, psychrophilic and psyschrotrophic gram negative bacteria such as Pseudomonads, *Aeromonas, Acinetobacter, Moraxella*, Enterobacteriaceae, yeasts and to a lesser extent lactic acid bacteria have been found to be present in jointed and minced meat (Nychas 1984; Nychas *et al.* 1992; Church *et al.* 1992; Sheridan *et al.* 1992; Stanbridge and Board 1992). It is well known (Mossel and Ingram 1955) that the spoilage of meat under particular conditions of storage is caused by only some of the mixed microbial contaminants mentioned above.

The final microflora of fresh meat (poultry, beef and lamb) and meat products (eg. ham, stuffed chicken) stored under vp/map at 0, 3, 5, & 10° C, comprised in decreasing order of magnitude, Lactic acid bacteria, *Brochothrix thermosphacta* and *Pseudomonas* spp and Enterobacteriaceae (Nychas *et al.* 1992; Church *et al.* 1992; Sheridan *et al.* 1992; Stanbridge and Board 1992). The degree of their contribution to the final flora depends on the percentage of carbon dioxide in the gas mix. The increased inhibitory action of CO_2 at lower temperatures is considered to be due to increased CO_2 solubility, as opposed to increased cell susceptibility. At 5 and 10° C there was little difference in the numbers of lactic acid bacteria between any of the atmopsheres. At 0 and 3° C the 50% CO_2/ 50% N_2 and the 100% CO_2 atmosphere increases the lag phase of the lactic acid bacteria but at the end of storage period numbers were similar. The numbers of *Br. thermosphacta, Pseudomonas* spp and Enterobacteriaceae generally increased faster, and attained highest numbers in the 20% O_2/ 20% CO_2 atmosphere, whilst in 100% CO_2 numbers remained static or gave a slight increase. Growth of *Br. thermosphacta, Pseudomonas* spp. and Enterobacteriaceae on vacuum - packed and 50% CO_2/ 50% N_2 stored fresh meat was intermediate between that on the 80% O_2/ 20% CO_2 and 100% stored meat. The 50% CO_2/ 50% N_2 atmosphere was somewhat more inhibitory to the Enterobacteriaceae and *Pseudomonas* spp. than vacuum packaging, whislt reverse was true of *Br. thermopshacta*. Similar microbiological results were obtained from normal and DFD minced beef (Erichsen and Molin 1981; Blickstad and Molin 1983; Gill 1986; Nychas and Arkoudelos 1990; Kakouri and Nychas 1994).

In general, it needs to be stressed that each of these atmospheres (map/vp) selects a microbial flora dominated by Gram-positive bacteria (principally lactic acid bacteria and *Brochothrix thermosphacta*) rather than the Gram-negative ones that develop on meat stored under chill conditions in a normal atmosphere (air). This is due to the inhibitory effect of CO_2 on Gram-negative aerobic psychotrophic meat bacteria (Huffman *et al.* 1975; Newton *et al.* 1977; Sheideman *et al.* 1979; Christopher *et al* 1980a, 1980b) whilst the Gram-positive species are more resistant to the effects of CO_2 (Sutherland *et al.* 1977; Siliker and Wolfe 1980; Stier *et al* 1981). Thus the preservative action of CO_2 is not so much due to the control of the total microbial population as to restriction of growth of those types of organisms which have the potential to cause most rapid deterioration. Indeed despite inoculation of meat with relatively high numbers of *L. monocytogenes, S.tymphimurium, S. enteritidis, St. aureus, Y. enterocolitica* at 0 and 5° C, no major differences were aparrent in the developing and dominant microflora or at either temperature, from that of the uninoculated packs (Sheridan *et al.* 1992; Church *et al.* 1992; Stanbridge and Board 1992; Nychas *et al.* 1992). Similar results were obtained in poultry samples (raw or cooked with microwave) inoculated with *Ps. fragi* at 10° C and stored in a 100% CO_2 atmosphere (Nychas *et al.* 1992; Nychas unpublished results). Moreover Gill and DeLacey (1991) reported similar numbers of spoilage flora in inoculated and uninoculated samples that had been packaged and stored identically.

The lactic acid bacteria that dominate throughout the storage in so packed samples could account for this effect. Indeed the selection of *Lactobacillus* spp. is considered by some to have an added benefit as many lactic acid bacteria are known to exert an antagonistic effect against other spoilage or pathogenic bacteria (Schroder *et al.* 1980).

It is worth to note that not only the contribution of each microbial group depends on the concentration of CO_2 in the gas mixtures but also the selection of strains finally grown on meat. Many investigations of the principal organisms in the microbial associations of chilled meat have used selective media without further characterisation of the isolates. The use of map/vp technology could affect the proportion of the specific strains present finally on meat. Of the 1012 strains isolated (Stanbridge and Board 1992) from pseudomonad selective medium, CFC, *Pseudomonas fragi*, was numerically dominant organism (Table 10) and *Ps. fluorescens* and *Ps. lundensis* were present as minor proportions of the flora. This group of organisms failed to grow to any appreciable extent other than on beef steak in 80% O_2/20% CO_2 with incubation at 5°C. In this case the growth of *Ps. fragi* was responsible for the increase in the size of the pseudomonad population.

Table 10

Identification (%) of CFC isolates from beef steaks (last sample) stored in modified atmopshere at 0° and 5° C

Organims	0°C				5°C			
	A*	B	C	D	A	B	C	D
Ps. fragi	85	42.5	60	62.5	32.5	18.8	56.3	35.0
Ps. fluorescens	-	10.0	12	12.5	4.9	8.7	17.5	6.5
Ps. lundensis	-	-	-	2.5	20.7	5	5	3.9
others	15	47.5	28	22.5	41.5	67.5	21.3	54.5

*; A: vacuum pack; B: 50%/50% N_2/CO_2; C: 80%/20% O_2/CO_2 D: 100% CO_2

Characterisation of 348 isolates from VRBG (selective for Enterobacteriaceae) led to the identification of *Hafnia alvei* (Table 11) as the dominant member of this group on beef stored in vacuum at 0° C or in atmospheres lacking oxygen at 5° C. Oxygen in a pack appeared to favour the development of *Serratia liquefaciens, Providencia alcalifaciens* and *Enterobacter aerogenes* occurred in low numbers on some occassions only (Table 11). These three species of *Pseudomonas*, *Ps. fragi, Ps. fluorescens* and *Ps. lundensis* have been reported also on red meats stored chilled in air (Shaw and Latty 1982, 1984 and Molin and Ternstrom 1982, 1986; Molin *et al.* 1986) and on British Pork sausages (Banks and Board 1983).

Table 11

Identification (%) of VRBG isolates from beef steaks (last sample) stored in modified atmopshere at 0° and 5° C

Organims	0°C				5°C			
	A	B	C	D	A	B	C	D
Hafnia alvei	90	15	10	-	100	90	15	100
Serratia liquefaciens	10	45	20	-	-	10	40	-
Providencia alcalifaciens	-	35	-	10	-	-	-	-
Enterobacter aerogenes	-	-	-	-	-	-	20	-
Pseudomonads	-	-	-	-	-	-	25	-

A: vacuum pack; B: 50%/50% N_2/CO_2; C: 80%/20% O_2/CO_2 D: 100% CO_2

Pantoea agglomerans (36.8%), *Aeromonas* sp. (23.7%), *Providencia alcalifaciens* (21.1%), *Serratia liquefaciens* (2.6%), *Esch. coli* (2.6%) and pseudomonads (10.5%) were found to be isolated from VRBG medium on day 0 on meat steaks

III. EFFECT OF VP/MAP ON PHYSICOCHEMICAL CHANGES OCCURRING IN MEAT AND MEAT PRODUCTS

A. SUBSTRATES AND MICROBIAL METABOLITES

1. Introduction

While the microbiogical changes are well established, the chemical ones accompanying the growth of bacteria on meat during storage, either aerobically or anaerobically, are not equally studied (Jay 1986; Dainty and Mackey 1992). In general, meat is a perishable commodity with the follow chemical composition: Water (75%), carbohydrates (1.2%), proteins (19.02%), fat (2.5%) nitrogenous compounds (1.65%) and inorganic compounds (0.65 %) (Lawrie 1985). It has a pH (*ca* 5.5-6.3) which readily supports the growth of microorganisms when stored under chill conditions.

The literature of 30 or so years ago leaves the impression that, simply because of the large amounts of proteins in meat, proteolytic activity would be the main feature of the meat spoilage. This contention was bolstered also by characteristic odors at the time of spoilage. It is now recognised that bacterial proteolysis is a late, spatially superficial, and relatively unimportant event in meat spoilage.

The levels of protein or fat do not change during the onset of rigor nor are they substrates for microbial attack prior to the onset of spoilage (Dainty *et al* 1975; Gill and Newton 1980). It is well established that spoilage starts when the bacterial numbers reach around 10^7-10^8 cfu/g (Dainty *et al.* 1975). Schmitt and Schmidt-Lorenz (1992a,b) and Nychas *et al.* (1992) found that breakdown of proteins could occur, in ckicken (skin, thigh or breast) during the earlier stages of storage regardless to microbial size. The protein breakdown could not only be attributed to the indigenous proteolytic meat enzymes (autolysis) but also to the microbial proteolytic activity. If only autolysis occurred during storage, a similar pattern of protein breakdown could be expected in all samples, irrespectively of the way of their storage or of the contribution of the various spoilage groups in the final composition of the microbial flora.

The HPLC and SDS-PAGE analysis of water soluble proteins from poultry samples stored under vp/map conditions, showed that the protein breakdown was more rapid and eventually more extensive in samples inoculated with *Ps. fragi* compared with the uninoculated samples (Nychas *et al.* 1992). Moreover Nychas *et al.* (1992), Schmitt and Schmidt-Lorenz (1992a,b) found that the profile of the proteins, analysed with HPLC or SDS-PAGE, were different among the samples (chicken; skin, thigh, breast) stored under identical conditions (vp/map) at the end of storage at 3, 4, 10, 25 and 37°C. As mentioned above the lactic acid bacteria dominated in samples stored under vp/map. These organisms can produce extracellular proteinases although concidered to be weakly proteolytic bacteria (Law and Kolstad 1983) compared with many other groups of bacteria, such as *Pseudomonas* spp.

The concept of microbial proteinase involvement must take into account also factors controlling the induction of such enzymes in proteolytic bacteria. It is well known (Harder 1979) that (i) organisms produce very low basal levels of extracellular enzymes in the absence of an inducer and (ii) that the regulation and extracellular production of proteinases are based on induction and end product and/or catabolite repression. For example it is reported that glucose inhibited proteinase production by a milk isolate of *Ps.fluorescens* (Juffs 1976), while Fairburn and Law (1986) found that, in continuous culture, a proteinase was produced by *Ps. fluorescens* under carbon but not under nitrogen limiting conditions. This was taken as evidence that proteinase induction ensured that an energy rather a nitrogen source was available to the organisms.

According to Gill (1976), so long as low molecular weight components (especially glucose) are available on meat the proteolysis is inhibited.

2. Low Molecular Weights

It was suggested that the critical physicochemical changes during spoilage take place in the aqueous phase (Schmitt and Schmidt-Lorenz 1992a,b). This water phase contains glucose, lactic acid, certain amino acids, nucleotides, urea and water soluble proteins which are utilised by almost all the bacteria of the meat microflora (Gill 1976; Nychas *et al* 1988; Kakouri and Nychas 1994).

Several studies have shown that bacteria grow on meat at the expense of one or more of the low molecular weight soluble components mentioned above. The order in which these substrates are attacked by the various groups of spoilage bacteria, under aerobic and anaerobic conditions, is reviewed by Nychas *et al.* (1988) and Lambert *et al.* (1991). Until spoilage is evident to the senses, the only detectable effect of bacterial growth is some reduction of the glucose concentration (Table 12), which does not alter the organoleptic qualities of the meat.

Table 12

Changes in glucose content in pork of low (5.8) or high (6.4) pH during storage at 3° C under different modified atmospheres

Day of Storage						
	Air	CO$_2$ low pH	N$_2$	Air	CO$_2$ high pH	N$_2$
1	110	110	110	60	60	60
2	85	120	122	32	62	45
4	45	100	86	-	-	-
5	-	-	-	28	70	51
7	-	-	-	16	40	38
8	25	90	56	-	-	-
10	10	85	34	8	24	22
12	0	45	7	-	-	-

Glucose (mg 100^{-1}); MA Conditions

Although the glucose depletion during aerobic or anaerobic conditions is well established (Gill 1976; Farber and Idziak 1982; Nychas 1984; Nychas and Arkoudelos 1990; Church *et al.* 1992) the pattern of the changes in the concentration of lactate varies. For example Nassos *et al.* (1985) have recommended the use of lactate as a spoilage index of ground beef. Indeed these workers have found an increase of lactate, during storage.

It needs to be mentioned that they analysed their samples with HPLC and for this were unable to distinguish between *D* and *L* lactate. On the other hand Nychas (1984), Nychas and Arkoudelos (1990), Borch and Agerhem (1992) Church *et al.* (1992), Nychas *et al* (1992) and Kakouri and Nychas (1994) have reported that *L*-lactate decreased during storage under aerobic or anaerobic conditions (Tables 13 & 14).

Table 13

Changes in *L*-lactate content in pork of low (5.8) or high (6.4) pH during storage at 3° C under different modified atmospheres

Day of Storage	Air	CO$_2$ pH 5.8	N$_2$	Air	CO$_2$ pH 6.4	N$_2$
1	260	260	260	220	220	220
2	230	270	270	200	230	170
4	190	260	230	-	-	-
5	-	-	-	180	210	190
7	-	-	-	100	190	123
8	160	230	200	-	-	-
10	110	225	140	90	170	143
12	70	230	95	-	-	-

L- lactate (mg 100^{-1}); MA Conditions

Table 14

Decrease of *L*-lactic acid during storage of meat and meat products stored under vacuum (vp) or 100% CO$_2$ (map)

Sampling day	map					vp				
	A[*]	B	C	D	E	A	B	C	D	E
1	220	350	582	650	682	220	350	582	650	682
last	170	280	484	450	425	90	265	407	420	507

type of meat

*; A: minced pork; B: stuffed chicken; C: pork steak; C:chicken breast; E: Beef steak

Gill and Newton (1978) reported that lactate was not exhausted at the meat surface when growth ceased. Church *et al*. (1992) and Nychas (unpublished results) analysed *D* and *L* lactate enzymatically, and found that *L*- lactate decreased while *D*-lactate increased during storage of meat under different gaseous conditions (Table 15). Similarly Dainty (1981), Ordonez *et al*. (1991) and de Pablo *et al*. (1989) found that *D*-lactate increased during storage under vp/map conditions. Obviously the *d*-lactic acid isomer found to increase during storage under different gaseous conditions (Church *et al*. 1992; Nychas unpublished results) does not come from either *Br. thermosphacta* or anaerobic glycolysis because in both cases only *l*-lactate could be produced (Ordonez *et al*. 1991). Therefore the increase of this compound is due to metabolism of lactic acid bacteria that could generate *D*, *L* or *DL* lactate (Kandler 1984).

Table 15
Changes in the concentration of *D*-lactic acid (mg/100g) produced in raw meat during the storage at different conditions

days of storage	vacuum pack	50%/50% N_2/CO_2	80%/20% O_2/CO_2	100% CO_2	vacuum pack	100% CO_2
		beef steaks			chicken breast	
0	0	0	0	0	0	0
1	0	0	0	0	-	-
7	191	-	175	249	-	-
9	-	-	-	-	229	138
13	-	-	-	-	243	225
16	-	-	-	-	276	249
28	437	395	-	-	-	-
35	-	432	-	419		

Table 16 shows the level of acetic acid (Nychas unpublished results) at various times throughout the storage period. Results in vp/ 100% CO_2 were similar. Acetic acid showed a clear rising trend from the beginning. This could be due to storage conditions. The storage of meat in vp/map not only select a microbial flora (lactic acid bacteria) on meat different from that stored in air, but also could possibly influence their metabolism (Nychas and Kakouri 1994). Indeed it is well known that the metabolism of lactic acid bacteria affected by environmental factors such as degree of aeration (low oxygen pressure), glucose limitation etc. (Marshall 1992). The increase of acetate (Tables 16) in meat samples stored under vp/map could be attributed either to switch of metabolism of lactic acid bacteria from homo to heterofermantative or to other predominant organism in such packaging system *Br. thermosphacta* (Nychas and Kakouri 1994).

Table 16
Changes in the concentration of acetate (mg/100g) produced in meat during the storage at at 3o C under different gaseous conditions

days of storage	vacuum pack	100% CO_2	vacuum pack	100% CO_2	vacuum pack	100% CO_2
	pork steaks		chicken breast		chicken thigh	
0	2	2	0	0	0	0
3	15.6	-	7.2	34.1		
5	41.6	12.1	-	-	5.95	12.3
7	87.3	47.6	68.8	50.2	14.6	52.4
10	-	179.2	-	-	-	-

Dainty 1981, de Pablo *et al*. (1989), Ordonez *et al*. (1991), Borch and Agerhem (1992) and Nychas and Kakouri (1994) reported similar results in beef, pork, chicken and meat products

The conversion of glucose to gluconate - a characteristic feature in meat exposed to a normal atmosphere during chill storage (Nychas *et al*.1988) - *via* the extracellular glucose dehydrogenase of pseudomanads provides them with a competitive advantage because gluconate is not so readily utilized by other members of the meat microflora (Farber and Idziak 1982; Nychas *et al*. 1988). The delay in the utilization of glucose (Table 12), as well as the delay in the production of gluconate (Table 17) during the early stages of storage of meat stored in carbon dioxide, was probably due to high pCO_2

or low pO_2 inhibiting the activity of glucose- dehydrogenase of pseudomonads (Mitchell and Dawes 1982; Nychas *et al.* 1988). The subsequent increase in the concentration of gluconate (Table 17) was associated with an increase in the size of the population of pseudomonads (Farber and Idziak 1982; Nychas and Arkoudelos 1990; Nychas *et al.* 1992) due to oxygen permeability of the film (Lambert *et al.* 1991) used. The limited changes in gluconate concentration in CO_2 stored meat contrast with those occurring in meat stored in either a normal atmosphere or nitrogen. In both of the latter instances the concentration of glucose diminished rapidly, and the transient peak in gluconate concentration occurred at an early stage of storage (Tables 12 and 17).

Similar results have been reported from Roberts (1989) and Nychas and Arkoudelos (1990) under aerobic and anaerobic conditions. In contrast Church *et al.* (1992) did not find any accumulation of gluconate in meat samples stored in 100% CO_2 and vacuum pack.

Table 17

Changes in gluconate content in pork of low (5.8) or high (6.4) pH during storage at 3° C under different modified atmospheres

Gluconate (μg 100^{-1})

Day of Storage	MA Conditions Air	CO_2 low pH	N_2	vp	CO_2 high pH	N_2
1	45	45	45	95	95	95
2	90	52	60	102	89	124
4	45	60	70	-	-	-
5	-	-	-	64	153	52
7	-	-	-	34	136	63
8	15	118	80	-	-	-
10	0	156	44	12	90	65
12	0	65	9	-	-	-

It was mentioned above that - off-odours are not produced until glucose and gluconate are exchausted (Gill 1976; Nychas *et al.* 1988). Gluconate is not connected directly with the production of off-odours, which are mainly caused by sulphur compounds, diacetyl, acetic acid and to, a lesser extent, short-chain branched fatty acids (eg. *iso*-butiric, *iso*-valeric) and to alcohols (e.g. 2,3 butanediol, 2-methylpropanol) (Dainty and Hoffman 1983; Dainty *et al.* 1985; Dainty and Mackey 1992).

The type of microbial metabolites detected in naturally contaminated samples of meat depends upon the composition of the microbial flora which is selected from the environmental storage conditions. For example the ethyl esters of acetic acid, propanoic, *n*-butanoic, isopentanoic and hexanoic acids were detected in meat stored chilled in air (Dainty *et al.* 1985) while none of the esters containing the branched chain alcohol and acid components were obsreved in meat stored in vacuum packs (Edwards and Dainty 1987).

As CO_2 inhibited growth - and presumably glucose assimilation by pseudomonads - the dairy/cheesy odours mainly found in samples stored in gas mixes with CO_2 and to a lesser extent in N_2 were no doubt produced by *Br. thermosphacta* and lactic acid bacteria both of which can produce diacetyl/acetoin and alcohols (Dainty and Hibbard 1983; Dainty and Hoffman 1983) exclusively from glucose under aerobic conditions or low O_2 tension (Blickstad and Molin 1984).

According to Church *et al.* 1992, alcohols (particularly ethanol and propanols) appear to be the most promising compounds as spoilage-indicators. Indeed in their study, a wide range of organic volatiles (Table 18) was detected in the headspace of meat packs. Many from these volatiles have reported also from Dainty *et al.* (1985) Nychas and Arkoudelos (1990) during refrigerated storage of naturally conatminated beef and pork. These alcohols (ethanol and propanols) are present at only trace levels at the beginning of storage, and their concentrations increase significantly prior to spoilage.

Experiments have shown that all three of these alcohols are able to diffuse through the packaging material used, and it may therefore be possible to use their production as the basis of a non-invasive spoilage sensor (Church *et al.* 1992). Dainty and Mackey (1992) reported that ethanol has been detected also consistently in their samples stored in vacuum packs. This compound is presumably a fermentation product of the heterofermantative leuconostoc and carnobacteria (Dainty and Mackey

1992)

Table 18
Volatiles present in packaged beef

ATMOSPHERE

Volatile	VP	20%/80% CO_2/O_2	50%/50% N_2/CO_2	100% CO_2
Ethanol	+	+	+	+
Acetone	+	+		+
Propan-2-ol		+	+	+
Dimethylsulphide	+	+		+
Propan-1-ol	+	+		
ethyl acetate	+	+	+	+
2,3 butandione	+	+		+
acetic acid	+	+	+	+
hexane		+	+	+
Heptane	+	+	+	+
pentanol			+	+
2 methylpropanol	+	+		
2 methyl butanol		+		+
pentanal		+		
heptadiene		+		
acetoin	+	+		+
3 methylbutan-1-ol	+	+		
2 methylbutan-1-ol	+	+		
dimethyldisulphide	+	+	+	
octane			+	+
2,3 butandiol		+		
3 ethyl pentane				+

Other less volatile compounds eg tyramine, putrescine and cadaverine have been detected in vacuum packs meat (Dainty *et al*. 1986; Edwards *et al*. 1987; Schmitt and Schmidt-Lorenz 1992a,b) mainly attributed to Gram-negative bacteria present.

Therefore spoilage of stored meat is not, as might be inferred from the findings of Gill (1976, 1986) and Gill and Newton (1977), exclusively associated with the onset of amino acid metabolism subsequent to glucose and gluconate depletion. Church *et al*. (1992) and Nychas *et al*. (1992) reported that glucose limitation does not appear to be the cause of total viable count becoming static or the reason for off-odour development. They have found that even the sample was considered spoiled organoleptically, glucose levels at the surface were found to be relative high compared to its initial concentration. Newton and Rigg (1979) found that the utilization of amino acids depends on oxygen concentration and under reduced conditions, amino acids were used even before glucose.

VI. CONCLUSIONS

Much of the data gathered in this EEC project, add considerably to the stock of knoweldge regarding spoilage and safety aspects of packaged meats. In general, vacuum pack and modified atmospheres have beneficial effects with respect to expansion of the self-life of meat and meat products. However, it is already clear that there are factors controlling the growth of certain spoilage (eg. *Ps. fragi*) and pathogens (*Y. enterololitica, L.monocytogenes*) in some meats and some conditions that are, as yet, inexplicable on the basis of current knowledge of the characteristics of those bacteria.

Although the factors that can select particular groups of organisms under vp/map are now well established in broad terms, we still need to explain why particular species eg. *Ps. fragi*, become the dominant organisms among the pseudomonads. In general the contribution of any group of micro-organisms to meat spoilage depends not only on their initial numbers in the spoilage bacterial population but also from the storage conditions. It has been shown that the important changes in microbial metabolites, and their possible role as indicators of incipient spoilage, are determined by storage methods. Indeed, each different type of storage condition (eg air, vp, CO_2, nitrogen) provided unique environments for the growth of particular groups of bacteria. For this reason it is desirable not only to inhibit or retard the total growth of all the initial flora present in meat, but to affect the growth of those microorganisms with the relative fastest rate of growth under these conditions which have the highest spoilage potential.

Although the presence of off-odours is the most rapid and universally accepted indicator of the end of shelf-life of meat stored in air, problems may arise with novel forms of gas packaging.
The potentail of certain compounds associated with microbial growth to indicate the status of the product within the package, seems evident, but methods of external detection and relations with spoilage have yet to be established.

REFERENCES

Anon Safety considerations for new generation refrigerated foods. *National Food Processors Association Dairy Food Sanitation* **8**, 5, 1988

Anon Synopsis of R & D Projects and Concerted Actions. XII/12520/91-EN, Commission of the Eyropean Communities Directorate General XII, Science, Research and Development, Brussels, 1992

Baker ,R.C.,Qureshi,R.A and Hotchkiss,J.H., Effect of an elevated level of CO_2 containing atmosphere on the growth of spoilage and pathogenic bacteria at 2, 7 and 13° C. *Poultry Science* **65**, 729, 1986

Banks, J.G. and Board, R.G. The classification of pseudomonads and other obligately aerobic Gram-negative bacteria from British pork sausage and ingredients. *Systematic and Applied Microbiology* 4,424, 1983.

Bergdoll, M.S. *Staphylococcus aureus*. In *Foodborne Bacterial Pathogens*, Doyle, M.P. (ed), New York, Marcel Dekker Inc. 1989

Blickstad,E. and Molin,G Carbon dioxide as a controller of the spoilage flora of pork, with special reference to temperature and sodium chloride. *Journal of Food Protection*, 46, 756,1983

Blickstad, E. and G. Molin. Growth and end-product formation in fermenter cultures of *Brochothrix thermosphacta* ATCC 11509^T and two psychrotrophic *Lactobacillus* spp. in different gaseous atmospheres. *Journal of Applied Bacteriology*, 57, 213, 1984

Borch,E. and Agerhem,H. Chemical, microbial and sensiory changes during the anaerobic cold storage of beef inoculated with a homofermantative *Lactobacillus* sp. or a *Leuconostoc* sp. *International Journal of Food Microbiology* 15,99, 1992

Borch,E., Berg,H. and Holst, O. Heterolactic fermentation by a homofermentative *Lactobacillus* sp. during glucose limitation in anaerobic continous culture with complete cell recycle. *Journal of Applied Bacteriology* 71, 265, 1991.

Buchanan,R.L and Klawitter, L.A., The effect of incubation temeprature,initial pH and sodium chloride on the growth kinetics of Escherichia coli O157:H7, *Food Microbiology*, 9,185, 1992

Buchanan, R.L., Stahl, H.G and Whiting, R.C. Effects and interactions of temperature, pH, atmosphere, sodium chloride, and sodium nitrite on the growth of *Listeria monocytogenes*. *Journal of Food Protection* 52, 844, 1989

Buchanan, R.L., Schultz, F.J., Golden, M.H., Bagi, K and Marmer, B. Feasibility of using microbiological ondicator assays to detect temeprature abuse in refrigerated meat, poultry and seafood products, *Food Microbiology*, 9,279, 1992

Clark, D.S., Lentz, C.P and Roth,L.A Use of carbon monoxide for extending shelf life of pre-packaged fresh fresh beef. *Canadian Institute of Food Science and Technology* 9, 114, 1976

Christopher, F.M., Smith, G.C., Dill, C.W., Carpenter, Z.L. and Vanderzant, C. Effect of CO_2-N_2 atmospheres on the microbial flora of pork. *Journal of Food Protection* 43, 268, 1980a

Christopher, F.M., Carpenter, Z.L., Dill, C.W., Smith, G.C. and Vanderzant, C. Microbiology of beef, pork and lamb stored in vacuum or modified gas atmospheres. *Journal of Food Protection*, 43, 259, 1980b

Chung,K.T., Dixon, J.S., and Crouse, J.D. Attachment and proliferation of bacteria on meat. *Journal of Food Protection*. 52, 173, 1989

Church, P.N., Davies, A.R., Slade, A., Hart, R.J., and Gibbs, P.A Improving the safety and quality of meat and meat products by modified atmopshere and assessment by novel methods. *FLAIR 89055 Interim 2nd Year Report*, EEC DGXII, Brussels, 1992

Dainty R.H Volatile fatty acids detected in vacuum packed beef during storage at chill temperatures. In *Proceedinds of 27th Meeting of European Meat Workers*, Vienna, p. 688, 1981

Dainty,R.H. and Hibbard,C.M Precursors of the major end products of aerobic metabolism of *Brochothrix thermosphacta*. *Journal of Applied Bacteriology* 55, 387, 1983.

Dainty,R.H. and F.J.K. Hoffman. The influence of glucose concentration and culture incubation time on end-product formation during aerobic growth of *Brochothrix thermosphacta*. *Journal of Applied Bacteriology* 55, 233, 1983

Dainty, R.H and Mackey, B.M The relationship between the phenotypic properties of bacteria from chill-stored meat and spoilage processes In *Ecosystems: Microbes: Food* Board, R.G D.Jones, Kroll, R.G Pettipher G.L, Eds, Society for Applied Bacteriology Symposium Series No.21, Oxford: Blackwell Scientific Publications, p.103S, 1992.

Dainty,R.H., Shaw,B.G., De Boer, K.A., and Scheps,S.J. Protein changes caused by bacterial growth on beef. *Journal of Applied Bacteriology* 39, 73, 1975

Dainty, R.H., Edwards R.A and Hibbard, C.M. Time course of volatile compound formation during refrigerated storage of naturally contaminated beef in air. *Journal of Applied Bacteriology*, 59,303, 1985

Dainty,R.H., Edwards,R.A., Hibbard,C.M and Ramantanis, S.V Bacterial sources of putrescine and cdaverine in chill stored vacuum-packaged beef. *Journal of Applied Bacteriology* 61, 117, 1986

de Pablo,O.B., Asensio,M.A., Sanz,B and Ordonez, J.A The D(-) lactic acid and acetoin/diacetyl as potential indicators of the microbial quality of vacuum-packed pork and meat products. *Journal of Applied Bacteriology* 66, 185, 1989

Edwards, R.A and Dainty, R.H Volatile compounds associated with spoilage of normal and high pH vacuum-packed pork. *Journal of the Science of Food and Agriculture* 38, 57, 1987

Edwards, R.A., Dainty, R.H.,Hibbard,C.M., and Ramantanis,S.V Amines in fresh beef of normal pH and the role of bacteria in changes in concentration observed during storage in vacuum packs at chill temperatures. *Journal of Applied Bacteriology* 63, 427, 1987

Eklund, T and Jarmund, T. Microculture model studies on the effect of various gas atmopsheres on microbial growth at different temperatures. *Journal of Applied Bacteriology* 55,119, 1983

Elliot,P.H and Gray,R.J.H., *Salmonella* sensitivity in a sorbate/modified atmosphere combination system. *Journal of Food Protection* 44, 903, 1981

Erichsen,I and Molin,G. Microbial flora of normal and high pH beef stored at 4°C in different gas environments. *Journal of Food Protection* 44, 866, 1981.

Fairbairn, D.J and Law, B.A. Proteinases of psychrotrophic bacteria: their production, properties effects and control. *Journal of Dairy Research* 53, 139, 1986

Farber, J.M. and E.S. Idziak. Detection of glucose oxidation products in chilled fresh beef undergoing spoilage. *Applied and Environmental Microbiology*, 44, 521, 1982.

Genigeorgis, C.A Present state of knowledge on staphylococcal intoxication. *International Journal of Food Microbiology*, 8, 143, 1989

Genigeorgis, C.A., Dutulescu, D. and Garayzabal, J.F Prevelance of *Listeria* spp. in poultry meat at the supermarket and slaughterhouse level. *Journal of Food Protection* 52, 618, 1989

Gill,C.O. Substrate limitation of bacterial growth at meat surfaces. *Journal of Applied Bacteriology* 41, 401,1976.

Gill, C.O. The control of microbial spoilage in fresh meats. In: *Advances in Meat Research Meat Poultry Microbiology*, Pearson A.M and Dutson T.R, Eds Macmillan, London, 1986

Gill,C.O Packaging meat under carbon dioxide:The Captech system. 34th International Congress of meat Science and technology. *Proceeding of Industry day*, Brisbane, Australia, 1988

Gill, C.O. and Newton, K.G. The development of aerobic spoilage flora on meat stored at chill temperatures. *Journal of Applied Bacteriology*, 43, 189, 1977.

Gill, C.O and Newton, K.G. The ecology of bacterial spoilage of fresh meat at chill temperatures. *Meat Science* 2, 207, 1988

GIll, C.O and Newton, K.G. Development of bacterial spoilage at adipose tissue surfaces of fresh meat. *Applied and Environmental Microbiology* 39, 1076, 1980

Gill, C.O. and Newton, K.G. Effect on lactic acid concentration on growth on meat of Gram-negative psychrotrophs from meatworks. *Applied Environmental Microbiology* 33, 1284, 1982.

Gill, C.O and Molin,G. Modified atmospheres and vaccum packaging, in *Food Preservatives*, Russell

N.J and Gould G.W., Eds, Blackie,Glasgow and London, 1991

Gill, C.O and Reichel, M.P. Growth of cold tolerant pathogens on high pH beef packaged under vacuum or CO$_2$. *Food Microbiology* 6, 223, 1989

Gill,C.O., and DeLacey,K.M Growth of *Escherichia coli* and *Salmonella typhimurium* on high-pH packed under vacuum or carbon dioxide. *International Journal of Food Microbiology* 39, 317, 1991

Gouet,P. Labadie, J.A and Serratore, C. Development of *Listeria monocytogenes* in monoxenic and polyxenic beef minces. *Zentralbaltt fur Bakteriologie, Parasitenkunde, Infektions Krankheiten und Hygiene I. Abteilung Originale B.*, 166, 87, 1978

Grau, F.H and Vanderlinde, P.B. Growth of *Listeria monocytogenes* on vacuum packaged beef. *Proceedings of the 34th International Congress of Meat Science and Technology, Brisbana*, Part B, Cannon Hill, Queensland:CSIRO Meat Research Laboratory, 1988

Gray, R.J.H., Hoover,D.G and Muir, A.M., Attenuattion of microbial growth on modified-atmopshere packed fish, *Journal of Food Protection* 47(7), 610,1983

Gray, R.J.H, Elliot, P.H and Tomlins, R.I. Control of two major pathogens on fresh poultry using a combination potassium sorbate/CO$_2$ packaging treatment. *Journal of Food Science* 49, 142, 1984

Harder, W., Regulation of the synthesis of extracellular enzymes in microorganisms. *Society for General Microbiology Quarterly* 6, 139, 1979

Hart, C.D., Mead G.C. and Norris, J. Effects of gaseous environment and temperature on the storage behaviour of *Listeria monocytogenes* on chicken breast meat. *Journal of Applied Bacteriology* 70, 40, 1991

Hintlian,C.B and Hotchkiss,J.H The safety of modified atmopshere packaging: a review. *Food Technolology* 40, 70, 1986

Hudson, W.R and Mead, G.C.*Listeria* contamination at a poultry processing plant. *Letters in Applied Microbiology* 9, 211, 1989

Huffman, D.L., Davis, K.A., Marple, D.N. and McGuire, J.A. Effect of gas atmospheres on microbial growth, colour and pH of beef. *Journal of Food Science* 40, 1229, 1975.

Jay, J.M. Microbial spoilage indicators and metabolites.In: *Foodborne microorganisms and their toxins: developing methodology*, Pierson M.D and Sterm N.J., Eds, Marcel Dekker Inc., New York and Basel, 1986

Johnson,J.L., Doyle,M.P., and Cassens, R.G. Survival of *Listeria monocytogenes* in ground beef. *International Journal of Food Microbiology* 6, 243, 1988

Jufss, H.S Effects of temperature and nutrient on proteinase production by *Pseudomonas fluorescens* and *Pseudomonas aeruginosa* in broth and milk. *Journal of Applied Bacteriology* 40, 23, 1976

Kakouri, A. and Nychas, G.J.E Storage of poultry meat under modified atmospheres or vacuum packs; possible role of microbial metabolites as indicator of spoilage, *Journal of Applied Bacteriology*, (in press), 1994

Kaya, M and Schmidt, U., Behaviour of *Listeria monocytogenes* on vacuum-packed beef. Fleischwirtschaft 71, 424, 1991

Kandler, O. Carbohydrate metabolism in lactic acid bacteria. *Antonie van Leeuwenhoek* 49, 209, 1984

Kleinlein, N. and Untermann, F. Growth of pathogenic *Yersinia enterocolitica* strains in minced meat with and without protective gas with consideration of the competatitive background flora. *International Journal of Food Microbiology* 10, 65, 1990

Lambert, A.D., Smith, J.P., and Dodds,K.L. Shelf life extension and microbiological safety of fresh meat - a review. *Food Microbiology* 8, 267, 1991

Law, B.A and Kolstad,J. Proteolytic systems in lactic acid bacteria. *Antonie van Leeuwenhoek*, 49, 225, 1983

Lawrie,R.A Meat Science, 3rd Ed, Pergamon Press, Oxford, 1985

Luiten, L.S., Marcello,J.A and Dryden, F.D. Growth of *Salmonella tymphimurium* and mesophilic organisms on beef steaks as influenced by type of packaging, *Journal of Food Protection*, 45, 263, 1982

Marshall, V.M. Inoculated ecosystem in a milk environment. In *Ecosystems: Microbes: Food* Board, R.G D. Jones, Kroll, R.G Pettipher G.L, Eds, Society for Applied Bacteriology Symposium Series No.21, Oxford: Blackwell Scientific Publications, p.127S, 1992.

Mitchell, G.C. and E.A. Dawes. The role of oxygen in the regulation of glucose meatbolism, transport and tricarboxylic acid cycle in *Pseudomonas aeruginosa*. *Journal of General Microbiology*,

128, 49, 1982

Molin, G. and Ternstrom, A. Numerical taxonomy of psychrotrophic pseudomanads. *Journal of General Microbiology* 128, 1249, 1982

Molin, G. and Ternstrom, A. Phenotypically based taxonomy of psychrotrophic *Pseudomonas* isolated from spoiled meat, water and soil. *International Journal of Systematic Bacteriology* 36, 257, 1986

Molin, G., Ternstrom, A. and Ursing, J. *Pseudomonas lundensis*, a new bacterial species isolated from meat. *International Journal of Systematic Bacteriology* 36, 339, 1986.

Mossel, D.A.A. and Ingram, M. The physiology of the microbial spoilage of foods. *Journal of Applied Bacteriology* 18, 233, 1955-68.

Nassos,P.S., King,A.D and Stafford, A.D. Lactic acid concentration and microbial spoilage in anaerobically and aerobically stored ground beef. *Journal of Food Science* 50, 710, 1985

Newton,K.G and Rigg,W.J., The effect of film permability on the storage life and microbiology of vacuum-packed meat. *Journal of Applied Bacteriology* 47, 433, 1979

Newton,K.G.,Harrison,J.C.L and Smith,K.M The effect of storage in various gaseous atmospheres on the microflora of lamb chops held at -1° C. *Journal of Applied Bacteriology*, 43, 53, 1977

Newton, K.G., Harrison, J.C.L. and Wauters, A.M. Sources of psychrotrophic bacteria on meat at the abattoir. *Journal of Applied Bacteriology* 45,75, 1978.

Nychas,G.J.E Microbial growth in minced meat. Ph.D Thesis, University of Bath,Bath, U.K, 1984

Nychas G.J.E *Staphylococcus aureus*: growth and toxin production in poultry meat and its control, In: *Other pathogens of concern (no salmonella and campylobacter)*. Nurmi, E., Colin, P., and Mulder, R.W.A.W, Eds. *Proceedings of a meeting held at Helsinki, Finland, June 11-14*, COVP-DLO Het Spelderholt, 1992.

Nychas G.J.E Survival/ihnibition of *Salmonella enteritidis* in poultry stored under map/vp. In. *Contamination with Pathogens in Relation with processing and marketing of products*. Mulder, R.W.A.W, Ed, *Proceedings of a meeting held at Fribourg, Switzeralnd, February 25-27*, COVP-DLO Het Spelderholt, 1993

Nychas,G.J.E and Arkoudelos,J.S. Microbiological and physicochemical changes in minced meat under carbon dioxide, nitrogen or air at 3°C. *International Journal of Food Science and Technology* 25, 389, 1990

Nychas,G.J.E, and Board, R.G Enterotoxin B production and physicochemical changes in extracts from different turkey muscles during the growth of *Staphylococcus aureus* S-6. *Food Microbiology*, 8, 105, 1991

Nychas,G.J.E, V.Dillon and Board, R.G. Glucose,the key substrate in the microbiological changes occurring in meat and certain meat products. *Biotechnology and Applied Biochemistry* 10, 203, 1988

Nychas, G.J.E, Kakouri, A., and Arkoudelos, J.S Improving the safety and quality of meat and meat products by modified atmposhere and assessment by novel methods. *FLAIR 89055 Interim 2nd Year Report,* EEC DGXII, Brussels, 1992

Ordonez,J.A, de Pablo,B., de Castro,B.P., Asensio, M.A and Sanz,B. Selected chemical and microbiological changes in refrigerated pork stored in carbon dioxide and oxygen enriched atmospheres. *Journal of Agricultural Food Chemistry* 39, 668, 1991

Palumbo,S.A, Is refrigeration enough to restrain foodborne pathogens?. *Journal of Food Protection* 49,1003, 1987

Patterson, M.F and Grant, I.R. Combination of irradiation with MAP. *Irradiation Symposium London*, IBC Technical Services Ltd, 1989

Price, R.J. and Lee, J.S. Inhibition of *Pseudomonas* species by hydrogen peroxide producing lactobacilli. *Journal of Milk and Food Technology* 33, 13, 1970

Roberts, T.A., Some recent findings in fresh refrigerated foods. Foodpuls VI-89. In *Proceedings of the 6th annual conference*, Lancaster Technomic Publishing p.419, 1989

Schmitt,R.E and Schmidt-Lorenz, W. Formation of ammonia and ammines during microbial spoilage of refrigerated broilers. *Lebensmittel-Wissenschaft und-Technologie* 25, 6, 1992a

Schmitt,R.E and Schmidt-Lorenz,W. Degradation of ammino acids and protein changes during microbial spoilage of chilled unpacked and packed chicken carcasses. *Lebensmittel-Wissenschaft und-Technologie* 25, 11, 1992b

Schofield,G.M. Emerging food-borne pathogens their significance in chilled foods In: *Other pathogens*

of concern (no salmonella and campylobacter). Nurmi, E., Colin, P., and Mulder, R.W.A.W, Eds, *Proceedings of a meeting held at Helsinki, Finland, June 11-14,* 1992, COVP-DLO Het Spelderholt.

Schroder,K.,Clausen, E., Sandberg, A.M. and Raa, J. Psychrotrophic *Lactobacillus plantarum* from fish its ability to produce antibiotic substances. In *Advances in Fish Science and Technology*, Conell, J.J.,Ed, Farnham, England:Fishing News Books, 1980.

Shaw, B.G. and Latty, J.B. A numerical toxonomic study of *Pseudomonas* strains from spoiled meat. *Journal of Applied Bacteriology* 52, 219, 1982.

Shaw, B.G. and Latty, J.B A study of the relative incidence of different *pseudomonas* groups on meat using a computer-assisted identification technique employing only carbon source tests *Journal of Applied Bacteriology* 57,59, 1984.

Sheideman, S.C., Smith, G.C., Carpenter, Z.L., Dutson, T.R. and Dill, C.W. Modified gas atmospheres and changes in beef during storage. *Journal of Food Science* 44, 1036,1979.

Shelef, L.A., Survival of *Listeria monocytogenes* in ground beef or liver during storage at 4 and 25° C. *Journal of Food Protection* 52, 379, 1989

Sheridan,J.J.,Doherty, A. and Allen, P. Improving the safety and quality of meat and meat products by modified atmopshere and assessment by novel methods. *FLAIR 89055 Interim 2nd Year Report,* EEC DGXII, Brussels, 1992

Silliker,J.H. and Wolfe,S.K., Microbiological safety consideration in controlled atmosphere storage of meats. *Food Technology* 34, 59, 1980

Stanbridge, L and Board, R.G Improving the safety and quality of meat and meat products by modified atmopshere and assessment by novel methods. *FLAIR 89055 Interim 2nd Year Report,* EEC DGXII, Brussels, 1992

Stern, N.J and Oblinger, J.L. Recovery of *Yersinia enterocolitica* from surfaces of inoculated harts and livers. *Journal of Food Protection* 43, 706, 1980

Stier, R.F., Bell, L., Ito, K.A., Shafer, B.D., Brown, L.A., Steeger,M.L., Allen, B.H., Porcuna, M.N. and Lerke, P.A. Effect of modified atmosphere storage on *Clostridium botulinum* toxigenesis and the spoilage microflora of salmon fillets. *Journal of Food Science* 46, 1639, 1981

Sutherland, J.P., Patterson, J.T., Gibbs, P.A. and Murray, J.G. The effect of several gaseous environments on the multiplication of organisms isolated from vacuum packaged beef. *Journal of Food Technology* 12, 249, 1977.

Tseng,C.P. and Montville,T.J. Enzymes activities affecting end product distribution by *Lactobacillus plantarum* in response to changes in pH and O_2. *Applied Environmental Microbiology* 56, 2761, 1990

Walker, S.J., Archer, P. and Banks, J.G Growth of *Listeria monocytogenes* at refrigeration temperatures. *Journal of Applied Bacteriology* 68, 157, 1990

Zee, J.A., Bouchard, C., R.E. Simard, B.Pichard and R.A Holley. Effect of N_2, CO, CO_2 on the growth of bacteria from meat products under modified atmospheres. *Microbiological Aliments Nutrition* 2, 354, 1984

DISCUSSION

ZOTTOLA: Do you have any explanations for the apparent stimulation of the growth of neurotoxigenic E. coli by the 80/20 atmosphere.

NYCHAS: No, we have not, at the moment, any serious explanation about this pathogen.

HENDRICKX: One of the project objectives suggests the use of time temperature integrating devices. Are these temperature recorders or TTI's?

NYCHAS: These were temperature recorders.

FARKAS: How do explain that *Listeria monocytogenes* grew well on lamb but not in minced beef under VP or gas-mixture MAP? Is it related to differences of growth of pseudomonas under the same conditions?

NYCHAS: We consider that this growth is related to the type of meat rather than the growth of pseudomonas. Actually these organisms were not able to grow under these conditions (VP-MAP).

GORMLEY: In your experiments when one partner obtains an unusual or unexpected result, is it possible to have the test repeated by another partner, i.e. in a different laboratory.

NYCHAS: We exchange samples in order to characterize isolates from different meat and meat products under different storage conditions, some of us duplicate the work done in order to confirm these unusual results.

DEBARDEMAEKER: Will the data collected under the FLAIR project eventually become available for modeling as kinetic equations rather than tables?

NYCHAS: Yes, these data could be used by the people doing modeling. We send our results to them and they do the modeling.

GORRIS: In many of you pathogen challenge experiment at 0-5 or $10^{o}C$, you found no growth of the inoculated cells, but they could survive very well (100%). Did you ever try to change temperatures from, for instance, 0 to $22^{o}C$ for some time to see whether the pathogens could grow, keeping in mind that initial and final microflora are very different?

NYCHAS: No, we do not change the temperature of storage, in order to examine if these organisms were able to grow.

FARKAS: (Comment) Approximately two-times higher dose seems to be optimal for strawberries than the low-dose clearance that FDA permits for fruits (2kGy, instead of 1kGy)
- Combination of MAP and irradiation may be more feasible for food of animal origin as shown by work of Dr. M. Patterson in the Queen's University of Belfast.

MODELING OF GAS EXCHANGE IN POLYMERIC PACKAGES OF FRESH FRUITS AND VEGETABLES

Jatal D. Mannapperuma and R. Paul Singh

I. INTRODUCTION

Fresh fruits and vegetables continue to respire after harvest. The rate of respiration is influenced by many factors including oxygen and carbon dioxide content in the storage atmosphere and temperature. In general the shelf life of fresh produce increases when its respiration rate is reduced. The design of modified atmosphere packages must allow for the respiration activity of the produce and its dependence on all the external factors.

When fresh produce is sealed in an impermeable package, then the respiration process causes depletion of oxygen and accumulation of carbon dioxide inside the package. This could eventually lead to anaerobic respiration with formation of ethanol, acetaldehyde and other organic compounds associated with undesirable quality. On the other hand, when the package is highly permeable to gases, then there is very little modification of the package atmosphere and little effect on shelf life.

A properly designed polymeric package balances the respiration rate of the produce and the permeability of the films to obtain a desirable modified atmosphere inside the package which results in extending shelf life.

In this study a simple mathematical model is used to illustrate the principles involved in the design of modified atmosphere packages. Another model is used to simulate the dynamic behavior of the package atmosphere. Experimental methods that can be used to determine the parameters of these models and some experimental packaging studies are also presented.

II. DESIGN OF MODIFIED ATMOSPHERE PACKAGES

The controlled atmosphere conditions that improve shelf life of fresh produce have been researched over a long period of time. A collection of atmospheric compositions recommended for storage of fresh produce can be found in Kader (1989) and Saltveit (1989).

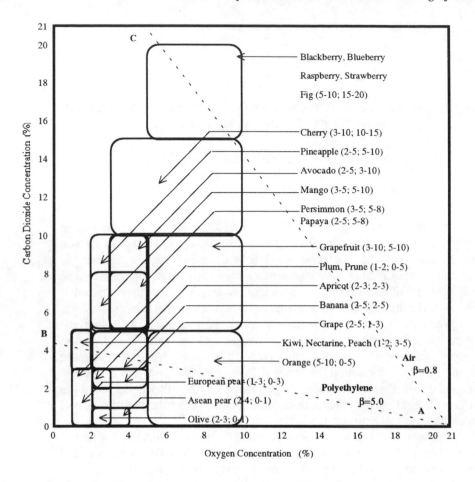

Figure 1. Recommended Modified Atmospheres for Storage of Fruits. Oxygen and carbon dioxide percentage ranges within parenthesis. A-B is the characteristic line for polyethylene film and A-C is the characteristic line for air.

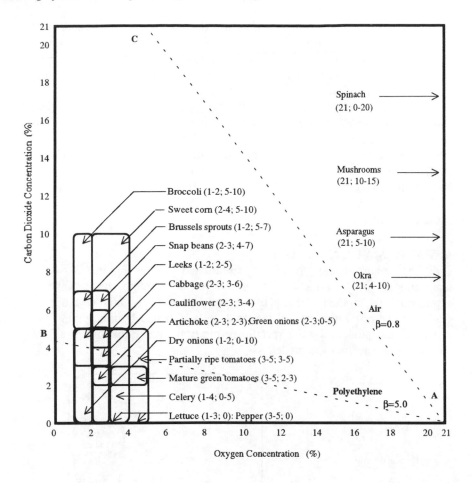

Figure 2. Recommended Modified Atmospheres for Storage of vegetables. Oxygen and carbon dioxide percentage ranges within parenthesis. A-B is the characteristic line for polyethylene film and A-C is the characteristic line for air.

The basic principles involved in the design can be illustrated very effectively by plotting the recommended modified atmospheric compositions on a two dimensional chart with oxygen concentration as the x axis and carbon dioxide concentration as the y axis. Figures 1 and 2 are such plots for some common fruits and vegetables.

Most of the recommended modified atmospheres contain lower oxygen concentrations and higher carbon dioxide concentrations compared to the ambient. In such cases these concentration gradients create a flux of oxygen into the package and a flux of carbon dioxide out of the package. Under steady state conditions these two fluxes should be equal to the oxygen consumption and the carbon dioxide generation by the produce in the package. The design parameters of the package are determined using these equalities.

A. MATHEMATICAL MODEL

The mathematical model is formulated using Figure 3 as a representation of the package. The gas flow rate through the package is expressed as the product of film permeability, film area and the concentration gradient. The respiration rate of the produce is obtained by multiplying the weight of the produce by the specific respiration rate. Equations 1 and 2 represent the mathematical model.

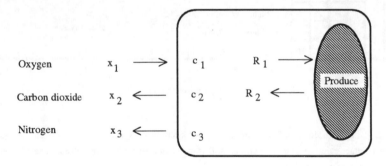

Figure 3. Model of the polymeric package of fresh produce

$$W R_1 = P_1 A \frac{(c_1 - x_1)}{b} \qquad (1)$$

$$W R_2 = P_2 A \frac{(c_2 - x_2)}{b} \tag{2}$$

This mathematical model contains 11 different variables. The effect of these variables on the package design can be illustrated by rearrangement of these equations together with the introduction of some new parameters.

B. PERMEABILITY RATIO

The gas permeabilities of commercially available polymers vary over a wide range. However, the ratio of the permeability of any two gases fall within a very narrow range for all the polymers (Stannett, 1968). For most of the polymers, the ratio of permeability of CO_2 to O_2 falls between 4 and 6. This ratio, denoted by β, imposes severe limitations in the design of polymeric packages for fresh produce. Equations 1 and 2 are combined and β is introduced as in equation 3 to illustrate this limitation.

$$x_2 = c_2 + \frac{1}{\beta} (c_1 - x_1) \frac{R_2}{R_1} \tag{3}$$

When the permeability ratio is close to 5, as for low density polyethylene (LDPE), equation 3 yields a straight line with a slope of 1/5 on the x_2-x_1 plot of recommended modified atmospheres. This line is shown as A-B in figures 1 and 2. The implication is that only the modified atmospheres falling along line A-B can be attained by LDPE film.

The recommended modified atmosphere windows for kiwi, nectarine, peach, prune, plum, orange, banana and avocado in figure 1 and windows for cauliflower, cabbage, leeks, and celery in figure 2 fall along the line A-B. Therefore LDPE film can create the recommended modified atmosphere for these fresh produce.

C. RESPIRATORY QUOTIENT

The ratio of carbon dioxide generation to oxygen consumption also affects the slope of line A-B. This ratio known as respiratory quotient is close to unity when substrate used in the metabolic process is carbohydrate and sufficient oxygen is available. It is less than one when the substrate is a lipid and greater than one when the substrate is an organic acid. Under anaerobic conditions respiratory quotient is greater than one even when the substrate is a carbohydrate. The effect of respiratory quotient on the package atmosphere can be illustrated by rewriting equation 3 by introducing the symbol α to this ratio.

$$x_2 = c_2 + \frac{\alpha}{\beta}(c_1 - x_1) \tag{4}$$

The effect of α is to oppose the effect of β. Therefore, respiratory quotients greater than unity will result in slopes greater than $1/\beta$ and vice versa.

D. PACKAGE DESIGN PARAMETERS

In the package design process, the type of produce determines R_1, R_2 and α. Produce also determines the recommended modified atmospheric composition x_1 and x_2, hence β. This selects the type of the film and hence P_1 and P_2. Therefore, the package designer is left with the choice of only three variables, thickness of the film, area of the package, and weight of the produce in the package. The criterion for the selection of these three variables can be illustrated by rewriting equations 1 and 2 by introducing a new parameter ϕ.

$$x_1 = c_1 - \frac{R_1}{P_1}\phi \tag{5}$$

$$x_2 = c_2 + \frac{R_2}{P_2}\phi \tag{6}$$

$$\text{where} \quad \phi = \frac{W\,b}{A} \tag{7}$$

The effect of increase in ϕ is to move the package atmosphere towards point B along line A-B in figures 1 and 2 and vice versa. Increase in ϕ can be achieved by decreasing A or by increasing W or b.

The respiration rates of fresh produce is affected by the composition of the package atmosphere. Usually the respiration rate increases with oxygen content and decreases with carbon dioxide content or as package atmosphere moves towards B. The respiration rate contours of green beans shown in figure 8 illustrates this effect.

This phenomenon tends to moderate the effect of ϕ. An increase in ϕ decreases x_2 and increases x_1. This will decrease respiration rates R_1 and R_2. Therefore, the final value of x_1 will be higher and x_2 will be lower than the predictions by equations 5 and 6 using constant R_1 and R_2.

The designer is benefited twice by this phenomenon. Firstly, any errors of respiration rate and film permeability data will result in smaller deviations of package atmospheres. Secondly, the effect of change of temperature on respiration rate and film permeability will be moderated to a large extent.

E. TEMPERATURE EFFECT

Temperature affects film permeability as well as produce respiration rate. Both these effects can be expressed as Arrhenius type relations (equations 8 and 9) in the temperature range encountered in modified atmosphere storage of fresh produce.

$$P = P_0 \exp\left|-\frac{E_P}{R\,T}\right| \qquad (8)$$

$$R = R_0 \exp\left|-\frac{E_R}{R\,T}\right| \qquad (9)$$

The effect of a change in temperature on package atmosphere can be illustrated by use of equations 8 and 9 between two temperatures T_{ref} and T.

$$\frac{R_1}{P_1} = \frac{R_{1,ref}}{P_{1,ref}} \exp\left|\frac{E_{R1}-E_{P1}}{R}\left\langle\frac{1}{T_1}-\frac{1}{T_{ref}}\right\rangle\right| \qquad (10)$$

The effective change in package atmosphere due to change in temperature is governed by the difference in activation energy of respiration and permeation. The activation energy of permeability of some films are listed in table 1. The activation energy of respiration rate of some fresh produce are listed in table 2. When the activation energies of respiration and permeation are close to each other the changes in temperature do not cause large changes in package atmosphere.

Table 1. Gas Permeability in Polymers, Water and Air

Medium	Gas	Permeability (m^2/s)	Activation Energy (J/mol)
Polyvinyl	N_2	0.0098×10^{-12}	69100
chloride	O_2	0.0376×10^{-12}	55600
(PVC)	CO_2	0.1302×10^{-12}	56900
High density	N_2	0.119×10^{-12}	39700
polyethylene	O_2	0.334×10^{-12}	35100
(HDPE)	CO_2	1.402×10^{-12}	37600
Low density	N_2	0.80×10^{-12}	41400
polyethylene	O_2	2.39×10^{-12}	40100
(LDPE)	CO_2	10.45×10^{-12}	38400
Water	N_2	0.0328×10^{-9}	15800
	O_2	0.0716×10^{-9}	15800
	CO_2	1.6470×10^{-9}	15800
Air	N_2	17.2×10^{-6}	3600
	O_2	20.3×10^{-6}	3600
	CO_2	15.7×10^{-6}	3600

Table 2. Respiration Rates of Some Fresh Produce (Data from Kader et al, 1989)

Produce	Atmospheric Composition	Respiration rate (ml/kg-hr)				Activation Energy (J/mol)
		0 °C	5 °C	10 °C	20 °C	
Broccoli	air	10.0	21.0	85.0	213.0	105000
(Green Valiant)	1.5% O_2; 10% CO_2	7.0	11.0	15.0	33.0	50950
Cabbage	air	1.5	-	4.0	10.0	63150
(Decema)	3% O_2	1.0	-	3.0	6.0	59800
Green beans	air	-	17.5	28.3	59.6	54900
(Blue Lake)	3 % O_2; 5 % CO_2	-	10.8	16.5	28.0	42200

F. LAYERS OF WATER ON THE FILM

The accumulation of a layer of water on the inside of the film due to condensation is observed frequently in fresh produce packaging. The effect of such water layers on the package atmosphere can be studied by converting the data on diffusion of gases in water and permeation of gases in polymers to the same basis using the convention proposed by Yasuda (1975). Table 1 contains such data for a number of polymers and water.

These values show that the permeability of gases in water is much higher than in polymers. Therefore, thin layers of water forming inside polymeric packages do not affect the package atmosphere significantly.

G. HOLES AND MICROPORES

Polymeric packages are susceptible to puncturing during handling. Also, microporous films and labels are used in polymeric packages. The presence of holes in the film allows the flow of gases by diffusion as well as by convection. The analysis of these phenomena is extremely complicated. However, the relative effect of diffusive flow through holes on package atmosphere can be studied by comparing the permeability of gases in air with permeability of the gases in polymers.

The permeability data listed in table 1 indicate that air is much more permeable than polymeric films. Therefore, even a very small hole in a polymeric package can affect the package atmosphere very significantly. However, this phenomenon can be used to the advantage of the designer through the use of carefully introduced micropores.

The permeability ratio β for air is about 0.8. This represents line A-C on figures 1 and 2. An otherwise impermeable package with a few small holes can be used to create atmospheres along this line, such as, the requirement for berries and figs in figure 1.

The atmospheres that lie between lines A-B and A-C can be created by introducing pinholes or microporous windows on LDPE packages. The results of a study by Burton et al (1987) on packaging of mushrooms are used in figure 4 to illustrate this possibility.

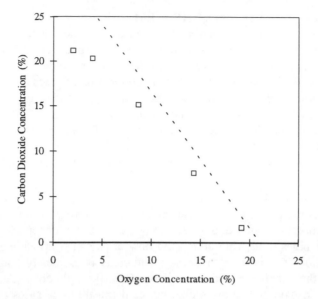

Figure 4. Packaging of mushrooms illustrating the effect of perforations on the package atmosphere (Data from Burton, Frost and Nichols, 1987). Characteristic line for air is shown as the dotted line. The point at top left corresponds to the smallest perforation area. The atmosphere moves towards bottom right as the area of perforations increases.

III. EXPERIMENTAL STUDY OF PACKAGES

The analysis of the packaging parameters presented in this work can be utilized effectively in planning experimental studies of packaging of fresh produce in polymeric films. This process is presented here using a recently completed experimental storage study with cauliflower florets as the example.

The first step is to determine the recommended modified atmosphere for storage of cauliflower. This is found in figure 2 as 2-3% oxygen and 2-5% carbon dioxide. Then the permeability ratio of the proper film is determined using the slope of a line A-B passing through this region. In this case a permeability ratio of about 5 is found to be suitable hence a 1,2-polybutadiene film meeting this requirement is selected.

If the respiration rate of cauliflower in an atmosphere with 2-3% oxygen and 2-5% carbon dioxide were available, together with gas permeability of the film then equations 5 and 6 can be used to determine the parameter ϕ. Equation 7 can then be used to determine W, A and b to suit other criteria.

However, when the respiration rates of cauliflower in the recommended modified atmosphere is not known, an experimental method can be used to determine the suitable value of ϕ. In this method a number of packages with different values of ϕ are prepared by placing different amounts of cauliflower in a series of identical bags. These bags are stored at 2 °C which is within the recommended temperature range of 0 to 5 °C, and the package atmosphere is monitored until equilibrium conditions are reached.

The results of an experiment where 150, 300, 450 and 600 grams of cauliflower were placed in 300 mm x 200 mm bags made of 1-2 polybutadiene film of 40 μm thickness is shown in figure 5.

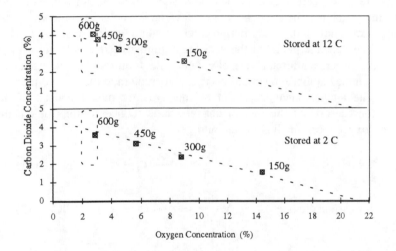

Figure 5. Experimental study of cauliflower packages illustrating the effect of produce weight and temperature on the package atmosphere. Storage at 2 C on top graph and storage at 12 C on bottom graph. Recommended modified atmosphere window is shown as dotted boxes. The characteristic line for the polymeric film is shown as dotted lines.

The bag containing 600 grams reached the target atmosphere while the bags containing lower weights had higher oxygen contents and lower carbon dioxide contents. The effect of temperature on the package atmosphere was studied by increasing the temperature to 12 °C and holding for an additional period of time. This increased the carbon dioxide content and decreased the oxygen content of all the bags. The bags with 450 grams also reached the target atmosphere while the bags with 600 grams moved further towards B.

This result is an indication that the temperature response of film permeability does not keep up with the temperature response of the respiration activity. Also it indicates that if a package is expected to suffer temperature abuse then a lower than optimum value of ϕ should be selected to avoid anaerobic conditions.

IV. DYNAMIC BEHAVIOR OF THE PACKAGE

The package design described earlier was based entirely on the steady state or equilibrium behavior of the package and its contents. However, the package atmosphere does not reach the design conditions for a certain period of time which may be a few hours or a few days. The gas composition inside the package during this early period can be studied using a dynamic model.

The dynamic model of the package requires the free volume inside the package as an additional variable. The time required by the package to approach the equilibrium conditions is closely related to the free volume.

The mathematical statement of the dynamic model consist of three equations describing the rate of change of the concentrations of the three gases, oxygen, carbon dioxide and nitrogen.

for oxygen
$$V \frac{dx_1}{dt} = \frac{P_1 A}{b} (c_1 - x_1) - WR_1 \quad (11)$$

for carbon dioxide
$$V \frac{dx_2}{dt} = \frac{P_2 A}{b} (c_2 - x_2) + WR_2 \quad (12)$$

for nitrogen
$$V \frac{dx_3}{dt} = \frac{P_3 A}{b} (c_3 - x_3) \quad (13)$$

In addition, the ideal gas equation relation has to be obeyed by the gases inside the package. This provides a restriction (Equation 14) on total pressure of the package.

$$p = (x_1 + x_2 + x_3) RT \quad (14)$$

The solution to this set of equation depends on the nature of the dependence of R_1 and R_2 on x_1 and x_2. When the relationship is simple an analytical solution can be obtained. A numerical technique such as Runge-Kutta or predictor-corrector method can be employed when the relationship is complicated.

Deily and Rizvi (1981) used the analytical solution of equations 11 and 12 obtained by assuming constant R_1 and R_2 in their work on peaches. Hayakawa et al (1975) assumed R_1 and R_2 to be piecewise multi-linear functions of x_1 and x_2 and obtained analytical solutions using Laplace transformation method. Both these studies do not include equations 13 and 14. Mannapperuma and Singh (1987) presented a number of models for packages with constraints on volume and pressure. These models included the gas concentrations inside the produce as a separate set of variables.

Equation 14 suggests that the sum of three gas concentrations remain invariant in spite of unequal changes in x_1 and x_2. This is possible only by the change of volume of the package. The contraction of polymeric packages with the establishment of the modified atmosphere is a consequence of this phenomenon. The change in volume imposed by equation 14 makes the system of equations non-linear requiring numerical methods to obtain solutions.

The solution of the dynamic model always converges to the steady state solution. However, the behavior during the initial stages depends on free volume in the package and the initial gas concentrations. The steady state is reached faster when free volume is smaller and when the initial gas concentrations are closer to the steady state concentrations.

Simulation of dynamic models can be used effectively to study the influence of these parameters. Figure 6 illustrates the effect of different free volumes on dynamic behavior of the package atmosphere using simulations of a package of 450 grams of cauliflower in a 300 mm x 200 mm bag made of 1,2-polybutadiene film of 40 μm thickness. The respiration rates and film permeabilities at 12 $^{\circ}$C were used in these simulations.

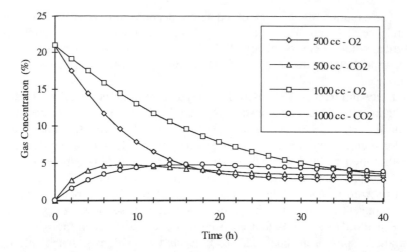

Figure 6. Dynamic simulation of package atmosphere illustrating the effect of free volume. The package with lower free volume approaches the target atmosphere faster.

V. RESPIRATION RATE MEASUREMENT

The respiration rate of fresh produce is time dependent. It starts at a high value soon upon harvest and gradually decreases until a relatively steady value is reached. Similar phenomenon is observed when fresh produce is mechanically damaged. The design of modified atmosphere packaging involves the steady state respiration rate. This is usually measured by the open system where a steady stream of gas is passed through a respiration jar. The closed system involving the unsteady state behavior of a closed respiration jar is some times used as a time saving measure.

The accuracy of both these methods depend on the accuracy of the gas analysis equipment used. The composition of oxygen, carbon dioxide and nitrogen can be measured using a gas chromatograph with a thermal conductivity detector and suitable columns. The use of infra red analyzer for carbon dioxide and zirconium oxide cell for oxygen is reported (Saltveit and Strike, 1989) to be much faster than gas chromatograph.

A. RESPIRATION RATE MEASUREMENT BY OPEN SYSTEM

In this method the produce is kept in a jar and a steady stream of the gas mixture of the required composition is passed through the jar until equilibrium is reached. The flow rate and composition of the gases leaving the respiration jars are measured.

The measurement of gas compositions can be automated by the use of a set of solenoids and a gas manifold. The solenoids and the detection equipment are operated in synchronization by a computer. Figure 7 shows such a system schematically. In the authors' laboratory a system with 12 respiration jars is used.

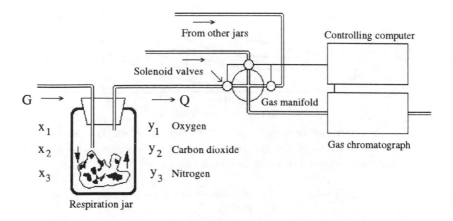

Figure 7. Experimental setup for the respiration rate measurement by an open system.

The respiration rates are calculated by performing mass balances on oxygen and carbon dioxide. A mass balance on nitrogen is used to calculate the gas inflow rate.

Oxygen balance $\qquad R_1 = G\,x_1 - Q\,y_1 \qquad\qquad (15)$

Carbon dioxide balance $\quad R_2 = Q\,y_2 - G\,x_2 \qquad\qquad (16)$

Nitrogen balance $\qquad G = Q\,\dfrac{y_3}{x_3} \qquad\qquad (17)$

The dependence of respiration rate on gas composition can be demonstrated by use of a contour plot (Figure 8). Figure 9 shows the effect of oxygen content and temperature on respiration rate of cauliflower. Reduction in oxygen is very effective at high temperatures while the reduction in temperature very effective at high oxygen contents.

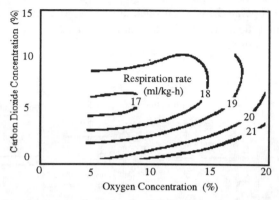

Figure 8. Respiration rate contours of greenbeans at 8 C illustrating the effects of oxygen and carbon dioxide concentrations. The lowest respiration rate is at about 5% oxygen and 5% carbon dioxide. It increases as the atmosphere moves away from this range in all directions.

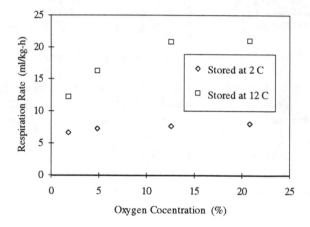

Figure 9. Respiration rate of cauliflower illustrating the effects of temperature and the oxygen concentration in the atmosphere.

The open system is a good method to measure steady state respiration rates. However, it is time consuming. Also it requires a number of gas mixtures. The closed system overcomes these disadvantages.

B. RESPIRATION RATE MEASUREMENT BY CLOSED SYSTEM

In the closed system method, the produce is kept in a closed container and the gas composition inside the container is analyzed over a period of time. The respiration activity causes oxygen content to decrease and carbon dioxide content to increase until all the oxygen is replaced by carbon dioxide. The respiration rates at any instant are calculated by mass balances on gas components expressed by equations 18 and 19.

$$R_1 = \frac{V}{W} \frac{dx_1}{dt} \tag{18}$$

$$R_2 = \frac{V}{W} \frac{dx_2}{dt} \tag{19}$$

The free volume inside the container, V, is determined by introducing a known small volume of an inert gas such as ethane into the container and then measuring its concentration once it is well mixed. The gas concentrations are usually measured by withdrawing samples at intervals. Mounting measuring electrodes inside the container (Cameron et al, 1989) allows the regular sampling without the need of an operator. At the authors' laboratory, a gas chromatograph is used in a closed loop with a peristaltic pump for unattended sampling. This scheme also can be extended for sampling multiple jars using two sets of gas manifolds and solenoid valves. The respiration rate of broccoli determined using this method is shown in figure 10.

This method, besides being fast, has the additional advantage of yielding the respiration rates in a series of different atmospheres. The composition of these atmospheres is determined by the respiration process itself and not under the complete control of the experimenter. However, it is possible to get a wide range of atmospheres by flushing the jars with different initial gas compositions.

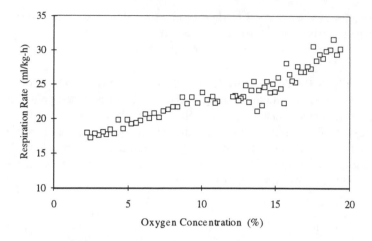

Figure 10. Respiration rate of broccoli measured by the closed system illustrating the effect of oxygen concentration.

VI. MEASUREMENT OF GAS PERMEABILITY OF POLYMERIC FILMS

In many instances the use of permeability values found in manufacturers' literature and other sources resulted in poorly designed packages. The method presented here was developed to enable testing films using the same equipment used in testing respiration rates and under conditions closer to those encountered in fresh produce packaging.

This method uses permeability cell with three chambers (Figure 11) in conjunction with a gas chromatograph and a soap bubble flowmeter to measure the permeability of oxygen, carbon dioxide and nitrogen in one experiment. The cell is placed in an incubator to control the temperature.

Two pieces of the film are placed separating the three chambers of the cell. Steady streams of the three gases are allowed to flow through the three chambers of the cell. The mixing of gases take place in proportion to permeability of the film to each gas. The outlet gas streams are analyzed for the presence of permeated gases once equilibrium conditions are reached.

The operation of this setup is also automated using solenoid valves and a gas manifold. The mass balances of permeated gases in each gas stream is used to evaluate the film permeabilities. The following equations use a double subscript notation for gas concentrations in the gas streams. The first subscript denotes the permeated gas while the second denotes the carrier gas.

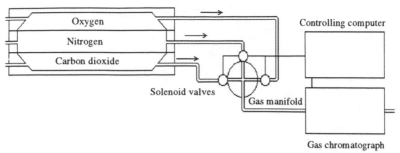

Figure 11. Experimental setup for the measurement of gas permeability of polymeric films.

The oxygen and carbon dioxide permeabilities are determined by writing the mass balances for these two gases in the nitrogen stream. Equations 20 and 21 for the gas permeabilities are derived by the rearrangement of these mass balances.

$$P_1 = \frac{Q_3 \, b \, x_{13}}{A \, [x_{11} + x_{12} - 2x_{13}]} \tag{20}$$

$$P_2 = \frac{Q_3 \, b \, x_{23}}{A \, [x_{21} + x_{22} - 2x_{23}]} \tag{21}$$

The nitrogen permeability can be obtained by similar mass balances for nitrogen in oxygen and carbon dioxide streams. This results in two different nitrogen permeabilities for the same film. When highly permeable films are being tested, significant concentrations of carbon dioxide is detected in oxygen stream and vice versa. In such cases, additional permeability values for oxygen and carbon dioxide are obtained.

The gas permeability of 1,2-polybutadiene film measured at three different temperatures is shown in figure 12. The figure illustrates the strong influence the temperature has on the permeability of all three gases.

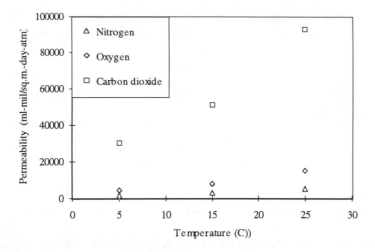

Figure 12. Permeability of 1,2 polybutadiene film for oxygen, carbon dioxide and nitrogen gases, illustrating the influence of temperature.

VII. DISCUSSION AND CONCLUSIONS

The simple mathematical model used in this study was quite effective in analyzing the influence of many parameters on the design of polymeric packages of fresh produce. The results of many experimental packaging studies were in good agreement with the predictions by the model.

The range of modified atmospheres obtainable by use of polymeric films is governed primarily by the permeability ratio of films. This parameter falls within a very narrow range for all the polymers. Careful introduction of pinholes or microporous windows can extend the range of attainable atmospheres considerably.

The atmospheres that lie outside this extended range cannot be attained by the equilibrium interaction of produce respiration and package permeability. Injecting the required atmosphere into a relatively impermeable package can be expected to maintain a favorable atmosphere for a considerable length of time. The commercial air shipment of strawberries utilizes this concept. The dynamic model presented in this study can be used to analyze this type of situations.

The temperature influences respiration rate and film permeability. However, the change in respiration rate is usually much greater than the change in gas permeability. This can result in anaerobic conditions inside optimally designed packages when subjected to temperature abuse. Suboptimal designs may be used to avoid such conditions.

The respiration activity is only one of the factors that influences the shelf life. A direct correlation between the cumulative respiration activity and the shelf life has been observed during storage of sweet cherries at different temperatures (Sekse, 1988). The possibility of extending this concept to modified atmosphere storage merits careful investigation.

Water vapor transfer plays a very important role during modified atmosphere storage of fresh produce. When the vapor pressure is too low it causes wilting of the produce. This is observed in packages where holes are made to allow for respiration. When the vapor pressure is too high condensation of water occurs. Antifog films prevents accumulation of droplets on the film but not the spoilage effects of free water. Control of water vapor transfer by selective permeation or by use of absorbers is possible. This aspect needs more attention than it has received.

NOMENCLATURE

A - area of the film (m^2)

α - respiratory quotient

β - permeability ratio

b - thickness of the film (m)

c - gas concentrations in ambient (mol/m^3)

E - activation energy (J/mol)

ϕ - a ratio

G - gas flow rate (m^3/s)

P - permeability (m^2/s)

p - pressure (Pa)

Q - gas flow rate (m^3/s)

R - respiration rate (mol/kg-s)

R - universal gas constant (J/mol-K)

T - temperature (K)

W - weight of produce in the package (kg)

x - gas concentrations in package atmosphere (mol/m^3)

        suffixes 1, 2 and 3 denote O_2, CO_2, and N_2 gases respectively.

REFERENCES

Burton, K. S., Frost, C. E. and Nichols, R., A combination plastic permeable film system for controlling post-harvest mushroom quality. Biotechnology Letters. 9(8):529, 1987.

Cameron, A. C., Boylan-pett, W. E. and Lee, J., Design of modified atmosphere systems: Modeling oxygen concentrations within sealed packages of tomato fruits. Journal of Food Science. 54(6):1413, 1989.

Deily, K. H. and Rizvi, S. S. H., Optimization of parameters for packaging of fresh peaches in polymeric films. Journal of Food Process Engineering. 5:23, 1981.

Hayakawa, K. I., Henig, Y. S. and Gilbert, S. G., Formulae for predicting gas exchange of fresh produce in polymeric film package. Journal of Food Science. 40:187, 1975.

Kader, A. A., A Summary of CA requirements and recommendations for fruits other than pome fruits. Sixth proceedings of the international controlled atmosphere research conference. June 15-17, Ithaca, NY, 1993.

Kader, A. A., Zagory, D. and Kerbel, E. L., Modified atmosphere packaging of fruits and vegetables. Critical Reviews in Food Science and Nutrition. 28(1):1, 1989.

Mannapperuma, J. D. and Singh, R. P., A computer-aided model for gas exchange in fruits and vegetables in polymeric packages. Presented at the 1987 International meeting of the American Society of Agricultural Engineers, Chicago, Il, December 15-18, 1987.

Saltveit, M. E., Jr. and Strike, T., A rapid method for accurately measuring oxygen concentrations in milliliter gas samples. HortScience. 24(1):145, 1989.

Saltveit, M. E., A summary of CA and MA requirements and recommendations for the storage of harvested vegetables. Sixth proceedings of the international controlled atmosphere research conference. June 15-17, Ithaca, NY, 1993.

Sekse, L., Storage and storage potential of sweet cherries as related to respiration rates. Acta Agricultura Scandinavia, 38:59, 1988.

Stannett, V., Simple gases in *Diffusion of Gases in Polymers* edited by Crank, J. and Park, G. S. Academic Press, London, 1968.

Yasuda, H., Units of gas permeability constants. Journal of Applied Polymer Science, 19:2529, 1975.

COMBINATIONS OF MINIMAL METHODS TO INCREASE SHELF-LIFE AND QUALITY OF TROPICAL FRUITS

G Campbell-Platt, JG Brennan and AS Grandison

INTRODUCTION

Tropical fruits are frequently wasted due to problems of deterioration during handling, transport and storage. This arises from a combination of senescence of the fruit, decay due to fungal pathogens and infestation with pests such as fruit fly or weevils. Improving the shelf-life of these fruits could benefit in terms of reducing post-harvest losses and improving the quality of produce. Treatments which could be used include the following:

Chilling - this is often inappropriate in tropical fruit due to the onset of chilling injury at different temperatures with diffcrent fruit. Chilling may also be prohibitively expensive in tropical countries.

Mild heat treatments - hot water dipping is effective in the control of disease and rots, although temperatures are limited due to damage to the fruit.

Low dose irradiation - is effective in some fruits in delaying ripening due to senescence, but has the additional benefit that it serves as a quarantine treatment against fruit flies and other pests. [Thomas, 1985] Irradiation has the problem that doses are limited due to irradiation damage, even at quite low doses in some fruit.

Packaging techniques - modified atmosphere packaging (MAP) can be used to control the carbon dioxide and oxygen levels in the atmosphere. This retards catabolic reactions in respiring fruit, and slows the growth of spoilage organisms. [Kader *et al*, 1989]

Coatings - can be used to delay senescence by reducing the respiration rate, with the additional benefit of reducing weight loss.

Combination treatments - may have the advantage that the severity of individual treatments could be reduced while the effectiveness of processing is maintained or even increased. Combinations may result in cumulative or even synergistic responses, and have been shown to be effective in the treatment of some tropical fruit. [eg Jessup *et al*, 1988; Padwal-Desai *et al*, 1973]

OBJECTIVES

To determine the effect of individual treatments on physico-chemical changes, and hence shelf-life of tropical fruit, and to optimise these treatments. To determine the effectiveness of combinations of these treatments.

MATERIALS AND METHODS

Fruits used were banana, plantain, avocado and starfruit.

Processing

Electron beam irradiation was carried out using a Van de Graaff electrostatic accelerator at an energy of 2MeV. Dose was controlled by altering beam current and the speed of the conveyor carrying the sample through the beam, and verified using Far West Optichromic detectors FWT-70-83M (Farwest Technology Inc, Goleta, California, USA).

Hot water treatments were carried out by immersing fruit in a water bath for controlled time/temperature combinations.

MAP (passive) employed a range of packaging films with different gas permeabilities. The composition of pack atmospheres was determined using Servomex oxygen and carbon dioxide analysers (Servomex Oxygen Analyser 574, and Servomex IR Analyser PA404, Servomex Controls Ltd, Crowborough, Sussex, UK). In a few cases, active MAP was used in which the atmosphere was created with a Multivac Gastrovac system (Multivac, Sepp Hagenmuller KG, Germany).

Semperfresh F is a commercial fruit coating material manufactured by Surface Systems International Ltd, East Challow, Oxfordshire, UK. It consists mainly of a mixture of sucrose esters and is approved by the major health bodies worldwide. Fruit were coated using different concentrations of Semperfresh, in the range 0.5 to 2%.

Samples were maintained under conditions of temperature and humidity appropriate to the particular fruit prior to analysis.

Analysis of Fruit

Fruit were analysed by a range of procedures to monitor changes in colour, texture, chemical properties and organoleptic properties during storage. However, for the purposes of this paper, results are restricted to the overriding factors affecting shelf-life ie. colour and external appearance of banana, plantain and starfruit; internal appearance of avocado.

RESULTS

Banana (*Musa paradisiaca*)

Bananas used in the trials were of the same variety of AAA group, Cavendish subgroup. Although they were obtained from different countries, untreated bananas behaved similarly during ripening, especially in terms of peel colour development. All experiments used green preclimacteric fruit at light full three-quarter grade. The end of shelf-life was defined as the time required to achieve colour stage 7 during storage at $21 \pm 1°C$; 85-95% RH. For untreated, non-ethylene initiated fruit this was 15-20 days.

The effects of electron beam irradiation, modified atmosphere packaging and hot water treatments were investigated both individually, and as combination treatments on unripened bananas. The work aimed to extend the shelf-life, and improve the quality of the ripened product. Another possibility with certain types of banana, is that the fruit could be harvested at a more mature stage, and subsequently delaying the ripening and senescence would lead to a more flavoursome product.

Shelf-life and quality were assessed on the basis of peel colour and texture, pulp texture, chemical analyses (including sugar/acid ratio and vitamin C content), and organoleptic assessment.

Irradiating at 0.1-0.3 kGy delayed the respiratory climacteric and hence delayed ripening for up to 3 days. Higher doses gave rise to darkening of the skin and fruit splitting, as well as loss of peel texture and reduced vitamin C content.

A range of hot water treatments were investigated, and it was concluded that the maximum treatment which could be tolerated without peel scalding was 50°C for 5 minutes. The primary reasons for hot water treatment are to control fungal infections and insect infestation. These effects were not tested in this study, but the treatment did give rise to a small (1 day) extension to shelf-life.

Bananas packaged in modified atmosphere retail packs ripened very well, with an extended shelf-life of up to 7 days. The optimum packaging material used was low density polyethylene (LDPE, thickness 25 μm), which produced an atmosphere containing 5% O_2 and 10% CO_2 after 4 days' storage.

Combinations of individual treatments (ie IR - Irradiation at 0.15 kGy, HW - Hot Water - 5 minutes at 50°C, and MAP - LDPE packaging) were found to be synergistic in some cases, but in others were additive, or less than additive. The effects of combinations on shelf-life extension are shown in Table 1.

Table 1
Shelf-life Extension of Bananas - Individual and Combination Treatments

	Treatment	Shelf-life extension (days at 21°C)
Individual	HW	0.8
	IR	2.0
	MAP	5.6
Combinations	HW + IR	2.8
	HW + MAP	9.6
	MAP + IR	5.4
	HW + MAP + IR	9.0

HW = hot water 5 min/50°C. IR = Irradiation at 0.15 kGy.
MAP = modified atmosphere packaging (passive) in 25μm LDPE

In summary, these results suggest that while electron beam irradiation gives some benefit in terms of extension of shelf-life, the treatment does not act synergistically with the hot water or MAP. On the other hand, MAP and hot water treatment in combination gave an enhanced effect on shelf-life, with an improvement of 3 days over the added effect of the individual treatments. Other studies have shown that gamma irradiation has a greater effect on shelf-life of bananas (up to 12 days), which suggests that the electrons used in this study were not penetrative enough to have the desired effect. However, it is possible that electron beam irradiation may have beneficial effects on surface fungi or insects, but this was not tested in this work.

Plantain (*Musa paradisiaca*) group AAB
Plantain fruits were subjected to the following treatments and combinations thereof:
(1) Hot water dip at 40, 50 and 60°C for 5 min.
(2) Electron beam irradiation at a dose level of 0.1 kGy.
(3) Coating in 1.5% Semperfresh.
(4) Packaging in LDPE, 4 gauges (20, 30, 40 and/μm).
Hot water dip: There was a decrease in the time taken by the fruit, stored at 27°C and 70-85% RH, to reach a ripe yellow colour (stage 6) as the water temperature was increased.

Irradiation: As the dose was increased from 0.1 to 0.5 kGy the fruits developed more and more dark brown patches, making them less acceptable.   The maximum dose level was set at 0.1 kGy.

Semperfresh: Coating the fruit with a 1.5% concentration of Semperfresh reduced the weight loss to 8.9% from 12.2% in the case of the control after 7 days storage at 27°C and 70-85% RH.

Packaging: Sealing in 30 μm thick LDPE reduced the weight loss to 0.35% after 7 days storage at 27°C and 70-85% RH.   The CO_2 content of the in-package atmosphere increased while the O_2 content fell.   In 30 μm film the atmosphere equilibrated at approximately 5.5% CO_2 and 10.5% O_2.   This thickness of film was the optimum in terms of shelf-life of the fruit.

Combinations: The inclusion of irradiation in any combination with the other treatments did not extend the shelf-life of the plantain.   The hot water treatment was similarly ineffective.   Sealing in polyethylene did significantly increase the shelf-life of the fruit which did not become fully ripe after 14 days storage.   The combination of coating in Semperfresh and sealing in polyethylene extended the shelf-life by another 2-3 days.   However, in some cases fruit treated in this way did not ripen to a uniform yellow colour.

Avocado (*Persea americana* Mill) Fuerte variety
Refrigerated storage

Unpacked samples were stored at 4, 7 and 10°C and 70-80% RH.   They were then moved to 20°C and 90-95% RH for ripening.   Time to ripen at 20°C was noted and also chilling injury.   Fruits stored at 10°C began to ripen after 3 weeks.   Fruits stored at 4 and 7°C did not ripen.   After transfer to 20°C for ripening, the longer the initial storage period, the shorter the ripening time.   Chilling injury occurred in fruits stored at 4 and 7°C.   This resulted in grey/brown discoloration of the mesocarp and the development of off-flavour.   These symptoms did not become apparent until the fruits ripened.   Fruits with slight internal injury did not show external discoloration.   Chilling injury occurred in fruit stored at 4°C for 3 weeks and at 7°C for 4 weeks.   After these storage periods the fruit took 8 days to ripen at 20°C.   No chilling injury occurred in fruits stored at 10°C.

Modified atmosphere (passive)

Fruits were packed in 7 packaging films with permeabilities ranging from very low to high and stored at 4, 7 and 10°C.   The in-package gas composition was monitored.   In all packs the CO_2 rose and the O_2 fell and a stage of equilibrium was reached after 3-5 days storage.   The higher the storage temperature the faster these changes occurred.   In the case of low permeability films, the O_2 permeability less than 65 cm^3 m^{-2} day^{-1} bar^{-1}, in-package atmospheres with less than 3% O_2 and more than 20% CO_2 developed.   Under these conditions spots or brown discoloration developed on the outer surface of the fruit.   However, when atmospheres containing 4-10% O_2 and 4-10% CO_2 developed in the package the shelf-life of the fruit could be extended by 2-4 weeks compared to the control.   Low density polyethylene and a microporous polypropylene were the most suitable of the films tested.

Modified atmosphere (active)

Experiments were carried out in which different initial gas concentrations were established in the packs prior to sealing them.   However, this method had no advantage over the methods outlined above when the initial gas concentration in the packs was that of the ambient atmosphere.

Irradiation

Avocado fruits were subjected to 0.5, 0.10, 0.21, 0.50 and 0.69 kGy of electron beam radiation. This resulted in an extension of the shelf-life of the fruit of up to 7 days at 20°C. However, at dose levels of 0.21 kGy and above, undesirable changes occurred in the fruit. With a dose of 0.21 kGy the fruit softened to a smooth normal consistency on ripening. However, brown vascular strands and discoloured patches appeared in the flesh. At higher doses the fruit softened unevenly and there was severe discoloration of the flesh and browning on the outer surface.

Combination treatments

Irradiation at a dose level of 0.1 kGy in combination with refrigerated storage did not significantly extend the shelf-life of the fruit as compared with refrigerated storage only. An extension of 1-2 days ripening time was noted. The irradiation seemed to intensify the symptoms of chilling injury in fruit stored for 5 weeks at 4°C.

Irradiation at a dose level of 0.1 kGy did not extend the shelf-life of fruit stored in a modified atmosphere at 3-5% CO_2 and 8.5-10.0% O_2, as compared with untreated fruit stored in a similar atmosphere.

Starfruit (*Averrhoa carambola* L) B10 variety

Samples of starfruit were sealed in two films. Film M is a polyamide/LDPE laminate with a gas transmission rate at 20°C, in cm^3 m^{-2} day^{-1} bar^{-1} as follows: CO_2 224.0, O_2 64.0. Film Q is a wax coated regenerated cellulose with gas transmission rates of 6.2 and 3.9 respectively. In each of these films some samples were sealed in air without modification (passive MA) while corresponding samples were packed in an initial atmosphere of 10% O_2 and 1% CO_2 (active MA). Samples were stored at 20°C and 90-95% RH, and 5°C and 70-80% RH.

Samples of the control fruit (unpackaged) had shelf lives of less than 1 week at 20°C, while those packaged in films M and Q had shelf lives of 2 and 1 week respectively. When stored at 5°C samples of the control and those stored in film Q developed chilling injury in the form of brown spots on the edges of the ribs of the fruit. Samples packaged in film M had shelf lives of up to 5 weeks with no or only slight signs of injury. Packaging in an initially modified atmosphere had no influence on the shelf lives of the fruit as compared with packaging in air.

Coating fruit with 1% Semperfresh extended its shelf life at 20°C from 1 to 2 weeks as compared with uncoated fruit. There was no difference in shelf life between coated and uncoated fruit when stored at 5°C.

CONCLUSIONS

In general, packaging techniques gave greater extension of shelf-life than the other processing methods investigated. Some benefit resulted in applying combination treatments including MAP, for example MAP and hot water dipping for banana; MAP and chilling for avocado.

Low-dose electron beam irradiation gave rather disappointing results. It was hoped that this technique may have given better results than gamma-irradiation, especially in terms of irradiation damage, due to the lower penetration of the radiation. While some extension of shelf-life was noted at very low dose for banana and avocado, there was no evidence of any advantage of incorporating electron beam as part of a combination treatment for banana, avocado or plantain.

REFERENCES

Jessup, A.J., Rigney, C.J. and Wills, P.A., Effects of gamma irradiation combined with hot dipping on quality "Kensington Pride" mangoes. Journal of Food Science *53*, 1486-1489, 1988.

Kader, A.A., Zagory, D. and Kerbel, E.L., Modified atmosphere packaging of fruits and vegetables. CRC - Critical Reviews in Food Science and Nutrition *28*, 1-30, 1989.

Padwal-Desai, S.R., Ghanekar, A.S., Thomas, P. and Sreenivasan, A.; Heat-irradiation combination for control of mould infection in harvested fruits and processed cereal food. Acta Alimentaria *2*, 189-207, 1973.

Thomas, P., Radiation preservation of foods of plant origin III. Tropical fruits: Bananas, mangoes and papayas. CRC - Critical Reviews in Food Science and Nutrition, *23*, 147-205, 1985.

Acknowledgement: The authors wish to acknowledge the major contributions to this work made by research students in this Department, as follows:

RA Rahman for work on banana

IB Falana for work on plantain

T Nagalingam for work on avocado

NA Azia for work on starfruit.

DISCUSSION

GORRIS: Bananas really benefit from being packaged in the polycthylene films with regard to color. Were sensory aspects (flavor, taste, and texture) beneficial as well?

CAMPBELL-PLATT: Yes, the fruit still had a good flavor and texture, as well as good color over the extended period of shelf life.

BRUHN: When a fruit is removed from the modified atmosphere package, what will be the effect on the shelf life? Does the advantage of extended shelf life go only to the food distribution system?

CAMPBELL-PLATT: Although it can depend on the type of fruit in general MAP-packed fruit still had similar shelf -life after removal from package. The benefits of extended shelf-life is to distributors, retailers and consumers. The packs can be left longer and wastage is reduced. An aim would be to harvest fruit at fuller maturity, with better flavor, and then keep good quality.

SHEWFELT: Your point is well taken that it is more important to extend shelf-life of improved quality fruits (e.g.. later harvest) rather than extending the life of currently inferior products.

CAMPBELL-PLATT: Yes, the advantages of improvements in shelf-life should be linked to higher quality to realize real consumer benefits.

POTENTIAL PRESERVATION OF PEACH FRUITS CV. BIUTI: EFFECT OF CALCIUM AND INTERMITTENT WARMING DURING COLD STORAGE UNDER MODIFIED ATMOSPHERE.

N. Holland; M.I.F. Chitarra and A.B. Chitarra

ESCOLA SUPERIOR DE AGRICULTURA DE LAVRAS

Department of Food Science

Cx. Postal 37. CEP 37200-000 Lavras, MG - BRAZIL.

SUMMARY

In order to verify the effects of calcium and intermittent warming on shelf-life the peach fruits (*Prunus persica* (L.) Batsch), harvested in mature-green stage were dipped in heated $CaCl_2$ solution 2% (w/v) at 49°C for 2,5 minutes. Then they were wrapped in PVC film (15 μ thickness) and placed in a constant cold storage condition (0°C, and relative humidity 90 to 95%). Half of the samples then were submitted to intermittent warming (48 hours at 21°C). Although calcium level had been high in treated fruits, it was not effective in delaying fruit ripening or in avoiding internal breakdown. Fruits stored under intermittent warming condition showed higher ripening degree (color development, increase in soluble pectin and loss of firmness) than did those at constant cold storage throughout 40 days. Nevertheless, the use of intermittent warming reduced the development of cold injury symptoms to almost 50% when fruits were stored for 4 days at room temperature.

INTRODUCTION

Peaches can be stored at low temperature (-1°C to 2°C) for two to six weeks, depending on the cultivar [Robertson & Meredith, 1988]. Nevertheless, some cultivars show cold injury symptoms which can limit the shelf-life. Damaged fruits usually have normal appearance, but they do not ripe satisfactorily at room temperature [Jackman et al., 1988]. Periodic interruptions in cooling by removing fruits to room temperature for short periods may decrease or delay cold damage in stone fruits, but it stimulates ripening [Anderson, 1982].

The use of modified atmosphere (MA) could be of value in delaying fruit ripening during storage and marketing periods. The MA is obtained by packing the product in either plastic bags or films with limited permeability to O_2 and CO_2 [Chitarra & Chitarra, 1990]. The reduction of O_2 level and increase of CO_2 level reduced respiration rate and the water loss [Hulbert & Bhowmik, 1987], slowing down changes in fruit acids, sugars, color and texture [Nakhasi et al., 1991].

The knowledge of both metabolic transformation and fruit physiology during storage is essential in order to control post harvest damages and avoid loss of quality. Glenn & Poovaiah [1975] showed that calcium would inhibit specific aspects of abnormal senescence and control physiological disorders by acting on the cell wall, through cross linking with pectic polymers. Scott & Wills [1975] found that the incidence of internal breakdown in apple during cold storage was reduced when calcium was used before the harvest as spray, or after the harvest for dipping. However, when infiltrated into the fruit, it increased skin damages [Wills & Mahendra, 1989]. Another limitation to the use of calcium in metabolic activity of fruit is the relatively low rate of uptake when it is applied by dipping fruits into solution [Wills & Tirmazi, 1982]. Brown rot (*Monilinia fructicola*) is the most important disease of peaches. The use of brief dips in hot water containing fungicides is recommended for decay control during storage at 0°C [Hardenburg et al., 1986].

Although numerous investigations have been carried out, physiological disorders still remain one of the major causes of fruit damages. Losses in quantity and quality of horticultural products occur between harvest and consumption, mostly in developing countries. Research and development efforts are aimed at improving the existing technology through the use of viable alternative procedures. Therefore, this work was carried out to verify the effects of calcium application by dipping fruits in heated calcium solution and the effects of intermittent warming on preservation of peach stored under low temperature and modified atmosphere.

MATERIAL AND METHODS
Plant material and treatments
Peaches (*Prunus persica* (L.) Batsch) cv. Biuti (crossbreeding of Halford 2 and Rubi) were obtained from Experimental Station of EPAMIG at Caldas, South of Minas Gerais State - Brazil. Fruits were harvested at mature-green stage, transported to the laboratory and then selected by size (medium), degree of maturation (firm texture and greenish color) and visual appearance (absence of damage). Calcium was applied by dipping the fruits for 2.5 minutes in 2% $CaCl_2$ solution heated at 49°C containing 0.06% (w/v) benomyl for decay control, and 0.025% (w/v) of Tween 80 as spreading adhesive. A similar treatment with hot water and benomyl was used as control. After drying, fruits from each treatment were randomly distributed into experimental units each containing 5 fruits, and placed in plastic trays (15 cm x 21 cm) and wrapped with a self-adhesive flexible PVC film (15 μ thick).

The experimental units were placed in a constant cold storage (CCS) chamber at 0°C, and relative humidity (RH) of 90 to 95%. The intermittent warming (IW) was applied after 10, 20, and 30 days of storage by removing half of the experimental units to room temperature for 2 days in air at about 21°C, and 60% RH. Following IW the fruits were returned to 0°C, and analyzed 8 days after the IW had been applied (20, 30 and 40 days of storage). Due to homogeneity of the experimental material, a randomized block design in a factorial arrangement 2 x 2 x 3 (calcium x IW x time) with 3 replications was used. Data were statistically analyzed by ANOVA and Test of Tukey at 5% level.
Physical measurements
Loss of weight was evaluated in 3 experimental units by the difference between initial weight after packaging and that after 20, 30, and 40 days of storage. The change in fruit firmness was determined on two opposite sides of 15 peaches using a Magness-Taylor Pressure tester with an 11-mm tip diameter. The mean of the two measurements was expressed in Newtons (N).
Chemical analyses
The same peaches used for physical measurements were used for chemical analyses. Calcium content was measured in slices of peeled fruit by atomic absorption spectrophotometry using the methods described by Sarruge & Haag [1974]. Soluble pectin was extracted according to the procedures used by Chitarra et al. [1989], and assayed by the method of Blumenkrantz & Asboe-Hansen [1973]. Cell wall material was derivatized into their alditol acetates [Albersheim et al., 1967] and analyzed by a Varian model 3.300 gas chromatograph (Varian, Palo Alto, CA) equipped with a flame ionization detector.
Subjective analysis
Skin color was determined by comparing the greenest area of the skin with standard color chips [Munsell, 1976]. The color index ranged from 1 to 5, that is, from yellowish-green to yellow-red. Fruits were taken out of storage conditions and kept 4 days at room temperature. Internal breakdown was evaluated daily through subjective scale of Bretch & Kader [1984], according to the internal browning intensity. The arithmetic mean of the four evaluations was designated as the average degree of internal breakdown.

RESULTS AND DISCUSSION

Because of their skin structure, peaches are rather sensitive to transpiration, so they need high relative humidity to avoid lossof weight . Fruits kept at CCS conditions showed around 2% total weight loss (Table 1). The value reached by IW fruit was two times greater and increased significantly (P < .01) during storage period (Table 1, Fig. 1A). However, no wrinkling of tissue was observed. The fruit appearance was found to be normal. Robertson et al. [1990] reported a 21% weight loss in peaches after six weeks of cold storage at 0°C, and 80 to 90% RH. According to Hardenburg et al. [1986] nectarines are more susceptible to shriveling than peaches. The symptoms become visible when weight loss reaches 4 to 5%. It was observed that the use of protective PVC film was a good supplementary way to reduce the transpiration rate of peach fruits during cold storage using intermittent warming.

Dipping fruits in 2% $CaCl_2$ heated solution caused an increase of about 37% in the calcium content (Table 1). It appeared that there was some calcium uptake through the fruit epidermis, as observed by Klein et al. [1990]. However, this tratment was not effective in delaying skin color development, as the color index values remained the same as those of control (Table 1). Thus, the pigment synthesis inhibition could be better due to the effect of low temperature during storage. Fruits submitted to IW showed higher values (P < .01) than those under CCS conditions. In all treatments there was a linear increasing tendency of increase of color index during the storage (Fig. 1B), mainly in IW-treated fruit, which had already shown visible ripening signs at the end of the storage period. This stimulus of normal metabolism by IW is important according to Dodd et al. [1986], due to the physiological activation of hydrolytic enzyme. Color changes during maturation are considered as a better indicator of edible quality after ripening than flesh firmness [Delwiche & Baumgardner, 1985].

Peach firmness as well as other quality parameters, are used as ripeness indices, since the tissues become soft as the fruits ripe. Dipping peach in heated calcium solution did not cause considerable changes in fruit firmness during 40 days of cold storage, since values remained identical to those of control (Table 1). The firmness of cv Biuti fruit was around 86 N (data not shown) decreasing to 82 N after storage under CCS condition. Tavares et al. [1992] found firmness values around 74 N in peaches cvs Talismã and Delícia from the same region, which decreased to 56 and 17 N, respectively, after 5 weeks of cold storage (0°C and 85 to 90% RH) under modified atmosphere. So, it should be stressed that Biuti cv. fruit showed a relatively firm texture, even when submitted to IW. In this case, fruit showed a significant loss of firmness (P < .01), which confirmed an enhanced ripening due to the temperature increase. There was no interaction found between calcium and IW treatments. Glenn & Poovaiah [1990], stated that the calcium concentration in the fruit improves its interaction with the cell wall through cross linkages with pectic polymers. This can reduce pectin solubility and consequent tissue softening.

Soluble pectin levels found in fruits under CCS conditions (Table 1), confirm the inhibitory action of low temperature upon ripening process rather than the effects of calcium as previously noticed in case of both color and firmness. Teixeira et al. [1983], found soluble pectin values of 'Biuti' fruits rising from 390 to 850 mg of galacturonic acid/100 g during maturation. The soluble pectin fraction was significantly increased (P < .01) by IW, especially in fruits dipped in heated calcium solution. According to Ferguson [1984] the inhibitoty effect of calcium upon ripening is extra-cellular at the beginning and occurs in both the cell walls and in the plasmatic membrane. The use of both heat in calcium treatment and IW during cold storage might have contributed to the evident ripening process. The breakdown of calcium cross-linkages between pectic polymers made the cell wall subtrates more accessible to hydrolytic enzymes, increasing the pectic soluble fraction.

The pectic polymers solubility leads to a decrease in the non-cellulosic neutral sugar content in the cell wall, and a loss of tissue firmness [Melford & Prakash, 1986]. Although the use of heat in calcium treatment has not apparently contributed to delayed fruit ripening,

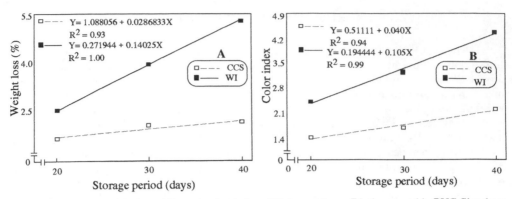

Figure 1. Average weight loss (A) and color index (B) in peach cv. Biuti wrapped in PVC film, kept at constant cold storage (CCS at 0°C) and under intermittent warming (IW at 21°C for 48 h).

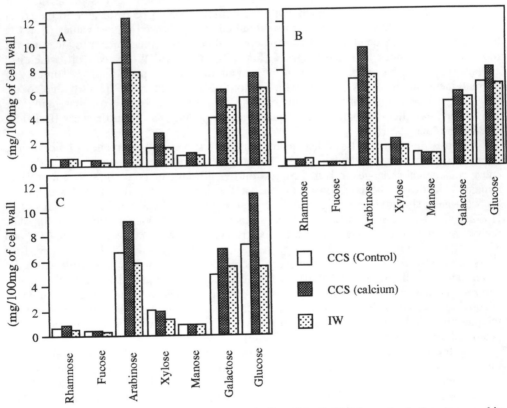

Figure 2. Neutral sugar of cell wall in peach dipped in CaCl 2 heated solution, wrapped in PVC film and stored under constant cold storage (CCS at 0 °C) and under intermittent warming (IW at 21 °C for 2 days). A = 20 days; B = 30 days and C = 40 days .

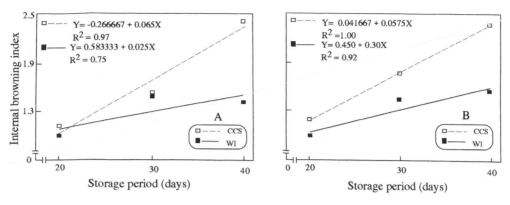

Figure 3. Average of internal browning in peach pipped in heated 2% CaCl$_2$ solution (A) and control (B), wrapped in PVC film and kept under constant cold storage (CCS at 0°C) and intermittent warming (IW at 21°C for 48 h)

TABLE 1

Changes in some quality parameters of peach dipped in 2% CaCl$_2$ heated solution wrapped in PVC fiml and kept at constant cold storage (CCS at 0°C) and under intermittent warming (IW at 21°C for 48 h).

Parameters	Time (days)	CCS Control	CCS Ca^{++}	IW Control	IW Ca^{++}
Weight loss	20	1.5^b	1.6^a	2.6^c	2.5^c
(%)	30	2.0^{ab}	2.0^a	4.0^b	3.9^b
	40	2.3^a	2.1^a	5.4^a	5.2^a
Calcium	20	73.0^a	124.0^a	75.0^a	104.0^a
(mg/100g)	30	64.0^a	144.0^a	66.0^a	99.0^a
	40	73.0^a	123.0^a	79.0^a	100.0^a
Color	20	1.4^b	1.3^b	2.5^c	2.2^c
(score 0 to 5)	30	1.7^{ab}	1.5^b	3.2^b	3.3^b
	40	2.3^a	2.1^a	4.5^a	4.4^a
Firmness	20	87.6^a	84.9^a	81.7^a	81.7^a
(N)	30	82.7^a	86.7^a	79.6^a	79.1^{ab}
	40	82.4^a	82.0^a	72.8^b	74.1^b
Soluble pectin	20	315.7^a	335.3^a	521.4^b	493.7^b
(mg galacturonic	30	338.9^a	354.1^a	581.3^{ab}	664.5^a
acid/100g)	40	386.4^a	411.4^a	669.0^a	704.7^a
Internal	20	1.2^c	1.1^c	1.0^b	1.0^b
browning	30	1.7^b	1.5^b	1.4^a	1.5^a
(score 1 to 5)	40	2.3^a	2.4^a	1.6^a	1.5^a

[abc] Means on the same column, separated by Tukey's Multiple Range Test after significant F ANOVA, having different superscripts differ ($P = .05$).

the levels of arabinose, galactose, glucose (P < .01) and xylose (P < .05) of the cell wall were higher than those of control (Fig. 2). A reduction of xylose and arabinose (P < .01) was found in fruits submitted to IW treatment. In peaches, arabinose is the main non-cellulose component of the lateral chains of pectic polysaccharides [Cross & Sams, 1984]. A remarkable reduction (50%) of this neutral sugar was observed by Ahmed & Labavitch [1980] during softening of pears. In the present work, the total loss of neutral sugar was low, even in fruits stored at IW. This fact could be due to a low hydrolytic action or concentration of pectic enzymes in cv. Biuti fruit, as confirmed by previous analysis on firmness and soluble pectin fraction.

Fruit maintained under CCS condition showed regular susceptibility to internal browning when they were taken out and kept at room temperature. The slight discoloration of the flesh around the stone was identical in both, calcium and in control (Table 1). IW was effective in reducing cold injury symptoms by reactivating normal metabolic pathways [Wang, 1982]. The increase of browning intensity over storage time was well correlated with ripening degree [Fig. 5].

CONCLUSIONS:

Dipping peach fruit cv Biuti in heated 2% $CaCl_2$ solution before cold storage was not effective either to delay ripening or to prevent cold injury symptoms. The use of intermittent warming (2 days at 21°C) during cold storage enhanced fruit ripening, especially if associated with calcium treatment. However, I.W. reduced cold storage injury which made it viable as an alternative method in the cold storage of 'Biuti' peach.

ACKNOWLEDGMENTS

We are grateful to Dr. Luiz Ronaldo de Abreu and to Dr. Prabir Kumar Chandra in the preparation of this manuscript and to CNPq for providing fellowship and research grants

REFERENCES

Ahmed, A. E. and Labavith, J. M., Cell wall metabolism in ripening fruit, *Plant Physiology*, 65,1009, 1980.

Albersheim, P.; Nevins, D. J.; English, P. D. and Karr, A., A method for the analysis of sugars in plant cell wall polysaccharides by gas-liquid chromathography, *Carbohydrate Research*, 5,340, 1967.

Anderson, R. E., Long-term storage of peaches and nectarines intermitentty warmed during controlled atmosphere storage, *Journal of the American Society and Horticultural Science*, 107, 214, 1982.

Blumenkrantz, N. and Asboe-Hansen, G., New method for quantitative determination of uronic acids, *Analytical Biochemistry*, 54, 484, 1973.

Brecht, J. K. and Kader, A. A., Description and postharvest physiology of some slow-ripening nectarine genotypes, *Journal of the American Society and Horticultural Science*, 109(5),596, 1984.

Chitarra, A. B.; Labavitch, J. M. and Kader, A. A., Canning-induced fruit softening and cell wall pectin solubilization in the 'Patterson' Apricot, *Journal of Food Science*, 54(4),990, 1989.

Chitarra, M. I. F. and Chitarra, A. B., *Pós colheita de frutos e hortaliças: fisiologia e manuseio*, ESAL/FAEPE, Lavras 1990, 320 p.

Delwiche, M. J. and Baumgardner, R. A., Ground color as a peach maturity index, *Journal of the American Horticultural Science*, 110(1),53, 1985.

Dodd, M. C., Hortmann, P. E. O. and De Kock, V. A., Influence of temperature manipulations on the storage quality of peaches and nectarines, *Deciduous Fruit Grower*, 36(12),517, 1986.

Ferguson, I. B., Calcium in plant senescence and fruit ripening, *Plant, Cell and Environment*, 7,477, 1984.

Glenn, G. M. and Poovaiah, B. W., Calcium mediated postharvest changes in texture and cell wall structure and composition in 'Golden Delicious' apples, *Journal of the American Society and Horticultural Science*, 115(6),962, 1990.

Gross, K. C. and Sams, C. E., Changes in cell wall neutral sugar composition during fruit ripening... a species survey, *Phytochemistry*, 23,2457, 1984.

Hardenburg, R. E.; Watada, A. E. and Wang, C. Y., The commercial storage of fruits, vegetables, and florist and nursery stocks, U.S. Department of Agriculture, Washington, D. C., 1986. (Agriculture Handbook # 66).

Hulbert, G. J. and Bhowmik, S. R., Quality of fungicide treated and individually shrink wrapped tomatoes, *Journal of Food Science*, 52, 1293, 1987.

Jackman, R. L.; Yada, R. Y.; Marangoni, A., Parkin, K. L. and Staney, D. W., Chilling Injury. A review of quality aspects, *Journal of Food Quality*, 11,253, 1988.

Klein, J. D.; Lurie, S. and Ben-Arte, R., Quality and cell wall components of 'Anna' and 'Granny Smith' apples treated with heat, calcium, and ethylene, *Journal of the American Society and Horticultural Science*, 115 (6), 954, 1990.

Melford, A. J. and Prakash, M. D., Postharvest changes in fruit cell wall, Advances in Food Research, 30,139, 1986.

Munsell, H. A., Munsell book of color, Mac Beth Division of Kollmorgen Corporation, Baltimore, 1976.

Nashasi, S.; Schlimme, D. and Solomos, T., Storage potential of tomatoes harvested at the breaker stage using modified atmosphere packaging, *Journal of Food Science*, 56(1),55, 1991.

Robertson, J. A. and Meredith, F. I., Physical, chemical and sensory evaluation of 'Flodakin' peaches stored under different conditions, *Proceedings of the Florida State Horticultural Society*, 101,272, 1988.

Robertson, J. A.; Meredith, F. I.; Horvat, R. J. and Senter, S. D., Effect of cold storage and maturity on the physical and chemical characteristics and volatile constituents of peaches (cv. Cresthaven), *Journal of Agricultural and Food Chemistry*, 38(3),620, 1990.

Sarruge, J. R. and Haag, H. P., Análise química de plantas, ESALQ, Piracicaba, 1974, 56p.

Scott, K. J. and Wills, R. B. H., Postharvest application of calcium as a control for storage breakdowm of apples, *HortScience*, 10(1),75, 1975.

Tavares, L. B. B.; Chitarra, M. I. F. and Chitarra, A. B., Using modified atmosphere in the storage of two peaches cvs. (*Prunus persica* (L.) Batsch): 2. Total sugar and pectic substances transformation. *Arquivos de Biologia e Tecnologia*, 35(2),203, 1992.

Teixeira, M. C. R.; Chitarra, M. I. F. and Chitarra, A. B., Some characterizations on peach fruit varieties (*Prunus persica* (L.) Batsch). II - pectin, calcium, tannins and color, *Revista Brasileira de Fruticultura*, 5, 81, 1983.

Wang, C. Y., Physiological and biochemical responses of plants to chilling stress, *Hortscience*, 17,173, 1982.

Wills, R. B. H. and Mahendra, M. S., Effect of postharvest application of calcium on ripening of peach, *Australian Journal of Experimental Agriculture*, 29,751, 1989.

Will, R. B. and Tirmazi, S.I.H., Effect of postharvest application of calcium on ripening rates of pears and bananas. *Journal of Horticultural Science*, 57(4),431, 1982.

THE INFLUENCE OF TEMPERATURE AND GAS COMPOSITION ON THE EVOLUTION OF MICROBIAL AND VISUAL QUALITY OF MINIMALLY PROCESSED ENDIVE[1]

F. Willocx, M. Hendrickx and P. Tobback

Katholieke Universiteit Leuven
Centre for Food Science and Technology
Unit Food Preservation
K. Mercierlaan 92, B-3001 Heverlee, BELGIUM

ABSTRACT

Minimally processed vegetables or 'grade 4' products refer to ready to eat vegetables and include fresh, washed, chopped vegetables ready for use and packed in a sealed polymeric film. Extension of the shelf-life is achieved by a combination of correct chilled storage, Modified Atmosphere Packaging (MAP), and good manufacturing and handling practices. The principal spoilage mechanisms affecting the prepared fresh vegetables are microbial growth, oxidation (enzymatic browning) and moisture loss.

The impact of the preservation technology on the microbial and visual quality evolution of minimally processed vegetables is evaluated in a case study of endive. In the case of 'grade 4' endive, *Pseudomonas marginalis* and *Lactobacillus plantarum* were selected as indicator microorganisms for spoilage. The effect of the extrinsic factors upon the growth under controlled conditions were measured and quantified. The bacterial growth was modelled with the logistic and Gompertz equation. The temperature dependence of lag time of both indicator microorganisms was modelled with an hyperbolic function whereas the Arrhenius equation was used to describe the temperature dependence of growth rate. Lag phase and generation time of *P. marginalis* was extended by CO_2, whereas the growth of *L. plantarum* was not affected.

In a Quantitative Descriptive Analysis (QDA), the visual quality deterioration of minimally processed endive was measured and quantified. The quality deterioration was described using a first order equation. The dependence of the quality deterioration rate constant on temperature (2-12°C) was described using the Arrhenius equation, whereas a linear relation with the CO_2/O_2 concentration was observed. It was concluded that temperature was the most important extrinsic factor for both microbial growth of the indicator microorganisms and the visual quality deterioration of minimally processed endive.

1 The following text presents research results of the Belgian Incentive Program 'Health hazards' initiated by the Belgian State - Prime Minister's Service - Science Policy Office. The scientific responsibility is assumed by its authors. The authors thank the Science Policy Office for financial support.

I. INTRODUCTION

Minimally processed vegetables or 'grade 4' products refer to ready to eat vegetables and include fresh, washed, chopped vegetables ready for use and packed in a sealed polymeric film or tray. This fourth form of trading - preceded by fresh, canned and frozen - originated in the 1980s as an answer to an emerging consumer demand for convenience products which are fresh like, less severely processed, preservative free and of high quality. Over the last years, the market of 'grade 4' vegetables is growing both in sales volume and assortment of vegetables e.g. vegetable salads of endive, lettuce, carrots, peppers, potatoes, onions, cabbages, soy beans, both as single vegetable and in mixtures, completed with sauces and dressings, as well as vegetable mixtures for soup or stew [1,2,3]

The extension of the shelf-life is achieved by a combination of correct refrigerated storage throughout the entire cold chain, Modified Atmosphere Packaging (MAP) and good manufacturing and handling practices. MAP is a food preservation technology whereby the composition of the atmosphere surrounding the product is different from the composition of air. In these minimally processed vegetables the gas composition in the package is modified through respiration of the vegetative tissue (passive modification). By the respiration activity, oxygen from the headspace is consumed and carbon dioxide evolved, usually in ratio of 1-1. After a time, an equilibrium modified atmosphere is created depending on the respiration activity of the produce and the permeability characteristics of the packaging material. Good hygiene and temperature control are very important factors in the extension of the shelf-life, but the addition of a suitable modified atmosphere can give an even greater extension [4].

The principal spoilage mechanisms affecting the minimally processed vegetables are microbial growth, oxidation (enzymatic browning) and moisture loss [4,5]. MAP is effective at inhibiting these spoilage mechanisms, as well as reducing respiration, delaying ripening and decreasing ethylene production, however MAP will not eliminate the need for refrigeration. In Belgium 'grade 4' products, as well as other refrigerated products, must be kept at a temperature of maximum 7 °C during transport, storage and display with a tolerance up to 10 °C in the warmest spot [6]. Retail display cabinets and domestic refrigerators are known to be critical points into the cold chain of refrigerated products [7,8,9,10], and temperatures of 12 °C or higher in these part of the cold chain are no exception. The actual shelf-life of the refrigerated minimally processed vegetables depend on the temperature conditions throughout the entire cold chain since the integrated effect of time and temperature allows the proliferation of pathogenic and spoilage microorganisms as well as organoleptic and nutritional quality to deteriorate. As a consequence the responsibility for safety and quality of these 'grade 4' products lies with the producer and distributor as well as the consumer.

The purpose of the study presented here was to evaluate microbial and visual quality evolution of minimally processed vegetables in a case study on endive. The impact of the preservation technology on the growth of indicator microorganisms (spoilage) and visual quality of minimally processed endive stored under controlled conditions of temperature and carbon dioxide, is measured and quantified. Models of microbial growth as a function of time, temperature and carbon dioxide concentration, together with the models for quality deterioration, will be helpful in the identification of critical control points in the cold chain of 'grade 4' products for domestic purposes. Validation of the models under dynamic storage conditions are currently under investigation.

II. MODELLING MICROBIAL SPOILAGE

A. PREDICTIVE MICROBIOLOGY

The traditional approach to evaluate the shelf-life of a food product has been to incubate it under conditions that are representative of those encountered during distribution, storage and use, and then determine whether the microbial load exceeded acceptable levels (challenge test). One of the problems with studying the growth of (spoilage) microorganisms on foods is the natural variability in food samples. For example for MA packed foods, it is difficult to ascertain if the effects noted are due to the atmosphere used, variations in the food composition, presence of the competitive flora, or a combination of all three. The use of a controlled microbial growth medium and only indicator microorganism strains will remove the effects due to competitors and sample composition variability, thereby allowing the effect of the parameter under consideration to be more easily determined. In addition, the magnitude of influence of the controlling factors upon microbial growth can be quantified. In predictive microbiology mathematical equations are used to estimate the growth, survival and death of indicator and/or index microorganisms as affected by the intrinsic parameters of the food or the extrinsic factors applied to the food. These models can be used to predict the microbial safety or shelf-life of the food product [11,12,13].

In the minimally processed vegetables, which fall into the low acid range category (pH 5.8 - 6.0), the high humidity and the large number of cut surfaces can provide ideal conditions for growth of microorganisms. The predominant microflora of fresh leafy vegetables are aerobic, psychrotrophic, Gram-negative rods, with *Pseudomonas* and *Erwinia* spp. being most numerous, with a count of approximately 10^5 cfu/g. During cold storage of 'grade 4' leafy vegetables pectinolytic strains of *Pseudomonas* are responsible for bacterial soft rot [13]. In the case of 'grade 4' endive, *Pseudomonas marginalis* has been selected as an indicator microorganism for spoilage. However, measurements of temperature in the so-called 'cold' chain revealed that temperatures above the minimum temperature of growth of lactic acid bacteria were frequently observed. Moreover, the growth of Gram-negative, aerobic, psychrotrophic bacteria is inhibited by the carbon dioxide concentration in the headspace of the package, whereas lactic acid bacteria are not. These temperature abuses allow a rapid development of the lactic acid population, accelerated by the inhibitory effect of the CO_2 upon the competitive Gram-negative, aerobic, psychrotrophic bacteria. A shift in storage temperature and gas concentration in the package, will shift the microflora towards lactic acid bacteria [14]. Variations of the storage conditions will alter not only the rate of growth of the spoilage microorganisms, but also the type of spoilage microorganisms that will predominate. For this reason, *Lactobacillus plantarum* has also been selected as indicator microorganism for spoilage in the case of minimally processed endive and the effect of the extrinsic factors (temperature and gas concentration) upon the growth were measured and quantified. The initial population density of lactic acid bacteria on 'grade 4' endive was approximately 10^3 cfu/g (Own measurements, not shown).

B. MATERIALS AND METHODS
1. Indicator microorganisms and growth medium

Pseudomonas marginalis pv. *marginalis* (ATCC 10844) and *Lactobacillus plantarum* (ATCC 14917) were obtained from the Belgian Coordinated Collection of Micro-organisms (LMG, RUGent). Cultures of *P. marginalis* were maintained on Nutrient Agar slopes (NA) (OXOID, CM3) and stored at 4 °C. Cultures of *L. plantarum* were maintained anaerobically at 4 °C on de Man, Rogosa, Sharp agar slants (MRS) (OXOID, CM361). Stock cultures were transferred monthly.

The bacterial growth curve is measured under controlled storage conditions in a highly nutritious medium and as a consequence, the substrate is not a limiting factor to growth until the maximum population density is reached. The model system used was Nutrient medium for *P. marginalis* and de Man, Rogosa, Sharp medium for *L. plantarum*. For *P. marginalis* Nutrient Agar (NA) was used as counting medium, and Nutrient Broth (NB) (OXOID, CM1) for growth. The pH of the broth (pH=6.8 $\pm$ 0.2) was slightly buffered (KH_2PO_4, 0.45 g.l^{-1} and Na_2HPO_4, 2.39 g.l^{-1}). Bacterial numbers of *L. plantarum* were determined with a pour plate (MRS agar) and MRS broth (OXOID, CM359) was used for growth (pH=6.2 $\pm$0.2). In order to reduce the foaming of the MRS broth, 50 ppm Antifoam A (Sigma, VEL, Belgium) was added.

2. Inoculum preparation and measurement of growth

In order to reduce the dependence of the lag phase on the history of the inoculum, the inoculum was taken from a stationary phase (72 h, 22 °C for *P. marginalis* and 37 °C for *L. plantarum*) after equilibration overnight (20 h) at the appropriate refrigeration temperatures. Hudson [15] showed for *Aeromonas hydrophila* that the duration of the lag phase was always smallest when the pre-incubation temperature of the inoculum matched the test culture temperature. For the growth experiments 250 ml incubator flasks were used, each containing 100 ml of the appropriate broth and inoculated with the test microorganism to give the demanded inoculum size. The incubator flasks were flushed with gas (100ml/min) entering the flask at the bottom. The sterility of this gas stream was maintained by a 0.2 μm teflon membrane (Millex FG50, Millipore, France). This hydrophobic membrane permits the passage of the dry sterilized gas into the flask and protects against leaking of the medium. The volume reduction of the broth by the drying force of the gas was measured to account for the increase in bacteria number by the concentration of the medium. Dehydration of the MRS broth was 1.05 ml per day, whereas 1.09 ml per day was calculated for the Nutrient broth. The experimental layout and the flow guarantee the necessary mixing of the growth medium. Since the oxygen (O_2) of the headspace of the 'grade 4' packages is consumed and the carbon dioxide (CO_2) is produced usually in a ratio of 1-1, and further more to cover the possible concentrations of O_2 and CO_2 in the headspace, the concentrations of the gas components were chosen according to the exchange of CO_2 for O_2. In table 1 the concentrations of O_2 and CO_2 in the different gas mixtures are given.

The flasks were incubated in refrigerators at the desired temperatures and temperature was measured every two minutes using an electronic temperature recorder with an accuracy of 0.1°C (0-80°C) and resolution of 0.1°C (Control One, VEL, Belgium). At appropriate time intervals (depending on the refrigeration temperature) samples of 1 ml were taken, diluted in sterile saline solution (NaCl, 0.7 g.l^{-1}), surface plated (0.1 ml on NA) in triplicate or triplicated on pour plates (1 ml in MRS). The NA surface plates were incubated at 22°C for 72 h before the colonies were counted whereas the MRS pour plates were incubated for 72 h at 37°C before enumeration. For the entire growth curve at least 15 data points were collected, evenly spread throughout each phase and sampling was ended when minimally three counts in the stationary phase were obtained.

Table 1
Gas mixtures and concentrations of components [volume %].

Gas mixture	conc CO_2		conc O_2		conc N_2
air	0.033	± 0.001	20.9	± 0.1	78.1
6% CO_2	5.4	± 0.4	15.5	± 0.4	Q.S.[a]
11% CO_2	10.6	± 0.1	10.2	± 0.1	Q.S.
16% CO_2	15.9	± 0.2	5.2	± 0.1	Q.S.
19% CO_2	18.4	± 0.2	2.1	± 0.1	Q.S.

[a]Quantum Satis

C. MODELLING MICROBIAL SPOILAGE

1. Kinetics of microbial growth

When plotting the bacterial population density [log(cfu ml^{-1})] against time for a set of controlled environmental conditions, the growth curve is sigmoidal and is characterised by several successive phases: the lag time, the exponential phase, the stationary phase and finally the death phase. The death phase is not considered in this study. The sigmoidal bacterial growth curve can be modelled with the modified logistic (Eqn 1) and Gompertz equations (Eqn 2) [13]:

$$\log(N) = A + C / [1 + \exp((4.r_{mL} /C).(\lambda_L - t) + 2)] \qquad (1)$$

where
N : bacterial population density at time t [cfu ml^{-1}]
t : time (h)
λ : lag time (h)
r_m : maximum absolute growth rate [log(cfu ml^{-1})/h]
A : population density at time t → -∞ [log(cfu ml^{-1})]
C : increase in population density at time t → ∞ [log(cfu ml^{-1})]

$$\log(N) = A + C. \exp [-\exp ((e.r_{mG}/C).(\lambda_G - t) + 1)] \qquad (2)$$

where the four parameters, A,C, λ and r_m are defined as in Eqn 1. The modified Gompertz and Logistic equations were fitted to the data of the growth experiments by nonlinear regression with a derivative free algorithm [16]. The algorithm provides the estimates for the lag phase (λ), the maximum absolute growth rate (r_m), the estimate of the initial inoculum level (A) and the asymptotic increase in population density (C).

When studying the effect of extrinsic factors upon the growth of key microorganisms, one of the problems encountered during the experiments is the variation in inoculum size. The inoculum is prepared from a stationary phase, where the population density is known from previous determinations (optical density or plate counting). Since the inoculum is taken from a stationary phase of only 72 hours, the exact concentration of microorganisms is not known, and by subsequent dilution, an approximately inoculum level of 10^5 (cfu/ml) of *P. marginalis* or 10^3 (cfu/ml) of *L. plantarum* is reached in the reactor. In order to verify the effect of the inoculum size on the kinetic growth parameters (lag time and maximum growth rate) a first experiment was set up. In figure 1, the growth data and growth curves (logistic equation) of *P. marginalis* with different inoculum levels in a range from 10^3 to 10^6 (cfu/ml) are shown.

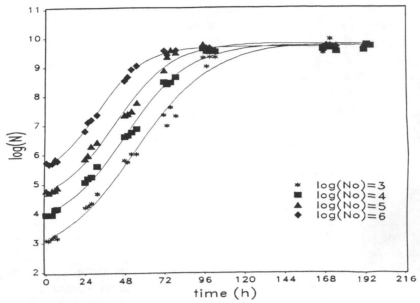

Figure 1. Growth data of *P. marginalis* (in air at 8.5°C) and growth curves modelled with the logistic equation. Range of the inoculum size from 10^3 (cfu/ml) to 10^6 (cfu/ml).

From the estimated growth parameters of lag time and maximum growth rate, it was concluded that the inoculum size did not affect the kinetic growth parameters of *P. marginalis* and as a consequence, growth curves with different inoculum levels in a range from 10^3 to 10^6 (cfu/ml) can be used to estimate lag time and maximum growth rate. Analogous results were obtained for *E. coli* by Jason [17] and for *Salomonella* by Mackey and Kerridge [18].

2. Modelling growth as a function of temperature

Various models have been proposed in literature to describe the relationship between growth rate, lag time and temperature whereas the maximum population density is considered independent of temperature [19]. In our study the temperature range is limited between the minimum (T_{min}) and optimum (T_{opt}) temperature of growth of the indicator micro-organisms. For *P. marginalis* 13 growth experiments were conducted in a temperature range from 1.9 to 18.1 °C and 10 experiments for *L. plantarum* from 12.8 to 18.7 °C. In table 2 the different models used in literature for the description of the temperature dependence of the growth rate (constant) and lag time are given.

These models were traditionally applied to the growth rate (constant). When taking the reciprocal of the lag time ($1/\lambda$), these models could also be used for the description of the temperature dependence of the 'lag rate' [20]. In order to compare the different models, no mathematical transformation of the growth rate (r_m) or lag time (λ) was carried out. Through the combination of these temperature dependence relations on both lag time and growth rate, and after introduction into the modified logistic (Eqn. 1) or Gompertz equation (Eqn. 2), the parameters of the different models were estimated by the combined multiple non linear regression and the Residual Sum of Squares (RSS) was used as selection criterion. Figure 2 shows the results of the global analysis for the growth curves of *P. marginalis* (in air) modelled with the hyperbolic temperature dependence relation of the lag time and the Arrhenius equation for the maximum absolute growth rate, together with the experimental data points of 5 growth curves.

Table 2
Temperature dependence relations for growth rate and lag time

Model	Equation	Reference
Linear relation	$r_T = r_0 \cdot (1 + c \cdot T)$	Spencer and Baines [21]
Arrhenius equation	$r_m = r_{mref} \cdot \exp[(E_a/R) \cdot (1/T_{ref} - 1/T)]$	
Square Root relation	$r_m = [b \cdot (T - T_{min})]^2$	Ratkowsky et al. [22]
Empirical relation	$\ln(r_m) = C_0 + C_1/T + C_2/T^2$	Davey [23]
Hyperbolic relation	$\ln(\lambda) = p /(T - T_{min})$	Zwietering et al. [19]

Where
T = temperature [°C or K]
T_{min} = theoretical minimum temperature for growth
T_{ref} = reference temperature
r_m = maximum absolute growth rate
r_T = rate of spoilage at temperature T (T=0 °C for r_0)
r_{mref} = maximum growth rate at reference temperature
E_a = activation energy or temperature characteristic [J/mol]
R = universal gas constant [8.314 J/mol,K]
b = Ratkowsky regression parameter [1/K]
c = constant for linear temperature response [1/ °C]
$C_{0,1,2}$ = regression parameters
λ = lag time [h]
p = hyperbolic regression parameter [K]

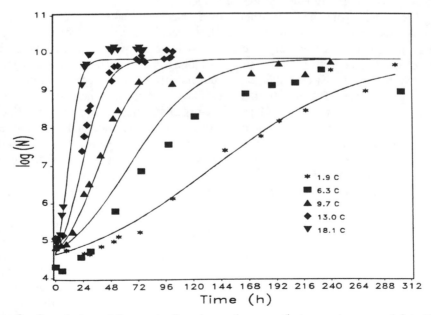

Figure 2. Growth data of *P. marginalis* and growth curves (in temperature range 1.9 to 18.1 °C, in air) modelled with the logistic equation (Eqn 1.) and temperature dependence relations of lag time (hyperbolic relation) and growth rate (Arrhenius relation).

The RSS for the Arrhenius equation was lower as compared to the Ratkowsky or linear Spencer and Baines relation. Due to instability, the empirical Davey model failed to converge for both logistic and Gompertz equation. Further more the Davey model consists of 3 (biological meaningless) parameters as compared to 2 regression coefficients for the linear, Arrhenius and Square Root model. Since microbial growth is a complex set of enzyme mediated biochemical reactions, microbial growth can be characterised by some overall activation energy (E_a) or temperature characteristic of the microorganism. The higher the (E_a) the more sensitive the growth rate is on temperature. The activation energy for *P. marginalis* was 100.3 [kJ/mol] modelled with the logistic equation (95 % confidence interval : 93.6 to 107.0 [kJ/mol]) and 100.0 [kJ/mol] modelled with the Gompertz equation (95 % confidence interval : 90.1 to 109.8 [kJ/mol]). The calculated temperature dependence (E_a) of the growth rate of *P. marginalis* is higher then the values reported for other *Pseudomonas* spp. in literature. Brocklehurst and Lund [24] found 83.0 [kJ/mol] for *P. fluorescens*, Stannard et al. [25] calculated 85.4 [kJ/mol] for pigmented and 84.5 [kJ/mol] for non-pigmented *Pseudomonas* spp. and Fu et al. [20] reported an E_a of 74.0 [kJ/mol] for *P. fragi*. However in literature the 95 % confidence interval on the activation energy is usually not stated. Beside the difference in microorganism, the temperature characteristic will depend on the determination method of the growth rate (constant) and the medium and the temperature range of the experiment.

For *L. plantarum* the Arrhenius equation also satisfactorily describes the temperature dependence of the maximum growth rate in a range from 12 to 19 °C. The temperature characteristic was calculated as 149.3 [kJ/mol] modelled with the logistic equation (95 % confidence interval from 124.4 to 174.3 [kJ/mol]) and 154.8 [kJ/mol] modelled with the Gompertz equation and the 95 % confidence interval from 130.2 to 179.4 [kJ/mol]). As compared to the activation energy of *P. marginalis*, the growth rate of *L. plantarum* is more sensitive towards temperature in the same temperature range.

3. Modelling growth as a function of gas composition

The bacteriostatic effect of carbon dioxide has been known for a long time and exploited for the preservation of packed food like meat, fish and poultry. The specific mechanism for this growth inhibitory effect of CO_2 is not known. [26]. Carbon dioxide has been shown to be effective for foods whose spoilage is dominated by gram-negative, aerobic, psychrotrophic bacteria [26,27]. The sensitivity to carbon dioxide varies according to the species. The overall effect of carbon dioxide is to increase both the lag phase and the generation time of spoilage micro-organisms but several factors influence this antimicrobial effect, specifically microbial load, gas concentration and temperature.

Enfors and Molin [28] reported an increased CO_2 resistance in the order: *Pseudomonas fragi, Bacillus cereus* to *Streptococcus cremoris*. As a consequence, preserving foods under high CO_2 concentrations selects the microflora towards lactic acid bacteria [29]. Some investigators have reported a maximum partial pressure of carbon dioxide above which a further CO_2 pressure increase has but minor additional inhibitory effect. Gill and Tan [30] indicate a maximum degree of inhibition of growth of *P. fluorescens* when an pCO_2 of 250 mm Hg was attained. In contrast, other investigators [28] observed inhibitory effects which were approximately proportional to the partial pressure of CO_2 over the entire CO_2 pressure range. At low oxygen levels, the combined growth inhibitory effect of oxygen deficiency (lower then 1%) and of the presence of CO_2 seems to be purely additive. [31,28]. No indication of a synergistic effect between CO_2 inhibition and oxygen limitation were noted. Several authors have stated that the effectiveness of the CO_2-inhibition increases at decreasing temperatures [30,29]. The increased efficiency of CO_2 at lower temperature could be related to the increased solubility of CO_2 into the medium with lowering of temperature. [27].

In this study, the effect of increasing carbon dioxide concentration (and complementary decrease in oxygen level) at constant temperature on the growth parameters of *P. marginalis* and *L. plantarum* was measured and quantified. For simplicity, only the CO_2 concentration of the gas mixture will be named, but the reader should bear in mind that the total concentration of CO_2 and O_2 equals 21 volume % in each gas mixture (see table 1). Different combinations of CO_2/O_2 dependence models (linear and exponential) for both lag time and maximum growth rate were investigated. An exponential dependence of the growth rate and a linear dependence of the lag time of *P. marginalis* with the CO_2 concentration in the gas mixture gave the lowest RSS, both for logistic and Gompertz equation. Figure 3 shows the growth data (at 11 °C) and growth curves modelled with the Gompertz equation (Eqn 1) and the CO_2 dependence relation of lag time (linear) and maximum growth rate (exponential) by the combined multiple non linear regression. As can be seen from figure 3, the effect of the increasing CO_2 concentration is primarily a decrease in growth rate and to a lesser extent an increase in lag time. The maximum population density (Y_{max} = A + C) was set independent of the concentration of CO_2 in the gas mixture.

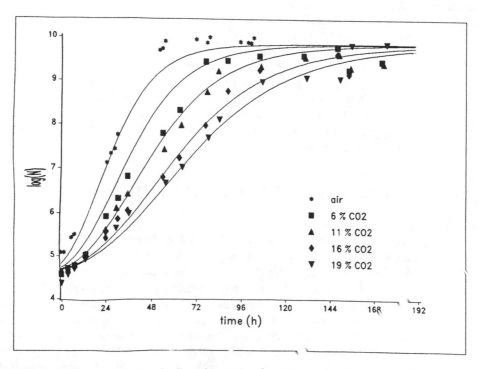

Figure 3. Growth data of *P. marginalis* and growth curves (temperature 11 °C) modelled with the Gompertz equation and CO_2 dependence relations of lag time (linear) and growth rate (exponential)

In table 3 the estimates of the growth parameters of *L. plantarum* modelled with the logistic and Gompertz equation are given for 5 gas mixtures (table 1). From table 3, it can be concluded that the concentration of CO_2 and O_2 in the gas mixtures in this study had no significant effect on both lag time and maximum growth rate of *L. plantarum*.

Table 3.

Estimates and 95 % confidence interval of lag time [h] and maximum growth rate [log(cfu ml⁻¹)/h] of *L. plantarum* (temperature 18.0 °C) for different gas compositions. Estimates from Gompertz equation (Eqn 2.).

Gas mixture	Lag time	95% interval	growth rate	95% interval
air	35.09	23.06 - 43.11	0.158	0.088 - 0.228
6% CO_2	35.12	31.96 - 38.29	0.154	0.137 - 0.171
11% CO_2	38.09	34.78 - 41.41	0.157	0.138 - 0.176
16% CO_2	36.33	33.44 - 39.21	0.158	0.141 - 0.174
19% CO_2	36.97	33.40 - 40.53	0.166	0.142 - 0.190

D. CONCLUSION

From this study it was concluded that the logistic and Gompertz equation are both useful in describing the bacterial growth curve as a function of time. The temperature dependence of the lag time of *P. marginalis* and *L. plantarum* were modelled with an hyperbolic function, whereas the Arrhenius equation was used to describe the temperature dependence of growth rate. At constant temperature, lag time of *P. marginalis* was extended linear with increasing CO_2 concentration (and complementary decreasing O_2 concentration), whereas the CO_2 dependence of the maximum growth rate was described using an exponential relation. The effect of the CO_2 concentration in the headspace will not affect the growth of *L. plantarum*. Preserving food under modified atmospheres like in 'grade 4' vegetables with CO_2 concentrations exchanged for O_2 in a ratio 1-1, will inhibit the aerobic flora (e.g. *Pseudomonas* spp.), but will not affect the growth of facultative anaerobic microorganisms (e.g. *Lactobacillus* spp.) and finally will select the spoilage microflora towards lactic acid bacteria, if the storage temperature is above their minimum temperature of growth.

III. MODELLING VISUAL QUALITY DETERIORATION

A. VISUAL QUALITY

The shelf-life of a food product can be defined as the period of time that the food product remains safe and wholesome under recommended production and storage conditions [5]. Safety and health are the implicit consumer needs of quality and are usually specified by legislation and rules. (e.g. absence of additives or pathogens). The term 'wholesome' for a food product reflects the combination of explicit consumer needs of quality such as nutritional and organoleptic quality attributes (appearance, flavour, taste and texture) of the food product or even service (like convenience and price). For refrigerated products, the organoleptic quality is as important for the shelf-life as the microbial load. But, because the consumer is unable to evaluate the microbial load at the moment of purchase or at the moment of consumption of the refrigerated products, appearance is the only quality attribute of the 'grade 4' products on which the consumer can and shall decide to buy and to consume.

The shelf-life of a food product is influenced by a wide variety of factors, either the intrinsic properties of the food (pH, aw, nutrient content, respiration rate,..) or the extrinsic factors applied to the food (processing and storage conditions: temperature, gaseous composition, hygiene,..). For minimally processed products the most important extrinsic factors are temperature and gas composition in the package. The quality characteristics of food products change as a function of the period and conditions of storage. The identification and description of the kinetics of the deterioration processes (and subsequent determination of the shelf-life) is important for both producer, distributor and consumer. The purpose of this study was to measure and quantify the visual quality deterioration evolution and to model the influence of the extrinsic factors of temperature and gas composition upon the visual quality of cut endive during storage.

B MATERIALS AND METHODS
1. Minimally processed endive

A tray of six Belgian endives were purchased from the vegetable mart of St.-Katelijne-Waver (Belgium) and transported to the laboratory. The time between purchase and initiation of the experiment never exceeded 1.5 h. The outer wrapper leaves were discarded and the endives were sliced with a sharp stainless steel knife into pieces of about 1 to 1.5 cm in width. The pieces were rinsed twice with cold tap water ($\pm$ 13 °C) thoroughly mixed and manually centrifuged to remove excess water. 100 gram of a mixture of white and green pieces (ad random selected) are placed on a perforated dish into a 1.5l jar. The jars are connected at the foot to the appropriate gas mixture (50 ml/min) (see table 3), with a vent on the top. This connection and gas flow permits the creation of a constant gas atmosphere inside the jar and copes with the modification of the atmosphere by the respiration activity of the endive. The gas flow was saturated with water before entering the jar, to avoid dehydration of the samples. In one experiment, the influence of two different gas mixtures at four different temperatures (2, 4, 8 and 12 °C) are measured. Two reference samples were also included to allow comparison of the quality deterioration between different experiments since processing and storage conditions are equal. The different jars are placed into four refrigerators at the appropriate temperature. The door of the refrigerators was modified with a transparent window (13 x 22 cm) and white light. This window allows the visual sensory evaluation without opening the refrigerators. Temperature was recorded with data loggers (Control One, VEL, Belgium) inside the jar during the whole experiment (12 days).

2. Visual quality evaluation

Each day, the visual quality was evaluated with a trained taste panel (minimally 15 persons). A Quantitative Descriptive Analysis test (QDA) was used to describe the sensory characteristics of the minimally processed endive, and to use these characteristics to quantify differences between products. The quality attributes of interest were discolouration (browning) of the cut surfaces, brown spots on the leaves and an overall quality evaluation. The judges indicate their perception of a quality attribute on a 100 mm unstructured line. Under the line the perception of a quality attribute is indicated as weak, moderate or strong (anchor points). The line scores of each judge are converted to numbers by the use of a template. By the use of this equal-interval scale, it is possible to calculate the mean and standard deviation of a quality attribute. The panel members were trained in a first experiment to get used to the descriptive test. Inconsequent panel members were removed from the statistical analysis. For more detailed information about the taste panelling, scaling method and statistical analysis is referred to the work of O'Mahony [32] and Meilgaard et al. [33].

C. KINETICS OF QUALITY EVOLUTION

1. Kinetic order of quality deterioration reactions

Traditionally, quality deterioration processes of foods stored under controlled environmental conditions are described with a zero order and/or first order rate functions [34,35,36]:

$$- dQ / dt = k [Q]^n \tag{3}$$

where Q : quality index
                    t : time
                    n : reaction order (0 or 1)
                    k : reaction rate constant

For many reactions in foods the reaction order $n=0$. The reaction can be assumed to be zero order mathematically although the mechanisms may be very complex. zero order can be applied to non-enzymatic browning reactions and to many quality losses in frozen foods [34]. For a zero order reaction a plot of the amount of quality left as a function of time should be a straight line:

$$Q_t = Q_0 - k_0 . t \tag{4}$$

where Q_t : amount of quality at time t
                  Q_0 : initial amount of quality
                  k_0 : zero order reaction rate constant [#quality units/day]
                  t : time [day]

Many other reactions leading to food deterioration follow a first order reaction ($n=1$). Labuza [34] reports that loss of vitamins A,B,C, unhibited lipid oxidation and microbial death follow first order reaction. In this case a semi-log plot of the 'amount' of quality left versus time gives a straight line:

$$Q_t = Q_0 . exp[- k_1 . t] \tag{5}$$

where Q_t : amount of quality at time t
                  Q_0 : initial amount of quality
                  k_1 : first order reaction rate constant [1/day]
                  t : time [day]

In much of the food science literature it is impossible to establish what order the data follow since few points are collected and errors in analysis may be large. Moreover, if the change in quality attribute is small (<30% of the initial amount), zero order and first order model give similar results.

In figure 4 the individual overall visual quality response, evaluated by the taste panel members of cut endive stored for 12 days at 11.7 °C under air is given. As can be seen from this figure, the numerical overall visual quality response value differs widely for each taste panel member. However, there is still a marked decrease in quality during the storage period. The numerical response values of the individual members are consistent during the whole experiment, but the use of the unstructured line scale differs according to the individual sensitivity. This problem generates a wide variability in the responses and this variation overshadows the variability between products.

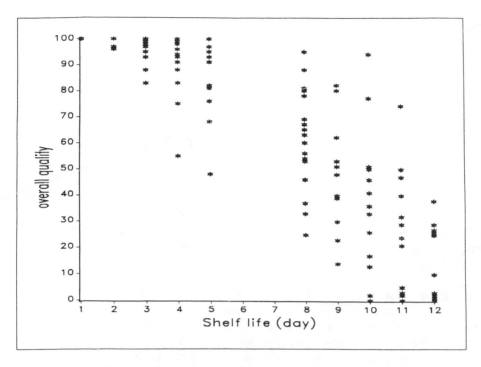

Figure 4. The individual overall visual quality response of each taste panel members for a sample of cut endive stored for 12 days at 11.7 °C under air.

A reduction in variability within a taste panel is obtained by a normalization method described by Weiss and Zenz [37] and is based on the hypothesis that the taste panel members behave in scaling like instruments working with different sensitivities (range of response, used scale range) and different zero setting (position of the range of each member on the scale. It is assumed that each subject works with a different but constant sensitivity (scale range factor, b_i) and with a different but constant zero setting (shift factor, b_i) during a session. The normalisation method used is, in the form $y = a + b.x$ a linear two-step transformation of scale range factors (b_i) and shift factors (a_i). In our experiment, a normalization of the shift factor (the zero setting for each member of the taste panel) was not necessary since the quality responses for each member was started at zero. The sensitivity or scale range factor (b_i) normalize the individual sensitivities (E_i) of each subject to the averaged sensitivity of the panel (E_p). The individual sensitivities of the members (E_i) is given by (Eqn 6):

$$E_i = [R_{i,max} - R_{i,min}] \, /c \tag{6}$$

where $R_{i,max} - R_{i,min}$: used scale range of the responses (R_i) of the i-th member
        c   : range of intensities of the stimulus or sample (usually not known)
        n   : number of taste panel members

The averaged sensitivity (used scale range) of the taste panel (E_p) is given by (Eqn 7):

$$E_p = \Sigma(E_i) \, / \, (c \cdot n) \tag{7}$$

The individual sensitivity factor of the taste panel member ($b_i = E_p/E_i$) from Eqn 6 and 7 is used to calculate the normalized responses of each taste panel member ($R_{i,norm}$):

$$R_{i,norm} = (b_i \cdot R_i) \tag{8}$$

In this experiment, the sensitivity factor (b_i) for the individual taste panel members varied between 0.85 and 1.31. After statistical analysis, it was concluded that a first order model was suitable for the description of the normalized visual quality deterioration rate. First of all, for the zero order model, the linear regression assumption of homoscedasticity was violated (homogeneous variance). A logarithmic transformation of the original data reduces the variance [32]. Secondly, an analysis of the residues indicate a non random distribution for the data modelled with a zero order model as compared to a (more) random distribution of the residues for a first order model. It was concluded that the quality deterioration rate constants, obtained with the first order model from the data points, are to be used for the analysis of the influence of temperature and gas composition.

2. Modelling visual quality as function of temperature and gas composition

As for microbial growth, quality deterioration rates of minimally processed vegetables are affected by temperature and gas composition during processing and storage. In literature, the temperature dependence of the rate constant is usually described by the Arrhenius equation. Taoukis and Labuza [38] report typical activation energy values (E_a) for food quality losses (table 4).

Table 4.
Activation energy (E_a) for food quality losses [kJ/mol]

Quality attribute	E_a	Quality attribute	E_a
Diffusion controlled	0 - 60	Nonenzymic browning	105 - 210
Enzymic	40 - 60	Microbial growth	80 - 250
Hydrolysis	60	Spore destruction	250 - 335
Lipid oxidation	40 - 105	Veg. cell destruction	210 - 630
Nutrient loss	80 - 125	Protein denaturation	335 - 500

The Q_{10} coefficient is also used to describe the dependence of quality deterioration with temperature. However, the activation energy is constant whereas the Q_{10} factor increases with decreasing temperature and the Q_{10} value is the result of only two measurements.

The influence of the gas composition, carbon dioxide and oxygen-concentration, on the quality deterioration rate constant is not quantified in literature.

In figure 5, the normalized mean quality responses for browning of the cut surfaces of three samples of cut endive stored under air at 1, 7 and 16 °C are given, evaluated in a period of 12 days. As can be seen from figure 5, samples stored at 1, 7 or 16 °C cannot be differenciated up to day 3 of storage period based on the standard deviations of the quality evaluation [16]. Figure 6 shows the normalized mean overall quality responses for samples of endive stored at 16 °C under air, 11% CO_2 and 19 % CO_2. As compared to figure 5, the taste panel members cannot differenciate samples stored at the same temperature but under different modified atmospheres. From this study it was concluded that temperature was the most important extrinsic factor for the visual quality evolution of minimally processed endive.

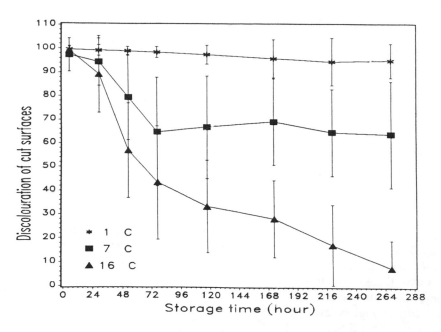

Figure 5. The normalized mean quality response (100-0) and standard deviation for browning of the cut surfaces of three samples of cut endive stored under air at 1, 7 and 16 °C. Storage period of 12 days.

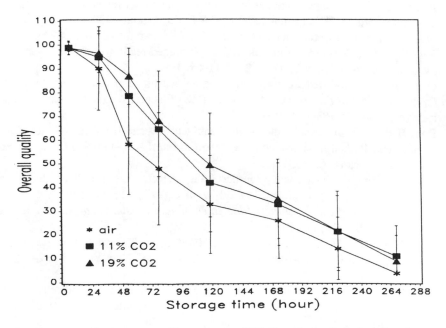

Figure 6. The normalized mean quality response (100-0) and standard deviation for overall quality of three samples of cut endive stored at 16 °C under air, 11% CO_2 and 19% CO_2. Storage period of 12 days.

The analysis of the influence of temperature and gas composition on the visual quality is based on the reaction rate constant modelled with a first order. The Arrhenius equation was used to describe the temperature dependence of the visual quality deterioration rate constant for discolouration of the cut surfaces and the overall quality evaluation. Browning of the cut surfaces was more sensitive towards temperature (E_a = 98.6 [kJ/mol]) then the overall quality evaluation (E_a = 58.6 [kJ/mol]), but independent from the gas composition. The calculated activation energy for the overall quality evaluation agrees fairly with values reported in literature for enzymatic reactions (table 4). For the description of the dependence of the quality deterioration rate with carbon dioxide concentration in the surrounding atmosphere, a linear relation was suggested.

D. CONCLUSION

Both temperature and gas composition exert an influence on the visual quality deterioration, but it was concluded that temperature is the most important factor. If the quality deterioration is independent of the history of the product, it is possible to predict the appearance of 'grade 4' endive under fluctuating temperature conditions based upon the results (kinetics) obtained under controlled storage conditions. Validation of this Time-Temperature-Tolerance hypothesis, first proposed for frozen foods by Van Arsdel [39] is currently under investigation.

IV REFERENCES

1   **Anon**. Guide de bonnes pratiques hygiéniques concernant les produits végétaux prêts à l'emploi, dits de "IV" gamma. *B.O.C.C.R.F.* du 13 août 1988.

2   **Scandella, D. and Leteinturier, J. (Ed.)** La 4$^\text{ème}$ gamme. *Ctifl*, 1989.

3   **Saracino, M., Pensa, M. and Spiezie, R.** Packaged ready-to-eat salads: an overview. *Agro-Industry Hi-Tech*, **2** (5):11-15, 1991.

4   **O'Beirne, D.** Modified atmosphere packaging of fruits and vegetables. In *Chilled foods. The state of the art*. (Ed. Gormley, T.R.) Elsevier Applied Sciences. London and New York Pp.183-199, 1990.

5   **Day, B.P.F.** Guidelines for the Good Manufacturing and Handling of Modified Atmopshere Packed food products. *Technical Manual N°.34* (1992) Campden Food and Drink Research Association.

6   **Anon**. Koninklijk Besluit van 4 februari 1980 betreffende het in de handel brengen van te koelen voedingsmiddelen. In *De Warenwetgeving. Verzameling van reglementen betreffende voedingsmiddelen en andere consumptieprodukten* **1.A** (dec. 1982) 13-14. Brugge, Die Keure.

7   **Bøgh-Sørensen, L.** The chill chain. In *Chilled foods. The state of the art*. (Ed. Gormley, T.R.) Elsevier Applied Sciences. London and New York, Pp. 245-267, 1990.

8   **James, S. and Evans, J.** Performance of domestic refrigerators and retail display cabinets for chilled products. In *Progress in the Science and Technology of Refrigeration in Food Engineering*. Proceedings of the meetings of Commissions B2, C2, D1, D2/3, September 24-28/1990, IIR, Paris (France): 401-410, 1990.

9   **James, S. and Evans, J.** Temperatures in the retail and domestic chilled chain. In *Processing and Quality of Foods Vol.3. Chilled foods: the revolution in freshness.* (Ed. P. Zeuthen, J.C. Cheftel, C. Eriksson, T.R. Gormley, P. Linko and K. Paulus), Elsevier Applied Science, London and New York, 3.273-3.278, 1990.

10   **Willocx, F., Hendrickx, M. and Tobback, P.** Temperature conditions and residence time distribution in 'grade 4' products retail display cabinets. *Int. J. Refr.* (submitted), 1993.

11   **Baird-Parker, A.C. and Kilsby, D.C.** Principles of predictive food microbiology. *J. Applied Bact. S. Suppl.* 43S-49S, 1987.

12   **Gould, G.** Predictive mathematical modelling of microbial growth and survival in foods. *Food Science & Technology Today* 3, 89-92, 1989.

13   **Willocx, F., Hendrickx, M. and Tobback, P.** Modelling the influence of temperature and carbon dioxide upon the growth of *Pseudomonas fluorescens. Food Microbiology*, 10, 159-173 1993.

14   **Manvell, P.A. and Ackland, M.R.** Rapid detection of microbial growth in vegetable salads at chill and abuse temperatures. *Food Microbiology*, 3, 59-65, 1986.

15   **Hudson, J.A.** Effect of pre-incubation temperature on the lag time of *Aeromonas hydrophila. Letters in Appl. Microbiol.* 16 (1993), 274-276, 1993.

16   **SAS Institute INC.** In SAS User's guide: Statistics, pp. 584. Cary, NC: SAS Institute Inc, 1982 Ed.

17   **Jason, A.C.** A deterministic model for monophasic growth of batch cultures of bacteria. *Antonie Van Leeuwenhoek J. Microbiol. Serol.* 49, 513-536, 1983.

18   **Mackey, B.M. and Kerridge, A.L.** The effect of incubation temperature and inoculum size on grwoth of Salmonellae in minced beef. *Int. J. Food Microbiol.* 6, 57-65, 1988.

19   **Zwietering, M.H., De Koos, J.T., Hasenack, B.E., De Wit, J.C. and Van 'T Riet K.** Modeling of bacterial growth as a function of temperature. *Appl. Environ. Microbiol.* 57, 1094-1101, 1991.

20   **Fu B., Taoukis, P.S. and Labuza, T.P.** Predictive microbiology for monitoring spoilage of dairy products with Time-Temperature-Integrators. *J. Food Sci.* 56, N°5 1209-1215, 1991.

21   **Spencer, R. and Baines, C.R.** The effect of temperature on the spoilage of wet white fish. *Food Technology*, 28, 175-179, 1964.

22   **Ratkowsky, D.A., Olley, J., McMeekin, T.A. and Ball, A.** Relationship between temperature and growth rate of bacterial cultures. *J. Bacteriol.* 149, 1-5, 1982.

23   **Davey, K.R.** A predictive model for combined temperature and water activity on microbial growth during the growth phase. *J. Applied Bact.* 67, 483-488, 1989.

24   **Brocklehurst, T.F. and Lund, B.M.** Properties of Pseudomonads causing spoilage of vegetables stored at low temperature. *J. Applied Bact.* 50, 259-266, 1981.

25   **Stannard, C.J., Williams, A.P. and Gibbs, P.A.** Temperature/growth relationships for psychrotrophic food-spoilage bacteria. *Food Microbiol.* 2, 115-122, 1985.

26   **Daniels, J.A., Krishnamurthi R. and Rizvi S.S.H.** A review of effects of carbon dioxide on microbial growth and food quality. *J. Food Prot.* 48, june 1985, 532-537.

27   **Lefevre, D.** The effect of modified atmospheres upon microorgnisms. In *International Conference on Modified Atmosphere Packaging 15-17th October 1990.* Conference Proceedings part 2. Campden Food & Drink Research Association.

28   **Enfors, S.O. and Molin, G.** Effect of high concentrations of carbon dioxide on growth rate of *Pseudomonas fragi, Bacillus cereus and Streptococcus cremoris. J. Applied Bact.* 48, 409-416, 1980.

29 **Blickstad, E., Enfors, S.O. and Molin, G.** Effect of hyperbaric carbon dioxide pressure on the microbial flora of pork stored at 4 or 14 °C. *J. App. Bact.* **50**, 493-504, 1981.

30 **Gill, C.O. and Tan, K.H.** Effect of carbon dioxide on growth of *Pseudomonas fluorescens*. *Appl. Environ. Microbiol.* Aug. 1979, 237-240.

31 **Dixon, N.M., Lovitt, R.W., Kell, D.B. and Morris, J.G.** Effects of pCO2 on the growth and metabolism of *Clostridium sporogenes* NCIB 8053 in defined media. *J. Applied Bact.* **63**, 171-182, 1987.

32 **O'Mahony, M.** Sensory evaluation of food. Statistical methods and procedures. Marcel Dekker Inc. New York and Basel. Pp. 487, 1986.

33 **Meilgaard, M.D., Civille, G.V and Carr, B.T.** Sensory evaluation techniques. CRC Press, Inc. (2nd Ed.) Pp. 281, 1988.

34 **Labuza, T.P.** A theoretical comparison of losses in foods under fluctuating temperature sequences. *J. Food Sci.*, **44**, 1162-1168, 1979.

35 **Labuza, T.P. and Riboh, D.** Theory and application of Arrhenius kinetics to the prediction of nutrient losses in foods. *Food Technology* **36**, 66-74, 1982.

36 **Meffert, H.F.Th.** Quality development of foodstuffs under time-temperature conditions in cold chains. In Progress in the Science and Technology of Refrigeration in Food Engineering. Commissions B2, C2, D1, D2/3, (Paris, September 24-28). 535-550, 1990.

37 **Weiss, J. and Zenz, H.** Reduction of panel variances by a simple two-step normalization procedure for grqphical line scale. *Acta Alimentaria*, **18**(3), 313-323, 1989.

38 **Taoukis,P.S. and Labuza, T.P.** Applicability of time-temperature indicators as shelf life monitors of food products. *J. Food Sci.*, **54**(4), 783-788, 1989.

39 **Van Arsdel, W.B.** The Time-Temperature Tolerance of frozen foods. I. Introduction - the problem and the attack. *Food Technology* **11**, 28-33, 1957.

CONSUMER PERCEPTIONS OF QUALITY

Christine M. Bruhn, Ph.D.

Overview

Quality depends on how a product meets the needs of the consumer (Benedict, 1986). It includes sensory properties of taste, texture, and color as well as psychological and other factors to which the food contributes. Quality food provides value in terms of sensory satisfaction, contributions to health, and compatibility with life-style conditions. As consumer lifestyle and nutritional needs change, quality judgments change. Hence quality is illusive and ever-changing.

Sensory quality and psychological attributes contribute to quality perception. An appropriate taste, odor, texture, mouthfeel, and appearance contribute to quality perception. Cultural and regional preferences formed during childhood influence quality perceptions. A product may be perceived as high quality if it offers nostalgic flavor experiences comparable to that enjoyed in the past. Some foods convey status in their use. Specific status foods differ by culture however they are usually expensive and difficult or time consuming to prepare. Some foods signify affection, security, comfort, or rewards. Specific examples vary by culture and personal background, however they could include home made bread, premium ice cream, or chocolate. Foods also create moods. Champagne and sparkling juices, for example, are celebration foods appropriate for festive occasions. Other foods convey body images, such as grapefruit for slimming and steak for athletic power. Some are thought to carry magical properties; oysters, for example, are consider aphrodisiacs. Appropriate sensory and psychological factors can enhance a product's quality image.

Today, freshly prepared food is desirable, however an increased number of households consist of two adults holding part or full time jobs. Demands from employment and other activities place a premium on time, including food shopping and preparation time. Despite the desire for fresh, a changing lifestyle places increased emphasis on convenient, partially prepared products. A quality processed product should require minimal preparation and/or cooking time and provide flavor and texture comparable to that of a freshly prepared counterpart.

Health aspects are of importance to an increasingly large segment of the population. A quality processed product should provide the nutrients comparable to the fresh counterpart without additional hazardous compounds added or formed during processing. Sensitivity to environmental needs places additional requirements on convenience foods.

Fresh convenient foods: What people want and what they buy

A review of convenience food purchases in the United States provides insight into factors which influence consumer behavior. In 1992, 42% of American consumers purchased at least one pre-cut, peeled, or mixed produce item (The Packer, 1993). Women were more likely than men to have purchased convenience items. The likelihood to buy convenience items decreases with age, with 60 years and older people substantially less likely than younger people. Households with three or more members, those with children under 18 years, and those with higher income purchased convenience items more frequently.

One might expect that consumers would purchase those convenience items that eliminate tasks that were difficult or time consuming. Consumers report that fruits and vegetables which take the most preparation times to prepare include melons, names by 18% of consumers, pineapple, named by 14%, potatoes, names by 30% and carrots, named by 26% (The Packer, 1992). Lettuce leads the list of items taking the least time to prepare. Consumer actual purchases reveal that preparation time was not as important as routine use. People purchased regular salad mix (16%), precut carrot sticks (15%), broccoli buds (14%), vegetable salad mix (13%), baby carrots (12%), and melon balls (12%) (The Packer, 1993). Therefore when considering potential market, the time consumption or difficulty of the task appears to be less important than the frequency of the task.

Consistent high quality is a key to attracting and maintaining convenience item customers. People who had one bad experience reported they were reluctant to try other convenience items. Most people who had tried convenience items seemed satisfied with them, however some problems were noted. Consumer's primary complaint was they would like to have convenience items stay fresh longer without the use of preservatives. Shelf life was perceived as too short. One consumer remarked, "They are fresh when they are first bought, but they get old quicker."

Quality: Anticipating Consumer Needs

Perceptions of quality are not limited to flavor. Quality also depend on how well the product package communicates to and meets the needs of the users.

The microwave oven has helped consumers cope with the increase time demands of contemporary lifestyles. In 1982 less than 30% of households in the United States owned a microwave, in ten years that percentage had increased to over 80% (Hollingsworth, 1993). The average use of a microwave in the US is four times a day (Anonymous, 1987). A sizable proportion of those who have no microwave at home use one at work. Analysts speculate that in the future, one microwave is not going to be adequate for many households because people don't like to wait for one meal while another is heating. Some speculate that the kitchen of the future will have a bank of microwaves offering the convenience of heating many different dishes at once. The number of food items prepared for the microwave is expected to grow. Consumers indicate there is room for improved quality, however, only a minority, 17%, would be willing to pay somewhat higher price for better tasting microwave frozen foods (Doyle, 1989).

Microwaves are used by children as well as adults. Simple instructions that specify steps make following cooking instructions easier. As international trade expands and nations include people with a variety of native languages, consumers could benefit from pictorial instructions. Consumers with poor eyesight would appreciate instructions they could read without their glasses.

Safety tips should also be included for children and adults. The possibility of burns from trapped steam should be stressed and proper handling suggested. Being clear about the use of correct utensils is important as is proper handling of packages that bend when they are hot. In a consumer survey, one parent wrote that packages have a long way to go to be considered "kid friendly" (Doyle, 1989). Reuse of packaging is also a consumer issue and can carry quality connotations.

Nutrition and health: Changing views, changing diets

Contributing to health is a potential quality factor for food. Nutritional awareness is at an all-time high. About 60% of men and 69% of women, including 74% of employed women, believe their diet could be at least somewhat healthier (Opinion

Research, 1993). This perception is found at all income levels, with the greatest percentage seeking improvement among those with higher incomes. People age 18 through 49 years also indicate a need for improvement.

Examining specific nutrients gives further indication of nutritional and food compositional concerns. In 1985, more consumers were concerned with sugar and salt than any other food ingredient, these ingredients were identified by 20% and 19% of consumers respectively (Opinion Research, 1993). Today these nutrients are still important, but the first nutritional concern is fat content, cited by 54% of the public. Cholesterol has been an important concern, cited by 44% in 1990, however the number mentioning this constituent has declined to 23% in 1993. Nutritional concerns in order are fat content 54%, salt 26%, cholesterol 23%, sugar 18%, and calories 15%.

People say they are making changes to improve their diet; 62% indicating they are eating more fruits and vegetables, 30% eating less meat, 26% eating less fats and oils, 15% more chicken and turkey, and 12% fewer snack foods (Opinion Research, 1993). Sales records confirm that people are eating more fruits, vegetables, and poultry and less red meat (Putman, 1991).

People are also changing purchases within food groups. In 1991 67% of Americans consumed low or reduced fat foods and beverages (Calorie Control Council, 1991). Lower fat foods are not just the choice of women, although women at 70%, were more frequent consumers than men at 61%. In 1993 consumers of light products increased to 78% of women and 66% of men (O'Connell, 1993). The most popular low fat products are reduced fat salad dressings, sauce and mayonnaise, consumed by 76% of respondents, followed by low fat cheese/dairy products consumed by 70% of adults, margarine, 69%, low-fat or skimmed milk, 67%, and meat products, 54% (O'Connell, 1993). Staying in better overall health, not weight reduction is the primary reason consumers chose reduced-fat foods.(Calorie Control Council, 1991).

The tendency toward lighter eating is not a uniquely American phenomenon (Calorie Control Council, 1992). In 1992 81 million adults including 74% of United Kingdom consumers, 48% of the French, and 69% of Germans chose light products. These percentages include both reduced fat and reduced sugar products, however reduced fat is the greater purchase. Like Americans, better health was the primary reason for European's choice. Other reasons include to reduce sugar or fat and to maintain current weight.

The emphasis on healthy eating implies that low fat and low calories are quality parameters. However taste, not nutrition is the primary factor in food choice (Opinion Research, 1993; Bruhn et al, 1992). Recently some lower fat products have recorded disappointing sales and a new wave of high flavor, highly indulgent, high fat foods have entered the U.S. market. The closer the food industry can come to providing premium flavor without the dietary penalty, the greater acceptance the product will find in the marketplace (Bruhn et al, 1992).

The relationship between diet and cancer has increased the importance of fruit and vegetable consumption to many consumers. The United States Department of Agriculture, National Cancer Institute, and various health groups have recommended consumption of five or more servings of fruit and vegetables daily. Label health claims will soon be permitted that associate diets low in fat and high in the nutrients found in fruits and vegetables with lower risks for heart disease and cancer. Promotion of these relationships has led to increased knowledge among consumers (Packer, 1993).

Consumer have associated nutrient attributes with specific fruits and vegetables (The Packer, 1993). Produce eaten to increase the fiber content of the diet include apples, 36%, broccoli, 27%, lettuce, 16%, and carrots, 15%. These same items in different order are eaten for cholesterol control: broccoli, 17%, apples, 15%, lettuce, 13%, and carrots, 12%. Oranges and bananas are most frequently mentioned as produce eaten for vitamin/mineral intake with 34% and 33% respectively, followed by broccoli, 27%, and apples and carrots, at 22% each. Produce eaten for calorie control include lettuce, 39%, apples, 28%, carrots, 24%, broccoli 17% and celery, 16%. Broccoli is a clear leader for cancer prevention, 48%, followed by carrots, 23%, and cauliflower, 19%. These associations suggest that some food products have a quality nutritional image. They also point to the central role of salads (lettuce) in contemporary life styles.

Food quality and chemicals: Consumer views of safety

Food safety has long been a consumer consideration. In the United States, 72% indicate food safety is a very important consideration in food selection (Opinion Research, 1993). Quality foods must be safe foods, but consumers have concerns about some manufacturing practices that are used to maintain safety and quality.

At one time concerns about the safety of additives and preservatives were cited by the greatest number of consumers, but today, bacterial contamination and pesticide residues lead concerns (McNutt, 1984; Center for Produce Quality, 1992, Opinion Research, 1993). When potential hazards are specified identified, 79% of consumers believe pesticide residues are a serious hazard, only 2% said it was no hazard; other concerns included antibiotics and hormones in meat, 55%, nitrites in food, 35%, irradiated foods 35%, additives and preservatives 23%, and artificial coloring 19% (Opinion Research, 1993). When consumers volunteered concerns, those related to microbiological contamination were mentioned most frequently. When consumers were questioned about a broad variety of potential safety areas, concerns about microbiological areas and pesticide residues are classified as significant hazards by an equal number of persons (Center for Produce Quality, 1992). Dutch consumers were most concerned about microbiological issues (improper processing) with 50% expressing concern, followed by pesticides, 45%, and preservatives, 34% (Cramwinckel and van Mazijk-Bokslag, 1989).

Supermarkets in the United States have used two market techniques to address pesticide residue concerns; certifying produce "residue free" and marketing "organic" products. In a survey in which tested produce was described as risk free and untested produce as having a specified risk of additional cancers per 50,000 consumers, 42% of consumers preferred tested produce, even though the items cost more (Ecom, 1992). Note this is an extreme situation and does not reflect the scientific perspective on actual risk in the marketplace. It is noteworthy, however, that consumers will pay to reduce risk, those with lower income or with fewer years of formal education were willing to pay more than higher income consumers or those with more formal education.

In response to safety concerns, many markets have added "organic" sections to their retail operations. Currently, there are no national standards or definition for organic, so consumers can not be assured of what they are buying (Frazao, 1992). Processed foods labeled organic could contain from 40% to 100% organically produced ingredients. Additionally consumers appear misinformed about the meaning of organically grown. Organic generally means grown without synthetic chemicals, however consumers have misinterpreted this to mean grown without any pesticides (Jolly et al, 1989). Although organic produce received considerable attention in 1989 after public concern about the growth regulator Alar, interest in organic has declined recently.

An alternative pest control approach which may become a quality indicator in the future is the use of integrated pest management (IPM) techniques to control pests. This is a biological based approach which considers the ecological impact of pest control. IPM has recently been endorsed by scientists and environmental groups as a consequence of a recent multi-year review by the National Academy of Science regarding the effect of pesticide residue on infants and children. Research with consumers demonstrated that the pubic has little knowledge of this approach, yet responds to it favorably (Bruhn et al, 1992). A method to certify foods produced using IPM practices may be proposed in the future.

Environmental impact, another facet of quality.

Consumers are increasingly concerned about the quality of the environment and the impact of their consumption decisions on environmental quality. An important component of concern about pesticide use is the impact of pesticides on the environment. In a study investigation the type of information consumers want to receive about pesticide use, people were equally interested in the environmental impact of pesticide use and personal health affects (Bruhn et al, 1992). Similarly a nationwide study found consumer concerns about the environmental impact of pesticide use increasing from 59% in 1991 to 66% in 1992 (Center for Produce Quality, 1992).

General environmental activism is also high. Two thirds of respondents in a national survey reported they had changed their life style in significant ways to protect the environment, almost half had stopped buying a product because it caused environmental problems, and one third had contributed money or time to an environmental cause (Hoban and Kendall, 1992). Almost sixty percent have sought products with recyclable packaging and 39% say they refused to buy products because they were not recyclable or used unnecessary packaging (Opinion Research, 1993).

Consumers have also modified their purchasing patterns because of company policies and procedures (Opinion Research, 1993). Fifty three percent of Americans report they refused to buy products manufactured by companies whose policies they do not agree with. Concern about ethical treatment of animals has caused 42% of consumers to refuse to buy specific products.

Quality products then must employ production, processing, and packaging methods which are environmentally benign. Packaging should be minimal to protect product quality and consumer health, and should be recyclable.

New technologies: Promises and potential, how people respond

New technologies like food irradiation and genetic engineering can help the food industry realize product characteristics and production methods consumer find desirable. Consumer acceptance however, will require informational programs that address consumer concerns and explain consumer benefits.

Food Irradiation

Food irradiation can enhance the microbiological safety of food, replace less safe post-harvest chemical fumigants, and enhance product shelf life. Food irradiation has received increased attention from public health officials in the United States because the process would have prevented recent deaths in the Pacific Northwest caused by hamburger containing the highly virulent bacteria, E-coli 0157:H7. Critical to application of this technology is the availability of treatment facilities and consumer acceptance.

Irradiation exposes food to high levels of energy from radioactive isotopes like cobalt or from X ray machines. The effect of this treatment varies by the amount of energy used. A high dose sterilizes spices and substitutes for the fumigant, ethylene oxide. Pathogens in raw poultry and meat can be reduced by as much as 99.9% by a lower "pasteurization" dose of radiation. Grain and produce can be disinfested by still lower doses, and natural decay of some fruit and vegetables can be retarded. Irradiation is an effective quarantine treatment for the Mediterranean fruit fly and other pests and can substitute for some uses of methyl bromide.

In all cases, irradiation is a cold treatment. The food maintains a raw, fresh-like character. Irradiation is considered safe by the World Health Organization, the American Medical Association, United States Food and Drug Association, and various health and safety authorities in over 30 countries. The process was recently endorsed by James Mason, Head of the US Public Health Service.

Early consumer attitude research in the United States found consumers were cautious regarding food irradiation. People knew little about the technology and wanted

nutritionists most credible, followed by farmers. University professors and environmental groups are comparable in generating trust. Government groups are lower in believability. (Center for Produce Quality, 1992; Hoban and Kendall, 1992; McNutt, 1985). It is appropriate therefore to seek assistance from those in the scientific and agricultural community to address consumer issues.

Delivering quality: Ethics in labeling

Labeling statements can be used to inform consumers about practices pertaining to quality in food production, processing, and packaging. Manufacturers may use ingredients from certain geographic areas with established reputations for quality. Many places in the world are well known for specialty products, such as French champagne, Russian caviar, Italian pasta, or Belgian chocolate. Manufacturers sometimes utilize consumer awareness of these products to enhance the image of their own line. Sometimes, however, the descriptive terms are not appropriately used. For example, an informal survey of markets in California and Hawaii found several products using geographical place names as part of the product name, with no relationship to the product ingredients or site of manufacturer.

A consumer survey has identified the public's perception of the appropriate use of terms associated with a geographic location. Half of the respondents considered the place of manufacture to have "some influence" on their purchase decision (Hodgson and Bruhn, 1992). Consumers associate specific geographic names with quality, price, and value. Almost three quarters of the consumers expected a relationship between a product and the name containing a geographical location. If a product was manufactured in an area but the ingredients did not come from that area, use of the name was considered inappropriate by 55% of the sample. When the major ingredients came from the area, the label was considered appropriate by 52% of consumers, even if the product was not manufactured at the area. The use of the United States Food and Drug Association approved term "style", meaning with flavorings or in the manner typical of the area, was not clearly understood by consumers. Almost half, 45%, of consumers considered using a geographic name with "style" was not appropriate, 34% considered it appropriate and 15% were uncertain. Fish caught in international waters could appropriately use a geographic name if the name referred to the vessel's home port or the products received significant processing in the geographic area.

Consumers also have perceptions of the appropriate use of a quality ingredient in a product. Consumers expect when a manufacturer uses an ingredient, then all of that particular ingredient should be of the type specified. For example, if the term "Kona Coffee" is used, all the coffee should have come from Kona (Hodgson and Bruhn 1993). If the label states "Kona Coffee Blend," more than 50% of the coffee should come from the area of Kona. When manufacturers proposed to adopt less stringent guidelines, permitting 10% Kona coffee in a Kona Coffee blend, consumer response varied from disappointment to outrage. People believed 10% was too small to justify using a descriptor term. Consumers felt that a product that followed the 10% guidelines would "look cheap." Similarly consumers would expect other ingredients to accurately reflect label declarations.

Conclusion

To meet consumer's demand for quality, manufacturers must recognize and respond to the way quality evolves over time. Quality does pertain to good taste attractive appearance, but it also includes cultural and emotional features, human and environmental safety, and accurate information. New technologies can help the food industry achieve these objectives, however efforts are needed in public education.

References

Anonymous, Microwave food, *Food Engineering*, 59(12), 29, 1987.

Baraldi, D., Technological tests at the pre-industrial level on irradiated potatoes, In *Food Preservation by Irradiation, Proceedings of FAO/IAEA/WHO Symposium*, 1977, As reported in Consumer acceptance of irradiated foods by Michelle Marcotte, Nordion, International, Inc. Ontario, Canada.

Benedict, J. and Steenkamp, E.M., Perceived quality of food products and its relationship to consumer preferences: Theory and measurement, *Journal of Food Quality*, 9, 373-386, 1986.

Bord, R.J. and O'Connor, R.E., Who wants irradiated food? Untangling complex public opinion, *Food Technology*, 43(10), 87-90, 1989.

Brand Group, Irradiated seafood product. A position paper for the seafood industry, Final report, Chicago, IL. 1986.

Bruhn C.M., Cotter, An, Diaz-Knauf, K., Sutherlin, J, West, E., Wightman, N, Williamson, E, and Yaffee, M., Consumer attitudes and market potential for foods using fat substitutes, *Food Technology*, 46(4), 81-86, 1992.

Bruhn, C., Diaz-Knauf, K. Feldman, N., Harwood, J., Ho, G., Ivans, E. Kubin, L., Lamp, C. Marshall, M. Osaki, S. Stanford G., Steinbring, Y., Valdez, I. Williamson, E. and Wunderlich, E., Consumer food safety concerns and interest in pesticide-related information, *Journal of Food Safety*, 12, 253-262, 1992.

Bruhn C.M. and Noel, J.W., Consumer in-store response to irradiated papayas, *Food Technology*, 41(9), 83, 1987.

Bruhn, C., Peterson, S., Phillips, P., and Sakovich, N., Consumer response to information on integrated pest management, *Journal of Food Safety*, 12, 315-326, 1992.

Bruhn, C.M., Schutz, H.G., Consumer awareness and outlook for acceptance of food irradiation, *Food Technology*, 43(7), 93-94, 97, 1989.

Bruhn, C.M., Schutz, H.G. and Sommer, R., Attitude change toward food irradiation among conventional and alternative consumers, *Food Technology*, (40(1), 86-91, 1986a.

Bruhn, C.M., Schutz, H.G. and Sommer, R., Food irradiation and consumer values, *Ecology of Food and Nutrition*, 21, 219, 1987.

Bruhn, C.M., Sommer, R., and Schutz, H.G, Effect of an educational pamphlet and posters on attitude toward food irradiation, *J. Industrial Irradiation Technology*, 4(1),1, 1986b.

Calorie Control Council, 2 out of 3 Americans take aim at dietary fat, Calorie Control Commentary, 13(1), 1,4, 1991.

Calorie Control Council, Europeans see the light, *Calorie Control Commentary*, 14(1), 1-2 1992.

Center for Produce Quality, Fading scares-future concerns: Trends in consumer attitudes toward food safety, Produce Marketing Association, Alexandria, Virginia, 1992.

Corrigan, James, personal communication, 1992.

Cramwinckel, A.B. and van Mazijk-Bokslag, D.M., Dutch consumer attitudes toward food irradiation, *Food Technology*, 43(4), 104, 109, 110, 1989.

Curzio, O.A. and Croci. C.A., Studies of pre-commercial scale irradiation of onions and garlic to control sprouting, Final Report, Research Contract No. 4453/R2/RB, Universidad Nacional del Sur, Bahia Blanca, Argentina, 1990.

Doyle, Mona, The shopper report, Philadelphia, PA, 1989.

Eom, Y.S., Consumers respond to information about pesticide residues, *Food Review*, 15(3), 6-8,10, 1992.

Fiszer, W., Status report on food irradiation in Poland. Submitted to Fifth Meeting of International Consultative Group on Food Irradiation, Vienna, As reported in Consumer acceptance of irradiated foods, by Michelle Marcotte, Nordion, International, Inc., Ontario, Canada, 1988.

Ford, N.J. and Rennie, D.M., Consumer understanding of food irradiation, *Journal of Consumer Studies and Home Economics,* 11, 305-320, 1987.

Frazao, E., Labeling of pesticide residues on produce, *Food Review*, 15(3), 9, 1992.

Hamstra, A.M., Biotechnology in foodstuff, toward a model of consumer acceptance, SWOKA research report no. 105, The Hague, The Netherlands, 1991.

Hoban, T.J. and Kendall, P.A., Consumer attitudes about the use of biotechnology in agriculture and food production, North Carolina State University, July, 1992.

Hodgson, A.S. and Bruhn, C.M., Geographical names on product labels: Consumer attitudes toward their use, *Food Technology*, 46(2), 83-89, 1992.

Hodgson, A.S. and Bruhn, C.M., Consumer attitudes toward the use of geographical product descriptors as a marketing technique for locally grown or manufactured foods, *Journal of Food Quality*, 16:163-174, 1993.

Hollingsworth, P., Convenience is king, Food Technology, 47(8), 28, 1993.

IAEA., Test market of irradiated strawberries in France, *Food Irradiation Newsletter*, 11(2), 45-46, 1987.

IAEA, Compilation of information on market trials (1984-89), *Food Irradiation Newsletter*, 14(1), 1990.

International Consultative Group on Food Irradiation, Paper submitted to ninth meeting of the ICGFI, Orlando, FL., 1992.

Johnson, F.C.S., Knowledge and attitudes of selected home economists toward irradiation in food preservation, *Home Economics Research Journal*, 19(2), 170-183, 1990.

Jolly, D., Schutz, H.G., Diaz-Knauf, K.V., and Johal, J., Organic foods, consumer attitudes and use, *Food Technology*, 43(11), 60-66, 1989.

Lustre, A.O. Ang, L., Dianco, A., Cabalfin, E.F. and Navarro, Q.O., Philippine's experience in marketing irradiated foods, paper presented at ASEAN Workshop on Food Irradiation, Bangkok, Thailand, November, 1985.

Marcotte, M., Irradiated strawberries enter the US market, *Food Technology*, 46(5), 80, 1992.

Martin, M., Bhuiya, A., Karim, A., Rahaman, S., Khatoom, J., Hossain, M. Islam, S., Islam, M., Amin, M. Hossain, A., and Siddiqui, A., Commercialization, storage, and transportation studies of irradiated dried fish and onions, paper presented at Asia Regional Cooperative Project Meeting, November, 1988. As reported in Consumer acceptance of irradiated foods, by Michelle Marcotte, Nordion, International, Inc., Ontario, Canada.

McNutt, K.W., Consumer acceptance of irradiated foods, paper presented to R&DA Irradiated Food Products Committee, Boston, Mass, 1985.

Nouchpramool, K. Charoen, S., Prachasitthisak, Y., Pringsulaka, V., Aduluatham, P., and Bunnak, J., Commercial storage and marketing trials of irradiated onions and garlic, *Food Irradiation Newsletter*, 14(1), 55, 1990 .

O'Donnell, C.D., Maximizing lowfat opportunities, *Prepared Foods*, 162(9), 49-51, 1993.

Opinion Research, Trends 93, consumer attitudes & and the supermarket 1993, Food Marketing Institute, Washington, DC., 1993.

The Packer, Fresh trends, a profile of fresh produce consumers, Lincolnshire, IL, 1992.

The Packer, Fresh trends, a profile of fresh produce consumers, Lincolnshire, IL, 1993.

Pohlman, A.J., Influence of education and food samples on consumer acceptance of food irradiation, Masters of Science Thesis, Purdue University, Indiana, 1993.

Prachasitthisak, Y., Pringsulaka, V., and Chareon, S., Consumer acceptance of irradiated Nham (fermented pork sausage), *Food Irradiation Newsletter*, 13(1), In Loaharanu, P. International trade in irradiated foods: regional status and outlook, *Food Technology*, 43(7), 77-80, 1989.

Pszczola, D.E., Irradiated produce reaches midwest market, *Food Technology*, 46(5), 89, 1992.

Putman, J.J., Food consumption, 1970-90, *Food Review*, 14, 2-12, 1991.

Resurreccion, A.V.A., Galvez, F.C.F., Fletcher, S.M., and Misra, S.K., Consumer attitudes toward irradiated food, results of a new study, paper presented at Institute of Food Technology Annual Meeting, Chicago, IL, 1993.

Schutz, H.G., Bruhn, C.M., and Diaz-Knauf, K.V., Consumer attitudes toward irradiated foods: Effects of labeling and benefits information, *Food Technology*, 43(10), 80-86, 1989.

DISCUSSION

SHEWFELT: Have you seen studies indicating whether consumers will sacrifice quality for less pesticides.

BRUHN: Attitude studies indicate a decrease in the number of consumers willing to buy blemished produce grown with less pesticides. (See CPQ, 1992 in reference). Organic produce has established a place in the market, however it represents less than 1% of produce sales. Consumers are concerned about the environmental impact of pesticides use, however, few want to sacrifice appearance.
A 1988 study in California investigated consumer willingness to buy scarred oranges. When told 50% less pesticide were applied to cosmetically scarred oranges, 63% of consumers said that they would prefer scarred product. This comparison, however is not an accurate description of the production practices in the state, hence the application of these findings to market behavior is questionable.

DEBARDEMAEKER: Is a local product and "small store" more appealing than a shipped product in large store? Does "brand naming" influence the consumer's decision?

BRUHN: Consumers believe a local product could be fresher, they are not willing to pay more for it however, and will not purchase local if it is of inferior appearance. Several studies have found consumers recognize brand names, but do not have a strong preference for branded products; they do not view branded products as superior to non-branded.

SASTRY: Regarding questionnaires listing foods safety concern, is it possible that the inclusion of an item on such a questionnaire (e.g. a harmless, intrinsic food constituent) be sufficient to cause consumers to list it as potentially unsafe?

BRUHN: Absolutely, experimenters can suggest concerns by the way a question is asked. In evaluating consumer surveys, it is important to review the questioning procedure and the exact wording of questions.

TANIGUCHI: There are usually some gaps between the answer of the consumer survey and their behavior. Sometimes, this gap will result in big issue on product development. Do you have any ideas or suggestion to reduce this issue?

BRUHN: Truly this does occur. Careful phrasing of questions and targeting the audience are a key components to preparing a survey that reflects a market behavior. Consumer research should be viewed as a first step that is followed by pilot testing in the market place.

FLAIR-FLOW: An Overview of Dissemination Research in Europe

T.R. Gormley
Teagasc, The National Food Centre,
Dunsinea, Castleknock, Dublin 15, Ireland.

I. SUMMARY

The FLAIR-FLOW EUROPE project of the European Community is an innovative international initiative, based on national networks in 16 countries whose aim is to disseminate R and D results from the FLAIR programme to food factories, health professionals and consumer groups. The project uses 1-page technical documents, workshops, and lectures at conferences and trade shows as its main vehicles for dissemination and also carries out research on the most suitable dissemination routes and on procedures for the quantification of feedback by the end users. To date the project has been highly successful with FLAIR results disseminated widely across Europe.

II. INTRODUCTION

Reaching the end users with results of research and development (R and D) is always a major challenge and the users of food R and D results are no exception. There is often a major imbalance between the amount of money spent (a lot) on carrying out R and D and that spent (very little) on ensuring effective and adequate dissemination of the results to end users. This chapter gives an overview of the ongoing FLAIR-FLOW EUROPE (F-FE) dissemination project and of its progress to date. FLAIR-FLOW EUROPE is a specialised dissemination project with the aims of redressing this imbalance and of helping to convey results from the other 32 international projects in the FLAIR (Food-Linked Agro-Industrial Research) programme, and also selected food-related items from other European Community (EC) programmes, to end users. The targeted end users are small to medium sized food enterprises (SMEs), health professionals, and consumer groups in the 12 EC states and in Austria, Finland, Norway and Sweden. FLAIR-FLOW is a cooperative project of the EC FLAIR and VALUE (Dissemination and exploitation of research results) programmes; the project commenced in 1990 and phase 1 will conclude in 1993. It will be followed in 1994 by a new (phase 2) three year dissemination project with the aim of disseminating results from the food R and D elements of the ongoing Agriculture and Agro-Industry Research Programme (AAIR) of the EC.

FLAIR-FLOW is in response to criticism that results from some EC programmes have been incompletely disseminated, i.e. results arrive at national organisations and institutions but do not filter down to the end users. The goal of FLAIR-FLOW, therefore, is to bridge this gap and to reach food SMEs, health professionals and consumer groups with tailored practical and useful information, written in layman's language, from the FLAIR programme. It should be stressed that FLAIR-FLOW is both a dissemination and a research project. The research relates to studying the effectiveness of an international network system as a dissemination tool and also to carrying out

Table 1
FLAIR Programme: Project Topics[1]

Concerted Actions
- assessment of food quality (125)[2]
- sensory analysis (150)
- predictive modelling (bacteria) (116)
- controlling poultry pathogens (180)
- HACCP & hurdle technology (595)
- toxicology & residues (68)
- food lectins (55)
- micronutrient availability (81)
- resistant starch (103)
- dietary intake, food composition (132)
- dissemination (FLAIR-FLOW) (322)

Shared-Cost Projects
- microwave & Joule heating (46)
- endogenous enzymes (6)
- transgenic food crops (17)
- limited shelf life products (29)
- fresh fruit mixes (35)
- virgin olive oil (44)
- SO_2 and wine quality (41)
- oxidoreductases (3)
- fruit juice quality (81)
- starters for wheat bread (22)
- dehydration technology (17)
- food intolerance (47)
- MAP of meat products (71)
- natural antimicrobials (24)
- 'late blowing' of cheese (25)
- food plant sanitation (81)
- raw milk - cheese safety (17)
- spoilage detection-methods (13)
- nutritious cereal products(44)
- in-pack thermal processing (51)
- functional fibres (45)
- probiotics for nutrition (17)

[1]NOTE: These are only "keyword" headings; more detail in the CEC booklet on the FLAIR programme (2nd edition, January 1992)
[2]number of requests for follow-up information: see section VI sub-section B

quantification of the penetration of the disseminated information and also its usefulness. However, FLAIR-FLOW is not the only dissemination route for FLAIR information as the researchers and technologists carrying out the research are also active in dissemination. It is also important to put the FLAIR programme in context in terms of its magnitude. It embraces circa 600 participants and 33 transnational projects in the areas of food quality, food safety and nutrition/wholesomeness. The budget is 25 million ECU with 8 million for the 11 concerted action projects (EC pays for coordination), 13 million ECU for the 22 shared-cost projects (EC pays up to 50% of the cost of the research) and 4 million ECU for administration. These 32 projects are generating results and these are being disseminated via FLAIR-FLOW in addition to other routes. The topics covered by the 33 projects are given in Table 1. The FLAIR programme is only one in a series[1] of food (and food related) R and D programmes which have been funded and/or administered by the EC over the last 16 years.

III. REACHING THE END USERS

The degree of difficulty in reaching end users with R and D results in user friendly form must never be underestimated. However, reaching the end user is only one part of the equation; the other part is the ability of food SMEs to understand and implement the results. This problem has been clearly recognised by FLAIR-FLOW personnel and potential remedial measures include setting up food SME focus groups or clusters and linking food SMEs to the newly established VALUE Relay Centres of the EC. The concept of an Integrated Networks Dissemination system has been developed and used by

FLAIR-FLOW in 1993 and is focusing on the synergy between four international networks. These are the FLAIR-FLOW national networks (see section IV below), the network of the 600 FLAIR participants, the network of the VALUE Relay Centres, and a collective fourth network embracing the BEUC (Bureau Européen des Unions de Consommateurs), the CIAA (Confédération des Industries Agro-Alimentaires de la CEC) and EUSJA (European Union of Science Journalists Associations).

The EC VALUE Relay Service and its component 29 VALUE Relay Centres is a recent initiative of the VALUE II programme for technology collaboration across Europe under two main headings:
- to enable organisations across Europe to benefit from the results of EC research and technological development (RTD) programmes
- to facilitate participation in EC RTD programmes.

For more information on the VALUE Relay Service contact J. Hernandez-Ros, VALUE Relay Service, Commission of the European Communities, DG XIII/D3, Bâtiment Jean Monnet, L-2920 Luxembourg.

IV. STRUCTURE OF FLAIR-FLOW

National networks are the kernel of the FLAIR-FLOW project and are operational in the 16 participating states; each national network has a leader and 15-25 members. Obviously the careful selection of both network leaders and network members is of key importance. The national networks have membership which is representative of a wide range of organisations (including persons from trade journals, the media, etc.) who are already disseminating information to the food industry or have the potential to do so. In many cases information from FLAIR can be included at little extra cost/difficulty with other ongoing disseminations, e.g. in newsletters. In some of the smaller countries a single person is in frequent contact with upwards of 80% of food SMEs; obviously the participation of such a person in a FLAIR-FLOW network results in extremely effective and efficient dissemination of FLAIR material. The collective disseminating power of the 16 FLAIR-FLOW networks via circa 320 active network members, and their downstream activities, is immense and runs into millions of potential contacts in Europe. Much of this potential has already been realised.

The 16 national network leaders together with the project leader, co-opted expertise (e.g. Bureau Européen des Unions de Consommateurs) and officials of the EC Commission comprise an international network which discusses and steers FLAIR-FLOW project policy, strategy and activities; the international network meets twice annually. The project leader and a sub-set of the above personnel also form a seven member project management group.

Each FLAIR-FLOW network is a dynamic entity which is in a continual state of change and improvement in accordance with requirements. There have been changes in network personnel and leadership in a number of countries over the duration of the project. Indications are that about 20% of network members are excellent performers with 20% of members non-performers. The remaining 60% are intermediate. Attendance by network members at national network meetings is about 46%. The affiliation of the 322 network members in the 16 countries is given in Table 2 with those from institutes/research associations and from academia predominating. Network members from food inspectorates/control are more active than expected (based on the whole international sample of network members) followed by those from consumer organisations, journals, and the media. Conversely, network members whose affiliations are academia or industry associations are less active (relatively speaking), and in the case

Table 2
Affiliation of 322 FLAIR-FLOW network members: collated data for the 16 networks

Affiliation	Percentage of network members	Affiliation	Percentage of network members
Institutes/Research Associations	19	Trade organisations	3
Academia	18	Food inspectorates/control	3
Assocs/federations/agencies	12	Confeds of industry	3
Journals/media/press	10	Health/nutrition orgs	2
Government depts	8	Research councils	1
Food industry (SMEs)	7	Quality control orgs	1
Consumer organisations	6	Others	7

Table 3
List of FLAIR-FLOW National Network Leaders

Country	Leader	Organisation	Fax
Austria	W. Pfannhauser	Technical University, Graz	+316-810599
Belgium	M.C. Donnet	Féd. de l'Ind. Agric. et Aliment	+32-2-7339426
Denmark	F. Holm	Food Group Denmark	+45-86-201222
Finland	K. Poutanen	VTT Food Research Laboratory	+358-0-4552103
France	J. Quillien	ADRIA, Quimper	+33-98-907328
Germany	W. Spiess	Federal Res. Centre for Nutrition	+49-7247-22820
Greece	Y. Totsiou	SPEED Ltd., Athens	+301-8225755
Ireland	G. Downey	The National Food Centre	+353-1-383684
Italy	C. Lerici	Universita di Udine	+39-432-501637
Luxembourg	G. Schlesser	LUXINNOVATION	+352-438326
Netherlands	H. van Oosten	Agricultural Univ. Wageningen	+31-8370-83342
Norway	H. Russwurm Jr.	Norwegian Food Res. Inst.	+47-9-970333
Portugal	T. Almeida	College of Biotech., Porto	+351-2-490351
Spain	J. Espinosa	Instituto del Frio, Madrid	+341-5493627
Sweden	B. Hedlund	The Swedish Food Institute	+46-31-832933
United Kingdom	S. Emmett	Leatherhead Food RA	+44-372-386228

of the former very much less active, than expected. Personnel in the food control/inspectorate area include environmental health officers, and various inspectors including veterinarians. Twelve network leaders operate an inner circle or core network, based on high performers, in addition to their full scale network. In the larger countries there is also a need for network members to be spread geographically in order to ensure the progress of FLAIR-FLOW in the regions. Currently the networks in Belgium, France, Italy and The Netherlands have network members in every region while those of Denmark, Luxembourg and Finland have network members in only one region. The other nine countries show an intermediate position with network members in most or some regions. The current list of FLAIR-FLOW national network leaders is given in Table 3.

V. INFORMATION FLOW

A. ONE PAGE DOCUMENTS

It was agreed at the outset that results and information from FLAIR for dissemination should be 'tailored' into 1-page documents, in layman's language, by the FLAIR-FLOW

Table 4
List of FLAIR-FLOW one-page and related documents
(complete to end September 1993)

1/91:	Comprehensive description of F-FE
2/91:	General description of FLAIR and F-FE
5/91:	Layman's description of FLAIR concerted actions
13/91:	Consumer attitudes to food quality
14/91:	Health aspects of food biotechnology
15/91:	Transporting chilled foods by air
17/91:	Controlling pathogens in poultry
18/91:	Sensors and sensor techniques
19/91:	Probiotics - fact or fiction?
20/91:	Precooked chilled foods in catering
21/91:	Fermented vegetables
22/91:	Pasta starch 'is best'
23/91:	The TTT-PPP concept for chilled foods
24/91:	Preparation of cheese analogues
25/91:	The frozen dough process in bread production
26/91:	Starter culture development
27/91:	Modelling for shelf life and safety
28/91:	Dough thawing by microwaves and baking by infrared
29/91:	Measuring minerals in foods and tissue
30/91:	Measuring vitamins in blood and tissue
31/91:	Resistant starch - the state of the art
32/91:	FLAIR/ECLAIR/FOREST technology days
33/91:	AAIR - what is it?
34/91:	FLAIR-FLOW EUROPE - the first year
35/91:	Controlling Salmonella in poultry
36/91 :	The Eurofoods - Enfant project
37/91 :	Predicting the growth and survival of bacteria in foods
38/92 :	FLAIR-FLOW technical documents in 1991 (pamphlet)
38A/92 :	FLAIR-FLOW technical documents in 1991 (booklet)
39/92 :	Sensing food quality
40/92 :	Testing for veterinary drugs in foods
41/92 :	FLAIR-FLOW EUROPE - the first year (6 page glossy)
42/92 :	Have you heard about lectins?
43/92 :	Rapid instrumental quality testing of foods
44/92 :	Food safety/quality and hurdle technology/HACCP
45/92 :	Hurdle technology
46/92 :	Hazard analysis critical control point
47/92 :	Sanitation of food processing equipment
48/92 :	Modifying the atmosphere in food packs
49/92 :	FLAIR newsletters
50/92 :	Dietary changes vs bioavailability
51/92 :	Enriching our food with dietary fibre
52/92 :	Toxicology : <u>in vitro</u> studies
53/92 :	Heat processing of foods containing particles
54/92 :	Coding our foods
55/92 :	Starch digestibility
56/92 :	Managing food composition data
57/92 :	Food safety : HACCP user guide
58/92 :	Quality of virgin olive oil
59/92 :	Treatment and utilization of whey from Greek cheese factories
60/92 :	Low-calorie low-fat cereal products
61/92 :	AAIR - The second call
62/92 :	Safe, high quality fruit products
63/92 :	Food intolerance : finding the answers
64/92 :	Sourcing information on HACCP

Table 4 (continued)

65/92 :   Optimising sulphur dioxide levels in wine
66/92 :   Assessing food intake from meals consumed outside the home : can you help?
67/92 :   Computer-aided process design procedures
68/92 :   AAIR - New food R and D projects
69/92 :   Authenticity and quality of fruit juices
70/92 :   Prevention of "late blowing" of cheese
71/92 :   Have you heard about FLAIR and FLAIR-FLOW? **(glossy)**
72/92 :   Food factories - ALERT!
73/92 :   Health professionals - ALERT!
74/92 :   Consumer groups - ALERT!
75/93 :   Sous vide - the state of the art
76/93 :   Poultry and poultry meat pathogens - prevention and control
77/93 :   Food quality - SENSory aspects
78/93 :   FLAIR-FLOW technical documents in 1992 **(booklet)**
79/93 :   FLAIR-FLOW EUROPE - the second year **(glossy)**
80/93 :   Resistant starch - the latest
81/93 :   Quality of low fat sausages
82/93 :   Food product safety through predictive modelling
83/93 :   Consumer acceptance of yoghurt
84/93 :   Quality and safety of raw milk cheeses
85/93 :   The new VALUE Relay Service - getting the information to you
86/93 :   Safety of transgenic food crops
87/93 :   Probiotics in fermented milk
88/93 :   Bacterial status of poultry meat in modified atmosphere packs
89/93 :   Natural antioxidants
90/93 :   Improved quality in vegetables and herbs
91/93 :   Novel heat treatments for fruit products
92/93 :   High quality is the key
93/93 :   European nutrition information management system (EuroNIMS)
94/93 :   Lactic acid bacteria in the food industry
95/93 :   Sourcing resistant starch
96/93 :   Ready-to-use fresh fruit salads
97/93 :   Fat content of meat and meat products in Germany
98/93 :   New near infra red technique for food analysis
99/93 :   Food balance sheets in the Nordic countries
100/93 :   Social and structural effects of FLAIR
101/93:   Natural antimicrobial systems
102/93:   Biosensors for detecting bacteria
103/93:   The attachment of bacteria to the gut

project leader/management team. Ninety one (Table 4) of these documents have been prepared to date and have been sent (in the language of each country) through the networks. Some of the documents are also sent to network leaders on disk and a number of network leaders have the documents on-line in their organisations. Each document has the contact address and phone/fax numbers for follow-up. In this way the person requesting the information can be put in direct contact with the individual researchers. The sequence of 1-page document distribution is from project leader to network leaders to network members, to SMEs, health professionals and consumer groups. Frequently, the dissemination is also via other intermediate persons and/or trade journals, scientific journals, the media, etc. The network members send the documents through their existing dissemination channels (e.g. monthly newsletters) or in some cases through new routes. The frequency of 1-page documents is three per month with a number being aimed at the consumer, i.e. 'consumer angle' documents. The 1-page documents for 1991 and 1992 have been collated in two booklets (F-FE 38A/92 and 78/93) and a

Table 5
Classification (percentage basis) of the 4535 direct-mail
addresses of the 16 FLAIR-FLOW networks

Small to medium sized food enterprises	81
Health professionals	2
Consumer groups	1
Academia and institutes	5
Journals and media persons	4
Government departments	2
Other	5

combined version embracing the three years (1991-1993) will be published at the end of 1993. The network leaders have a combined total of circa 4500 direct mail addresses for the 1-page documents in addition to those sent to network members. Almost half of the total is in Sweden (2146), followed by Belgium (1322), Holland (214), Germany (178) and then the remaining 12 countries (675). A classification (percentage basis) of the direct mail addresses is given in Table 5 with over 80% of them relating to SMEs. These data show that direct mail is a significant part of the overall FLAIR-FLOW dissemination process for the 1-page documents.

The 1-page documents represent Phase 1 information in that they may be the 'first wave' of information to arrive at the end user. However, follow-up information is equally important i.e. where end users request more information. As FLAIR-FLOW is only an alerting mechanism, most of the responsibility for follow-up information lies with the coordinators of the other 32 FLAIR projects. Arising from the 1-page documents over 1000 technical articles (see Table 7) on FLAIR results have been published (FLAIR-FLOW Phase 2 publications) in trade and other journals Europe-wide and have been collated in two volumes (F-FE 100A/93 and 100B/93). This represents a very high level of dissemination as confirmed in a pilot study by the UK network leader as follows: the multiplier effect of 15 journals x number of issues annually x circulation list x a readership of 1.7 persons (a used statistic) per copy amounts to a potential readership of 636,000 people for the UK alone; this assumes a FLAIR item in every issue of each journal. The 1-page documents are also being abstracted monthly in Food Science and Technology Abstracts.

B. FLAIR-FLOW WORKSHOPS

FLAIR-FLOW also operates by initiating national workshops on FLAIR topics. The workshops involve bringing together national personnel involved in FLAIR projects with representative SMEs, consumer groups and other end users. Thirty such workshops have taken place to date with a predicted total of 50 for the 16 countries before the end of 1993. Some of the workshops were INCLUSION workshops and are seen as very powerful dissemination tools. The concept is to include FLAIR workshops as components of food conferences/meetings/trade shows being organised by other parties (e.g. a FLAIR half day as part of a 2-day conference of environmental health officers). The FLAIR session is chaired by the national network leader and the speakers are national FLAIR participants and/or coordinators/contractors from the FLAIR programme. A list of active participants in each of the other 32 FLAIR projects has been circulated to FLAIR-FLOW network leaders. Collectively the participants constitute a network of FLAIR researchers and are a major national and international resource as disseminators of FLAIR results via lectures and participation in FLAIR-FLOW and FLAIR workshops. Special-initiative FLAIR-FLOW workshops, aimed at attracting large numbers of food SMEs, are planned in the autumn of 1993 in Spain, Greece, Portugal and Ireland. These are seen as major information transfer events and are likely to be linked to other SME activities thus attracting a large audience.

C. OTHER ROUTES

Virtually all national network leaders and also the project leader, have given lectures/posters on FLAIR and on FLAIR-FLOW at conferences, food fairs and trade shows. These are seen as key routes for FLAIR information and many more presentations are envisaged. The newly established Information Relay Centres of the EC, mentioned already above, are seen as key dissemination routes for FLAIR results and also potentially as providers of backup expertise to aid implementation of R and D results at end-user level.

Apart from a small number of interviews, radio and television have not been used by network leaders for creating an awareness of FLAIR or FLAIR results. However, this situation may change in the future as a greater attempt is made to disseminate suitable FLAIR results to consumers (see section VI, subsection B). Food equipment and ingredient suppliers have not been used extensively to date to disseminate FLAIR results despite the fact that SMEs, and especially very small enterprises, get most of their technical information from these sources. There appears to have been a reluctance by network leaders to involve food equipment and ingredient suppliers in the dissemination process but attitudes are changing and they may be used more frequently on a selective basis in the future.

The inclusion of FLAIR information in technical databases containing food R and D information has not been actively pursued to date as it is felt that many SMEs, and especially very small enterprises (VSEs), would not have access to the information. However, SMEs at the larger end of the size range would have access and so future policy will be to include the information in any suitable databases that will accept it free or at a nominal charge. To date the 1-page information is in the CORDIS database of the EC and also in databases in the organisations of some of the network leaders.

VI. QUANTIFICATION AND FEEDBACK

The FLAIR-FLOW networks are vehicles for a two-way flow of information and so it is important to assess feedback up through the network system. Each network leader is assessing feedback via the number of queries for further information and also from the phase 2 publications arising from the 1-page technical articles. Quantification of the feedback is difficult and only a fraction of it is being documented for at least two reasons. Firstly, network leaders may not be aware of a request for follow-up information if it is made to a network member or directly to the researchers in the individual FLAIR projects. Secondly, it is difficult for network leaders to remember to log all queries especially if they are made by phone.

A. COUNTRY STATISTICS

The network leaders were asked to quantify the numbers of food SMEs, consumer groups, and other groupings of relevance to the FLAIR-FLOW project. The cumulative totals for all 16 countries were as follows: food SMEs ($>$ 130,000), consumer groups (126), food databases (63), food trade journals (320), food newsletters (187) and food related TV/radio programmes (55). It is important to stress that these are (at best) estimates. The figure for food SMEs is probably a large underestimate as it is very difficult to define what is and what is not a food SME; for example how is an in-store (supermarket) bakery classified? Suffice to say there is a vast number of food SMEs in Europe and collectively they have an immense sociological impact. For this reason it is imperative that network leaders search out as many of these as possible and ensure that they are receiving the FLAIR-FLOW information. While reaching food SMEs with information is a major challenge, helping them to use it and apply it is even a greater one. This will be tackled in the new FLAIR-FLOW II project by setting up focus groups of food SMEs in the 16 countries.

B. ENQUIRIES FOR FOLLOW-UP INFORMATION (Cumulative to June 30, 1993)

Enquiries logged from SMEs numbered 1476 and from consumer groups 117; these were made by 817 SMEs and 73 consumer groups respectively. The largest number of queries came from the UK (322), France (240), Sweden (168) and Austria (166) with the least from Portugal (3) and Norway 11. Enquiries from other sources amounted to 2134 with the breakdown shown in Table 6.

Table 6
Number[1] of enquiries for follow-up information from non SME and non consumer sources (cumulative to 30 June, 1993)

Enquiries from :	Number
National network members	330
Confederations of industry	85
Institutes	353
Universities/Polytechs	343
Marketing organisations	49
Industry support organisations	215
Trade journals	127
Newspapers/news agencies	99
Radio/TV	20
Government departments	170
Medical/nutritionist	69
Other	274
Total	**2134**

[1]accumulated over countries

These figures are likely to be large underestimates of the real situation as referred to already above. The response to, and the participation in FLAIR-FLOW by consumer groups has been disappointing, i.e. minimal. This is surprising in view of the vocal stance usually taken by consumer groups to food and related issues. Enquiries for follow-up information on FLAIR projects reflect the fulfilment of the real goal of FLAIR-FLOW, i.e. to bridge the gap between researchers and the end-users of research and to bring them together on a one to one basis.

The enquiries for more information have been broken down on a FLAIR project by project basis (see numbers in brackets against each project topic in Table 1). The data show that the HACCP/hurdle technology project is the most requested at 595 queries followed by FLAIR-FLOW itself (322), and poultry pathogens (180). The most requested shared-cost projects were those on fruit juice quality (81), food plant sanitation (81) and the modified atmosphere packaging of meat products (71).

Table 7
Number and accuracy of phase 2 FLAIR-FLOW publications

Articles in:	Total No.	Accuracy	
		No. accurate	No. inaccurate
Newsletters	269	267	2
Trade journals	373	368	5
Scientific journals	254	254	0
Consumer journals	38	38	0
Newspapers	64	28	36
Membership letters	6	6	0
Bulletins	16	16	0
Total	1020	977	43

C. FLAIR-FLOW PHASE 2 PUBLICATIONS

The FLAIR-FLOW one page documents have been reproduced extensively across Europe as 1020 technical articles (Table 7) in trade and scientific journals, and in other media. Virtually all the 1-page documents were reproduced accurately (often in shortened form) with the exception of 36 articles (on the same topic) in Danish newspapers; these arose from an inaccurate/distorted article written by one journalist which 'snowballed'. This situation highlighted a fear often expressed by the national network leaders of the wisdom of feeding FLAIR-FLOW results to newspapers and or radio/TV. The extent of the FLAIR-FLOW phase 2 publications is very satisfying and represents a major dissemination effort by any standard as discussed in section V, subsection A above.

VII. CONCLUSIONS

- the FLAIR-FLOW project is an innovative, international initiative based on national networks in 16 countries whose aim is to disseminate R and D results from the FLAIR programme to the end users

- FLAIR-FLOW uses one-page technical documents on FLAIR results; phase 2 publications in trade and other journals based on the 1-page documents; workshops on FLAIR results, and lectures and presentations at conferences/trade shows, etc., as vehicles for dissemination

- the success rate of FLAIR-FLOW to date is excellent with FLAIR results being disseminated widely throughout Europe

- research is continuing on new routes and procedures for dissemination and also for logging quantification and feedback

- FLAIR-FLOW EUROPE I finishes in December 1993 but will be followed by a new three year (1994-96) dissemination project, FLAIR-FLOW EUROPE II, under the EC AAIR programme.

VIII. ACKNOWLEDGEMENTS

Sincere thanks is due to the national network leaders, and the project management team, for their unstinting work in bringing national networks on-stream, for operating them so effectively, and for their overall support for the FLAIR-FLOW EUROPE project. The financial support from the FLAIR and VALUE programmes is gratefully acknowledged as is the support and advice of EC Commission staff.

IX. REFERENCE

1. Gormley, R., FLAIR-FLOW EUROPE: a dissemination route to the food industry and consumers, *Trends in Food Science and Technology*, 3(5), 103, 1992.

DISCUSSION

ZOTTOLA; Any hard evidence that the SMEs utilize the information received from FLAIR project in their every day processing?

GORMLEY: Yes, a considerable number of introductions have been achieved via FLAIR-FLOW where researchers and end-users have met on a one-to-one basis for the transfer of information and technology.

BRUHN: I commend the activities of FLAIR-FLOW. Does the one-page information sheets include a section that describes the importance of the project to the public?

GORMLEY: Yes! One in every four of the FLAIR-FLOW technical documents is called a "consumer angle" document. These are specially written in simple language to inform the public of the benefits (to them) of the research results in question.

GORRIS: In an EC-Project, by contract there is some room left for holding back findings that may be patentable. for instance, for a participant's industrial partner. Does this not happen in case of FLAIR-FLOW results?

GORMLEY: Some commercially sensitive results from a few of the FLAIR shared-cost projects are confidential and will not be disseminated via FLAIR-FLOW. However, these are the exception rather than the rule and have little overall impact on FLAIR-FLOW dissemination.

SASTRY: Is there any measure of the extent of damage caused by inaccuracies in the media?

GORMLEY: Not really. The one media "incident" experienced in FLAIR-FLOW peaked and subsided over a relatively short time span which suggests that it did not cause any lasting damage.

Index

INDEX

A

Accelerated testing, 194
Acetaldehyde, 175
Acetic acid treatment, 233–238, 428
Acid treatment, 233–238, 267, 428
Activation energy
   thermal death time constant and, 355
   time-temperature indicators, 322, 323, 393
Additives, consumer attitudes, 496
Aeromonas spp., 425
   processing factors affecting growth, 203
Aeromonas hydrophila, 52, 58, 64
   meat under MAP, 418
   pre-incubation temperature and, 478
Agitation, osmotic dehydration and, 79–80
Airline food, 300
Alcohols, as spoilage indicators, 429
Amino acids, 174
Amylases, 6–7, 326, 399
Anthocyanins, 172
Antifog films, 455
Antimicrobial hurdles, 43, 52
Appearance, 172–173, 188, See also Color
   temperature and gas composition effects, 475, 484–490
Apple
   high pressure treatment, 9
   osmotic dehydration, 78, 88–104
   pulsed vacuum osmotic dehydration, 110
Arabinose, 470
Archimedes number, 154
Area fraction, 25
Aroma, 174, 179, See also Off-odors
Aromatic compounds, 174
Ascorbic acid
   blanching-related losses, 267
   measurement, 179
   meat preservation and, 44, 52
   storage and, 177
Ascorbic acid oxidase, 136
Aseptic processing, 268
   fluid fraction quality, 353, 359–360, 362, 364–366
   microbiological TTIs and, 325
   process setting, 198
   residence time distribution, 153–165
Aspergillus oryzae, 7
Avocado, 462–463
A_w, See Water activity

B

Bacillus amyloliquefaciens, 399–401
Bacillus cereus, 58, 63, 64, 482
Bacillus coagulans, 324
Bacillus licheniformis, 326, 399–401
Bacillus stearothermophilus, 10, 324

Bacillus subtilis, 7–8, 10
Bacteria, See Pathogens; Spoilage bacteria; specific types
Bacterial bioluminescence, See Bioluminescence
Bacterial growth, See Microbial growth; Predictive microbiology
Bacterial recognition database, 200
Bacteriocins, 65–66, 72
Bahia-Blanca model, 272
Banana
   microwave-freezing process, 138–148, 151
   minimal treatment combinations, 460–461, 465
   ripening, 142
Beans, 369–385
Bean sprouts, 63–64
Beef, See Ground beef; Meat products
Beet, ohmic heating, 18–19
Betaine, 59–60
Betalains, 172
Bigelow model, 355
Biological Indicator Unit (BIU), 324
Biological time-temperature integrators, 324–326
Bioluminescence, 213–223
Biopreservation, 57, 65–66, 72
Biotechnology
   bioluminescence gene, 213
   consumer attitudes, 498–499
Biot number, 287, 338, 343, 348, 351
Bitterness, 176
Blanching, 87
   enzyme indicators, 136–137
   high pressure systems, 9
   in situ evaluation, 317
   microwave applications, 135–148
   transport phenomena, 265–267
Blueberries, respiration rate, 411
Boiling, ohmic heating systems, 23
Brand names, 504
Broccoli, respiration rate, 442
Brochothrix thermosphacta, 424, 429
Browning, 142, 145–147, 172, 173, 265

C

Cabbage, respiration rate, 442
Calcium chloride treatment, 467–470
Cancer, consumer concerns, 496
Candying effect, 74
Canning, transport phenomena, 268–270
Capillarity, 263
Capillary pressure, vacuum osmotic dehydration, 107–108
Carbon dioxide, 10–12, 475, 477, 482–484
   endive quality and, 475
   endogenous microflora and, 65
   gluconate production and, 428–429
   internal package dynamics, 446
   meat spoilage and, 419–424, 429

2 Symptoms and Ailments

Many ailments can cause discomfort or even distress. When sickness strikes you or a family member, refer to this part of the book for advice on the natural remedies you can use to help alleviate the condition or symptom. Choose the treatments that fit your case best and that you feel most comfortable with. Also use this section to learn about steps you can take to prevent disease. Remember that while some minor problems can be treated at home, you should always seek your doctor's opinion for serious symptoms or when self-care isn't working.

Abdominal Pain

Most of us are familiar with a stomach ache that results from anxiety or from eating too fast or too much. Natural remedies and lifestyle changes can ease discomfort in many such cases. More severe or persistent abdominal pain, however, requires medical attention, since it can be an indicator of a serious underlying condition.

Abdominal pain is frequently related to a problem in the digestive system. There may be infection (as with gastroenteritis or a peptic ulcer), inflammation (as with ulcerative colitis, diverticular disease, or appendicitis), or an obstruction (as with constipation or a stuck gallstone). An imbalance in the intestinal flora—the bacteria and other organisms normally present in the intestines—may occur, often as a result of treatment with antibiotics or steroids. In some cases pain may result from the nerves supplying the intestines being oversensitive to stress, or to distension caused by wind (as with irritable bowel syndrome). Food allergies and other sensitivities can also involve irritation of the digestive tract.

Prevention

Once your doctor has ruled out any serious underlying cause of your pain, you may be able to avoid a recurrence by making changes in your diet and lifestyle. To prevent digestive problems in particular, take the following steps:

- Eat a wholefood, high-fibre diet, avoiding rich, fatty, and spicy foods.
- Have regular meals, and take the time to eat in a relaxed way. If you are in the habit of eating too quickly, slow down, chew thoroughly, and avoid gulping air as you swallow. Also limit the amount that you drink while eating.
- Try to identify any food sensitivity, getting professional help, if necessary. Some people experience abdominal pain after eating certain foods or additives; the flavour enhancer monosodium glutamate (MSG) is a common cause of this problem.
- Drink enough water-based fluids to relieve thirst, prevent constipation, and produce pale-coloured urine. This helps improve digestive efficiency. Too much alcohol, tea, coffee, and smoking can interfere with digestion and can make many types of abdominal pain worse.

Nondigestive causes

Abdominal pain can sometimes occur as a result of the following problems, which are unrelated to the digestive organs:

- Urinary-tract infection.
- Strained muscles in the abdominal wall.
- Kidney stones.
- Hernia.
- Gynaecological disorders, such as dysmenorrhoea (painful periods), pelvic inflammatory disease, fibroids, and endometriosis (in which cells of the type that line the uterus develop in the abdomen and cause painful cysts).
- Pressure or irritation from a disorder outside the digestive tract (as with certain tumours, pneumonia, and swollen lymph nodes surrounding the bowel).

■ Do what you can to reduce unnecessary stress in your life. Use effective stress-management strategies (see "Treatment"), and be sure to get enough sleep and exercise. Make mealtimes as leisurely as possible.

Treatment

Stress management: When you are stressed, blood levels of stimulating hormones, such as adrenaline, increase and make the muscles in your stomach and intestinal walls tense, leading to the sensation of "butterflies in the stomach" or actual pain. Stress-relieving strategies include regular exercise, yoga, and relaxation classes and tapes. If these don't work, consider taking a stress-management course or seeing a therapist to help you work out better ways to cope with unavoidable stress.

Heat and cold: Abdominal cramps, indigestion, and painfully strained abdominal wall muscles may be eased with a safely wrapped hot-water

Relaxation through yoga

A daily half-hour yoga session is believed to balance the body's flow of energy, helping to reduce the harmful effects of feeling stressed, such as the muscle tension that is associated with many digestive disorders.

You can join a yoga class or buy a book or video that will teach you some simple yoga techniques. The yoga-based relaxation exercise described below can help dissipate tension and relieve related abdominal pain.

1 Lie on your back with back and shoulders relaxed, your legs comfortably apart and your arms away from your body and with the palms upward. If this is uncomfortable for your lower back, bend your knees.

2 Close your eyes and focus on your bodily sensations. Be aware of the contact your body makes with the ground and of your abdomen rising and falling as you breathe slowly and easily.

3 Try to allow extraneous or unwanted thoughts to pass through your mind quickly without attracting your attention, so your awareness is centred on the movement of air in and out of your body.

4 Count your breaths, noting the sensation as the air enters your lungs and as it leaves. When you inhale, think of the air as incoming energy, making you feel light. Then, when you exhale, feel your abdomen sink back down. Continue this exercise for five minutes.

bottle or a hot compress (see p. 295) over the painful area. You may want to put a thick, folded towel over the bottle or compress to hold the heat in longer.

Another method is to leave the bottle or compress in position for three minutes, then substitute a bottle filled with cool water, or a cold compress, for one minute. Continue alternating hot and cold applications in this way for about 20 minutes, ending with a hot one.

Aromatherapy: Add two drops of peppermint oil and two drops each of bitter orange and caraway oils to a teaspoon of sweet almond oil, olive oil, or other cold-pressed vegetable oil, and smoothe this over the painful area of your abdomen (see box, below). If you prefer, you can

Healing oils
A massage using essential oils of bitter orange, peppermint, and caraway can soothe abdominal pain caused by indigestion or wind.

add two drops each of bitter orange and caraway oils to a pint of hot water and inhale the vapour. Do not use caraway oil if you are or might be pregnant. Avoid sunshine on skin treated with bitter orange oil.

Abdominal massage

A circular abdominal massage can sometimes relieve pain. You can do it yourself, but it's easier for a partner or friend to do it for you. With this massage, try using some of the essential oils suggested under "Aromatherapy".

1 Lie on your back with your partner kneeling beside your hips.

2 Your partner should then put one hand gently on your abdomen and stroke firmly, smoothly, and slowly, with a clockwise circling movement, around your abdomen. The circle should begin just inside one hipbone, move up the side of your abdomen, go across the lower border of your rib cage, and then back down your other side to the other hipbone.

3 Continue the massage for several minutes or as long as is comfortable.

Herbal remedies: Herbs have been used for centuries to ease abdominal pain. It's worth trying several to discover which works best for you.

■ Chamomile is especially good for pain from tension or sluggish digestion, since it may relieve mental stress, help the intestinal muscles relax, and encourage food residues to pass through the intestines. It can also soothe inflammation and promote healing. Drink two or three cups of chamomile tea a day.

■ Parsley, eaten raw, may reduce pain from gas and indigestion. Parsley-seed tea relaxes

the intestinal muscles and mitigates cramping.

- Peppermint tea or capsules may alleviate intestinal cramps; however, be careful not to take too much, as this herb can irritate the lining of the stomach.

Caution: For safety concerns, see pp. 34–37.

Homeopathy: Several homeopathic remedies may help abdominal pain caused by digestive problems. Depending on your symptoms, take:

- Nux vomica: if you have eaten or drunk too much and you feel queasy—especially if the meal was rich and you drank a lot of alcohol.
- Arsenicum: if you have a burning sensation in your stomach and you feel restless and chilly.

A remedy from ancient times
Cinnamon tea (made by simmering the bark in boiling water) has maintained its age-old reputation as an effective treatment for abdominal pain resulting from wind, diarrhoea, or nausea.

- Pulsatilla: if you've eaten too much rich food and it leaves an aftertaste, and you feel better in the fresh air.
- Carbo vegetabilis: if you are bloated but feel better after belching or passing wind, and you feel chilly but nevertheless prefer to be in the open air.

Acupressure

Apply thumb pressure on the point (St 36) four finger-widths down from the lower edge of the kneecap and one finger-width from the crest of your shinbone on the outside of your leg. Make firm, circling movements with the thumb for about two minutes on each leg. This point is said to regulate the functioning of the stomach and spleen.

When to get medical help

- Pain doesn't ease in two to three days.
- Pain recurs within a month.
- You have other symptoms, such as a fever, severe diarrhoea, vomiting, menstrual or urinary problems, or general malaise.
- You have red or black blood in your stools.
- The usual frequency of your bowel movements changes with no obvious cause, such as a change in diet.
- You lose weight for no apparent reason.

Get help right away if:

- Pain is severe or continues to worsen.
- You vomit and see blood or what looks like coffee grounds in the vomit.

See also:
Constipation, Diarrhoea, Diverticular Disease, Flatulence, Gallbladder Problems, Indigestion, Irritable Bowel Syndrome, Nausea and Vomiting

Abscesses and Boils

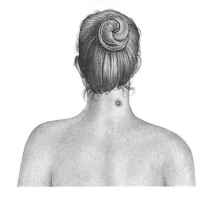

Lumps containing pus, abscesses and boils are generally somewhat painful and tender. They are usually caused by bacterial infection. These disorders, especially if they occur repeatedly, may be a sign that your immune system is weak.

An abscess can develop in any organ and in soft tissue beneath the skin anywhere on the body. Common sites include the breasts, the gums, the armpits, and the groin. Boils develop in the skin, usually around a hair follicle, and common sites include the back of the neck, the armpits, and the groin. A carbuncle is a large boil or a cluster of boils. Carbuncles occur less frequently than regular boils, often on the neck or the buttocks.

Boils and carbuncles are readily visible, whereas an abscess is usually invisible or apparent only as a tender swelling. Most boils subside or come to a head and discharge through the skin. An abscess may subside, grow, burst inside, become a sac of uninfected pus, or discharge via a long track through the skin.

Prevention
Strengthen your body's resistance to infection by eating fewer foods containing sugar, white flour, and saturated fats, and more fresh vegetables and fruit. If you feel run-down, make sure that you are getting enough rest, fresh air, and exercise, and consider having a medical checkup.

Treatment
General measures
- Don't burst a boil yourself; there is a risk that this might spread the infection.
- For boils that have not burst, wring out a clean piece of cloth in warm water and place it over the affected area until the cloth cools. Do this every two hours to speed healing.
- For a boil on the torso or legs, add Epsom salts to your bathwater. Use two handfuls of salts, and soak for 15 minutes daily. (This treatment is unsuitable for the very young, the elderly, and those with high blood pressure.)
- If your boil, abscess, or carbuncle has burst, take a shower instead of a bath to reduce the risk of the infection spreading.

Herbal remedies: Hot herbal poultices encourage boils and superficial abscesses to come to a head. You can use various herbs, including marshmallow, slippery elm, and burdock (see box, facing page). Cabbage poultices also help draw pus from boils: Dip green cabbage leaves in boiling water for about a minute, until they have just

Natural antibiotic
Garlic, especially when raw, fights many types of bacteria.

wilted. Cool and wrap in gauze. Place this poultice over your boil as with a herbal poultice. Apply a fresh poultice each day.

Garlic and echinacea, taken orally, help strengthen the immune system and clear infection. Use several cloves of fresh garlic a day in your food, or take garlic capsules or tablets as recommended on the bottle. Take echinacea as a tincture or capsules, or as a tea made from two teaspoons of dried root added to one cup of water: Bring to a boil and simmer for 10 minutes; cool and strain. Drink three cups a day.
Caution: For safety concerns, see pp. 34–37.

Aromatherapy: Some essential oils have significant anti-infective properties.
- Add three drops of geranium oil or two drops of tea tree oil to a cup of warm water. Gently apply the mixture with cotton wool. Use a new piece each time you wipe to avoid reinfection.
- When a boil ruptures, bathe the area in an antiseptic solution made with one tablespoon of tea tree oil or distilled witch hazel added to a pint of warm water. Apply and cover with a sterile bandage. Leave the bandage in place for two to three days, or until a scab forms and swelling and redness subside.

Homeopathy: Try the remedy given below that most closely matches your case. It can safely be used in addition to any treatment your doctor may recommend.
- Arsenicum: when a boil or abscess is very red and pain is eased by a hot compress.
- Belladonna: when a boil or abscess is red, hot, painful, and throbbing.

Making a herbal poultice

1 Moisten two tablespoons of dried herbs with hot water and mix well to make a smooth paste.

2 Add a few drops of lavender, thyme, or eucalyptus oil. Soak a piece of gauze in the mixture and place it over the affected area.

3 Secure it with a bandage for about an hour, applying a hot-water bottle to keep it warm.

- Hepar sulphuris: when pus has formed and the lump is very sensitive to touch.
- Tarentula cubensis: when a boil or abscess develops rapidly after a slow incubation, when it feels very hard and looks bluish, and when the pain is agonizing and burning.
- Silica: when a boil or abscess is slow to clear up.

When to get medical help
- A boil or abscess does not heal within a few days.
- You don't feel well.
- The skin is extremely tender.
- You have swollen or tender lymph nodes (glands).
- You suffer from recurrent boils or abscesses.

Get help right away if:
- Inflammation or red streaks appear around a boil.
- You have a fever, chills, or night sweats.

See also:
Acne, Skin Problems

Acne

Pimples on the face, neck, back, and/or shoulders characterize acne. The disorder often begins during adolescence, the very time when appearance becomes especially important. Although it can cause considerable distress, acne poses no risk to health. The malady is more prevalent in boys than girls and affects 80 percent of teenagers in Western countries.

The underlying cause of acne is an oversensitivity to normal levels of the sex hormone testosterone. This oversensitivity makes the sebaceous (oil-producing) glands in the skin produce an abnormally high level of sebum, the oily secretion that lubricates and protects the skin.

At the same time, an abnormal reaction of the lining of the sebaceous ducts makes the lining cells too sticky. Instead of being shed when they die, these cells build up and block the duct. The dammed-up sebum solidifies, forming a blackhead or whitehead. Bacteria that are normally present on the skin can then easily multiply in the trapped sebum, causing the typical redness and swelling of an acne

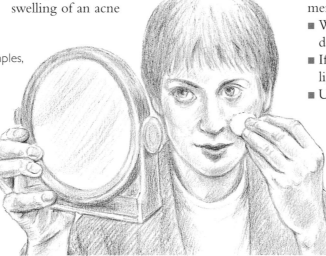

pimple. In severe cases these pimples, or pustules, may develop into hard lumps or fluid-filled cysts.

Certain medications—for example, some types of the contraceptive Pill, corticosteroids, and drugs for epilepsy—can make acne worse. In some women, acne is a symptom of polycystic ovary syndrome (see p. 274), a hormonal disorder that can also cause irregular periods, infertility, excess body hair, and weight gain.

Prevention

The following general advice is helpful for anyone who tends to get acne, and it may be all you need to do to control mild acne. For more severe cases, try some of the treatments recommended on the next page.

- Wash your face with an oil-free soap that doesn't over-dry the skin.
- If you need a moisturizer, use a product that is light and nongreasy.
- Ultraviolet rays can help relieve acne, so expose affected skin to outdoor daylight for at least half an hour each day, but not when the sun is at its most intense (usually 10 A.M. to 3 P.M. in summer).
- Choose your sunscreen with care. Buy products that are free from PABA (para-amino-benzoic acid) and do not contain ingredients that may clog pores.

Avoiding acne scars
Do not pick or squeeze pimples, as this can lead to scarring. Make sure your hands are clean before applying herbal teas or essential oil remedies with a fresh cottonwool ball.

- Avoid getting very sweaty or spending long periods in a humid environment. Many people find that this makes their skin condition worse.
- Iodine-containing food supplements and iodized salt may exacerbate acne, so avoid these.
- Apply tea tree oil to blemishes with a clean fingertip or a piece of gauze once or twice a day. If you prefer, dilute the oil with water.
- Keep your hair off your face, especially if it tends to be oily, and try to avoid touching your face too often.
- If you use hair gel or mousse and have acne on your forehead, do without these products for a couple of weeks to see if this helps.

Treatment

Aromatherapy: Some essential oils help regulate sebum production. These include juniper berry (avoid during pregnancy) and Atlas cedarwood (avoid during first 20 weeks of pregnancy). Lavender and geranium have antiseptic and healing properties, while Roman chamomile and petitgrain help reduce skin inflammation. Try the following treatments:

- Add two drops each of juniper berry and Atlas cedarwood oils to half a cup of water. Use a cotton wool ball to apply this solution to your skin every two hours during the day.
- Add two drops of juniper oil to a tablespoon of jojoba or sweet almond oil. Mix this and apply gently to the problem areas as above.
- Add two drops each of petitgrain and Atlas cedarwood oils to five teaspoons of jojoba oil. Also add these essential oils to five teaspoons of an unscented skin lotion. Use the oil each night, and the lotion each morning.

Acne and diet

Most surveys show that food has no effect on acne. However, it may be worth keeping a food diary—a record of what you eat each day—to see if any foods appear to cause breakouts. For example, some women find that eating chocolate just before menstruation is associated with acne. People with acne tend to have lower than normal levels of vitamin A in the blood, so ensure that your food provides enough of this vitamin, or of beta-carotene, which the body makes into vitamin A. Do not take a supplement except on medical advice.

Herbal remedies

- Combine equal amounts of dried dandelion root, burdock, nettle, yarrow, cleavers, and echinacea, and make an infusion in the usual way. Drink a cup once or twice a day.
- Boil two to three teaspoons of dried basil in a cup of water. Cool and apply with a cotton wool ball to clean skin.
- Clean your skin once a day by steaming your face over a bowl of very hot water for 5 to 10 minutes with a towel over your head. Dry your skin, then apply rosewater, elderflower water, distilled witch hazel, or calendula tea. Alternatively, use warm water containing a few drops of calendula tincture.

Caution: For safety concerns, see pp. 34–37.

Herbs for the skin
Dandelion root, burdock, nettle, yarrow, and echinacea help reduce acne, possibly by making the skin hostile to bacteria.

When to get medical help

- Natural remedies don't improve your acne.
- You suspect that your acne is a side effect of a prescribed medication.

See also:
Skin Problems

Addictions

Compulsive behaviour resulting from a physical or emotional dependence is commonly known as addiction. You may be dependent on a substance, such as a drug or tobacco, or an activity, such as overeating or gambling. Despite your cravings, simple treatments may help you break your habit. However, severe addictions often require professional help.

If you associate an activity with pleasure or with the relief of stress or uncomfortable feelings, you can fall into the habit of doing it whenever times are tough or when certain events in your everyday life easily lead to it. Some addictions, such as smoking and excessive drinking, are always physically damaging. However, any habit that becomes the dominant feature of your life can take a toll on your physical or emotional health and may even be dangerous to your family and others.

Such substances as narcotic drugs and the nicotine in tobacco are highly addictive; most of those who consume them become physically dependent on them. Research indicates that some people are genetically predisposed to addiction because they have a particular balance of chemicals in the brain. Depression or anxiety, for example, can be associated with an imbalance of neurotransmitters (such as serotonin) and hormones (such as adrenaline) that makes the cravings leading to addictive behaviour more likely. Nevertheless, social, emotional, and behavioural influences are equally important. With determination, persistence, and support, patterns of addictive behaviour can often be changed.

Increasing self-awareness

Buy a notebook and fill it with lists of alternative, life-enhancing activities—such as phoning a friend, having a scented bath, or going for a walk—which you can substitute for your habit. Then any time you feel that your addictive behaviour is about to break through, take out the notebook, write down exactly what you are feeling, and choose one of the alternative activities from your list.

Prevention

If you have a tendency towards addictive behaviour, try to recognize the warning signs that a new cycle is about to begin, and find alternative, less self-destructive ways of behaving. For example, if you react to depression or stress by reaching for a drink, a cigarette, or a biscuit, you need to find a better way of meeting your underlying needs and managing stress (see p. 347).

Treatment

Support groups: Find out if there is a local support group for your type of addiction. The well-known 12-step Alcoholics Anonymous programme, for example, is echoed by similar groups for people who have other addictions.

Yoga and meditation: Studies show that yoga exercises, meditation, and visualization can help break addictions and enable you to attain a more positive attitude towards your health. A daily session of yoga postures, breathing exercises, and relaxation (see p. 69) may calm you down, reduce withdrawal symptoms, and increase willpower. Exercise of any type tends to lift mood and improve self-esteem, thereby providing a valuable weapon against addiction.

Massage away cravings

Self-massage may help reduce some withdrawal symptoms and ease tension when you are trying to resist a substance or activity.

1 Sit somewhere comfortable and give each foot a firm massage, using a tablespoon of sweet almond oil mixed with two or three drops of lavender, geranium, neroli, or jasmine oil.

2 Using the same blend of oils, stroke your lower legs, working with long, slow sweeps from your ankles up to your knees.

3 Gently smooth moisturizing cream over your face and neck with your fingertips.

Diet: If you are a compulsive drinker, include in your diet foods rich in beta-carotene, magnesium, chromium, zinc, and vitamins B, C, and E to help meet your body's increased need for vitamins, minerals, and essential fatty acids, and to reduce cravings. Addiction to alcohol or tobacco may be associated with a low blood-sugar level. To combat this eat regular, small meals and keep sugar-containing foods to a minimum. Follow a diet high in protein and complex carbohydrates, such as wholegrain bread and pasta and brown rice. Drink plenty of nonalcoholic fluids; dehydration can be a trigger for addictive behaviour.

Other therapies: Seek help from a counsellor who uses cognitive techniques—ways of helping you understand what you are doing and why, and enabling you to discover other ways of achieving the same ends or meeting the same needs. Even when stress is not the trigger for addictive behaviour, therapy can help. Acupuncture, hypnotherapy, and biofeedback can all be effective.

Replacing nutrients
Wholegrain cereals contain high levels of the vitamins and minerals which are often lacking in those addicted to alcohol.

When to get medical help
- You realize that a habit is out of control.
- Addictive behaviour causes emotional or physical problems.
- Addictive behaviour starts to affect your relationships adversely.

Get help right away if:
- You have taken an overdose of an addictive substance.
- Addictive behaviour makes you a danger to others or yourself.

See also:
Anxiety, Depression, Eating Disorders, Overweight, Stress

Aging

There is as yet no wonder drug or other treatment to prevent all the physical symptoms of aging or to make us live longer. There are, however, many safe, natural methods of reducing some of the problems associated with aging and of helping ourselves to stay as fit and well as possible so that we can enjoy our natural life span to the fullest.

Aging is associated with increasingly slow cell division, mechanical wear and tear, the continued loss of brain cells, genetic damage from various environmental factors, and oxidative damage by free radicals. These unstable oxygen molecules are created through the normal process of generating energy within cells, but their levels are raised by such things as smoking, an unhealthy diet, and anxiety. Their effects on the body as it ages include loss of elasticity in the skin, blood vessels, and tendons, as well as a progressive decline in organ and joint function. If you lay the foundations of good health in youth and middle age, however, you are much more likely to maintain it into old age.

Prevention

You may not be able to stop your hair from turning grey or your skin from losing the smoothness of youth, but you can do much to avoid a pot belly, mental confusion, dimming eyesight, stiff joints, and many of the other possible problems of aging. You can also keep your immune system working well and take steps to prevent age-associated diseases, such as Alzheimer's disease, arterial disease, arthritis, and osteoporosis.

It is important to stay active in mind and body. The principal ways of doing this are to maintain your interests and social activities, sustain close relationships, exercise regularly, eat a

Always active
Keep physically and mentally lively in your later years. Those who enjoy friendships and explore interests tend to stay fitter.

healthy diet, and not smoke. You should also expose your skin to daylight so that you make enough vitamin D to keep your bones strong, but be careful to avoid overexposure to ultraviolet rays at their most intense (10 A.M. to 3 P.M.).

Mental well-being: Such activities as engaging in stimulating discussion, reading a book or newspaper, doing a crossword puzzle, and playing chess or bridge help prevent the gradual loss of nerve connections in the brain.

Relationships: Fostering good relationships with family and friends is life-enhancing and life-lengthening, probably because it reduces stress, boosts the levels of natural body chemicals that foster a feeling of well-being, and provides another reason for living. Even having a pet can prolong life by reducing stress and lowering blood pressure.

Stress management: High levels of stress encourage more rapid aging. Stress can also lead to arterial disease and trigger such illnesses as infections, arthritis, and migraine. Minimize the pressures in your life and learn how to manage your reactions to those that remain (see p. 347).

Sex: Those who enjoy regular sex tend to live longer, perhaps because of the emotional intimacy and physical exercise of sex, the stress relief of cuddling, and the increased levels of hormones that are released during orgasm, which widen the blood vessels.

Exercise: At minimum, get half an hour of moderately vigorous exercise five days a week. This will make you feel happier, increase your mental powers, and improve the health of your heart, lungs, skin, eyes, and other tissues.

   Consult your doctor if you are uncertain as to what type of exercise is suitable in your case,

Youthful skin

Help slow the aging of skin by taking the following measures:
● Protect yourself from excessive sun exposure.
● Help prevent skin dryness and wrinkles by applying a moisturizer with an SPF of at least 15 every day.
● Smooth on a little antioxidant oil each day. Make it by adding one drop of geranium or thyme oil to a teaspoon of sweet almond oil. (If you make this up fresh each time, no preservative is necessary.)

especially if you have a health problem. Popular forms of exercise for older people include swimming and low-impact aerobics. Weight-bearing forms of exercise, such as walking, tennis, and dancing, help ward off osteoporosis. Yoga and tai chi improve flexibility, and most types of exercise strengthen muscles to some extent. Regular yoga practice has been shown to reduce aches and pains in old age.

Diet: Well-nourished people who are not overweight tend to age more slowly and have fewer ailments. A poor diet—especially when combined with lack of exercise—can lead to the accumulation of fat around the abdomen. This form of weight gain carries an increased risk

Good fats
The omega-3 fatty acids in oily fish (such as sardines, mackerel, and tuna) can help keep your blood vessels in good condition. Try to have three three-ounce servings a week.

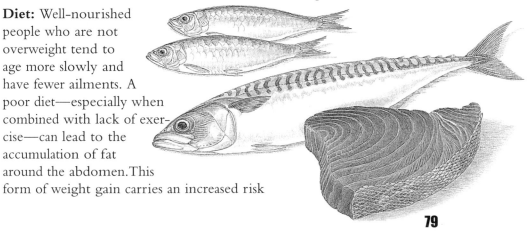

of diabetes and arterial disease, which may reduce quality of life and life expectancy.

As you grow older, your digestive system becomes less efficient, so you need a more nutritious diet, with fewer refined, fatty foods and more foods rich in antioxidant nutrients (beta-carotene, vitamins C and E, selenium, and zinc) and plant pigments (such as flavonoids). Avoid large amounts of alcohol. However, if you do drink, a daily glass of wine will stimulate appetite and may reduce your risk of heart disease.

Herbal remedies
- Garlic: To help immunity and circulation, include a crushed raw clove in your food each day, or take garlic capsules.
- Ginkgo biloba: To aid mental functioning (especially memory) and circulation, drink a daily cup of tea made from this herb.
- Ginseng: To protect against loss of sex drive or general debility, take some ground ginseng root each day, either in tea or tablet form.
- Green tea: For its antioxidant and blood-thinning effects, have at least a cup a day.

Caution: For safety concerns, see pp. 34–37.

From the chemist: Consider taking supplements of vitamin B_{12} and the antioxidants selenium, beta-carotene, and vitamins C and E.

Massage and aromatherapy

The overall benefits of massage, which include improved circulation and relaxed muscles, make it an excellent way to help maintain good health as well as to treat indigestion, arthritis, and aches and pains associated with aging. Ask a friend or partner to give you a massage from time to time. You can increase its therapeutic benefits by adding essential oils to your massage oil or lotion. Choose the oils according to the effect you want.

Massage oil for relaxation
Two drops of geranium oil, two of lavender, two of sandalwood, and one of ylang ylang added to two teaspoons of an unscented lotion or of sunflower or sweet almond oil.

Massage oil for revitalization
Two drops of clary sage oil, two of juniper berry, and two of rosemary in two teaspoons of lotion or oil as above. (Avoid this mix of oils during pregnancy.)

When to get medical help
- You feel depressed about growing older.
- You have symptoms of disease that aren't responding to natural treatments.

See also:
Arterial Disease, Arthritis, Diabetes, Eyesight Problems, Hearing Loss, High Blood Pressure, Memory Loss, Menopausal Problems, Osteoporosis

Allergic Rhinitis

Airborne allergen
Pollen in the atmosphere is the most common cause of seasonal allergic rhinitis. Pictured above are pollen grains from the artemisia plant enlarged 1,500 times.

The runny nose, frequent sneezing, and watery, itchy eyes of allergic rhinitis are triggered by exposure to an allergen (allergy-causing substance), such as pollen or dust. You can take steps to avoid exposure, and natural remedies will provide some respite, but the best way to increase resistance is to strengthen your immune system.

The symptoms of allergic rhinitis are similar to those of the common cold, but they are not caused by infection. Instead they represent an inappropriate response to a normally harmless substance. Your immune system reacts to the substance as if it were a dangerous invader, leading to inflammation and irritation.

The most familiar form is seasonal allergic rhinitis, also called hay fever or pollen allergy. Tree pollens are the main culprit in spring, grasses in summer, and weeds in autumn. If you are allergic to more than one kind of pollen, your symptoms may last for several months each year. The year-round symptoms of perennial allergic rhinitis result from such triggers as mould, animal dander, dust-mite droppings, certain foods, and environmental toxins.

Prevention

Bolster your immune system by eating a healthy diet, with plenty of foods rich in vitamins B and C as well as flavonoids (see p. 12). Vitamin C supplements may be helpful.

Seasonal allergic rhinitis: Pollen grains are so plentiful and minuscule that it is difficult to avoid them completely, but you can reduce the impact of exposure to them. You should take the following measures in the three months before your hay fever usually begins:

- Eat a teaspoon of locally produced, non-heat-treated honey containing wax cappings from the honeycomb cells. This is thought to strengthen resistance to local plant pollens.
- Drink a daily cup of ginseng or echinacea tea to strengthen your immune system.

During hay fever season
- Stay indoors with the windows closed whenever feasible, especially in the early evening, when pollen levels often peak.
- Damp-dust regularly.
- Clear pollen from the air with an ionizer or a high-efficiency particulate air (HEPA) filter.
- Wear wraparound sunglasses and/or a cyclist's face mask when you go out.
- Trap pollen by applying petroleum jelly in and around your nostrils.
- Keep car windows and air intakes closed.
- Avoid city centres and smoky rooms, since polluted air traps pollen.
- Put on clean clothes when you come home, and launder clothes frequently.
- Stay indoors before a thunderstorm and for two to three hours after it's over. The high humidity that precedes a storm makes pollen grains swell and burst, releasing particles of pollen starch.
- Use a mask when cleaning dusty, mouldy, or extremely dirty areas.

DID YOU KNOW?

A healthy diet which includes beta-carotene, vitamins B_5, B_6, C and E, selenium, zinc, plant pigments and a good balance of essential fatty acids can help reduce the symptoms because it provides natural antihistamines, anti-inflammatories, and immunity boosters.

If you're pregnant

Pregnancy can bring about both iron-deficiency anaemia and folic-acid-deficient megaloblastic anaemia. Try to prevent these by eating foods rich in the following nutrients: vitamins A, B_6, C, and E, folic acid, pantothenic acid, flavonoids, iron, manganese, zinc, and essential fatty acids (see p. 12). Take any supplements your doctor recommends.

The role of stomach acid

Some anaemic people make too little stomach acid and as a result don't absorb iron as well as they should. Carbohydrates may temporarily lower your stomach acid level. To improve iron absorption, therefore, avoid eating high-carbohydrate foods (such as bread, pasta, sugar, and rice) in the same meal as iron-rich, high-protein foods (such as meat, fish, and eggs). Protein requires the presence of adequate amounts of stomach acid for optimum digestion and iron absorption.

many months, untreated haemorrhoids, or bleeding in the digestive tract (due, perhaps, to irritation by painkillers or to cancer of the colon)—you need an especially iron-rich diet.

- Enhance your meals with liver, lean red meat, fish, egg yolks, dried fruits, soy products, and treacle.
- Eat plenty of onions, garlic, beans, peas, nuts, seeds, green leafy vegetables, and herbs (such as watercress, parsley, chives, nettles, coriander and dandelion leaves).
- Consume foods high in vitamin C, which boosts iron absorption. For example, have a glass of orange juice with your meals.
- Whole grains are a good source of iron, but they also contain phytates, which hinder iron absorption, so eat whole grains separately from other iron-rich foods.
- Limit consumption of spinach and rhubarb; their oxalic acid also reduces iron absorption.

- Eat foods that contain copper, which assists iron absorption. These include cheese, egg yolks, seafood, liver, whole grains, green vegetables, apricots, cherries, and dried figs.
- Avoid tea, coffee, cocoa, cola, and wine at mealtimes, since the tannins in these drinks block iron absorption from food.

Treatment

Your doctor may suggest an iron supplement, but it should only be taken under medical supervision; too much iron can be dangerous. The following measures, in addition to those described under "Prevention", may be helpful for iron-deficiency anaemia:

- Consider kelp supplements, which are rich in minerals and may boost iron absorption.
- Take a daily dose of the homeopathic remedy Ferrum phosphoricum.
- Try teas or tinctures of yellow dock, which contains iron, and of dandelion or burdock, both of which may aid in absorbing iron.
- If you feel tired, drink a tea or take a tincture made from wild oats or liquorice.

Caution: For safety concerns about the use of herbs, see pp. 34–37.

When to get medical help
- In all cases of suspected anaemia.
- For regular checkups if you're under treatment.

Get help right away if:
- You are short of breath.

See also:
Fatigue, Menstrual Problems

Anal Problems

With symptoms ranging from itching and soreness to bleeding and pain, anal problems are surprisingly common. The good news is that they are generally easy to treat or correct by using simple self-help measures. Bleeding, however, should always be checked out by your doctor to rule out any more serious underlying disorder, such as cancer of the rectum or colon.

Most anal conditions are linked to intestinal problems, particularly those that cause constipation, diarrhoea, or haemorrhoids. Poor hygiene, yeast (candida) infections, threadworm infestations, food sensitivities, and skin allergies may also contribute to such problems.

Prevention

You can avoid many of these conditions by eating ample amounts of fruits, vegetables, and whole-grain foods, and by drinking plenty of liquids. This provides fibre and fluid to help prevent constipation and supplies the nutrients needed to keep the intestines, the anus, and the skin of the anal area healthy. Avoid any foods that act as irritants, such as coffee and spicy foods.

Keep to the following guidelines to protect against anal problems:

- Clean the anal area gently but thoroughly after a bowel movement. (Women and girls should wipe from front to back to avoid bringing intestinal bacteria or threadworms near the vulva or the entrance to the vagina.)
- Wear cotton underwear, and change it daily.
- If you have sensitive skin or are prone to allergies that make your skin itchy and/or red, use nonbiological soap powder (available from most large supermarkets) to launder your clothes, and wash with unscented soap or a soap-free cleanser. Avoid scented bath products.

Treatment

Itching and soreness in the anal area can usually be soothed by the following simple measures:

- Keep the skin around the anus clean and dry.
- Don't scratch the anal area.
- Apply calendula (marigold) cream twice a day.
- Soak the area for 10 to 15 minutes at least once a day by sitting in a shallow bath or bidet of warm water containing a tablespoon of colloidal oatmeal or three drops of lavender oil.
- Take alternate hot and cold sitz baths (see p. 49) unless the skin is broken. Finish the sequence with cold water.
- Boost resistance to infection and inflammation by taking vitamins A and C with flavonoids, and zinc (with copper to aid absorption).
- If itching mainly occurs soon after going to bed at night, you may have threadworms and should treat this problem (see p. 314).

Anal fissure

This is a split in the lining of the anal canal that is usually caused by straining to pass a hard stool when constipated. Symptoms of a fissure are pain on defecation and bleeding, as the fissure is opened further. Treatment includes a high-fibre diet and plenty of fluids to soften the faeces. A fissure usually heals after a few days, but persistent or recurrent splits may need medical treatment.

When to get medical help
- There is no improvement within a few days.
- You notice anal bleeding.
- You have an unusual discharge or severe pain.

See also:
Constipation, Diarrhoea, Haemorrhoids, Pinworms (Threadworms)

Anxiety

Distress or tension is a normal response to an especially difficult situation, but the feeling should pass once your problem is resolved. During anxiety-producing times, use stress-management strategies and try natural remedies. If you're seeking an alternative or supplement to prescribed drugs, such as tranquillizers, check with your doctor first.

Anxiety is the body's "alarm response" to a perceived physical or psychological threat. A state of acute anxiety may occur before an important or difficult event, such as an exam or a job interview, and is usually short-lived. For some people, however, anxiety becomes an almost permanent state that seriously affects their ability to cope with everyday life. This condition is known as chronic anxiety.

Symptoms

Acute anxiety can be either a vague or a focused feeling of foreboding. It may be accompanied by physical symptoms, such as stomach cramps, a dry mouth, a racing heartbeat, sweating, diarrhoea, and insomnia. If you suffer from chronic anxiety, you may feel agitated without knowing why. Some people experience panic attacks, seemingly without warning. The symptoms, which may include a feeling of suffocation, chest pain, shaking, tingling in hands and feet, faintness, and terror, can be so extreme that sufferers —and onlookers—may believe they are having a heart attack.

Whatever form your anxiety takes, the symptoms can be distressing and disabling, and in the long run can damage your physical health. You need to find ways of dealing with the underlying cause. Consider seeking professional help from a therapist or support group.

Prevention

During periods of stress, the body uses up nutrients faster than normal, and unless these are replaced, the nervous system becomes progressively depleted, resulting in an anxious state. It is therefore important to eat a healthy diet—one high in complex carbohydrates, such as whole-wheat bread and brown rice, which may have a calming effect. Be sure to include essential fatty acids (for example, whole grains, nuts, seeds, and vegetables), vitamins (especially B complex), and minerals to nourish your nervous system. Eat

Restful foods

The amino acid tryptophan has a soothing effect on the brain. This is because tryptophan is converted in the brain to serotonin, a chemical messenger that exerts a calming action. Most protein-containing foods contain tryptophan. However, absorption of this substance is improved when it is taken with carbohydrates. Good sources include:
- Milk with biscuits.
- A turkey or cheese sandwich.

small amounts often during the day to keep your blood-sugar level steady. A good balance of rest, exercise, and recreation will also help you feel better physically and thus more positive about life in general.

Treatment

If you are a very anxious person, you may need professional counselling, but there is also a lot you can do to help yourself.

Self-understanding: Many experts believe that some of us are naturally more fearful and fretful than others, but it may be that something in your past has triggered a tendency to become especially anxious. You might reflect on the possible origins of your anxiety as a first step to overcoming it. For example, overly protective parents, constantly warning you of risks and dangers inherent in normal life, may have passed along their own fears, or you may have had a traumatic experience that set off your anxiety. Anxious people often have a highly developed

imagination, too, so that they quickly foresee all the possible unpleasant or disastrous consequences of any action and can't help dwelling on them. If you have a phobia, such as an abnormal fear of flying, this may be a way of focusing your fear on something specific, even though that is no easier to handle than generalized anxiety. You

Yoga breathing for calm

Yoga is excellent for people who often feel anxious, because it can relax both mind and body, encourage steady breathing, and help overcome negative emotions. This breathing exercise is designed to strengthen and relax the chest and abdominal muscles, and to harmonize the flow of life energy (*prana*). Take five breaths at each stage.

  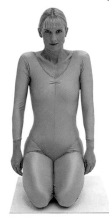

1 Kneel with one hand on your abdomen and the other palm placed downwards on your thigh. Feel your abdomen expand as you inhale and retract as you slowly exhale.

2 Place your palms on either side of your rib cage. Lift and lower your rib cage as you breathe, pushing inwards with your hands to help expel the last of the air as you exhale.

3 Tighten your abdomen. Lift your shoulders and upper chest as you inhale, and then let them drop as you exhale and release the muscles of your abdomen.

Hyperventilation

When you are anxious, and especially if you experience a panic attack, your breathing becomes fast and shallow. This disrupts the body's oxygen and carbon-dioxide levels. To correct this overbreathing, or hyperventilation, sit with one hand over the top of your abdomen and breathe in and out slowly so that your hand moves outwards each time you inhale. This "abdominal breathing" helps slow and deepen your breathing.

Appetite Loss

Eating is one of life's necessities and also one of its greatest pleasures. Yet there are many reasons why you may not feel like eating from time to time. Most of these are minor and pose few problems. Natural remedies and changes in lifestyle often help. However, if you experience loss of appetite for a prolonged period, seek medical advice, especially if you have also lost weight.

Many people lose their appetite for a short period of time, for example, when suffering from indigestion or a fever. Anxiety or depression may also reduce the appetite (though some depressed people want to eat more). Once the underlying cause is over, the appetite usually returns to normal, with no harm done.

Drinking too much alcohol on a regular basis supplies so many calories that it usually removes the desire for proper nourishment. And anything that makes you nauseated—including motion sickness, gastroenteritis, early pregnancy, treatment with certain drugs, and exposure to pesticides, lead, and other poisons—will keep you from wanting to eat.

A continuously poor or nonexistent appetite can suggest a more serious disorder, such as infectious mononucleosis, rheumatoid arthritis, high blood pressure, stomach cancer, tuberculosis, hepatitis, or severe depression.

Surprisingly, perhaps, people who suffer from anorexia nervosa, which can involve life-threatening weight loss, don't often lose their appetite. While victims of this eating disorder consume only small amounts of food, most continue to feel very hungry.

Prevention

If your appetite loss is a symptom of illness, you need to treat the underlying cause. Otherwise, there are many ways in which you can encourage a good appetite.

- Eat small, regular meals, making them as nutritious and appealing as possible.
- Try a pre-meal appetizer containing a traditional, bitter-tasting appetite booster. Examples are olives, watercress, young dandelion leaves, rosemary, and chicory.
- Avoid excessive alcohol intake. However, if you drink, a bitter aperitif may whet your appetite before a meal. Having a small glass of wine with a meal may also encourage you to relax and you may therefore eat more.
- Take your time over meals and make them as stress-free as possible.
- Opt for meals composed of several smaller courses rather than one or two larger ones.

Appetite-boosting juices

Sipping a cup of carrot and watercress juice half an hour before a meal can be an excellent appetite booster. Juice four carrots and a bunch of watercress, then add an equal volume of water. Increase the proportion of carrot juice if you find the mixture too bitter.

When a child doesn't eat
If your child's appetite is poor for no clear reason, find out whether there's a problem at home or at school. Refusal of school lunches, for example, may not be because of the food but instead for social reasons.

- Say no to foods containing white flour and added sugar, as these lack the nutrient quota of wholefoods. If refined foods really are the only ones you feel like eating, take a daily multiple vitamin and mineral supplement.
- Drink coffee and tea that contain caffeine only after meals. Caffeine can suppress appetite. Excessive fluid intake during a meal can fill the stomach and hamper digestion.
- Exercise regularly to stimulate hunger.
- If you smoke cigarettes or use other tobacco products, get help to stop or at least cut down. Any tobacco use can inhibit appetite.
- Ask a friend to give you a relaxing massage, using five drops of lavender oil in two table-spoons of sweet almond oil, if stress is suppressing your interest in food.

Treatment

Your choice of treatment for appetite loss depends on the underlying cause. See, for example, advice on indigestion (p. 252), morning sickness (p. 318), and motion sickness (p. 282).

Nausea and tension at mealtime: Give your-self a soothing, gentle aromatherapy massage. Add two drops each of peppermint and black pepper oils, and one drop of rose otto oil, to five teaspoons of sweet almond oil. Use this mixture to massage your abdomen with slow, circular, clockwise strokes.

Anxiety or depression: Get expert assistance to address the source of the problem. In the meantime, care for yourself by preparing attractive, nutritious meals of foods that you normally like, and take a daily multiple vitamin and mineral supplement if you are still unable to eat much.

Helpful hors d'oeuvre
Chew a little fresh ginger or take it as a tea before meals. Or add a few drops of the essential oil to five teaspoons of olive oil for an appetite-stimulating abdominal massage.

When to get medical help

For children aged 1 to 3 years
- Reduced appetite lasts longer than a few days or is accompanied by other symptoms.

For older children and adults
- Appetite loss lasts longer than seven days, or there is also vomiting, diarrhoea, a cough, abdominal pain, rapid weight loss, or other symptoms.

Get help right away for babies under 12 months if:
- Poor appetite is accompanied by fewer than five very wet nappies in a 24-hour period.
- Your baby refuses all feedings for longer than 24 hours.
- He or she vomits, cries uncontrollably, or shows other symptoms.

See also:
Anxiety, Breast-feeding Problems, Depression, Eating Disorders, Indigestion, Nausea and Vomiting, Pregnancy Problems, Weight Loss

Arterial Disease

Endemic in the Western world, arterial disease plagues the lives of many people and causes a high proportion of premature deaths. Cardiovascular disease claims the lives of over 124,000 people in the UK each year. Lifestyle changes and other natural approaches can reduce—or even reverse—artery damage and make a significant difference to your well-being.

Diseased arteries can reduce the delivery of blood to the tissues and cause internal bleeding. They may also contribute to fluid retention. Arterial disease underlies most cardiovascular conditions, including the following: high blood pressure; angina and heart attacks; pain in the legs when walking; abnormally cold hands, feet, and nose; Raynaud's syndrome; migraine; and strokes and mini-strokes (transient ischaemic attacks). The disorder can also cause dizziness, memory loss, and confusion. Less common manifestations include a group of inflammatory conditions known collectively as vasculitis or arteritis.

Causes

Many factors can lead to arterial disease. Often several factors are present, acting together to cause the disorder.

Irritation of the artery lining: This encourages the accumulation in the arteries of "bad" cholesterol (oxidized low-density lipoprotein, or LDL). The oxidized LDL collects as a layer of yellowish material, called atheroma, in the artery lining. This narrows the artery and encourages turbulence in the blood, leading to tiny tears in the artery lining that form scars as they heal. Scarred arteries can, at worst, tear through at their weakest points, resulting in haemorrhages that damage surrounding tissues. The roughened interior of a scarred artery promotes the formation of blood clots and further blockages.

Known irritants include:
- A high level of oxidized blood fats.
- A high blood-sugar level.
- Smoking and passive smoking.
- Autoimmune disorders (in which the damage is caused by a person's own immune system).

Possible irritants include:
- High blood levels of the amino acid homocysteine (the result of such things as insufficient folic acid and vitamins B_6 and B_{12}, smoking, inactivity, and genetic flaws).
- Viral or bacterial infections, such as long-term lung or gum disease.

Hardening of the arteries: This reduces the efficiency of the circulation. The artery walls may lose elasticity as a result of high blood pressure, smoking, inactivity, or atheroma.

Thickening of the blood: This encourages blood clots to form in arteries and can be caused by inactivity, infection of any kind, cold weather, and/or a diet that is low in nutrients and high in saturated fats.

Oversensitivity of the artery muscles: This leads to unexpected changes in artery diameter and blood flow. The condition can be caused by

low blood levels of magnesium and calcium, and also by stress, cold weather, smoking, changing hormone levels, insufficient intake of vitamin C, or a high fat intake.

Prevention

There are a number of established, preventable risk factors for arterial disease. By avoiding them, and by practising other healthy habits, you can help keep your arteries strong.

Give up smoking: Smoking tends to narrow the arteries and to increase oxidation of fats in the blood. Stopping reduces your risk of serious arterial disease, and the longer you are a non-smoker, the lower your risk becomes. If you find quitting impossible, at least cut down, and eat plenty of foods rich in vitamins C and E. Each cigarette you smoke destroys about 25 milligrams of vitamin C, and you need both it and vitamin E to protect against heart disease.

Avoid inactivity: A half hour of exercise—strenuous enough to boost your circulation and make you feel warm (for example, a two-mile walk)—five or more days a week is good for your heart and circulation. Exercise increases the ratio of protective HDL (high-density lipo-protein) cholesterol to the potentially damaging LDL type, strengthens the heart, and promotes weight loss. Always warm up before exercise (see p. 13), especially after eating a fatty meal.

Ease tension: Make time each day to relax both mind and body. Good ways to do this include seeing friends, watching movies or TV, cooking

Reduce tension with yoga

By helping you deal with stress, yoga practice can make a significant contribution to the prevention of arterial disease. One relaxing yoga exercise is the standing forward bend: Stand with your feet together and your weight slightly forward. As you inhale deeply, raise your arms, stretching them above your head. Hold this position for 20 to 30 seconds. Then exhale while slowly bending from the waist and stretching your hands and arms down towards the floor. While breathing normally, hold the position for up to half a minute, then straighten slowly.

creatively, gardening, walking the dog, enjoying humour and sex, and playing with your children. Or you can try progressive muscular relaxation (see p. 245), meditation, and yoga.

High stress levels trigger the release of adrenaline, cortisone, and other hormones. These raise blood pressure and increase oxidation of fat in the blood. Even slight depression raises stress hormone levels and makes the blood more sticky and prone to clot. Moreover, if you feel stressed or depressed, you are less likely to take care of yourself and more likely to indulge in such habits as smoking and compulsive eating—especially of sugary and fatty foods.

Stress and depression management skills include not only looking after your physical health, but also recognizing and dealing with

Risks dating from birth

Low birth weight and/or premature birth may make high blood pressure, heart attacks, and strokes more likely. If this applies to you, be especially careful to take steps to prevent arterial disease.

your feelings—by, for example, learning safe techniques for handling anger and other difficult emotions. This is especially important if you often find yourself hiding emotional distress, if you're argumentative, or if you've recently experienced the death of a loved one.

Choose a healthy diet: Avoid eating too much, and especially limit foods containing animal fats (as these are mainly saturated fats) and trans-fatty acids (present in many commercially hardened vegetable fats). A large, fatty meal creates a surge of fat in the blood, which encourages the formation of atheroma. Limit your intake of animal products—meat, eggs, butter, and other high-fat dairy products. Some people have an increased risk of arterial disease if they eat too much salt, because this can raise blood pressure.

Certain nutrients are good for the arteries and the heart. They include antioxidants, such as beta-carotene, the mineral selenium, and vita-

mins C and E, which reduce the oxidation of LDL cholesterol. Omega-6 fatty acids (from vegetables, nuts, whole grains, seeds, and olive oil), omega-3 fatty acids (for example, from oily fish and pumpkin seeds), plant oestrogens (for example, from soy products) and soluble fibre (in oats and apples, for example) help combat the effects of damaging fats in the blood.

These nutrients help maintain the health of the artery linings, strengthen the immune system, and counter danger from saturated fats and other artery irritants. Fruits and vegetables also contain natural salicylates, which help prevent blood from becoming too thick. Omega-3 fatty acids from fish also help prevent abnormal thickening of the blood.

An occasional glass of wine is safe for most people and may even be good for the arteries—partly because it dilates them, thereby increasing the blood supply to the heart muscle and other tissues, and partly because red wine (as well as some other alcoholic drinks) contains antioxidants such as flavonoid plant pigments. Grape juice is an alternative source of flavonoids for those who need or prefer to avoid alcohol.

Manage your weight: Keep your weight down, especially if you are accumulating fat around your waist and abdomen and becoming "apple-shaped". Laying down fat in these areas makes a heart attack more likely.

Stay out of the cold: Living in a cold climate makes the blood thicker, increasing the possibility of clots. If you live somewhere cold, wet, and windy and also have additional risk factors,

Foods to counter artery damage
Consumption of onions, leeks, garlic, tomatoes, apples, oats, and oily fish is associated with a lower risk of heart disease because of their blood-thinning and/or antioxidant effects. Eating half a raw onion a day may help lower your blood level of LDL cholesterol.

consider a move to somewhere warmer. If this isn't feasible, avoid the cold and cover up sufficiently when you must be outdoors.

Watch your breathing: Avoid overbreathing (hyperventilating) when stressed. Breathing too fast makes the blood relatively alkaline, which means that it is harder for red blood cells to supply enough oxygen to the cells of the artery walls. Try the breathing exercises on page 87.

Herbal remedies: Warming spices—such as ginger, cinnamon, cloves, and cayenne—added to recipes or used in a tea, aid arterial health by stimulating the circulation and dilating the blood vessels that supply the arms and legs.

- Drink up to three cups a day of ginkgo biloba tea to help counter artery-wall inflammation and prevent the formation of damaging oxidized cholesterol.
- Eat raw garlic or take garlic tablets or capsules regularly. This can reduce high blood pressure and keep the blood flowing smoothly. Garlic lowers the level of cholesterol and reduces oxidation of LDL cholesterol.

Caution: For safety concerns, see pp. 34–37.

Supplements: There is evidence to suggest that certain food supplements, available at a chemist or health-food store, can lower the risk of developing arterial disease.

- The pectin in powdered grapefruit fibre may help reduce high blood cholesterol and open up blocked arteries.
- The antioxidants beta-carotene, vitamins C and E, and selenium are thought to play an impor-

Dress warmly
Being cold causes the blood to thicken, making it more likely to clot.

tant part in protecting the circulatory system.
- Bromelain, an enzyme derived from pineapple stalks, reduces blood stickiness.
- The amino acid L-arginine helps form nitric oxide, which relaxes artery walls.

Treatment

If you develop arterial disease, remember that it is never too late to take steps to reduce or reverse the damage to your arteries. In consultation with your doctor, consider all of the preventive measures outlined above, as well as the following natural remedies:

- Hawthorn, which is particularly valuable in treating high blood pressure caused by hardening of the arteries, angina, and arterial spasm. Use a teaspoon of the flowers or the crushed fruit to make one cup of tea, and sip at regular intervals throughout the day.
- A daily dose (90–300 milligrams) of the antioxidant coenzyme Q_{10}, which may reduce the frequency of angina attacks.

When to get medical help

- You suffer pain when walking or unexplained dizziness or faintness.

Get help right away if:

- You experience pain in your chest, arms, or neck.
- You become unexpectedly short of breath, have sudden vision loss, or black out.
- Your arms or legs suddenly lose sensation or become very pale or bluish.

See also:
Cold Hands and Feet, Depression, Diabetes, Dizziness, High Blood Pressure, Memory Loss, Migraine, Overweight, Palpitations, Raynaud's Syndrome, Stress

Arthritis

More than 100 diseases fall under the heading of arthritis. The symptoms they have in common are joint stiffness and pain, which in some cases can severely limit movement. Arthritis is a chronic condition affecting about 8 million people in the United Kingdom. If you have arthritis, you may be able to relieve your symptoms by using a variety of natural methods.

Pain in a joint may be caused by an injury, a condition such as rheumatoid arthritis, food sensitivity, or infection. Natural remedies can be used for pain that has not been directly caused by an injury. The main forms of arthritis are described in the box below.

Prevention

Although mainly associated with damaged cartilage in the joints from the wear and tear of aging, osteoarthritis may have a genetic component. To reduce your likelihood of suffering from joint damage, avoid becoming overweight, since this puts additional strain on the joints. Eat a healthy diet, containing plenty of fruits and vegetables, to provide the nutrients your body needs to repair damaged cartilage.

To help prevent rheumatoid arthritis, try to discover and avoid any trigger factors. Possibilities include smoking, stress and certain foods, such as (in order of likelihood) wheat and other cereal grains, red meat, sugar, animal fats, salt, coffee,

---

The main types of arthritis

Osteoarthritis occurs when the cartilage lining a joint—usually a weight-bearing joint, such as the knee or hip—wears away. Excess fluid can then accumulate in the joint, causing swelling, pain, and reduced mobility. Osteoarthritis is more common in people over 60. It affects almost three times as many women as men.

Rheumatoid arthritis results from inflammation of the membrane lining the joints and eventually the cartilage, too. The affected joints become swollen, stiff, and painful. In people with this arthritis, the normal immune response that is designed to protect against infection turns against the joint lining. The disorder usually begins in the fingers, wrists, and toes. You may inherit a tendency to rheumatoid arthritis, and one or more of several triggers may play a part. Some people with this condition have additional symptoms, such as fatigue, anaemia, poor circulation, and trouble with the tendons, eyes, and thyroid gland. The disease generally begins in young or middle-aged adults and is two to three times more common in women. Women who have taken the oral contraceptive Pill seem to have a lower risk of developing rheumatoid arthritis.

Infective arthritis occurs when inflammation is caused by bacterial infection. This may be the result of germs entering a wound or may be linked to infection elsewhere in the body, as with, for example, tuberculosis, gonorrhoea, or a urinary-tract infection.

Gout causes acute attacks of pain in the joints. It results from high levels of uric acid (a waste product) in the blood, which causes crystals of uric acid to form in a joint. High uric acid levels may be due to inefficient kidney function. Often only one joint is affected, frequently that of the big toe. In an attack the affected joint becomes red, hot, swollen, and extremely painful. Gout is 10 times more common in men. It is generally found in women only after they have gone through the menopause.

dairy foods, and potatoes. Tomatoes, peppers, and aubergines may promote arthritis in some people. Try excluding these foods to see if this eases your condition. It may be that pesticide poisoning and certain medications can also act as rheumatoid arthritis triggers.

You may be able to prevent attacks of gout by reducing alcohol consumption. Although doctors used to suggest that people with gout should avoid eating liver and other organ meat, poultry, peas, and beans, this advice is no longer considered helpful.

Treatment

Most arthritis treatments aim to relax muscles and ease pain by reducing inflammation. Only a few are aimed at correcting the underlying causes.

Heat and cold: Heat can help relieve pain, aid circulation, and ease stiffness; cold is better for a hot and swollen joint. You may find relief by using one or more of the following:

- A compress made by dipping a small towel in hot or cold water and wringing it out. Place this over the affected joint(s).
- A pack of frozen peas wrapped in a towel and applied to the painful area.
- Gloves, socks, and other warm clothing.
- Warm (not hot) melted beeswax or paraffin wax. Immerse arthritic hands for 10 minutes. The wax peels off easily when cooled. It can be reused for further treatments.
- An Epsom salts bath for the discomfort of

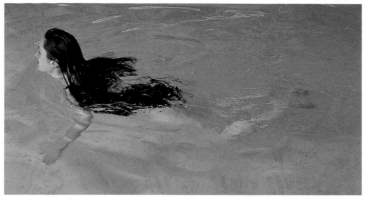

Stress-free exercise
Swimming is an ideal activity for those suffering from arthritis, because it helps strengthen affected joints without jarring them. Choose the stroke you find most comfortable.

rheumatoid arthritis. Add two handfuls of Epsom salts to a hot tub and soak for 15 to 20 minutes. (This treatment is unsuitable if you are very elderly or have high blood pressure.)

Rest and exercise: A joint that is very painful or inflamed should be rested. When it feels better, exercise daily for a half-hour to warm the joint without stressing it. Do some muscle-strengthening exercises as well, to keep the muscles around the joint supple and to increase the production of joint-lubricating fluid.

Diet: Supply your joints and your immune system with the nutrients they need by eating a healthy diet, with five servings of vegetables and fruit daily. The essential fatty acids in oily fish (such as herrings, sardines, and salmon), nuts, seeds (especially linseeds), and whole grains are particularly helpful for reducing inflammation. Some people find pineapple beneficial, possibly because it contains bromelain, an enzyme that aids digestion and increases the absorption of nutrients. Bromelain can reduce inflammation in

A spicy remedy
Turmeric contains strong anti-inflammatory substances that ease sore joints. You can add either the fresh or powdered root to curries and other dishes.

Relaxation with yoga

Ask your doctor if it would be appropriate for you to try yoga to ease your discomfort. Daily yoga exercises will not only help keep you and your muscles relaxed and your joints supple, but may also allow you to reduce your dosage of painkilling drugs. The gentle movements can, in addition, boost the immune system and help

stimulate the circulation of blood and lymph. If you think this form of exercise may suit you, try joining a yoga class. Some exercises which are beneficial for stiff joints are shown below. Practise them slowly and carefully, and avoid overstretching. Do not exercise joints which are red and painful.

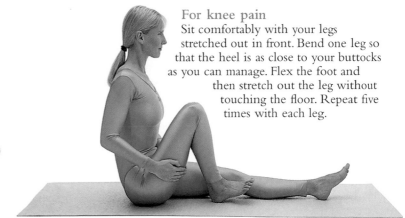

For knee pain
Sit comfortably with your legs stretched out in front. Bend one leg so that the heel is as close to your buttocks as you can manage. Flex the foot and then stretch out the leg without touching the floor. Repeat five times with each leg.

Shoulder rotation
Rotate arms one at a time, from the shoulder. Go forwards a few times, then backwards. Try it first with your arms straight, then bent.

Foot rotation
Use your hands to rotate the upper foot each way, then rotate the foot without using your hands. Wriggle your toes. Repeat with the other foot.

For hip pain
Lie on the floor with your lower back relaxed and feet together. Inhale slowly and bring one knee as close to your chest as is comfortable. As you slowly exhale, gently return your leg to the ground. Do this five times with each leg. Repeat the exercise, holding the bent knee and moving it in circles. Finally, repeat the exercise moving the bent knee from side to side.

Hand rotation
Slowly rotate each wrist a few times, first in a clockwise, then a counter-clockwise direction. Follow by wriggling your fingers.

An effective supplement

Glucosamine, a cartilage-building compound, is one of the best remedies for osteoarthritis. Take one 500-milligram capsule three times a day. Chondroitin sulphate, another cartilage-building substance, can be taken along with it. Expect the treatment to take a few weeks to begin working.

Anti-arthritis foods
To help alleviate symptoms, include in your diet foods containing beta-carotene, such as yellow and orange produce; omega-3 fatty acids, such as oily fish and nuts; and proanthocyanidins, such as cherries.

the body when taken between meals and is also available as an extract. Blueberries can also help combat arthritis. They are an excellent source of powerful antioxidants called proanthocyanidins, which can ease osteoarthritis.

Special advice for gout: In addition to the advice given above, try the following measures to relieve joint pain resulting from gout:

■ Omega-3 fatty acids reduce inflammation, so take fish oil daily, which is a rich source of two such fatty acids: eicosapentaenoic acid (EPA) and docosahexanoic acid (DPA).

■ Eat plenty of foods with high levels of beta-carotene (orange and yellow vegetables and fruit), vitamin C (citrus fruit), and vitamin E (vegetable oils, nuts, seeds, beans, and egg yolks—or from a vitamin E supplement), and selenium (whole grains, fish, and nuts).

■ Eat cherries, and berries, such as blueberries. They are an excellent source of proanthocyanidins, substances that are helpful for reducing the joint inflammation of gout.

■ If you drink alcohol, reduce your intake or even cut it out completely.

Aromatherapy: German chamomile and lavender can ease inflammation. Juniper berry and cypress may reduce swelling. The warming properties of black pepper and sweet marjoram help relax muscles and relieve aching. Roman chamomile and cajuput may alleviate severe pain. Do not use cypress and cajuput oils in the first 20 weeks of pregnancy. Juniper berry oil should be avoided throughout pregnancy.

■ Take a warm, scented bath: Sprinkle two drops of lavender oil into the bath water. You can add two drops of either cajuput or Roman chamomile oil if you are suffering from severe pain.

■ Make a warm compress by sprinkling two drops of lavender oil and two drops of juniper berry oil into a bowl of warm water. Place a piece of cloth over the surface of the water to pick up the oil film, then lay it over the affected joint. Cover with a towel, put a hot-water bottle on top, and leave for 30 minutes.

■ Prepare some massage oil by mixing two drops each of juniper berry, black pepper, and Roman chamomile oils and five drops of lavender oil with 10 teaspoons of olive or jojoba oil. Smooth this scented oil over the affected joints each day.

Herbal remedies: Depending on your condition, choose from:

- Buckbean (bogbean) tea and celery-seed tea for rheumatoid arthritis. Drink one small cup three times a day. You can also add celery seeds to soups, casseroles, salad dressings, and pizzas.
- Devil's claw tea for rheumatoid arthritis. Add one teaspoon of the herb to a cup of water; simmer for 15 minutes. Drink a cup three times a day. Or buy devil's claw tablets from a health-food store or chemist. Devil's claw can cause indigestion; stop taking it if you experience this problem.
- Wild thyme tea if stress is making your arthritis worse. Have one small cup three times a day.
- Nettle or coriander tea for gout. These herbs help the kidneys get rid of uric acid.
- Two or three dandelion roots, boiled in two pints of water in a covered pan for one hour. Drink this tea three times a day before meals.

Caution: For safety concerns, see pp. 34–37.

Breathing: Some people find that their joint pain is worse if they breathe too fast or deeply. Exhaling too much carbon dioxide makes the blood more alkaline and prevents red blood cells from releasing oxygen. The lack of oxygen results in a need to breathe even more deeply or quickly. Practise breathing more slowly. Consult your doctor before trying breathing exercises if you have a disorder of the heart or lungs.

Kitchen-cabinet remedies: Some people find the following traditional treatments effective:

- A mixture of four parts olive oil, eight parts spirit of camphor, and one part cayenne pepper, used to massage the joints.
- Two large teaspoons of apple-cider vinegar in a glass of hot water flavoured with a teaspoon of honey, taken twice a day.
- Tea made by adding boiling water to two teaspoons of powdered ginger.

Joint-soothing herbs
Teas made from wild thyme, devil's claw, nettles, or coriander contain natural anti-inflammatory substances.

When to get medical help
- Pain is severe.
- You also have a fever or don't feel well.
- Self-help treatments don't improve your symptoms.
- Your arthritis is getting worse, or discomfort impairs sleep or everyday functioning.

Get help right away if:
- The joint swells suddenly or severely.

See also:
Back Pain, Pain

Asthma

Recurrent attacks of wheezing and breathlessness—known as asthma—can develop at any age, but they usually begin early in life and sometimes lessen or clear up completely by adulthood. Asthma is becoming more common; according to estimates, it now affects 5.1 million people in the UK, including 1.4 million children.

Asthma occurs when the small airways in the lungs become overly sensitive, reacting to one or more substances that have no effect on non-asthmatic people. This reaction makes the lining of the airways swell and produce mucus, reducing the space available for incoming air. Principal symptoms include breathlessness (with particular difficulty in exhaling), wheezing, coughing, and tightness in the chest. Attacks vary in severity, from mild breathlessness to life-threatening respiratory failure. Frequent asthma attacks can interfere with everyday life.

Some people have an inherited tendency to get asthma, while many develop it for no known reason. Asthma is caused by an allergic reaction in 1 in 8 children and in 1 in 13 adults who are over 30 years old. A wide variety of other things can trigger the oversensitivity of the airways that underlies the condition (see box below).

There are several possible reasons for the increase in asthma. The improved living conditions of people in developed countries mean that they have fewer early childhood infections and parasitic infestations. Because their immune systems don't have the early stimulation derived from exposure to foreign organisms, they may overreact not only to these organisms, but also to certain other substances later. The

---

Asthma triggers

Allergens (inhaled or eaten), activities, or conditions may set off an asthma attack. They include:
- Dust-mite or cockroach droppings.
- Pollen.
- Dander (hair and skin scales) shed by pets.
- Certain drugs (including aspirin).
- Certain foods (such as peanuts, shellfish, and eggs).
- Mites in food (cereals, flour, and other dried foods) well past the expiry date.
- Mould spores.
- Respiratory infection.
- Cigarette smoke.
- Vehicle exhaust fumes.
- Vapour from air fresheners, perfumes, finishing chemicals in new clothes, dry-cleaning solvents, paints, glues, shoe-waterproofing sprays, and pesticides.
- Nitrogen dioxide in burnt gas fumes from gas appliances.
- Industrial chemicals and other substances (for example, dusts from paper, wood, and flour; fumes from welding and soldering; vapour from hardening agents and isocyanates).
- Cold air or a sudden temperature change.
- Exercise.
- Strong emotion.

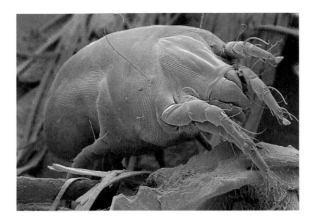

Tiny troublemaker
The dust mite, shown above at 245 times life-size, thrives in the modern super-insulated, heated home. Soft furnishings and fitted carpets provide its ideal environment. Dust-mite droppings are one of the most common asthma triggers.

widespread use of antibiotics to treat relatively minor childhood infections also makes asthma more likely. This may be because antibiotics disturb the development of normal immunity by altering the population of bacteria in the intestine. As children grow older, they are more likely to overreact to viral infections and to substances in the environment. Environmental pollutants—from, for example, nitrogen dioxide, sulphur dioxide, ozone, and airborne particles in vehicle exhaust emissions—can also be to blame.

Prevention

Try to identify your asthma trigger and take steps to avoid it. In cold weather, wind a scarf loosely around your nose and mouth so that you breathe warmer air. Air pollution is often at its worst when it's hot and humid, so you may want to stay inside on those days. Don't use spray products, and make your home a no-smoking zone.

If allergy tests reveal a sensitivity to dust mites, consider replacing fitted carpets or rugs with hard flooring, such as tiles or wood. Vacuum upholstery and curtains frequently, and clean them regularly. Wash bedclothes often, at as high a temperature as the fabric can withstand. Put your pillow and mattress in allergen-proof covers. Buy a vacuum cleaner or air purifier with a high-efficiency particulate air (HEPA) filter, or install such a filter in your heating system. Dust mites thrive in moist conditions, so purchase a dehumidifier if indoor humidity is high.

Once a week, kill dust mites in soft toys by laundering washable ones and putting others in the freezer overnight. If a pet causes asthma, you may have to give it away, though bathing it weekly may help.

Diet and supplements: A healthy diet (see pp. 9-12) reduces the risk of wheezing. A magnesium supplement may help, and vitamins C and E may lessen the frequency of asthma triggered by smoke or air pollution. Garlic tablets may reduce production of mucus, which can contribute to narrowing of the airways. Take measures to identify and exclude any foods to which you might be allergic or otherwise sensitive (see p. 214).

Exercise: Improving your general fitness with a daily half hour of strenuous exercise will boost your sense of well-being and your lung fitness.

Breathing and yoga: Regular breathing exercises, such as those taught by Buteyko instructors (see p. 60) and yoga practitioners, can increase your ability to relax your chest muscles. They may also improve your control over anxiety

---

Spotting danger signs early

You may want to buy a peak flow meter and use it each morning and evening to measure how fast you can expel air. If your performance decreases, you can step up preventive actions.

and other stress responses that can narrow the airways. Together these skills will make you less susceptible to frequent or severe asthma attacks.

Frequent episodes of feeling overstressed can readily instill the habit of rapid breathing. This eliminates too much carbon dioxide from the body, reducing the blood's acidity and thus diminishing the oxygen available to cells. All this leads to "air-hunger", a desire to get more oxygen by breathing even more rapidly.

Herbal remedies: Certain herbs may help relax airways and expel mucus, including echinacea, coltsfoot, elecampane, liquorice, hyssop, and

Better breathing

When resting, always breathe deeply enough to expand both the lower and upper parts of your lungs. As your diaphragm descends, it presses against the abdominal organs, causing your abdomen to expand visibly. Doing the exercises described below once a day may reduce the severity and frequency of asthma attacks by helping to improve your breathing.

Rapid abdominal breathing
This relaxes the chest wall and helps reduce the severity of attacks.

1 With the neck, shoulder, and facial muscles relaxed, exhale strongly by contracting your abdominal muscles, then relax your abdomen as you inhale. Continue to breathe in this way, moving your abdomen with a slow rhythm, taking one breath every two seconds.

2 Gradually increase the speed until you are taking two breaths a second. Start with three sequences of 10 breaths each, then each week add another 10 until you do 30 in each sequence. Take a 20-second break between each sequence.

Breath-holding after exhaling
If you can hold your breath for no more than 30 seconds after exhaling, you may be in the habit of hyperventilating. This exercise allows you to measure and increase your breath-holding ability. Set aside a half-hour each day to practise. Aim to increase the length of time you can comfortably hold your breath to 40–60 seconds. This may take several months of regular practice. To help you achieve this goal, be sure to get some form of aerobic exercise daily (see p. 13) and take steps to manage the unavoidable stress in your life effectively (see p. 347).

1 Use a stopwatch or a clock with a second hand. Sit in a relaxed manner but with your back straight, and keep your mouth closed throughout the exercise.

2 Exhale normally, then hold your nose and count how many seconds you can hold your breath. Then inhale slowly and smoothly. Reduce your breath-holding time if you feel faint or dizzy or get "pins and needles" in your fingers.

thyme. Vervain and chamomile are especially good if you feel tense. Drink one to two cups of one of these teas daily.

Caution: For safety concerns, see pp. 34–37.

Treatment

As soon as you feel an attack coming on, remove yourself, if possible, from any trigger, stay in a warm room, and summon help.

Breathing exercise during an attack: Sit comfortably upright, put one hand on your chest and the other on your stomach, and concentrate on relaxing, taking slow, easy breaths through your nose and filling your chest by using your abdominal muscles. The hand on your stomach should move farther up and down with each breath than the one on your chest.

Homeopathy: The following remedies may relieve a mild attack of asthma. Choose the one that most closely matches your symptoms.

- Arsenicum: if attacks often occur at night and make you feel restless and anxious.
- Ipecacuanha: if you feel that your lungs are congested with mucus and you have coughing along with nausea.
- Natrum sulphuricum: if attacks are brought on by damp conditions.
- Pulsatilla: if you feel better in the fresh air.

Other therapies: Acupuncture can help some asthma sufferers.

Massage to avert an attack

If you sense that you may be on the verge of an asthma attack, ask a friend to perform the back massage described below to relieve chest tension. You should lie on your front with your head to one side, while your friend does the following:

1 Kneel at a right angle to the person's back and place the heels of your hands on the far side of the upper back, avoiding the spine.

2 Leaning forwards and pressing firmly, slide your hands over the ridge of muscle running alongside the spine from top to bottom.

3 Repeat this action, moving the starting point slightly down the spine towards the waist each time, until reaching the bottom rib. Do the same on the other side.

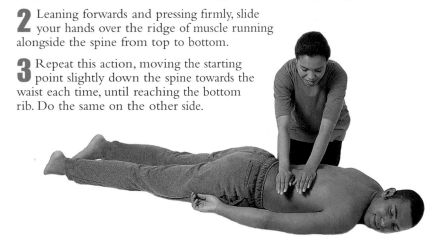

When to get medical help
- You need to find the cause of your asthma.
- Preventive measures don't stop attacks.
- You have an attack along with a worsening upper respiratory infection.

Get help right away if:
- You feel very breathless or dizzy, turn pale or bluish, or have chest pain during an attack.

See also:
Anxiety, Food Sensitivity

Back Pain

The health of your back is affected mainly by how you carry and use your body, but also by diet, digestion, and the way you cope with stress. Back pain affects most people at some time in their lives and is the leading cause of disability in the United Kingdom. Many natural therapies can be used to provide relief from back pain, even when orthodox therapies fail to help.

Back pain may be a slight problem lasting only a few days, or a serious, long-lasting disability. Pain can result from any weakness in the back or from injury. Carrying heavy objects, exercising without due care, using a poor lifting technique, twisting or turning awkwardly—all these can cause damage. Bad posture, lack of exercise, sitting in one position for long periods, pregnancy, and being overweight make back pain more likely and may worsen existing pain. Stress is another major cause of back pain. Less commonly, backache results from a condition unrelated to the back, such as a urinary-tract infection.

Causes

Pain can result from problems with any part of the complex structure of the back (see illus-

tration, below). Often it is caused by a combination of factors, such as:

- Strained muscles or ligaments.
- Damage to a facet joint in the spine.
- Muscle spasm, which occurs as a protective measure to immobilize the spine following injury, strain, or stress. The muscles surrounding the damaged area become tense and painfully stiff, often preventing all movement.
- Herniation of an intervertebral disc ("slipped disc"). This is a less frequent cause of back pain than was once thought. The pain results from the soft centre of a disc bulging out and pressing on a nerve or irritating nearby tissues. Discs tend to dry out in later life, making them more vulnerable to injury. Inactivity also increases the risk of disc problems.
- Sciatica, a sharp, severe pain in the buttock that shoots down the back of the leg. The most common causes of this are a herniated disc pressing on the sciatic nerve, and osteoarthritis.
- Osteoarthritis, which is one of the most common causes of back pain among those over 50.
- Osteoporosis, which can weaken the vertebrae so that they are easily crushed or fractured.
- Spinal stenosis, which is the narrowing of the spinal canal, causing pressure on the spinal cord or the nerves joining it.

The structure of the spine

Your spine contains 33 firmly linked vertebrae. The first 24 are separated by natural shock absorbers consisting of cartilage discs filled with gel. The spine is held together by ligaments (thick bands of fibrous tissue), muscles, and the facet joints of each vertebra.

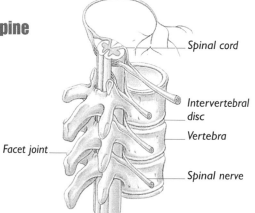

Spinal cord

Intervertebral disc

Vertebra

Facet joint

Spinal nerve

Prevention

The most important way to prevent back pain is to maintain your back's strength and flexibility with daily exercise. It is also essential to avoid excessive or awkward movements that can cause stress and strain. Take the following precautions to protect your back:

■ Never carry a heavy bag in one hand or on one shoulder. Instead, divide the contents between two bags, held one on each side. Better still, carry a backpack (use both straps), or sling a long-handled bag over the opposite shoulder.

■ To lift a heavy object, get as close to the object as possible, bend your knees, then lift by straightening your knees while keeping your back as straight as you can.

■ Don't twist your back while holding something heavy; move your feet to turn your whole body instead.

■ Sit with your back straight, making sure that your lower back is well supported. When driving, recline the seat to increase the thigh-to-trunk angle.

■ Wear comfortable shoes with good support that will not cause an abnormal gait. Shoe alterations prescribed by a chiropodist or other podiatrist can often reduce back strain.

■ If you work for long hours at a computer, make certain that you sit correctly (see photograph, left). Use a footrest if you can't place your feet flat on the floor.

■ Never stoop to do a job such as weeding. Kneel instead.

■ Check that your mattress is firm enough to support your back, but not overly hard. If it is too soft, you can make it firmer by placing a board between the mattress and base.

Treatment

If stress is a contributing factor, try to reduce the sources of stress in your life, and use better stress-management techniques (see p. 347).

Heat and cold: This form of hydrotherapy improves circulation and promotes healing.

■ To ease the pain of strained muscles or sciatica, apply a towel that has been dipped in hot water and wrung out.

■ Apply a hot towel (see above) for two minutes and then a towel wrung out in cold water for one minute. Alternate the two for about 15

The supine twist

This gentle exercise can ease back stiffness (stop if it causes pain).

Lie on your back. Place your left foot on your right knee, and lower your left knee to the right, twisting from the waist. Rest your right hand on your left thigh. Stretch your left arm out to the left with palm up and look towards it. Relax in this position for one minute and then repeat on the other side.

Avoiding strain
When working at a computer for any length of time, be sure that your knees are bent at a 90-degree angle and that your lower back is well supported by a correctly adjusted chair.

Exercises for your back

Prolonged bed rest is no longer recommended for back pain. Instead, get up and about as soon as possible and, after checking with your doctor, begin to do a few minutes of gentle exercise each day. Build up a regular exercise routine to strengthen both your back and abdominal muscles and to increase your flexibility. The following exercises can help to relieve strained muscles or ligaments, irritated facet joints, and sometimes can help ease sciatica. As with any exercise, you should stop immediately if you feel pain.

1 Lie on your back on the floor, with a small cushion or rolled towel under your neck. Bend your knees, then press the small of your back to the floor, tilting your pelvis slightly upwards. Placing both hands around one knee, gently pull your thigh towards your chest for 10 seconds, then lower your foot to its original position. Do the same with the other leg, and repeat the sequence 10 times.

2 Lie on your back on the floor, with legs straight and feet flexed. Rest your linked hands on your lower abdomen. Breathe slowly and deeply, and as you inhale, slowly raise your straight arms—keeping your hands linked—until they are vertically above your face. Without stopping the movement, start exhaling, moving your arms down behind your head to rest on the floor. Reverse this procedure, inhaling as you raise your arms, and exhaling as you lower them to rest your linked hands on your lower abdomen. Press down firmly with your hands. Repeat the whole exercise five or six times.

3 Lie on your back with your knees bent to your abdomen and your hands holding your lower thighs. Curl your back slightly, with your weight on your lower back, then rock from side to side several times.

4 Lie face down with your palms on the ground just below shoulder level.

5 Raise your head and shoulders as far as is comfortable for you, using the muscles of your back and without pushing with your arms. Hold for a few moments and then lower yourself gently to the floor. Do this 10 times.

minutes. This is especially helpful if your pain results from muscle strain. Use just the cold towel for pain resulting from inflammation.

Herbal remedies: These can help relieve pain and stiffness, as well as stimulate circulation.
- Rub equal amounts of glycerin and tincture of cayenne into the skin over the painful area.
- Apply frequent hot compresses of cramp bark, valerian, chamomile, or ginger to the back.
- For sciatica, rub the skin with St. John's wort oil (see p. 192).
Caution: For safety concerns, see pp. 34–37.

Homeopathy
- Arnica: if back pain has followed an injury, take orally twice a day.

- Hypericum: to relieve back pain resulting from nerve irritation, take orally twice a day.

Other therapies: Chiropractic and osteopathy may help relieve persistent back trouble.

When to get medical help
- The pain does not improve after a few days or worsens at any time.

Get help right away if:
- The pain is sudden and severe.
- You lose sensation or strength in a limb.

See also:
Arthritis, Neck and Shoulder Pain, Osteoporosis

Massage for back pain

Gentle massage of the muscles on either side of the spine can often ease pain. Try these special massages to alleviate sciatica (below left) and stretch the spine (below right).

Stretching the spine can help some back pain by relaxing tense bands of muscle on either side of the spine. Lie on a mat or on folded blankets or towels.

Buttock massage
This simple massage can ease sciatica. Lie face down, with your partner kneeling beside your hips. He or she should then lean over and knead your opposite buttock using slow, firm, circling movements of the whole hand. Repeat on the other side.

Spinal stretch
Kneel on the floor with your forehead touching the ground, your arms pointing back, and your hands near your feet. You may need a cushion between your calves and buttocks for comfort. Have your partner kneel beside you with one hand at the top of your spine and one at the base. He or she should then push down with each hand pressing slightly away from the other, so as to stretch out your spine.

Bad Breath

Unpleasant breath, or halitosis, usually results from eating strongly flavoured foods, such as garlic or onions, smoking cigarettes, or drinking alcohol. Your breath may also be affected by disorders involving the mouth, lungs, or digestive tract. Sweeten your breath by using herbs, spices, and other kitchen-cabinet items. Paying attention to oral hygiene is also vital.

Mouth freshener
Add two drops of peppermint oil to a cup of warm water and gargle.

Persistent bad breath is frequently caused by poor oral hygiene, tooth decay, or gum disease. Fasting or eating too much food of animal origin may also be to blame. Occasionally, bad breath is a symptom of sinusitis, a respiratory disorder, anaemia, diabetes, or a fever. In addition, if you produce insufficient stomach acid and eat fruit after a protein meal, the fruit may ferment in the stomach, producing gases that cause bad breath.

Prevention

Eat plenty of green leafy vegetables, because these contain the plant pigment chlorophyll, a natural breath freshener. Limit your intake of refined carbohydrates, coffee, alcohol, and dairy products, since they encourage bad breath.

If you also suffer from poor digestion or flatulence, you may be producing insufficient stomach acid. To boost your digestion, add vinegar to meals of meat, fish, or eggs. Apple-cider vinegar provides acetic acid in roughly the same strength as that of normal gastric acid. Some people benefit from gargling (do not swallow) with a solution of two teaspoons of this vinegar in half a pint of warm water each day. Brush your teeth meticulously twice daily, then use dental floss or tape. If you are not sure how to do this correctly, consult your dentist. Brushing the tongue also helps prevent bad breath.

Treatment
Herbal remedies

- Chew parsley leaves to counteract the scent of garlic on the breath.
- Chew aniseed or fennel, dill, cardamom, or caraway seeds to prevent bad breath at any time.
- Drink fenugreek or peppermint tea to sweeten breath after a meal.
- A daily mouthwash of echinacea tea helps treat gum infections that cause bad breath. Add two teaspoons of echinacea root to one cup of water. Simmer, covered, for 10 minutes, then cool before using.

When to get medical help
- Your gums bleed or you have mouth ulcers.
- Bad breath persists despite self-help treatment.
- Bad breath is a new problem and does not stop within a week.

See also:
Candida Infections, Colds, Gum Problems, Indigestion

Bloating

Distension of the abdomen is a very common problem. It often results in discomfort, especially if the swollen abdomen makes clothes too tight. Bloating has a variety of possible causes, including fluid retention and excess wind, most of which can successfully be treated with a range of natural remedies and lifestyle changes.

The most likely cause of abdominal distension is an overfull stomach and intestines. Overeating—especially of foods high in refined carbohydrates and fats, such as biscuits, cakes, and pies—may result in excess wind production, leading to bloating. Other causes include fluid retention, gastroenteritis, food sensitivity, constipation, irritable bowel syndrome, coeliac disease (intolerance to gluten, a cereal protein), lactose intolerance (inability to digest lactose, the sugar in milk), inflammation or ulceration of the stomach or intestinal lining, and candida infection. Bloating may also result from a disruption in the population of bacteria in the intestines, leading to an insufficient number of "friendly" bacteria, such as *Lactobacillus acidophilus*.

Many women experience bloating just before their period, when hormonal changes lead to fluid retention. Occasionally, heart, kidney, or liver disease, an ovarian cyst, or another condition or complication causes distension.

Prevention

Bloating from fluid retention: Fluid retention can be associated with a salty diet. Many commercially processed foods contain high levels of salt and other sources of sodium (including bicarbonate of soda and monosodium glutamate, or MSG), so always read labels. Minimize your sodium intake in the following ways:

How you eat and what you eat
Choose your foods carefully, and take your time during meals to relax and chew well.

- Avoid processed meats (such as bacon, ham, sausage, and corned beef), stock cubes, yeast extract, and salted snacks.
- Buy low-salt foods. Look for low-salt bread, cereals, canned and frozen vegetables, butter and spreads, sauces, pickles, and soup.
- Don't cook vegetables in salted water.
- Don't add salt or sodium products to home-cooked recipes or to the food on your plate.

Bloating from digestive problems

- Eat small, frequent meals.
- Chew each mouthful well and avoid talking while eating. Powerful enzymes in the mouth aid digestion.
- Avoid fizzy drinks.
- Don't drink during meals. Fluid dilutes stomach acid and digestive enzymes, which may slow the passage of food through your stomach. In particular, any fruit eaten is then more likely to ferment and produce wind.
- Avoid eating starchy foods, such as pasta and potatoes, in the same meal as protein or fruit.
- Avoid excessive quantities of legumes, fatty foods, and raw fruit and vegetables, as they can increase the volume of wind in the intestines.
- Curtail a high fibre intake, especially of insoluble fibre, such as wheat bran.
- Take some papain, a papaya enzyme preparation from health-food shops, with each meal.
- If you suspect a food sensitivity, follow an exclusion diet to identify the cause (see p. 214).
- Try for one day eating only finely grated or puréed apples and drinking only mineral water or peppermint tea. On the next day, add some vegetable soup or steamed vegetables to the midday and evening meals, and then expand your diet over the next two days to include savoury rice and live yogurt. After that, continue with a normal healthy diet.

Bloating from abnormal intestinal flora

- Eat foods, such as yogurt, containing *Lactobacillus acidophilus*. If you're milk-sensitive, take an *acidophilus* supplement, widely available in health-food stores.

Premenstrual bloating: Try the following for the three days before bloating usually begins:
- Eat more foods rich in vitamins B_6 and E, magnesium, and essential fatty acids (see p. 12).
- Cut out caffeine and added sugar and salt.
- Mix three drops of juniper berry oil and two drops each of rosemary and lavender oils with five teaspoons of grapeseed or wheat-germ oil. Omit juniper berry and rosemary oils if you are trying to conceive. Ask a friend to use this mixture to massage the backs of your legs (with upward strokes), lower back, and abdomen (with clockwise circular movements).
- Take 500 to 1,500 milligrams of evening primrose oil capsules twice a day, in this case starting one week before your period.

Treatment

Herbal remedies

- For bloating that occurs with indigestion or wind, pour a pint of boiling water over an ounce of cut angelica root, cover, and steep for 10 minutes. Take two tablespoons of this tea three times a day before meals. Or, chew some raw angelica leaves or root.

A fragrant compress
To prevent premenstrual bloating, use a compress. Place a towel in a bowl of warm water to which you've added three drops each of rosemary and juniper berry oils and two drops each of lavender and cypress oils. Wring out and apply to the abdomen and chest for as long as is comfortable. Do not use this treatment if you are trying to conceive.

- Have tea made from wild yam, peppermint, ginger, cinnamon, fennel, lemon balm, or chamomile after a meal to prevent wind-related bloating. If preferred, drink half a cup of warm water containing a few drops of a tincture made from one of these herbs.
- Prevent wind by adding warming spices—such as cayenne, ginger, and cardamom and caraway seeds—to your meals.
- For bloating resulting from premenstrual fluid retention, drink a cup of celery seed or horse-tail tea twice daily.

Caution: For safety concerns, see pp. 34–37.

Aromatherapy: For bloating with indigestion, gas, or fluid retention, try the following treatments with essential oils (but not if you are or might be pregnant):

- Add three drops of black pepper oil to three tablespoons of sunflower oil. After a meal, smooth gently onto your abdomen, using clockwise movements.
- Add four drops of peppermint oil and two drops each of juniper berry and black pepper oils to half a teaspoon of

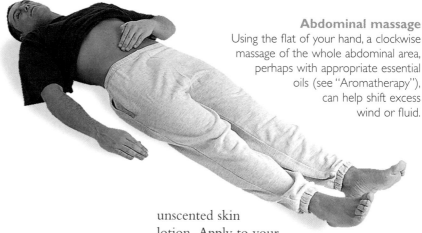

Abdominal massage
Using the flat of your hand, a clockwise massage of the whole abdominal area, perhaps with appropriate essential oils (see "Aromatherapy"), can help shift excess wind or fluid.

unscented skin lotion. Apply to your abdomen with slow, clockwise movements.

- Soak in a bathtub of warm water to which you have added three drops of black pepper oil and three drops of fennel oil.
- Massage your abdomen with two teaspoons of grapeseed or wheatgerm oil to which you have added two or three drops of cinnamon, ginger, clove, or peppermint oil.

Plant remedies for fluid retention
Cucumber, parsley, nettles, yarrow, elderflower, and meadowsweet have all been used to remove excess fluid. Include them in your diet or make herbal teas.

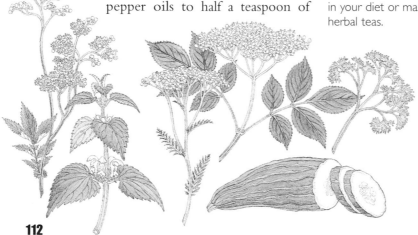

When to get medical help
- You have persistent or worsening pain.
- You have vomited or you have a fever.
- Your ankles are swollen.
- You have continued bloating.
- You have unexpected bloating when pregnant.

See also:
Candida Infections, Constipation, Diarrhoea, Flatulence, Food Sensitivity, Indigestion, Irritable Bowel Syndrome, Premenstrual Syndrome

Breastfeeding Problems

Your milk is the best food for your baby. It provides all the nutrients he or she needs as well as protection against infection and allergy. Breastfeeding also gives you and your baby a special sense of closeness, and it has health benefits for you. Any difficulties can usually be overcome with proper information and the support of family, friends, and health-care professionals.

Breast milk is nutritionally perfect for babies and contains enzymes that make it easy to digest, as well as substances that boost immunity and enhance growth. Nourishing your baby with breast milk alone for the first six months and continuing to breastfeed for at least another six months protects against many infections. Breastfed babies rarely develop gastroenteritis and are less likely to suffer from lung, ear, and urinary-tract infections or to develop such allergic conditions as asthma and eczema. Breast-feeding may also reduce the risk of sudden infant death syndrome (SIDS), or cot death.

Almost every woman is able to breastfeed, although some, especially first-time mothers, experience problems. Common challenges include sore or cracked nipples, engorged breasts, blocked ducts, and a poor milk supply.

Prevention

Diet: To maintain your energy level and ensure an adequate milk supply, increase your calorie intake by 400 to 600 Calories a day, concentrating on nutritious foods. You will need to maintain a calcium intake of about 1,200 milligrams a day (see "Calcium-rich foods", p. 298). Eat small amounts often, with a meal or snack between each feed. Drink at least eight glasses of caffeine-free fluids a day. Be aware that most substances you consume will be present in

Vital protection
In the first few days after giving birth, your breasts produce a special milk called colostrum. This contains antibodies to help protect against infection.

your breast milk. Therefore ask your doctor before taking any medications. Don't have beer, wine or other alcohol for at least two hours before breastfeeding.

Everyday routine: Your baby will feed frequently at first, so delegate other tasks whenever possible. Get as much rest as you need, especially in the first few weeks following the

birth, when your milk supply is becoming established. Breastfeeding triggers the release of hormone-like substances called endorphins, as well as prolactin (a hormone that causes milk glands to produce milk), all of which can induce a feeling of calm and well-being.

Relaxation: Try to be relaxed when you feed your baby. If you feel tense or anxious, you may fail to release enough oxytocin, the hormone that triggers the initial flow of milk. Your hungry baby will then become frustrated, making you more tense. Simple measures, such as sitting quietly with a warm drink or watching TV, may be enough to make you feel relaxed. If this does not work, try using the relaxation techniques, such as deep breathing exercises, that you learned in ante-natal classes.

Treatment

It is important to know how to tackle the common breastfeeding challenges that can beset any mother. If you can't solve a problem yourself, get advice from a trained breastfeeding counsellor or other health-care professional.

Engorged breasts: When the mature milk comes in a few days after the birth, your breasts may feel tight, swollen, and painful. This makes it difficult for your baby to "latch on"—to take hold of the nipple and areola properly—and this in turn can make the nipples sore.

■ Feed frequently, day and night, to prevent too much milk from accumulating in your breasts.

■ To relieve some of the pressure and allow your baby to take the breast more easily, express a

Collecting breast milk

If you are going to be away from your baby, allow time to collect some milk in advance by hand expressing or pumping your breasts. Your partner or babysitter can then give your milk to the baby, from a cup, ideally, or from a bottle. If you will be collecting milk only at home, a full-size electric pump may be appropriate. These can often be hired. If you need to use a pump outside the home, choose a smaller type—either manual or electric. Remember to follow the instructions on sterilizing the pump and to use sterilized containers or milk-collection bags (for freezing).

little milk before a feed, either by hand or with a breast pump (see box, above).

■ If your breasts are so full that expressing is impossible, bathe them in hot water for several minutes or place a hot compress on each breast before a feed to help the milk flow. Place a cold compress on each breast between feeds to reduce discomfort.

Sore or cracked nipples: If your nipples hurt when your baby feeds, try the following:

■ Place your baby so that he or she takes the whole areola into the mouth. Change the position in which you hold your baby at each feed to ensure that no one part of the nipple takes too much of the force of your baby's sucking action. Your counsellor can advise you.

■ Treat any engorgement as soon as possible (see previous section).

- Encourage your milk to flow—before you put your baby to the breast—by relaxing and making yourself comfortable.
- Breastfeed your baby frequently, but if your nipples are very sore or cracked, limit the length of each feed for a day or two, and express any remaining milk.
- Offer the less sore nipple first.
- Don't use soap on your nipples.
- After a feed, dry your nipples, then apply some breast milk or calendula (marigold) ointment.
- Allow your nipples to air-dry as much as you can. If possible, expose your nipples to sunlight for just a few minutes each day.
- If leaking is a problem, keep your nipples dry by wearing breast pads (not plastic-lined), changing them frequently.

Blocked ducts: A tender, red lump in your breast may be a sign of a clogged-up milk duct. Try to clear the duct as soon as possible to avoid infection. The following measures will help:

- Empty your breasts thoroughly each time your baby feeds. Feed your baby from the affected breast first.
- Feed your baby more often, and express between or after sessions, if necessary.
- Gently but firmly massage the lump toward the nipple during a feed.
- In the bath or shower, soap the area of the affected duct and then gently run a wide-tooth comb over it to stimulate milk flow and help clear the blockage.
- Do some arm-swinging exercises (see right).
- Ensure that your bra fits well and isn't pressing too hard and causing the blockage.

Boosting circulation
Prevent and treat blocked milk ducts by swinging your arms in big circles—forwards and backwards—for five minutes every hour or two.

- Vary your feeding position at each nursing.
- To relieve pain, place a hot, wet compress or a covered hot-water bottle on the breast every hour. Before a feed, splash the breast or immerse it in hot water for 5 to 10 minutes.
- Increase your intake of vegetables, whole grains, oily fish, and vegetable oils. A supplement of vegetable lecithin may also help.

Poor milk supply: Only rarely is a woman unable to produce a sufficient amount. Breastfeed frequently and for as long as your baby wants, since this will stimulate the milk to flow.

- To try to increase your milk supply, drink a cup of tea once or twice daily of chasteberry, nettle, fennel, vervain, raspberry leaf, cinnamon, blessed thistle, or marshmallow. Caraway, coriander, cumin, sunflower, sesame, celery, and fenugreek seeds are also said to help.

Caution: For safety concerns about herbs, see pp. 34–37.

Homeopathy: Calcarea carbonica, Pulsatilla, and Urtica are believed to promote milk flow. Silica may help cracked nipples.

When to get medical help
- A breastfeeding problem worries you or doesn't clear up in a few days.
- A lump or a red area persists or worsens.

Get help right away if:
- Your baby has not been feeding for 24 hours.

Bruises

Most bruises are the result of minor bumps or falls and disappear after only a few days. Even so, they may cause pain, especially if the site of the bruise is pressed or knocked. You may be able to keep a bruise from developing, or at least reduce its severity, if you carry out self-help treatment of the affected area as soon as possible after the injury.

A bruise is a discoloured area of skin that forms usually after a blow damages underlying small blood vessels (capillaries). At first the blood leaking from the capillaries makes a bruise appear black and blue. Then, as the blood breaks down, the bruise turns yellow, green, or purple.

Treatment

Immediate action: Apply a cold compress as soon as possible after the injury. Soak a cloth in ice-cold water and place over the area for 10 minutes, or use a pack of frozen peas wrapped in a towel. Take one or two pilules (tiny tablets) of the homeopathic remedy Arnica 30c as soon as possible to minimize bruising. Repeat the dose every 15 to 60 minutes, depending on the severity of your injury, and continue for several doses. Also apply herbal arnica ointment or cream if the skin is not broken.

Herbal remedies: A number of herbs are effective in reducing the severity of bruising.
- If the skin is not broken or scraped, make a compress of comfrey (also known as bruise-wort) by soaking some gauze in comfrey tea and placing over the injured area for an hour. Make the tea by pouring a pint of boiling water over one ounce of dried—or two ounces of fresh—leaves, infuse for 10 minutes, then strain. (Do not drink the tea.)

- A cabbage-leaf poultice is also soothing. Take the greenest leaves of a cabbage and discard the ribs. Warm the leaves in hot water, then drain them and flatten with a rolling pin. Place several leaves over the bruise, and secure them with a bandage or taped cling film. Change the poultice every few hours.
- Calendula ointment and distilled witch hazel solution are traditional remedies for bruising, as both help stop bleeding under the skin.

Caution: For safety concerns, see pp. 34–37.

Aromatherapy: Mix five drops of sweet marjoram oil and two drops each of myrrh (omit during the first 20 weeks of pregnancy) and German chamomile oils with five teaspoons of calendula (marigold) oil or lotion. Apply to unbroken skin as soon as possible after the injury and repeat hourly until the pain subsides.

When to get medical help
- Bruises appear for no apparent reason.
- Skin discoloration is accompanied by severe pain and swelling.
- A bruise does not fade after a week.

See also:
Strains and Sprains

Burns

Heat, friction, and chemicals can all burn or scald the skin. Many burns result from accidents at home, and prompt first aid can mean the difference between rapid recovery without scarring or further damage to the skin. After applying standard first aid, try natural treatments to soothe discomfort and promote healing.

Fast action is essential for both major and minor burns. You can safely self-treat small burns that affect only the top layer of skin. Carry out first aid as described in the box below. Deeper or more extensive burns and those caused by electricity need emergency medical attention.

Treatment

After first-aid treatment, you can let minor burns heal by themselves, but the following may help.

Aromatherapy: Apply lavender oil to the burn and cover it with gauze, secured at the edges with tape. Reapply the oil every two hours for 24 hours, without removing the gauze. If the burn has not then healed, apply six drops of lavender oil and two of geranium oil mixed with one teaspoon of grapeseed or sweet almond oil. Repeat four times daily until healed.

Herbal remedies: Apply one of the following:
- Aloe vera gel.
- St. John's wort oil.
- Calendula (marigold) ointment.
- A compress soaked in diluted distilled witch hazel or cooled tea made from chamomile, marigold, elderflower, or chickweed (or use a flannel soaked in whole milk).
- Elderflower lotion or ointment.
- Crushed yellow dock leaves.

Homeopathy
- Apply Urtica or Hypercal (ointment or tincture) to burns that do not blister.
- Take Cantharis by mouth every hour for burns that blister.

Flower essences: Take Rescue Remedy to counter the emotional shock of the burn.

First aid for minor burns

- As soon as possible, immerse the area in cold water for at least 10 minutes. This cools the skin, stops burning, and relieves pain. If water is unavailable, use any cold, non-irritating liquid, such as milk or iced tea.
- Remove jewellery or clothing that may constrict the area should swelling occur.
- Cover the burn with a sterile, non-adherent dressing. If unavailable, use any clean, dry, absorbent cloth.
- Don't break blisters.

When to get medical help

Get help right away if:
- A burn covers more than two inches.
- A burn is deeper than the top layer of skin.
- A severe burn affects the mouth or throat.
- There is pus or increasing swelling or redness.
- The burn was caused by an electric shock or by caustic chemicals.

Candida Infections

A yeast-like fungus, *Candida albicans* usually lives in the intestines, along with many other varieties of microorganisms. These organisms are known collectively as the intestinal flora. The balance among them normally prevents overgrowth of any one type. However, sometimes the balance is upset, allowing candida to multiply unchecked, causing infection.

Also called thrush or a yeast infection, a candida infection can affect any area of mucous membrane or moist skin. In the vagina it causes a thick, whitish, curdlike discharge and, frequently, itching and soreness. A candida infection of the mouth—oral thrush—causes sore, raised, creamy-yellow patches in the mouth and on the tongue. Candida can also affect the nipples and the folds of skin at the sides of the nails. Some babies develop a rash infected with candida; this is typically bright red with white patches. Generalized infection of the body with candida is a rare but serious condition.

Some people in very poor health may develop an overgrowth of candida in the intestines which can cause a variety of symptoms, including poor digestion, wind, bloating, and even fatigue, headaches, and other aches and pains.

Candida infection can be triggered by anything that upsets the balance of microorganisms in the body—for example, poor general health, pregnancy, stress, and taking antibiotics, oral steroids, or the contraceptive Pill. Other possible causes include uncontrolled diabetes, reduced stomach-acid production, and a diet high in yeast-containing foods, alcohol, and refined carbohydrates, such as sugar and white flour.

Prevention

Diet: If you are vulnerable to candida infections, there are a number of adjustments you can make to your diet to boost your resistance.

- Cut out, or at least limit, foods containing added sugar and refined carbohydrates.
- Reduce your intake of foods containing natural sugars, such as milk, fruit, and wine.
- Drink no more than three cups of caffeine-containing tea, coffee or cola a day.
- Eat two crushed or finely chopped cloves of raw garlic three times a day, perhaps in a salad dressing or added to soup at the last minute. Alternatively, take garlic capsules or tablets.
- Choose more foods containing B vitamins, particularly biotin and vitamins B_6 and B_{12}.
- Eat some yogurt with live *Lactobacillus acidophilus* cultures each day. These bacteria help

Anti-candida foods
Include in your diet foods that discourage candida, such as yogurt with live cultures, leafy green vegetables, garlic, and olive oil.

maintain a beneficial balance of micro-organisms in the intestines. You can also take acidophilus in tablet, capsule, or powder form.

■ Eat a healthy diet containing plenty of iron, magnesium, selenium, zinc, flavonoids, and vitamins A, C, and E (see p. 12).

Preventing vaginal yeast infections: If you are prone to vaginal candida, try the following:

■ Wear cotton rather than nylon underwear, and stockings rather than tights. Avoid tightly fitting trousers. Good circulation of air in the genital area discourages infection.

■ Make the water in your bath or shower as cool as is comfortable for you.

■ Change your towel frequently. Don't dry your genital area too roughly; sore skin is more prone to infections.

■ Avoid perfumed soaps, bubble baths, and other bath additives. Do not use scented vaginal deodorants, as these may irritate the skin, making it more vulnerable to infection.

■ Use a lubricant during sexual intercourse if you feel dry, since too much friction can cause soreness, which promotes candida infection.

■ When using the toilet, wipe from front to back to prevent intestinal microorganisms from entering your vagina.

Treatment
From the chemist

■ For a yeast infection of the vulva, apply boracic acid solution: dilute one teaspoon of boric acid crystals in a quart of tepid water.

■ For intestinal candida, take caprylic acid (coconut extract) in capsule form.

Herbal remedies: The following are known to inhibit the growth of intestinal candida: black walnut seed hull, goldenseal, and oregano oil. These can be taken as capsules or tinctures. Caution: For safety concerns, see pp. 34–37.

Aromatherapy: The antifungal action of lavender, sweet thyme, and tea tree oils may help clear up vaginal yeast infections.

■ Kneel in bathwater to which you have added essential oils (see illustration, right) and splash the water around the external genital area several times before sitting down. Remain in the bath for at least 10 minutes.

■ Mix two drops of sweet thyme oil and four of lavender oil with 2½ teaspoons of a cold-pressed oil, such as sweet almond or olive oil. Smoothe over your vulva to soothe soreness.

■ Apply two drops of tea tree oil to the top of a slightly dampened tampon and insert into your vagina. You should leave it there for 3–4 hours.

A healthy soak
To treat a yeast infection, sprinkle six drops of tea tree oil and two of sweet thyme oil into your bathwater, or add a few dandelion leaves.

When to get medical help

● A candida infection does not clear up after a few days of self-treatment.
● A baby or young child is affected.
● You develop a mouth infection.
● You suspect you may have a candida infection of the nipples.

See also:
**Diabetes,
Diaper (Nappy) Rash,
Vaginal Problems**

Cellulite

Up to four out of five Western women have lumpy, dimpled, fatty areas known as cellulite. Occasionally men are affected, too. The belief that the fat in cellulite is fundamentally different from fat elsewhere is popular among non-scientists, but incorrect. There is no special health risk associated with cellulite, unless you are also significantly overweight.

Certain parts of the body, such as the thighs and buttocks, are especially prone to developing dimpled fat. This cellulite can be hard to shift and usually appears when the reproductive hormones are in a state of flux—for example, at puberty, during pregnancy, around the menopause, and when a woman is on the Pill.

How cellulite forms

The fat of cellulite is separated by inelastic bands of connective tissue. These bands run between the skin and the deeper tissues, where they are firmly secured. Many people with cellulite are overweight. As a person gains weight, the fat cells swell but the connective bands stay the same length. The fat then bulges the only way possible—towards the skin—and as this happens, the bands pull at the skin, creating the characteristic dimpled appearance of cellulite. The problem often becomes more apparent with age.

Poor circulation of lymph may contribute to cellulite formation. Lymph is a milky liquid that carries excess fluid, waste products, and toxins from the tissues to the bloodstream via the lymphatic vessels. Poor circulation of lymph—because of excess fat, lack of exercise, or fluid retention—may allow these wastes to build up in the tissues. This may be one reason that some women who are not overweight have cellulite.

Prevention

Try to keep your weight at a normal level and avoid rapid increases in weight. Exercise regularly to improve your body tone, reduce fat, build muscle, and boost the circulation of lymph and blood. Avoid sitting or standing for long periods. Eat five servings of vegetables and fruit each day. Limit your intake of alcohol and caffeine. Drink enough water-based fluids each day to relieve thirst and produce pale-coloured urine.

Treatments

A number of natural therapies can help improve the appearance of the affected areas temporarily.

Massage and skin brushing: A firm massage improves the circulation, and massaging with a blend of essential oils may be especially helpful. Stimulation of the skin surface contributes to an improvement in the appearance of the skin.
- Mix four drops each of cedarwood and rosemary essential oils, and three drops each of cypress and patchouli oils, into five teaspoons

Exercise for cellulite control
Working out tightens the muscles and improves the appearance of areas of cellulite. Ask a fitness trainer about exercises for the parts of the body that are affected by cellulite in your case.

of cold-pressed vegetable oil, such as olive or sweet almond oil. Avoid cedarwood, cypress, and rosemary oils in the first 20 weeks of pregnancy. Massage the areas of cellulite twice a day.

■ Gently brush the affected parts with a soft, dry bristle brush. Start at the feet and work towards the heart. This boosts the circulation, removes dead skin cells, and may improve the appearance of the skin.

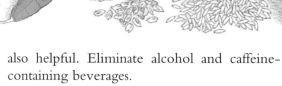

Herbal remedies: Dandelion contains substances that enhance the liver's ability to break down waste products and toxins. It also aids elimination from the blood of water and waste by the kidneys.

■ Add dandelion leaves to salads, or cook them as you would spinach.

■ Drink a cup of dandelion tea each day, using two ounces of fresh leaves to a pint of water.

Caution: For safety concerns, see pp. 34–37.

Detoxifying diet: This type of diet may help clear excess fluid, waste products, and toxins from the body. However, seek advice from your doctor or a nutritional therapist before embarking on such a programme.

■ Eat a high percentage of your food in the form of fresh, raw fruit and vegetables, as well as brown rice, seeds, and bean sprouts.

■ Drink plenty of fluids, especially spring water. Fruit and vegetable juices and herbal teas are also helpful. Eliminate alcohol and caffeine-containing beverages.

■ Avoid refined, processed foods, and limit your intake of animal products, such as meat and dairy products.

When to get medical help

Although you may not find cellulite attractive, there is little treatment other than a healthy diet and exercise that your doctor can suggest. If you are considering cosmetic surgery to remove excess fat, ask your doctor to recommend a reputable surgeon and take the opportunity to discuss the very real risks inherent in such a procedure.

See also:
Bloating, Overweight

Firming foods
Fresh fruit and vegetables are the mainstay of a detoxifying diet, since they are rich in nutrients, are easily digested, and contain a lot of fibre.

Chapped Lips

The lips may easily become chapped and sore, especially after exposure to hot, cold, or windy weather—for example, when taking part in outdoor sports. They are particularly vulnerable to chapping because, unlike skin elsewhere, they have no sebaceous (oil-producing) glands and therefore lack a protective oily barrier.

Hot or windy weather dries out the skin, while cold weather reduces the circulation in exposed areas. These adverse weather conditions can lead to sore, cracked, rough skin on the lips. Constantly wetting the lips is also likely to cause chapping. It is often tempting to lick sore lips to soothe them, but this will give only short-term relief. Not only is the prolonged wetting of the skin likely to dry it out and worsen the chapping, but as saliva dries, it can irritate the lips and increase soreness.

Prevention

Protect your lips against dry, cold or windy weather, especially if they are already dry. This means applying a barrier, such as lip balm or salve, when you go outdoors. Reapply frequently, especially after eating, drinking, brushing your teeth, or washing your face.

On sunny days, especially during the summer, use a sunscreen or sun block formulated for use on the lips. Lipstick can act as a barrier, but some types, especially those designed to last all day, can be drying, so buy only those containing moisturizers. In winter, use a humidifier or place bowls of water on or near radiators to counteract the drying effect of central heating on the skin and the lips.

Lip care
Apply lip balm at frequent intervals when chapping occurs.

Treatments

- Massage dry lips with a generous dab of petroleum jelly. Then leave it on for two minutes to soften the skin. To remove the petroleum jelly together with any loose flakes of dry skin, rub your lips with a warm, damp flannel or gently brush them with a soft toothbrush (kept for the purpose) dipped in warm water.
- Apply a little olive oil, which has soothing properties and makes the skin more supple.
- Combine rosewater with the same quantity of glycerine, which also has soothing properties and can reduce irritation and inflammation. Apply two or three times a day to chapped lips. Or mix two drops of rose oil with a teaspoon of cold-pressed vegetable oil, such as olive or sweet almond oil, and smoothe a little of this mixture onto your lips several times a day.
- Puncture a vitamin E capsule and apply the oil directly to the skin.

When to get medical help
- A sore on or near your lips fails to heal within a week.
- The sore area is encrusted or oozes.

See also:
Cold Sores, Dry Skin

Childbirth

A woman about to give birth needs not only good midwifery or obstetric care but also practical and emotional support. There are many ways in which you and your partner can prepare for the event and a wide variety of natural techniques you can use during labour to promote a trouble-free delivery and a positive start to your new relationship with your baby.

Childbirth in the United Kingdom generally takes place in the hospital under close supervision, sometimes with the assistance of a battery of equipment. However, even in this high-tech environment it is possible, in consultation with your birth attendants, to use natural therapies to relieve discomfort and promote a healthy labour.

Many women and their partners choose to attend ante-natal classes to learn how to use Lamaze and similar techniques for breathing and relaxation to ease childbirth. Most experts agree that preparation for the birth helps reduce anxiety. The presence of a supportive, encouraging midwife and, perhaps, a trained birth companion can also ease childbirth.

Positions for labour

Until relatively recently, British women could expect to give birth lying on their backs, often with their feet up in stirrups. Today, however, many professional birth attendants encourage women to adopt the position they find the most comfortable, and one in which the contractions can be aided by the force of gravity. Upright and semi-upright positions have the advantage of widening the pelvic outlet.

In the early stages, walking around the room can relieve discomfort and promote a speedier labour. When you feel the need to rest, try the positions suggested in the box on page 124. You

The three stages of labour

1 Intense and regular contractions of the uterus begin, and the cervix starts to dilate. The cervix opens wider with each contraction, until it is fully dilated. A transitional stage occurs at this time.

2 The second stage lasts from the end of stage one, when the mother feels the urge to push, until the delivery of the baby is complete.

3 Some minutes after the birth of the baby, the placenta is expelled, along with the amniotic membrane that enclosed and protected the baby in the uterus.

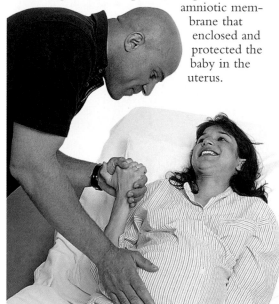

Has labour started?

In the following circumstances, labour may have started or be about to start. Call your doctor or midwife if:

- Your waters break.
- A plug of mucus (possibly blood-stained) is expelled from your vagina.
- You are having contractions, felt in your abdomen, back, or thighs, at least every 10 minutes.

may find it easier to go through labour on a sheet on the floor rather than on a bed, because it is easier to move and change position. Even when you need to be on a bed—for example, if your attendants wish to monitor the baby—you can return to a more comfortable position later.

Fluid and energy levels

Delivery rooms are often very warm, and you will lose fluids through sweating during labour.

You will need to drink to avoid dehydration and thus help maintain your strength and energy.

- Take frequent sips of water or fruit juice (if allowed) throughout labour.
- You may prefer to suck on a face flannel dipped in iced water or to suck crushed ice.
- Add Rescue Remedy, a standard blend of flower essences, to your drinking water to help you remain calm and in control.
- Homeopathic Arnica is said to help control

Positions for labour

Some popular and effective positions are shown here. Practise them during pregnancy, and during labour alternate among those that help you feel most comfortable.

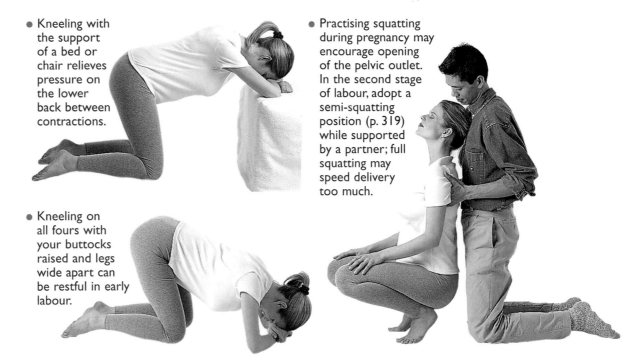

- Kneeling with the support of a bed or chair relieves pressure on the lower back between contractions.

- Kneeling on all fours with your buttocks raised and legs wide apart can be restful in early labour.

- Practising squatting during pregnancy may encourage opening of the pelvic outlet. In the second stage of labour, adopt a semi-squatting position (p. 319) while supported by a partner; full squatting may speed delivery too much.

pain. Take three drops or one tablet every two hours, if your doctor agrees.

- In the early stages of labour, eat one or two small snacks to maintain your energy level—as long as your medical attendants agree that you are not expected to need an anaesthetic. Suitable foods include cereal bars and bananas and other fruits. Vegetable broth, honey, and even glucose tablets also provide energy.

- You will almost certainly feel hungry once your baby is born, so ask your partner to bring a picnic or some soup.

Coping with contractions

Breathing and relaxation: Instead of tensing and holding your breath as you feel the pain of a contraction begin, concentrate on exhaling slowly through your mouth while keeping your face, neck, and shoulder muscles relaxed. Once your lungs have emptied, you automatically inhale—do so deeply, lowering your diaphragm towards your abdomen and breathing through your nose. This slow abdominal breathing will help you through the first stage of labour and ensure that you and your baby receive plenty of oxygen. When your baby is about to be born, resist the urge to hold your breath. Instead, breathe quickly and lightly, inhaling and exhaling from your chest rather than your abdomen. Your birth partner can help by reminding you of the breathing techniques you have learned.

Ante-natal class teachers encourage focusing on a particular point, sound, or sensation to help their clients relax through the contractions. Some

Massage strokes for your partner

Lie against a large pillow or on your side, whichever is more comfortable, while your partner tries these massage techniques to help ease the pain while you're in labour. You may find some are more comfortable than others. These massage strokes can be learned and practised any time during pregnancy.

1 Long strokes down your spine and around your hips and thighs.

2 Fairly firm pressure applied with the fists or a massage roller in the small of the back or the buttocks. This may be especially useful for backache.

3 A very light and rapid stroke—the so-called butterfly massage—at the base of your spine, just above the crease between the buttocks.

women like to have a photograph or illustration to focus on. Many other relaxation techniques can assist you through labour. For example, you may want to try a type of meditation or visualization that you have practised during your pregnancy (see pp. 46–47). Soft music can also be calming during labour, although some women prefer the room to be quiet.

Massage: Besides being emotionally soothing, massage can relieve backache or abdominal pain during labour. Your partner should be well-practiced in the techniques involved, which can be learned at ante-natal classes. He can use either his hands or, since massaging for long periods can be tiring, a wooden massage roller. Apply talcum powder, oil, or cream to lubricate the

Acupressure for easing labour

If you would like your partner to use acupressure during labour and delivery, identify the points beforehand, so that he can work quickly and easily when labour begins.

However, none of the points below should be pressed before labour has started. The pressure exerted during labour should be only as firm as is comfortable.

- To relieve pain and help make contractions effective, use the point (LI 4) on the back of your hand, at the base of the crease formed when you hold your thumb and index finger together. Your partner should support the palm of your left hand in his or her right, and then use the right thumb to apply pressure towards your wrist.

- An alternative point to promote effective contractions is Sp 6 at the side of the calf, about four finger-widths above the inside ankle bone and just behind the shinbone (tibia). Your partner should press very gently over the point or slightly up towards your knee.

- Have your partner apply pressure with both thumbs to the sacral (lower back) area to help relieve pain.

skin, choosing one that is not strongly scented; you may be extra-sensitive to smells while in labour. Experiment with the massages shown on page 125. Some strokes may not be soothing for you; tell your partner which help most. You may no longer want to be touched during the "transition" stage of labour—the stage just before your cervix dilates fully.

Hot and cold therapy: A well-covered hot-water bottle or a hot pack can help ease the discomfort of contractions early in labour and can be especially useful if backache is predominant. Or try rubbing an ice pack or a bag of ice cubes quickly over your back. Refresh yourself during labour by spraying mineral water on your face or wiping it with a cool, damp sponge or flannel.

Aromatherapy: Many essential oils can have a calming effect during labour. They're best used in a vaporizer. Experiment with different mixtures in advance to discover which you prefer. For example, try mixing four drops each of lavender and sandalwood oils with two drops of geranium oil and sprinkling eight drops of this mixture into the vaporizer bowl. This scents the air in a gentle, rather than overpowering, way.

Homeopathy: These remedies should present no risk to you or your baby, but consult your doctor before taking them at this time.
- Arnica: to reduce bruising and soreness after the birth. It is especially effective if you start

taking it as soon as you go into labour and then for several days after the birth.
- Caulophyllum: to help promote effective contractions, especially when they are short or stop altogether. Take a daily dose if your baby is overdue.
- Coffea: to reduce pain.
- Kali carbonicum: for backache.
- Pulsatilla: for weak or ineffective contractions, or for when you feel very discouraged or especially emotional.

Flower essences: The following relate especially to the difficulties of childbirth:
- Olive: if you are exhausted during or after childbirth.
- Walnut: if you need help adjusting to the changes in your body afterwards.

Traditional tonic
Some herbalists recommend raspberry leaf during pregnancy and childbirth for its alleged strengthening and toning effects on the uterus. Check with your doctor before taking.

See also:
Pregnancy Problems

127

Childhood Viral Infections

Chickenpox, measles, mumps, and rubella (German measles) are viral illnesses that can occur in childhood. Immunization has made most of these diseases much less common than in the past. Symptoms may include a fever, a rash, and swollen lymph nodes ("glands"). Natural therapies provide a range of treatment options that can alleviate symptoms.

The symptoms of the childhood viral infectious diseases are distinct, but treatment of these conditions at home is similar. If you suspect such an infection, seek medical diagnosis to exclude the possibility of a more serious disorder and to prevent complications, as with measles.

Chickenpox: Most non-immunized children catch chickenpox, the most common of the childhood viral infections. Two to three weeks after exposure to infection, an itchy rash appears, first on the trunk and then on the rest of the body. The rash consists of crops of pink spots that quickly become fluid-filled blisters. After about five days, they form crusty scabs, which gradually disappear within about 14 days. There is also usually a fever for the first few days.

Measles: This disease is now rare in the United Kingdom. It starts with a fever, runny nose, cough, and inflamed eyes. After three or four

Rest and fluids
A few days in bed are usually all that's required to recover from a childhood viral infectious illness. Your child may not feel like eating much, but be sure to provide plenty of drinks.

days, a rash of brownish-pink spots begins behind the ears. This spreads to the rest of the body, usually lasting five to seven days. The child may feel ill and have small white spots inside the cheeks. When the rash stops spreading, the temperature falls, and the child begins to feel better. Possible complications of measles include ear and chest infections and, more seriously, encephalitis (inflammation of the brain).

Mumps: This illness produces swelling and tenderness of one or both of the salivary glands below each ear. There is often a fever as well as earaches and headaches. Chewing may be painful. The illness usually lasts about a week.

When infections strike adults

These illnesses are rare in adults, but when they do occur, they often produce more severe symptoms than in children. Even if she has had the infection, a pregnant woman should avoid contact with chicken-pox, measles, and, in early pregnancy, rubella, since any of these diseases may harm the baby. Mumps is more likely to cause inflammation of the testes and ovaries when it occurs in adulthood.

Adolescent boys and men may develop painfully swollen testes, and girls and women may have abdominal pain from swollen ovaries. Mumps occasionally leads to fertility problems.

Rubella: Also known as German measles, this is usually a mild disease that causes a light red rash, sometimes accompanied by slight fever and swollen lymph nodes. The rash lasts between one and five days, and joint pain may occur after the rash has faded and continue up to 14 days. Rubella may make your child headachy and irritable.

Prevention

These viruses are spread by airborne droplets. Sufferers are highly infectious during the incubation period (usually two to three weeks, depending on the infection), which means that they can pass on the disease before symptoms appear. Immunization is the preferred way of preventing these diseases (see p. 130), since it also limits the spread of the infection in the community, but an attack of any of these infections usually provides natural immunity for life.

Treatment

General measures: Treatment of all these diseases includes rest, reducing any fever, and general nursing care. Keep an infected child, or one who may be incubating an infection, away from young babies and others who are at particular risk, such as pregnant women, the elderly, and the chronically sick.

Chickenpox: Discourage your child from scratching the spots, which could damage the

Recognizing symptoms

Chickenpox: The blistery rash normally first appears on the trunk and later spreads to other parts of the body. The photograph below shows a typical blister.

Measles: The dark, brownish-red rash, shown below, typically starts behind the ears and then spreads to the trunk and abdomen.

Mumps: The hallmark of this infection is swelling of the parotid (salivary) glands, located below each ear.

Rubella: The light red rash is usually first seen on the face and chest. It may later affect the limbs, as in this photograph.

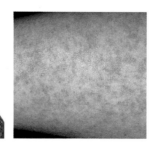

Chronic Fatigue Syndrome

This disorder is characterized by attacks of unexplained and disabling tiredness over a period of more than six months. The condition has been previously known as myalgic encephalomyelitis (ME), chronic fatigue and immune dysfunction syndrome, post-viral fatigue syndrome, Epstein-Barr virus syndrome, persistent virus disease, neurasthenia, and "yuppie flu".

About four out of five people with chronic fatigue syndrome are women, and the condition most often strikes people in their 20s and 30s. Diagnosis is made by excluding other disorders, such as infectious mononucleosis, that could account for the symptoms. Chronic fatigue syndrome can be disabling as well as lengthy, lasting months or even years, with symptoms varying in severity. Relapse is common, but many people have only minimal symptoms after five years, and others make a complete recovery.

The cause of chronic fatigue syndrome is unknown. Sometimes it follows a viral infection (with the flu virus or with the Epstein-Barr virus, which causes infectious mononucleosis); however, some experts think such infections are coincidental or the result of lowered immunity in the early stages of the syndrome. Recent research has found that people with the condition have higher than normal levels of melatonin, a hormone produced by the pineal gland and thought to influence daily body rhythms. Other suggested causes include pesticide or lead poisoning, and a low oestrogen level.

Psychological factors are often significant. One in two sufferers feels depressed, and one in four has some other mental health problem, but it is unclear whether this is a cause or an effect of the disorder. While studies show that the strongest risk factors for the condition are

Regaining your strength
Exercise is a vital part of any treatment programme for this condition, but build up your fitness gradually.

depression, anxiety, and stress, these are also an understandable result of the experience of having the illness, especially if it lasts for months or years. Mind and body are intimately linked by hormones, neurotransmitters, and other body chemicals whose levels can be altered by both "physical" and "mental" illness.

Symptoms

Besides overwhelming weariness, chronic fatigue syndrome can cause headaches, nausea, sleep problems, and weak, aching, or twitchy muscles. Some people have poor concentration and memory, difficulty keeping warm, or swollen lymph nodes ("glands"). Other possible symptoms include dizziness, slurred speech, sore throat, fainting, breathlessness, and an abnormal sensitivity to sound, light, touch, and smells.

Prevention

Improve your lifestyle with a good diet and regular exercise. Try to reduce sources of stress in your life, and learn and practise more effective stress-management strategies (see p. 347), especially if you've already had one or more episodes of chronic fatigue syndrome.

Treatment

Experts recommend a graduated programme of physical exercise. People who are also depressed may benefit from cognitive-behavioural therapy. This attempts to help you change the way you think about your illness and to deal with your symptoms.

Rest and exercise: When your symptoms are at their worst, a daily rest is essential, perhaps in the form of an afternoon nap. However, since too much rest can lead to loss of muscle bulk and therefore increased weakness, step up your physical activity gradually when you feel well

Stimulating oil
Add one tablespoon of fresh rosemary, which can relieve fatigue, to a bottle of olive oil. Use this fragrant oil for cooking or for salads.

Instant relaxation
Sit comfortably with your eyes closed. Take slow, regular breaths and focus your mind on a peaceful scene. Let any intrusive thoughts pass through your mind.

enough. Start with gentle exercise, such as walking or swimming. As you get better, increase the intensity and duration of exercise, but stop before you feel tired. If your symptoms come and go, take particular care to rest as soon as you feel them coming on. With luck, this will shorten the duration of an attack.

Stress management: Reduce stress by asking others to limit the demands they make on you, and learn to say no when necessary. Listening to relaxation tapes or practising gentle yoga postures and breathing techniques on a regular basis can have a calming effect.

Diet: A healthy diet can boost your immunity and enhance your ability to manage stress well. You should avoid eating too many foods containing added sugar, especially if your appetite is poor. A reliance on sugary foods can cause a succession of rapid peaks in your blood glucose level, with each peak being followed by a dip below normal that can make you feel tired as a result. You should also cut down on caffeine and alcohol, and ensure that you drink plenty of fluids.

Sensitivity to one or more foods may coexist with chronic fatigue syndrome and this can make the symptoms worse. If you suspect a food sensitivity, take steps to discover any foods (see p. 214) that may provoke symptoms, and then avoid them. Common trigger foods include milk, wheat, and fermented foods, such as vinegar, cheese, pickles, and soy sauce.

Cold Hands and Feet

Hands and feet feel cold when they don't receive an ample supply of warm blood containing oxygen and nutrients. The most common reason for this problem is exposure to cold air, especially from a draught or the wind. An underlying health problem may be responsible, but whatever the cause of the condition, there is much you can do to relieve symptoms.

Although cold extremities can be uncomfortable or even painful, the problem is usually relatively minor. When you are inadequately protected from the cold—especially if you also smoke or feel tired, faint, or anxious—your peripheral arteries become narrower. This restricts the circulation of warm blood to your hands and feet with the purpose of keeping the rest of your body warm. Cold extremities can also result from hormone fluctuations before menstruation, or a lack of circulating nutrients, as when a person is on a very strict diet or is suffering from an eating disorder. In addition, they may occur during the incubation period before an infection.

Other causes of cold hands and feet include Raynaud's syndrome and circulatory problems associated with such conditions as chronic bronchitis and arterial disease. Prolonged restriction of the blood supply may lead to chilblains— shiny red or purple lumps on the fingers or toes that can be painful and itchy.

Prevention

Prevent cold hands and feet and the development of chilblains in four simple ways:

- Dress warmly in cold weather.
- Stop or reduce smoking.
- Get exercise that raises your pulse rate for about 20 minutes every day.
- Eat regular, nutritious meals to fuel your body so that it raises the metabolism and creates heat. Small, frequent meals are better than one or two large meals a day.

Treatment
Keeping warm

Protect yourself from the cold by wearing several layers of thin fabric, since this traps heat more effectively than just a single thick layer. Close any gaps in clothing where cold air can enter, especially around the neck, wrists, and ankles. Wear shoes with thick soles, and insert insulating boot liners.

Winter clothing
Keep warm by wearing a hat (a large proportion of body heat is lost from the head), scarf (to close gaps), wind-proof warm coat, gloves or mittens, sturdy shoes, and thick socks.

Diet: Avoid heavy, fatty meals, which divert blood from the extremities to the stomach and intestines for several hours afterwards. Eat three servings of oily fish each week to encourage the production of prostaglandins (hormone-like

substances that help keep the blood flowing smoothly through the veins and arteries).

Herbal remedies: To improve your circulation and keep the blood from becoming too thick and slow-moving:

■ Add warming spices, such as ginger, mustard, and cayenne, to food and drinks.

■ Eat three cloves of fresh garlic, or take garlic tablets or capsules, each day.

■ Take a daily dose of ginkgo biloba, a traditional remedy for boosting the circulation.

■ Drink a cup of hawthorn tea either alone or with ginger, cinnamon, prickly ash, or dong quai twice a day.

■ Soothe chilblains with ointment containing calendula (marigold) or cayenne. (Do not use cayenne on broken skin.)

■ Soak your feet until they feel warm. Add to a basin of hot water one tablespoon of dried—or two of fresh—thyme, marjoram, or rosemary; two teaspoons of powdered ginger or crushed black pepper; or one tablespoon of mustard powder.
Caution: For safety concerns, see pp. 34–37.

A stimulating massage
In the bath or shower, rub your arms and legs with a loofah or bath brush. Cover up well afterwards. This may warm your hands and feet for several hours.

A circulation-booster
Hawthorn (*Crataegus oxyacantha*) has the ability to dilate blood vessels, thus making the blood flow more quickly.

Aromatherapy: Massage with stimulating essential oils can boost the circulation in the hands and feet.

■ Mix three drops each of rosemary and black pepper oils into a tablespoon of warm almond or olive oil, and massage your hands, arms, feet, and calves with the mixture. Use a firm stroke as you sweep your hand up your leg or arm, and a lighter one as you sweep down towards your hand or foot. (Avoid rosemary oil in the first 20 weeks of pregnancy.)

Hydrotherapy: To help the circulation before you go outdoors:

■ Place a hot compress on your feet or hands for three minutes, then a cold compress for one minute. Repeat several times, ending with a cold compress. Dry yourself briskly.

Homeopathy: To relieve chilblains, apply a thin layer of calendula or Hypercal ointment. The following remedies, taken by mouth, may be useful for chilblains:

■ Agaricus: if chilblains are worse when cold.

■ Pulsatilla: if they are worse when hot.

When to get medical help
● The condition occurs often or suddenly.
● Self-help remedies have no effect.

See also:
Anxiety, Arterial Disease, Eating Disorders, Raynaud's Syndrome

Cold Sores

A cold sore not only looks unsightly and feels unpleasant, but the condition is also highly contagious. You therefore need to be very careful to avoid passing on the infection to others. If you take action as soon as you start to feel the tell-tale tingling or hot sensation that heralds the advent of a sore, you can sometimes prevent a fresh outbreak.

Cold sores are small blisters or ulcers that occur on or near the lips or around the nose. There are usually some warning signs that a sore is about to erupt. The area may suddenly itch and tingle or feel hot, sore, and irritated. A day or two later, a blister appears. It enlarges and bursts to form an open sore that gradually crusts over and dries up. Most cold sores disappear in about a week with or without treatment.

Cold sores are usually caused by *Herpes simplex* type I viruses. You can become infected by direct contact with a sore, the fluid from a sore, or the saliva of someone with a sore. There are usually no outward signs of the first infection, but children occasionally become very ill. Newborn babies are particularly at risk.

Once infected, you either develop immunity to further infection or the virus lies dormant in the skin until something reactivates it. The most common triggers of a new outbreak are:

- Exposure to very hot or cold temperatures
- Strong sunlight
- A cold (which is why they're called cold sores)
- Stress
- Fatigue and exhaustion
- Menstruation
- Deficient nutrition.

Don't spread infection

Cold sores are contagious, so don't kiss anyone when you have a cold sore. Change your towel, flannel, and pillowcase daily, and don't share them. Try not to touch a sore, but if you do, wash your hands as soon as possible. Avoid touching your eyes, since this could lead to a potentially serious eye infection.

Prevention

You may be able to prevent further cold sores by boosting your immunity. Eat a healthy vitamin-rich diet (see pp. 9–12), get enough sleep, and get some exercise each day. Herbal remedies containing echinacea root are a traditional immune system tonic. You can take echinacea in tablet or tincture form or use the roots to brew a tea. Garlic, liquorice, and ginseng are believed to strengthen the immune system, but avoid liquorice root if you have high blood pressure.

The amino acid lysine can be a deterrent, although its effect is not immediate. Take 500 milligrams twice a day for several weeks. You can also boost your lysine intake by eating more lysine-rich foods (meat, potatoes, milk, yogurt, fish, beans, and eggs), while avoiding foods rich in another amino acid, arginine (chocolate, peanuts, nuts, seeds, and cereal grains).

Treatment

Start treatment as soon as possible.

Kitchen-cabinet remedies: These traditional treatments are all reported to help cold sores heal. As soon as you feel the initial symptoms, apply a wet Earl Grey tea bag or, with a cotton bud, cooled black coffee or an alcoholic spirit (such as gin, vodka, or whisky). An ice cube applied to the affected area may also be effective.

Aromatherapy: Geranium, bergamot, eucalyptus, tea tree and lavender oils are all astringent and antiseptic, and geranium contains substances that are thought to help fight viral infections.

■ Add four drops each of bergamot, eucalyptus, and geranium oils to two and a half teaspoons of calendula (marigold) oil or lotion. Protect this mixture from bright light by keeping it in a brown glass bottle, and apply a little to the affected area several times a day with a cotton bud. Because this mixture may increase sensitivity to light, avoid exposure to bright sunlight for at least 30 minutes after use.

■ Combine 10 drops of thyme or tea tree oil with two teaspoons of lavender tincture and four teaspoons of water. Apply with a cotton bud two or three times a day.

Marigold power
Remedies containing marigold (calendula) are antiseptic, helping prevent the broken skin of a cold sore from becoming infected with bacteria. The yellow pigments are also thought to promote the healing effect of light.

Genital herpes

Infection with another strain of the herpes virus, *Herpes simplex* type II, leads to blisters and sores in the genital area. The virus can be sexually transmitted and also be passed from mother to baby during childbirth. Do not rely on self-treatment; see a doctor.

Herbal remedies

■ Dab some diluted marigold tincture or cooled tea onto the area with a cotton bud.

■ Crush some marigold flowers and apply the liquid that oozes out directly onto your cold sore with a cotton bud.

■ Apply lemon balm (melissa) cream to the affected area.

Caution: For safety concerns, see pp. 34–37.

Homeopathy: Natrum muriaticum or Rhus toxicodendron can sometimes help to speed the healing of a cold sore.

Other remedies: Some people find that covering a cold sore with a greasy substance speeds healing. Apply vitamin A, or E, oil, or dab on some petroleum jelly or zinc cream.

Healing vitamin
Break open or puncture a capsule of vitamin A or E oil and apply a little of the oil to your cold sore.

When to get medical help
● The cold sore doesn't heal within a week.
● You have frequent or severe sores.
● You develop eye pain or light sensitivity.

See also:
**Mouth Ulcers,
Skin Problems**

Colds

Most of us are familiar with the classic symptoms of the common cold: a sore throat followed by sneezing and a runny, stuffy nose. Fatigue, headache, and a slight fever may also occur. In some cases, the lymph nodes (glands) in the neck become swollen. Although there is no cure, natural remedies can offer plenty of help to reduce discomfort and speed recovery.

More than 200 viruses can cause the annoying common cold. These are generally kept at bay by our immune systems, but when our resistance is lowered, colds may develop. Most adults catch a cold once or twice a year, usually in winter. Young children may get colds more frequently, as they have not yet built up sufficient resistance to the many strains of cold viruses.

Prevention

The key to preventing colds is to build up and maintain a healthy resistance to infection. To do this, you need a well-balanced diet that is high in vitamin C and flavonoids and low in sugar (see p. 12). You should also avoid factors that reduce immune function, such as stress, lack of sleep, and excessive consumption of caffeine and alcohol. Wash your hands frequently, since cold viruses are often spread by hand-to-hand contact.

Treatment

At the first sign of a cold: If you think you may be developing a cold—for example, if you notice a dry tickle in the back of the throat—you can stave off or lessen the severity of the infection by taking one or more of the following:

- Twenty drops of echinacea tincture in water every two hours (for up to two weeks).
- Tea from an appropriate herb (see facing page) every two hours.
- The homeopathic remedy Arsenicum.
- 250 milligrams of vitamin C with added flavonoids every two hours, but limited to 2,000 milligrams a day.
- 25 milligrams of zinc, in the form of lozenges, four times a day (for one week maximum).
- Half a clove of garlic or two garlic tablets or capsules every two hours.

General cold relief: Make yourself more comfortable by taking the following steps.

- Stay in a warm but well-ventilated environment. Keep the air from becoming too dry by

Acupressure to prevent colds

1 Place both hands over the point (GB 20) behind the ears at the top of your neck and rub your palms back and forth about 30 times.

2 Then apply pressure to the point (LI 4) on the hands in the fleshy web between the thumb and index finger. Press towards the bone at the base of the index finger for several seconds, until it aches. Repeat on each hand several times. Do not use this point if you are or might be pregnant.

placing bowls of water near heat sources, such as radiators, or by using a humidifier.

- Eat light, vitamin-rich foods—vegetable and chicken soups are ideal.
- Drink plenty of fluids, excluding drinks containing sugar, caffeine, or alcohol.

Herbal remedies

- Take elderberry extract according to the instructions on the packet.
- Drink tea made from fresh or dried ginger, sweetened with a little honey, if preferred.
- Steep one teaspoon of dried oregano or thyme

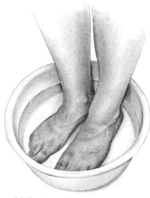

Winter foot-warmer
A mustard footbath, prepared by adding two teaspoons of mustard powder to two pints of hot water, is said to draw blood to the feet, thereby relieving congestion in the head and lungs.

Clearing congestion
Add a few drops of essential oils (see "Aromatherapy", above right) to a bowl of hot (but not boiling) water to make an inhalation. Position your face over the bowl and drape a towel over your head to capture the rising vapour. Inhale through your nose.

in a cup of hot water. Drink this brew three times a day while symptoms persist.
Caution: For safety concerns, see pp. 34–37.

Aromatherapy: Clear congestion with a steam inhalation to which you have added eucalyptus, lemon thyme, clove, and/or tea tree oils (see illustration, below left), or add up to 10 drops to your bathwater. Do not use clove oil at any time during pregnancy.

Homeopathy

- Aconite: if symptoms strike after a chill.
- Allium: if you have a very runny nose that leads to soreness of the nostrils and upper lip.
- Ferrum phosphoricum: if you feel generally run down. Repeat the remedy daily for a week or more, if necessary.

When to get medical help

- You have facial pain, an earache, or a significant fever that lasts for more than two days.
- You have severe coughing with phlegm and tightness in the chest.
- You have yellow-green phlegm along with a feeling of pressure in the head or chest lasting more than two days.
- You have a severe sore throat, and other cold symptoms do not develop within a day or so.

See also:
Allergic Rhinitis, Coughs, Earache, Fever, Headache, Influenza, Sore Throat

Colic

Some 25 percent of newborns suffer from unexplained bouts of inconsolable crying—often called colic. This disorder begins in the first few weeks of life and is often at its worst at about six weeks. Nearly all colicky babies grow out of the problem within three months. In the meantime, various treatments may mitigate your child's discomfort.

Living with colic

If nothing seems to help your baby's colic, you will have to resign yourself to a challenging few weeks until the problem resolves itself. In the meantime, minimize other stresses in your life, and try to rest whenever your baby sleeps.

A baby with colic may cry for several hours a day, often mainly in the early evening. The infant's knees may be drawn up to the chest, and the abdomen may feel hard and look distended. When the baby does eventually fall asleep, he or she may wake up screaming in an hour or so.

Most babies with colic are otherwise healthy. Babies cry for many reasons, but the crying of colic is distinctive, and the infant is obviously in pain, rather than simply tired or hungry. The pain is believed to arise from tense muscles in the intestinal wall, but the underlying cause is unknown. Possible explanations include:

- Trapped wind.
- A feeding problem.
- A reaction to breathing tobacco smoke or drinking nicotine-tainted breast milk.
- A sensitivity to formula milk or to something in a breastfeeding mother's diet.
- Immaturity of the nerves in the digestive tract.
- Having solids before the age of three months.

Treatment

Diet: If you are breastfeeding, give your baby plenty of time to drink the hind-milk—the milk released (let down) from the breast only after the first few minutes of a feed. If you switch to the second breast too soon and then take your infant off that breast before he or she has been able to drink the hind-milk, the baby may seem satisfied

Comforting cuddles
When your baby is colicky, walk around with the infant held securely against your chest. The warmth of your body is calming and makes your baby feel safe.

but may actually be filled only with fore-milk— the lower-fat milk released from the breast first. Without the presence of hind-milk, fore-milk quickly passes from the stomach into the intestine. Its lactose (milk sugar) may then ferment in the intestine, producing wind. If you have a large

amount of milk, breastfeed your baby from only one breast at each feed.

Adjust your own diet, if necessary, to exclude possible irritants that may enter your milk. Dairy foods can cause problems for some breastfed babies, and others are sensitive to substances in eggs, bananas, apples, oranges, strawberries, tomatoes, chocolate, coffee, cola, or alcohol.

Traces of foods that produce wind in the intestines may also enter breast milk. Possible culprits include cauliflower, broccoli, Brussels sprouts, peppers, onions, garlic, beans, rhubarb, and cucumber. If you eat any of these and your baby cries after the next feeding or two, try excluding the suspected food. Add soothing herbs and spices to your meals, such as ginger, caraway, dill, thyme, fennel, and cinnamon.

If your baby is formula-fed, don't change to a different formula unless advised to do so by a health professional.

Rocking to sleep
Many colicky babies are soothed by the gentle motion of a stroller or by a car ride.

Herbal remedies: Ask your paediatrician if it would be appropriate to try herbal treatments. If so, give your baby small sips of warm herbal tea from a cup or spoon before each feeding. Fennel, chamomile, ginger, lemon balm (melissa), or peppermint may help relax the stomach and intestines and prevent the painful spasms that are probably the cause of colic. Give no more than a quarter of a cup of herbal tea daily.
Caution: For safety concerns, see pp. 34–37.

Homeopathy: Magnesia phosphorica is an effective all-round remedy if your baby draws the knees up to the chest and is generally better with warmth and gentle massage. If your baby appears to be extremely upset, try Chamomilla or Colocynthis.

Other therapies: Cranial osteopathy seems to provide relief for some babies.

Massage for baby

Each day, when your baby is calm, lay him or her on your lap or on a towel on the floor or bed. Gently massage the abdomen with a slow, clockwise movement. As you do the massage, lubricate the skin with a teaspoon of warmed olive or jojoba oil combined with a drop of sweet marjoram, Roman chamomile, or mandarin oil. All of these oils—when their vapour is inhaled—encourage the baby's tense stomach and intestines to relax.

When to get medical help
- You find it hard to cope or find yourself becoming angry with your baby.
- Your baby continues to cry for long periods after four months of age.

Get help right away if:
- Your baby vomits, passes unusual stools or blood, or has a fever, a rash, or a rigid abdomen.
- Your baby feeds less or is unresponsive.

See also:
Breastfeeding Problems

Constipation

This disorder, which is increasingly common among people living in the Western world, involves infrequent bowel movements, with stools that are difficult or painful to pass. Constipation may make you feel generally out of sorts, and it can lead to other problems, such as haemorrhoids. It may be a symptom of an underlying intestinal disease.

Working out
Inactive people who have sedentary jobs are especially susceptible to constipation. Make sure you build plenty of exercise into your daily routine.

The last stage in the digestive process occurs in the colon (the major part of the large intestine), which is five feet long and populated by billions of bacteria, yeasts, and other microorganisms, collectively known as the intestinal flora. Some of these help detoxify waste and guard against infection. A good balance of microorganisms helps prevent constipation as well as diarrhoea.

Food residues and other waste material move through the bowel by peristalsis—involuntary rhythmic movements. The healthy transit time from eating to having a bowel movement is 12 to 48 hours, though up to 72 hours is acceptable if bowel movements are regular. People with healthy intestines generally have at least one movement a day, and their stools are smooth, soft, and easily passed. As we get older, our intestines may become more sluggish.

The usual cause of constipation is a poor diet. Highly refined, low-fibre foods and insufficient fluids almost inevitably slow down the digestive system. Other possible factors include too little exercise, long-term stress, pregnancy, and, very occasionally, a more serious underlying disorder, such as an underactive thyroid gland or colon cancer. Certain medications—including some painkillers, antidepressants, antibiotics, and the contraceptive Pill—may produce constipation. Paradoxically, regular use of laxatives can cause constipation by making the intestines weak and sluggish.

Prevention

Most of us can easily avoid constipation if we eat a high-fibre diet and avoid too many refined, processed, and fatty foods. However, adding fibre in the form of extra wheat bran to your food is not recommended; it can be irritating to the intestines. Oat or rice bran is a gentle alternative.

Naturally fibre-rich
Fruit, vegetables, and wholegrain foods (such as whole-wheat bread and brown rice) help waste matter move quickly through the digestive tract.

Check that you eat plenty of foods rich in magnesium (shellfish, nuts, seeds, beans, whole grains). If this is a problem, consult your doctor about taking a magnesium supplement.

Drink at least six glasses of water-based fluids every day to make your stools softer, bulkier, and easier to pass. Increase your intake if you are large or active or if the weather is hot. Fruit and vegetable juices can actively stimulate the intestines. You can buy many juices in the supermarket, but it's better to make your own juice and drink it right away (see box, right).

Treatment

Compresses: Putting alternately hot and cold compresses on the abdomen can stimulate the intestines. Start with a hot one. Wring out a small towel in very hot water, fold it over the abdomen, and leave for three minutes. Remove and replace with a towel wrung out in very cold water and leave in place for one minute. Continue to alternate the hot and cold compresses for 10 to 20 minutes.

Psychological factors: Just as some people react to stress by getting a migraine or eczema,

---

Juices for regularity

These recipes, which are especially recommended for alleviating constipation, each make one large glass of juice. Drink up to three glasses a day.

Spinach juice
Juice a handful of fresh spinach leaves, a third of a cucumber, and two tomatoes. Dilute with the equivalent volume of uncarbonated (still) bottled or filtered water. You can substitute watercress for the spinach if you prefer.

Apple and grape juice
Juice two apples and six ounces of grapes. You can substitute pears, papaya, or pineapple for the apples for variety.

Massage for constipation

With a partner

A friend can give you the following gentle abdominal massage, but ask to stop if you feel any discomfort: Rest both hands on the right side of the lower abdomen, then make big, slow, circling movements, moving the hands slowly upwards, beneath the lower left ribs, then down to the inside of the left hipbone. Slide the hands across to the starting point and repeat, moving in rhythmic circles for up to 10 minutes.

Self-massage

Make massage oil by adding two drops each of rosemary and sweet marjoram oils plus two drops of Roman chamomile oil to five teaspoons of sweet almond or a cold-pressed vegetable oil. (Omit rosemary oil in the first 20 weeks of pregnancy.) Massage the oil into your lower abdomen and then the small of your back—using firm, gentle, clockwise strokes—for several minutes every day.

others develop diarrhoea or constipation. If you feel pressured, try to find healthy ways of managing your reaction to stress (see p. 347).

Diet: Try the following measures in addition to those outlined under "Prevention":

- Drink the juice from five or six stewed prunes each morning, and eat the prunes themselves late at night. Figs are good alternatives.
- Eat yogurt with live cultures, which help re-establish the balance of intestinal flora if this has been upset by antibiotics or gastroenteritis.
- Take one teaspoon of molasses daily.
- Soak one to two teaspoons of linseeds or psyllium seeds in a cup of hot water for two hours. Add lemon and honey to taste and drink the

Yoga to start the day

Try this exercise if you tend to be constipated in the morning. (Do not attempt it, however, if you have high blood pressure.) Ask a yoga teacher to check that you are doing it safely.

1 Start by drinking two glasses of tepid water. Lie on your back and raise your legs as you inhale, bending your knees, if necessary.

2 Exhale, lifting your legs higher, so that your hips come off the floor and are supported with your hands. Your trunk should remain at an angle of 45 degrees and your legs should be as vertical as possible.

3 Exhale forcibly from your abdomen, then inhale and relax. Continue breathing like this for 40 breaths, then put your legs down and rest. Do this 40-breath sequence twice more. Then sit on the toilet with your feet raised to encourage the bowels to open.

Acupressure for constipation

Apply thumb pressure to the point (TH 6) four finger-widths above the wrist on the back of the forearm. You should do this for two minutes each day while the problem persists.

liquid and seeds at bedtime. While using this remedy, make sure you drink extra fluids.

Herbal remedies

- Make a tea of liquorice, ginger, dandelion root, yellow dock root, and burdock. Drink a cup of this mixture three times a day.
- Take three garlic capsules each night for up to one week to help rebalance intestinal flora.

Caution: For safety concerns, see pp. 34–37.

When to get medical help

- Onset is sudden and self-care measures do not work within three days.
- Your stools are black or blood-stained.
- You have abdominal pain or a fever.
- You suffer from long-term constipation.
- New medication may be the cause.

See also:
**Abdominal Pain,
Anal Problems, Diarrhoea,
Diverticular Disease,
Haemorrhoids,
Irritable Bowel Syndrome**

Convulsions

A sudden episode of violent involuntary contractions of the muscles, often with loss of consciousness, is known as a convulsion, or seizure. It results from chaotic electrical activity in the brain. Recurrent attacks can be caused by a form of epilepsy. An isolated convulsion may be provoked by a very high fever or a reaction to a medication.

One person in 130 has recurrent convulsions—known as epilepsy—but one in three affected people eventually grows out of the condition. The two main types of epilepsy are grand mal and petit mal seizures. A person having a grand mal convulsion may become unconscious for a few minutes, falling to the ground. The muscles may twitch and jerk in an uncontrolled way. Petit mal epilepsy results in momentary loss of consciousness but no convulsions. It mainly affects children.

Causes of epilepsy include head injury, birth trauma, brain infection or tumour, withdrawal from alcohol or drugs, stroke, and metabolic disturbance. Often there is no apparent cause.

Prevention

In a minority of people with epilepsy, convulsions are triggered by flashing or flickering lights, as from a television or computer screen, a failing fluorescent tube, or strobe lighting in a disco.

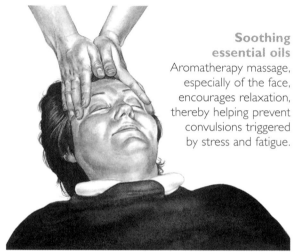

Soothing essential oils
Aromatherapy massage, especially of the face, encourages relaxation, thereby helping prevent convulsions triggered by stress and fatigue.

Some individuals appear to be sensitive to certain wavelengths of light.

- Stay away from risky environments. When you watch TV, sit several feet away from the screen in a well-lit room. If a fluorescent tube flickers, get it fixed or replaced. And don't go to places with strobe lighting.
- Ask your doctor or optician about the possibility of obtaining glasses with specially coloured lenses, tailored for the individual. There is some evidence that this might help reduce the frequency of convulsions.

Diet: Some people experience convulsions due to a food sensitivity that affects their brain chemistry. Use an elimination diet to check for foods

---

Trigger factors

If you are prone to convulsions, try to identify and avoid anything that may bring them on. Common triggers include:
- Flashing or flickering lights.
- Stress or fatigue.
- Changing hormone levels before a period or during pregnancy.
- Food sensitivity.
- High fever—especially in babies and young children (see box, p. 148).

Fluids: For all types of cough, drink plenty of fluids—choose hot ones in particular. Fruit juice, warm water, or blackcurrant tea will not only quiet a dry cough but also help loosen any phlegm. Avoid drinks containing caffeine and alcohol, however, as these are diuretics—flushing fluid from the body via the urine—and fluid loss makes mucus harder to cough up. It's also wise to eat a healthy diet (see pp. 9–12) with five servings a day of vegetables and fruit.

From the vegetable bin: Many vegetables form the basis of traditional cough remedies, such as juices, teas, and poultices.

Croup in children

A child with croup has a harsh, barking cough and difficulty breathing as a result of inflammation and congestion of the voice box and windpipe. Croup is often worse at night and is most prevalent in winter. Although distressing, this condition, caused by a viral or bacterial infection, is not usually dangerous. Steam is the best home remedy. Create a steamy atmosphere by keeping a kettle of water safely boiling, running a hot shower or bath, or using an electric vaporizer. The homeopathic remedy Aconite may be effective. Belladonna can also be used for croup.

Seek emergency medical help if treatment is ineffective and the child is struggling to breathe and getting bluish around the lips.

- Make a juice from carrots, which help shift phlegm, or turnips, which boost immunity and have an antiseptic effect on the respiratory system. To draw out the juice from a turnip, slice and cover with sugar for a few hours.
- Squeeze boiled leeks through a clean cloth to extract their juice. Drink the juice sweetened with honey according to taste.
- Slice a raw onion, drizzle honey over it, and leave overnight. The next day, take two teaspoons of the juice from the onion every two hours. Or substitute three or four cloves of finely sliced garlic for the onion, and take a teaspoon every two hours.
- Drink cabbage tea or apply a cabbage-leaf poultice (see p. 116) to the chest.

Herbal remedies: Certain herbs help soothe inflamed breathing passages, break up mucus, counter infection, or bolster immunity. Remedies containing echinacea can bolster immunity, and those containing elderberry *(Sambucus nigra)* help counter the viral infections that are the underlying causes of many coughs.

Valuable vegetables
The juices from carrots, onions, leeks, and garlic may reduce or stop coughing.

➤ continued, p. 152

Strengthen your resistance to coughs

Asian exercise therapies can help overcome susceptibility to coughs. Consider learning yoga, tai chi, or chi kung. With practice, all these gentle forms of exercise encourage proper breathing and improve posture. They also relax the muscle tension in the shoulders, chest, and abdomen that in many cases accompanies a cough, and they help you cope with stress. Try the chi kung exercise, known as supporting the sky, described below.

1 Stand relaxed, with your feet shoulder–width apart, knees slightly bent, and your head held lightly and freely.

2 Hold your hands out in front of you, palms upward, as if you were holding a large, light object, such as a ball.

3 While inhaling, slowly raise your arms above your head, palms upward. Stretch your arms and look up.

4 Exhale as you lower your arms out to your sides, and bend your knees slightly. Repeat six times.

- For a cough with phlegm, drink hot teas made from a herb that helps liquefy and bring up the mucus, such as thyme, marshmallow, mullein, or hyssop.
- For a dry cough, try a tea made from a soothing and antispasmodic herb, such as lungwort.
- Astragalus and reishi mushrooms, in the form of capsules or tinctures, or as teas or used in cooking, can also provide relief from a variety of types of cough.

Caution: For safety concerns, see pp. 34–37.

Aromatherapy: Essential oils may relieve the persistent cough of such long-term conditions as chronic bronchitis.

- Try steam inhalations scented with a few drops of eucalyptus oil and/or peppermint oil.
- Sprinkle a few drops of the above oils on a handkerchief or tissue and inhale the vapour.
- For a cough caused by a cold or a sore throat, soothe irritation at night by putting two drops of eucalyptus oil on a handkerchief under your pillow. Alternatively, put a few drops of the oil in a vaporizer (see photograph, p. 29).
- Add oils of cedarwood (three drops), peppermint (two drops), and cajuput (one drop) to two teaspoons of unscented lotion and massage onto your throat and chest. Do not use cedarwood or cajuput oil if you are in the first 20 weeks of pregnancy.

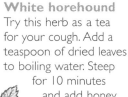

White horehound
Try this herb as a tea for your cough. Add a teaspoon of dried leaves to boiling water. Steep for 10 minutes and add honey.

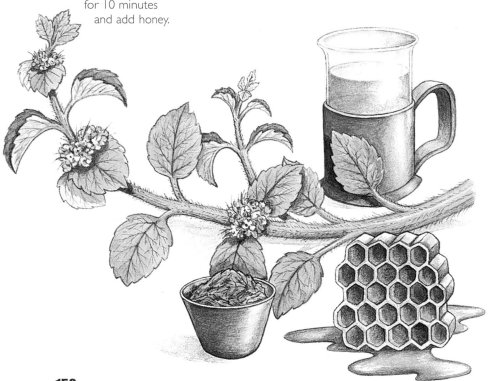

When to get medical help
- A mild cough is no better after seven days or gets worse after three days.
- A dry cough lasts longer than a month.
- A chronic cough gets worse.
- Any change occurs in a smoker's cough.
- Your cough is painful or produces green or yellow phlegm throughout the day.
- You have a fever.
- The cough develops after taking medication.

Get help right away if:
- You cough up blood.
- You have chest pain.
- You become breathless.
- A baby has a cough.

See also:
**Asthma, Colds,
Sore Throat, Voice Loss**

Cuts and Scrapes

An inevitable part of an active childhood and an occasional occurrence for adults in the kitchen and garden, minor cuts and scrapes can nearly always be dealt with effectively at home by using traditional first-aid remedies. The purpose of treatment is to stop any bleeding, protect the broken skin against infection, and promote healing.

Severe bleeding

If a wound is bleeding profusely and nothing is embedded in it, raise the injured part, apply a clean dressing, and press with your hand to control the bleeding. Wrap a bandage snugly over the dressing.

Cleaning a scrape Wash away any dirt in and around a scrape with a cotton wool ball dipped in an antiseptic solution.

Cuts to the skin and underlying tissue from a sharp object, such as a knife or piece of glass, will bleed, sometimes profusely, if blood vessels are severed. Scrapes, or abrasions, are superficial wounds that occur when the skin is grazed. They may contain embedded particles of dirt or grit.

Treatment

Immediate action: The following steps are designed to remove dirt, stop bleeding and protect against infection:

- Wash hands with soap and warm water before touching the wound.
- Gently rinse the wound under tepid running water to remove any dirt.
- Pat dry with a sterile gauze pad or clean cloth, and press the wound gently for a few minutes.
- Hold the edges of a gaping cut together with thin strips of surgical tape.

Cover a small cut or scrape with an adhesive dressing, and a larger one with a sterile gauze dressing, held in place with bandages.

Preventing infection: The following natural remedies reduce the risk of infection.

- Bathe a cut or scrape in a bicarbonate of soda solution or cooled cinnamon tea, sage tea, or parsley-leaf tea—all of which are antiseptic. Ordinary black tea, onion juice, and garlic juice are also antiseptic as well as helpful in reducing bleeding. Honey applied to a cut helps prevent infection and promotes healing.
- Apply a solution of distilled witch hazel on gauze to the wound. This rapidly stops bleeding, helps relieve pain, reduces swelling, and promotes healing.
- Clean a wound with a solution of 10 drops of the homeopathic remedy Hypercal in half a cup of water. If the wound is very painful, apply a temporary dressing soaked in a fresh batch of the same solution.
- Make a mixture of two drops of geranium essential oil, two of lavender oil, and one of peppermint oil, and sprinkle it on the clean dressing before you cover the wound.
- Apply calendula (marigold) ointment to a scrape before putting on the dressing.

When to get medical help

Get help right away if:
- Bleeding is severe or doesn't stop within a few minutes.
- A cut does not form a scab within a week or shows signs of infection (swelling, redness, pus).
- A foreign body is embedded in the wound.
- A cut was caused by a dirty object and the person has not had a tetanus immunization within the past ten years.

Depression

Most people feel "down" once in a while, for short periods of time. A true depressive illness causes persistent sadness, pessimism, and feelings of anxiety and hopelessness. It has both physical and behavioural effects. While severely depressed people require medical attention, natural therapies may help those with mild depression and can also support any treatment prescribed by a doctor.

An occasional low mood is a normal part of life. Triggers include stressful events and unresolved problems or disputes. Deeper "situational" depression may follow a significant loss, such as a marital breakdown, a bereavement, or a business failure. Some people who experience such a loss are unable to recover because they suffer from an imbalance of brain neurotransmitters. An imbalance of neurotransmitters can provoke depression even in the absence of a distressing event. In these cases, professional treatments that deal with both emotional and physical problems usually work best.

More women than men are diagnosed with depression. This is partly because women are more likely to seek professional help for their symptoms and because they may experience hormone-related depression.

Causes

Depression can have many causes. These include:

- A genetic and/or biochemical tendency to have fluctuating or depressed moods.
- Depression before a menstrual period—possibly because of changing hormone levels.
- Postnatal depression as a result of hormonal changes and other problems, such as lack of a supportive partner or other close person, isolation, and the loss of work status and income.
- Difficult childhood experiences, such as loss of a parent, leading to depression in later life.
- A food sensitivity. Sufferers may also have other symptoms (see p. 213).
- Sensitivity to lack of bright daylight in the winter, leading to seasonal affective disorder (SAD). Sufferers may crave sugary, starchy foods, gain weight, and feel tired.

Are you seriously depressed?

Listed below are possible symptoms of an illness requiring professional care:

- Inability to work or do everyday tasks.
- Little pleasure from activities once enjoyed.
- Difficulty in relating to people close to you.
- Lack of interest in sex.
- Poor sleep and early waking.
- Constant fatigue.
- Feeling cold.
- Anxiety, agitation, or irritability.
- Excessive eating or alcohol or drug abuse.
- Reduced appetite.
- A feeling of worthlessness.
- Lack of interest in personal appearance.
- Inability to make decisions.
- Forgetfulness and poor concentration.
- Generalized feelings of guilt.

Prevention

Since stressful situations make you more vulnerable to depression, reduce their impact by using stress-management strategies. These include:

- Arranging or accepting help and support.
- Eating a healthy, nutrient-rich diet to optimize brain function (see p. 9 and the box below).
- Limiting your alcohol intake (see "Alcohol limits", p. 230), and limiting your caffeine consumption to the equivalent of one or two cups of coffee a day. A high intake of either of these substances can affect mood adversely over time (see "Treatment").
- Being kind to yourself, allowing yourself to enjoy life's pleasures, and not setting your standards too high.
- Getting enough sleep every night (at least six hours for most people).
- Practicing yoga, meditation, or another relaxation technique regularly.

Treatment

It is not helpful to be told to count your blessings or snap out of it. But there is much you can do to lighten a low mood or to complement professional therapy for a more serious depression.

Support from others
Confiding your worries and sad, fearful, or angry thoughts to someone you trust can be comforting.

Nutrients for mind and body

Make sure your diet includes plenty of foods containing these vitamins and minerals, which promote a healthy balance of mood-enhancing chemicals in the brain and ensure that the brain cells are able to operate normally.

Nutrient	Food sources	Benefit for depression
Calcium	Milk products, green leafy vegetables, legumes, nuts, seeds	Activates enzymes needed for normal brain cell activity
Folic acid	Dark-green leafy vegetables, such as cabbage	Promotes production of serotonin, a mood-lifting chemical in the brain
Inositol	Brewer's yeast, fruit, vegetables, legumes, meat, milk, whole grains	Helps regulate mood swings
Iron	Meat, fish, egg yolks, beans, dark-green leafy vegetables	Boosts the production of a range of chemical transmitters in the brain
Magnesium	Shellfish, beans, whole grains, dark-green leafy vegetables, nuts	Promotes normal brain cell activity
Potassium	Whole grains, vegetables, fruit (especially bananas)	Redresses the low levels of this mineral commonly found in depressed people
Zinc	Meat, shellfish, egg yolks, peas, beans, whole grains, root vegetables, nuts	Enhances the release of energy from brain cells, which may help prevent depression
Vitamin B_6	Meat, fish, egg yolks, whole grains, bananas, avocados, nuts, seeds, dark-green leafy vegetables	Helps convert tryptophan to mood-lifting serotonin in the brain
Vitamin C	Fresh vegetables and fruit, especially citrus	Enhances iron absorption (see above)

Exercise: Exercise makes you focus on your body, providing relief from difficult feelings and thoughts. If you work your body hard enough, it releases chemicals called endorphins, which may act in the brain to help lift your mood.

If you don't exercise already, you may want to start simply, with a brisk half-hour walk at least three times a week. Once in the habit of exercising, consider something more strenuous, such as aerobic dance. Activities that you share with other people, such as tennis or team games, may be preferable to solitary forms of exercise, such as swimming, which leave your mind free to dwell on negative thoughts. Vary the type of exercise you do, and don't choose something you don't enjoy at all. Be careful not to overdo exercise, or you will feel exhaustion instead of increased energy and improved well-being.

Diet

- You may find it easier to eat several small meals daily than three larger meals.
- Counter a craving for one sort of food—such as sweets or cheese—by eating small, frequent, balanced meals. Cravings may result from a low level of the neurotransmitter serotonin. To increase serotonin levels, eat protein-containing foods that are rich in the amino acid tryptophan.

Boost tryptophan levels

Tryptophan is needed for the production of mood-elevating serotonin. Found in many protein foods, it is best absorbed when eaten with carbohydrates. Good snack choices include milk, turkey or chicken sandwiches, dates, and hazelnuts.

- Occasional treats like chocolate ice cream or French fries and a milk shake may cheer you up temporarily. Indeed, chocolate contains a compound that is thought to have a positive effect on mood. However, don't eat such sugary or fatty foods too often, and avoid them altogether if you are sensitive to them.
- Don't use alcohol to "drown your sorrows". It may help you relax for a short while, but it also depresses the central nervous system, making depression worse and reducing your ability to deal with problems.
- Limit caffeine consumption. Excessive intake may make depression worse.
- If you suspect that a food allergy is causing your depression, try to identify the food (see p. 214) and then avoid it.

Stress reduction: Practise strategies for minimizing stress. For example:

- Make sure that you and those around you understand that there are limits to what you can do when you don't feel well.
- If possible, delegate tasks that you find too difficult—for example, business trips or organizing the school PTA. People are often willing to assist if you tell them what you need.
- You may find it helps to break down large, daunting tasks into small, manageable steps, listing each one and checking it off when you have completed it.
- Take the time to reflect on your problems and practise activities that will help you cope—for example, yoga, meditation, or prayer.

Light therapy

Exposure to bright light may diminish winter depression, or seasonal affective disorder (SAD), by reducing the levels of the sedative brain chemical melatonin and boosting those of serotonin, which is stimulating. You can increase your exposure to light in the following ways:

- Go outside in the middle of the day for a half hour.
- If your doctor approves, try sitting in front of a high-intensity light box for a half hour to two hours every day. Do not use if there is a flickering fluorescent tube in the box, as this could provoke light-sensitive migraine or epilepsy.
- Install a bright, full-spectrum, fluorescent tube in the room where you spend most of your waking hours.
- Make the best use of daylight by adjusting your wake-sleep pattern to dawn and dusk. Alternatively, buy a bedroom lamp specially designed to turn on at dawn and gradually increase in brightness until you rise.

Herbal remedies

- St. John's wort is the best-known herbal remedy for mild to moderate depression. Take it according to the manufacturer's instructions.
- Tea made from a combination of wild oats, vervain, and ginseng may help.

Caution: For safety concerns, see pp. 34–37.

Aromatherapy

- Take a daily warm bath containing a few drops of oils of lavender, chamomile, bergamot, rose, or clary sage oil. Don't use clary sage if you are in the first 20 weeks of pregnancy.
- Ask a friend to give you a massage. Use two drops each of lavender or geranium oil, and one of Roman chamomile oil, in a tablespoon of sweet almond or grapeseed oil.

Flower essences

- Gorse: for deep pessimism, when you feel that nothing can help you.
- Larch: for feelings of failure and worthlessness.
- Mustard: for depression that comes on for no apparent reason.

Other therapies: A counsellor or psychotherapist can work with you to find the causes of your depression and ways of overcoming it. Laughter therapy can also be effective for some people.

When to get medical help

- You feel incapable of trying any of these approaches.
- Your depression lasts longer than two weeks, prevents you from sleeping for more than a few nights, or significantly interferes with your work or relationships.

Get help right away if:

- You have suicidal thoughts.

See also:
Addictions, Anxiety, Eating Disorders, Emotional Problems, Fatigue, Grief, Menopausal Problems, Premenstrual Syndrome, Sex Drive Loss, Sleeping Difficulties, Stress

Treatment

Diabetes treatment aims to keep the blood-sugar level within the normal range. Type I requires daily injections of insulin. If you have type II, you are unlikely to need insulin injections, but you may need oral medication. Whichever type you have, you can help manage your condition by improving your diet, losing any excess weight, and exercising regularly.

Diet: Space your food intake evenly throughout the day. Four or five small meals enable the body to better regulate blood sugar than three large ones. Follow the dietary guidelines outlined under "Prevention". Carbohydrates should come mainly from legumes, whole grains, fruit, and vegetables. Limit your intake of saturated fats (found mostly in meat and full-fat dairy products), but, if necessary, increase your intake of omega-3 and omega-6 fatty acids (found in oily fish and nuts and seeds and their oils) as well as monounsaturated fats (such as olive oil) to help keep your arteries healthy. Eat three servings of oily fish a week, a few nuts and seeds every day, and five daily servings of vegetables and fruit (to supply antioxidants). Special "diabetic foods" are unnecessary. Using your knowledge of the glycaemic index of the foods you eat can help stabilize your blood-sugar level. Keep sugary foods to a minimum: they cause fluctuations in blood-sugar level and encourage weight gain.

You may be able to drink some types of alcohol in moderation, but check with your doctor first. If you do drink,

Preventing hypoglycaemia
Keep some sweets or a sugary drink with you in case you experience a dangerous drop in blood sugar (hypoglycaemia), and tell friends how to recognize this and what to do in such a situation.

always eat something at the same time, especially if you are on insulin treatment or take medication to lower blood sugar.

Exercise: Regular exercise helps keep weight down and therefore can work to control type II diabetes. If your diabetes results from insulin resistance, you may find that exercise also increases your sensitivity to insulin and helps improve blood-sugar control. Exercise increases the amount of sugar taken up by the muscles for up to 72 hours. If you are sedentary or have stopped exercising for a while, get advice from your doctor before you start.

If you are on insulin or another drug treatment, guard against a sudden drop in your blood-sugar level by checking it before and after each exercise session and by learning to balance exercise with your food intake. You may need to adjust your dose and/or food intake to allow for the energy expenditure during exercise and to prevent a drop in blood sugar.

Smoking: Since the risk of heart disease and other complications increases with diabetes, and smoking further increases that risk, it is especially important not to smoke.

Herbal remedies: The following therapies are no substitute for orthodox treatment, but they may improve your sense of well-being. Check with your doctor, however, before taking herbs.
- Eat raw garlic, cooked onions, or onion juice daily to enhance circulation and help reduce your blood-sugar level.
- Try to drink a cup of dandelion-root tea daily.

Dandelion root is believed to stimulate the pancreas and encourage insulin production. Burdock leaves and roots are another traditional remedy for an elevated blood-sugar level, as is bilberry-leaf tea.

■ Take Siberian ginseng. This is an "adaptogenic" herb, which means it helps the body regulate certain aspects of metabolism, perhaps including a fluctuating blood-sugar level.

■ Evening primrose oil, rich in gamma linolenic acid, may reduce pain and tingling from nerves damaged by diabetes.
Caution: For safety concerns, see pp. 34–37.

From the chemist: Certain vitamins and minerals may help your body cope with diabetes. These can usually be obtained from food, but ask your doctor whether you should also take supplements.

■ Vitamins A, B, C, and E and

Bilberry benefits
Bilberry works in two ways for people with diabetes: it strengthens the blood vessels, which are often weakened by the disease, and helps prevent related eyesight problems.

flavonoids may help you avoid long-term complications of diabetes.

■ Chromium helps boost the action of insulin. As a supplement, chromium picolinate may be beneficial in regulating the blood-sugar level.

■ Magnesium may decrease insulin resistance and lead to better blood-sugar control.

■ Manganese is often lacking in those with diabetes. This mineral can improve arterial health.

■ Those with type II diabetes may need to increase their intake of zinc.

Aromatherapy: Eucalyptus and lemon oils are believed to influence the pancreas. Add these to your bathwater, individually or in combination, or add five drops to four teaspoons of grapeseed or sweet almond oil for a massage.

When to get medical help
● You have any of the symptoms described on page 163 or slow-healing wounds, blurred vision, or numbness or tingling in the hands or feet.

Get help right away if:
● You have diabetes and develop persistent light-headedness, repeated vomiting, prolonged fever, severe shortness of breath, or acutely worsening pain or ability to think clearly.
● A person with diabetes loses consciousness.

See also:
Arterial Disease, High Blood Pressure, Overweight, Pregnancy Problems

Diaper (Nappy) Rash

Keeping a baby's skin clean and healthy when it is covered most of the time with a nappy is a real challenge. Most babies—however well cared for— occasionally experience soreness. Nappy rash can mean anything from slight redness to severe inflammation with infected sores, but the condition is usually mild and easily treated at home.

The most common type of nappy rash results from ammonia, a skin irritant formed by the action of intestinal bacteria on the urine. The wet nappy chafes the skin and makes it sore, and the ammonia makes the soreness worse. Some babies develop psoriasis in the nappy area; others have dermatitis, seborrhoea (overly oily skin), or an allergy (for example, to the rubber in elastic). Broken skin encourages infection with yeasts (candida), bacteria, or viruses.

Prevention

The best way to prevent nappy rash is to keep your baby's skin as dry as possible by changing a nappy as soon as it's wet or dirty. Also:

- When changing a nappy, wash your baby's bottom with water and dry it thoroughly, then smooth on a thick, waterproofing layer of ointment or cream. Use a zinc cream, or stir a few drops of lavender oil into a tablespoon of unscented cream. Don't use talcum powder; fine particles can irritate the lungs and prevent the umbilicus from healing properly. If you want to use a powder, choose a product made from cornflour.
- Plastic pants retain moisture and may harbour infection. If you use them, wash frequently.
- Sterilize, wash, and rinse cloth nappies thoroughly. Use the highest temperature possible and the minimal amount of nonbiological washing powder (available from supermarkets).
- Avoid fabric conditioner.

Treatment

- Use the preventive methods above, but instead of your usual cream, apply calendula (marigold) or chamomile cream at each nappy change, or use a soothing oil made from four drops of lavender oil, two drops of Roman chamomile oil, one drop of sandalwood oil, and 10 teaspoons of calendula oil.
- Give your baby a long soak, morning and evening, in a warm bath to which you have added two drops of lavender oil. This is calming and stimulates the growth of new skin cells.
- When washing cloth nappies, add six drops of lavender oil to your washing machine during the rinse cycle.

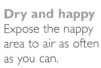

Dry and happy
Expose the nappy area to air as often as you can.

When to get medical help

- Mild nappy rash persists after trying the above treatments for a few days.
- The skin is very inflamed or broken.

See also:
Dermatitis, Fungal Skin Infections

Diarrhoea

Attacks of frequent and loose or runny bowel movements affect most people at some time. They occur when food residues travel too rapidly through the digestive tract, most commonly because of an infection in the tract. In healthy adults, bouts of diarrhoea generally clear up with rest and fluids. There is usually no serious underlying problem.

DID YOU KNOW?

Stewed apple, ripe banana, and brown rice are among the easiest foods to digest after an attack of diarrhoea.

Poisonous plants
Yew, deadly nightshade, and laburnum are among the plants that may cause diarrhoea if ingested.

Acute attacks of diarrhoea, lasting from a few hours to seven days, are most often the result of food poisoning caused by viral or bacterial infections. The microorganisms (or the toxins they produce) cause gastroenteritis, inflammation of the lining of the stomach and intestines. Diarrhoea resulting from infection may be accompanied by vomiting, abdominal cramps, bloating, wind, and a slight fever. Sometimes diarrhoea is caused by a food sensitivity (such as an intolerance to lactose, or cow's-milk sugar) or it may be a side effect of drugs—for example, antibiotics. Poisoning with lead, pesticides, or certain plants can also cause attacks of diarrhoea.

Long-term and recurrent diarrhoea may indicate a chronic problem, such as irritable bowel syndrome, inflammatory bowel disease, diverticular disease, an overactive thyroid gland, or a stress-related disorder. In rare cases, diarrhoea is a symptom of intestinal cancer.

Prevention

Personal hygiene: Paying attention to the basic rules of hygiene is the best way to prevent diarrhoea caused by infection.
- Always wash your hands thoroughly with soap after using the bathroom and before preparing food or eating.
- In public washrooms, dry your hands with a clean paper towel rather than under a hot-air drier, which may harbour germs.

Kitchen hygiene: Care in choosing, using, and storing food is important for avoiding diarrhoea from food poisoning.
- Use only fresh eggs (check the date label) and discard cracked ones. Cook thoroughly.
- During pregnancy, prevent diarrhoea due to infection with listeria bacteria by avoiding soft and mould-ripened cheeses; unpasteurized milk; cheeses made with unpasteurized milk; soft ice cream; precooked, refrigerated foods, unless thoroughly reheated; precooked poultry; pâtés; deli meats; rare or undercooked meat; and prepackaged salads, unless washed thoroughly.
- Don't buy cans that are swollen or dented at the rim or seam. Don't buy any food that is past its "sell by" date.
- Consume food before any "use by" date.
- Cook raw meats and reheat leftovers thoroughly. Be especially careful with chicken.

Cooking implements Choose utensils that are easy to clean. Allow wooden chopping boards time to dry between uses.

Stress: Stress can readily disrupt the digestive system and cause diarrhoea. If you feel too pressured, find effective ways of preventing stress or managing your reaction to it (see p. 347).

Treatment

Preventing dehydration: One of the dangers of diarrhoea is dehydration, especially if you are also vomiting. Particularly at risk are young children and people who are frail, elderly, or have a weakened immune system. Take the following measures to prevent this problem:

- Drink plenty of water or water-based fluids, but avoid alcohol and caffeine. If vomiting is one of your symptoms, take frequent sips to help keep some fluid down.
- Replace the salts and sugars lost through diarrhoea, especially if you can't eat or keep much food down. Make a specially balanced drink by squeezing out the juice of two fresh oranges, adding half a teaspoon of salt and two teaspoons of honey, and then adding water until you have a pint. Drink a glass every half hour until your symptoms improve.

- Defrost frozen food completely in the refrigerator, not the kitchen counter, before cooking, unless package instructions advise otherwise. Do not refreeze.
- Store cooked and uncooked meats on separate refrigerator shelves. Make sure that raw meat does not come into contact with other foods.
- Regularly clean your refrigerator, freezer, kitchen surfaces, and utensils.
- Wash dishcloths, sponges, and dishwashing utensils frequently, and disinfect them with a bleach solution (with the bleach diluted as directed on the bottle).
- Don't handle or prepare food for others if you have gastroenteritis.

Avoiding traveller's diarrhoea

In some foreign countries, you may be vulnerable to local bacteria. When in doubt, take these measures:

- Use only chemically sterilized, boiled, or bottled water for drinking or brushing your teeth.
- In restaurants ask for bottles of water to be opened in front of you (some establishments may refill bottles with tap water).
- Avoid ice.
- Peel all fruit and avoid raw vegetables and salads.
- In restaurants eat only cooked foods served hot. Don't eat food that has been kept warm for a long time.

Acupressure to relieve diarrhoea

Use the point (St 37) eight finger-widths below the lower border of the kneecap, one finger-width outwards from the crest of the shinbone. Apply thumb pressure for two minutes on each leg.

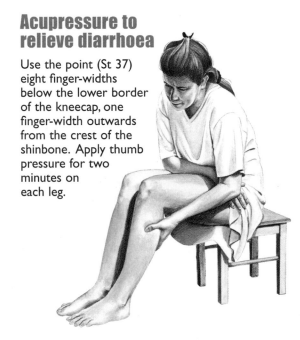

- Drink herbal teas such as chamomile, thyme, ginger, peppermint, and fennel. These soothing herbs are mildly antiseptic and help relieve the cramping abdominal pains that often accompany diarrhoea caused by infections of the digestive tract.
- Drink unsweetened blackcurrant or elderberry juice to reduce inflammation in the intestines. The vitamin C in these juices also helps fight infection.
- Mix one tablespoon of arrowroot with a little water to make a smooth paste, then add a pint of boiling water and stir as it thickens. Flavour with honey or lemon juice. Drink at regular intervals throughout the day to thicken the bowel contents.

Getting back to normal: Once you start feeling better, you can gradually resume a normal diet if you've been eating little or nothing. Start with rice, which is non-irritating and helps bind the bowel contents. Eating some yogurt with live cultures each day will help restore the balance of microorganisms in your intestines, and a daily vitamin and mineral supplement will help restore your body's nutrient balance.

Homeopathy: If you still have diarrhoea after 24 hours, try a homeopathic remedy.
- Arsenicum: for food poisoning involving vomiting and nausea.
- Colocynthis: for diarrhoea accompanied by severe abdominal cramps.

Aromatherapy: Astringent (drying) oils, such as ginger and geranium, can be helpful for the relief of diarrhoea, as can oils, such as peppermint and Roman chamomile, that calm muscular contractions in the intestines. Tea tree oil is also good, because it is an antiseptic.
- Sprinkle three drops each of geranium and ginger oils and two of peppermint oil into your bathwater. Relax in the bath for 15 to 20 minutes. Be sure to keep the water warm.
- Make a massage oil with three drops of tea tree oil, two drops each of peppermint, geranium, and sandalwood oils, and five teaspoons of sweet almond oil, olive oil, or another cold-pressed vegetable oil. Warm the oil (see illustration, right) and use it to massage

Warm massage oil
Heat your aromatherapy massage oil by immersing the container in hot water for a few seconds. This helps release the healing vapours and makes the oil feel pleasant on the skin.

your abdomen, as described on page 70. Repeat the massage every few hours.

Yoga: Deep relaxation while lying on the floor (see p. 69) can reduce the mental and physical tension that can sometimes provoke attacks of stress-related diarrhoea.

Special advice concerning young children: Infants and toddlers are at an increased risk of dehydration from fluid loss. If your baby or young child has diarrhoea:

■ Check whether the child could have eaten anything poisonous. Phone NHS Direct on 0845 4647.

■ Check the child's diet for irritating foods. Hot spices, peppers, onions, tomatoes, rhubarb, and too much fruit may provoke diarrhoea. Reduce your own intake of these foods if you are breastfeeding.

■ Give your breastfed baby as much time at the breast as desired, and offer more feeds than usual both day and night. Your milk supply will increase naturally; drinks of cooled boiled water are needed only rarely. Give your bottle-fed baby extra diluted formula or cooled boiled water.

■ Give a weaned child extra drinks, including pasteurized apple juice or well-diluted orange juice that you've squeezed yourself. The water in which rice has been boiled is also suitable.

■ Wash your hands thoroughly after touching your child and before eating, serving, or preparing food.

Extra fluids
Bottle-fed babies who are suffering from diarrhoea may need additional drinks of cooled boiled water.

When to get medical help

Babies and young children
● Mild diarrhoea has lasted longer than 24 hours or severe diarrhoea longer than 12 hours.

Get help right away if:
● The child is vomiting and can't keep fluids down or won't nurse or eat.
● The child seems dehydrated, with a dry mouth, sunken eyes, loose skin, and fewer wet nappies than you usually expect. Dehydration may also cause the soft spot (fontanelle) on top of a baby's head to become depressed.
● The child won't stop crying or is unusually sleepy or listless.
● There is blood in the stools, or the child has a fever or other symptoms.
● You think your child may have eaten or drunk something poisonous.

Older children and adults
● You are normally fit and healthy and have mild diarrhoea for more than three days or severe diarrhoea for more than two days.
● You are elderly or frail.
● You suffer from frequent infections or are taking steroids, antibiotics, or any other medicine that may be a possible cause.
● Others who share your home are affected by similar symptoms.
● You have severe diarrhoea and are taking the oral contraceptive Pill.
● There is red or black blood in the stools.

Get help right away if:
● You have signs of dehydration—including drowsiness, glazed eyes, loose skin, and very small amounts of dark urine or no urine at all.

See also:
Abdominal Pain, Diverticular Disease, Food Sensitivity, Irritable Bowel Syndrome, Nausea and Vomiting

Diverticular Disease

Commonly occurring in people over 50, diverticulosis is a disorder in which many small pockets (diverticula) form in the wall of the large intestine. When the lining of one or more of these pockets becomes inflamed, the condition is called diverticulitis. Both forms of the condition are less likely to develop if you eat plenty of fibre.

Most cases of diverticular disease probably stem from a diet low in fibre and high in refined carbohydrates. The condition is rare before the age of 20, but in the United Kingdom 40 percent of the over 50s are affected, and 50 percent of the over 70s. Many people with diverticular disease also have symptoms of irritable bowel syndrome.

When you eat a refined diet, the muscles of your colon have to work harder than normal to push along sticky, dry food residues, and they may temporarily go into spasm. Pressure builds up, causing the colon to dilate and forcing pockets of its lining (known as diverticula) out from the stretched wall. These distended segments usually cause little trouble, but they can lead to constipation or diarrhoea, together with a nagging ache—usually on the left side and sometimes relieved by passing wind. Possible complications include diverticulitis (when a pocket becomes inflamed or infected), bleeding from a burst pocket, and an obstructed intestine.

Prevention

Diverticular disease is rare in countries in which the normal diet is high in fibre. You are unlikely to develop the condition if you adopt similar eating habits. This means eating five servings of vegetables (including peas and beans) and fruit daily, as well as some nuts and seeds (chewed

Early habits that pay off
A high-fibre diet starting in childhood can reduce susceptibility to diverticular disease in later life.

well). You should also choose foods made with whole grains rather than white rice or flour. However, do not add wheat bran or other insoluble fibre to your diet, as it can be irritating to the intestines. A high-fibre diet also requires plenty of fluids—you should drink at least eight glasses of water-based fluid a day.

Another important way of preventing the disease is to exercise every day. This massages the intestines and helps to prevent constipation, which is a major risk factor for diverticulitis. Taking fresh garlic (preferably raw) or garlic tablets three times daily counters abnormal bacterial activity and may therefore also help to prevent diverticulitis.

a persistently sore throat, abdominal pain, bloating, digestive disturbances, and various metabolic disorders. Excessive use of laxatives can disturb the muscle action of the intestines, leading to chronic constipation. It may also cause fluid, electrolyte, and nutritional deficiencies.

Compulsive eating: People with this disorder, who are likely to be overweight, have the urge to eat even when they are not hungry. Binge eating is a type of compulsive disorder that involves eating an enormous amount of food at one sitting, often in a very short time. The sufferer feels completely out of control while eating and eats rapidly—but fails to feel full. Some binge eaters spend a great deal of time planning meals and buying food.

However, they feel embarrassed and guilty about their bingeing, and they usually eat secretly.

Helpful diversions
To overcome a compulsive eating disorder, look for new, pleasurable ways of relieving stress. Listening to music, for example, can provide a valuable alternative to using food as a source of comfort.

Treatment

Stress management: Since stress and anxiety play a large part in many eating disorders, finding ways to deal with these problems can be very helpful. You need emotional support from someone with whom you can talk openly about your condition. Some people find that creative activities, such as keeping a journal, painting, and sculpting, help them express feelings they can't verbalize. If you are eating too much, make a list of interesting and enjoyable things to do instead, whenever you crave food. For example, have a relaxing, fragrant bath, go for a walk, or phone a friend to arrange a visit.

Flower essences: To help counter negative feelings about yourself and your body, choose one or more essences that most closely match your emotions. Take four drops in a little spring water, four times a day on an empty stomach. The following essences may be especially appropriate:
- Crab Apple: if you dislike your appearance.
- Rock Water: if you are very hard on yourself—for example, by denying yourself food.

Crab Apple
This flower essence may help those who are overly concerned with cleanliness and hate the way they look.

Diet: Compulsive eaters often eat high-sugar, starchy foods, which make their blood-sugar level rise quickly. The pancreas responds by

producing excess insulin, which can make the blood-sugar level drop too low. This can trigger a craving for yet more sugar or refined carbohydrates. Try to reduce your intake of these types of foods, and also do the following:

- Increase the fibre in your diet.
- Keep blood sugar stable by eating small amounts several times during the day, rather than two or three big meals.
- Eat more chromium-rich foods (see p. 12). Chromium helps regulate blood-sugar levels.
- Reduce your intake of alcohol and caffeine-containing drinks, since these can produce energy dips that encourage food cravings.
- Identify any food sensitivity that may be to blame for food cravings and weight fluctuation. Try an elimination diet to identify the foods that may be responsible (see p. 214), but consider professional help in doing this, especially if your weight is low.

Vitamins and minerals: Appetite is controlled by neurotransmitters in the hypothalamus. The efficiency of these natural substances is influenced by the blood levels of sugar, fatty acids, and hormones, which, in turn, are affected by anxiety, depression, and deficiencies of vitamin B or zinc. To help keep your neurotransmitters in balance, eat foods rich in the following nutrients (or take them in the form of a multiple vitamin and mineral supplement):

- Calcium (milk, cheese, nuts, seeds, green leafy vegetables, legumes)
- Magnesium (shellfish, nuts, whole grains, legumes, green leafy vegetables)
- Manganese (tea, nuts, seeds, whole grains,

organic green leafy vegetables, pineapple, raisins, blueberries)
- Potassium (fruit, vegetables)
- Selenium (fish, whole grains, organic fruits and vegetables)
- Vitamin B complex (lean cuts of meat, milk, whole grains, fresh vegetables)
- Zinc (lean cuts of meat, poultry, shellfish, egg yolks, nuts, seeds, whole grains, hard cheeses, root vegetables)

Yoga: Regular yoga practice can relieve tension and depression, and so helps you to meet emotional needs in other ways besides eating. Yoga works by:

- Giving you more self-control, which will allow you to master the urge to gratify your desires instantly or in an unhealthy fashion.

Fresh and nutritious Foods carefully selected for freshness and variety are likely to provide the balance of nutrients you need to regulate appetite and control blood sugar.

Colour breathing for inner harmony

This technique helps you relax, clear your mind of negative thoughts, and feel at peace with the world around you. As you breathe, you focus on each of the body's so-called energy centres (known as *chakras* in Ayurvedic medicine), the colours they are associated with, and the emotions and qualities they are said to govern.

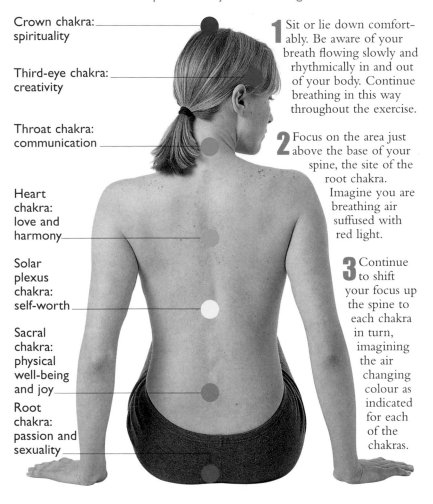

Crown chakra: spirituality

Third-eye chakra: creativity

Throat chakra: communication

Heart chakra: love and harmony

Solar plexus chakra: self-worth

Sacral chakra: physical well-being and joy

Root chakra: passion and sexuality

1 Sit or lie down comfortably. Be aware of your breath flowing slowly and rhythmically in and out of your body. Continue breathing in this way throughout the exercise.

2 Focus on the area just above the base of your spine, the site of the root chakra. Imagine you are breathing air suffused with red light.

3 Continue to shift your focus up the spine to each chakra in turn, imagining the air changing colour as indicated for each of the chakras.

- Helping you feel relaxed.
- Providing an alternative to abnormal eating as a way of feeling in control of your life.

Emotion culturing: This technique aims to cultivate a positive state of mind and to control the urge to binge without inducing a sense of guilt when around food. When you feel a craving for food, say to yourself, "I am going to eat, but why hurry? Be slow, be easy—let me enjoy this food." Then take five deep breaths and start to eat slowly, savouring the flavours in each morsel.

Other therapies: Ask your doctor to refer you to a psychiatrist or psychotherapist who specializes in eating disorders. A dietician or other nutritional therapist can provide advice on eating behaviours and on choosing a good balance of foods to help combat your eating problem.

When to get medical help
- Your eating pattern has been abnormal for longer than four weeks.
- You are continuing to lose or gain a significant amount of weight.
- You are abusing laxatives, diuretics, or emetics.

Get help right away if:
- You develop bulimia or stop eating completely.

See also:
Anxiety, Appetite Loss, Depression, Emotional Problems, Food Sensitivity, Overweight, Stress

Emotional Problems

Psychological distress, whether arising from personality traits, biochemical imbalances, or difficult or uncomfortable circumstances or events, can have a profound effect on your physical well-being and quality of life. Sympathy and support from a trusted person, together with home remedies and lifestyle changes, can often ease difficulties.

DID YOU KNOW?

The willow tree, which has many healing properties, provided one of the original flower essences used by Dr. Edward Bach to combat the negative emotions of resentment and bitterness.

The constant stress of living with unresolved emotions—such as sadness, anger, fear, and anxiety—can affect you physically, because it changes levels of hormones and neurotransmitters (chemicals that convey messages between nerves). These changes can make you sluggish and tired and may depress your immune system, increasing your susceptibility to infection and illnesses, such as rheumatoid arthritis, that are triggered by attack from your own immune cells.

The way you respond emotionally to problems in life is shaped by your personality, experience, and circumstances. You may become ill as a result of a problem that someone else copes with relatively easily. However, you can learn how to better manage emotional problems.

Prevention

Your customary responses to difficult situations are not unchangeable. You may have emotions that are uncomfortably close to the surface, last too long, and disrupt your well-being and relationships; however, you can learn to deal with situations in a different way. Or if you are emotionally restrained, bottling up your feelings, you can learn to become more open.

- Aim to develop self-knowledge, accept your strengths and weaknesses, and recognize and commit yourself to your primary goals in life.
- If decision-making upsets you, practise clarifying your options and objectives. Don't be afraid to recognize your sense of loss or other feelings about the possible life choices you have rejected.
- Remember that total control of all

Sharing your feelings
Let others around you help you through difficult times. You may benefit from practical help they can offer, but it is equally important to have someone tell you they understand how you're feeling.

aspects of your life is neither possible nor desirable, and adjust your expectations accordingly.

Treatment

Mind games: A growing body of evidence suggests that optimism gives some protection against mental and physical ailments. If you tend to see only the worst aspects of situations, make an effort to take a more balanced view. For example, if a relationship breaks down, the disadvantages may be all too clear, but there are potential benefits, too: you might have the opportunity to move to a different area or follow interests that your partner didn't share.

Express yourself

Emotional distress sometimes results from relationship problems. Try the techniques described below to improve your communication skills.

Problem-solving

If a misunderstanding with another person is causing you anxiety or distress, try the following:
- Arrange a meeting at which both people take uninterrupted turns to say how they feel.
- State your case in the first person. For example, "I feel hurt when you do that", rather than "You make me feel hurt when you do that". By doing this, you take responsibility for your response to the other person's actions. You also comment on the behaviour, but not on the person. The person can then—if able or willing—take responsibility for the behaviour and modify it in the future.
- Next, listen to the other person's response (see below).

Empathic listening

This type of listening focuses on the other person's emotions. It helps to be aware of your own feelings, so you should get into the habit of listening carefully to yourself. Empathic listening has three stages:
- Put your own feelings to one side for a while.
- Try to identify the other person's emotions by using your ears, eyes, and intuition.
- Tell the person what you think those emotions are. This generally has the powerful effect of imparting a feeling of understanding. At the least, the person will know you're trying to see things from his or her point of view.

Even if you are unable to see things in a positive way, you may find a situation more bearable if you try to move on, accepting that nothing stays the same in life, rather than dwelling on your emotional pain. Losing your job may feel like a disaster, but if you acknowledge your feelings of anger and bitterness without letting them overwhelm you, it may then be possible to consider future plans more constructively.

Communication: One of the main reasons we have emotional problems is that we sometimes find it difficult, if not impossible, to express our feelings clearly, especially during stressful times. Instead of explaining what we are feeling and why—as well as listening to the other person— we may shout, argue, accuse, or stop talking completely. Try to communicate your emotional responses clearly. The more uncomfortable or embarrassing this is, the more important is good communication. It is often better to deal with upset feelings sooner rather than later. (See also the box on the facing page.)

- Don't wait until anger or resentment has built up and both parties have become entrenched in their positions.
- Before discussing a problem, make a list of the points you want to raise, so you don't lose track of them.
- Tell the other person clearly and in a non-confrontational way what you want, and listen when it's his or her turn to do the same.
- Be careful to avoid falling into the trap of inappropriate sharing of emotions. Choose your confidant and the time and place for such confidences with care.

Improving self-understanding: Gaining a good understanding of your emotional responses and behaviours is a key to overcoming emotional difficulties. Some people find it useful to keep a diary to record their feelings. Try noting each day how you have felt, the events or circumstances that you think contributed to your feelings, and how you coped with any challenging emotions. Gradually you may begin to see patterns of feelings and behaviour and may then be able consciously to alter any habitual negative responses. For example, if you always respond to insults or criticism by becoming upset and blaming yourself, it may be helpful to try focusing instead on other possible explanations for the person's behaviour. You thereby avoid taking responsibility for the behaviour of others (see also "Affirmation", p. 88).

Aromatherapy
- Put a few drops each of clary sage and juniper berry oils in a warm bath, to clear thoughts and promote a good night's sleep. (Do not use these oils if you are or might be pregnant.)
- When you feel unhappy, scent your room by using six

Time of serenity
Try meditation to achieve a clearer, calmer view of yourself and any distress you're feeling.

drops of clary sage (omit if you're pregnant), three of rose otto, and one of sandalwood oil in a vaporizer (see p. 29).

Flower essences: These are used to treat mental and emotional problems, and each flower essence is said to help alleviate a particular negative emotional state. Choose one or more remedies based on your personality or the way you normally respond to difficulties. Useful remedies include the following:

- Cherry Plum: if you find it difficult to control your temper.
- Holly: if you feel jealous or full of hate.
- Larch: if you lack belief in your own abilities.
- Mimulus: if you are shy or anxious.
- Pine: if you have a tendency to blame yourself for past actions.
- Scleranthus: if you are not able to make decisions easily.
- Star of Bethlehem: if you are mourning the loss of a loved one.

- Willow: if you are feeling sorry for yourself or are bitter and resentful.

Homeopathy: You should ideally take into account your personality type, as well as your symptoms, when choosing the most appropriate remedies. The aim of treatment is not to suppress your emotions, but to help you achieve a state of emotional balance so that you can express your feelings without being overwhelmed by them.

- Arsenicum: for extreme anxiety, exhaustion, and restlessness.
- Ignatia: for emotional upsets, sadness and grief, hysteria, mood swings, excessive weeping, or inability to cry.
- Nux vomica: for irritability and explosive anger when you cannot take any more stress.

Other therapies: A trained counsellor or psychotherapist can help. Many other therapies can be of benefit, including biofeedback, cranial osteopathy, and art and creative therapies.

Flower power

Flower essences are extracted from many common plants, including (left to right): willow, star of Bethlehem, holly, pine, and scleranthus.

When to get medical help

- Your emotional problems are affecting your relationships or are otherwise disrupting your usual lifestyle.

Get help right away if:

- You have suicidal thoughts.

See also:
Addictions, Anxiety, Depression, Eating Disorders, Grief, Sex Drive Loss, Stress

Eye Irritation and Discharge

Conjunctivitis, or pink eye, is the most common form of eye irritation, causing soreness, redness and/or discharge. Blepharitis produces similar symptoms, and both conditions may be a result of injury, infection, a foreign body in the eye, allergy, or exposure to irritating chemicals, such as those in cigarette smoke and vehicle exhaust fumes.

Conjunctivitis involves inflammation of the conjunctiva, the transparent membrane covering the whites of the eyes and lining the lids. The eyes look red, feel gritty and itchy, and produce a discharge that in infective conjunctivitis may cause the eyelids to stick together during sleep. In allergic conjunctivitis—often associated with allergic rhinitis (hay fever)—the discharge is clear, and the eyelids are often swollen. There may be abnormal sensitivity to light.

Blepharitis, inflammation of the eyelids, causes redness, irritation, and scaly skin at the lid margins. There may be crusty beads of dried discharge on the lids and lashes. The condition, which is sometimes associated with dandruff or eczema, tends to recur.

Prevention

Reduce your chances of catching eye infections by never sharing towels or flannels and by encouraging any household member who is affected to seek treatment. Protect your eyes with goggles when performing tasks that generate dust or smoke and when swimming in pools if you are sensitive to chlorine.

If you are susceptible to eye infections, build up your resistance by eating five servings of fruit and vegetables daily, including berries and yellow and orange produce. Avoid foods made with white flour and sugar.

Foreign body in the eye

Dust particles are usually removed by blinking. Try removing larger objects using the methods below. Seek medical help if a foreign body is on the pupil, or embedded in the eye, or if first aid fails.

1 Lower lid: Try flushing out the object with water or lifting it out by touching only the white part of the eye with the dampened corner of a clean cloth or tissue.

2 Upper lid: Holding the lashes, gently pull the lid outwards and downwards over the lower lid, or try flushing the object out by blinking under water.

Treatment

If you have an eye infection, you should avoid touching your eyes, and ensure that you wash your hands often. You should also wash towels, flannels, and bedsheets frequently to prevent reinfecting yourself. If you have both blepharitis and dandruff, the dandruff can be treated at the same time.

Diet: Persistent eye irritation that may result from an allergy may improve if you exclude dairy products, tea, and coffee from your diet.

From the chemist: To help fight infection and soothe irritation, take supplements of evening primrose oil, vitamins A, B, and E, and flavonoids, as directed on the bottle. For infectious conjunctivitis, try a boracic-acid eyewash.

Homeopathy

- Aconite: if eyes feel hot, dry, and gritty and look red and inflamed, with swollen lids. The symptoms may have been caused by cold wind.
- Apis: if eyelids are red and puffy.
- Belladonna: if eyes are dry, bloodshot, and sensitive to light, and symptoms develop quickly.
- Euphrasia: if eyes water and burn. Bathe the eyes with a few drops of the tincture in an eyebath of sterilized water. Use fresh solution for each eye.
- Pulsatilla: if you have a yellowish discharge that does not burn or irritate, but the eyelids are sore and, perhaps, sticky.

Herbal remedies: Bathing the eyes in eyebright tea soothes many symptoms. Simmer the tea for 10 minutes to sterilize the liquid, then cool and strain (through a sterile gauze pad or cheesecloth) into a sterile eyebath. Use fresh tea for each eye and do not store any leftover tea.

To boost your immune system, drink tea made from a combination of eyebright, echinacea, cleavers, burdock, and liquorice. Drink a cup once or twice daily for two weeks.
Caution: For safety concerns, see pp. 34–37.

Acupressure for eyes

Soothe irritated eyes by pressing one or more of the points described below steadily for about two minutes. Release and repeat every few breaths.

- The point (*tai yang*) one thumb-width away from the outer bony margin of the eye, level with the top of the ear.

- The point (B 1) immediately above the inner corner of the eye, at the inner end of the eyebrow.

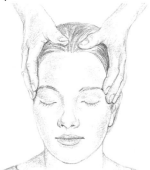
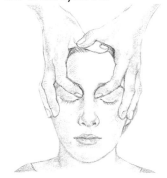

- The point (Liv 3) on the top of the foot between the first and second toes, along with the point (GB 37) on the shin five thumb-widths directly above the outer ankle tip, in front of the bony ridge on the outside of the leg.

When to get medical help

- Your eye problem is causing pain.
- There is a profuse, coloured discharge.
- Your vision is gradually affected.
- Your symptoms fail to improve within two days.

Get help right away if:

- Your eye has been injured.
- Your vision is acutely affected.

See also:
Allergic Rhinitis, Eyesight Problems, Hair and Scalp Problems

Eyesight Problems

When your eyes are unable to focus images accurately, the result is blurred vision. Wearing glasses or contact lenses is the conventional way of correcting the problem, but simple home remedies, dietary measures, and exercises may be able to arrest the rate of deterioration and in some cases even improve your existing eyesight.

The causes of blurred vision include too long or too short an eyeball (shortsightedness and longsightedness) and discrepancies in the curvature of the surface of the eye (astigmatism). In middle age the lenses start to harden (presbyopia), so that it becomes more and more difficult to bring close objects into focus. Other conditions that cause increased blurring of vision include a clouded lens (cataract), excess fluid in the eyes (glaucoma), and changes in the retina resulting from aging (macular degeneration).

Prevention

Some eyesight problems are inherited, so prevention is impossible, but for others you can safeguard the health of your eyes.

- Wear sunglasses in strong light to reduce the risk of cataracts from cumulative ultraviolet damage.
- Protect your eyes with goggles when doing hazardous jobs.
- Nourish your eyes and the blood vessels and nerves that supply them by eating a diet containing plenty of vegetables and fruit. Those shown below (see illustration, left) and dark-green leafy vegetables are especially beneficial.
- Limit your consumption of saturated fats and stay away from cigarette smoke to help prevent free radicals forming in the retina and encouraging macular degeneration.

➤ *continued, p. 191*

Vision vitamins
Red, orange, and yellow fruit and vegetables are rich in antioxidant nutrients, such as beta-carotene and vitamins C, and E, that counter eye damage.

Yoga-based eye exercises

Practised daily, the exercises below help relax and strengthen the eye muscles. The first sequence, known as candle gazing, is also good for steadying the mind.

The second sequence helps train the eyes to make the adjustments needed during everyday activities, such as reading, driving, and working at a computer screen.

1 Position a candle three feet away from your eyes at eye level.

2 Gaze at the flame for 10 seconds, then palm your eyes for 30 seconds (see "The Bates method" on the facing page), while observing the after-image of the flame.

3 Next, gaze at it with one eye at a time, then with both eyes while turning your head from side to side. Palm between each stage to avoid eye strain.

4 Repeat, gazing first for 10, then 20, then 30 seconds. Increase the times by 10 seconds each week, until you are gazing for one, two, and three minutes.

1 With your forefinger, touch the spot between your eyebrows. Gradually move your finger away, focusing on it as it moves and continuing until your arm is fully extended. Hold the position for a few seconds, then bring the finger back between your eyebrows. Cover your eyes with your palms, then repeat the whole sequence.

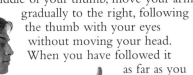

2 Extend your right arm in front of you, with the thumb pointing upwards. Focusing on the middle of your thumb, move your arm gradually to the right, following the thumb with your eyes without moving your head. When you have followed it as far as you can, hold the position for 15 to 30 seconds. Move your arm slowly to the front again, following your thumb with your eyes. Cover your eyes with your palms. Repeat with the left arm, and palm once more.

3 Hold both arms out in front of you, with the thumbs pointing up. Gaze at both thumbs and gradually move your arms apart. Continue until you are about to lose sight of the thumbs. Hold the position for one minute, then bring your arms slowly back to the starting position. Cover your eyes with your palms.

4 Hold out your right arm, pointing with your forefinger to the left. Without moving your head, slowly raise your arm while focusing on the finger. Move your arm as far as you can without losing sight of the finger. Hold for 30 seconds, then slowly lower your arm to eye level. Cover your eyes with your palms. Repeat, this time moving your arm below eye level. Repeat both stages five times, then cover your eyes with your palms for one minute.

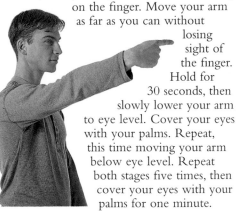

Treatment

Although natural therapies may not cure problems with vision, there is much you can do to slow deterioration of eyesight.

- Do close work, such as reading, in good light. Position yourself so that the light comes over your left shoulder if you are right-handed and over your right shoulder if you are left-handed.
- When reading a newspaper, start with the largest type, such as headlines, before reading the small print.
- If you have a cataract, take bilberry extract. Research has shown that this herbal remedy, which contains antioxidant pigments called proanthocyanidins, can help slow the development of cataracts.

Homeopathy: Use Ruta for tired eyes when it's hard to focus—for example, after spending too much time in front of a computer screen.

The Bates method: This technique aims to "re-educate" the eye muscles and improve the

Palming
Covering your eyes with your palms between exercises allows your eyes to rest.

eyesight. Bates teachers use such exercises as the following (do not wear glasses or contact lenses while you are doing these exercises):

- Every morning, close your eyes and splash them 20 times with warm water, then 20 times with cold. At bedtime, repeat the exercise, starting with cold water.
- Throughout the day, blink frequently— about 15 times per minute.
- Cover your eyes with your palms to rest them completely three or four times a day. Sit with your elbows resting on a table and place the palms of your hands over your eyes, with the base of your little fingers on the bridge of your nose. Do this for several minutes. Listening to music will help you relax.
- Hold your forefinger 6 to 10 inches from your eyes. Move it from one shoulder to the other, and follow it with your eyes, keeping your neck relaxed and allowing your head to move with your finger. Focus on the finger, but be aware of the moving background too. Reverse the movement and repeat several times.
- Do the same exercise, but focus on the moving background instead.

Focusing
Exercises in which you follow your finger with your eyes are a key element of the Bates method.

When to get medical help

- Your eyesight deteriorates.
- You experience eye pain.
- You have a vision change in one eye only.

Get help right away if:

- You lose vision suddenly.
- You have headaches along with loss of vision.

See also:
Eye Irritation and Discharge

Facial Pain

Many people experience aching or other pain in the face at some time. Facial pain can range from a continuous dull ache to a sharp, intense spasm, depending on the cause. Simple home remedies often make all the difference, but some causes of facial pain, such as an inflamed temporal artery, are potentially serious and require medical attention.

Soothing oil
For nerve pain, add three drops of St. John's wort oil to two teaspoons of olive or sweet almond oil. Smooth onto the painful area every three hours.

Understanding why your face hurts can help you plan what to do. Possible reasons include dental problems, migraine, sinusitis, upper respiratory infections, and mumps. Another cause is a dysfunctional jaw joint (temporomandibular joint syndrome, or TMJ), which may be triggered by dental problems, stress, poor posture, or injury.

An inflamed temporal artery (temporal arteritis) creates persistent pain in one or both temples. This serious condition, affecting mainly people over 50, requires prompt medical treatment. A damaged trigeminal nerve (trigeminal neuralgia), also more common in older people, brings about attacks of brief but severe pain in the face.

Shingles, an infection of the facial nerves caused by the varicella zoster virus, produces discomfort and sensitivity in the affected side of the face. Several days later, a blistery rash occurs. Pain may persist after the rash has disappeared.

Prevent facial pain by practising good dental hygiene, reducing stress, and treating any infections promptly.

Treatment

Home treatments can help relieve pain, boost immunity, and manage the stress that may underlie the condition or result from long-term pain.

Heat or cold: Place a hot-water bottle wrapped in a towel or other cover over the painful area,

Gentle massage
Stroke both sides of the person's face with the fingers. Use only light pressure, following the direction of the arrows shown above. Start at the centre of the forehead and finish with sweeping movements along the jawline and up towards the ears.

or apply a hot, damp flannel repeatedly. If you find the application of cold more soothing, use a flannel wrung out in cold water instead.

Massage: Massage encourages the release of natural painkillers called endorphins, and a gentle facial massage also releases tension in the muscles. You can either give yourself a massage or ask a friend to do it (see illustration, above). Use five drops of lavender or peppermint oil in two

tablespoons of sweet almond oil or another cold-pressed vegetable oil. If the pain is so severe that you cannot bear your face to be touched, massage the tops of both big toes instead. Reflexologists call this the pituitary reflex; the pituitary gland affects many parts of the body.

Acupressure: The point (LI 4) between the base of your forefinger and thumb, used for treating colds (see p. 140), can also be helpful for relieving facial pain. Press this point several times on each hand. Do not use this point if you are pregnant. Alternatively, press with both thumbs along the base of your skull. Use small circular movements and work from the centre outwards.

Immunity-boosting remedies: Increasing immunity may help relieve sinusitis and shingles. Eat a healthy diet, with five daily servings of vegetables and fruit. Include some fresh garlic, or take garlic capsules. Consider supplements of vitamins B and C and flavonoids. Make a herbal tea of echinacea, goldenseal, and liquorice; drink a cup three times a day. Or take echinacea and astragalus as capsules or tincture.
Caution: For safety concerns about herbal remedies, see pp. 34–37.

Stress management: Continual stress makes pain worse and depresses immunity. Use stress-management strategies (see p. 347).

For TMJ: Eating soft foods puts less stress on your jaw. Keep your jaw muscles relaxed. If you work at a desk, check your posture, and don't lean forwards. Avoid propping up your chin

with your hands or cradling a telephone between your shoulder and chin. Do not carry heavy shoulder bags; instead use a backpack.

Finding the cause of your facial pain

Symptoms	Possible cause
• Pain in forehead, nose, ears, cheekbones, or behind eyes	Sinusitis
• Aching cheeks	Stress-induced tension
• Pain and a rash on the upper half of one side of face	Shingles
• Pain in jaw and mouth	Dental problems
• Sensitivity on one side of face and, possibly, in one eye, followed by severe pain	Migraine
• Intense shooting pain on one side, often provoked by even a light touch	Trigeminal neuralgia
• Swelling and tenderness in temple, possibly with headache and, at worst, vision problems	Temporal arteritis
• A stiff, clicking jaw joint and dull ache or pain in the jaw muscles	TMJ

When to get medical help
- You also have an earache, eye pain, headache, toothache, or a facial rash.
- The pain is severe.
- The pain has lasted longer than a week or is worsening after one or two days.

Get help right away if:
- You have had a head injury.
- You have a throbbing ache in your temple, vision problems, or a rash near your eye.

See also:
Colds, Headache, Migraine, Toothache

Fainting

When insufficient oxygen reaches the brain, a brief loss of consciousness, known as fainting, can result. Recovery is usually quick, because falling to the ground places the head at the same level as the heart, thus restoring blood flow and oxygen to the brain. Prolonged loss of consciousness—for more than a minute—requires emergency treatment.

Fainting is caused by a sudden reduction in the blood supply to the brain, which leaves it short of sugar and other nutrients, as well as oxygen. Simply being in a hot, stuffy room is enough to make some people pass out, but fainting can also indicate a serious medical condition. Other causes include the following:

- Standing still for a long period, which makes the leg veins dilate and fill with blood, leaving less circulating blood available for the brain.
- Hunger, which can lead to a low level of sugar in the bloodstream.

Restoring blood flow
If you feel you are about to faint, place your head below the level of your heart as quickly as possible. If it is difficult to lie down, sit with your head between your knees.

- Breathing too fast, or hyperventilating, because of anxiety. This reduces the blood's carbon-dioxide level and the amount of oxygen available to the brain cells.
- A shock—for example, from witnessing an accident or hearing bad news—which affects the nerve that controls blood pressure.
- Severe anaemia, fever, uncontrolled diabetes, heart and circulatory disease, and certain other medical disorders.
- New medication.

Fainting is often preceded by such unpleasant sensations as dizziness, sweating, clamminess, and nausea. Other warning signs or symptoms include repeated yawning, feeling hot, shaking, pallor, and breathlessness. You may feel that you are suffocating and need fresh air, and some people experience an overwhelming sense of impending disaster.

Most people feel well again quickly after fainting, unless they are injured from falling against a hard object. If you have hit your head hard enough to cause a cut or a large bump, have the injury examined by a doctor; you may have a concussion or a fracture.

Prevention

- Eat regularly. If you miss a meal, at least have a nutritious, high-fibre snack, such as a whole-grain sandwich, a cereal bar, an apple, a banana,

or some nuts to prevent low blood sugar. Most important, don't skip breakfast.

- Eat a well-balanced diet, with plenty of fibre-rich foods and limited amounts of refined carbohydrates, such as white flour and sugar. Over the years this protects your pancreas—the gland that produces insulin, the blood-sugar regulating hormone—from becoming overstimulated by excess sugar. An overstimulated pancreas can react to the intake of refined carbohydrates by producing too much insulin. This leads to a sudden dip in the blood-sugar level, which makes fainting more likely.
- Open a window if a room is hot or stuffy or, if you feel that you are well enough, go outside.
- Check your breathing. Hyperventilation is a common response to stress. Practice slow, deep breathing every day (see p. 87).
- When standing still for a long time, stimulate your circulation by rocking back and forth on your heels and the balls of your feet, or by alternately tightening and relaxing your calf muscles—just as sentries do on guard duty.
- To avoid feeling faint after rising from a lying-down position, sit upright for a few moments, then get up slowly.
- If your blood pressure is low, drink a daily cup of hawthorn tea to help normalize it, but first see pp. 34–37 for safety concerns.

Treatment

Posture: Occasionally, fainting results from a sudden compression of the blood vessels in the neck—for example, from a sudden or violent sideways or backward jerk of your head. If you think this compression may have happened,

When someone has fainted

If the person is breathing normally and no serious injury has occurred, raise the legs above head level for a few moments, and loosen any tight clothing. This allows more blood to get to the brain. Discourage people from crowding around. When the person regains consciousness, advise him or her to sit or lie quietly until fully recovered.

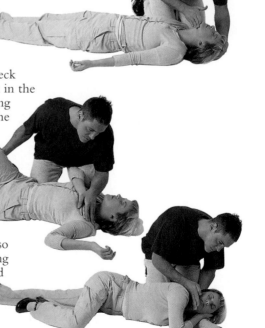

For prolonged unconsciousness

If the person does not revive after a minute, summon medical help, and follow the steps below.

1 Make sure that the unconscious person is breathing easily, and tilt the head back slightly to keep the airway open.

2 If you saw the person faint and know that there is no neck or spinal injury, place the victim in the recovery position: While kneeling down beside the person, bend the leg that's close to you and fold the arm on the same side across the body. Raise the other arm above the head.

3 Gently roll the person over so that his or her cheek is resting on the hand of the bent arm and the upper leg stabilizes the body. Lift the chin to improve the airway.

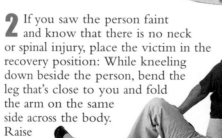

stand up straight and imagine that the top of the back of your head is attached by a taut string to the ceiling. Allow your head to rise up out of your neck, and keep your shoulders relaxed.

Breathing: If you are hyperventilating (breathing too rapidly), try to take slow, deep breaths into your abdomen, or cup your hands over your mouth and nose as you breathe; this raises your carbon-dioxide level and helps restore alertness. Or breathe in and out of a paper bag (see photograph).

Diet: Don't accept any medicinal alcohol, which could make you choke or vomit, and inhale the vomit because of a reduced level of consciousness. Once the faintness passes, sip some cold water. Upon regaining full consciousness, have something sweet—sweets, a biscuit, a teaspoon of honey, fruit juice, or sweetened tea or coffee. As soon as possible, eat something more nutritious and fibre-rich, or you may start to feel faint again when the raised blood-sugar level induced by the sugary snack passes its peak.

Homeopathy: Take your chosen remedy every 10 minutes while feeling faint or after reviving, then use as necessary. Helpful remedies include:
- Arnica: for an emotional shock.
- Aconite: for fright that remains after a shock.
- Gelsemium: if you feel weak and shaky after an emotional shock.

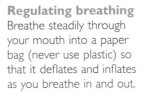

Regulating breathing
Breathe steadily through your mouth into a paper bag (never use plastic) so that it deflates and inflates as you breathe in and out.

An instant lift
The application of lavender oil to the temples can provide rapid relief from faintness. You can also add a few drops of this oil to sweet almond or wheat-germ oil for a reviving massage after an episode of faintness.

- Ignatia: if you have had bad news or another shock and you feel you are losing control.
- Pulsatilla: after an emotional shock, when you are tearful and feel better from being comforted or getting fresh air.

Aromatherapy: Smelling salts (ammonium carbonate crystals) are a traditional remedy. You can achieve a similar result by inhaling the aroma from certain essential oils if you are feeling faint.
- To revive yourself and clear your head, rub a few drops of undiluted lavender oil into your temples and over the backs of your hands.
- To calm yourself and restore vitality, pour a few drops of lavender oil onto a tissue, and inhale the vapour for several minutes.

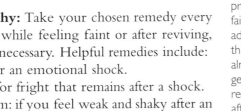

Herbal remedies: The following teas help speed recovery after fainting. Do not drink anything if you feel you are about to faint, because this could make you choke.

- After regaining consciousness, sip tea made from ginger, rosemary, or elderflowers.
- If fainting resulted from emotional shock and you continue to feel tense or stressed, have up to four cups daily of tea made from lemon balm and chamomile.

Caution: For safety concerns, see pp. 34–37.

Flower first aid
The combination of the flower essences Star of Bethlehem, Rock Rose, Cherry Plum, Impatiens, and Clematis—known as Rescue Remedy—helps you recover after fainting. Place a few drops on your tongue, or take them in a little cold water. Repeat every few minutes until you feel better.

Acupressure for a fainting victim

Use these points to help revive a person who has fainted, but not if there is a head, neck, or spinal injury.

- Press firmly with your thumbnail on the point (GV 26) in the furrow between the nose and upper lip. Continue until the person revives, keeping your other hand on the forehead just above the hairline.

- As the person is reviving, apply firm pressure to the point (Sp 6) four finger-widths above the tip of the ankle bone on the inside of the calf, just behind the shinbone. (Do not use this point on a pregnant woman.)

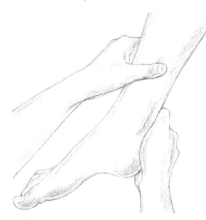

When to get medical help

Get help right away if:
- You experience first-time or unaccustomed fainting.
- Fainting results in a head injury or another possibly serious injury.
- The victim remains unconscious for longer than one minute.
- You have confusion, numbness, or loss of vision or movement upon recovery.

See also:
Anaemia, Arterial Disease, Diabetes, Dizziness, Shock, Stress

important to maintain a good blood supply to the ovaries: poor circulation can in some cases lead to reduced fertility. However, be aware that excessive exercise can reduce a woman's fertility.

Contraception: At least three months before you plan to conceive, stop taking the oral contraceptive Pill, or have your intrauterine device or hormone implant removed. Use a barrier method of contraception, such as a diaphragm, until you wish to conceive.

Ovulation and sex: Find out the time in the month when you normally ovulate by using an ovulation prediction kit or by taking your temperature each morning and noting the slight rise that accompanies ovulation. Then each month, during the week before you expect ovulation and for a day or two after your temperature rises, have sexual intercourse once a day, ideally just after waking in the morning. At other times, have sex whenever you wish.

Smoking: Smoking reduces fertility and increases the risk of miscarriage. Smoking during pregnancy can also affect the health of your baby. If you can't stop immediately, try cutting down, and seek help in quitting.

Stress management: Feeling stressed can suppress ovulation and reduce sperm count. Practise stress-management strategies and relaxation techniques (see p. 347). You may also want to try yoga or meditation.

Concerns at work: The workplace can be hazardous if you are hoping to conceive.
- Check that your employer enforces any necessary safety precautions, for example, those related to exposure to chemicals or radiation.
- Sitting for very long periods (especially with your legs crossed) or being exposed to direct heat may overheat the testes, depressing sperm production. If low sperm production is a problem, consider whether your occupation—

Advice for men

Men with fertility problems should:
- Avoid beer, since its natural oestrogen content may lower the sperm count.
- Reduce consumption of meat, dairy products, beans, and peas, which may also contain oestrogens.
- Avoid hot baths. Sperm production is more efficient when the testes remain cool.

Precautions for women

Tell your doctor or pharmacist that you are trying to conceive before accepting any medications; some can damage a newly conceived baby. Before agreeing to an X ray, inform your doctor, dentist, and X-ray technician that you are trying to become pregnant so that appropriate precautions can be taken, or so that the X ray can be rescheduled.

Drugs
Avoid all unnecessary drugs, including non-prescription medicines (except on medical advice) and so-called "recreational" drugs, such as marijuana.

Immunization
If possible, women should avoid immunization with live vaccines, such as those for measles, mumps, rubella, polio, and yellow

fever, during pregnancy and in the six months before conceiving. Tell your doctor if you become pregnant soon after such an immunization.

Herbal remedies, essential oils, and supplements
Always seek the advice of a qualified practitioner before using these remedies while trying to conceive or when pregnant.

for example, bus driver or welder—may be the cause. If so, it may be worthwhile to change to another type of work, if possible, until your partner conceives.

- Some occupations are statistically linked to lowered female fertility. Examples include being a dental assistant, and jobs involving exposure to textile dust.
- Some office buildings have poor supplies of fresh air; elevated levels of extra-low-frequency electromagnetic radiation (from computers and other electrical equipment); and heightened concentrations of chemicals in the air, such as formaldehyde, which vaporize ("outgas") from synthetic wall and floor coverings. Such an environment could theoretically affect fertility. The same may also apply to some homes. If you live or work in such conditions, spend as much time outdoors as possible.

House and garden: Many household products contain chemicals that can inhibit fertility.

- Don't breathe organochlorine pesticide vapour —for example, from pet flea sprays—or touch any pesticide-containing product.
- Don't inhale smoke from burning plastic that contains polyvinyl chloride (PVC).
- Don't strip old paint that may contain lead.
- Don't inhale the vapour from solvents, glues, felt-tip markers, and paints.

Avoiding infection: Because the mother's state of health can affect a developing baby, it is important to try to avoid contracting infections of all kinds when trying to conceive and throughout pregnancy.

- Both partners should stay away from people who have colds, coughs, the flu, childhood infections, or other viral illnesses.
- At least six months before attempting to conceive, a woman should check her immunity to rubella and be immunized, if necessary.
- Consult your doctor about precautions against toxoplasmosis, a protozoal infection that can damage a developing baby. If you own a cat, ask someone else to clean the litter tray. If you must do it yourself, line the tray with newspaper, wear rubber gloves, and empty used litter directly into a plastic rubbish bag. Wash your hands after handling raw meat and eat only well-done meat.
- Take measures to avoid infection with listeria, an organism that can harm a foetus and may cause miscarriage. Don't eat pâté; soft, mould-ripened cheeses, blue-veined cheese, and feta cheese; soft ice cream; unpasteurized milk; precooked poultry; prepacked salads (unless washed again); and—unless thoroughly reheated—precooked, refrigerated foods.

When to get medical help

- You are a woman who has been trying unsuccessfully to conceive, using the above guidelines, for one year if under 35, or six months if over 35.

See also:
Fibroids, Food Sensitivity, Overweight, Pelvic Inflammatory Disease, Stress

Fever

Having a fever—a higher than normal body temperature—is generally a sign that your body's immune system is doing its best to combat an infection. But if your temperature is so high that it makes you uncomfortable, you'll benefit from using some simple home remedies. You will also need to find out what is causing the rise in temperature.

Normal body temperature varies from person to person, ranging from 96.5°F to 99.5°F. Body temperature is affected by food, drink, exercise, sleep, time of day, and the menstrual cycle. A fever is an abnormally raised body temperature. It can be a sign of any of several common ailments, including influenza, tonsillitis, and the childhood infectious illnesses, such as chicken-pox, as well as rarer infections, such as malaria and typhoid fever. Heatstroke caused by prolonged exposure to heat is another possible cause.

A rise in temperature may be heralded by bouts of shivering and by feeling alternately sweaty and chilled. Once a fever begins, it may also be accompanied by a headache and rapid breathing. In certain illnesses, such as malaria, the episodes of shivering are so severe that they are described as rigors. The feverish stage of an infection usually lasts no more than three days.

Treatment

If you feel unexpectedly cold and shivery, it is likely that you are developing a feverish illness. It is best to rest, but it is not necessary to go to bed unless you want to. A lukewarm bath may also make you feel better, but take care not to become chilled.

Reducing a fever: There is no need to take special action to bring down your temperature unless it is above 100°F or is hard to tolerate. However, if you do want to do this, turn room heat down or off, open windows and doors, if necessary, and turn on an electric fan if you have one. Dress in light clothes, and if you feel too hot in bed, remove blankets or replace a duvet with a single blanket or sheet. Most people are more comfortable when lightly covered than with no bed covers at all. Clothes and bed covers made of natural fabrics allow sweat to evaporate more easily and aid cooling.

Compresses: You can use cold, wet compresses on your forehead, back of the neck, wrists, and calves. Keep the rest of your body covered.

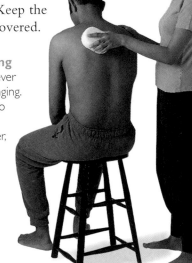

Cooling by sponging
An adult with a high fever may benefit from sponging. Use lukewarm water to which you have added a few drops of lavender, Roman chamomile, or eucalyptus oil. Treat the skin a little at a time. If necessary, cover parts of the body that aren't being sponged to avoid becoming too cold.

Change the compresses as they become warm. Continue until your temperature falls and you feel better.

Body wrap: A body wrap (see p. 50) is an effective means of bringing down a high temperature, but take care not to reduce the temperature too much. Wrap the feverish person firmly in a cold, wet sheet or several cold, wet towels, and then with a dry woollen blanket. Change this wrap every 15 to 20 minutes, until the person feels comfortably cool.

Replacing fluid: Consume enough liquids to enable you to pass plenty of pale urine. Choose water, and drinks with a high vitamin C content, such as blackcurrant, orange, or lemon juice.

Herbal remedies: Drink teas made from white willow bark, lime tree flower, lemon balm, elderflower, echinacea, ginger, or peppermint. Echinacea tincture may also be helpful.
Caution: For safety concerns, see pp. 34–37.

Homeopathy

For a high fever
- Aconite: if thirst and sweating are pronounced, perhaps due to a sudden chill.
- Belladonna: if you feel a dry, burning heat and you have a red face.

For a slowly developing low fever
- Bryonia: if you are noticeably irritable and experience intense thirst.
- Gelsemium: if fever is accompanied by marked shivering and shaking.
- Pulsatilla: if a child is clingy as well as feverish.

Fever reduction in children

A child with a fever should not be sponged or bathed in cool or lukewarm water. This may result in an excessive drop in temperature, leading to shivering, which may cause a further rise in temperature. If your child has a fever, remove all but a light layer of clothing and make sure the room is cool and well-ventilated. Offer plenty of fluids.

When to get medical help

- You have vomited, coughed up phlegm or blood, or passed blood during bowel movements.
- You have a condition, such as heart disease or diabetes, that requires medical monitoring.
- You have a temperature of 101°F or 102°F that has lasted more than 72 hours.

Get help right away if:
- You have a fever of 103°F or more.
- You also have a severe headache, a stiff neck, a rash, or a sensitivity to bright light.
- You also have severe abdominal pain or urinary problems.
- You experience confusion, unusual drowsiness, irritability, and/or laboured breathing.
- A baby under three months has a temperature of 100°F or more.
- A baby of three to six months has a fever of 101°F or more.
- A child older than six months has a fever of 103°F or more.

See also:
Childhood Viral Infections, Colds, Convulsions, Coughs, Influenza

Fibroids

A benign tumour of the uterus, a fibroid consists of an abnormal collection of muscle and fibrous tissue that grows slowly in the uterine wall. Fibroids may be as small as a pea or as big as a grapefruit. They may occur singly, or several may develop within the uterus. Fibroids pose few serious risks to health and usually produce no symptoms unless they are quite large.

DID YOU KNOW?

In the Middle Ages the herb *Calendula officinalis* was dedicated to the Virgin Mary, hence its common name marigold. Tradition has it that this herb has an affinity with the female reproductive system.

One in four Western women has fibroids. They are most common in childless women over the age of 35 and before the menopause. The specific cause of fibroids is unknown, but it is thought to be related to an abnormal response to the hormone oestrogen. Oestrogen stimulates fibroids to grow larger, so they may become more troublesome during pregnancy, when oestrogen levels are higher than normal. Fat cells produce oestrogen, which may account for fibroids being more likely in obese women. The association with oestrogen also explains why most fibroids shrink and disappear after the menopause, when oestrogen production falls, unless a woman takes HRT (hormone replacement therapy). Surgery to remove fibroids is necessary only if the growths cause severe symptoms and if other treatments fail.

Prevention

Keep your weight within recommended limits (see p. 302). Eat relatively little saturated fat and animal protein. Instead, consume vegetable protein and fibre. Such a diet may discourage fibroids by lowering your oestrogen level.

Anti-fibroid foods? Some researchers believe that plant hormones help counteract the high oestrogen level that encourages fibroids. They are present in beans, peas, seeds, whole grains, and most fruit and vegetables.

Do you have fibroids?

Fibroids may be discovered only when you have a physical examination, and if they produce no symptoms, they need no treatment.

If a fibroid grows, it may erode the uterine lining and cause prolonged or heavy menstrual periods or bleeding between periods. If you lose a lot of blood month after month, you may eventually become anaemic and experience fatigue and shortness of breath.

Other possible symptoms include:
- Severe cramps and a dull ache or feeling of uncomfortable pressure in the lower back and thighs during menstruation.
- Constipation or a need to urinate more often than usual (a fibroid may be pressing on the intestines or bladder).
- Very light menstrual flow (a large fibroid near the cervix may be partially blocking it).
- Pain during sexual intercourse.

Treatment

Aromatherapy

■ A gentle abdominal massage may soothe painful symptoms and reduce tension. Make a massage oil with four drops each of clary sage and lavender oils, and two of true melissa or rose otto oil, to one tablespoon of sweet almond or a cold-pressed vegetable oil.

■ To ease abdominal pain or backache, add four drops each of clary sage and sweet marjoram oils and three of Roman chamomile oil to a bowl of hot water. Use this to make a warm compress (see p. 295) to hold against the painful area. Do not use clary sage oil in the first 20 weeks of pregnancy.

Exercise: If fibroids cause cramps or heavy bleeding during menstruation, you may not feel like being active. However, exercise at other times of the month stimulates uterine circulation and may therefore make pain from fibroids less likely.

Herbal remedies

■ For fibroids with heavy periods, drink a cup of tea twice daily—or 15 drops of

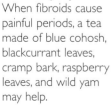

Herbs for severe symptoms
When fibroids cause painful periods, a tea made of blue cohosh, blackcurrant leaves, cramp bark, raspberry leaves, and wild yam may help.

Improving circulation with hydrotherapy

Blood flow in the wall of the uterus may be sluggish if you have fibroids. Increase uterine circulation by sitting in cool water up to your hips for two or three minutes every morning. This may reduce the severity of your symptoms.

tincture in a little water—made from agrimony, beth root, chasteberry, and raspberry leaves. You can also add nettles, which are rich in vitamin C and iron, to soups, stews, and salads, and drink a cup of nettle tea twice daily.

■ For fibroids with painful, heavy periods accompanied by blood clots, drink a cup of tea twice daily—or 15 drops of tincture in a little water—made from the herbs shown at left.

Caution: For safety concerns, see pp. 34–37.

When to get medical help

● You feel excessively tired or weak.
● You have irregular or problematic periods.
● You're due for your gynaecological checkup—to ensure that any growth is not cancerous.

Get help right away if:

● You have any unaccustomed severe pain.

See also:
Abdominal Pain, Anaemia, Menstrual Problems

Flatulence

Having excessive wind in the digestive tract can be annoying as well as embarrassing. Many people are able to overcome or at least reduce the problem by avoiding wind-producing foods and improving their eating habits. Flatulence may be accompanied by indigestion or constipation. It can be a sign of underlying disease, but this is rare.

Flatulence is often accompanied by a bloated sensation. This discomfort is generally relieved when the wind is expelled, either via the mouth—belching—or the anus. Belching is more likely when you're standing or sitting, and expelling wind from the anus is more likely when you're lying down. Excessive wind production may lead to obvious distension of the abdomen.

Flatulence occasionally results from gastritis (an inflamed stomach lining), irritable bowel syndrome, gallbladder disease, or a peptic ulcer. It can also occur as a result of long-term stress.

Prevention

Adjusting your diet usually eases the problem. Flatulence results from the action of bacteria and fungi in your intestines causing certain digested foods to ferment and produce wind. Common culprits include cabbage, onions, peas, and beans. You may find that fatty foods, fizzy drinks, sugar, drinks containing caffeine, or uncooked vegetables and fruits make the problem worse. Flatulence can also result from sensitivity to a particular food or foods, or from combinations of certain foods, notably starch with protein or fruit. Examples are pastry with fruit (as in apple pie), and bread with meat (as in a ham sandwich).

Eating habits: It isn't just *what* we eat but *how* we eat that can cause flatulence. Some of us, without realizing it, swallow air along with our food, especially when we are nervous or under stress. This causes wind to accumulate in the stomach and intestines. The same thing can happen if you overeat, eat too fast, don't chew

Abdominal massage

Massage your abdomen with a slow, clockwise movement. For extra benefit, massage with oil made by mixing four drops of peppermint oil and two drops each of juniper berry and caraway oils (omit these oils if you are pregnant) in two and a half teaspoons of sweet almond oil.

The effect of antibiotics

You may develop flatulence as a result of taking antibiotics, which temporarily alter or destroy the normal balance of microorganisms in the intestines. This population of microorganisms, known as the intestinal flora, is essential for healthy digestion. To help restore a healthy balance of intestinal flora, take a supplement of *Lactobacillus acidophilus* or eat yogurt that contains live acidophilus cultures at least once a day.

well, or talk or drink while eating. The best way to avoid these hazards is to eat slowly and to chew food thoroughly. Try not to leave too much time between meals. Do not eat on the run or at erratic hours, and avoid having a late evening meal.

Treatment

Herbal remedies: Chamomile, fennel, lemon balm, wild yam, ginger, or angelica root can ease flatulence. Take the herb in a tea or diluted tincture after meals. For angelica root tea, pour one pint of boiling water over one ounce of chopped root, steep for 10 minutes, and take two teaspoons three times a day before meals.
Caution: For safety concerns, see pp. 34–37.

From the chemist or health-food store
- Take charcoal tablets according to the instructions on the label.
- Take an acidophilus supplement each day for

Acupressure for flatulence

Thumb pressure on the point (St 4) on the inside arch of the foot just behind the ball of the foot can regulate digestion. Sitting comfortably, press this point for 5 to 10 seconds once or twice a day.

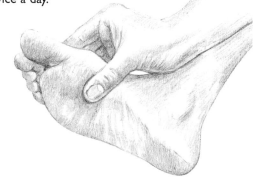

two to four weeks. Choose a dairy-free product that contains more than one billion organisms in each dose.

Wind-reducing juices: Pineapple juice contains the enzyme bromelain, and papaya juice contains an enzyme called papain. Both aid digestion. Juice half a papaya or a third of a pineapple and add an equal amount of uncarbonated water. Drink three times a day for up to three days. Some people find that sauerkraut juice also helps.

Hot and cold compresses: Direct heat sometimes soothes a windy abdomen and relieves distension. Place a covered hot-water bottle on your abdomen and leave for as long as it feels comfortable or until it starts to cool. Or try hot and cold compresses one after the other: put a hot towel (dipped in hot water and wrung out) over your abdomen for three minutes, then a cold towel (dipped in cold water and wrung out) for one minute, and repeat several times.

When to get medical help
- Flatulence does not respond to natural remedies within seven days, or sooner if it is accompanied by abdominal pain, fever, vomiting, or severe nausea or diarrhoea.

See also:
Abdominal Pain, Bloating, Constipation, Indigestion, Irritable Bowel Syndrome

Herbs and spices
Adding herbs and spices to your food can help reduce flatulence by inhibiting the growth of wind-forming bacteria. Try some of the following: parsley, dill, fennel, cayenne, ginger, and cardamom and caraway seeds. Add winter or summer savory when you cook beans to reduce their wind-forming effect.

Food Sensitivity

An allergic response or an intolerance to a food or food ingredient is called a food sensitivity. Such reactions can cause a variety of unpleasant symptoms, including digestive upsets, headaches, and bloating. In most cases, recognizing and avoiding the offending food—or at least cutting down on it—is the only effective means of treating a food sensitivity.

Contrary to popular belief, not all food sensitivities are allergies. Food sensitivities are grouped into two types: allergies and intolerances. Allergies are the rapid-onset reaction of the immune system to a particular food. If you have a food allergy, your immune system over-reacts to certain foods (allergens, or "culprit" foods), treating them as if they were harmful (see box below).

Intolerances are delayed-onset, non-allergic reactions to food, which may or may not involve the immune system. Gluten intolerance, which is the cause of coeliac disease, is an example of a delayed-onset immune-system reaction. However, lactose intolerance (inability to digest sugar in cow's milk) does not involve the immune system, but is instead caused by the failure of the body to produce an enzyme necessary for digesting this food substance properly. Common intolerance triggers include monosodium glutamate, or MSG, which can produce faintness, flushing, headache, and abdominal pain. Certain artificial food colourings, especially orange and yellow, can trigger eczema or asthma, and caffeine can lead to insomnia and shaking. Some individuals develop abdominal pain after eating fruit in the same meal as fatty foods. The stomach is slow in emptying such foods, allowing time for bacterial fermentation to produce gas, which distends the stomach and intestines.

It is thought that damage to the intestinal lining—for example, by irritation from certain foods, infection, or antibiotics—can cause a "leaky gut" and allow traces of undigested foods to enter the blood, triggering a food allergy or other immune-system reaction in susceptible people.

Symptoms

Allergy: In this form of food sensitivity, symptoms appear within 1 to 2 hours of eating even a small amount of the culprit food. The immune system produces inflammatory substances and a type of antibody known as IgE. Blood tests and skin tests for IgE can identify immediate-onset allergy. Symptoms may include:

- Urticaria (nettlerash or hives).
- Allergic rhinitis (hay fever).
- Asthma.
- Swelling of the lips, mouth, and the lining of the respiratory tract.

The immune response

When you have a food allergy, or other immune-system related sensitivity, your immune system reacts by producing IgE antibodies which attack the culprit food. Their attack releases chemicals such as histamine and leukotrienes, which cause inflammation. Your immune cells interpret this inflammation as a warning sign of invading bacteria or viruses and in response they produce harmful particles called free radicals. These free radicals can cause damage if left unchecked because they interfere with the healthy functioning of body cells.

Non-allergic immune response: Symptoms appear up to 72 hours after eating the culprit food. They're usually provoked by eating a large amount of it or having it frequently. The immune system produces IgG (and, perhaps, IgE) antibodies and in some cases inflammatory substances as well. Identifying the culprit can be difficult because symptoms tend to be vague, take time to appear, and often result from foods that are common in the diet (see "Common culprits," p. 214). You may be more susceptible to this form

Hot tip
If you have food-related symptoms, experiment by avoiding paprika, cayenne pepper, and chilli pepper. Evidence suggests that these spices irritate the intestines, encouraging food sensitivity.

A healthy start
Reduce your baby's risk of developing food sensitivities later in life by selecting first foods with care. Fruit and vegetable purées are the best starter foods.

> ### Anaphylactic shock
> This is a potentially fatal allergic reaction that includes breathing difficulty, a rapid fall in blood pressure, and, in some cases, loss of consciousness. If you experience shortness of breath or feelings of faintness after eating a suspected food allergen, seek urgent treatment. An adrenaline injection could save your life.

of reaction when physically or mentally stressed, and you may crave the very food that makes you ill. In addition to the symptoms listed for immediate-onset allergy, possible symptoms include:
- Flushing.
- Fatigue.
- Muscle weakness, aching, and stiffness.
- Eczema.
- Nausea and vomiting.
- Diarrhoea.
- Abdominal pain.
- Joint pain.
- Palpitations.
- Bloating.
- Weight fluctuation.

Prevention
- Help prevent damage from immune reactions to foods by eating a healthy diet that includes foods containing flavonoids, copper, iron, magnesium, selenium, zinc, essential fatty acids, and vitamins A, B (especially B_6), C, and E (see p. 12).
- If you are sensitive to wheat, don't use biscuits and cakes as "comfort foods" unless you know that they are wheat-free.

■ If you have an attack of gastroenteritis, you may suffer from temporary intolerance to gluten or sugar, so take care as you recover to reintroduce only gradually foods containing wheat, barley, oats, rye, or added sugar into your diet.

■ To reduce your baby's risk of an allergy or other immune-system related food sensitivity developing in later life, breastfeed for at least a year, and wait until the infant is four to six months old before introducing other foods.

■ Withhold those foods most frequently implicated in sensitivities (see illustration below) until a baby is at least six months old—longer if there is eczema or asthma in the family.

Common culprits
Wheat, milk, soy, bananas, eggs, fish, shellfish, nuts, seeds, beans, peas, lentils, tomatoes, citrus fruit, yeast, chocolate, and food additives are among the most likely causes of food sensitivity.

Treatment

Diet: Try to identify your culprit foods. People with a food allergy generally react to one or two foods, while those with a delayed-onset immune reaction may react to up to 15.

■ Keep a food diary for three months, recording everything you eat and any symptoms. This may allow you to identify an obvious relationship between a food and an adverse reaction.

■ If you suspect a certain food, don't eat it for three weeks, then reintroduce it to see if it causes problems. If so, stop eating it and then reintroduce it again. Repeat this challenge twice to be sure. If this single-food elimination doesn't work, exclude all the most likely culprits for three weeks. Then try a small amount of one of these foods every four days to see if it triggers a reaction. However, if you are not knowledgeable about nutrition, don't attempt this exclusion diet without medical supervision, and always get help if a child is involved.

■ Be aware that when you give up a food to which you are sensitive, you may experience temporary withdrawal symptoms, such as headaches, fatigue, and irritability.

■ Once you identify your culprit foods, you can omit them from your diet altogether, in which case all your symptoms should disappear within three to six months. Or, if you are not severely allergic, see whether you can eat a small amount once every four days or more without trouble. You may need to exclude the culprit foods completely for six months before you gradually reintroduce them in this way.

Supplements

- If you have digestive problems, take *Lactobacillus acidophilus* tablets (as directed on the packet), along with vitamins A, C, and E, as well as flavonoids and zinc, to help heal the lining of the intestines.
- If your food sensitivity results from a lack of a specific digestive enzyme, try taking a supplement of that enzyme.

Herbal remedies: Herbs cannot cure an allergy or other immune-system reaction, but some may help soothe symptoms while you are searching for culprit foods. They may also aid in healing after withdrawing the problem food from your diet.

- Drink a cup of a herbal tea or a few drops of a tincture made from echinacea, liquorice, and red clover three times daily.
- Drink a cup of yarrow or chamomile tea every four hours—or combine both herbs in a tea. Yarrow and chamomile contain substances that act as natural antihistamines and therefore help counter food allergies.

Caution: For safety concerns, see pp. 34–37.

Anti-allergy salad

Nettles are especially rich in vitamins A and C and iron and other minerals, and their astringent properties can reduce the inflammation associated with allergy. Add nettle leaves to salads, stews, and soups. Or drink a cup of nettle tea twice a day or half a teaspoon of tincture daily.

When to get medical help

- You are uncertain about the cause of your symptoms.
- You suspect lactose intolerance or coeliac disease.
- Symptoms from a suspected food allergy do not improve after self-help measures.

Get help right away if:

- You suffer a severe allergic reaction, such as a rash, shortness of breath, fainting, or swelling of the mouth, tongue, or throat.

See also:
Abdominal Pain, Allergic Rhinitis, Arthritis, Asthma, Bloating, Convulsions, Depression, Dermatitis, Diabetes, Diarrhoea, Fatigue, Headache, High Blood Pressure, Indigestion, Irritable Bowel Syndrome, Itching, Migraine, Nausea and Vomiting, Overweight, Stress

Foot Problems

Corns, calluses, bunions, and ingrown toenails are not only unsightly and uncomfortable, but they also make it difficult to buy shoes that fit. These problems are preventable, and many natural home remedies can ease them. By caring for your hard-working feet, you make sure that walking and other forms of exercise remain pain-free and pleasurable.

Prolonged friction or pressure from ill-fitting shoes is the main cause of most foot problems. The following descriptions are of the most common foot conditions:

- Corns are raised areas of thickened skin, usually on the toes. Each corn has a hard centre formed from a cone of tightly packed dead skin that points downward into the underlying tissue. If you press a corn, this hard point puts pressure on the nerve endings beneath the skin, causing pain. Corns between the toes are usually softened by sweat, and corns on the soles are generally very small—the size of a grain of rice. People with high arches are most likely to be affected, because when they walk, they put greater pressure on their toes, increasing the risk of rubbing against the shoes.

- Calluses are raised patches of thickened skin on the sole. Unlike a corn, a callus is insensitive. Calluses may be caused by any disorder of the foot or leg that prevents the sole from accepting the body's weight evenly. (Calluses can also occur on the hands, as a result of excessive pressure or friction.)

- A bunion is a fluid-filled pad (bursa) on the side of the big toe, overlying the joint between the toe and the foot. It may become red and tender. The natural shape of the big toe joint in some people makes it more likely to be damaged by shoes that compress the toes or don't match the shape of the feet.

- An ingrown toenail develops if pressure on the nail—usually that of the big toe—forces its growing edges into the adjacent skin, leading to pain, inflammation, and sometimes infection. The condition is most likely to occur if the nails are cut incorrectly or if the skin is continually damp from sweating—for example, because of shoes or socks made from synthetic materials that do not allow moisture to escape or that are too tight.

Reviving footbath

A regular herbal footbath not only helps keep your feet clean and fresh, but also improves circulation. Add a tablespoon of dried rosemary leaves (for their antiseptic and stimulating properties) to some hot water in a large basin—big enough to allow you to immerse your feet—and let stand for five minutes. Fill another basin with cold water, and place both basins on the floor by a chair. Sit with your feet in the hot, herb-scented water for one minute, then put them into the cold water for about 20 seconds. Repeat several times, ending with a brief plunge of the feet into the cold water. Pat your feet dry.

Prevention

Go barefoot as often as you can. In cultures in which this is the custom, there are few deformed feet and no corns or bunions. Wear shoes made from materials that allow the feet to "breathe", such as leather, canvas, or certain synthetic products. Make sure the shoes fit well, with space for the toes to move but without so much room that the feet slip. The heel should be well supported. Socks should not be too tight.

Wearing properly fitting shoes is crucial during infancy and childhood, when growing feet are so flexible that they feel no pain if shoes don't fit correctly. The wrong shoes can, over the years, lead to foot problems and even deformity. Babies don't need anything on their feet until they begin to walk outdoors. Children's feet should be measured for length and width by a trained fitter. They should be remeasured and shoes replaced (if necessary) every three months in the first seven or eight years, then less often as foot growth slows.

Make certain that you cut your toenails correctly (see illustration, left). Before cutting, let your nails grow long enough to allow their edges to clear the skin. Avoid digging into the sides of the nail to clear away dead skin and other debris.

Pedicure pointers
Prevent ingrown toenails by cutting the nails straight across. Do not cut into the corners, but be sure to remove any protruding spikes at the nail edges.

Treatment

General measures

■ Remove any source of pressure or friction on your feet and toes by buying

Aromatherapy foot massage

Self-massage with aromatherapy oil is relaxing, reduces a tendency toward excessive sweating, and suppresses the bacteria that act on sweat and cause foot odour. Add three drops of lemon, tea tree, peppermint, or rosemary oil to two teaspoons of soya or sweet almond oil, and rub the mixture into your feet. (Avoid rosemary oil during the first 20 weeks of pregnancy.) Use slow, smoothing strokes and/or firm kneading, depending on what feels most comfortable to you.

better-fitting footwear. Avoid high-heeled shoes. If necessary, stretch tighter-fitting leather shoes. Ask your shoemender to do this, or buy shoe-stretching liquid or an expandable wooden shoe-stretcher from a shoemender, department store or specialist shop. Consider wearing shoes that are designed for walking or running; these tend to give more support.

■ Prevent fungal infection by keeping your feet clean and dry.

■ For an ingrown toenail, make a small, V-shaped cut in the centre of the nail to ease the pressure at the sides.

➤ *continued, p. 219*

Foot massage

A massage brings relief for tired and aching feet. You can massage your feet yourself (see p. 217), or have a friend or partner do it for you while you lie on your stomach, with one lower leg lifted from the knee. Do not massage the foot if the joints are swollen or inflamed, and stop any massage that causes pain or discomfort.

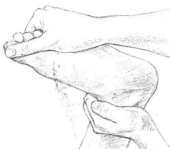

1 Holding the toes and ankle, slowly rotate the whole foot first one way, then the other.

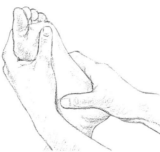

2 Press down on the heel with one hand, and with the other push the front of the foot back for about 15 seconds.

3 Hold either side of the Achilles tendon (just above the heel), then with the other hand press downwards on the ball of the foot, while pushing the heel upward.

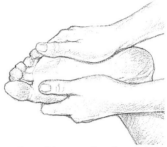

4 Holding each side, push the foot towards one side, then the other, several times.

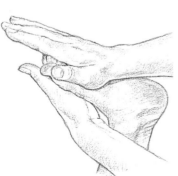

5 Roll the ball of the foot firmly between the heels of your hands, working across the foot, below the toes.

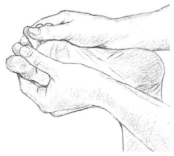

6 Stretch each toe in turn by grasping the two adjacent toes and pulling them slowly apart, stretching the skin between them.

7 Rotate and pull each toe by holding the foot in one hand, grasping each toe in turn between your thumb and forefinger, and gently rotating it several times, then pulling it gently for a few seconds.

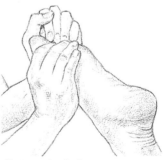

8 Stretch all the toes at once by grasping the four small toes with one hand and holding the big toe in the other, then lifting the leg up and shaking it slightly. Perform this sequence on both feet.

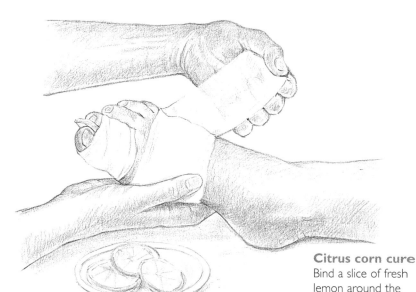

Citrus corn cure
Bind a slice of fresh lemon around the affected toe with a bandage before you go to bed. Remove the next morning, and repeat each night for a week or two.

From the chemist

- For corns or bunions, apply felt or sponge corn pads or rings to ease pressure and pain.
- For a callus, soften by soaking in water or with a product containing salicylic acid. Then use a pumice stone or callus file to remove the thickened skin. Repeat once or twice a week.
- Castor oil also softens corns and calluses, as does pure lanolin (do not use if you're allergic to wool).
- For smelly feet, try bathing them daily in a solution of sea salt and water.

Kitchen–cabinet remedy: For an ingrown toenail, soak your foot in a basin of warm water to which you have added about three tablespoons of salt. Do this for five minutes twice a day. Cut the nail only when inflammation has subsided.

Aromatherapy

- Rub a corn with a few drops of lemon oil, then cover in cling film for 20 minutes. Repeat this procedure daily for a week or two.
- For an ingrown toenail, soak the affected foot for five minutes twice a day in a basin of warm water to which you have added six drops of tea tree oil. This will help prevent infection.

Herbal remedies

- For an ingrown toenail, soak your foot in two pints of warm water containing two teaspoons of marigold (calendula) tincture. Because marigold has antiseptic properties, this solution will also help cleanse and soften the nail. Or, instead of marigold, add a tablespoon of distilled witch hazel to the water.
- Thuja may help clear corns or verrucas. Apply thuja ointment to the sore area, and cover with an adhesive bandage overnight.

When to get medical help

- You have pain, a sore, or swelling that persists longer than seven days.
- You have a fluid- or pus-filled swelling.
- No shoes are comfortable.
- Calluses remain painful.
- You experience numbness or tingling.
- An ingrown nail remains embedded.

See also:
**Cold Hands and Feet,
Fungal Skin Infections,
Strains and Sprains,
Warts**

Haemorrhoids

About half the population in Western countries suffers at some time from haemorrhoids, which means that the spongy. bloody-vessel-rich pads of tissue that line the anus are swollen. Otherwise known as piles (from the Latin *pilae*, meaning "balls"), they can often be prevented by eating the right foods and, unless severe, they can usually be alleviated through natural therapies.

Haemorrhoids may be internal—that is, within the anal canal—or external, when they may be felt as little knobs or balls around the anal opening. Prolapsed haemorrhoids protrude outside the anus. Symptoms may include painful defecation, itching, and rectal bleeding. Pain may be severe if a large haemorrhoid is inside the anus. Haemorrhoids are usually caused by constipation, when straining to pass hard stools increases the pressure in the anal veins, making them dilate. Constipation also makes existing haemorrhoids worse, since passing hard stools abrades dilated veins and the strain of trying to have a bowel movement puts pressure on the anus. Haemorrhoids often occur during pregnancy and immediately after childbirth. They are more likely if you are overweight or inactive, or if you eat a diet low in fibre and high in refined foods.

prevent constipation. Drink at least six extra glasses of water daily to help soften stools. Exercise regularly to ensure good circulation. Be careful about reading in the bathroom; sitting too long on the toilet seat increases pressure on the rectum.

Potent fruits
Berries and cherries are rich in vein-strengthening antioxidant flavonoids called proanthocyanidins.

Prevention
Lose any excess weight. Eat a well-balanced, high-fibre diet (see pp. 9–12) to provide the nutrients needed to keep veins strong and healthy and to

Treatment
During flare-ups of pain, take daily breaks (see photograph, left). Don't stand still for long periods, and exercise vigorously for at least 20 minutes on most days to improve circulation.

Diet: Eat plenty of fresh fruit and vegetables, whole grains, nuts, and seeds. These contain fibre, which helps treat constipation, as well as vitamin C, silica, and flavonoid pigments that help make vein walls strong and flexible. Berries, cherries, and buckwheat contain especially beneficial flavonoids.

Rest position
When haemorrhoids are painful, rest for two half-hour periods every day by lying or sitting with the feet above hip level.

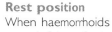

From the chemist: Strengthen vein walls with supplements of vitamin C, proanthocyanidins, rutin, and silica.

Herbal remedies: Gently wash and dry the anal area, then apply one of the following to soothe and shrink swollen veins:

■ A cold, astringent solution made from a tablespoon of distilled witch hazel and four drops of marigold (calendula) tincture in a cup of water.
■ Cooled tea made by boiling a teaspoon of grated bistort root in a cup of water for 10 minutes in a covered pan.
■ Pilewort (lesser celandine) ointment. Apply in the morning, at bedtime, and after washing.
■ A preparation available from the health-food store containing extracts of horse chestnut and black haw. Or apply a paste made by mixing two teaspoons of one or both of these powdered herbs with one to two teaspoons of walnut oil at night; remove by spraying with cold water in the morning.

Caution: For safety concerns, see pp. 34–37.

Homeopathy
■ Apply homeopathic cream containing Aesculus and Hamamelis to soothe the affected area.
■ Sulphur and Nux vomica, alternated on a daily basis, can often be helpful.

Aromatherapy: To strengthen vein walls, add six drops of lemon, lavender, rosemary, cypress, or juniper berry essential oil to your bathwater. Avoid juniper berry oil throughout pregnancy and rosemary and cypress oils during the first 20 weeks of pregnancy.

Acupressure for haemorrhoids

To relieve haemorrhoids, try pressing with your thumb the point (B 57) in the middle of the back of your calf at the junction of the muscle and tendon. Do not use this point during pregnancy.

Hydrotherapy
■ To ease itching, wash with unscented soap and water. Rinse the area with cold water.
■ Take alternate hot and cold sitz baths each day (see p. 49) to improve circulation.
■ If your haemorrhoids are very painful, apply crushed ice to the area for a few seconds or spray with iced water.

When to get medical help
● You notice blood in the stools or rectal bleeding.
● You experience unexplained diarrhoea or constipation for more than two weeks.
● Symptoms fail to improve with treatment.
● You feel ill or lose weight.
● You have protracted rectal pain or excessive discomfort during bowel movements.

Get help right away if:
● Rectal bleeding is severe.

See also:
Constipation, Diarrhoea, Overweight, Pregnancy Problems, Varicose Veins

Hair and Scalp Problems

Most hair and scalp problems respond well to home remedies. Dandruff and hair loss may result from poor nutrition, stress, illness, or incorrect or harsh treatment. They may also occur as side effects of some medications. Thinning hair may have hormonal causes that may arise from a temporary condition or from long-term age-related changes.

The oiliness of your hair reflects your genetic make-up and hormone balance; the scalp's oil glands are particularly active during adolescence. The excessive skin flaking found with dandruff represents a high turnover of scalp cells and, possibly, a yeast infection. Temporary hair thinning may follow stress, shock, childbirth, illness, or treatment with certain drugs. Stress-related muscle tension tightens the scalp and prevents enough nutrient-rich blood from reaching the hair follicles. The starved hair roots then shrink and the hair may fall out prematurely. Patchy hair loss may be due to a fungal infection, such as ringworm of the scalp (*Tinea capitis*).

Prevention

The condition of your hair and scalp is often directly related to your lifestyle and the way you take care of yourself. To prevent problems from arising or worsening, eat foods rich in protein, carotenes, essential fatty acids, iron, silica, zinc, and vitamins B (especially biotin), C, and E. Cut down on fizzy drinks, which acidify the blood and may lead to mineral loss from the hair. Avoid losing weight too quickly with an unhealthy diet, and, if necessary, use stress-management techniques (see p. 347). Get regular exercise to improve the circulation to your scalp and supply the hair follicles with plenty of nutrients.

Treatments

Aromatherapy: Choose one of the following recipes, according to your specific problem. Apply the formula to the affected areas and leave on overnight. Shampoo your hair the next morning, and use a final rinse made by stirring the same blend of essential oils into a jug of warm water. Repeat every one or two days at first, then twice a week as your scalp improves.

- Red, itchy scalp: add two drops each of geranium and lavender oils and one of sandalwood

Scalp savers

Diluted with oil or water, such essential oils as geranium, lavender, and juniper can help alleviate scalp problems.

Tips for healthy hair

- Choose suitable products for your hair type. Consider the degree of oiliness or dryness, and previous treatments, such as permanents.
- Avoid using a hair dryer, heated rollers, or curling rods for too long, and don't blow hot air too near your hair. Let dry or damaged hair dry naturally; for extra bounce or curl, use thermostatically controlled, steam-producing heated rollers.
- Bleach, perms, and some tints can harm the hair and scalp, so always follow the instructions. Have treatments done professionally if your hair is in poor condition. If your hair is dry and you want to colour it, choose such herbal products as henna and other vegetable dyes; these can strengthen hair and make it shine.

essential oil to two and a half teaspoons of olive or sweet almond oil.

- Dry, flaking scalp: Mix three drops of lavender, three of geranium, and two of sandalwood oil with five teaspoons of olive oil, sweet almond oil, or another cold-pressed vegetable oil.
- Oily, scaly scalp: Mix three drops of Atlas cedarwood, two of rosemary—omit these if you're pregnant—and two of lemon oil with five teaspoons of unscented body lotion.

Massage: A herbal or aromatherapy scalp massage relieves stress-induced muscle tension and improves circulation. Be sure to move the skin of the scalp over the skull, rather than just moving the hair around. Use nettle or burdock

Spices that go to your head
Fenugreek seeds contain an enzyme called triogonelline, which increases the blood supply to the scalp and allows more nutrients and oxygen to reach the hair follicles and nourish the hair roots. Its chemical structure is similar to that of minoxidil, a topical drug that promotes hair growth in some people. Drink a cup of fenugreek tea every day for best effect.

tea or a tonic made by mixing three drops each of rosemary and ylang ylang oils, two of Atlas cedarwood oil, half a teaspoon of vodka, and five teaspoons of orange-flower water. Omit rosemary and Atlas cedarwood oils if you're pregnant.

Kitchen-cabinet remedies

- To remove excess oiliness, rinse your hair with distilled water to which you have added a splash of lemon juice.
- Smooth a tablespoon of castor oil into the hair with your fingers two hours before shampooing. This makes hair shine and seals split ends.
- Boil four tablespoons of dried thyme in two cups of water and allow to cool. Apply to your clean, damp scalp to treat dandruff.

When to get medical help

- The problem fails to clear up after home treatment.
- Hair loss results in bald patches.
- You have other symptoms, such as fatigue or weight loss.

See also:
Fungal Skin Infections, Head Lice

Hangover

You may have, at some point in your life, woken up with the throbbing headache that usually signals a hangover. This is not a serious condition in itself, and it generally disappears within a few hours. However, a hangover is a sign that you have been drinking too much and without due care. A range of natural remedies can help your symptoms.

DID YOU KNOW?

Women who are pregnant or trying to conceive should either drink no alcohol, or have no more than one or two units once or twice a week. Alcohol may also be hazardous if you are taking certain medications.

Hangovers vary in severity, depending on the quantity and type of alcohol consumed. Symptoms include headache, nausea, and dizziness. Alcohol dilates the blood vessels, and too much dilatation can cause a headache. It also has a dehydrating effect because it encourages the kidneys to remove more water than usual from the body.

Several factors influence the speed at which alcohol is absorbed from the digestive system into the bloodstream. Food in the stomach can slow absorption. Research also suggests that the more you weigh, the more time it takes for you to absorb alcohol. A woman tends to have a higher level of alcohol in the blood than a man of the same weight who has drunk the same amount. This is partly because women produce less of an enzyme that breaks down alcohol.

Frequent or persistent hangovers indicate that you should reduce the amount of alcohol you drink. Regularly drinking large amounts of alcohol is likely to permanently harm your health—for example, by damaging your liver or raising your risk of developing certain cancers.

Prevention

Besides simply not drinking too much, there are several strategies to prevent a hangover:
- Always eat something before you drink and/or while you drink to slow alcohol absorption.

Alcohol limits

Drink no more than the officially recommended alcohol limit. The UK Department of Health recommends no more than two or three units of alcohol a day for women, three or four for men. It also recommends at least two alcohol-free days a week. Consider also your size, weight, health and metabolism. Your ideal limit may be less than the standard guidelines. A unit of alcohol is half a pint of average-strength beer, one measure of spirits, or a small (four to five fluid ounce, or 125ml) glass of 8–9% proof wine.

- Reduce your alcohol intake by sipping drinks slowly so they last longer and by alternating non-alcoholic drinks with alcoholic ones.
- Don't mix drinks, and avoid those to which you react badly. Some people find that a hangover is more likely after drinking inexpensive red wine (because of the high additive content) or a fortified wine, such as sherry (because of the high content of natural flavourings and colourings called congeners).

Cleansing cabbage
Raw cabbage, a traditional remedy for hangovers, has been used for centuries because it is believed to detoxify the liver. Its protective effect on the lining of the digestive tract may also counter the adverse effects of alcohol on the stomach and intestines.

- Willow bark tea—made by simmering the bark in a covered pan for 10 minutes—contains natural aspirin-like substances that can soothe a hangover headache.
- Chamomile tea soothes irritation of the stomach and intestines.

Caution: For safety concerns, see pp. 34–37.

Homeopathy: Take the following remedy every 30 minutes for up to six doses.
- Nux vomica: if you feel nauseated but are unable to vomit.

Stimulating shake
A banana milkshake (made from half a glass of milk, a banana, and two tablespoons of honey) replaces potassium lost in dilute urine, and the honey raises your blood-sugar level, which is lowered by excessive alcohol intake.

- Avoid dehydration by drinking plenty of water before and after drinking alcohol.

Treatment
- First drink a large glass of water for immediate replacement of lost fluids.
- Replace lost nutrients, such as magnesium and potassium, with a banana milkshake (see illustration, far right).
- Don't have another drink (the "hair of the dog"). It will worsen or prolong symptoms.
- Don't drink coffee. It won't do any good and may make you jittery. It's also dehydrating.

Herbal remedies: Drink a cup of one of the following teas every hour until you feel better.
- Rosemary tea relieves headaches and is said to help the liver detoxify alcohol.
- Milk thistle or dandelion tea may also aid detoxification of the liver and blood.

When to get medical help
- You cannot remember what happened during a drinking session.
- You have regular hangovers.

See also:
Addictions, Headache

Head Lice

The head louse is a brownish-grey, wingless insect that lives in human hair. Smaller than a match head, the head louse lays its eggs (called nits) along the base of the hair shaft, close to the scalp. Because of concern about the safety of some insecticides, natural methods of eradicating the problem are becoming increasingly popular.

Head lice feed by sucking blood from the scalp. Their bites may itch severely and sometimes become infected. Their brownish eggs hatch within eight days of being laid. The pearly-white, empty egg cases, which look somewhat like dandruff, are carried along the hairs as they grow, so that the egg cases are often found further from the scalp than the lice themselves.

Head lice are easily transmitted by head-to-head contact or by sharing combs, brushes, hats, or towels. Infestations are common in school-children and those who work with children.

The problem magnified
The head louse, seen here enlarged about 75 times, has three pairs of legs that enable it to cling firmly to the hair shafts.

Prevention

Inspect your child's hair and scalp regularly for lice and nits, paying particular attention to the hairline under a fringe, the nape of the neck, and the area above the ears. If there is an outbreak of head lice at your child's school, check even more thoroughly every few days. The best way to do this is to comb your child's hair using the wet comb method (see "Treatment") and see if any lice fall out.

To repel lice, comb the hair twice a week with a comb dipped in a mug of warm water containing 10 drops of tea tree oil.

Treatment

If your child gets head lice, he or she should use separate towels and flannels, and should keep them away from those belonging to the rest of the family. After treating the head lice, wash your child's clothing, sheets, and pillowcases in hot water, and rinse brushes and combs thoroughly. Check all other family members, and treat them in the same way, if necessary.

The "wet comb" method: This is one of the best ways to get rid of head lice without using chemical insecticides. Wash the hair, then apply a lot of silicone-based conditioner evenly through it. Comb the hair first with a wide-toothed comb to remove any tangles. Then, working systematically, use a very fine-toothed comb (one made for the purpose) to remove the lice from one section of the hair at a time. Inspect the comb for lice after each stroke, and rinse them away. Rinse out the conditioner, and comb through again. You won't be able to dislodge the unhatched eggs with the comb because they stick firmly to the hair, but by repeating this treatment every other day for two weeks, you will gradually catch all the lice that hatch from the most recently laid eggs before they reproduce.

Careful combing
Whether you use herbal treatments, essential oils, or conditioner alone, thorough combing with a specially designed "nit comb" is the key to successful natural treatment of head lice.

the same mixture as a final rinse when washing hair, or a blend of six drops each of sweet thyme and rosemary oils with one pint of warm water.

Herbal remedies: Quassia bark and tansy are reputed to have insecticidal properties. Make a double-strength tea from either quassia bark chips (boil the chips in the water for 20 minutes) or tansy. Mix a cup with your usual amount of conditioner, and use as a rinse after shampooing.
Caution: For safety concerns, see pp. 34–37.

Gentle insecticides
Rosemary, geranium, lavender, eucalyptus, and lemon oils are excellent lice deterrents. During an outbreak, add two drops of eucalyptus and one drop each of lavender and geranium oils to shampoo. After shampooing, add two drops of any of these oils to conditioner or warm water and use as a final rinse.

Aromatherapy: To kill lice and soothe itching, add two drops of eucalyptus and one drop each of lavender and geranium oils to one teaspoon of unscented body lotion. Massage into the scalp and leave for half an hour. Run a fine-toothed comb through the hair before shampooing out the lotion, then rinse well. Next, apply an antiseptic rinse made by stirring two drops each of eucalyptus, lavender, rosemary (omit if you're pregnant), and geranium oils as well as two and a half teaspoons of vinegar into one cup of water. Rinse the entire head, and let the hair air-dry. Repeat daily. To prevent further infestation, use either

An all-round anti-infestation treatment
As well as repelling head lice, tansy is said to combat scabies and intestinal worms. However, it should be taken internally only on the advice of a qualified medical specialist.

When to get medical help
- Treatment is ineffective.
- The scalp becomes inflamed or infected.

See also:
Hair and Scalp Problems

Headache

A symptom, not an illness, a headache is rarely a sign of serious disease, but is instead most often caused by muscle tension resulting from a state of mental stress. Pain is produced by hard strands of contracted muscle fibres pressing on nerves or obstructing the flow of blood, lymph, tissue fluid, or—according to Asian medicine—energy.

Headaches range in severity from mild, short-lived discomfort to a raging pain that makes activity impossible. The site of the pain varies from behind the eyes to the temples, forehead, back of the head, or even the whole head. You may be able to feel the hard, taut muscle fibres responsible for a tension headache under the skin. Migraine is a recurrent severe headache that may be accompanied by additional symptoms. Almost everyone has occasional headaches, but you are likely to have fewer as you get older.

Causes

There are many causes of headaches, including:
- Stress.
- Dehydration.
- A lack of nutrients to the muscles, blood vessels, and nerves.
- Flu or other viral infection.
- Sinus infection.
- Excessive alcohol intake.
- Environmental factors, such as a smoky atmosphere or poor lighting.
- Prescription and over-the-counter medications.
- Hormonal changes, as with premenstrual syndrome (PMS).
- Dental abnormalities.
- Temporomandibular joint (TMJ) syndrome.
- Eyestrain.
- A food sensitivity or allergy.

- Blood vessel contraction or dilation (vascular headaches), which can lead to migraine.
- Osteoarthritis or a misalignment in the bones of the neck.
- In rare cases, brain tumour, meningitis, high blood pressure, or a stroke.

Prevention

A healthy diet reduces susceptibility to headaches by increasing resistance to infection, improving the condition of the muscles, nerves, and blood vessels, and helping prevent food sensitivity. Drink at least six glasses of water daily to prevent dehydration. Try to respond to stress in as positive a way as possible (see p. 347). Make any necessary changes to your environment, such as avoiding cigarette smoke, moistening dry air with houseplants or a humidifier (be sure to clean thoroughly and regularly), and adjusting the levels and angles of ambient lighting. Some people find an ionizer—a device that removes positively charged ions from the air—helpful.

Treatment

Herbal remedies: Take the following herbs as a tea (up to three cups daily) or a tincture (five drops diluted in a little water) drunk during or after eating—every two hours, if necessary, but no more than three times a day.
- Take feverfew tincture for a maximum of two

weeks at a time. Or you can eat feverfew leaves, putting them in a sandwich to disguise their bitterness. Be aware that feverfew leaves cause mouth ulcers or a sore tongue in some people.

Feverfew
This staple of the herbal medicine cabinet has a marked effect on headaches caused by dilation or contraction of the blood vessels.

- To prevent a headache from worsening, drink a cup of tea or take some tincture made from passionflower, rosemary, or wood betony—or a combination of one or more of these herbs— together with meadowsweet.
- To relax tight muscles, drink a cup of tea or take a tincture made from valerian or cramp bark, with meadowsweet and rosemary. If you feel stressed, include chamomile, vervain, wild oats, or dried pasqueflower.
Caution: For safety concerns, see pp. 34–37.

Aromatherapy: Fragrant essential oils can relieve pain, ease tension, and clear congestion. Put them in your bath water, in hot water for an inhalation, or in oil for a massage.

- For a tension headache, try lavender, sweet marjoram, and chamomile oils, all of which have relaxing properties.
- If your headache is caused by nasal or sinus congestion, perhaps from a cold or allergic rhinitis, use eucalyptus or peppermint oil.

➤ *continued, p. 239*

Massage away tension

Smoothing, kneading, or pressing taut muscles in the shoulders, neck, face, and scalp is one of the most effective ways of dealing with a tension headache. Ask a friend to try the following massages, while you are lying down.

Forehead massage
1 Kneel at the person's head and place your thumbs at the centre of his or her forehead, just above the eyebrows, with your fingers at each side of the head.

2 Draw your thumbs out towards the temples, then lift them off when you reach the hairline. Repeat the movement, starting a fraction higher each time, until you have massaged the whole forehead.

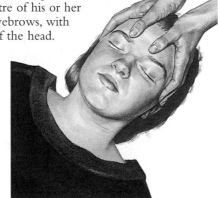

Massaging the temples
1 Place your fingertips on the temples, and press firmly for 10 seconds with the flat ends of your fingers.

2 Release the pressure gradually, and make slow, circling movements with your fingers over the temples.

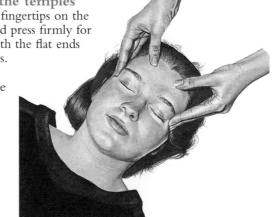

Yoga for tension

Simple exercises based on yoga techniques may relax tense muscles and thus relieve headaches.

Five-minute relaxation technique

1 Lie on your back on a firm surface, with a cushion under your neck, if necessary, for comfort. To minimize the space between your lower spine and the ground, lift your knees to your chest, then slowly lower them by sliding your feet along the floor. Place your feet 8 to 12 inches apart, and your hands about 18 inches from your sides, with palms upward. Close your eyes.

2 Focus your attention on your body and your breathing. As you breathe, be aware of the movement of your abdomen as it rises and falls. Continue slowly breathing in and out, concentrating on how it feels as the air fills your lungs, and on how your abdomen expands with each inhalation. Notice your abdomen relaxing and sinking towards the floor as you exhale, and feel your whole body sinking into the ground. As you continue this deep breathing, notice the feeling of increasing relaxation as you exhale and of lightness and energy as you inhale.

Neck rolling

This exercise is not suitable if you have neck pain or other neck problems.

1 Get on your hands and knees, and place your hands flat on the floor under your shoulders.

2 Bend your arms until the top of your head touches the floor, then as you exhale, roll your head gently forward. Stop when you feel a gentle stretch down the back of your neck, and hold the position for a few seconds.

3 As you inhale, slowly roll your head in the other direction until your forehead touches the floor. Repeat this sequence slowly 20 times.

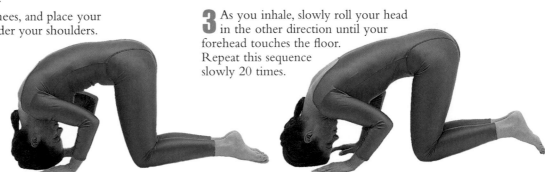

Aid in a nutshell
Eating almonds may help your headache. These tasty nuts contain pain-relieving chemicals.

■ Gently massage two drops of lavender oil along the base of your skull at the back of your neck, on your temples, and behind your ears. Keep the oil well away from your eyes.

■ Sprinkle two drops each of sweet marjoram, lavender, and peppermint oils on a tissue. Inhale deeply three times.

Diet: Reduce your caffeine intake. If you are a heavy consumer, do this gradually—for example, by combining decaffeinated and caffeinated coffee. Be prepared for a withdrawal headache that may last a few days. Eat foods rich in calcium, magnesium, and essential fatty acids (see p. 12). Avoid foods to which you might be sensitive. Some people get headaches from nitrites and nitrates (in cured meats), monosodium glutamate (MSG), tyramines (in fermented foods, certain red wines, processed meats, aged cheeses, beer), sulphites (in dried fruits and relishes), or salicylates (in tea, vinegar, and many fruits).

Hydrotherapy: Apply alternating hot and cold compresses to the nape of the neck. Fold a small towel two or three times, wring it out in hot water, and leave it in place for two minutes; then replace it for one minute with a towel wrung out in cold water. Repeat for 15 to 20 minutes.

Homeopathy: For isolated headaches, try one of the following:
■ Belladonna: for throbbing, hammering headaches that are worse for light and noise.
■ Bryonia: for bursting headaches that are worse for the slightest movement.
■ Ignatia: for headaches arising from acute emotional distress.

Other therapies: Cranial osteopathy, biofeedback, self-hypnosis, meditation, reflexology, and acupuncture may also be helpful.

Acupressure for headache relief

1 Place one hand across the front of your partner's head, while using your other thumb to press the point (GB 12) just behind the bony prominence at the hairline at the back of the ear (see illustration). Direct gently pulsing thumb pressure towards the eye for about two minutes.

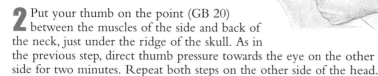

2 Put your thumb on the point (GB 20) between the muscles of the side and back of the neck, just under the ridge of the skull. As in the previous step, direct thumb pressure towards the eye on the other side for two minutes. Repeat both steps on the other side of the head.

When to get medical help
● You have repeated, unexplained headaches.

Get help right away if:
● You have a sudden, severe headache of a kind not previously experienced.
● You experience vomiting, a stiff neck, fever, rash, dislike of bright light, confusion, and/or vision problems (with a severe headache).

See also:
Hangover, Migraine, Stress

Hearing Loss

Deterioration of hearing can make effective communication a real challenge and it reduces the amount of useful information we receive from our environment. Luckily, hearing loss is rarely total. Several of the causes of this problem, such as wax blockage, are often easy to prevent, and some others are treatable with natural remedies and therapies.

There are two types of hearing loss: conductive and sensorineural. Conductive hearing loss occurs when sound waves from the outside world fail to reach the inner ear, usually as a result of a blockage in the outer ear canal or problems in the eardrum or middle ear. In sensorineural hearing loss, sounds reach the inner ear but fail to reach the brain because of damage to the inner ear or the acoustic nerve. The chief causes of such permanent hearing loss are the effects of age, arterial disease (which can reduce the blood supply to the acoustic nerve), and prolonged exposure to excessive noise levels (above 85 decibels). Sudden or prolonged noise—for example, from a loud workplace or rock concert—can damage the hairlike endings of the acoustic nerve. Researchers now believe that such damage may be partly caused by a surge of unstable oxygen molecules, known as free

What is Ménière's disease?

Attacks of vertigo, nausea, ear discomfort, and tinnitus (ringing and other noises in the ear), combined with progressive hearing loss in one ear, are usually symptoms of Ménière's disease. This results from fluid build-up in the inner ear, which puts pressure on the hearing nerve endings. Factors that may trigger Ménière's disease include poor circulation, premenstrual fluid retention, and, possibly, food sensitivity. One in three sufferers also has migraine.

radicals, that follows exposure to loud noise. Some prescribed drugs can also affect hearing.

Common causes of conductive hearing loss include inflammation of the middle ear, a boil in the outer ear, and impacted wax in the ear canal. Other causes include thick fluid remaining in the middle ear after an infection, Ménière's disease (see box), and aging. Temporary hearing loss experienced when flying occurs because of the inability to equalize air pressure in the middle ear and the throat during takeoff and landing. This is more likely if the eustachian tubes are blocked due to allergy or infection.

Prevention

Measures to prevent hearing loss include boosting resistance to infection and allergy,

Inside the ear
The ear is divided into three areas: the outer ear, the middle ear, and the inner ear. Blockage of the outer or middle ear leads to conductive hearing loss, which is usually reversible. Damage to the inner ear can cause sensorineural hearing loss, which is more likely to be permanent.

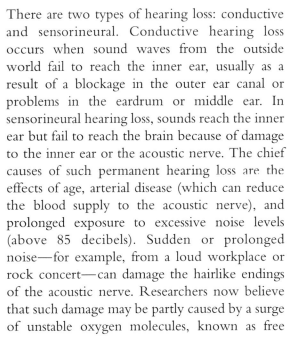

Outer ear

Middle ear

Inner ear

Eustachian tube

stabilizing the body's fluid balance, combating arterial disease, and addressing the causes of premature aging. This means eating a healthy diet (see p. 9), with plenty of whole grains, fruit, and vegetables, and limiting your intake of sugar, refined carbohydrates, and salt. Get some exercise daily, and if you smoke, find a way to stop. If you are suffering from allergic rhinitis or Ménière's disease, try to identify and deal with any underlying triggers.

Avoid sudden or prolonged loud noise. Signs that the decibel level may be high enough to damage hearing include being unable to hear someone speaking directly to you and experiencing ringing in the ears during exposure and muffling of ordinary sounds afterwards. If you cannot avoid loud noise, wear efficient earplugs or other protectors and take a supplement of antioxidants for a few days in advance. These nutrients will help neutralize any damaging effects of free radicals on the ear.

Treatment

Age-related hearing loss: Help slow down the progression of this condition by following a diet that preserves the health of the arteries and nerves supplying the ears (see "Prevention"). Take supplements of antioxidants and magnesium for two weeks after exposure to loud noise.

Nasal congestion: You can tackle hearing loss resulting from mucous congestion in several ways.
- Add a teaspoon of red sage tincture to half a cup of warm water. Use as a gargle.
- Drink plantain tea, which helps liquefy and loosen thick mucus.

Removing earwax

To soften wax so that it will work its way out of the ear canal, place a few drops of warm olive, almond, or mullein oil in your ear, then lie on your side with that ear uppermost for 15 minutes. Repeat several times a day for two to three days. Caution: Don't put anything in your ear if you have a discharge or pain—your eardrum may be perforated.

- Drink a cup of tea made with one or more of the following herbs two or three times a day: chamomile, cleavers, echinacea, elderflower, goldenseal, and liquorice.
- Put a few drops of eucalyptus or lemon oil on a tissue and inhale the vapour. Or add one or two drops of one of these oils to a teaspoon of cold-pressed vegetable oil and smooth over your throat and around your ears.
- Add lemon balm leaves to salads and soups.

Caution: For safety concerns, see pp. 34–37.

When to get medical help
- You have a severe earache or a discharge from the ear.
- You feel dizzy.
- Home treatment does not help within a few days.
- The hearing loss is sudden.

See also:
Aging, Allergic Rhinitis, Arterial Disease, Colds, Earache, Migraine, Sore Throat

Hiccups

Involuntary intakes of air, known as hiccups, are provoked by repeated spasms of the diaphragm, the sheet of muscle separating the chest and abdomen. Most attacks of hiccups, which include making an uncontrollable noise with each intake, last only a few minutes. Although irritating, hiccups are not serious, but you can try a variety of home remedies to stop an attack.

Hiccups often begin after having a large hot drink or a large meal, when an overfull stomach presses on the diaphragm. They also frequently follow a prolonged bout of uncontrollable laughter. But there may be no obvious cause. Attacks of hiccups usually stop of their own accord, rarely lasting more than 20 minutes. If they persist or make you uncomfortable or embarrassed, any of several simple traditional remedies may be effective. In rare cases, continuing hiccups signify an underlying disorder.

Prevention
Get into the habit of eating and drinking slowly. Avoid ingesting too much at one time, always chew thoroughly, and avoid stress at mealtimes.

Rhythmic drinking

Slowly sipping a glass of cold water or sucking on an ice cube can sometimes stop a bout of hiccups. A variation is to bend your head forward over a glass of water and slowly sip from the other side of the rim.

These remedies may work because most people breathe relatively little as they sip, thereby increasing their body's carbon dioxide level. The rhythmic contractions of the oesophagus induced by slow sipping and swallowing also override the spasms of the diaphragm.

Treatment
The following two remedies work by raising the body's level of carbon dioxide, which seems to relax the diaphragm.
- Take a breath and hold it for a while, but exhale before you get light-headed or dizzy.
- Hold a paper bag—never use plastic—over your nose and mouth and breathe in and out several times (see photograph, p. 196).

Acupressure: Apply firm pressure with finger and thumb on the point (GB 20) on each side of the neck under the base of the skull (see p. 239) for two to three minutes while taking short, shallow breaths. This calms the nerve that controls contraction of the diaphragm.

Herbal remedies: The antispasmodic and relaxant effects of peppermint on the digestive tract may help relieve hiccups resulting from an overfull stomach. Have the tea after main meals. Caution: For safety concerns, see pp. 34–37.

When to get medical help
- Your hiccups last longer than a day.
- You suffer from frequent attacks of hiccups.
- You also have difficulty breathing, chest pain, or light-headedness.

242

High Blood Pressure

Arise in pressure exerted by circulating blood on the artery walls is a normal response to stress and physical activity. However, if this pressure remains persistently high, it can overwork your heart and arteries, making arterial disease, heart attacks, and strokes more likely. Simple lifestyle changes can help reduce high blood pressure, or hypertension.

More men than women have hypertension, and the disorder is most common after middle age. No obvious cause for high blood pressure is found in about 90 percent of those affected; this type is called essential hypertension.

Because hypertension is usually symptomless, it is often discovered only during a medical examination. It must be taken seriously, because it increases your risk of a stroke, "mini-stroke" (transient ischaemic attack, or TIA), heart attack, and other problems, such as kidney disease.

Causes

Smoking, a high alcohol intake, obesity, a sedentary lifestyle, and stress all increase the likelihood of hypertension. The stiffening and loss of elasticity of artery walls that accompany aging and certain forms of arterial disease can also elevate blood pressure. High blood pressure during pregnancy (pre-eclampsia) makes hypertension more likely later in life. Genetic factors can also underlie elevated blood pressure.

You may be at risk of high blood pressure if your mother did not eat a nutritious diet during pregnancy. This can adversely influence the long-term elasticity of the arteries, as well as encouraging low birth weight. Babies of low birth weight are more likely as adults to develop high blood pressure, along with obesity, elevated blood fats, and oversensitivity to insulin (which can precede diabetes). These symptoms are known collectively as the metabolic syndrome, or syndrome X.

Prevention

Help prevent high blood pressure by reducing all possible risk factors.

- Get regular, brisk exercise.
- Don't smoke.
- Eat a healthy diet (see pp. 9–12).
- Avoid excessive alcohol intake (see p. 232).
- Keep off excess weight.

Losing weight is especially important if you are "apple-shaped", with excess fat carried around your middle, rather than your hips and thighs. A waist measuring more than 37 inches in a man, or 32 inches in a woman, slightly increases the risk of hypertension; more than 40 inches

---

How blood pressure is measured

Blood pressure is recorded as two values, the systolic (the pressure of the blood as it enters the aorta from the heart) and the diastolic (the pressure when the heart's ventricles relax between beats). Normal blood pressure (BP) is usually defined as less than 140/90. However, a person's BP varies with age, and you can work out what your normal systolic pressure (the higher number of the two) should roughly be by adding 100 to your age in years (though this is not so accurate in later years). BP also varies with diet, weight, activity level, emotions, and degree of relaxation at the time of testing.

in a man, or 35 inches in a woman, increases it more significantly.

Because the adrenaline released by a high level of anger, anxiety, or other stressful emotion encourages high blood pressure, learn stress-management techniques to keep you calm. Conversely, feeling happy and relaxed tends to reduce high blood pressure.

Treatment

High blood pressure requires monitoring and, often, medical treatment. If you have only mild hypertension, you and your doctor may find that lifestyle changes and, perhaps, natural supplements will reduce your blood pressure enough so that you have no need for prescription drugs. In other cases where medication is required, many of the following natural remedies can work to support that treatment.

Diet: Recent research has identified the benefits of certain minerals in the treatment of high blood pressure. The following tips will help you maintain your intake of the nutrients you need to keep your blood vessels in good condition while also helping you control your weight.

- Eat at least five daily servings of vegetables (not including potatoes) and fruit. These supply nutrients—such as vitamins B_6 and C, magnesium, and potassium—necessary for healthy arteries and heart function. Vegetables and fruit also supply fibre and natural salicylates, which may—like their relative, aspirin (acetyl salicylic acid)—reduce the risk of a heart attack resulting from hypertension.

Dietary helpers
Potassium can work to lower blood pressure. The richest sources of this mineral include bananas, tomatoes, and peas.

- If you need to lose weight, follow a sound weight-control diet (see p. 303), combined with daily exercise.
- Boost fibre intake by choosing wholegrain breads and cereals rather than refined products.
- Eat oily fish three times weekly; they contain omega-3 fatty acids which may lower blood pressure.
- Reduce your overall intake of fat, especially saturated fat, but eat enough foods containing essential fatty acids (oily fish, nuts, seeds, and cold-pressed olive, walnut, or sunflower oil).
- Eat more calcium-rich foods, such as dairy products. Choose low-fat types.
- Keep your sodium intake low. Don't add salt (sodium chloride) to food or eat commercially processed foods containing added salt. Between 70 and 80 percent of our salt intake comes from such processed products as breads, breakfast cereals, crisps, and other snacks.

Testing for salt sensitivity

Salt encourages hypertension only in certain "salt-sensitive" people, but until a test to detect salt sensitivity becomes available, the only way you can detect it is to eat a low-salt diet for a month and have your blood pressure checked before and afterwards.

- Cut down on sodium from foods containing sodium bicarbonate, fish or meat cured with sodium nitrite, and foods containing the flavour enhancer monosodium glutamate, or MSG.
- Use a salt substitute from the supermarket or chemist. Such products are high in potassium and magnesium (which may help lower blood pressure), but low in sodium.
- Minimize your intake of caffeine, since it may contribute to raised blood pressure.
- Keep your alcohol intake low.
- Have some garlic (raw or as capsules or tablets) every day. A substance, called allicin, in garlic dilates blood vessels.

From the chemist: Supplements are sometimes used along with or instead of prescription drugs for very mild hypertension. However, consult your doctor before trying any of them to make sure they are safe for you and to find out the proper amounts and combinations. Also keep in mind that high blood pressure requires careful monitoring. Supplements may include:

- Calcium, magnesium, and/or potassium.
- Fish oil (for its omega-3 fatty acids).
- The amino acids arginine and taurine.

Herbal remedies: To help reduce mild high blood pressure, use remedies containing hawthorn or calcium elenolate (derived from a bitter antioxidant flavonoid called oleuropein in olive leaves). Drink a cup of tea or take a few drops of tincture made from one or more of these remedies once or twice daily for up to two months. Do not use these remedies in addition to the supplements described in "From the

Progessive muscular relaxation

Relaxation exercises are among the most effective natural means of combating high blood pressure. Try the following method every day:

1 Lie down on your back with your legs slightly apart and your hands palms upward at your sides. Breathe deeply and steadily throughout the exercise.

2 Starting at your toes and slowly working up to your face, tighten and then consciously relax the muscles of each part of the body in turn.

3 When you have worked on all your muscles, lie still and enjoy the feeling of relaxation for about five minutes.

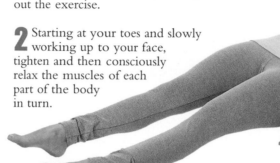

chemist" (left). Consult your doctor first if you're already taking prescribed medication.
Caution: For safety concerns, see pp. 34–37.

Stress management: Reduce the demands of work and home when you can. For the unavoidable stress in your life, find ways to cope (see p. 347), if necessary by working with a counsellor.

- Relax at regular intervals, using meditation and/or progressive muscular relaxation (see box). Consider also visualization (p. 47).

Exercise: Get daily aerobic exercise, such as brisk walking or cycling. The combination of

➤ *continued, p. 247*

Aromatherapy massage for relaxation

The calming actions of some essential oils add to the relaxing and therefore blood-pressure-lowering benefits of massage. Ask a friend to follow these instructions to give you a peaceful, soothing, whole-body massage to reduce stress-induced hypertension. Make a massage oil from three drops each of ylang ylang, lavender, and marjoram oils, and one tablespoon of jojoba oil. Warm the oil to body temperature in a container of hot water.

1 Sitting as shown, oil your hands and rest them gently on the centre of the upper back. Leaning forwards from the hips, slide your hands down along each side of the spine, circling back up the sides. Repeat several times.

2 Glide one hand across the upper back, away from the direction your partner is facing. Follow with a stroke of the other hand. Repeat, alternating strokes of each hand. Then ask your partner to face the other way and work in the other direction.

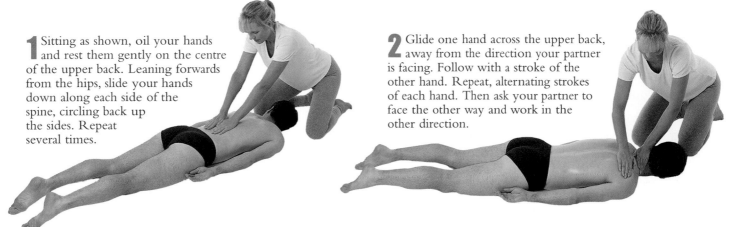

3 Place your thumbs on one side of the neck. Glide your thumbs outwards along the shoulder between the top of the shoulder blade and the shoulder muscle. Repeat, gradually moving a little farther towards the top of the shoulder. Work on the other shoulder in the same way.

4 Facing your partner's head, place your oiled hands on the lower back. Make counterclockwise circles with your right hand and clockwise circles with the left. Increase the pressure a little as you move towards the centre of the back.

5 Place both hands on the back of the leg below the calf. Glide slowly up to the top of the thigh. Move your hands apart, allowing one to circle the hip and the other to move to the inner thigh, avoiding the genitals. Slide both hands back to the ankle and down the foot. Repeat several times on each leg.

this and weight loss lowers blood pressure more than either alone. However, avoid lifting heavy weights. This type of physical exertion causes a rise in blood pressure that might burst a blood vessel, leading to a nosebleed or bleeding in the retina or in the brain (a stroke). Gentle forms of exercise, such as yoga (see below), can promote relaxation, which helps lower blood pressure.

Yoga: Many studies have shown that the regular practice of certain yoga postures and breathing exercises can have a marked effect on hypertension, partly by teaching you how to keep calm.
- Practise the corpse pose with breathing exercises (see p. 54) for 30 to 60 minutes daily.
- Don't do any exercise that involves holding your breath, and avoid the shoulder-stand and other inverted postures. The half shoulder-stand can bring high blood pressure down, but should be done only under the supervision of an experienced yoga teacher.

Aromatherapy: Soak in a warm bath to which you have added a total of five to six drops of bergamot, German or Roman chamomile, or frankincense oil.

Light therapy: Exposure to the ultraviolet light of the sun may help reduce blood pressure. It elevates the level of vitamin D, which alters the

Soothing herb
Lime tree flowers aid relaxation of mind and muscles, and so have beneficial effects on high blood pressure caused by nervous tension. Drink a cup of this tea once or twice a day.

body's use of calcium and relaxes tense arteries, and which may in turn encourage a healthy blood pressure. Indeed, hypertension appears to be less common in areas that have more daylight. Expose your skin to daylight every day for 15 to 20 minutes, but avoid too much strong midday sun (usually 10 A.M. to 3 P.M. in summer) to reduce the risk of skin cancer.

Hydrotherapy: Alternate hot and cold footbaths are said to be useful for hypertension.
- Fill one bowl with hot water and another with cold. Put your feet in the hot water for three minutes, then in the cold for one minute. Repeat this process three or four times.

Other therapies: Studies suggest that acupuncture and autogenic training can reduce high blood pressure. A biofeedback teacher or counsellor can teach you how to lower blood pressure elevated by stress-induced muscle tension.

When to get medical help
- You haven't had your blood pressure checked in the past 12 months.

Get help right away if:
- You have high blood pressure and feel poorly for any reason.
- You experience dizziness, faintness, unusual fatigue, or unusual headaches.
- You have shortness of breath, confusion, or chest or arm pain.
- You notice a sudden change in strength or feeling in a limb.

See also:
Arterial Disease, Diabetes, Overweight, Stress

Hives

This type of itchy rash is also known as urticaria or—because one cause is contact with stinging nettles—nettle rash. It is usually caused by an allergic reaction to something you have touched, such as a plant, or eaten, such as strawberries. About 20 percent of the population develop hives at some time. Women are more prone to the condition than men, but the reason for this is unclear.

Common culprits
Among the most prevalent dietary causes of hives are eggs, milk, nuts, shellfish, and strawberries. Skin contact with nettles can produce similar symptoms.

When an allergic reaction causes certain skin cells to release histamine, local small blood vessels dilate and their walls become permeable. Clear fluid called serum then leaks into the surrounding tissues and causes the characteristic inflammation and itchy, raised lumps of hives. This rash develops within minutes of eating or coming into contact with the offending substance. The lumps vary considerably in size, and larger ones may merge to form irregular patches known as weals. Initially a weal is red; later it becomes white at the centre, leaving a red rim. It lasts for about a day, and then disappears.

Hives often result from an allergy to a food, stinging plant (such as stinging nettle), medication (such as aspirin or penicillin), or insect bite. Sun exposure, temperature extremes, and stress may also cause hives.

Prevention

You can prevent a recurrence of hives if you are able to identify and avoid the cause. Sometimes it is easy to relate the symptom to a particular food, but more often the link is not obvious. To pin it down, keep a diary in which you list everything that enters your mouth—not only food and drink, but also supplements, medications (including over-the-counter drugs), toothpaste, and mouthwash.

If your symptoms reappear, you may be able to determine a connection between something you have ingested and your hives. If this is unsuccessful, try an exclusion diet, eliminating one or two foods at a time for a period of three weeks each (see illustration, left). If you have no further symptoms during this period, reintroduce the foods into your diet one at a time, every three weeks. If the hives recur, the most recently reintroduced food is probably causing the symptoms and therefore should be avoided.

To lower your risk of allergic reactions to food, avoid highly refined foods, which can reduce the efficiency of the immune system, and eat more immunity-boosting foods, such as those rich in essential fatty acids and vitamins A and C (see p. 12). Vitamin C also has powerful antihistamine properties. Be aware that certain factors may weaken your

immune system and make it react unusually to a normally harmless food. These include infection, stress, a lack of sleep, digestive problems, air filled with tobacco smoke or other pollutants, certain drugs, and pesticides and other chemicals contained in foods.

If you have had a baby, reduce his or her risk of developing a food allergy or other sensitivity by breastfeeding exclusively for a minimum of four months. While you are pregnant or breast-feeding, eat a wide range of foods to reduce your child's risk of developing allergies.

Treatment

Herbal remedies

- For hives induced by anxiety or stress, drink a cup of valerian or chamomile tea twice daily.
- Apply aloe vera gel to soothe the rash.
- Wash with tea made from chamomile, chickweed, or elderflowers.
- Drink a cup of tea made from nettles, burdock, echinacea, myrrh, and/or marshmallow twice a day for their antihistamine action.
- Apply cream made from calendula (marigold), chickweed, comfrey, elderflowers, or plantain.

Caution: For safety concerns, see pp. 34–37.

Kitchen-cabinet remedies

- Relieve itching by adding three tablespoons of bicarbonate of soda to a warm bath and soaking for 10 to 15 minutes. Or add five tablespoons of coarse (colloidal) oatmeal to the water.
- Add nine cups of vinegar to your bathwater, or add a teaspoon of vinegar to a tablespoon of lukewarm water and apply with cotton wool to the affected area. Vinegar helps prevent

Protecting your baby
Breastfeeding a baby helps prevent allergies, which can cause hives, in later life. If possible, breastfeed your baby for at least a year.

itching by acidifying the skin.
- Apply extracted cucumber juice to the affected area, or lay cucumber slices over it.

From the chemist: Take 500 milligrams of vitamin C with flavonoids twice a day.

Homeopathy

- Urtica: if hives are very itchy.
- Rhus toxicodendron: if the rash is itchy, red, and sore, and you feel very restless.

When to get medical help

- The rash persists longer than a day or recurs.

Get help right away if:

- Your mouth and throat are affected.
- You experience faintness, shortness of breath, or a rapid pulse.
- The hives occur after taking medication.

See also:
Food Sensitivity, Insect Bites and Stings, Itching, Sunburn

Incontinence

Urinary incontinence, the involuntary passing of urine, is often a result of injury or disease of the urinary tract. This loss of bladder control may be frustrating and potentially embarrassing, but in many cases the problem can be improved significantly with pelvic-floor exercises, devised to strengthen pelvic-floor muscles, and various changes in routine.

In women, who are mainly affected, incontinence usually results from weakness of the bladder-neck muscle (urethral sphincter) around the top of the urethra, which helps close the bladder, or from weakness of the pelvic-floor muscles—the muscles that support the bladder and uterus and also help close the bladder. Weakness of these muscles may accompany aging, any disorder (such as a stroke) affecting the nerves supplying these muscles, or injury to the muscles or their nerves. Other causes of incontinence include irritation of the bladder lining and over-sensitivity of the bladder-wall (detrusor) muscle, making it contract unexpectedly. In men, incontinence may also stem from prostate problems.

Types of incontinence

Stress incontinence: The most common result of weakness of the bladder-neck and pelvic-floor muscles, stress incontinence is leakage of urine that occurs because of raised pressure in the abdomen, often caused by laughing, coughing, lifting, or jumping. The condition is widespread among women, especially during pregnancy, when the growing uterus puts pressure on the bladder neck and pelvic floor, and after childbirth, when the bladder neck and pelvic floor may be stretched or damaged. Standing or straining may provoke stress incontinence in men with an enlarged prostate or other prostate disorder. People of either sex may suffer from this problem if they are very overweight.

Urge incontinence: This sudden, irresistible urge to empty the bladder is often triggered by a change in position, for example, from sitting to standing. It may also occur during the night. One cause—often responsible for incontinence in older people—is overactivity of the bladder-wall muscle. This may happen as a result of stress, a full bladder, certain drugs (including diuretics, antidepressants, tranquillizers, and high blood pressure medication), or for no apparent reason. Other causes of urge incontinence include irritation of the bladder lining by a urinary-tract infection, over-concentrated urine from low fluid intake, nicotine, certain food colourings, sugar (with untreated diabetes), caffeine, and alcohol. Some researchers believe that fluoride-containing toothpaste is a possible cause.

Overflow incontinence: This occurs as a result of chronic urinary retention, when the bladder is unable to empty and is always full, leading to constant dribbling of urine. Possible causes

What are the pelvic-floor muscles? Located within the pelvis, these vital muscles (shown in red) provide a supporting web for the female reproductive and urinary organs and their outlets.

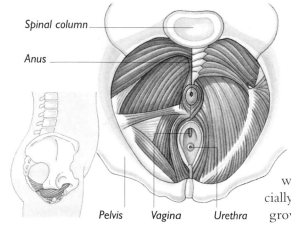

Spinal column

Anus

Pelvis   Vagina   Urethra

include an enlarged prostate, a prolapsed uterus, and certain medications.

Prevention

Doing pelvic-floor exercises, especially during pregnancy and after childbirth, strengthens and tones the pelvic floor, and helps prevent stress incontinence. If you have done these exercises regularly, you will more easily be able to adjust the speed and ease of the baby's descent down your vagina in the second stage of labour. A controlled descent helps prevent subsequent urinary incontinence caused by overstretching of

Bladder irritants
Eat foods containing oxalate (such as strawberries, rhubarb, and spinach), which increase the frequency of urination, only in the morning.

the pelvic-floor muscles. These exercises are also beneficial for men and older women.

Treatment

If you have stress incontinence:
- Do pelvic-floor exercises every day.
- Women can also try vaginal muscle "weight-training". Cone-shaped vaginal weights of various sizes to insert and hold in the vagina are available from some pharmacies.
- Lose any excess weight, since fat deposits put pressure on the bladder and pelvic floor.

If incontinence results from bladder irritation:
- Keep the genital area clean and dry to avoid irritation of the urethra from infection.
- Avoid caffeine-containing drinks and alcohol in the evening; these increase urine output and may irritate the bladder lining.
- Cut out artificially coloured foods, as some food colourings may irritate the bladder lining.
- Stop smoking.

Pelvic-floor exercises

To recognize the pelvic-floor muscles, try to stop your urine flow midstream. Then, at least five times daily, sit or lie with your knees slightly apart, and tighten these same muscles for two seconds at a time, relaxing for two seconds in between. Repeat the cycle up to 10 times. Gradually work up to 10-second muscle tensing. Practise anywhere, standing or sitting. Check your progress each week by trying to stop your urine flow midstream.

When to get medical help
- Lack of bladder control is preventing you from undertaking normal activities.
- Urination is painful.
- You think a prescribed medication may be responsible.

Get help right away if:
- You suddenly lose bladder control.

See also:
Prostate Problems, Urinary-tract Infections, Urinary Difficulties

Influenza

Popularly known as flu, influenza is caused by a number of similar viruses. It occurs more often in winter and sometimes reaches epidemic proportions. It spreads quickly, especially in schools and institutions. Most otherwise healthy people recover quickly from flu without need for medical intervention, but those in poor health should see their doctor.

Keeping infection at bay
At the first sign of flu, gargle morning and night with two drops each of antiseptic tea tree and geranium oils in half a glass of warm water.

Flu usually starts suddenly with chills, fever, aching muscles, and sneezing. Soon you may develop a sore throat, a dry cough, sensitive skin, painful eyes, weakness, and a headache. You probably won't feel hungry. The fever accompanying flu generally lasts from three to five days. It is common to feel run-down for some weeks after these symptoms have passed.

Prevention

Try to improve your resistance to infection before the winter flu season gets underway. If you still get the flu, you may get it more mildly.
- Eat a healthy diet (see pp. 9–12).
- Get some aerobic exercise each day.
  - Take a daily multiple vitamin and mineral supplement (including beta-carotene, vitamins C and E, flavonoids, and the minerals selenium and zinc).
    - Try not to get too tired or stressed.
  - Don't expose yourself to crowds when there's flu around.
    - Take echinacea tea, tincture, or tablets two or three times a week in winter. Take a dose every day if you are in close contact with someone who has the flu.
      - Eat garlic or take garlic tablets daily.
      - Every three weeks throughout the winter consider taking flu nosode, a preventive homeopathic remedy.

Cold or flu?

Both colds and flu can cause a sore throat, cough, and runny nose. Although there's usually no mistaking flu because of the severity of symptoms, the only sure way of distinguishing a bad cold from flu is for your doctor to send nose and throat swabs to a lab. However, this is rarely necessary.

Treatment

Treat flu carefully because it can cause serious complications. Stay at home so that you don't spread infection, and stay in bed because you need your strength to fight the flu viruses. Follow the recommendations for reducing fever on pages 206–207. Be wary of over-the-counter medicines that may mask unpleasant symptoms and make you think you can be up and about when you're actually still sick. Stay home for at least one day after your temperature has returned to normal.

Diet
- Drink plenty of non-alcoholic, caffeine-free fluids, including water, fruit juice, barley water (see illustration, p. 358), and herbal teas (see facing page). Sipping blackcurrant tea may soothe your sore throat as well as providing extra vitamin C, which helps fight infection.

Herbal remedies

■ Drink echinacea tea three times a day, or take 200 milligrams of the supplement five times a day.

■ A tea made from elderflower, peppermint, and yarrow, may bring down a fever and reduce aches and pains.

■ Elderberry extract may help stop flu viruses from multiplying.

Caution: For safety concerns, see pp. 34–37.

Aromatherapy

■ Cajuput oil helps reduce fever and ease aching muscles. Add two drops of this essential oil to a tablespoon of jojoba oil and smooth behind the ears, over the forehead, on each side of your nose, and on your chest four times a

Nature's medicine chest
The elder tree provides many remedies. The flowers and berries may ease colds and flu, the leaves may soothe bruises, and the inner green bark has been used as a purgative.

day. Or add six to eight drops of cajuput oil to a warm (not hot) bath. Do not use this oil if you are in the first 20 weeks of pregnancy.

■ Soothe congestion with steam inhalations, scented with a few drops of tea tree or eucalyptus oil (see p. 141).

Homeopathy

■ Aconite: if you suddenly become feverish, especially after catching a chill.

■ Eupatorium: if the aching feels as though it is penetrating your bones.

■ Gelsemium: if shivering and shaking are predominant symptoms.

■ Oscillococcinum: as an overall remedy (take within the first 36 hours of the onset of your flu-like symptoms).

Should you get a flu jab?

Immunization can shorten flu and help prevent complications. It is offered to people who are more likely to become seriously ill with flu, including:

● Those aged 65 or over.
● Those with long-term heart, lung, liver or kidney disease, diabetes, or sickle cell anaemia.
● Those on corticosteroid or immuno-suppressant drugs.
● Those with no spleen.
● Those who live in residential care or nursing homes. Flu vaccine is also recommended for the staff in these homes and those who work in hospitals and clinics.

When to get medical help

● You are no better after a week or your symptoms worsen after three or four days.
● Symptoms remain after four weeks or recur.
● You cough up lots of green or yellow phlegm.
● You are among those for whom flu immunization is especially recommended (see box).

Get help right away if:

● You experience chest pain or breathlessness.
● You have a stiff neck, a severe headache, an aversion to bright light, a rash, confusion, or severe joint pain.
● You cough up blood.

See also:
Colds, Coughs, Fever, Headache, Sore Throat

Insect Bites and Stings

Blood-sucking insects, such as mosquitoes, inflict tiny puncture wounds when they bite. The stings from bees, wasps, and hornets contain venom. People vary in their reactions to bites and stings, but in most cases symptoms last only a day or two.

An allergic response to an insect's saliva or venom causes pain or itching, redness, and swelling at the site of a bite or sting. Some people experience a severe reaction to insect stings—a life-threatening swelling of the airways. Certain ticks living in wild country areas transmit a bacterial infection called Lyme disease and mosquitoes cause malaria in some parts of the world.

Prevention

Keep your skin covered if you are in an area where insect bites are likely. Screen windows and doors. Repel insects by adding five drops of citronella oil to a cup of water and dabbing on exposed skin. Eat garlic and take a daily supplement of vitamin B_1 (thiamine) and zinc.

Treatment

General measures: Distilled witch hazel and calamine lotion are effective for soothing pain and itching from mosquito bites and other bites and stings. An ice cube, aloe vera gel, or onion juice are also good remedies.

Bee sting: Remove a bee's sting sac by pressing it out sideways with a thumbnail. Afterwards, press out any poison.

Pain-relief paste
For a bee sting or ant bite, add bicarbonate of soda to a little water and apply to the painful area.

Wasp sting: Apply lemon juice, vinegar, or cinnamon tea as soon as possible after being stung. Repeat if necessary.

Tick bite: If the tick is clinging to the skin, you can dislodge it by covering it with oil or petroleum jelly. Then you should gently twist it out using tweezers.

Homeopathy: Apply Hypercal cream and take apis (if the bite is swollen and red), or Cantharis (if burning pain is predominant).

Aromatherapy: For any sting, apply one drop each of lavender and tea tree oils hourly.

When to get medical help

- With a suspected tick bite, if you have localized redness or a circular rash, flulike symptoms, or joint pain.

Get help right way if:

- The bite or sting is on the face or in the mouth or throat.
- You have hives, breathing difficulty, nausea, or vomiting.

See also:
Itching

Irritable Bowel Syndrome

The most common of all intestinal disorders, irritable bowel syndrome may affect as many as one in three people in the Western world at some time. The condition tends to be recurrent, although symptoms can subside for long periods between attacks. Orthodox medicine offers no certain cure, but there are many natural ways of managing and easing symptoms.

Irritable bowel syndrome, or IBS, consists of a number of related symptoms, including intermittent abdominal pain and irregular bowel movements. The disorder is caused by a disturbance in the normal muscle movements of the wall of the large intestine. This creates problems in the way food moves through the digestive tract, which leads to diarrhoea or constipation and/or pain from intestinal muscle spasms. The nerves in the intestinal wall may also overreact to painful stimuli, such as intestinal distension. IBS affects twice as many females as males. In some cases the disorder may result from too much or too little stomach acid or inadequate digestive enzymes (see "Indigestion", pp. 252–255, and "Food Sensitivity", pp. 212–215).

Symptoms

Different individuals have different combinations of symptoms, which can also vary in severity from time to time in any one person. Most people who suffer from IBS have more than one of the following symptoms:

- Lower abdominal pain, eased temporarily by having a bowel movement or passing wind.
- Bloating after a meal.
- A rumbling stomach.
- Excessive flatulence.
- Diarrhoea, especially early in the morning.
- Constipation.
- Alternating constipation and diarrhoea.
- A sense that the bowels are never completely empty, even after defaecation.
- Mucus in the stools.

Headaches, fatigue, and depression are also common with IBS. Other associated symptoms include pain in the back or thighs, heavy or painful menstrual periods, pain during sexual intercourse for women, and a frequent or urgent need to urinate.

If you have IBS, your intestines may react adversely to one or more triggers. Possibilities include smoking, antibiotics, certain foods, a changing level of oestrogen, excessive exercise, and anxiety, depression, and other forms of stress.

Several disorders produce symptoms that can be confused with those of IBS, including lactose intolerance, gluten intolerance, pelvic inflammatory disease, and endometriosis (a condition in which uterine lining cells stray into the abdomen, where they may cause pain, diarrhoea, and other symptoms by bleeding within the abdomen).

Many people find IBS much easier to live with once they have a diagnosis and realize that their symptoms are not the result of a serious underlying disease.

Prevention

This condition occurs less frequently in those who adhere to a high fibre diet. Therefore, a

Jet Lag and In-flight Health

Flying through different time zones either shortens (going east) or lengthens (going west) a traveller's day. This disrupts the normal sleep–wake cycle and disturbs the body's usual hormonal rhythms, leading to jet lag. Flying can also have other ill effects on your well-being as a result of fluid depletion and long periods of immobility.

Gaining and losing hours

When travelling east to west, you lengthen your day. For example, after a five-hour flight from New York to San Francisco the local time of your arrival is only two hours later, so you "gain" three hours. When travelling west to east, you shorten your day. After a seven-hour flight from New York to Rome, the local time of your arrival is 13 hours later, so you "lose" six hours.

Throughout the 24-hour day, many of our bodily functions, such as alertness, temperature, sexual interest, and sleepiness, vary according to predictable rhythms. This happens in response to environmental factors, including time, temperature, and light levels. The hormonal activity responsible for these "biorhythms" is orchestrated by the hypothalamus and the pituitary and pineal glands in the brain. The pineal gland, for example, responds to the level of ambient light by producing a hormone called melatonin. Production of melatonin increases as light falls in the evening, continues during the hours of darkness, and ceases as day dawns. Your body's level of melatonin influences your sleep pattern, determining whether or not you are ready for sleep.

All these biorhythms are disrupted by a long-haul flight, producing the symptoms of jet lag. The symptoms, similar to those of a hangover, include fatigue, a desire to sleep during the day, and difficulty in sleeping during your new night-time. Memory and concentration may also be impaired. An eastward flight tends to produce more severe jet lag than a westward journey.

Flying entails other health threats besides jet lag. Breathing dry, recycled air on long flights can lead to dehydration, dry skin, headaches, and infections. Sitting still for long periods can cause swollen feet and ankles and an increased risk of developing a potentially serious blood clot in a vein deep in the leg. Some people also suffer from stress as a result of a fear of flying.

West from Denver	Local time
Denver	12 noon
Honolulu	9 am
Tokyo	4 am
Beijing	3 am
Bangkok	2 am

San Francisco 11 am · Denver 12 noon · New York 2 pm

East from Denver	Local time
Denver	12 noon
London	7 pm
Rome	8 pm
Moscow	10 pm
Karachi	Midnight

Prevention
Before you go

- Have several good nights' sleep before travelling, so you don't start your flight already tired.
- When travelling east, begin to accustom yourself to what will be your new bedtime by going to bed an hour or two earlier for several nights.
- When travelling west, begin to accustom yourself to what will be your new bedtime by going to bed an hour or two later for several nights.

During the flight

- Set your watch to your destination's local time as soon as you board. Then, on the flight, begin to adjust your sleep–wake cycle to this time to reduce the adjustment your body has to make when you arrive.
- Prevent dehydration by drinking plenty of water or soft drinks. Avoid alcohol and caffeine, which can encourage fluid loss, and fizzy drinks, which may cause bloating.
- What you eat may also affect how you feel after flying. Avoid having foods on the plane that you don't usually eat. Take some fruit with you for a healthy snack.
- Walk around for 5 to 10 minutes at least every two hours. This helps prevent your ankles and feet from swelling and makes a blood clot in the legs less likely. Contracting your calves frequently and doing other leg exercises helps too.
- Don't cross your legs while sitting, as this encourages ankle swelling.
- Promote sleep with breathing exercises (p. 87) and progressive muscular relaxation (p. 245).
- Counteract tension with a neck and shoulder massage or by sprinkling a tissue with a few drops of lavender or geranium essential oil and inhaling.

Treatment

On arrival during the day

- Go outdoors immediately, and stay there for at least an hour. Exposure to bright light helps the body clock readjust.
- Try to make yourself stay awake (or take only a very short nap). Go to bed at the local time.
- Get some exercise to help you stay awake.

A wake-up bath
After you've been on a long flight, add a few drops of peppermint or eucalyptus oil to a bath. These oils are stimulating and will help revive you.

Skin saver
Apply moisturizer to your face and hands at regular intervals during the flight to counteract the drying effects of recirculated cabin air.

On arrival at night

- Go to bed at the normal hour, even if you don't feel sleepy. If you can't sleep, you could try practising progressive muscular relaxation (see p. 245).
- To help you sleep, use a few drops of a relaxing essential oil, such as lavender or geranium oil, in a bath.

When to get medical help

- Severe jet lag lasts longer than a week.
- You have recurring, disruptive jet lag.
- See your doctor before flying if you are on hormone replacement therapy or the Pill, are pregnant or newly delivered, have cancer, or have recently had a stroke, or surgery.

Get help right away if:

- On arrival, you experience a bad headache or any other internal pain.

See also:
Sleeping Difficulties

Kidney Stones

When urine is too rich in certain mineral salts, a stone may form in the kidney. Stone formation occasionally follows a urinary-tract infection. If the stone moves out of the kidney and lodges in the ureter, the duct to the bladder, it causes waves of intense pain, known as renal colic. Kidney stones affect men three times as often as they do women.

Most kidney stones result from excess oxalic acid, which produces stones composed of calcium oxalate. Less commonly, stones result from too much uric acid. Stones are less likely to develop if you drink plenty of water-based fluids to dilute the urine; limit your consumption of animal protein, fat, added sugar, and alcohol; and lose any excess weight. Don't smoke, since cadmium from smoke encourages stones.

Treatment

Diet
- Drink six to eight glasses of water a day.
- If you have calcium oxalate stones, eat fewer foods containing oxalates, such as beetroot, celery, cucumber, grapefruit, parsley, rhubarb, spinach, strawberries, sweet potatoes, nuts, chocolate, tea, and cola. Eat more foods rich in calcium, magnesium, potassium, and vitamin B_6 (see p. 12). You may want to consult a dietitian.
- If you have uric acid stones, cut down on foods containing purine—red meat, fish, shellfish, whole grains, beans, cauliflower, peas, spinach.

From the chemist
- Daily supplements of vitamin B_6 (40 milligrams), magnesium (300 milligrams), and potassium citrate (150 milligrams) may help.

Herbal remedies: Drink two or three cups daily of corn silk, buchu, or couch grass tea. Caution: For safety concerns, see pp. 34–37.

The effect of calcium

Calcium in the diet may help prevent calcium oxalate stones—probably because calcium combines with oxalate in the intestine and so prevents the absorption of pure oxalate. Foods rich in calcium include milk, cheese, and yogurt (these can be low-fat), sardines, dark-green leafy vegetables, nuts, seeds, and dried fruit. Taking calcium supplements during or just after meals may have a similar effect; however, taking them between meals can increase the risk of stones. You may want to ask your doctor about dosage and timing.

When to get medical help
- You have intense intermittent pain, probably starting in the back and moving to the groin.

Get help right away if:
- You have blood in your urine.

See also:
Urinary-tract Infections

Leg Cramps

Sudden muscle spasm can cause pain severe enough to make a person cry out. This usually strikes the calf muscle but sometimes affects the foot, arm, or hand. For most people cramp is only an occasional nuisance, but for others it is a frequent occurrence. Home remedies can help, and simple lifestyle changes may prevent the problem from recurring.

Stretch for relief
When you have a cramp, lean forwards against a wall, placing the affected leg out behind you with the foot flat.

The pain of a cramp results from excessive, prolonged, and unusual muscle contraction. Possible causes include a build-up of lactic acid in an overused muscle, nerve damage from repeated movement, sitting or lying in an awkward position, and poor circulation.

Prevention

- Eat fresh green vegetables (especially watercress and parsley), fresh fruit, whole grains, and low-fat dairy products to provide your muscles with the calcium, iron, magnesium, potassium, and zinc they need to function properly. Make sure you drink plenty of fluids.
- Get moderate daily exercise, but if this involves a repeated movement, such as when running or swimming, stretch and relax your muscles for five minutes every half hour.
- Take a warm bath before retiring and keep warm in bed.
- Avoid either too much or too little salt, since both extremes can cause cramp.
- Drink a cup of tea made from cramp bark, chamomile, and vervain before going to bed.

Treatment

Ease a cramp by stretching the contracted muscle (see photograph). Then knead and squeeze the muscle to disperse any remaining tension.

Acupressure

With your thumb, press the point (B 57) in the middle of the back of your calf where the muscle and the tendon meet. Begin with a brushing action, and build up to more force. Then press the point (B 54) in the middle of the crease at the back of your knee for a few seconds. Don't do this if you're pregnant.

Aromatherapy: Apply a hot compress made with lavender, marjoram, or ginger oil. Mix one or two drops of oil into a bowl of hot water. Lay a cloth on the surface of the water. Gently wring out and apply to the leg, oily side down.

Homeopathy: Apply arnica cream to the affected muscle.

When to get medical help

- You suffer from frequent leg cramp.
- You experience leg cramp when walking.
- A leg cramp persists longer than an hour.
- A leg cramp is associated with circulation problems.

Memory Loss

We all experience forgetfulness from time to time, especially when preoccupied or under stress. Short-term memory loss also becomes more frequent during middle age. If you often forget where you've left something or have difficulty remembering people's names, you may need to learn ways to improve your memory. Simple lifestyle changes can make a big difference as well.

The computer cannot compare with the miracle of the human brain, but there are some similarities. Both receive, store, and recall information, and each can fall prey to problems with these processes. Just as you can enhance a computer's memory, so too can you optimize your own.

Memory loss among older people may be the result of poor blood flow to the brain coupled with lack of mental stimulation. However, there are other causes of impaired memory—anaemia, an underactive thyroid, depression, and chronic fatigue syndrome—that may need to be ruled out. A poor diet and excessive alcohol intake may also be factors. Alzheimer's disease is also a possible cause of memory loss in the elderly.

Prevention

Reduce your risk of memory loss from poor circulation with a half hour's daily exercise, and by not smoking. Brain function is affected by nutrition, so be sure to eat a healthy diet (see p. 9). This is especially important as you get older, because you need fewer calories to meet your energy needs, and you absorb nutrients less efficiently.

Optimizing your memory

Try to present your brain with information in the way it prefers. Some people memorize best by seeing, others by hearing, and still others by the use of other senses. For example, if you learn best via your ears, record information on tape.

Jog your memory
- Use written memory aids, such as lists and diary entries.
- Mentally run through the alphabet when searching for a word.
- Refresh your long-term memory regularly with photographs, videos, letters, and other memorabilia.
- Use mnemonics—for example, "Every Good Boy Deserves Favour (or Figs)" for the notes on the lines of the music staff (E, G, B, D, F).

Train your brain
Use these memory-enhancing tricks:
- Keep repeating a new fact to yourself.
- Recall each day's events before falling asleep.
- Help to remember names by creating pictures from them. Mr. Lightfoot, for example, could be visualized as a lightbulb with a foot.

Stay interested
- Try to keep up with family, community, national, and world events.
- Keep your brain alert by reading, engaging in stimulating conversation, and doing crossword and other puzzles.
- Take up a new intellectual endeavour—for example, learning a foreign language.
- Make new friends and enjoy old relationships.

Treatment

Even if you have been absent-minded for some time, it is never too late to take steps to improve your memory or slow memory loss.

Diet

▪ Eat frequent small meals to maintain a normal blood-sugar level, which is important for optimum mental function.

▪ Eat plenty of whole grains, nuts, seeds, fruit, vegetables, and legumes. These foods contain high concentrations of beta-carotene, folic acid, calcium, copper, iron, iodine, magnesium, manganese, selenium, zinc, essential fatty acids, plant pigments, and vitamins B, C, and E. These nutrients are vital for maintaining the health of the brain and its blood supply.

▪ Eat three servings of oily fish weekly. These are rich in brain-friendly omega-3 essential fatty acids, including DHA (docosahexaenoic acid) and phospholipids (important constituents of all membranes).

▪ Don't drink too much alcohol.

From the chemist: If you think your diet may be lacking in essential nutrients, consider taking the following supplements:

▪ A multiple vitamin and mineral preparation.

▪ Vitamin E.

▪ Fish oil.

Healthy body, healthy mind
A half hour's daily vigorous exercise improves the brain's blood supply for more than 24 hours. It will also help you sleep soundly. Interrupted or inadequate sleep can contribute to forgetfulness.

▪ Phosphatidyl serine. The brain uses this phospholipid during memory recall, and supplements can slow age-related memory loss.

Breathing: The rapid, shallow breathing that accompanies anxiety reduces the brain's oxygen supply and, if prolonged, can interfere with memory. Use special strategies to handle the stress in your life (see p. 347). Yoga-based breathing exercises (p. 54), regular meditation, and/or other relaxation exercises may help improve memory.

Herbal helper
Ginkgo biloba may improve brain functioning and may prevent further short-term memory loss in people with Alzheimer's and other types of dementia. Take it as a tea or as a supplement.

When to get medical help
● You have severe or rapidly worsening memory loss.
● You are depressed.
● You have just started taking, or have increased the dose of, a prescribed drug.

Get help right away if:
● You suffer sudden memory loss following a head injury or loss of consciousness.

See also:
Aging, Anaemia, Anxiety, Arterial Disease, Chronic Fatigue Syndrome, Depression, Sleeping Difficulties, Stress

Menopausal Problems

When menstruation ceases—usually between the ages of 45 and 55—a woman's natural capacity to bear children also ends. The menopause is the start of a new phase of life, which may bring exciting opportunities. However, this transition may also trigger physical and emotional problems. Many of these respond to simple measures.

The timing of the menopause is determined not only by genes, but also by certain medical conditions and by such lifestyle factors as smoking, diet, and stress levels. It is caused by reduced production of hormones by the ovaries, as the number of eggs in the ovaries dwindles. The supply of the hormone oestrogen begins to fall before the menopause, drops faster for a while afterwards, then stabilizes. The same thing happens with progesterone, the other sex hormone, but most experts think it is the reduced oestrogen level that causes most of the signs of the menopause.

Especially if fertility has been at the heart of your sense of worth, you may experience a lowering of self-esteem at this time. Together with other possible major life changes—such as children leaving home, divorce, and elderly parents growing frail or ill—the menopause can create a

A new beginning
For many women, the menopause signals the start of a time in life to concentrate on their own interests.

sense of multiple losses. If you are also dissatisfied with your life—struggling with a demanding job or other burdens—you may find this a hard time. On the other hand, you may welcome the release from menstruation and worries about contraception and the possibility of pregnancy. You may also enjoy increased freedom from responsibilities and discover a new zest for life.

Signs and symptoms
Apart from the cessation of periods, the most common signs of the menopause—experienced by four out of five women and sometimes predating the end of menstruation—are hot

Hormone replacement therapy

Many women are prescribed oestrogen and progestogen supplements to counteract the adverse effects of the menopause—not only such symptoms as hot flushes but also the greater risk of heart disease and osteoporosis. Such hormone replacement therapy (HRT) can be very effective, but it also carries possible long-term risks—in particular, breast cancer. Some evidence suggests that an HRT programme involving treatment breaks may be the best option. Discuss the issues with your doctor, including whether natural therapies, such as consuming soy products, may offer a sound alternative to HRT in your case.

flushes and night sweats. These signs may continue for weeks, months, or even years. Many women find them to be little, if any, problem, but about 25 percent seek help because they interfere with sleep or are embarrassing or uncomfortable during the day. Vaginal dryness, leading to pain during intercourse, may occur. Some women experience headaches, fatigue, insomnia, and depression. The skin may become thinner and drier.

Other body changes take place around this time as part of the natural aging process. The bones become less dense, and the risk of osteoporosis (weak, fragile bones prone to fracture) increases. Heart disease and strokes become more likely after the menopause, largely because of the reduction in the protective effect of the sex hormones.

Prevention

You will have a lower risk of osteoporosis if you have had plenty of calcium, magnesium, and other bone-building nutrients in your diet throughout your life. Regular weight-bearing exercise, such as walking or dancing, also boosts bone-mineral density. To protect against arterial disease, do an aerobic exercise, such as jogging or swimming, three times a week. Choosing the right foods can

Fit and active
Exercising regularly throughout the menopause is one of the best ways of feeling good. There is evidence that aerobic exercise can boost oestrogen levels.

help ward off heart attacks, strokes, and other forms of arterial disease (see pp. 92–95), as well as hot flushes and other menopausal signs.

Stress management is useful at any time, but after the menopause it has the additional benefit of preventing a stress-induced fall in oestrogen. This, along with regular sexual activity, reduces the risk of developing vaginal dryness.

Treatment

Diet: Foods that supply calcium, essential fatty acids, plant hormones, and vitamins C and E help maintain strong bones and the proper balance of various blood fats. They also help guard against hot flushes, night sweats, and vaginal dryness.

A "Japanese" diet
For menopausal health, eat foods that contain plant hormones, including soy products, celery, fennel, rhubarb, alfalfa, fruit, whole grains, and linseeds. The fatty acids found in oily fish can help too.

Women who have a diet that includes soy products—such as tofu, tempeh, soy flour, and soybeans—have fewer hot flushes and night sweats. Research suggests that the plant hormones found in high concentrations in soy products mimic the action of human oestrogen and thereby compensate for falling hormone levels. Most important, plant hormones can also help block the action of a woman's oestrogen when fluctuating hormone levels create an imbalance in the ratio of a woman's oestrogen to progesterone.

Foods rich in essential fatty acids are critical for hormone production, as well as for healthy skin and nerves. Eat nuts, seeds, and whole grains; three helpings a week of oily fish (such as salmon, mackerel, tuna, and sardines); and cold-pressed vegetable oils.

Acupressure for symptoms

To alleviate hot flushes, night sweats, and anxiety, treat the points shown for two minutes each, every other day.

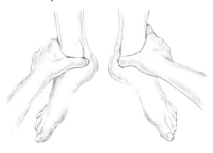

- Apply pressure with your thumb or fingertip, using a small, firm circling motion, on the point (H 6) half a thumb-width above the wrist crease, on the little-finger side of the forearm.

- Apply firm stationary pressure with thumb or fingertip to the point (K 6) a thumb-width below the tip of the inside ankle bone.

Drink at least six glasses of water a day. If hot flushes are a problem, reduce your intake of hot drinks, especially tea and coffee, and avoid spicy foods and alcohol. Wean yourself off caffeinated drinks slowly by blending them with increasing amounts of decaffeinated coffee or tea. Stopping immediately can lead to headaches.

Exercise: Regular exercise helps relieve depression, possibly by raising the level of endorphins, hormone-like substances that lift mood. It may also reduce hot flushes and, when combined with a healthy diet, make weight control easier and more permanent by raising your metabolic rate. Exercise also helps prevent or slow osteoporosis.

From the chemist: Useful nutritional supplements include:
- Vitamin C and flavonoids: for hot flushes.
- Vitamin E: for hot flushes and vaginal dryness.
- A vitamin supplement designed for the menopause, especially if you are not eating well.
- Vitamin B complex and magnesium: for mild depression or anxiety.
- Fish oil: for hormone production (if insufficient in your diet).

Herbal remedies: Many herbs rebalance hormone levels, often by acting as mild oestrogens. For an all-round benefit, take red or purple sage and/or dong quai. You can also try linseed oil or evening primrose oil. For specific menopausal symptoms, try the following:
- Teas or supplements of chasteberry, black cohosh, or motherwort for hot flushes and/or night sweats.

An aromatic relaxant
Ask a friend to give you a back massage using this mixture: two drops each of rose otto and sandalwood oils and three drops each of neroli and cypress oils in five teaspoons of sweet almond oil.

- Peppermint, which has cooling qualities, is useful for treating hot flushes. Drink peppermint tea regularly, and carry a bottle of peppermint oil with you to relieve symptoms when they occur: sprinkle a few drops of the oil onto a tissue and inhale deeply.
- St. John's wort for sleeping difficulties and depression.
- Motherwort for anxiety.
- Motherwort or dong quai for vaginal dryness.

Caution: For safety concerns, see pp. 34–37.

Herbal helpers
Peppermint, chasteberry, black cohosh, motherwort, St. John's wort, red clover, and dong quai may all alleviate menopausal signs. Yarrow may help regulate excessive menstrual flow in the months preceding cessation of menstruation, but avoid it if you have hot flushes.

Aromatherapy: Essential oils that may help regulate hormone production include neroli, sandalwood, lavender, clary sage, rose otto, and geranium. To prevent a recurrence of hot flushes or other symptoms, add three drops of clary sage and two drops each of rose otto and geranium oils to a daily bath. A few drops of lavender oil in a vaporizer can help promote relaxation.

Flower essences
- Larch: for lack of confidence.
- Mustard: for depression with no apparent cause.
- Scleranthus: for mood swings.
- Walnut: to help you cope with change.

Homeopathy
- Sepia: for hot flushes, anxiety, painful sex, and/or a decline in libido.
- Lachesis: for hot flushes, heavy bleeding, night sweats, uterine cramps, and/or irritability.
- Calendula ointment: for vaginal dryness.
- Pulsatilla: for hot flushes, overwhelming emotions, or mood swings.

When to get medical help
- You have menopausal signs that are severe or suddenly change.

See also:
Aging, Anxiety, Arterial Disease, Depression, Menstrual Problems, Osteoporosis, Sex Drive Loss, Stress, Vaginal Problems

Menstrual Problems

The delicate hormonal balance of the monthly menstrual cycle can be easily upset by such factors as stress, a change in weight or diet, and too little or too much exercise. Menstrual problems include painful or heavy periods, and irregular or infrequent periods. It is often possible to relieve symptoms by normalizing hormone levels naturally, without drugs.

Pain during a period (dysmenorrhoea) is the most common menstrual problem, especially among teenage girls and young women. It often abates after the age of 25 and after childbirth. The pain, which usually starts just before a period and lasts up to 12 hours, may be severe enough to interfere with everyday life.

Menstrual pain, felt as a cramping in the lower abdomen, often comes in waves and may be accompanied by low back pain and nausea. The cramps are caused by contractions of the uterus and are associated with high levels of hormone-like substances called prostaglandins. An unhealthy diet and a lack of exercise can make cramps worse.

If you start to experience such pain in your 40s or if it recurs after years of pain-free periods, it may be due to an underlying disorder, such as endometriosis, a condition in which patches of uterine-lining cells stray into the pelvic cavity and settle on other organs. Each month these cells swell and bleed, and they may inflame the underlying tissues.

Heavy periods are most likely soon after the onset of menstruation and in your late 30s and 40s. Possible causes include a hormonal imbalance from an unhealthy diet, stress, pelvic inflammatory disease, fibroids, and endometriosis.

Irregular or infrequent periods are common during the first few years after the start of menstruation, until regular ovulation is established. Periods may also be irregular in the few years before the menopause. Disrupted periods and bleeding between periods may also be associated with difficulty in conceiving. The disruption may be caused by a hormonal imbalance due to stress or polycystic ovary syndrome, in which multiple cysts on the ovaries—often due to weight gain—disrupt hormone production.

Other causes of menstrual irregularity include uterine polyps and cancer of the cervix or uterus. Spotting of blood between menstrual periods may occur if you are taking the contraceptive Pill or undergoing hormone replacement therapy. Intensive exercise, weight loss, and eating disorders can also result in disruption of the normal pattern of menstruation.

Exercise away pain
Swimming, as well as other activities that work muscles throughout the body, can help prevent menstrual cramps.

Treating pain

Diet: In the week before you expect your period, take special care to eat a healthy diet; this helps prevent an imbalance of prostaglandins in the wall of the uterus. Choose foods rich in essential fatty acids, calcium, magnesium, zinc, and vitamins B, C, and E (see p. 12). Prevent constipation by drinking plenty of fluids and by eating vegetables, fruit, and whole grains daily. Avoid foods made with white flour, sugar, and saturated fats.

Massage: A daily massage of your abdomen and back in the week before your period helps prevent cramps by aiding muscle relaxation. Massage your abdomen yourself and ask someone else to massage your back (see photograph, below left). Use sunflower or sweet almond oil alone or with three drops of lavender oil added to each tablespoon.

Exercise: Get regular exercise that works your whole body. Try swimming, or join a yoga or

> ### Sexual healing
>
> Having an orgasm can help reduce menstrual pain and bring it to an end faster. It may be that the contractions of the uterus that occur during orgasm help dispel congestion of blood and tissue fluid as well as painful muscle spasms.

aerobics class. If you have cramps, kneel on all fours, then flex and stretch your lower back quickly and repeatedly. This pushes your hips up and down, which exercises the pelvis, boosts blood circulation to it, and helps prevent congestion in the uterine wall (see photograph, p. 312).

Heat: Take a warm bath, or rest in bed with a covered hot-water bottle over your lower abdomen or the small of your back. Alternatively, put a hot compress over your lower abdomen for two or three minutes, then replace it with a cold one for half a minute. Repeat two or three times. For severe pain, apply to your lower back and abdomen a compress soaked in hot water to which you have added four drops each of clary sage and sweet marjoram oils, and three of Roman chamomile oil.

Herbal remedies: For cramping pains combined with relatively light bleeding, drink tea made from prickly ash, cramp bark, blue cohosh, black cohosh (for intense pain), chasteberry,

Back massage
Curl up on your side, and have a friend circle one hand counter-clockwise over your lower back, with the other hand placed gently on your abdomen.

Helpful hip-rocking
Lie on your back, and have a friend straddle your legs and use his or her hands to rock your hips rhythmically from side to side. Once the person has a rhythm going, a light touch is all that's needed.

chamomile, and ginger, taken singly or in combination. Drink a cup three times daily.
Caution: For safety concerns, see pp. 34–37.

Homeopathy: For persistent problems, consult a homeopathic practitioner. Otherwise, choose the remedy that most closely matches your case.

■ Belladonna: for violent pain along with bright red, clotted blood and a feeling of heat.

■ Colocynthis: for pain relieved by warmth or doubling up, and for irritability.

■ Lachesis: for menstrual cramps that start before the period, are worse on the left side, increase when pressure is applied, and improve when the flow starts.

■ Magnesia phosphorica: for shooting pain that improves with heat or gentle massage.

■ Nux vomica: for cramps, low back pain, irritability, and constipation.

■ Pulsatilla: for pain when you also feel extremely emotional, weepy, and needy.

■ Sepia: for pain accompanied by irritability and indifference to those close to you.

Treating heavy periods

Get medical advice if you have heavy bleeding. Your doctor may need to test for anaemia and such disorders as fibroids and endometriosis.

Diet: Eat foods rich in vitamin C and flavonoids (see p. 12) to strengthen blood vessel walls, and those rich in iron to prevent and treat anaemia that may result from excessive blood loss.

Aromatherapy: During menstruation, massage your abdomen each night with a blend of the following: one drop each of rose otto, Roman chamomile, and clary sage oils with two drops of sweet marjoram oil and one tablespoon of sweet almond or olive oil.

Herbal remedies: Take a tea or tincture made from equal amounts of beth root, blue cohosh, agrimony, goldenseal, and raspberry leaves.

Acupressure for menstrual problems

● To dissipate pain that builds up just before a period and lasts for the first day or two, apply pressure first to the point (Liv 3) in the furrow between the first and second toes where the bones meet on the top of the foot.

Then try the point four finger-widths above your inner ankle bone, just behind the shinbone (Sp 6). Press each point firmly for two minutes, one at a time, with your thumb. Use small circular movements.

● Two points are useful for heavy periods: the point (Sp 1) on the outer corners of your big toenails, and the point (CV 4) four finger-widths below your navel. Press

each point firmly with your thumb or fingers in a downwards direction for about two minutes.

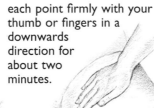

Yoga for pain relief

Practising yoga can help reduce cramps and other symptoms. The following exercise promotes healthy circulation in the pelvic region.

1 Sit with your back straight and knees bent, so that your soles touch and your heels are close to your body.

2 Hold your feet, then gently raise and lower your knees several times.

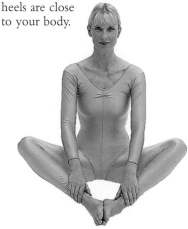

3 Next, lean forwards slowly, bending from the hips and keeping your back straight. Hold this position for two minutes, feeling the stretch in your legs.

4 Relax and lean forwards a little more as you exhale. Hold this position as you inhale. Repeat as necessary.

Dong quai and chasteberry are also effective. Take as tea or tincture three times each day.
Caution: For safety concerns, see pp. 34–37.

Treating irregular periods

Diet: Try to maintain a normal weight for your height (see p. 302) and, most important, avoid crash diets and binge eating. Too low a body weight prevents ovulation, which may make periods irregular. Too high a weight can over-stimulate the ovaries, which may contribute to multiple ovarian cysts and disrupted periods. Get the professional help you need to lose excess weight and/or manage your eating disorder.

Exercise: Get regular exercise, but if your periods become irregular or stop, reduce the exercise time by 10 to 20 percent, and have one or two exercise-free days each week.

Other therapies: Acupuncture may be helpful in regulating hormonal imbalances.

When to get medical help
- You have heavy or irregular periods.
- You have severe menstrual pain lasting several days a month.
- You have fertility problems.
- You have bleeding between periods.
- Your periods have stopped.

See also:
Anaemia, Eating Disorders, Fertility Problems, Fibroids, Premenstrual Syndrome

Migraine

A recurrent severe headache, usually accompanied by other disturbing symptoms, is known as migraine. It may last up to three days and be extremely disabling. The best way to manage a tendency towards migraine is to learn which factors trigger your attacks, so that you can try to avoid them. If a migraine does develop, various self-help measures may bring relief.

A clear head with yoga
Yoga can help relieve stress and reduce the frequency of migraines. Exercises that release tension in the upper back, shoulders, and neck may be especially beneficial. Ask a yoga teacher about suitable postures.

Around eight percent of people suffer from migraine. The problem occurs three times more often in females than males, largely because of changing hormone levels before and during menstruation, pregnancy, and the menopause (although for some women, the menopause brings relief). A first attack of migraine usually occurs in the late teens or twenties; in some cases it happens at an even younger age. It is rare to have your first migraine over the age of 50.

Attacks often become less frequent and severe as people get older: they are much less common in those over 65 years of age.

Migraine results from some sort of trigger (see box) making certain arteries in the brain first constrict, then dilate. Serotonin levels in the brain are low between attacks and high during them. Levels of other neurotransmitters may also be disrupted during attacks, as may calcium and magnesium levels.

Symptoms vary from one individual to another, but the common factor is a fierce, throbbing pain in one side of the head.

Migraine triggers

One or more of a large number of factors may set off a migraine. They include:
- Certain foods, especially cheese, chocolate, red wine, fried foods, and citrus fruits.
- Low blood sugar, brought on by hunger or excessive intake of refined carbohydrates.
- Dehydration.
- Stress, shock, or worry.
- Lack of sleep.
- Bright light or certain colours of light.
- Loud noise.
- Weather or climate changes.
- A dry atmosphere or a warm, dry wind.
- Hormonal changes.

You may experience one or more strange sensations—called an aura—that precede an attack and last up to an hour, such as flashing lights, zigzag lines, or a blind area in your field of vision. Once the headache begins, many people feel nauseated or vomit, and they may become sensitive to light and sound. Other possible symptoms are vertigo, tingling, and numbness.

Prevention

The simplest way to prevent attacks is to try, with experience and the help of a diary, to recognize your triggers, so that you can take steps to avoid them or minimize their effect. For example, if

both stress and premenstrual hormonal changes tend to bring about migraines, don't plan demanding events, such as a job interview, in the week before a period. If you know that motion sickness can precipitate a migraine, take steps to prevent the condition from arising, for instance by avoiding a large meal before a long journey. A diary may reveal a cycle of regular attacks—for most people with migraine, occurring every 10 to 40 days. Being aware of such a pattern helps you know when to be especially careful about staying away from triggers.

Diet: Prevent any nutrient shortage that might make migraine more likely by eating a well-balanced diet, including foods rich in vitamin B_2, or riboflavin, such as whole grains, egg yolks, milk, spinach, and lean meat. To avoid large blood-sugar swings, eat smaller, more frequent meals, choosing slowly digested complex carbohydrates, such as whole grains and beans, rather than foods containing white flour or added sugar. Sugary, refined foods cause a rapid rise in blood-sugar level, followed by a steep fall (see p. 199). Such changes in blood-sugar level may provoke a migraine attack in susceptible people.

To help you discover whether any food triggers a migraine, keep a detailed food diary for several weeks. Some people are unable to tolerate foods containing tyramine, an amino acid that can cause migraine by affecting blood vessels in the brain. Tyramine-containing foods include

Culprit foods
Keep a diary of the foods you have eaten in the 24 hours preceding an attack. Chocolate, cheese, oranges, caffeine, and milk are possible migraine triggers.

Protective foods
Eating oily fish, such as tuna and sardines, three times a week may protect against migraine because of their omega-3 fatty acids. Emphasizing whole grains over refined carbohydrates can also be beneficial.

chocolate, cheese, beef, liver, eggs, beer, red wine and other fermented foods, and some fruit and vegetables (bananas, oranges, plums, broad beans, spinach, tomatoes).

Another possibility is a food allergy. Common culprits include wheat, milk, cheese, tomatoes, oranges, and potatoes. Some people develop a migraine after drinking tea, coffee, cola, or other caffeine-containing drinks.

However, there is one special difficulty in knowing whether any particular food is a migraine trigger. Before the headache itself begins, the already disrupted neurotransmitter levels may lead to a craving for certain foods. The foods commonly eaten before a migraine may not, therefore, actually cause the migraine, although symptoms

tend to occur after you eat them. The real trigger for your migraines may be another factor.

Light sensitivity: The glare from water, snow, and other surfaces can set off migraines in light-sensitive people. Wearing polarized sunglasses with good-quality lenses should help. If a very bright picture on a computer screen is a problem, adjust the contrast and brightness levels, and position the screen so as to avoid reflections. Ask your optician or ophthalmologist about trying tinted lenses to see if these help you. Certain blue or green tints help prevent migraines in some people.

Air quality: Very dry air increases the proportion of positively charged ions in the atmosphere, which raises the body's level of serotonin, a neurotransmitter whose level increases during a migraine. Negative ions make the symptoms less severe and less long-lasting. Raise the concentration of negative ions in your home by opening windows and doors, having plenty of house-plants, and using a humidifier and/or an ionizer.

Stress: Stress is one of the most common triggers, so learning to handle it may reduce your migraine risk. Relaxation therapy can be very helpful, as can massage, yoga, meditation, and aromatherapy. Some people cope with stress while it lasts, but succumb to migraine afterwards. If you recognize this pattern for your migraines, try to adjust your lifestyle so as to manage stress more effectively (see p. 347).

Supplements: There is evidence that the amino acid 5-HTP (5-hydroxytryptophan), a form of tryptophan, can help to prevent migraines. Take 100 milligrams three times a day. Some people experience nausea with this supplement. Seek medical advice first if you are taking antidepressant medication. Supplements of magnesium and calcium may also be beneficial. Take vitamin B_2 (riboflavin) if you find it hard to increase your dietary intake of this vitamin.

Herbal prevention: Substances in the bitter leaves of feverfew help combat inflammation and relax narrowed blood vessels. If you suspect that a migraine is starting, add two or three leaves of feverfew to a sandwich, which disguises the taste and helps prevent any possible irritant effect of the leaves on the mouth. If you prefer, take feverfew capsules or tablets. Feverfew can also be taken on a daily basis for the long-term prevention of migraines. Do not take feverfew if you

Acupressure for migraine

- Press with your thumbs between your eyebrows (the *yintang* point) for 7 to 10 seconds. Relax and breathe deeply as you apply pressure. The points recommended on page 239 may also help.

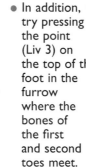

- In addition, try pressing the point (Liv 3) on the top of the foot in the furrow where the bones of the first and second toes meet.

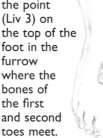

Cool relief
Rest your head against a hot-water bottle filled with cold water. Alternatives are putting your head under a shower of cold water, and pouring a basin of cold water over your head.

forehead and a hot one where your neck meets your skull, and switch them every two minutes. Repeat up to six times.

Other therapies: A cranial osteopath or practitioner of craniosacral therapy may be able to help with migraine. Utilizing biofeedback methods and the Alexander technique may also help reduce your susceptibility to this condition.

are pregnant or breastfeeding, or if contact with the plant gives you a rash.

Treatment

Rest and sleep: As soon as you feel you are about to get a migraine, lie down in a well-ventilated dark room. By doing this, you may cut short the attack or you may even succeed in staving it off completely.

Herbal remedies: Rosemary, lemon balm, and peppermint teas all have antispasmodic properties that may reduce the effects of a migraine.
Caution: For safety concerns, see pp. 34–37.

Aromatherapy: If you feel an attack is imminent, add one drop each of peppermint and lavender oils to two teaspoons of sweet almond oil, and rub a little gently into your temples and the back of your neck.

Heat and cold
■ Place a cold compress against your forehead or neck. If this does not work, use hot and cold compresses. Start with a cold one on your

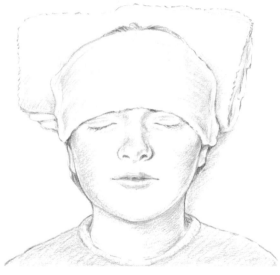

Aromatherapy compress
During stressful times, use daily compresses on your temples and forehead. Make these by wringing out a cloth in water to which you have added a few drops of lavender or marjoram oil.

When to get medical help
- Migraines are more than an occasional problem or are severely disrupting your life.
- You are taking the oral contraceptive Pill.
- The pattern of migraines changes.

Get help right away if:
- You experience a debilitating headache for the first time.

See also:
Headache, Motion Sickness, Nausea and Vomiting, Stress

Motion Sickness

Some people experience nausea and/or vomiting when travelling by car, boat, train, or plane. This condition is especially common in children, whose balance mechanism is more sensitive than that of adults, and most people suffer less as they grow older. The sickness usually clears up quickly at the end of the journey.

Motion sickness occurs when movement disturbs the semicircular canals, the balance mechanisms in your inner ear. Although your eyes adjust to the motion, your ears do not. In its mildest form, the condition causes slight discomfort or a headache; more severe cases produce nausea, sweating, and vomiting, which continue until the motion stops. A stuffy atmosphere, a full stomach, or the sight or smell of food makes the condition worse.

Prevention

Avoid large meals before travelling. Choose the front seat in a car or above the wing in a plane, and stay amidships on a boat. Get some fresh air—for example, by opening the car window. Look straight ahead at the road, or at the horizon if at sea. Avoid reading or any activity that involves focusing on nearby objects. Keep away from people who are smoking, since breathing the smoke may make you feel nauseated. If your child is prone to motion sickness, don't increase anxiety by talking about it before or during the journey, as this makes sickness more likely.

Treatment

Homeopathy: Take one dose of an appropriate remedy just before you set out, and then as necessary during the journey.

Spicy relief
Take ginger for motion sickness two hours before departure and every four hours after that. Chew fresh ginger or drink ginger tea or a few drops of ginger tincture in warm water. You can also nibble crystallized ginger or a ginger biscuit.

Acupressure

Press the point (HP 6) between the tendons two thumb-widths above the crease at the front of your wrist to relieve nausea and anxiety. Some chemists and travel agents sell acupressure wristbands; these work in a similar way.

- Cocculus: for all symptoms of motion sickness.
- Tabacum: if the slightest movement brings on extreme nausea and vomiting, especially if you are pale or are sweating a lot.

When to get medical help
- You have a fever, severe headache, or feel faint or dizzy.
- Symptoms don't lessen within 24 hours.

Get help right away if:
- You have severe abdominal or chest pain.

See also:
Dizziness, Nausea and Vomiting

Mouth Ulcers

Small white, grey, or yellow sores can occur singly or in clusters anywhere in the mouth, including the tongue. Often very painful, mouth ulcers affect about 20 percent of the population at any one time. They are most likely in people who have a poor diet or an underlying infection, or are under a lot of stress.

Salt water mouthwash
Add one teaspoon of salt to a glass of warm water and rinse with this solution once or twice a day. This may cause stinging when the ulcer is new.

The many causes of common mouth ulcers, called aphthous ulcers, include a vitamin deficiency, digestive upset, food sensitivity, infection, injury (such as from biting the tongue or cheek or wearing poorly fitting dentures). Mouth ulcers are much more likely to occur if you are exhausted or under stress. Occasionally they may be a sign of an underlying condition, such as tuberculosis, *Herpes simplex* infection, coeliac disease (intolerance of gluten, a cereal protein), Crohn's disease, anaemia, or leukaemia.

Prevention
Ensure that you eat a healthy diet (see pp. 9–12). Boost your resistance to infection by eating garlic, onions, foods rich in vitamin C and flavonoids, and those containing zinc, and by limiting your intake of refined carbohydrates. Avoid using toothpaste containing a detergent such as sodium lauryl sulphate, as this can destroy the protective mucin in oral mucus.

Treatment
Kitchen-cabinet remedies:
Soothe mouth ulcers with a once- or twice-daily mouthwash or gargle of:
- Cold tea for its astringent effect.
- Sage tea for its antiseptic, healing, and astringent properties.

Aromatherapy: Antiseptic essential oils, such as geranium, lavender, lemon, myrrh, and tea tree, work well on ulcers caused by viral or bacterial infection. Avoid myrrh oil if you are in the first 20 weeks of pregnancy.
- Mix five drops of tea tree oil, three of lemon oil, and two of myrrh oil with two and a half teaspoons of grapeseed oil. Apply to the ulcers with your finger every two hours.
- Add one drop each of tea tree, geranium, and lavender oils to half a glass of water. Use as a mouthwash three or four times a day.
- Add a few drops of tea tree, geranium, or lavender oil to a cup of warm water and use as a mouthwash three times a day.

When to get medical help
- A mouth ulcer fails to heal within two weeks or increases in size over one week.
- The problem recurs frequently.
- You suspect the problem is caused by a tooth or dentures (see your dentist).
- You also have a cough, diarrhoea, or a tendency to get recurrent infections elsewhere in your body.
- You are taking new medication.

See also:
Candida Infections, Cold Sores, Gum Problems

Muscle Aches and Stiffness

Tender, stiff, aching muscles from too much exercise, heavy lifting, or other overuse—or from alterations in body chemistry—are a common complaint. The various remedies and therapies used to relieve them are a part of many people's healing repertoire, but these extra tips may provide relief the next time the problem arises.

Muscle-warming oil
Dried chilli infused in sunflower oil makes a warming massage oil, but test it on a small area first.

Stiffness and pain in the skeletal muscles (the muscles that control movement of the body) often occur as a result of strains, cramps, and injuries. The affected muscles may go into a state of prolonged tension—or spasm—as a reflex to prevent further damage through movement.

Muscle tension may also be a response to injury or pain in an adjacent part of the body. If certain groups of fibres within a muscle become especially tense, they can be felt as taut, hard strands beneath the skin; these are most likely at the edges of the shoulder muscles. Sometimes tender lumps (fibrositic nodules) appear in strained muscles in the back, neck, and shoulders.

Muscle tension impedes local blood circulation, which inhibits the healing process in an injured muscle or other part. It also triggers the release of chemicals called cytokines and prostaglandins, which are responsible for the pain. If muscle tension is a response to pain from a compressed nerve (as in some back pain), the tension further compresses the nerve and increases pain.

One special type of muscle problem is described in the box below. Additional causes of aches and stiffness include:

- Stress, anxiety, or depression.
- Exposure to cold air, as with a draught.
- Insufficient nutrients, especially minerals.
- Food sensitivity.
- Cramps.
- Chronic fatigue syndrome.
- Tension headaches.
- Repetitive strain injury.
- Poor posture.

Fibromyalgia: when you ache all over

Formerly called fibrositis or muscular rheumatism, this condition involves widespread muscle pain, morning stiffness, and tenderness over particular points where a muscle is attached to a bone. These are known as myofascial trigger points, or, in traditional Chinese medicine, "ah-shi" points, because when pressed, they tend to make a person cry out "ah shi!" (the equivalent of "ouch!"). It is thought that each skeletal muscle has a trigger point, and that this type of tenderness can result from any damage to that muscle or its protective coating of connective tissue. For a firm diagnosis of fibromyalgia, at least 11 of the 18 points must be tender.

People who suffer from fibromyalgia usually can't achieve deep sleep and are prone to depression, headaches, menstrual pain, restless legs, and Raynaud's syndrome. There is considerable overlap between the symptoms of fibromyalgia and those of chronic fatigue syndrome, but the cause of fibromyalgia is unclear.

Prevention

To protect muscles when doing physical work or exercising, follow some basic rules. Always warm up first, increasing the circulation to the muscles with some light whole-body exercise, because muscle aches and stiffness are more likely when muscles are cold. Once the body is warm, stretch each of the main groups of muscles. When you have finished your workout, cool down by gradually decreasing the intensity of the exercises, then stretch the muscles again.

Lifestyle factors can affect your susceptibility to muscle stiffness. Getting enough rest to balance the amount of exercise you get, and vice versa, helps prevent muscle problems and their sources, such as aching shoulders and back, fibromyalgia, cramps, restless legs, and tension headaches. Maintaining good posture so that no one group of muscles remains tense for too long is also important for preventing aches and stiffness. Two of the most important ways of doing this are keeping your head in line with your spine and not hunching your shoulders.

A healthy diet provides the nutrients your muscles need to perform well and to recover quickly from strain. Such nutrients include amino acids, calcium, magnesium, potassium, selenium, vitamins B and C, and flavonoids.

Warm-up stretches
To avoid stiff, aching muscles after your workout, start your routine with a programme of stretches. Learn them from a fitness trainer or an exercise video that you know to be good.

Drinking enough fluids is important. Stress management (see p. 347) can help prevent stress-induced muscle tension and fibromyalgia.

Treatment

The most important aspect of treatment is to protect the muscles from further damage while healing the affected tissue. Relaxing the muscles helps alleviate symptoms and allows injured tissue to heal more quickly.

Rest: Rest stiff, aching muscles that result from overuse or strain, but resume normal activity after two or three days (long periods of inactivity are not advisable because this can result in shortening of the muscle fibres). Gradually resume normal exercise.

Exercise: After a few days' rest, start to gently stretch your stiff muscles. Exercise for a maximum of five minutes at first. Repeat several times daily, doing a little more activity every day, but stopping if you feel pain. After several days, weeks, or months—depending on the problem—you should be able to extend the muscle fully. Chronically stiff and aching shoulder and back muscles may take months to extend fully.

Aromatherapy: Having a massage or bathing with certain essential oils can provide pain relief.
- Mix 10 drops of rosemary oil, or two drops of German chamomile, with a tablespoon of sweet almond or soybean oil, and rub gently over the sore area. Don't use rosemary oil in the first 20 weeks of pregnancy.
- Add three drops of lavender oil and three of

cypress to comfortably hot bathwater, and soak in it. Omit cypress oil if you are in the first 20 weeks of pregnancy.

■ Add three drops of a warming spice oil, such as ginger, to a hot compress and apply to the affected muscle.

Herbal remedies: Gently rubbing a herbal ointment over painful, stiff muscles three or four times daily may help.

■ Apply arnica ointment.

■ Make a warming oil to rub into aching muscles by infusing an ounce of dried chilli in a pint of sunflower oil. Apply to a tiny area of skin first, since it makes some people blister. Alternatively, buy an ointment containing capsaicin (a chemical derived from capsicum peppers) from a chemist or health-food store.

Caution: For safety concerns, see pp. 34–37.

Heat and cold: Stiff, aching muscles respond well to heat, whereas a torn or overstretched muscle generally feels better with a cold compress for the first 24 hours, then heat.

■ Warm stiff muscles with an infrared bulb placed at a distance of 20 to 30 inches for a half hour twice daily.

■ Place over a sore muscle a covered hot-water bottle, a gel-filled pack heated in the oven or microwave, or an electrically heated pad.

■ Have a long soak in warm (not hot) bathwater.

■ Apply a hot compress for three minutes, then a cold one for one minute, and repeat two or three times. Do this several times daily until you feel better. For added benefit, first soak the hot compress in half a pint of water containing

Massage away the aches

One of the best therapies for aching muscles is massage. Take a warm bath first, and have the massage in a warm room to encourage the muscles to relax. Ask a friend to follow the steps below.

1 Massage the sore muscles with gliding and kneading movements.

2 Encourage tense strands or points to relax by moving your fingers across the muscle fibres without sliding on the skin (cross fibre friction).

3 Press tender points firmly with a thumbtip.

two tablespoons of distilled witch hazel.

- Try placing an ice pack (such as a covered freezer pack or package of frozen peas) over painful muscles resulting from injury.

Diet

- Eat magnesium-rich foods (see illustration, right). Alternatively, take 300 milligrams of magnesium daily in the form of a supplement.
- Consider a calcium supplement (250 milligrams twice a day).
- Reduce your intake of animal protein (for example, meat, cheese, and eggs) and of foods containing white flour and added sugar.

Homeopathy

- Arnica: for muscle stiffness after overstrenuous or unaccustomed exercise; for muscles that feel more painful when you move, and for related restlessness and irritability.
- Bryonia: for aching muscles made worse by movement or by a dry, cold wind.
- Rhus toxicodendron: for muscle stiffness after overuse that improves with gentle movement.

Other therapies: If long-standing aching and stiffness result from poor posture, you may benefit from physiotherapy, Alexander technique lessons, or yoga classes. An osteopath, chiropractor, or shiatsu therapist may also be helpful.

Muscle-friendly magnesium
This mineral has muscle-relaxing properties. Include in your diet a variety of magnesium-rich foods, such as wholegrain cereals, nuts, and beans.

Acupressure for stiff muscles

To relax your muscles, treat the following points for two minutes at a time:

- Apply constant thumb pressure to the point (GB 34) in the depression below the outside of the knee joint at the top of the fibula (shinbone).

- Apply pressure with rotating movements of the thumb to the point (TH 5) two thumb-widths above the wrist crease on the back of the forearm.

When to get medical help

- Symptoms persist longer than a week.
- Symptoms worsen.
- Other symptoms, such as fever, headache, diarrhoea, pain, bruising, and numbness, accompany muscle aches or spasms.

See also:
Anxiety, Back Pain, Chronic Fatigue Syndrome, Headache, Leg Cramps, Neck and Shoulder Problems, Pain, Repetitive Strain Injury, Restless Legs, Strains and Sprains, Stress

Nail Problems

Healthy nails are smooth, evenly coloured, and strong, with a pale pink or flesh-coloured nail bed. A number of disorders may affect their appearance. Your nails can therefore be an indicator of your underlying state of health. There's much you can do at home through diet and externally applied treatments to improve the look of your nails.

The nails and surrounding skin can be infected by bacteria and such fungi as tinea and candida. These conditions can make the nails soft, discoloured, thickened, and misshapen. Paronychia is a bacterial or fungal infection of the fold of skin at the side of the nail.

Certain skin diseases can also affect the nails, including psoriasis, which may cause thickening, pitting, or even separation of the nail from the nail bed. The patchy hair loss known as alopecia areata is sometimes associated with ridged, pitted, rough nails. Other, more serious conditions leading to nail problems include excessive production of the thyroid hormone (thyrotoxicosis), disorders of the blood-clotting process, and inflamed heart valves (endocarditis). Blueness of the nails may result from severe asthma, heart disease, emphysema, or bronchitis, while yellowing and excessive hardening of the nails may be a sign of bronchiectasis (damaged airways in the lung) or lymphoedema (accumulation of lymphatic fluid in the tissues). Nail discoloration can also result from smoking or regular use of nail polish. Toenails may become ingrown—for example, because of the pressure from ill-fitting shoes.

Several nutritional deficiencies may also show up in the nails:

■ Iron-deficiency anaemia may cause the nails to appear pale and become thin, brittle, ridged, and easily cracked or broken.

Tell-tale shape
Iron-deficiency anaemia can cause the profile of each nail to become spoon-shaped.

Looking after your nails

- Have manicures and pedicures regularly.
- Massage in cuticle cream at bedtime.
- Dry your hands and feet thoroughly after washing.
- Wear protective gloves for housework and other manual work.

■ Zinc deficiency leads to white spots and/or brittleness of nails.

■ Severe protein deficiency makes the nail beds appear white.

■ Lack of linoleic acid (an essential fatty acid) may cause the nails to split and flake.

Treatment
Diet: Adapt your diet according to your specific nail problem.

■ For nail problems resulting from iron deficiency, follow the advice on pages 83–84.

- For fungal or bacterial nail infections, you can boost immunity by eating garlic and onions, foods rich in zinc (nuts, root vegetables, shellfish), vitamin C and flavonoids (fruit and vegetables), and omega-3 fatty acids (nuts, seeds, dark-green leafy vegetables, oily fish, rapeseed oil).
- For splitting and breaking nails, increase your body's level of vitamin A by drinking carrot juice and eating more eggs, milk, and liver. Eat three servings of oily fish each week, and consider taking a fish oil supplement.
- For white spots in the nails, increase your intake of zinc and B vitamins by eating more poultry, seafood, and whole grains.
- For nail problems caused by psoriasis, eat more oily fish and other foods with essential fatty acids (whole grains, nuts, seeds).

Aromatherapy: Several essential oils are useful.
- For fungal infections, tea tree oil is particularly beneficial. Dab the oil directly onto the affected nail and

Anti-infective oils
Keep your selected mix of essential oils in a dark bottle and use to massage around the nail bed three times daily while symptoms persist.

Nutrition for nails
If your nails are brittle or flaking, give them a boost by increasing your intake of key nutrients, such as calcium (in milk), omega-3 fatty acids (in oily fish), vitamin C (in most fresh fruit and vegetables), and zinc (in seafood and poultry).

cuticle, or mix five drops with the same amount of marigold (calendula) oil into two tablespoons of sweet almond oil, and massage into the nail.
- For other infections, mix five drops each of eucalyptus and patchouli oils with 10 drops each of tea tree and marigold (calendula) oils, and two tablespoons of sweet almond oil.

Homeopathy
- Silica can often strengthen weak nails. Try one pill a day for a month.

When to get medical help
- Your nail problem continues to affect new nail growth in spite of treatment.
- You experience pain, swelling, inflammation, or pus around the nail.
- You suspect that your nail problem is a symptom of an underlying condition.

See also:
Anaemia, Foot Problems, Fungal Skin Infections, Hair and Scalp Problems, Skin Problems

Nausea and Vomiting

Each of us is familiar with the unpleasant sensation of nausea, with its accompanying sweating, pallor, and faintness, and the feeling that we may vomit. In some cases, actual vomiting (the "throwing up" of stomach contents) soon follows these warning signs. Many simple natural remedies can help you feel better or stop you from feeling nauseated in the first place.

Nausea and vomiting most often result from a digestive upset caused by overeating, too much rich food, or food to which you are allergic, intolerant, or otherwise sensitive. Irritation of the stomach lining caused by too much alcohol or by infecting microorganisms or their toxins (food poisoning) may also produce these symptoms. Infection of the digestive tract may lead to diarrhoea as well.

Other triggers of nausea and vomiting include anxiety, fear or shock, and migraine. Disturbance of the inner ear's balance mechanism that leads to dizziness—as with motion sickness, an inner-ear infection, and Ménière's disease—may also cause nausea and vomiting. Nausea, especially first thing in the morning ("morning sickness"), is a common occurrence in early pregnancy. Young children often vomit as a result of a feverish illness.

Nausea and vomiting are occasionally a sign of an underlying disorder, such as a stomach ulcer, uncontrolled diabetes, jaundice, gallbladder disease, a heart attack or cancer. Certain drugs, including some anticancer and anaesthetic agents, may also cause nausea and vomiting.

Prevention

Always eat slowly and chew thoroughly to avoid nausea after meals. Avoid very large meals. You may suspect a food sensitivity if symptoms consistently occur after eating a particular type of food. In this case, avoiding the food is the simplest way of averting the problem.

Prevent digestive upsets from infection by paying careful attention to personal and kitchen hygiene (see p. 167). To help prevent sickness during pregnancy, eat frequent light meals, and have a ginger biscuit or drink a small cup of ginger tea, sweetened with a little honey, as soon as you get up in the morning.

To avoid feeling nauseated when you are stressed or excited, practise breathing and other relaxation exercises so that you can more easily relax your mind and body when necessary.

Treatment

Maintain an adequate fluid intake if you are suffering from repeated bouts of vomiting, especially if accompanied by diarrhoea.

- Take frequent sips of water or fruit juice throughout the day.
- To replace salts and fluids, make drinks using oral rehydration salts from the chemist, or make

Rehydration fluid
Mix eight level teaspoons of sugar (or four heaped ones of honey) and one teaspoon of salt into two pints of water.

your own (see illustration, facing page).

■ Experiment to see whether warm drinks are easier to keep down than cold ones.

■ Avoid alcohol, as this boosts urine production and encourages dehydration. It also irritates the stomach and may therefore slow recovery.

Herbal remedies

■ Ginger counters nausea and vomiting. Chew a piece of fresh or crystallized ginger, take ginger tablets, or sip ginger tea.

■ To help settle nausea associated with anxiety, sip chamomile tea, which has a calming effect as well as digestive properties.

Caution: For safety concerns, see pp. 34–37.

Kitchen-cabinet remedies: Warming spices, such as cloves, cinnamon, and cardamom, assist digestion and facilitate the elimination of toxins via the bowel. Use one or more of these spices to make a tea to sip when you feel nauseated. Do not use this remedy if you have a stomach ulcer.

Diet: After a bout of nausea and vomiting, give your system time to recover by returning to a normal diet gradually. Choose bland, easily digested foods at first, such as rice, clear soup, low-fat yogurt, wholegrain toast, and apple purée. Avoid coffee, tea, and fatty foods.

Homeopathy

■ Arsenicum: for acute gastrointestinal ailments, with diarrhoea and burning stomach pains, exhaustion, and chilliness.

■ Ipecacuanha: for constant nausea. Other symptoms may accompany the nausea.

Acupressure

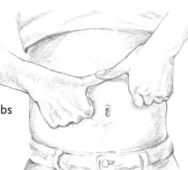

● Use the point (HP 6) described on page 282.

● Or if the risk of vomiting has passed, lie down and ask someone to press gradually and gently with the thumbs on the point (CV 12) four thumb-widths above your navel, using small circling movements, for up to two minutes.

■ Nux vomica: for nausea when you can't vomit but wish you could.

■ Sepia: for unrelieved nausea triggered even by the thought of food.

Flower essences: When nausea results from anxiety, try mimulus if you know what you are anxious about, or aspen if you don't.

When to get medical help
● You have been suffering from bouts of unexplained nausea and vomiting for longer than 12 hours, or you have eaten undercooked food.
● A medication may be the cause.
● You have recently been to a tropical country.

Get help right away if:
● Vomit contains blood or black material resembling coffee grounds.
● You have a severe headache, dizziness, fever, severe abdominal pain, drowsiness, aversion to bright light, or chest pain.
● A young child or an elderly person is experiencing repeated vomiting.

See also:
Abdominal Pain, Anxiety, Arterial Disease, Eating Disorders, Gallbladder Problems, Hangover, Indigestion, Motion Sickness, Pregnancy Problems

Neck and Shoulder Problems

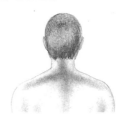

Pain and stiffness in the neck and shoulders can range from slight discomfort that hinders full movement to severe pain that prevents even minimal movement. Fortunately, the problem can often be prevented or relieved by reducing strain and tension in the muscles and joints.

In most cases, neck and shoulder pain is caused when the muscles go into spasm—that is, they contract and become rigid so that the joints are unable to move normally. This can result from the cumulative stresses of long periods of being in the same position, as when driving, sitting at a computer, or doing repetitive factory work.

"Whiplash" injuries sustained in automobile accidents and sports injuries (especially among those unaccustomed to strenuous activity) are common causes of shoulder pain. Before you try to treat any such injury, check the box on page 295 for danger signs. In addition, pain may recur at the site of an old injury.

Persistent pain and stiffness often result from osteoarthritis, in which bony outgrowths from the vertebrae in the neck pinch nearby nerves and put pressure on the muscles and ligaments. There are many other causes of neck and/or shoulder pain that may require medical attention, including a herniated ("slipped") disc, tendinitis, bursitis, spinal disorders, and Lyme disease.

Muscular tension in the neck and shoulder area is a common physiological manifestation of psychological and emotional stress. In this case, the underlying source of the problem needs to be identified and addressed.

Prevention

Avoid strain by paying attention to how you stand, sit, lift, and carry (see advice in "Back Pain", pp. 105–108). Learn to recognize the signs of stress, and offset any negative physical effects by dealing with it properly (see p. 347). Always warm up and stretch before exercise; the overuse of cold muscles can cause pain and stiffness.

Treatment

For pain from minor strain or tension, try the following natural remedies:

- Ask a friend to give you the shiatsu massage described on the facing page. Enhance its soothing effects by adding a few drops of a relaxing essential oil, such as lavender or geranium, to your basic massage oil or lotion.

➤ continued, p. 294

Carrying bags

Regularly carrying a heavy shoulder bag (left) is a common cause of shoulder and neck pain. A backpack (right), which distributes the weight evenly on both shoulders, is a healthier option.

292

Shiatsu massage for painful neck and shoulders

This sequence will help relieve pain and stiffness when symptoms are not severe and do not include redness. The person being treated should sit on a chair, stool, or floor cushion. The instructions are for the masseur. Omit step one if the person being treated is pregnant.

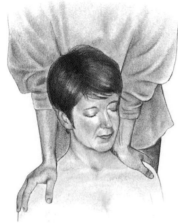

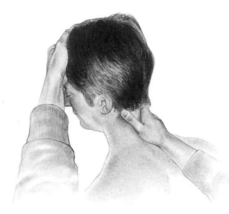

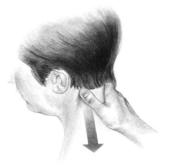

3 Apply firm but careful thumb pressure into the hollow between the neck muscles at the base of the skull. Work downwards. Change sides and hands, repeating on the other side of the neck.

2 Move your hands so you can work on one side. Support the forehead with one hand and, with the other hand, squeeze the muscles at the back of the neck, working downwards from the base of the skull.

1 Stand behind the person and lean straight down on the shoulders with open palms. Start with gentle pressure and gradually lean more heavily.

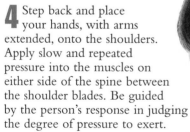

4 Step back and place your hands, with arms extended, onto the shoulders. Apply slow and repeated pressure into the muscles on either side of the spine between the shoulder blades. Be guided by the person's response in judging the degree of pressure to exert.

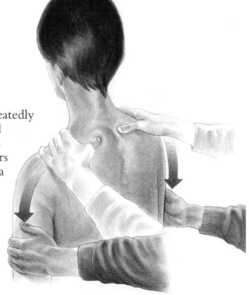

5 Finish by repeatedly squeezing and stroking the arms from the shoulders to the elbows in a series of quick movements to dispel any remaining tension.

Symptoms and Ailments

■ Place a hot compress (see facing page) over the affected area. Or apply alternate hot and cold compresses for a total of 20 minutes. Leave each hot compress in place for three minutes and each cold one for one minute.

Posture: Much neck and shoulder pain results from muscle tension caused by poor posture, especially when driving or working at a desk.

■ Be sure your desk and chair at work are properly adjusted for your height (see p. 106).

■ Take care to adjust your car seat, especially before long journeys. Your back should be well supported along its length. It is especially important for shorter people to make sure that

A simple solution
If you are prone to neck pain or if you have to sit in the cold or a draught for more than a few minutes, prevent neck stiffness by wearing a scarf or a turtleneck sweater, which will help keep the muscles warm and relaxed.

Neck-release exercises

Try the sequence of exercises described below to release tension in the neck and shoulders. Sit comfortably in an upright chair, and breathe slowly and evenly throughout. Repeat each exercise six times before going on to the next. Stop if you feel dizzy.

1 Allow your head to drop forwards until your chin rests on your chest. Raise your head slowly.

2 Tilt your head to one side and then the other, while keeping your shoulders level.

3 Turn your head to face left and then turn slowly to face right.

4 Lift your shoulders up towards your ears and roll them forwards and then backwards.

the seat is high enough to prevent the need to strain the neck to see clearly over the steering wheel. Use a cushion if your car seat is not fully adjustable.

Herbal remedies: Pain and stiffness in the neck and shoulders are often due to muscle tension resulting from emotional tension. Herbs with relaxing properties may help.

- Take a bath in warm water to which you have added tea made from rosemary or lavender. Avoid rosemary if you are in the first 20 weeks of pregnancy.
- Try chamomile, hops, passionflower, or valerian. Take as tea before bedtime.

Caution: For safety concerns, see pp. 34–37.

Homeopathy

- Arnica: for pain and stiffness arising from a recent injury. Take the remedy as tablets or apply it topically as ointment.
- Rhus toxicodendron: for long-standing pain and stiffness when symptoms are eased by gentle movement, warmth, and massage.

Other therapies: If you don't have any of the symptoms listed under "When to get medical help" (right), seek advice from a qualified practitioner of a manipulative therapy, such as osteopathy or chiropractic. Acupuncture may also help. Long-standing neck and shoulder problems may benefit from improved posture achieved through the Alexander technique, the Feldenkrais method, or similar bodywork therapies. Practising tai chi and yoga can also help prevent posture-related problems.

Making a neck compress

Applying a hot compress may relieve neck and shoulder problems resulting from muscle stiffness. Here's how to make one:

1 Soak a towel in hot (not boiling) water. Fold the towel and wring out well.

2 Unfold the towel and place over the back of the neck and shoulders. Cover with a dry towel. Leave in place for up to 10 minutes.

When to get medical help

- You also have a headache, a fever, dizziness, faintness, or sensitivity to bright light.
- You have swollen lymph nodes ("glands") in the neck, or difficulty in swallowing.

Get help right away if:

- You also have difficulty in moving a limb, loss of bladder or bowel control, tingling or numbness in a limb, breathlessness, chest pain, or shooting pains in one or both arms.
- The pain follows an injury.
- You have difficulty moving your neck.

See also:
**Arthritis, Back Pain,
Muscle Aches and Stiffness**

Nosebleeds

Relatively common in childhood, when the tiny blood vessels of the nasal lining may be fragile, nosebleeds are usually insignificant. They occur less often in healthy adults but may become more frequent again during old age. You can easily treat most nosebleeds at home, using traditional techniques to stem bleeding and allow healing.

A nosebleed occurs when one or more blood vessels inside a nostril ruptures. This may happen after a blow to the nose or head, repeated sneezing, or picking or blowing the nose. An upper-respiratory infection also makes nasal blood vessels more fragile. Indoor heating can dry out the mucous membrane, affecting the vessels and causing a nosebleed. In rare cases, nosebleeds are a sign of an underlying condition, such as high blood pressure or a blood-clotting disorder.

Prevention

If you suffer from recurrent nosebleeds, eat more foods containing vitamin C and flavonoids (see p. 12) to strengthen capillary walls.

Treatment

Herbal remedies

- Apply to your nose and the back of your neck cold compresses soaked in dilute witch hazel.
- Hold a cotton wool ball soaked in marigold (calendula) tincture under your nose.

Aromatherapy: Cypress and helichrysum oils help stop blood loss. Put a few drops on a cotton wool ball and hold under your nose. Do not use cypress oil in the first 20 weeks of pregnancy.

Homeopathy

- Ferrum phosphoricum: use daily for a month if you are prone to nosebleeds.
- Phosphorus: for sudden heavy bleeding.

First aid for nosebleeds

- Sit up and lean slightly forwards. Breathing through your mouth, pinch your nostrils together for 10 to 15 minutes.
- If you find it hard to maintain pressure, take two tongue depressors and place a rubber band around them (about one-third of the way down). Then position them over the nose so that a depressor is on either side (see illustration).
- Slowly release your nostrils.

If the bleeding has not stopped, pinch for 10 minutes more.
- Once the bleeding stops, gently clean away dried blood with luke-warm water.

When to get medical help

- Bleeding lasts longer than 30 minutes.
- Blood loss is severe.
- You suffer from frequent nosebleeds.
- You have high blood pressure.

Get help right away if:

- A nosebleed follows a head injury.
- You get a headache along with a nosebleed.

See also:
High Blood Pressure

Osteoporosis

As people age, the density of their bones naturally decreases because of the gradual loss of minerals from the skeleton. Severe mineral loss, a condition called osteoporosis, causes the bones to become weak and increasingly susceptible to fractures. The disorder occurs most frequently in older women due to hormonal changes after the menopause, but men can get it too.

About three million people in the United Kingdom suffer from osteoporosis. Four out of five are women who have gone through the menopause, though osteoporosis can also affect men and younger women. Each year in the United Kingdom an estimated 200,000 fractures—primarily of the spine, hip, and wrist—result from this condition. By the age of 80, one in two women have had at least one fracture caused by osteoporosis.

Osteoporosis often produces no symptoms until a fracture occurs after a fall. However, there may be back pain if the vertebrae (spinal bones) become weakened and collapse. Other indications of osteoporosis are loosening of the teeth, a loss of height, and kyphosis (an excessive curvature of the upper spine, also known as dowager's hump). A bone density scan can reveal the severity of osteoporosis.

Risk factors

Besides being postmenopausal, important risk factors in developing osteoporosis include being small-boned, underweight, and inactive. Excessive alcohol consumption, a poor diet, and smoking also increase the risk. All these factors can make mature bones less dense and can prevent a young person's bone density from

How osteoporosis develops

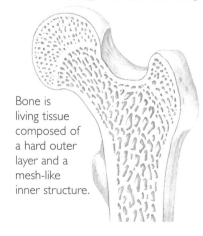

Bone is living tissue composed of a hard outer layer and a mesh-like inner structure.

Bone density increases throughout childhood and adolescence and for some years afterwards. New bone cells are continually being produced and old ones broken down. The bones are usually at their most dense and resilient in the late 20s. Osteoporosis—which means "porous bones"—occurs when bone tissue breaks down faster than new bone is formed. As a woman nears her menopause, more bone is destroyed than is produced, and her bones start to become less dense. The onset of osteoporosis may result from the naturally falling level of oestrogen, which assists the body in absorbing calcium. After the menopause the rate of bone mineral loss increases the risk of osteoporosis. In contrast, men's risk of osteoporosis is more gradual as they get older.

The amount of calcium available also plays a major role in determining the rates of bone formation and resorption. If insufficient calcium is provided by the diet, and levels in the blood are inadequate for the needs of other vital body parts, such as the heart, nerves, and muscles, cells called osteoclasts (bone cell destroyers) release calcium from bones. This results in more porous bones that become increasingly light and fragile.

reaching its potential peak. If bone density does not develop fully, the natural demineralization that accompanies aging takes effect sooner and makes a person prone to osteoporosis at a younger age.

Osteoporosis may also have a genetic link. Experts suspect that a gene can interfere with the body's ability to use vitamin D, which is crucial for calcium absorption. People with small bones—such as, generally speaking, Caucasians and Asians—are particularly prone to osteoporosis. Other risk factors include diabetes, thyroid disease, and some prescribed drugs, notably cortico-steroids and anticonvulsants.

Oestrogen boosters
Soybeans and soy products, such as tofu, miso, and soy milk, contain hormone-like substances that may mimic the bone-protecting properties of oestrogen.

Prevention

You can do much to prevent osteoporosis from developing. It is never too early to begin. Encourage children to eat foods rich in calcium and other bone-building nutrients (see chart) to help ensure optimal bone density. Whatever their history, adults too can make their bones stronger or at least slow the rate of bone loss by improving their diet and making lifestyle changes.

■ Eat more fruit and vegetables, including soy products. When a person's natural level of oestrogen falls, plant hormones may take its place by locking onto cell receptors in the bone. This is thought to confer benefits similar to those of your own oestrogen.

■ Eat a diet that will provide the minerals required to keep bones healthy. In addition to calcium, the most abundant mineral in bone, these include boron (for calcium absorption and retention), copper (for bone production), magnesium (for the efficient use of calcium), manganese (for strengthening connective tissue in bones), silicon (for bone resilience), and zinc (see facing page). Most experts recommend that an adult woman consume at least 1,000 milligrams of calcium daily.

■ Teenage, pregnant, breastfeeding, and post-menopausal women, as well as older men, need at least 1,200 milligrams of calcium a day. Sources include dairy products (milk, cheese, yogurt—choose low-fat varieties), dark-green

---

Calcium-rich foods

The following foods are good sources of calcium. The approximate amount of the mineral present in an average serving of a selection of foods is shown below.

Hard cheese (2 ounces) ....300 mg
Milk (8 ounces) ....250 mg
Yogurt (4 ounces).....................................250 mg
Sardines (2 ounces, canned)....................250 mg
Tofu (bean curd, 4 ounces)......................150 mg
Green cabbage (4 ounces, raw)...............50 mg
Baked beans (4 ounces)...............................50 mg

leafy vegetables (kale, spring greens, broccoli), beans, carrots, almonds, and fish with edible bones (sardines, canned salmon, anchovies).

■ If you are not eating well or have a high risk of osteoporosis, take a daily calcium supplement (with milk to increase absorption). Calcium supplementation can decrease bone loss by 40 percent after the menopause.

■ Increase your intake of zinc-rich foods (shellfish, nuts, seeds, root vegetables). Zinc encourages the production of bone protein and gastric acid, which is needed for the optimal absorption of calcium. Reduce your intake of saturated fat and sugar, since both can reduce acid production. If you suspect that your acid production is low (for example, if you produce

Strength from the sun
The body synthesizes vitamin D, a nutrient required for strong bones, in response to exposure to ultraviolet light. Being outdoors with bright light on your face for 15 minutes a day is usually sufficient. Avoid overexposure, which can cause skin cancer.

Feed your bones
Fruit, vegetables, dairy products, nuts, and oily fish contain many of the nutrients essential for healthy bones.

a lot of wind and suffer from indigestion and a feeling of fullness for a long time after a meal), consider taking a health food product, such as betaine hydrochloride, that contains acid to enhance mineral absorption. Don't take it if you have an ulcer.

■ Cut down on alcohol; drinking can accelerate the rate of loss of minerals from your bones.

■ Avoid fizzy drinks, as these encourage bone-mineral loss, and limit your caffeine intake, because it can interfere with calcium absorption. However, some studies indicate that moderate caffeine consumption (defined as about

two cups of average-strength coffee a day) has little or no effect. Adding milk to tea and coffee helps replace the calcium being lost.

- Cut salt intake, since salt may reduce bone density by increasing the amount of calcium that is lost in the urine.
- Limit your intake of meat, which can promote calcium loss. Research shows that meat eaters are more likely to develop osteoporosis than those who follow a vegetarian diet.
- Eat plenty of foods rich in vitamin A (dairy products, eggs, yellow and orange fruit and vegetables, green leafy vegetables) for bone protein production, and vitamin C (most fresh fruit—especially citrus—and vegetables) for the production of collagen, which helps keep bone's connective tissue strong.
- Eat foods rich in vitamin D, such as oily fish and fortified dairy products.
- Get some direct daylight—unfiltered by windows—on your skin every day. Take a daily vitamin D supplement if you do not get much sunlight, live in a region with little sunlight for much of the year, or don't eat a balanced diet. Most people need 400 IU a day. Be aware that vitamin D can be toxic in large doses, so do not exceed 600 IU a day.
- Eat foods rich in vitamin K (leafy green vegetables); one in three people with osteoporosis has too little of this vitamin in the blood.
- Stop smoking. Smoking increases the rate of bone resorption, probably because nicotine reduces calcium absorption in the intestines.
- Exercise regularly. Weight-bearing exercise is especially good for bones.
- Use stress-management techniques (see p. 347)

when you feel emotionally overwhelmed. High levels of adrenaline and other hormones produced during periods of stress deplete the body of magnesium and other minerals necessary for strong bones.

Treatment

If you have osteoporosis, take the preventive measures above to minimize further bone loss. It is never too late. Take care with exercise, however: while regular exercise is important in treating the condition, seek medical advice first. Prolonged or strenuous exercise may make the problem worse or even cause a fracture.

Building bone
Weight-bearing exercise increases bone density as well as muscle bulk, but you don't have to lift weights to gain this benefit. Forms of exercise in which you bear the weight of your body—such as walking, jogging, and dancing—also help build strong bones.

When to get medical help
- You have symptoms indicating osteoporosis, such as back or hip pain and loss of height.
- You have a condition or are taking a medication that increases your risk of developing osteoporosis.
- You are nearing your menopause and have had many risk factors for much of your life.

See also:
Aging, Back Pain, Eating Disorders, Menopausal Problems

Overweight

Carrying a lot more body fat than you should for your height and build increases the risk of health problems. Some people put on weight more easily than others, but with motivation and know-how, most overweight people can lose pounds and thereby improve their well-being. Scientists increasingly see regular aerobic exercise as one of the key factors to weight control.

More than two in three women and one in two men in the United Kingdom are too fat. The problem is more common in older people. Being overweight increases the risk of serious illness. Obesity—a body mass index of 30 or above (see p. 302)—increases the risk of encountering disorders such as arterial disease (heart disease, strokes, and high blood pressure), type II diabetes, osteoarthritis, digestive troubles (gallstones and constipation), certain cancers (for example, of the breast and uterus), infertility, pregnancy problems, varicose veins, and menstrual problems.

Prevention

Maintaining a healthy balance between energy expenditure and energy intake from an early age is the best insurance against excess weight gain. You need to get plenty of exercise (see p. 13) and to fuel your capacity for it by consuming all the nutrients you need for energy production and cell growth and renewal (see p. 12). Monitor your weight and adjust your exercise level and food intake to keep your weight within the optimum range for your height and build. Ensure your intake of calories is balanced: approximately 45 percent from complex carbohydrates, 25 percent from protein, and 30 percent from fats (with the emphasis on unsaturated fats).

Treatment

Whatever weight-loss programme you use, aim to lose an average of no more than two pounds per week. If you lose any more than this, you are losing muscle, not fat. A dramatic drop in the first week usually results from fluid loss.

Diet: The only sure way of arriving at and maintaining a healthy weight is to change your eating habits permanently. Crash diets that promise rapid weight loss do not provide lasting results. On any one day in the United Kingdom, three in five women and two in five men are trying to lose weight. However, 98 percent regain their lost weight—or more—within five years.

Causes of excess weight gain

Scientists are gradually learning why some people are especially prone to gaining weight. Proven factors include:
- The time of day at which food is eaten.
- How food is eaten.
- Genetic makeup—governing, for example, the levels of body chemicals that control appetite.
- Activity level.
- The body's metabolic rate—the rate at which it burns energy, which is influenced by genetic factors, activity level, types of food eaten, thyroid hormone levels, body weight, stress level, and food sensitivity.
- Increasing age. Many people also become less active as they get older, resulting in a lower metabolic rate.

➤ continued, p. 303

Are you carrying too much weight?

The most common way of judging if you are overweight is through the body mass index, which relates weight to height. Another useful indicator is your waist-to-hip ratio, since research now shows that those who accumulate fat in the abdominal area have a higher risk of encountering health problems. A body fat percentage calculation distinguishes between those who are heavy owing to muscle bulk and those whose excess weight is the result of fat deposits.

Body mass index (BMI)

BMI is calculated by multiplying your weight in pounds by 700, then dividing the result by the square of your height in inches. A BMI of:
- Under 20 suggests you are underweight.
- 20–24 is healthy.
- 25–29 means that you are overweight, but the risk to your health is low.
- 30 or more means you are obese, and your weight may be harming your health.

Use the chart (right) as a quick guide to your BMI.

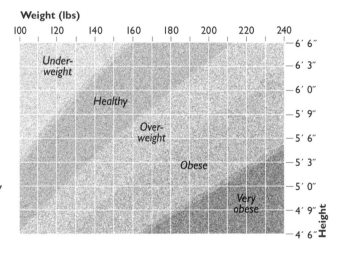

Waist-to-hip ratio

Measure your waist, with your stomach relaxed, and divide this number by your widest hip measurement. If the result is 0.85 or more (women) or 0.95 or more (men), you have an increased risk of health problems. The chart (right) provides a quick reference guide.

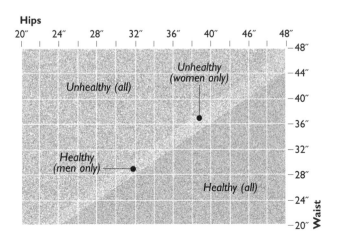

Body fat percentage

This measure of fat deposits needs expert assessment. Ask your doctor or fitness adviser to test you. Or you can measure it yourself on a special weighing scale that has this facility. It's natural for women to carry more body fat than men. The healthy range is 15 to 20 percent body fat for men and 17 to 23 percent for women.

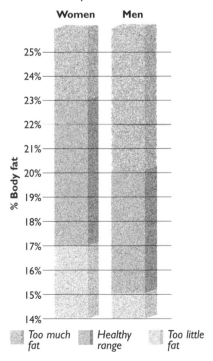

To not only lose weight but also keep it off, follow these suggestions:

■ Eat a healthy diet with enough nutritious foods. If you choose a high proportion of refined foods (those made from white flour and sugar) and foods containing a lot of saturated fats but devoid of essential fatty acids, you are wasting a good part of your caloric intake, and your body may not get enough nutrients.

■ Choose high-fibre foods, such as brown rice, nuts, legumes, seeds, fruit, vegetables, whole-grain bread, cereals, and pasta. Fibre fills you up, prevents excessive blood-sugar swings, and converts to fat much more slowly.

■ Eat foods rich in essential fatty acids—oily fish, nuts, seeds, and whole grains. These help raise your metabolic rate.

■ Eat five or six small meals daily, rather than only one or two large ones. When you are hungry, your metabolic rate slows and you burn fewer calories to conserve energy, since your body expects famine. Eating small amounts frequently helps prevent this.

■ Have fresh foods whenever possible, because many processed foods have had their essential fatty acids destroyed to increase shelf life. They also tend to contain high levels of sugar and salt.

■ Cut down on fatty foods. These include butter, margarine, whole milk, cream, cheese, and foods made with these ingredients, as well as fatty cuts of meat. Instead, choose fish, poultry, lean cuts of meat (trim off any visible fat); buy semi-skimmed (1.7 percent fat) or skimmed milk; and avoid fried foods.

■ Have a snack, such as a piece of fruit, about 90 minutes before a meal to prevent hunger from making you overeat during the meal.

■ Eat peppers and other spicy foods occasionally. These increase the metabolic rate for three or four hours afterwards.

■ Drink green tea: studies suggest it can promote weight loss.

■ Eat less food with added sugar, such as cakes, sweets, biscuits, many canned vegetables and fruit, many canned or bottled drinks, and sweetened breakfast cereals. Take special care to avoid foods containing high levels of both sugar and fat, since eating sugar encourages your body to store fat.

■ Keep your alcohol intake low. Alcohol contains a lot of calories, and drinking alcohol before a meal is a well-known appetite stimulant.

■ If you suspect that food sensitivity is behind your weight problem, try an elimination diet (see p. 214). A food sensitivity can lead to fluid

Chromium power
Foods such as meat, dairy products, egg yolks, and whole grains contain chromium, which assists weight loss because it helps regulate the release of energy from body cells.

retention because once the offending food proteins are captured by antibodies, they irritate the blood-vessel walls, which allows clear fluid to seep from the blood into the tissues. It can also trigger a craving for sweet, starchy food, and, perhaps, slow the metabolic rate. Wheat, milk, nuts, and shellfish are common causes of food sensitivity reactions.

From the chemist

■ The antioxidants in multiple vitamin and mineral supplements help reduce the health risks arising from an increase in the level of fats in the blood, which is apt to occur when fat stores are utilized while you are losing weight.

■ Evidence suggests that taking a chromium supplement helps obese people with a high risk of diabetes to lose weight. Foods rich in chromium may not always provide enough of this mineral. Chromium supplements may also help prevent blood-sugar fluctuations, which can encourage binge eating. Consult your doctor before taking this mineral if you are receiving insulin treatment.

Yoga for regulating weight

Practising yoga can help to restore and maintain healthy body weight by toning muscles and enhancing body awareness and willpower. The following exercises also help prevent or reduce a distended abdomen.

1 Lie flat on the floor with your legs together and your arms down by your sides. Inhale, raising both legs as high as possible. Then exhale, bringing your legs down again. Repeat this up to 10 times. If your abdominal or back muscles are weak, press your palms down as you lift your legs, and slightly bend your knees. Or try raising one leg at a time, but only as far as is comfortable. Your lower back should remain flat on the floor. Avoid this posture if you have lower back pain.

2 Stand with knees bent, your legs wide apart, and your hands on your thighs. Exhale completely, then, without inhaling, pull the abdomen in and up, expanding the chest at the same time. Then relax your abdomen. Suck in the abdomen again, and pump it in and out (aim for 10 to 18 times) until you need to inhale. Take a normal breath, then exhale and repeat.

Exercise: Physical activity is vital to any weight-loss programme, for a variety of reasons. Most important, it burns calories that would otherwise be stored as fat. An hour-long workout burns 200 to 400 calories and speeds up your metabolic rate for 24 to 48 hours afterwards. Exercising with weights builds muscles, and muscle cells burn more energy, even when you are inactive, than fat cells. Strenuous physical activity also enhances your sense of well-being, partly by raising the level of mood-lifting chemicals, such as endorphins, and so helps you avoid eating too much to make yourself feel better.

Get your doctor's approval before starting an exercise programme, especially if you are new to exercise, have a health problem, or are considerably overweight.

- Exercise four or five times weekly for about 30 minutes each session. Include aerobic exercise, such as brisk walking, swimming, cycling, or aerobics classes.
- Put your exercise sessions in your diary to make sure you save time for them.
- Experiment with different sports and classes, such as tennis lessons, salsa dancing, and swimming, to find types of exercise that you enjoy.
- Ask a friend to join you. You will benefit from the stimulus of having company, and you'll be less likely to drop out if you know this will disappoint someone.
- Take every opportunity to be more active throughout the day. For instance, climb stairs instead of using the lift, and walk to the store instead of taking the car. The extra calories you burn will soon add up.

Walk away pounds
You don't have to do a vigorous workout to start burning fat. A regular brisk walk will help you lose weight.

Stress management: If food is your main source of comfort or security, you may find that you undermine your desire to lose weight at stressful or anxious times by overeating. Make a list of alternative, enjoyable activities to take the place of eating when times are tough. You might include having a scented bath, phoning a friend, and taking a walk (which is also good exercise).

Aromatherapy: Massage with essential oils helps to keep the skin smooth and supple during and after weight loss. Useful oils include celery, fennel, juniper berry, lavender, lemon, orange, oregano, and rosemary. Avoid rosemary oil in the first 20 weeks of pregnancy, and fennel, juniper berry, and oregano oils throughout pregnancy.

- Add six drops of your chosen essential oil to a warm bath.
- Alternatively, massage your feet, legs, and abdomen—or ask a friend to give you a massage—using four drops of your chosen oils in a tablespoon of sweet almond or grapeseed oil.

When to get medical help
- You have a body mass index of 30 or above.
- The weight–reduction strategies suggested here do not work for you.
- Carrying excess weight is or may be creating health problems for you.
- You have sudden, unexplained weight gain.

See also:
Bloating, Diabetes, Eating Disorders, Food Sensitivity, Stress

Pain

Never ignore pain; it is a sign that something is harming you. The cause may be an external injury, such as a burn, or internal damage or disease. Once you know the source of the pain, you can often use natural remedies to ease the discomfort, while seeking medical advice, if necessary, to address the underlying cause.

DID YOU KNOW?

There is a scientific reason why the pain goes away when a parent rubs a child's "hurting place". The pain signals from the injured area are blocked by the pressure signals produced by massage.

Without pain to warn us about dangerous situations, human beings would probably not survive for long. For example, as children we need to learn that the pain from a wound is a sign that in the future we need to avoid the situation that caused it. Similarly, the discomfort from a sprain helps us understand that we need to rest the injured part. We therefore learn to prevent pain by protecting ourselves from its sources.

The perception of pain

Everyone's pain threshold—the point at which a stimulus becomes painful—is the same, but each person's tolerance of pain varies. What seems a minor discomfort to one individual may be experienced as troublesome, or even agonizing, by another. People's responses also vary: someone brought up to suffer in silence will probably admit to less pain than someone from a background in which expressing feelings freely is accepted behaviour.

You may also perceive the same level of pain differently at different times. A toothache that seems unbearable in the middle of the night may be unnoticed while you are enjoying a movie or cheering your football team. This may be either because your mind is occupied or because your enjoyment or excitement raises your body's levels of natural painkillers called endorphins (see "Natural painkillers", facing page). Similarly, if you are in a car accident, you may not notice any pain from your injuries until later, when you are clear of the crash scene. The brain can suppress the sensation of pain until a crisis has passed and you are better able to cope.

Treatment

Addressing the underlying cause is the key to alleviating pain, so it is important to seek your doctor's advice about your symptoms. However, with many conditions it may take time for any type of treatment to begin to have an effect on your symptoms. In such cases, as with conditions that have no reliable cure, self-help measures can ease discomfort, help alleviate the mental stress associated with pain, and promote healing.

Stress and pain

Whereas acute stress resulting from trauma may temporarily block pain (see "Natural painkillers", opposite), chronic physical or emotional stress can cause pain and also affect your ability to cope with it. By heightening your response and lowering your tolerance to pain, stress makes existing discomfort worse. For example, if you are worried that pain results from a serious disease, it may be hard to bear, but once you know that this is not the case, the pain may seem much more tolerable. Long-term stress also causes physical tension, which makes you prone to headaches and minor injuries, such as strained muscles.

➤ continued, p. 308

Pain pathways

The mechanisms enabling us to sense pain are complex. What you feel depends on many factors, including how active you are and whether or not you are already feeling stressed.

Painful messages

Skin receptors report a painful stimulus by sending electrical messages along sensory nerve fibres to the spinal cord. From there they travel to the brain, where the signals are interpreted as the sensation of pain. Sometimes you can feel pain in one part of the body even though the area affected is actually elsewhere. This is called referred pain, and one example is angina—pain originating from oxygen starvation of the heart muscle but felt in the upper arm, shoulder, or neck.

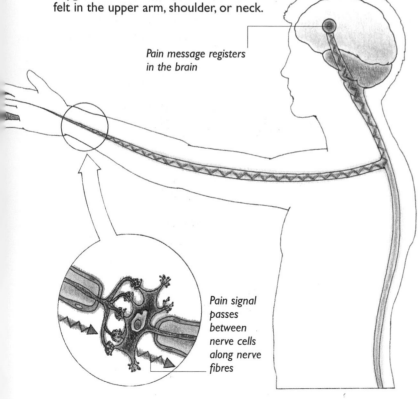

Pain message registers in the brain

Pain signal passes between nerve cells along nerve fibres

Natural painkillers

At times of acute stress, the brain produces chemicals, called endorphins, that block the action of the chemical messengers that transmit pain signals between nerve cells. This reduces the perception of pain.

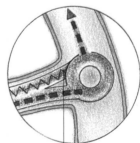

Endorphin blocks pain signal in brain

The "gate control" theory

Pain may be partly or completely blocked by a process known as "gate control". Nerve messages carrying information from the skin about other sensations can sometimes bar the transmission of pain messages. This happens when nerve fibres carrying "non-pain" messages prevent nerve fibres carrying pain messages from relaying these signals up the spinal cord. It explains why therapies involving touch, pressure, heat, and cold can relieve pain.

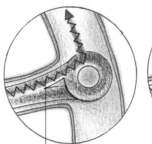

Normal pain signal passing through "pain gate" in spinal cord

Pain signal blocked by other sensations

Relaxation: Counter the muscle tension that can exacerbate chronic pain by learning some relaxation techniques. Relaxation classes and cassettes, breathing exercises, yoga, meditation, visualization, and self-hypnosis can all help. Massage and aromatherapy are good for relaxing tense muscles; if you can't consult a professional, ask a friend or relative if he or she would be willing to learn the basic techniques.

Exercise

- A daily half hour of brisk exercise raises your body's levels of endorphins (natural painkillers in the blood) for several hours. To spread the effects of a raised endorphin level over a longer time, exercise for 20 minutes in the morning and another 20 minutes in the late afternoon.
- To lessen localized pain, exercise the affected part to increase local circulation. For example, gently bending and stretching a knee that is painful because of arthritis can assist the removal of inflammation-producing chemicals in the bloodstream.

Diet

- Limit your intake of animal proteins, which can increase the production of pain-promoting prostaglandins (hormone-like substances).
- Eat more foods containing salicylates, the family of natural painkillers from which aspirin was first made. These include most fruits (preferably unpeeled) and many vegetables, spices, nuts, and seeds. Some people are sensitive to salicylates, so seek medical advice if you notice unusual symptoms following a change in diet.

Pain-relieving pepper
A topical preparation containing capsaicin, a substance in cayenne pepper, can relieve arthritis or nerve pain. When first applied to the skin, it burns—an indication that it's working. Check with your doctor before using.

Healing touch
The contact of caring hands—for example, a friend giving you a massage or holding your hand—can be comforting if you are in pain.

Naturopathy

- Injuries to muscles, ligaments, or other soft tissues—such as a sprained ankle—can be improved by applying a cold compress to reduce swelling and inflammation and ease pain. Make a compress by wringing out a small towel or cloth in cold water. Lay this over the painful area and bind it in place with a bandage. Leave the compress in place for a half hour at a time. During the first two days after an injury, repeat as often as necessary to relieve pain.
- Some other types of pain, including menstrual cramps, may respond better to heat. A hot pack improves blood flow to the affected area, which helps clear away the inflammatory chemicals that contribute to pain. You can buy

a pack to heat in the oven or microwave, or use a towel wrung out in hot water or a hot-water bottle wrapped in a towel.

Herbal remedies: Drink one or two cups daily of the following herbal teas, according to your symptoms:

- Black cohosh, wild yam, or meadow-sweet: for pain from muscle tension.
- Chamomile: for relaxation and relief of pain caused by tension.
- Passionflower and St. John's wort: for back-ache resulting from tension.
- Peppermint tea: for pain arising from muscle spasm.
- Yarrow, wild yam, and meadowsweet: for pain arising from inflammation.

Caution: For safety concerns, see pp. 34–37.

Homeopathy

- Arnica: for sprains, and aching muscles.
- Belladonna: for throbbing pains, especially in the head, and when the area looks red.
- Bryonia: for pain that worsens with the slightest movement and for pain on coughing.
- Hypericum: for nerve injuries, especially when there are shooting pains.
- Magnesia phosphorica: for neuralgia and abdominal pains, such as cramps, that are better with warmth and gentle rubbing.
- Rhus toxicodendron: for sprains and strains with stiffness that eases after gentle movement.

Flower essences: Long-term pain can lead to depression, and this can add to your difficulties. Choosing flower essences may help relieve

A natural analgesic
Meadowsweet has pain-relieving properties similar to those of aspirin. However, whereas aspirin can irritate the stomach, meadowsweet has a soothing effect.

emotional distress. Choose the essence according to your state of mind. For example:

- Olive: for those who feel drained by a long period of illness.
- Sweet Chestnut: for those in despair, who are at the limits of their endurance.
  - Willow: for those who feel bitter or resentful about their illness.

Other therapies: Acupuncture, reflexology, biofeedback, meditation, yoga, and such healing therapies as reiki, can help those with chronic pain, as can counselling for stress. Some types of pain, such as lower-back, respond well to osteopathy and chiropractic.

When to get medical help

- You do not know the cause of pain.
- Pain continues despite home treatment.
- You have swelling or redness at the site of the pain, breathlessness, fever, or difficulty moving the affected part.

Get help right away if:

- You have severe pain of any kind.
- The pain is associated with numbness, tingling, or muscle weakness.

See also:
Abdominal Pain, Arthritis, Back Pain, Facial Pain, Headache, Menstrual Problems, Muscle Aches and Stiffness, Neck and Shoulder Pain, Raynaud's Syndrome, Sore Throat, Strains and Sprains, Toothache

Palpitations

The sensation of your heart pounding, beating fast, missing a beat, or being out of rhythm can be alarming. Because it can indicate a serious heart disorder, this symptom always needs to be brought to your doctor's attention, but there is often a simple cause. Once your doctor has eliminated an underlying disorder, using appropriate natural remedies can help control palpitations.

Most of the time we are unaware of our heartbeat. However, when the heart works harder than normal, its pumping action becomes readily apparent. During and after exercise it beats faster to pump more blood around the body; this provides more oxygen and nutrients to the working muscles and removes carbon dioxide and other waste substances. Anxiety also increases the heart rate, because high levels of adrenaline prepare the body for a possible "flight or fight" response to assumed danger.

Other causes of a racing heart with a regular beat include fever, caffeine-containing drinks (which raise the adrenaline level), and food sensitivity. Certain drugs (for example, those used for allergic rhinitis), an overactive thyroid gland, or arterial disease may also be responsible. Only in rare cases is a rapid heartbeat a sign of a heart disorder that may be life-threatening.

One type of very irregular heartbeat, atrial fibrillation, affects one in 50 people over the age of 65. It may be caused by thyroid disease or arterial disease and can be triggered by alcohol. This condition is a medical emergency. Another less serious cause of an irregular heartbeat is an incompetent (prolapsed) mitral valve in the heart. This condition is present in up to one person in 20, and most commonly affects women. It requires expert assessment, but no specific treatment is usually needed.

Prevention

To diminish the risk of palpitations and at the same time reduce the likelihood of heart disease, keep fit with regular exercise, don't smoke, use stress-management strategies (see p. 347), eat a well-balanced diet, and maintain a proper body weight for your height (see p. 302).

Treatment

Palpitations manifested as an occasional rapid pulse or skipped beat usually need no treatment, unless they indicate underlying disease or are accompanied by dizziness or certain other symptoms (see box, facing page). But in all cases, it is important to get a medical opinion promptly.

Exercise: Aerobic exercise that provokes an uncomfortably rapid heartbeat indicates that your heart and lungs cannot supply your body with enough oxygenated, nutrient-enriched blood. An appropriate, graduated programme of aerobic exercise over a period of months should boost lung capacity and heart strength. Be sure to consult your doctor before intensifying your exercise level, especially if you have any cardiovascular risk factors, such as being overweight or a smoker.

Stress management: If you are suffering from stress-related palpitations, make time for regular relaxation and massage. Using stress-management

Nourish your heart
Folate, magnesium, and vitamins B$_6$ and B$_{12}$ are essential for healthy heart rhythm. A diet that includes meat, shellfish, beans, peas, dark-green leafy vegetables, and whole grains will supply these nutrients.

Herbal remedies: Do not take herbal remedies for palpitations except on the advice of a doctor experienced in the use of herbs. One remedy that he or she might recommend is hawthorn, which improves the pumping ability of the heart. Caution: For safety concerns, see pp. 34–37.

Aromatherapy: Use calming oils, such as lavender, sweet marjoram, bitter orange, or neroli oils. Sprinkle a few drops onto a handkerchief and inhale, or use in a vaporizer. Rub two drops of bitter orange oil on your chest.

Other therapies: A qualified fitness instructor can work out a progressive exercise programme to increase your cardiovascular tolerance to exertion, if lack of fitness is at the root of your problem. Biofeedback can help you manage your physical responses to stress, if anxiety is the cause. Regular yoga practice, with deep breathing and relaxation exercises, or meditation may also help regulate the action of the heart.

strategies will spare your heart from the excessive stimulation of a constantly high adrenaline level in the blood (see p. 347).

Diet
- Eat foods rich in folate, magnesium, and vitamins B$_6$ and B$_{12}$ (see illustration).
- Lose any excess weight. If you reduce the amount of body tissue the heart has to supply with blood, your heart will have to work less.
- If you suspect food sensitivity, take steps to identify the culprit food so that you can omit it from your diet (see p. 214).
- If drinking caffeine makes your heart race and prevents you from sleeping, try decaffeinated coffee and tea instead. Alternatively, drink chamomile tea, which is a calming, caffeine-free, beverage.
- If alcohol causes palpitations, you should cut your intake.

When to get medical help
- You have unexplained palpitations.
- You want to begin an exercise programme and have a history of palpitations.
- You intend to use herbal remedies for palpitations.

Get help right away if:
- You have chest pain, a feeling of faintness, dizziness, a sudden change in vision, nausea, shortness of breath, or confusion.
- Your palpitations are very irregular, change in some way, or are associated with a pulse faster than 140 beats per minute.

See also:
**Anxiety,
Arterial
Disease,
Stress**

Pelvic Inflammatory Disease

This disorder, also known as PID, is a potentially serious infection of the internal female reproductive organs, and its damaging effects on the fallopian tubes can cause female infertility. However, there is much you can do to prevent it and, if you have the disease, to increase the effectiveness of the treatment prescribed by your doctor.

Pelvic inflammatory disease may involve the cervix, uterine lining, fallopian tubes, ovaries, or the tissues surrounding the uterus, bladder, and bowel. Many kinds of bacteria, including chlamydiae and gonococci, can cause the disease. PID may develop after sexual intercourse with an infected partner, or, less frequently, after child-birth, miscarriage, or abortion. Those who use an intrauterine device (IUD) have a higher than average risk of PID, as do young, sexually active women, especially those with multiple partners. Sometimes the cause cannot be found. Without adequate treatment, the infection can produce recurrent pain and can block the fallopian tubes—leading to an increased risk of infertility, ectopic pregnancy, or premature birth.

Symptoms
The infected areas are sometimes tender and inflamed, but symptoms may be nonexistent, minimal, or vague. You may have painful, heavy, or irregular periods, bleeding between periods, an abnormal vaginal discharge, backache, fever, or nausea. You may also experience lower abdominal pain or a dull ache, whether intermittent or steady. Sexual intercourse may be painful.

Prevention
You can reduce the chances of catching any sexually transmitted disease by having an exclusive sexual relationship with one healthy partner. There is no risk if you were both previously celibate. Otherwise, you can reduce your risk of infection by using a condom in addition to any other contraceptive method. This is especially important if you have recently had a miscarriage or an abortion. If you do not use condoms, arrange an annual test for chlamydia infection and gonorrhoea (as well as HIV). If you have an IUD, you may want to consider an alternative method of contraception.

Back arching
Based on a yoga posture, this exercise alleviates back pain, tones the pelvic region, and strengthens the uterus. Arch your back upwards, as shown, for a count of five, then inhale and slowly drop your back down, with your head stretched back. Repeat several times. Do not do this exercise if you have back pain from spine, muscle, or joint problems.

Treatment

Both you and your sexual partner should seek medical treatment and should abstain from unprotected intercourse until you are both free from infection. Use the following natural remedies alongside any prescribed treatment.

Heat: If you experience abdominal pain, rest until symptoms subside. Hold a covered hot-water bottle or an electrically heated pad against your abdomen to boost the circulation of blood and lymph and thus ease pelvic congestion of blood, lymph, and tissue fluid. Alternatively, take a warm bath to which you've added six to eight drops of rosemary, cypress, or peppermint oil. Later, when you feel better, try sitting in a warm bath, then a cold bath, and then a warm bath for 10 minutes each time.

Herbal remedies: Both echinacea and astragalus enhance the functioning of the immune system. Take either one as a tea, tincture, or tablet. Chamomile and hyssop teas may have relaxing and pain-relieving effects.
Caution: For safety concerns, see pp. 34–37.

Diet: Increase your resistance to infection with a healthy diet that includes foods rich in zinc, folic acid, flavonoids, and vitamins A, B_6, D, and E (see p. 12). Avoid refined foods, and if your consumption of alcohol and caffeine is high, cut down, since both substances can depress your immune system and your levels of B vitamins and zinc. If you are taking antibiotics, eat live yogurt daily to help prevent adverse effects on the digestive system.

Strengthening the abdomen

Lie on your back with knees bent and feet apart. Put your hands over your ribs, then inhale and exhale deeply. Now stretch your hands out at your sides, clench your buttocks tight and draw in your abdominal muscles. Continue until the small of your back is pressed against the floor and then relax for a few moments before repeating. Perform this exercise five times running, twice a day, building up to 20 times running, twice a day.

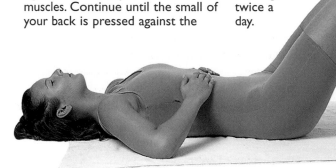

Exercise: If you feel well enough, do aerobic exercise regularly to help strengthen the immune system and stimulate the pelvic circulation, which allows more infection-fighting cells to reach the infected areas. Exercises for the abdomen (see panel, above) and back help reduce congestion in the pelvis. "Crunches" are effective, as are yoga exercises.

When to get medical help
- In all cases in which symptoms suggest pelvic inflammatory disease.

See also:
Abdominal Pain, Fertility Problems, Menstrual Problems

Pinworms (Threadworms)

Infestation of the intestines by these tiny worms is a common cause of anal irritation among children. Although the idea of having these parasites is unpleasant and, perhaps, embarrassing, threadworms pose little risk to health. Since the infestation is easily transmitted, all members of the family should be treated at the same time.

Testing for threadworms

You can often diagnose the condition by using the transparent tape method: Secure a piece of tape sticky-side up just outside the anus at bedtime. In the morning you may see the tiny white worms stuck to the tape.

Threadworms look like half-inch lengths of white cotton thread. The females emerge to lay their eggs at night, and their movements cause the characteristic night-time itching. The eggs are too small to see with the naked eye, but if affected children scratch, they pick up eggs under their fingernails. If they later put their fingers to their mouths, the eggs readily travel from there to the intestines, where they mature, hatch, and start the cycle again. Dislodged eggs can live for some time away from the body, in bedding and on floors or other surfaces.

Treatment

General hygiene: Take these measures to prevent reinfestation if you have a case, and see that other affected household members do so as well.
- Wear cotton gloves in bed. If you scratch while half-asleep, the gloves will keep you from picking up eggs under your nails.
- Keep the nails cut short.
- Wash your hands and scrub your nails after using the bathroom and before meals.
- Wash nightclothes, gloves, and bed linen daily, at as high a temperature as possible.

Eliminating infestation: Make a pint of herbal tea using one part of wormwood, one part of peppermint, and one part of aniseed. Sweeten with honey, treacle, or fruit juice. Drink a cup of this mixture before breakfast, then another cup two or three times during the day, before meals. Use the treatment for a week, then repeat after a break of two weeks. Do not use this treatment if you are pregnant. Consult a medical herbalist about the dose for a child.
Caution: For safety concerns, see pp. 34–37.

Dietary measures
- Grate a carrot and eat it mixed with a tablespoon of ground pumpkin seeds for breakfast.
- Eat raw onions, apples, and coconut, or add cayenne pepper and fresh or dried thyme to your meals to help kill worms.
- Add one or two cloves of crushed garlic to a little warm milk or a teaspoon of honey, and eat half an hour before breakfast.

Symptomatic relief: Apply a salve to the anal area at bedtime. Use calendula (marigold) ointment. Or mix two drops of lavender, eucalyptus, sweet thyme, or tea tree oil with two ounces of warmed petroleum jelly, and allow to cool.

When to get medical help
- The problem doesn't respond to the above treatments within a few days.

See also:
Anal Problems

Pregnancy Problems

Most women remain healthy throughout pregnancy, but the enormous changes taking place in the body are capable of causing numerous problems and discomforts. Natural approaches, in addition to your regular antenatal care, are especially useful at this time, when only essential medications advised by your doctor should be taken.

The nine months of pregnancy have profound effects on a woman's body that can sometimes cause problems. The enlarging uterus may press on other internal organs, possibly leading to such disorders as heartburn, constipation, haemorrhoids, varicose veins, and stress incontinence. The growing weight of the baby, placenta, amniotic fluid, increased blood volume, and extra fat stores can cause fatigue and back pain. The baby's requirement for iron, calcium, and other minerals can result in maternal anaemia and an increased risk of dental problems.

In addition, the body's changing metabolic, hormonal, and immunological states make diabetes, candida infection, fainting, itching, nausea and vomiting, gum problems, and leg cramps more likely.

A serious condition called pre-eclampsia may cause elevated blood pressure, migraine, fluid retention (which results in puffiness of the hands, face, and ankles), and protein leakage into the urine. If untreated, it can affect the supply of oxygen and nutrients to the baby and may cause miscarriage.

Prevention

Many minor problems of pregnancy can be averted or minimized by ensuring that you and

Pampered pregnancy
When you are expecting, focusing on your health is not self-indulgence but a necessity. Getting enough rest and gentle exercise and eating a nutritious diet are among the most important ways of giving your baby a good start in life.

your partner are in optimum health before you conceive. Follow the advice on pages 202–205 for improving your health at this time. Once you know you are pregnant, adhere to the following guidelines for a naturally healthy pregnancy.

Diet: Continue to eat a healthy diet, with plenty of foods rich in folic acid, calcium, iron, essential

Antenatal care

It is essential that you have regular checkups by an obstetrician or by another appropriate specialist to ensure that any problems with you or your baby can be identified and treated as early as possible.

fatty acids, and vitamins B_1, B_2, C, and D. The average woman needs about 200 extra calories per day—roughly equivalent to a small tomato sandwich. Well-balanced meals and snacks will increase the chances of your baby having a normal birth weight, and, research suggests, will reduce the likelihood of the child suffering from diabetes, obesity, high blood pressure, and arterial disease as an adult. A good diet will also help protect you from deficiency of essential nutrients.

- Eat at least five helpings daily of fruit and vegetables. Many of these contain natural salicylates, which are chemically similar to aspirin and make the blood less likely to clot. It's thought that a high intake of salicylates might reduce your risk of pre-eclampsia and miscarriage. Some studies already suggest that a daily dose of aspirin (which should be prescribed by a doctor) lessens the possibility of miscarriage in a few women at risk of pre-eclampsia and in women with antiphospholipid syndrome (APS), an immune-system disorder that interrupts the baby's blood supply by causing blood clots to form in the blood vessels of the placenta.
- Avoid or limit consumption of liver, liver products, and vitamin-A-enriched foods. Too much vitamin A may cause birth defects.

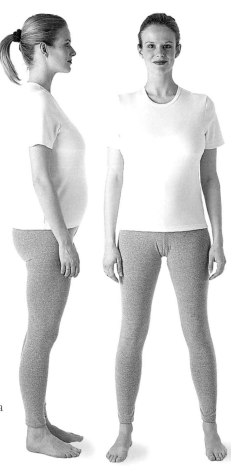

Avoiding back trouble
To reduce the risk of back strain and other aches and pains when standing, keep your feet shoulder-width apart, your shoulders relaxed, and your buttocks tucked in.

Abdominal pain

Mild tightening of the abdomen commonly occurs from about the fifth month of pregnancy onward. However, this should not cause more than slight discomfort. Abdominal pain at any stage of pregnancy should be taken seriously, especially if accompanied by vaginal bleeding. It may indicate a problem requiring urgent medical attention.

- Don't eat any food you are sensitive to. Also avoid having a large amount of any protein food at one time, as this might increase a baby's risk of developing a sensitivity to that food.

From the chemist
Protect your nutrient stores from becoming depleted by taking a multiple vitamin and mineral supplement designed for pregnancy. Ask your doctor for a specific recommendation.

Exercise: Exercise moderately every day.
- If you have already had a miscarriage or premature baby, or are at risk this time, don't play contact sports or engage in other jarring exercises, such as high-impact aerobics and jogging.
- Do pelvic-floor exercises (see p. 251) several times daily. These make urine leaks less likely, give you more control of the baby's descent through the vagina during the second stage of labour, and lower the risk of postnatal stress incontinence.

Rest: Get sufficient rest, and take maternity leave sooner rather than later. Research indicates that

Yoga-based exercise programme

Specially modified yoga exercises help reduce the normal aches and pains of pregnancy and prepare your body for childbirth. If you have back pain, check with your doctor before trying them.

Pelvic tuck-in
This exercise releases tension in the lower back.

1 Kneel on all fours with your knees and shoulders about 12 inches apart.

2 Drop your head and tuck in your buttocks, allowing your back to arch. Hold for a few seconds and relax. Repeat 5 to 10 times a day.

Pelvic release
This exercise releases stiffness in the pelvic area and helps widen the pelvic outlet.

1 Sit on your heels with your knees apart and toes pointing inwards. Keeping your shoulders relaxed, raise your arms above your head. Breathe steadily.

2 Bend forwards from your hips and rest your forearms on the floor or on a large pillow. Keep your buttocks as close to your heels as possible. Hold for 30 seconds.

Inner thigh stretch
This exercise releases tension in the inner thighs and helps reduce ankle swelling.

1 Lie down so that your buttocks are touching a wall and your legs are up against the wall. Breathe deeply and allow your legs to drop apart and your lower back to relax towards the floor. Bring your arms over your head and rest them on the floor. Relax and breathe deeply.

2 Bend your knees and bring your heels together as close to your body as you can without strain. Press your knees gently towards the wall. Hold for 30 seconds, then slowly roll onto your side to come up.

women who remain at work during most of their pregnancy have a higher risk of premature labour. Try to avoid exerting yourself and standing for long periods, since both can induce miscarriage.

Light: Exposure to bright sunshine boosts the body's production of vitamin D (which protects the bones) and serotonin (a "feel-good" neurotransmitter that helps you cope with stress). Go outside for 10 minutes at midday—longer at other times—to get sunlight on your skin, but take care not to burn.

Smoking: Stop smoking, or at least cut down. Chemicals in tobacco diminish the flow of blood to the placenta and reduce the supply of oxygen and vital nutrients to the baby. The more you smoke, the more likely your baby is to be born prematurely and the higher the risk of sudden infant death syndrome (SIDS, also called cot death). Be aware that low-tar cigarettes produce higher levels of carbon monoxide, which further reduce a baby's oxygen supply.

Tooth and gum problems: Both are more likely during pregnancy, so take extra care to protect yourself with good oral hygiene (brushing twice a day along with a session of flossing) and an adequate intake of calcium-rich foods.
Avoiding infection: Do your best to ward off infections, which can harm the baby.
- Flu: Keep away from infected people and crowded places during an outbreak. Wash your hands often.
- Toxoplasmosis: Wear rubber gloves when

Vaginal bleeding

Slight blood loss occurs in 1 out of 10 normal pregnancies without serious cause. However, vaginal bleeding can be a symptom of an impending miscarriage, an ectopic pregnancy, or a blood disorder. In all cases of bleeding, even spotting, notify your doctor—do this immediately if you bleed heavily, have abdominal pain, or feel faint. If you have noticed bleeding of any kind, avoid physically strenuous activities, including lifting heavy weights, at least until a few days after the bleeding stops, and don't stand for long periods.

---

gardening and handling raw meat. Wash fruit and vegetables well. Don't eat undercooked meat, and don't empty cat litter (have someone else disinfect the litter tray daily) or handle a sick cat.
- Listeria bacteria: Don't eat pâtés and deli meats, blue-veined and soft, mould-ripened cheese; feta cheese; soft ice cream; precooked poultry that's been kept warm for long periods; prepackaged salads (unless well washed); store-bought refrigerated meals (unless reheated thoroughly).

Treatment
Always consult your doctor about any troubling symptoms that you experience during pregnancy, and mention any home remedies that you are considering.

Sound sleep
If you find it difficult to fall asleep because you can't get comfortable, try lying on your side with the upper knee resting on a pillow.

Semi-squatting
If you use this position during labour it can help to open your pelvis and ease the baby's passage. You should practise it regularly even if you aren't sure you'll use it. Stand with your feet turned out, about 18 inches apart. Bend your knees to lower your body as far as is comfortable. If necessary, hold on to a chair or have your partner support you, as shown.

Nausea and vomiting: This is one of the most common problems in pregnancy. It typically diminishes after about 12 weeks, but these unpleasant symptoms continue longer for some women. Prevent attacks by eating smaller, more frequent meals and staying well rested. Avoid fatty foods and other foods or drinks that seem to provoke attacks. When nausea or vomiting occurs, try the following:

- Eat a little bread or a cracker.
- Sip some water.
- Drink a cup of ginger or chamomile tea, sweetened with honey, or chew fresh ginger.
- Press the acupressure point HP 6 (see p. 282).

Fatigue

- Rest with your feet up for at least a half hour daily, or take a daily nap.
- Get some whole-body exercise each day; lack of activity will make you more tired than a good balance of exercise and rest.

Carpal tunnel syndrome: This causes numbness and tingling in the middle three fingers, generally as a result of pressure on the nerves in the wrist due to fluid retention.

- Don't dangle your bent wrist in an attempt to relieve discomfort. If you do this when asleep, put on a wrist splint before going to bed.

- Take a five-minute break every half hour if your symptoms arise from working at a keyboard (see also the box on p. 325).
- Every hour, exercise your arms and wrists for two minutes to increase blood circulation.

Other therapies: Seek advice on backache and other postural problems from an Alexander technique teacher.

When to get medical help

Ask your doctor about all unexplained symptoms. In particular, report the following:
- Severe nausea or repeated vomiting.
- Itching lasting more than a few days.
- Abnormal vaginal discharge.
- A bad cold, the flu, diarrhoea, or a fever.

Get help right away if:
- You have abdominal pain, vaginal bleeding, or sudden and severe swelling of the hands, feet, or ankles.
- You have dizziness, severe headaches or shortness of breath, chest pain, or a sudden increase in frequency of urination.
- You have yellowing of the skin or the eyes.
- You have rubella, chickenpox, or a genital herpes infection.

See also:
Anaemia, Back Pain, Candida Infections, Childbirth, Constipation, Fatigue, Gum Problems, Haemorrhoids, High Blood Pressure, Incontinence, Indigestion, Leg Cramps, Migraine, Nausea and Vomiting, Swollen Ankles, Varicose Veins

Premenstrual Syndrome

During the second half of the menstrual cycle, many women experience a number of physical and/or emotional changes known collectively as premenstrual syndrome (PMS). Symptoms usually begin at or after ovulation (up to about 14 days before the next period) and gradually increase in severity until the onset of menstruation.

DID YOU KNOW?

The folk name of *Vitex agnus-castus*, a herb commonly used for menstrual problems, is chasteberry, which suggests that its effects on the reproductive system have long been known.

About 80 percent of ovulating women suffer from premenstrual syndrome, or PMS, at some time in their lives. It may be masked by taking contraceptive pills that contain both oestrogen and progestogen, which override the body's normal hormonal cycle. Although the condition is extremely common, only one in 20 women is so severely affected that PMS greatly disrupts her life. As many as 100 symptoms have been ascribed to PMS. They fall into four categories:

- **Type A (for Anxiety):** anxiety, tension, irritability, and mood swings.
- **Type C (for Cravings):** cravings for foods containing sugar and other refined carbohydrates, increased appetite, faintness, dizziness, and headaches.
- **Type D (for Depression):** tearfulness, forgetfulness, insomnia, and confusion.
- **Type H (for Hydration):** fluid retention, with weight gain, swollen ankles and fingers, bloating, and breast tenderness.

PMS was long thought to result from an abnormal balance between oestrogen and progesterone, but it now seems more likely that some women are simply oversensitive to the normal hormonal fluctuations of the menstrual cycle. Another possible cause of PMS is that premenstrual hormonal changes lead to changes in diet, resulting in nutritional deficiencies which trigger the symptoms.

Prevention

Help even out mood swings by exercising daily and practising relaxation techniques. Avoid or limit alcohol and caffeine-containing drinks, since they may worsen tension and depression. Also limit your intake of processed foods, animal fats, and sugar, but increase consumption of fruit, vegetables, and complex carbohydrates, such as wholegrain bread and pasta. Eat plenty of foods containing calcium, magnesium, and vitamins B_6 and E (see p. 12).

Treatment
Diet

- Eat more foods that contain helpful plant hormones (see illustration).
- Eat frequent small meals, based on whole grains, fruit, and vegetables, to help stabilize the blood-sugar level and reduce mood swings and cravings.
- Reduce your salt intake a few days before symptoms usually begin.

The soy factor
Soybeans and soy products, such as tofu and soy milk, contain plant isoflavones, which can help balance hormonal activity and ease PMS symptoms.

Supplements

- Two days before the onset of symptoms until the second day of your period, take supplements of vitamins B_6 and E, magnesium, and calcium, or a multiple vitamin and mineral formula specifically for PMS.
- Boost chromium intake (200 milligrams a day) to help regulate blood-sugar level and possibly control cravings for sweet foods.

Herbal remedies

- Chasteberry, along with dong quai and black cohosh, can help regulate hormone levels for all types of PMS. Take these herbs as tablets, tincture or tea. Supplements containing combinations of herbs for PMS are widely available.
- Teas made from herbs that help prevent a high level of oestrogen by improving liver function, such as dandelion root or milk thistle, may also be helpful.
- Take evening primrose oil (1,000 milligrams twice a day) two days before the usual onset of symptoms until day two of your period, to help regulate any hormonal imbalance and relieve breast tenderness.
- For Type A, the calming herbs recommended for anxiety (see p. 89) may alleviate symptoms.
- For Type D, try St. John's wort tablets.
- For Type H, take horsetail and celery seed tea (two cups daily).

Caution: For safety concerns, see pp. 34–37.

Homeopathy

- Nux vomica: if you feel irritable.
- Pulsatilla: for tearfulness and a feeling of not being loved.

- Sepia: for moodiness, irritability, and a general feeling of exhaustion.
- Lachesis: if you feel jealous and quarrelsome.

Flower essences: These remedies may help dispel the emotional symptoms of PMS. Choose one or more remedies that match your symptoms and personality type (see p. 33).

Aromatherapy: Soak in a warm bath with three drops each of Roman chamomile (for its calming properties) and geranium oils and two drops of lavender oil (see illustration).

Light therapy: If depression is your main symptom, exposure to two hours of bright light in the late afternoon may help. In the winter months, a full-spectrum light box can provide an alternative to daylight (see p. 157).

When to get medical help
- You are suffering from extreme depression.
- Your symptoms are so severe that they interfere with daily life and/or relationships.

See also:
Anxiety, Bloating, Depression

Soothing scents
Essential oils of geranium and lavender are said to help balance the emotions—a property that may enable them to curtail the mood swings often associated with PMS.

Prostate Problems

The prostate gland, which lies just below the bladder in men, is normally the size of a walnut. It secretes a fluid that enables the sperm to swim and reach the cervix after ejaculation. Problems may arise if the prostate becomes enlarged or inflamed. The gland is also a common site for cancer in older men.

The prostate gland begins to enlarge in most men around the age of 50. This condition, known as benign (that is, not inflamed or cancerous) prostatic hypertrophy, requires no treatment unless it produces symptoms. If the prostate swells enough to compress the urethra, which passes through it, it will obstruct urine flow. This can lead to a desire to urinate frequently, day and night, and the flow may become increasingly slow and hesitant. You may be unable to pass urine except in small amounts, yet you may suffer from incontinence.

Inflammation of the prostate is known as prostatitis. It may be a result of infection, usually transmitted sexually, or occur after jarring exercise (such as jogging) with a full bladder. Symptoms include a heavy feeling just behind the scrotum and pain in the lower back and abdomen. You may also experience pain while

Tomato power
A high intake of lycopene, a flavonoid found in large amounts in cooked tomatoes, can help prevent cancer of the prostate.

Protect your prostate

The following are best avoided, especially if you have an enlarged prostate:
- Delayed urination despite a full bladder.
- Jarring exercise with a full bladder.
- Constipation.
- Smoking.
- Unprotected sexual intercourse, unless in a long-standing, monogamous relationship.

Science meets tradition
Saw palmetto is a remedy long used by Native Americans for prostate trouble. Research now shows that this plant contains substances that help block the production of hormones, especially testosterone, that stimulate prostate cell growth.

urinating. Your urine may be cloudy, contain blood, and smell fishy, and you may have a fever.

The risk of prostate cancer begins to rise in the early 40s. The cancer may be symptomless. A PSA blood test helps indicate the likelihood of cancer, but is not reliable enough to be used for screening.

Prevention

Eat foods rich in zinc, beta-carotene, flavonoids, and vitamins C and E (see p. 12). Eat more beans, nuts, and seeds, and have three helpings of oily

Yoga for prostate health

The following exercise stimulates the circulation to the prostate and nearby areas. It may also reduce any inflammation.

1 Lie on your back with your arms by your sides. Raise your knees and move your feet close to your buttocks.

2 Put the soles of your feet together —or, if you find this difficult, just let the inner edges of your soles touch. Relax your knees and allow them to sink towards the floor. Hold the position for about five minutes.

fish every week. Their omega-3 fatty acids boost immunity, thus guarding against infections and other causes of prostatitis. Daily exercise helps the prostate by stimulating pelvic circulation. Regular ejaculation is also good for the prostate. Going out in bright daylight for half an hour each day may help prevent prostate cancer.

Treatment

Prostatitis

This disorder generally requires prompt medical treatment. To promote healing, eat plenty of vegetables and fruit, as well as foods containing essential fatty acids, selenium, and zinc. Try to avoid physical and psychological stress.

Benign enlargement

Diet

- Reduce your consumption of dairy products and refined carbohydrates.
- Eat more foods containing zinc (including shellfish, root vegetables, and pumpkin seeds).
- Have more soy-based foods, such as tofu. Soybeans contain substances called isoflavones, which may protect against prostate disease.
- Increase your intake of foods containing beta-carotene, vitamins C and E, flavonoids (see "Tomato power", facing page), magnesium, fibre, and essential fatty acids.
- Eat lentils, nuts, and sweetcorn regularly. These contain an amino acid called glutamic acid, which may help reduce prostate enlargement.
- Drink fewer caffeinated drinks.

Hydrotherapy: Take alternating hot and cold sitz baths (see p. 49) every day or two to stimulate the circulation to the prostate gland.

Herbal remedies: Take the recommended daily dose of a remedy made from rye grass pollen, saw palmetto, or *Pygeum africanum*. Linseed oil (one tablespoon a day) contains essential fatty acids that may help combat the symptoms of prostatitis and an enlarged prostate.
Caution: For safety concerns, see pp. 34–37.

When to get medical help

- You need to urinate frequently and/or the stream is weak.
- You have pain when urinating.
- You have a fever, possibly with lower back and abdominal pain.
- You have blood in the urine or a discharge.

Get help right away if:

- You are unable to urinate or have lower abdominal swelling or severe pain.

See also:
**Incontinence,
Urinary-tract
Infections**

Raynaud's Syndrome

This disorder of the blood vessels causes the small arteries supplying the fingers and toes to contract suddenly, usually upon exposure to cold. The nose and ears may also be affected. The resulting reduction in the circulation of the blood produces a succession of colour changes in the skin, accompanied, perhaps, by pain, numbness, and tingling.

Maintaining circulation
When sitting still, keep the feet moving, and stimulate the blood flow to the hands by repeatedly squeezing a small rubber ball.

The symptoms of Raynaud's syndrome occur when the muscles surrounding the tiny blood vessels in the fingers and toes constrict, cutting down the blood supply to those areas. A cold environment, working with vibrating power tools, or certain drugs (including beta-blockers) can cause an attack. The skin turns white from a lack of blood, then blue as the blood flow starts to return; finally, the skin reddens and may hurt. Raynaud's syndrome sometimes occurs with rheumatoid arthritis or arterial disease.

Prevention

Avoid known triggers. Smoking constricts the arteries, so quit or at least cut down. Keep hands and feet warm by wearing gloves and socks. Also wear several layers of clothing. Such fabrics as wool, silk, and polypropylene help conserve heat.

Exercise and massage: Improve your circulation by getting daily vigorous exercise that raises your heart rate. Regular massage may also help.

Diet: You may need extra magnesium, flavonoids, vitamins B and E, and essential fatty acids (see p. 12). Consider a daily multiple vitamin and mineral supplement. Limit your intake of alcohol and caffeine since these are triggers for some people.

Herbal remedies: Add warming spices such as cayenne, ginger, coriander seeds, cloves, and cinnamon to food, or drink one or two cups of tea made with one of these spices each day. Caution: For safety concerns, see pp. 34–37.

Treatment

Heat: Warm yourself up, for example, by taking a comfortably warm, but not hot, bath. A mustard footbath (see p. 141) or hand bath may also be helpful during an attack.

Exercise: During an attack, hold your hands above your head for a minute, then whirl them around for half a minute to boost circulation. Rubbing your hands together for one to two minutes may also help alleviate symptoms.

When to get medical help
- The above measures produce no improvement.
- You notice damage to the skin.
- You also have a leg ulcer, a rash, or arthritis.
- White or bluish skin does not return to normal.

See also:
Arterial Disease, Cold Hands and Feet

Repetitive Strain Injury

Injury to the muscles, ligaments, tendons, or nerves of the wrists, hands, arms, or shoulders that results from repeated movements is known as repetitive strain injury, or RSI. This includes such conditions as carpal tunnel syndrome, tenosynovitis, and tendinitis. In severe cases RSI can be painful and disabling, but there is much you can do to lessen the symptoms.

Repetitive strain injury—sometimes called work-related upper limb disorder—results from performing the same movement repeatedly throughout the day. This commonly occurs in the workplace, as a result of working on a computer keyboard for long hours or doing assembly-line production. Musicians are also susceptible to RSI. The damaging effects of the repeated action are encouraged by incorrect posture, poor working conditions, a lack of rest breaks, and the failure to perform muscle-stretching exercises.

Symptoms

- Pain or aching in the hands, wrists, arms, or shoulders, aggravated by movement.
- Restricted movement.
- Swelling over the back of the hand or wrist.
- Fatigue and weakness in the hands, wrists, arms, or shoulders.
- Weak hand grip and/or a lack of strength in the wrists.
- Tingling and numbness in the fingers.

Prevention

If your work involves small repeated movements for long periods of time, make sure that you take a regular five-minute break every hour. Use this break time to move around and stretch your hands, arms, and shoulders.

Avoiding RSI at the computer

- Adjust the height of your desk and chair as described on page 106.
- Choose a chair with arms that support your elbows.
- Arrange your position so your forearms are parallel to the floor.
- Position the screen at a right angle to the window or strongest light to avoid glare.
- Make sure that your eyes are at the same height as the screen, which should be about 25 inches away from your eyes.
- Sit directly facing the keyboard and screen.
- Use soft but firm wrist supports, so you don't need to flex your wrists upwards.
- Use a stand that holds documents at eye level to minimize neck bending while typing.

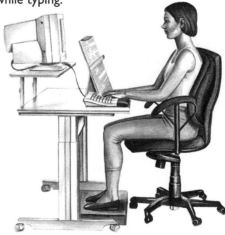

Stretch out
Yoga postures that release tension in the neck, shoulders, and upper back may help relieve the symptoms of RSI.

Treatment

Self-help measures reduce the risk of a problem becoming chronic and causing serious and permanent damage.

■ Rest the affected area as much as possible. If the condition is already serious, you may need to rest it completely for several weeks.

■ Wear a splint to protect and support an affected wrist or arm. You may need to wear this while working to prevent the condition from becoming worse.

■ Get regular upper body exercise to maintain mobility. Include frequent stretching exercises, such as yoga postures (see photograph), to improve the range of motion of the affected part.

Diet: A deficiency of vitamin B_6 has been linked to carpal tunnel syndrome, a type of RSI in which swelling of the soft tissues in the wrist pinches a nerve that runs between the hand and the arm. Increase your intake of foods rich in this vitamin (lean meat, fish, whole grains, peanuts, beans, avocados, and bananas).

Hydrotherapy

■ To reduce pain and swelling, apply either a cold compress (or ice pack) or a hot compress (or wrapped hot-water bottle), according to preference, to the affected area every half hour, as needed. Leave in place for as long as is comfortable—except for an ice pack, which should be applied for a maximum of three minutes.

■ Avoid very hot baths. Although a soak in a hot bath may temporarily soothe aching muscles, too much heat may ultimately increase inflammation and therefore worsen RSI.

Herbal remedies: Comfrey is a traditional herbal remedy for inflammatory conditions. Apply comfrey ointment over a sore tendon or muscle. Do not use this herb if the skin is broken, and do not take it by mouth.

Aromatherapy: Certain essential oils help reduce inflammation and muscle stiffness.

■ Mix 10 drops each of peppermint, lavender, and eucalyptus oils with three tablespoons of sweet almond oil or other cold-pressed vegetable oil. Use this mixture to massage the affected area morning and evening.

Homeopathy

■ Ruta: for most types of RSI, especially of the hand and wrist.

■ Rhus toxicodendron: for pain relieved by gentle movement.

■ Arnica: for any injury when there is a sensation of bruising, Arnica cream massaged into the affected area may be beneficial.

When to get medical help
● Pain and/or stiffness is severe.
● Discomfort continues even during rest.
● Your range of movement is restricted.
● You notice wasting of muscle, especially near the thumb.
● You have numbness or tingling in the hand.

See also:
Back Pain, Pain, Strains and Sprains

Restless Legs

This disorder is characterized by unpleasant aching, prickling, itching, or tickling sensations in the legs. The legs may twitch or feel tired, heavy, or tense, as if swollen. The most effective way to relieve the discomfort is by moving the legs. A few simple changes in your routine can reduce the frequency and/or the severity of symptoms.

Restless legs affect up to one in six people, and symptoms usually occur after sitting for a prolonged amount of time in the evening or while in bed at night. The condition sometimes runs in families. It affects mainly middle-aged women but is also common toward the end of pregnancy. One in four people with rheumatoid arthritis develops the disorder, and it is also sometimes associated with iron-deficiency anaemia, varicose veins, diabetes, and lung disease. The arms may also be involved.

The exact cause of restless legs is unknown, but possible triggers include smoking, fatigue, stress, alcohol, caffeine, and a lack of iron, folic acid, or vitamin E.

Cold comfort
Relieve affected muscles with a covered cold pack—a package of frozen peas or a picnic ice pack. Leave in place for up to 30 minutes (10 over a bony area), but remove sooner if your toes go numb or blue.

Prevention
- Eat more foods rich in iron, magnesium, vitamin E, and folic acid (see p. 12).
- Take a daily multiple vitamin and mineral supplement.
  - If you think you might have iron deficiency anaemia, ask your doctor to arrange a blood test.
    - Don't smoke.
    - Avoid alcohol in the evening.
    - Reduce your caffeine intake.
      - Take some moderately strenuous exercise every day.
    - Get enough rest and sleep.

Treatment
When symptoms appear, massage your calves with firm, kneading movements. If you can, take a walk and/or do some leg-stretching exercises.

Aromatherapy: Add 5 drops each of basil (avoid during first 20 weeks of pregnancy) and marjoram oils to three teaspoons of sweet almond oil and use this to massage your legs.

Hydrotherapy: Try the following therapies to see which works best for you.
- Apply a hot compress, such as a towel wrung out in hot water, for two minutes, and then a cold compress for one minute.
- Similarly, apply a hot-water bottle for two minutes and then a cold pack for one minute.
- Have a cool leg bath in a bath of shallow water.
- Alternate a hot leg bath for two minutes with a cold for one minute.

When to get medical help
- Your symptoms persist or worsen.
- Your sleep is seriously interrupted.

See also:
**Muscle Aches and Stiffness,
Sleeping Difficulties**

Sex Drive Loss

The level of desire for sex, also termed libido, changes throughout life in response to both emotional and physical factors. These include fatigue, stress, illness, childbirth, and relationship difficulties. Understanding the reason behind any fluctuation is the first step in dealing with it. Natural therapies can help by boosting your general health so that you feel your best.

As is the case with most other aspects of human behaviour, we all differ in our need for sex; there is no rule about what is "normal". A low sex drive is no cause for concern in itself, provided that you and your partner are content with the amount of sex in your relationship. However, reduced interest in sex can indicate an underlying problem. Depression and other psychological difficulties can lower libido, as can physical illnesses such as arthritis and diabetes.

Several medications, including certain antidepressants and blood-pressure lowering drugs, may reduce sex drive. Some older men have difficulty in becoming aroused or maintaining an erection. Women who are past their menopause may not want to have intercourse because of discomfort resulting from vaginal dryness or other hormone-related changes.

Treatment

Aromatherapy: Ylang ylang and clary sage oils can encourage feelings of relaxation and sensuality. Rose otto oil is traditionally associated with a heightening of sexual desire, while sandalwood and geranium are said to help stimulate desire subdued by depression. Avoid clary sage in the first 20 weeks of pregnancy.

- To restore calm and harmony—a prerequisite for resolving sexual difficulties—place a blend of four drops each of ylang ylang and clary sage oils, or of sandalwood and geranium oils, in a vaporizer, or use four drops of rose otto oil alone. Or sprinkle a few drops of selected oils on your bedsheets.
- Add two drops each of ylang ylang and clary sage oils and three drops of geranium oil to two and a half teaspoons of sweet almond oil or other cold-pressed vegetable oil, and use this mixture to massage your partner's back, legs, and abdomen.
- To boost energy levels, add two drops each of ylang ylang and clary sage oils and three drops of geranium oil to an evening bath.
- Avoid smoking, and drinking too much alcohol, as both can reduce libido.

General help for loss of libido

- Share your concerns with your partner and involve him or her in your treatment plan.
- Try not to worry about your lack of sexual energy, as anxiety may aggravate the problem.
- Practise stress-management strategies, such as deep breathing, creative visualization, or other relaxation techniques.
- Exercise regularly to increase general, as well as sexual, energy. This will also help relieve anxiety and increase your awareness of—and confidence in—your body.
- Get enough sleep. Fatigue is one of the major causes of a lack of interest in sex.

Easy closeness
Relieve anxiety and strengthen the bond between you and your partner by giving each other a gentle massage, perhaps with fragrant oils.

Specially for men
Damiana (*Turnera diffusa*) has an ancient reputation as an aphrodisiac. Modern herbalism confirms its action as a tonic for the nerves and male hormonal system.

Herbal remedies:

Certain herbal teas or tinctures may be useful.

- Women should drink a cup of raspberry-leaf tea in the evening or take a few drops of raspberry-leaf tincture in half a cup of water. Chasteberry (*Vitex agnus castus*) is another useful herb, but avoid this if you are on hormone medication. Tea or a few drops of tincture made from a combination of dong quai, liquorice, nettles, prickly ash bark, and wild oats can increase vitality.
- To aid relaxation, men and women may benefit from a daily cup of tea made from a soothing herb, such as chamomile or vervain.
- Improve male libido with a cup of tea or a few

drops of tincture made from damiana, saw palmetto, cinnamon, liquorice, and ginger.
Caution: For safety concerns, see pp. 34–37.

Diet: A poor diet probably accounts for many people's low sex drive.

- Eat foods rich in vitamins A, B, and E, magnesium, manganese, zinc, essential fatty acids, and protein (see p. 12). These nutrients are important for sex hormone production and the maintenance of healthy reproductive organs.
- Eat raw fruit and vegetables every day to ensure maximum vitamin and mineral content.
- Avoid too much alcohol; it tends to impair sexual performance.

Other therapies: Couple counselling can provide valuable help for those whose sex drive loss originates in a relationship problem or creates stress in the relationship. Psychotherapy may also be effective.

When to get medical help

- Your relationship with your partner is adversely affected.
- You suspect you have—or you are taking medication for—an underlying condition.
- You have other symptoms, such as pain during or after intercourse.
- You are depressed.
- The problem comes on suddenly.

See also:
Anxiety, Arterial Disease, Depression, Menopausal Problems, Stress

Shock

Emotional shock is the mental distress that may occur after a sudden traumatic or upsetting event. Although it may require medical attention, it is quite different from clinical, or medical, shock, which follows serious injury or illness and always requires urgent attention. Natural remedies apply only to mild or moderate emotional shock.

Emotional shock may occur after such experiences as receiving news of bereavement or job loss or being physically attacked. The shock may be strong enough to affect the nerves that control blood pressure, leading to fainting (see p. 194).

Clinical shock is a potentially severe physical reaction to an injury or another medical emergency. It follows a sudden fall in the body's blood flow and blood pressure which leads to a reduction in the amount of oxygen reaching the brain and other vital organs. Untreated, it can damage these organs and may even be life-threatening. Causes of clinical shock include a heart attack, serious accident, burns, severe allergic reaction (anaphylactic shock), acute pain, low blood-sugar level and internal or external bleeding. When this type of shock is suspected, seek emergency medical help. After either type of shock, it is possible to develop a constellation of symptoms known as post-traumatic stress disorder. This may include flashbacks, sleep disturbance, depression, and relationship problems.

Treatment

The following treatments should be used only in cases of emotional shock that do not require emergency help.

First aid for clinical shock

1 Call for medical assistance immediately.

2 If you are able, treat any obvious cause, such as bleeding.

3 Lay the person down and raise their legs higher than the heart.

4 Loosen tight clothing, and cover the victim with a blanket or extra clothing, to prevent heat loss.

5 Do not move a person in shock after a serious injury or if unconscious, except into the recovery position (see p. 195).

6 Do not give anything to eat or drink, though you can wet the person's lips with dabs of water.

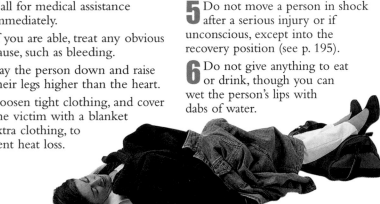

Homeopathy

- Arnica: the primary remedy for emotional shock. Take doses of arnica and Rescue Remedy (see "Flower essences", facing page) every few minutes until you feel better.

- Aconite: if the shock is accompanied by feelings of terror and impending doom.

- Ignatia: if you feel faint or hysterical from an emotional upset.

- Gelsemium: if the shock is accompanied by weakness, trembling, diarrhoea, or frequent urination, or occurs in anticipation of an event.

- Carbo vegetabilis: if the shock is accompanied by a feeling of icy coldness and the need for fresh air.

Symptoms of clinical shock

- Rapid or weak, irregular pulse.
- Pale, grey skin.
- Clammy, cold, sweaty skin.
- Weakness and giddiness.
- Rapid, shallow breathing.
- Fainting.
- Anxiety.
- Nausea and possibly vomiting.
- Thirst.
- Visual difficulties.

Aromatherapy

■ Massage your legs, arms, or face, or ask a friend to give you a massage. Mix a massage oil from one drop of lavender essential oil, two of geranium, and three of lemon, plus two teaspoons of sweet almond or jojoba oil. Or use other calming oils, such as chamomile, valerian, rose, and palma rosa.

■ If an emotional shock makes you feel faint, rub lavender oil on your temples and inhale its scent from your hands.

Flower essences: Take Rescue Remedy. Place four drops on your tongue every few minutes, or put four drops in a little water and sip it over a quarter of an hour.

Herbal remedies

■ Sip tea made from rosemary or elderflowers.
■ For lingering shock associated with faintness, sip ginger tea.
■ For feelings of stress after a shock, drink a cup of tea—or take a few drops of tincture—made

from lemon balm and chamomile. Do this four times daily.
■ To help stimulate appetite after an emotional upset, drink one or two cups daily of tea made with fenugreek seeds, clover flowers, or yarrow.
Caution: For safety concerns, see pp. 34–37.

Diet: Shock can affect your appetite.
■ Try to eat small meals of foods you especially like.
■ Emphasize foods containing zinc and vitamins B$_1$ and B$_5$, which may improve mental function and help counter the effects of stress.

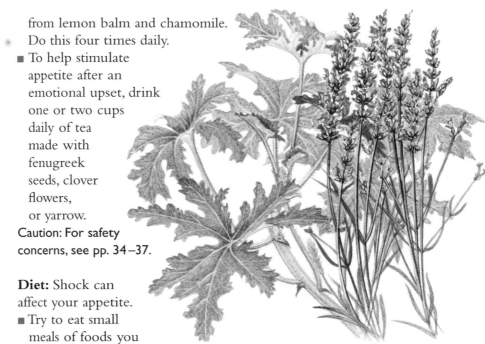

Gentle restoratives
Such essential oils as geranium and lavender can help induce a sense of calm in those who have experienced an emotional trauma.

When to get medical help

- The after-effects of emotional shock prevent you from coping with ordinary life.

Get help right away if:

- You suspect clinical shock.

See also:
Abdominal Pain, Anxiety, Asthma, Emotional Problems, Fainting, Food Sensitivity, Grief, Pain

Skin Problems

The skin is your body's first line of defence against damage from infection, extremes of light and temperature, pollution, and physical injury. Its health can be affected by internal disorders as well as external factors. Look after your skin carefully—with natural therapies, when necessary—and the odds are that it will continue to serve you well for a lifetime.

Essential skin protection
Contact with detergents can cause dryness, inflammation, and cracking of the skin. Protect your hands by wearing rubber gloves whenever possible.

The most common skin problems in adults include oily or dry skin, acne, psoriasis, and eczema and other forms of dermatitis. For skin to remain healthy, smooth, and supple, it needs an adequate production of sebum (the oily substance secreted by glands attached to the hair follicles); good hydration; a healthy circulation of blood and lymph; sufficient nutrients; hormonal balance; the ability to repair and renew itself; and the absence of irritation from detergents, ultraviolet sun rays, and other potentially damaging agents.

Prevention

Encourage sebum production by managing stress effectively, eating a healthy diet, maintaining a normal body weight, and avoiding temperature extremes, strong winds, and contact with harsh soaps and detergents. Apply a moisturizer each day, and hydrate your skin from within by drinking at least six glasses of water every day. Boost your circulation and help prevent fluid retention with regular brisk exercise and a sound diet. Protect your skin from wind and sun (use a sunscreen with an SPF of 15 or more). Wash your face and body with a mild soap containing vegetable extracts, such as palm-kernel and coconut oils, or with a soapless cleansing bar containing synthetic ingredients to help maintain the skin's natural acidity. Encourage the repair and renewal of your skin cells by getting enough sleep and by being a nonsmoker (see box, below).

Diet: Prevent premature aging of skin by eating plenty of raw fruit and vegetables. Limit your intake of saturated fats, refined foods, and alcohol, all of which tend to speed skin aging. Only about 13 percent of your calories should come from animal protein foods, such as meat, eggs, and cheese. Try to lose excess weight: slim yet well-nourished people tend to age more slowly.

Vitamin C, flavonoids, and zinc help keep skin supple, so make sure your diet contains foods rich in these nutrients (see p. 12). You also need foods

Aging effects of smoking

- Toxins from inhaled smoke restrict the blood flow to the skin, thereby depriving it of oxygen. This reduces the ability to heal, and to regenerate new skin.
- After years of regular smoking, the skin may become much thinner than that of a nonsmoker, resulting in deeper wrinkles.
- Smoking reduces the collagen content of the skin so that wrinkles develop more quickly.

rich in carotenoids; vitamins B (especially biotin) and E; and the minerals copper, manganese, and selenium. Include foods containing essential fatty acids, together with those that contain magnesium, zinc, and vitamin B_6, which are needed to metabolize them.

Treatment

Aromatherapy: The balancing, cleansing, and regenerative qualities of certain essential oils are excellent for improving the appearance and health of the skin. Used daily, they can improve the skin's texture, especially that of the face and hands—the areas most exposed to the elements.

Create your own range of aromatherapy skin products by mixing the oils listed opposite with sweet almond or jojoba oil or an unscented body lotion. Apply your mixtures to problem areas, or use for massage. You can also add diluted oils to your bathwater or mix them with water for a compress, but don't apply undiluted oils, other than tea tree and lavender, to the skin.

- Clary sage is soothing and anti-inflammatory. It also helps preserve moisture in dry or mature skin. Do not use this oil during the first 20 weeks of pregnancy.
- Cypress, which is astringent and soothing, may help regulate the production of sebum in oily skin. Do not use this oil during the first 20 weeks of pregnancy.
- Eucalyptus is cooling and antiseptic. Use it to treat boils and pimples.
- Geranium has astringent and balancing properties. It helps cleanse and tone the skin, reduces inflammation, and can soothe acne, eczema, and minor wounds.

- Juniper berry is astringent and cleansing. It may be beneficial for acne, oily skin, and oozing eczema. Do not use this oil at any time during pregnancy.
- Lavender has antiseptic and anti-inflammatory properties. It can soothe eczema, sunburn, and insect bites, and it also promotes cell growth and healing, so may therefore help to minimize scarring.
- Peppermint oil cools and cleanses the skin. It can soothe itchy skin, though too much can make itching worse.
- Roman chamomile is soothing and antiseptic, and good for sensitive or dry skin. It helps heal acne and dermatitis and reduces other types of skin inflammation.
- Sandalwood helps soften dry, mature, or wrinkled skin. It may also reduce irritation from sunburn, hives, and other rashes.
- Tea tree oil has antiseptic, anti-inflammatory, and antifungal properties. It is good for boils and rashes, as well as for countering skin infections of all kinds.

Fragrant skin care
Essential oils may help a wide variety of skin conditions. The plants from which some of the most effective are obtained are (clockwise from far left) eucalyptus, cypress, peppermint, clary sage, juniper berry, chamomile, and lavender.

Symptoms and Ailments

Food sensitivity: Such skin problems as roughness, hives, and eczema may result from a food sensitivity. Sometimes it is easy to relate a skin problem to a particular food, but often the link is not obvious. Identifying a culprit food may be difficult if it is one that you eat frequently or if your symptoms are vague and long-standing.

You may want to try an exclusion diet, avoiding the suspected foods one by one for up to three weeks each. If your symptoms settle while avoiding a food, challenge whether it really is to blame by eating it again. If your symptoms return, stop eating it. Repeat this challenge twice to be sure. This process is essential for a reliable diagnosis of food sensitivity.

You may be able to boost your immunity—thereby lowering your risk of food sensitivity—by avoiding refined foods and eating more foods rich in essential fatty acids, such as most cold-pressed vegetable oils, nuts, and seeds. Supplements of omega-3 fatty acids (found in fish oil) and evening primrose oil may be beneficial. Eating foods rich in vitamins A and C may also be useful in preventing food sensitivity.

Herbal remedies: Many herbs are excellent for healing skin conditions. They can ward off infection, relieve pain, and reduce scarring. Use them either in a compress, made by soaking a pad of gauze in herbal tea and binding it to the damaged area, or in cream or ointment form.
- Aloe vera gel softens and nourishes dry skin and encourages skin regeneration after injury. It also helps prevent "barber's rash"—the crop of tiny pimples that may appear after shaving.
- Burdock seeds are cleansing. They're good for

treating eczema, dermatitis, psoriasis, boils, abscesses, and acne.
- Chamomile's antiseptic oils have a soothing and anti-inflammatory effect. They help stimulate skin repair.
- Chickweed soothes irritating skin conditions such as eczema and psoriasis. It also softens the skin and alleviates itchy rashes and eruptions.
- Elderflower, an anti-inflammatory, is a traditional remedy for ulcers, burns, cuts, and wounds. Distilled elderflower water makes an excellent facial toner and cleanser.
- Marigold (calendula) is a good first-aid remedy for cuts, burns, and bruises, since it combats infection and reduces inflammation.
- Rosewater cleanses and tones the skin, smoothes wrinkles, and can help clear up acne. It cools the skin by acting as an astringent, aids skin repair, and reduces swelling and bruising.
- Witch hazel, widely available in chemists, is a traditional home remedy. It is strongly antiseptic and healing: its astringent tannins stop bleeding and lessen inflammation and scarring.

Help from the herb garden
Herbal teas and ointments are gentle treatments for skin conditions caused by infection, inflammation, or allergy. Many of these herbs have been used for centuries.

It is also effective for bruises and irritated varicose veins. Apply neat distilled witch hazel to unbroken skin, but dilute with water if applying to broken skin (any stinging will stop quickly).

Caution: For safety concerns, see pp. 34–37.

Kitchen-cabinet remedies

- Bicarbonate of soda applied to a weepy, allergic rash will help keep it dry.
- Beetroot and cabbage juices are good blood cleansers and helpful for oily skin that is prone to acne. Drink half a glass of one of these juices once or twice a day for 10 days.
- Carrots are antiseptic and help speed wound healing. Apply cooled carrot broth to your skin to soothe chapping, roughness, and itching.
- Cucumber soothes inflamed, itchy rashes, such as eczema, and cucumber water also makes an ideal cleanser and toner for oily skin: cut a cucumber into cubes and boil in two pints of water for 15 minutes. Strain and press through a cloth or sieve.

Versatile vegetable
Raw potato juice is a traditional remedy said to soothe inflammation and help heal dermatitis, wounds, and ulcers. To treat cracked skin, use grated potato mixed with olive oil.

- Linseeds are rich in omega-3 essential fatty acids, which aid skin health. Apply linseed tea to dry skin, or add it to bathwater to soften skin.
- Honey has antiseptic properties; smooth a little over infected skin.

When to get medical help

- A skin condition worsens or fails to clear up after home treatment.
- The problem is close to the eyes.
- A baby or young child has a skin problem other than typical nappy rash or cradle cap.

Get help right away if:

- A rash occurs along with a raised temperature, a headache, drowsiness, a stiff neck, breathlessness, or faintness.

See also:
Abscesses and Boils, Acne, Aging, Candida Infections, Chapped Lips, Cold Sores, Cuts and Scrapes, Dermatitis, Diaper (Nappy) Rash, Dry Skin, Food Sensitivity, Fungal Skin Infections, Hair and Scalp Problems, Hives, Insect Bites and Stings, Itching, Sunburn, Warts

Sleeping Difficulties

A good night's sleep regenerates the body and refreshes the mind, but nearly everyone complains of difficulty sleeping at some time or another. Anxiety about your lack of sleep can make the situation worse, but using natural techniques (known collectively as sleep hygiene) may be all you need to start slumbering deeply once again.

The average adult in the United Kingdom sleeps between six and a half and eight and a half hours a night. Women report more problems with sleep than men, and sleeping difficulties become more common as we get older. If you do not sleep well, you may feel tired and generally unwell. A chronic lack of sleep reduces physical and mental performance during the day and may lead to anxiety, depression, or other problems.

Prevention

Sleep hygiene: This is a way of reorganizing your routine to create conditions that are conducive to sleep. Paying attention to sleep hygiene can help prevent problems with sleeping and can even help overcome established insomnia. You

Dream land
Keep your bedroom tidy and free of clutter as a restful setting can help to encourage sleep.

are unlikely to get a good night's sleep if you are uncomfortable, hungry, thirsty, or too hot or cold, so it is important to ensure that these factors do not interfere with your sleep.

- Make sure that your bed is comfortable. If the mattress is more than 10 years old, you probably need a new one. Choose one firm enough to provide support, but not so hard as to put too much pressure on your hips and shoulders.
- Check that the temperature in your bedroom is comfortable—research indicates that around 65°F is right for most people. Your bedroom should also be dark and quiet. If it is not, consider using an eye mask and earplugs.

Types of sleeping problem

- Difficulty getting to sleep often occurs because a person is unable to stop thoughts and worries from whirling around in their mind. The cause is often anxiety or a strong emotion, such as anger, that may not have been dealt with during the day. Anticipatory excitement can also make it difficult to drop off. Physical reasons include indigestion, restless legs, and alcohol-induced agitation.

- Waking up early and being unable to get back to sleep may be a symptom of depression and/or anxiety. It can also occur if a person has drunk too much alcohol or caffeine.
- Waking up frequently during the night and being unable to drop off again immediately is more common in older people and is sometimes linked to depression and/or anxiety.

- Choose a restful decor, with warm, soft colours.
- In the hour or two before going to bed, do not smoke, eat large or indigestible meals containing a lot of fat and refined carbohydrates; do not drink alcohol or caffeine-containing drinks; and do not watch exciting or disturbing TV programmes, read stimulating articles or books, or do strenuous exercise. Have a warm bath just before bedtime.
- Eat a light, sleep-inducing snack (see "Treatment"). Hunger can keep you from getting to sleep or can make you wake up during the night or in the early morning.
- Ideally, your bedroom should be a peaceful haven associated with sleep. Do not use it for such activities as working and ironing.
- If you stay awake for more than half an hour after going to bed, get up, go to another room, and do something else, such as yoga or light reading, until you feel sleepy.
- Get into the habit of going to bed and getting up at the same times each day. Set your alarm to help you wake up, if necessary.
- Do not nap during the day.
- Exercise daily for at least 30 minutes.
- Try—unless you are required to do shift work—to take full advantage of the bright early-morning light by being awake then, rather than sleeping in a darkened room until mid-morning. The natural variation of light intensity throughout the 24-hour day has powerful effects on the brain's pineal gland, which helps regulate our daily rhythms. Going to sleep when it is dark and staying awake when it is light can help reinforce a healthy pattern of sleep and waking.

Treatment

Soporific snacks: A light, easily digested snack before bedtime can often prevent hunger pangs from disturbing your sleep. Moreover, certain foods or food combinations can promote sleep. Such foods are high in carbohydrates and rich in vitamin B, calcium, magnesium, essential fatty acids, and the amino acid tryptophan, which is used by the brain to make sleep-inducing chemicals (see p. 86). This combination of nutrients allows the tryptophan to reach the brain and may help you feel sleepier within about half an hour. Snack suggestions include:

- A wholegrain bread and lettuce sandwich.
- Boiled potato and cauliflower mashed with a little hazelnut or walnut oil.
- Sliced banana with chopped dates.
- Warm milk and biscuits.

If you suspect that certain foods disagree with you, cut them out of your diet for a couple of weeks to see if this improves your sleep. If you find that cheese causes nightmares, for example, eat it only early in the day.

Herbal remedies: Herbs are often a gentle alternative to prescription sleeping pills.

- Drink a cup of celery-seed tea before going to bed. Add two teaspoons of the crushed seeds to a cup of boiling water and steep.
- If your sleep problem is long-standing, opt for a cup of passionflower or valerian tea before bedtime. You can have another cup of this tea if you wake during the

The sleep hormone

Melatonin, the body's sleep-inducing hormone, can be bought in many countries, but not in the UK because there is no proof of its safety.

Snooze foods
Eating snacks made from wholegrain bread, bananas, lettuce, cauliflower, and dates, may be better than counting sheep.

Symptoms and Ailments

course of the night and find it hard to fall asleep again.

■ Help yourself unwind by adding essential oils or herbal tea to your bathwater in the evening. Lime tree flower, lavender, Roman chamomile, frankincense, neroli, and rose oils are all suitable, as are teas made from lime tree flower,

chamomile, catnip, lemon balm, and hops.

■ St. John's wort tincture, taken according to the manufacturer's instructions, can aid sleep.

■ Put a mixture of your favourite soothing and sleep-inducing herbs inside your pillowcase, or take a small fabric bag, stuff it with herbs, and put it near your head so you breathe in the

Yoga as sleep preparation

These simple yoga techniques can help if you have difficulty "switching off" at bedtime.

1 Stand with your arms straight out in front of you at shoulder height, and your palms facing each other, with fingers stretched forwards.

2 Move your arms out to each side as you inhale. Then return your arms to their original position as you exhale. Breathe in time with your arm movements. Repeat 10 times.

3 Stand with the palms of your hands on your chest and your fingers linked. As you inhale, move your arms out in front of you, at chest height, with palms outward and fingers

stretched forwards. Bring your hands back to their original position as you exhale. Do this three times. Then repeat, but this time lift your stretched arms to form an angle of 45

degrees to your body. Bring them back to your chest. Then repeat, this time raising your arms above your head. Repeat six times.

scent during the night. Suitable calming herbs include lavender, lime tree flower, chamomile, catnip, lemon balm, and hops.
Caution: For safety concerns, see pp. 34–37.

Homeopathy: Try the remedy that most closely matches your symptoms:

- Arsenicum: if anxiety is preventing you from sleeping.
- Coffea: for an overactive mind, crowded with unwanted thoughts.
- Passiflora: for uneasy sleep.

Flower essences: These are gentle remedies designed to address emotional problems and are particularly appropriate for the treatment of sleep disorders. Choose a flower essence that seems to suit your emotional state—for example, Rock Rose for terror, Mimulus for fear, or Agrimony for worry (see p. 33). Add two drops each of your chosen remedies to a one-ounce, dark-glass vial three-quarters full of spring water, and top off the bottle with brandy as a preservative. Use glycerine instead of brandy if consuming alcohol creates problems for you.

Take four drops of your mixture, either directly on the tongue from a dropper or added to spring water. Do this four times a day before a meal or snack. If this doesn't help after four weeks, add up to three more remedies to the mixture and continue taking as before.

Stress management: Lessening the stress in your life will probably improve your sleep. Some stress is unavoidable, but there are positive ways of handling it. Classes in relaxation, meditation, visualization, yoga, or assertion training may help

you to manage stress. Self-hypnosis can also be effective. These techniques are especially good if you have been using food, cigarettes, or alcohol or other recreational drugs as a way of dealing with pressure. If you can't handle stress on your own, consult your doctor or a trained counsellor.

Exercise: Regular exercise reduces anxiety, improves the circulation, and burns up the adrenaline and other stimulating hormones you produce when you're feeling stressed. However, you should do serious exercise during the day rather than in the evening, or you may be too stimulated to sleep.

Herbs for sleep
Chamomile, passion-flower, or hops can be used to make a soothing, caffeine-free bedtime drink.

When to get medical help

- Your sleep problems last longer than a week or two.
- You fall asleep or cannot function properly during the day (especially important if your work requires you to be alert, physically strong, or well-coordinated).
- You fear that tiredness is affecting your family relationships.
- You suspect that an illness, such as depression or severe anxiety, underlies your sleep problem.
- You also have symptoms such as night sweats, fever, chills, abdominal pain, and neck ache.

See also:
Anxiety, Depression, Indigestion, Restless Legs, Stress

Snoring

People who snore are usually unaware of the noise they make, but the disorder can easily become the bane of their partner's life by seriously disturbing their sleep. There is no certain cure for snoring, but you can take several measures that may reduce or even prevent this annoying problem as well as improve your general health.

Sleep on your side
Sew a tennis ball into the back waistband of your pyjamas to keep you from rolling onto your back and allowing the tongue and uvula to block the throat.

During the waking hours, the activity of the throat muscles keeps the airway open. During sleep, these muscles relax and in certain people allow the airway to collapse. Their breathing then needs to deepen to force the airway open and allow air to enter the lungs. This deeper, more forceful breathing through the mouth causes a loud vibration and rattling of the soft palate and uvula (the small flap at the back of the soft palate), which produces the noise of snoring. Several factors may temporarily cause or increase snoring: a cold, allergic rhinitis ("hay fever"), polyps in the nose, and enlarged adenoids. The following groups are most likely to snore:
- Men—the problem affects three times as many men as women.
- People of middle age and older.
- Overweight people. Fat beneath the lining of the throat reduces the size of the airway and encourages breathing through the mouth. The larger you are, the more likely you are to snore.
- Smokers. Smoke irritates the nose and throat and swells the lining of the passages. This constricts the airway and encourages breathing through the mouth.
  - Those who have consumed alcohol within the last few hours before bedtime. Alcohol relaxes the throat, face, and jaw muscles, encouraging breathing through the mouth.

Treatment
To prevent or reduce snoring:
- Lose excess weight.
- Don't smoke.
- Reduce your alcohol intake.
- Eat your evening meal at least three hours before you go to bed.
- Go to sleep and get up at about the same times every day.
- Try using a different number of pillows, or a neck pillow.
- Use books or blocks to elevate the head of your bed by three to four inches.
- During a cold or an attack of allergic rhinitis, gargle before bedtime with one drop of peppermint oil in cold water. This remedy enlarges the airway by shrinking the swollen lining of the nose and throat.

When to get medical help
- The measures described don't help.
- You or your partner experiences daytime fatigue and sleepiness.
- Your partner reports that you sometimes stop breathing while snoring (sleep apnoea).

See also:
Overweight, Sleeping Difficulties

Sore Throat

This common complaint may be caused by a viral or bacterial infection, allergy, dry air, or by inhaling smoke and other airborne pollutants. A sore throat is often the first sign of a cold, flu, laryngitis, or infectious mononucleosis ("glandular fever"). Choose from a number of natural remedies to soothe your symptoms and fight infection.

A sore throat, which may look red, feels raw and rough, making it painful to swallow. It may be accompanied by a fever and congestion. In most cases, the symptoms clear up quickly.

Prevention
- Keep warm.
- Drink at least six glasses of caffeine-free, non-alcoholic fluids daily.
- Don't smoke, and avoid polluted environments.
- Eat foods rich in folic acid and vitamins A, B, C, D, and E, as well as flavonoids, iron, magnesium, and zinc (see p. 12). Increase your intake of raw fruit and vegetables, and limit that of refined carbohydrates, caffeine, and alcohol.
- Include several cloves of crushed raw garlic in your food daily.

Treatment
Herbal remedies
- Gargle three to six times daily with tea made from thyme, goldenseal, or myrrh, or a few drops of tincture in a little water. Or use red sage tea with a teaspoon of apple-cider vinegar. These herbs are both antiseptic and soothing.
- To boost your immunity, drink a cup of tea made from echinacea, wild indigo, and sage every two hours. Add soothing mullein, marshmallow, or coltsfoot if your throat is very sore or you have a cough. If you have congestion or a fever, add elderflower, yarrow, and peppermint.
- The Chinese herbal supplements astragalus and reishi mushrooms have a powerful strengthening effect on the immune system. Add them to teas or use them in cooking.
Caution: For safety concerns, see pp. 34–37.

Air quality
Moisten dry air with a humidifier, a bowl of water placed near a heat source, and house plants.

Aromatherapy: Sandalwood and lavender are soothing; lemon, geranium, pine, and tea tree help fight infection; and eucalyptus, peppermint, and Atlas cedarwood help clear congestion. Choose from these treatments:

- Every two hours, gargle for several minutes with two drops each of sandalwood and lemon oils, or two drops each of Atlas cedarwood and eucalyptus oils, in a glass of warm water, then spit the water out.
- Gargle with warm water to which you have added three drops of geranium or lemon oil.
- Massage your face and chest with a blend made with three drops of sandalwood oil, two drops of eucalyptus oil, and one drop of peppermint oil in two and a half teaspoons of sweet almond oil or unscented lotion or cold cream.
- Inhale the steam from a bowl of very hot water containing two drops of eucalyptus or pine oil and two drops of peppermint oil. Do this for 5 to 10 minutes twice daily.

Homeopathy

- Aconite: for soreness that begins suddenly, begins or worsens at night, and makes your throat hot and dry. Swallowing is difficult, although you are thirsty.
- Apis: if your throat is red, stinging, and puffy, and cold drinks help.
- Belladonna: if you have a sudden dry, burning, throbbing soreness, with bright red throat.
- Hepar sulphuris: if your throat feels as if a fish bone were stuck in it. You feel bad-tempered

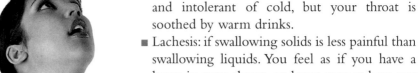

Acid action
A gargle made with a teaspoon of lemon juice or apple-cider vinegar in a glass of warm water is a traditional remedy for a sore throat. The acid in these ingredients is hostile to the bacteria and viruses that most often cause the problem.

and intolerant of cold, but your throat is soothed by warm drinks.

- Lachesis: if swallowing solids is less painful than swallowing liquids. You feel as if you have a lump in your throat, and you cannot bear any constriction around the throat. The pain is often on the left side, or it begins on the left and moves to the right.
  - Lycopodium: if the right side of your throat feels worse or the pain moves from right to left. Warm drinks help, and you feel worse in the late afternoon or early evening.
  - Phytolacca: if your throat looks dark or bluish-red, and the pain is especially bad on swallowing. You feel as if you have a hot lump in your throat, and your body aches.

Kitchen-cabinet remedy: Mix one teaspoon of salt with warm water. Gargle or use as a mouthwash up to six times daily.

When to get medical help

- Your symptoms last longer than a few days, or if severe, a day.
- You have a high fever, a rash, dizziness, a severe headache, enlarged glands in the neck, or extreme difficulty in swallowing.

Get help right away if:

- You have concurrent breathlessness, chest pain, or palpitations.

See also:
Childhood Viral Infections, Colds, Coughs, Fever, Influenza, Swollen Glands, Voice Loss

Splinters

A small splinter of wood, metal, glass, or other hard substance that becomes embedded in the skin can be surprisingly painful. If not removed, such a foreign body may cause inflammation and tenderness and may become infected. Splinters are often quick and easy to take out, however, using simple first-aid techniques.

The hands, feet, and knees are the areas most likely to come into contact with splintery materials. Although splinters may seem insignificant, don't ignore them. If left alone, they can be a source of infection that could, at worst, spread to other parts of the body.

Treatment

Aromatherapy: Before and after removing a splinter, apply one drop of lavender, tea tree, or myrrh oil (avoid myrrh during first 20 weeks of pregnancy). These essential oils encourage healing and, because of their antibacterial properties, help prevent infection from taking a hold.

Herbal remedy: To draw out a splinter, apply slippery elm ointment under a sticking plaster or a slippery elm poultice for 24 hours.

Kitchen-cabinet remedies
- To help bring a deeply embedded splinter to the surface, apply a warm poultice (see p. 37) of bread or bran several times throughout the day. Then apply comfrey ointment and cover with a light bandage for 24 hours.
- Draw out a stubborn splinter by covering it overnight with a sticking plaster spread with honey or kaolin (from a chemist).

Homeopathy: Silica, taken orally, may help bring a small splinter to the skin surface.

Removing a splinter

- Clean the skin around the splinter with soap and warm water.
- If the splinter projects from the skin, use tweezers to grasp the end of it, as close to the skin as possible, and gently pull it out. If the splinter lies under the skin, sterilize the tip of a sewing needle by holding it in the clear part of a flame for a few seconds. Allow it to cool, then use it to ease the splinter out.
- When the splinter is out, squeeze the wound to encourage a little bleeding, apply a drop of tea tree oil, then cover with a dressing.

When to get medical help
- The splinter is large or cannot be removed.
- The splinter lies over a joint.
- The area surrounding the splinter is red, swollen, or hot.
- The wound around the splinter looks dirty.
- Pus has formed in the area of the splinter.
- You have not had a tetanus immunization within the past ten years.

See also:
Cuts and Scrapes

Strains and Sprains

A strain is damage to a muscle and its tendon, and a sprain is damage to a ligament, which connects and supports the bones on either side of a joint, and the fibrous capsule that encloses the joint. Both types of injury are caused by a sudden pulling or twisting of the affected part that overstretches or tears the tissues. They are a frequent consequence of falls and sports accidents.

The symptoms of strains and sprains are pain, swelling, and bruising. A strained muscle is most likely to occur with unaccustomed exertion or when exercising or lifting carelessly. An ankle sprain often happens after turning the ankle on uneven ground. The tissues around the finger joints and the facet joints of the spine (see p. 105) are other common sites of sprains.

To prevent sprains, watch your step, wear appropriate shoes for sports, and discard shoes that permit your ankles to turn over easily. To prevent strains, warm up and stretch before exercising, and cool down and stretch afterwards.

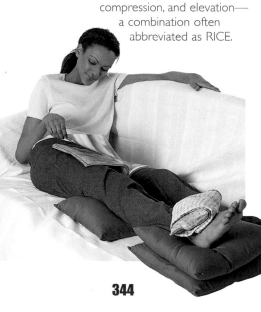

RICE remedy
Treat sprains and strains with rest, ice (or a cold pack), compression, and elevation— a combination often abbreviated as RICE.

Treatment

- Apply a well-wrapped ice pack or a cold compress for 10 minutes several times daily for the first two days. After that, alternate a hot compress or heated pad with a cold compress. Apply an elastic bandage. If the skin becomes numb or blue, remove the source of cold. Get medical advice if normal sensation and colour do not return within a few minutes.
- Elevate the injured part to reduce swelling.
- Rest the joint or muscle for at least 48 hours. Then gradually start moving the affected part.

Diet
- Add turmeric to your food. This spice is a potent anti-inflammatory agent.
- Eat pineapple. It contains bromelain, an enzyme that helps reduce inflammation.
- Eat foods rich in calcium and magnesium (see p. 12). These are key nutrients for muscles.

Herbal remedies
- To encourage healing, apply comfrey or arnica ointment over the injury before bandaging. Do not use either remedy on broken skin.
- To reduce inflammation, make a tea or take a supplement of the Indian herb boswellia.

Caution: For safety concerns, see pp. 34–37.

Homeopathy
- Arnica: for immediately after injury.
- Rhus toxicodendron: for severe strains, when the pain is eased by gentle motion.

When to get medical help
- Your symptoms are severe or don't improve within two or three days.

Get help right away if:
- The part appears misshapen and/or you cannot move it.

See also:
Back Pain, Muscle Aches and Stiffness

Stress

People experience stress in response to various physical, mental, or emotional stimuli. Some stress in our lives can bring out the best in us, but if the level of tension is too high or lasts too long, we may lose our ability to cope and may even become ill. Everyone needs to learn effective ways of dealing with unavoidable stress.

The challenge of stress can be exciting, stimulating, or energizing. Some individuals thrive under continual stress, but most people find that they can cope for only so long before developing physical problems, or "distress".

Stress hormone levels normally fall once the stress is over and you can relax. The hormone levels may stay elevated if the stress continues or recurs frequently, or if you get into the habit of reacting to any stress, however minor, with distress. Research indicates that as much as 75 percent of disease is stress-related.

Stress-related illnesses include high blood pressure, heart attack, stroke, depression, anxiety, chronic fatigue syndrome, irritable bowel syndrome and other digestive disorders, obesity, migraine, and respiratory problems. Feeling continually stressed also disturbs the immune system, making you more likely to develop infections, cancer, and autoimmune disease, in which

Causes of stress

Common causes of continual stress include:

- Loss or bereavement.
- Poor relationships.
- Money worries.
- Unemployment.
- Poor time-management.
- Insufficient recreation and relaxation.
- Boredom.
- Ill health.

the immune system turns against the body's own cells. Examples of autoimmune disorders are rheumatoid arthritis, lupus, thyroid disease, and certain types of anaemia and fertility problems. Smoking, overeating, and other common addictions are often stress-related.

Prevention

It is not always possible to prevent the events that cause stress, but you can modify your response to stimuli. Learn stress-management strategies (see p. 347) so that the next time you feel stressed, you can choose one that is appropriate for you. This will allow your stress hormone levels to fall and help you cope with whatever life brings.

Symptoms of stress

The prolonged presence of high levels of stress hormones may produce one or more of the following mental and physical symptoms:

- Raised heart rate.
- Rapid, shallow breathing.
- Dry mouth.
- Sweating.
- Shaking.
- Weight loss or gain.
- Indigestion.
- Recurrent infections.
- Tearfulness.
- Headaches.
- Difficulty in concentrating.
- Irritability.
- Sleeping difficulties.

Many things influence your response to stress, including whether you are a first-born, or a second, middle, youngest, or only child, your age, gender, education, experience, personality, expectations, and health.

Treatment

General measures: The first part of stress management is to take note of what or who makes you feel stressed, and how you react. The second is to reduce avoidable stresses by, for example, leaving an unsuitable job, living within your means, and making your workplace more comfortable. The third element of stress management is to use strategies to help you respond to pressures in a constructive way (see facing page).

Diet: During periods of stress, the body uses up nutrients faster than usual. This may lead to deficiencies and the consequent lowering of immunity unless the nutrients are replaced through the food you eat or by supplements.

- Eat regularly, and aim for relaxed mealtimes.
- Eat foods rich in vitamins A, B, C, and E, flavonoids, calcium, magnesium, selenium, and essential fatty acids (see p. 12).
- Reduce your caffeine and alcohol intake.
- Eat more fruit, vegetables, and wholegrain cereals. Consume less animal protein and fewer refined foods made with white flour or added sugar, which tend to reduce immunity.
- Take a daily multiple vitamin and mineral supplement.

Aromatherapy: Use relaxing essential oils—for massage, in the bath, or for inhaling from a vapor-

---

Relax with acupressure

To reduce feelings of stress, use your thumb to press the following points firmly and repeatedly for two minutes each:

- The point (LI 4) in the web between the thumb and index finger on the back of your hand (see p. 88). Do not use this point during pregnancy.
- The point (Liv 3) in the furrow on the top of the foot between the first and second toes, where the bones merge (see p. 276).

---

izer or a tissue. Massage yourself or, even better, ask someone else to do so. Use a massage oil made by mixing a total of eight drops of lavender, neroli, and/or ylang ylang oil into one tablespoon of sweet almond or grapeseed oil. You can also use this oil for a shiatsu neck and shoulder massage (see p. 293), to help relieve tension.

Meditation: The calming effects of meditation —feelings of being centred and at peace—can help you feel more detached about the causes of stress. You can learn basic meditation techniques by joining a class (see also pp. 46–47). Take time to practise meditation every day.

Herbal remedies: Teas made from calming herbs are helpful at especially stressful times. Drink a cup of chamomile, valerian, lime tree flower, or clover blossom tea once or twice daily. Caution: For safety concerns, see pp. 34–37.

➤ continued, p. 348

Exercise troubles away
A daily 30 minutes of brisk exercise reduces distress by "burning off" excess levels of stress hormones and by raising blood levels of mood-lifting chemicals such as endorphins.

Stress-management strategies

Relaxation
The many forms of relaxing include:
- Taking regular work breaks.
- Going on holidays.
- Enjoying hobbies.
- Doing breathing exercises (see p. 87).
- Keeping a journal.
- Watching film comedies.
- Enjoying pets.
- Talking to friends.
- Praying or meditating.
- Soaking in a warm, scented bath.
- Visiting a spa or beauty salon.

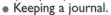

Balance and perspective
You may need to:
- Accept that you can change only your behaviour, not that of others.
- Accept that you cannot have everything you want.
- Recognize what really matters.

- Challenge "shoulds", "oughts", and "musts" in your thinking.
- Forgive yourself and others.
- Try to see stressful situations as opportunities.
- Learn to say no.
- Reject perfectionism.
- Sift information to avoid overload.
- Resist negative thinking.
- Don't magnify or belittle problems.
- Accept the inevitable.

Self-management
Improve your ability to tackle stressful situations:
- Be assertive, rather than passive or aggressive. Take assertion training if you find this hard.
- Define problems, choose goals, and work out how to achieve them.
- Listen to your own needs and concerns.
- Try to stick with a decision once you've made it. Accept that every choice involves giving up something.

Time management
To feel more in control of your life, organize your time efficiently. Use such techniques as:
- Prioritizing.
- Setting realistic deadlines.
- Doing important or difficult tasks when you feel freshest.
- Anticipating stressful times and planning ahead.
- Making time for yourself daily.
- Not taking on too much.
- Delegating.
- Doing one thing at a time.

Relationships
Other people can reduce or bring on stress.
Try to:
- Nurture intimate relationships.
- Improve your communication skills.
- Recognize other people's feelings and separate them from your own.
- Deal with relationship problems.
- Keep in touch with friends.
- Offer encouragement and support to those around you.

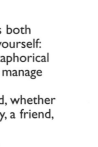

Help and support
Realize that you have resources both within yourself and outside of yourself:
- Every day, give yourself a metaphorical pat on the back for trying to manage stress more effectively.
- Ask for any support you need, whether from a member of your family, a friend, or a counsellor.

Reflexology for stress

Since the principal effect of reflexology is relaxation, you can relieve stress with a reflexology workout on your hands, or by asking a friend to do one on your feet (see pp. 52–53). The reflex points shown below are especially good to work on. Massage each of the areas shown two or three times in each direction. If any part is very sensitive, gently work the area again. (See "Giving treatment", p. 52.)

1 With your left thumb, work around the base of your right thumb.

2 Use your left thumb to work from the little finger side of the base of your right palm across to the thumb side, then up the outer margin of the thumb.

3 With your left thumb, work all over the area between halfway up the fleshy pad at the base of the right thumb, and the V formed by the thumb and index finger.

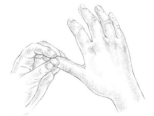

4 Use your left middle finger to work on the outside edge of your right hand, in front of the wrist bone.

5 Use your left thumb to work the top of the right thumb.

6 With your right thumb, work in horizontal lines from the little finger side over your left palm.

Flower essences: Choose a flower remedy whose description closely matches your state of mind and your personality (see p. 33). Relevant essences include:

- Rescue Remedy: for acute distress following a sudden shock.
- Olive: for exhaustion following a period of prolonged stress.

Other therapies: A counsellor or psychotherapist can help you manage stress and examine the reasons why you may find coping with stress difficult. Acupuncture and many forms of therapy that involve exercise or physical manipulation (such as tai chi, massage, and yoga) can ease stress-related symptoms.

When to get medical help
- You feel overwhelmed by stress to the extent that your work or relationships are adversely affected.
- You have physical symptoms of stress, such as palpitations, breathlessness, or headaches.

See also:
Addictions, Anxiety, Appetite Loss, Chronic Fatigue Syndrome, Depression, Eating Disorders, Emotional Problems, Grief, Headache, High Blood Pressure, Indigestion, Irritable Bowel Syndrome, Migraine, Overweight, Pain, Palpitations, Shock, Sleeping Difficulties

Styes

A small abscess in the eyelid, usually at the base of an eyelash, a stye makes the lid painful and swollen. Styes are most likely to occur when you are run-down or tired and therefore especially prone to infection. Home remedies can ease pain and speed up the healing process.

Styes, which are normally not serious, result from infection in an eyelash follicle. They generally heal either by bursting and releasing their pus or by slowly disappearing. Left untreated, a stye usually lasts 7 to 10 days. Most treatments aim to encourage a stye to diminish or burst, stop the infection from spreading, and build up immunity to further infection.

A soothing compress
Put some cotton wool into the bowl of a wooden spoon, and wrap with a bandage. Dip the spoon in a suitable herbal tea (see "Herbal remedies") or hot water and apply to the eye for 5 to 10 minutes.

Prevention

Enhance your resistance to infection with regular exercise, a healthy diet, and stress management. Styes are contagious, so anyone affected should not share towels or flannels with other people, and they should change their flannel daily. If you live with someone who has a bacterial infection of any kind, take care not to touch your eyes after touching the person; wash your hands first.

Treatment

Do not attempt to squeeze pus out of a stye. If it bursts spontaneously, carefully wipe away any discharge.

Herbal remedies

- Eat two cloves of raw garlic daily to boost immunity.
- Make a hot tea from one of the following: eyebright (which helps soothe any surrounding inflammation), burdock, or marigold (both of which counter infection). Prepare and apply a compress as described in the illustration. Repeat every two hours.
- In addition, drink two cups of tea daily made from echinacea, burdock, goldenseal, cleavers, and peppermint, to boost immunity.

Caution: For safety concerns, see pp. 34–37.

From the chemist: Take daily supplements of vitamin C, flavonoids, and fish oil (for omega-3 fatty acids) to help overcome the infection.

Homeopathy

- Pulsatilla: for the onset of a stye, and for recurrent styes with yellow pus.
- Silica: to bring the stye to a head if pulsatilla doesn't help.

When to get medical help

- The stye has not burst or diminished after three days.
- You suffer from recurrent styes.
- The stye grows large.
- The eye itself is red or has a discharge.

> *See also:*
> **Abscesses and Boils**

Sunburn

Although most common in fair-complexioned people, whose skin produces only small amounts of the protective pigment melanin, sunburn can affect anyone who has been excessively exposed to the sun's ultraviolet rays. Prevention is the best approach, especially for children, because severe childhood sunburns have been linked to malignant melanoma in later life.

Safe in the sun
When you have to go out in strong sun, a wide-brimmed hat helps shade your face.

Repeated overexposure to ultraviolet A (UVA) rays leads to premature wrinkling and an increased risk of skin cancer. Overexposure to UVB rays causes sunburn, which damages the top layer of skin and also contributes to the cancer risk. The redness and tenderness of sunburn usually increase for several hours after exposure and last for up to a week. The skin may blister.

Prevention

Limit exposure to strong sunlight to no more than 15 minutes a day at first, increasing the time gradually, if you wish. Before going into the sun, apply a broad-spectrum sunscreen of factor 15 or higher, depending on your skin type. Reapply every few hours and after swimming. For further protection, wear tightly woven clothing, and try to avoid strong sunlight from 10 A.M. to 3 P.M. Keep babies under six months out of the sun; use shading devices on buggies. When out in the sun, drink plenty of water to keep your skin hydrated. Take the antioxidants beta-carotene, vitamins C and E, and selenium daily. After sun exposure, apply an antioxidant moisturizer.

Treatment

- Don't peel burned skin or prick blisters.
- Soothe sunburned skin with cool compresses or a cool bath or shower, or apply aloe vera gel or calamine lotion.
- Ease dryness and flaking with a moisturizer.
- Reduce the discomfort of mild sunburn with a dusting of cornflour.
- Protect burned skin from further sun exposure.

Homeopathy

- Urtica, calendula, or Hypercal cream: for stinging, unblistered skin.
- Urtica, calendula, or Hypercal tincture (1 part diluted with 10 parts water): for any sunburn, applied to the skin liberally and immediately.
- Cantharis: for severe sunburn, taken hourly.

Flower essences: A cold compress made with a few drops of Rescue Remedy helps cool the skin.

When to get medical help

Get help right away if:
- You have severe redness and/or blistering.
- You experience fever, chills, dizziness, headaches, or vomiting.
- A young child is badly burned.

See also:
Burns

Swollen Ankles

An abnormal accumulation of fluid in the soft tissues can cause swelling, especially of the ankles. This condition, known as oedema, may have an underlying cause that requires medical treatment, such as heart disease. Nevertheless, lifestyle changes and home remedies can often play an important role in reducing fluid retention and easing symptoms.

Ankle swelling unrelated to injury usually occurs because the normal balance between fluid levels in the blood and the body cells is disrupted. This can happen when higher than normal pressure in the small blood vessels, the capillaries, forces water into adjoining tissues. Long periods of standing or sitting, a high salt intake, and hormone fluctuations before menstruation and during pregnancy may all make swollen ankles more likely. The disorder sometimes indicates an underlying condition, such as varicose veins, a kidney problem, heart or other arterial disease, high blood pressure, or pre-eclampsia (high blood pressure and fluid retention during pregnancy).

Treatment

General measures

- Exercise daily for a half hour. This boosts the circulation and discourages fluid from pooling in the lower limbs.
- To strengthen capillary walls, eat foods rich in vitamin C and flavonoids (see p. 12).
- Reduce your salt intake.
  - If you sit or stand still for a long time, contract your calf muscles

frequently. This squeezes blood up through the calf veins, which helps prevent ankle swelling.

Aromatherapy

- Take a warm bath to which you have added six drops of geranium oil, six drops of lemon oil, or three of rosemary oil. (Avoid rosemary oil for the first 20 weeks of pregnancy.)

Herbal remedies

- Drink a daily cup or two of lukewarm-to-cool tea made from dandelion leaves and yarrow.

Caution: For safety concerns, see pp. 34–37.

DID YOU KNOW?

Drinking lemon juice may help relieve ankles swollen from fluid retention because it increases urine production. Other diuretics include carrots, leeks, cucumbers, onions, and turnips.

Put your feet up
Lie with your feet raised above your hips for at least 20 minutes twice daily to allow excess fluid to drain away from the ankles.

When to get medical help

- Your symptoms worsen or fail to improve.
- Your symptoms begin after you take a new medication.
- The swelling is in only one ankle.
- You also have swelling or tenderness in the groin or lower abdomen.
- You are short of breath.
- You are pregnant.
- You experience numbness or tingling in the leg or foot.

See also:
Bloating, Pregnancy Problems, Premenstrual Syndrome, Varicose Veins

Swollen Glands

The lymph nodes—commonly but mistakenly referred to as "glands"—are collections of cells along lymph vessels that play an important role in the body's defences. Swollen glands are a sign that you are actively fighting an infection or other invasion by harmful organisms or cells. There's a lot that you can do to help the lymph nodes protect your body.

The lymph nodes manufacture infection-fighting antibodies and white cells. They also trap infecting microorganisms, cancer cells, and other invading particles. When they are swollen due to an infection, you feel them as tender, painful, slightly warm lumps, usually under the skin of the neck, armpits, or groin. During a localized infection, as from an infected cut, lymph nodes near the affected area may become enlarged.

Swollen lymph nodes are most often caused by viral infections, such as colds, flu, and certain types of sore throat. They are one of the main symptoms of infectious mononucleosis ("glandular fever"), a viral disease. Swelling can also result from bacterial infection, as of the throat. In rare cases, swollen lymph nodes result from cancer of the blood or lymphatic system.

Prevention

You can take steps to reduce the likelihood of the infections that cause swollen lymph nodes.
- Eat plenty of immunity-enhancing foods—those rich in folic acid, flavonoids, copper, iodine, iron, magnesium, selenium, zinc, essential fatty acids, lecithin, beta-carotene, and vitamins B complex, C, D, and E (see p. 12).
- Limit your intake of refined carbohydrates, such as sugar and white flour.
- Get regular exercise to boost immunity.
- Avoid alcohol and tobacco.

Treatment

Follow the advice given under "Prevention" and try these additional measures:
- Get enough rest and sleep.
- Eat some raw garlic daily.
- Take the herbal remedy echinacea and/or astragalus or shiitake mushrooms, all of which are immune-system boosters, and elderberry extract or goldenseal to fight infection. For safety concerns about the use of herbs, see pp. 34–37.
- Apply cold compresses (see p. 50) to the swelling for the first one to two days. Then apply heat, such as from a covered hot-water bottle, at regular intervals until symptoms clear.

When to get medical help
- Mildly swollen lymph nodes remain swollen for more than one week.
- Your lymph nodes increase in size or become more tender after three days.
- You have a high or persistent fever.
- You have a headache, aching muscles, abdominal pain, jaundice, a rash, or a bacterial infection, such as tonsillitis.

See also:
Abdominal Pain, Childhood Viral Infections, Coughs, Earache, Fever, Sore Throat

Toothache

An aching tooth (or teeth) is usually a sign of decay or infection or other inflammation inside the tooth or the gum. These conditions always require treatment by a dentist, but you can use a variety of natural home remedies to relieve pain while waiting for dental treatment. You can also reduce the risk of future dental problems.

The main cause of tooth decay is plaque, a sticky film of food residue, saliva, and bacteria that accumulates on teeth. As the plaque bacteria consume sugars and starches from food particles, they produce an acid that can destroy enamel, the protective surface of the tooth. If your food contains added sugar, the amount of acid increases sharply, and its level stays high for about 25 minutes after eating unless removed, for example by brushing. If you eat sugary foods frequently, you run a high risk of suffering from tooth decay.

Advanced decay, with deep cavities that allow bacteria to enter the pulp well within the tooth, will eventually lead to a painful dental abscess.

Prevention

Keep your teeth healthy by brushing at least twice daily, and preferably after every meal. Use dental floss daily to remove food particles and plaque from between your teeth.

- Minimize your consumption of refined carbohydrates (such as added sugar and white flour). Such foods promote plaque formation and raise acid levels in the mouth.
- Reduce excess acidity in your mouth after eating by brushing your teeth or by ending your meal or snack with naturally antacid foods, such as nuts and cheese. Simply swilling water around in your mouth can also remove food residue and counter acidity.
- Drink tea, which is a good source of fluoride, a decay-fighting mineral. Tea contains tannins which can also help to combat bacterial activity.
- Discourage bacterial activity by rinsing your mouth daily with a few drops of myrrh tincture in a glass of water.
- Ensure that you visit a dentist regularly for checkups.

Pain-relieving herbs
Tea made from hops, valerian, or wild lettuce may ease toothache, because these herbs have a soothing effect on the nervous system.

Teething pain

Many babies experience discomfort as their first teeth erupt through the gums, usually starting at about six months. Your child may cry more, have a "dribble rash" around the mouth, have difficulty sleeping, and be irritable or clingy. A teething ring to chew on may help. You can also try massaging the gum with your fingertip or using one of these homeopathic remedies:

- Chamomilla: for a cross, irritable child, who has one cheek redder than the other. Give this remedy as granules, which dissolve easily in the mouth.
- Pulsatilla: for a clingy child whose pain is helped by cool, fresh air.

Treatment

If you have a toothache or painful swollen gums, see your dentist as soon as possible. To reduce the pain of an abscess, apply a hot compress for five minutes every half hour and wash out your mouth with warm salt water afterwards. This may help release the pus.

Aromatherapy: Rub a few drops of clove or cajuput oil on the gum above or beneath the aching tooth. Repeat several times daily. Avoid clove oil throughout pregnancy and cajuput oil in the first 20 weeks of pregnancy.

Herbal remedies

- Apply a cotton wool ball soaked in echinacea tincture to the affected tooth.
- Chew prickly ash bark, or apply a paste of powdered bark mixed with a little water.

Caution: For safety concerns, see pp. 34–37.

Homeopathy

- Chamomilla: for unbearable toothache made worse by cold air and warm food and drink.
- Calcarea carbonica: for toothache that begins when you eat and is made worse by cold air and cold drinks and food.
- Magnesia phosphorica: for toothache like little electric shocks, that is relieved by warmth.
- Belladonna: for an abscess that develops quickly and looks red, hot, and swollen. The tooth throbs and your mouth feels dry.
- Mercurius solubilis: for sore, bleeding gums. The pain is worse at night; you feel thirsty; your mouth has more saliva than usual; and you have bad breath.

Acupressure: With your thumb, press the point LI 4 on the back of the hand opposite to the side of the mouth with toothache, in the web of skin between the thumb and index finger (see p. 88). Apply pressure to the point for about two minutes. Do not use this point if you are pregnant.

A natural anaesthetic
Soak a cotton wool ball in clove oil and apply to the affected tooth, or chew a clove. Cloves contain eugenol, a local anaesthetic often used by dentists.

Tooth-friendly foods
After meals, eat nuts or cheese, which counteract acidity, or a fibrous food, such as celery, which removes plaque because of its slightly abrasive effect.

When to get medical help
- See your dentist in all cases.

See also:
Abscesses and Boils, Gum Problems

Urinary Difficulties

Pain when urinating, frequent urination, and urinary retention (inability to empty the bladder properly) are the main difficulties of this type. These problems often require medical attention or even surgery. However, in many cases, self-care with natural remedies can support and enhance the effectiveness of the treatments prescribed by your doctor.

Most people pass urine between four and six times daily, depending on the volume of fluids they drink and how active they are. The problem of frequent urination may result from a large fluid intake, a urinary-tract infection, irritation from stones in the bladder, an enlarged prostate gland, urge or overflow incontinence (see p. 250), diabetes, or anxiety. Pain during urination, usually described as burning or scalding, may be caused by urinary-tract infection or other inflammation, a vaginal yeast (candida) infection, inflammation of the penis, or kidney or bladder stones.

A slow or weak stream of urine is common in older men because of prostate enlargement. This symptom may also occur in older women, possibly because of weakness in the bladder-wall (detrusor) muscle. Inability to pass urine (urinary retention) most often affects men. It may result from an obstruction (such as a stone or a congenital anatomical abnormality) in the bladder or urethra, a tight foreskin, or an enlarged prostate. It occasionally happens in women because of pregnancy or fibroids in the uterus, both of which can put pressure on the urethra. Some urinary problems arise from a long-standing nervous-system disorder, such as multiple sclerosis.

Prevention

- Drink plenty of water. Avoid caffeinated drinks, such as coffee, tea, and cola; these have a diuretic effect, creating large amounts of urine.
- Eat a healthy diet, with plenty of foods rich in flavonoids, selenium, zinc, plant hormones, and vitamins A, C, and E (see p. 356). These help boost your resistance to infection, ensure that your muscles, nerves, and other tissues are in good condition, and make prostate enlargement and fibroids less likely.

Flush away problems
Drink plenty of water to keep your urinary tract in good health. For most people this means six to eight glasses of water-based fluid a day.

- Eat ample fibre-rich foods, so as to prevent pressure on the bladder from constipation.
- Lose any excess weight, since deposits of fat in the abdomen can put pressure on the bladder.
- Empty your bladder regularly to prevent infection from developing in stagnant urine.
- To avoid anxiety, practise stress-management techniques, such as yoga, deep breathing, and other relaxation exercises (see p. 347).
- Women of all ages should practise pelvic-floor exercises (see p. 251) to maintain the strength of the muscles that support the urinary and reproductive organs.
- Be a non-smoker, or, if you smoke, at least cut down, since cadmium from cigarette smoke can accumulate in the kidneys and

encourage the formation of stones. Smoking is also a significant risk factor for certain types of bladder cancer.

Treatment

Appropriate treatment depends on the under-lying problem. But anyone with urinary difficulties should follow the advice given under "Prevention" along with any prescribed treatment for an underlying disorder.

Herbal tonics for the urinary system
Take cleavers or couch grass in the form of a tea or tincture when a gentle diuretic effect is beneficial, as with cystitis or kidney stones.

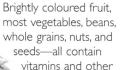

The right nutrients
Brightly coloured fruit, most vegetables, beans, whole grains, nuts, and seeds—all contain vitamins and other substances that help prevent urinary difficulties.

When to get medical help

- Urination is painful, or you have a discharge.
- Copious urination is accompanied by unusual thirst.
- Symptoms are severe or continue for more than a few days.
- Symptoms occur along with pain in the lower abdomen or, for men, the perineum.
- Urine is pink, cloudy, or bloody.
- A baby or child has urinary problems.

Get help right away if:

- You are unable to urinate.

See also:
Abdominal Pain, Anxiety, Candida Infections, Diabetes, Fibroids, Incontinence, Kidney Stones, Overweight, Prostate Problems, Urinary-tract Infections

Urinary-tract Infections

Inflammation and/or infection of the lower part of the urinary tract (bladder and urethra) affects four in five women at some time. Such conditions are not usually a risk to general health. However, it is important to treat them promptly to avoid the risk of the kidneys becoming affected—a more serious problem. You can use natural remedies along with prescribed medication.

Most urinary-tract infections are caused by *Escherichia coli (E. coli)* bacteria. These are normally present in the large intestine, but they can spread from the anus to the urethral opening and up to the bladder. If conditions allow, the bacteria can settle into the bladder lining, multiply, and cause infection and, perhaps, inflammation. Conditions that make this more likely include urinary-tract injury, kidney stones, anatomical abnormality, concentrated or acidic urine, and certain food constituents.

Symptoms

If your urinary tract is inflamed, you may notice some of the following:

Sites of inflammation or other infection
This can occur anywhere in the urinary tract, from the kidneys (pyelonephritis) to the bladder (cystitis) or urethra (urethritis).

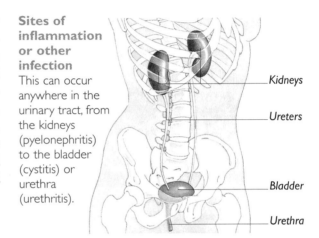

Kidneys

Ureters

Bladder

Urethra

- Burning or stinging before, during, or immediately after urination.
- A frequent urge to urinate.
- Low abdominal pain (from the bladder).
- Pain in the back above the waist (from one or both kidneys).
- Tenderness above the pubic bone.
- Blood in the urine, making it pink or cloudy.
- Pus in the urine, making it cloudy.
- Strong—perhaps fishy—smelling urine.
- A fever.
- A sudden and irresistible urge to urinate.

Why women?

Several factors make cystitis more common in women than in men:
- The relatively short urethra in women allows bacteria easy access to the bladder.
- During intercourse the vagina, the urethra, and the base of the bladder may be bruised. Bacteria may also be pushed into the urethra.
- A tampon may press against the upper vaginal wall and irritate the base of the bladder.
- The rim of a diaphragm may press against the upper vaginal wall and irritate the bladder base.
- Spermicidal foams and gels may irritate the urethra.
- During pregnancy, the pressure of the enlarged uterus can make complete bladder emptying difficult, allowing infection to become established.
- After the menopause, the urinary-tract lining may gradually become thinner, less elastic, and more vulnerable to irritation.

Prevention
- To flush out bacteria and other irritants, drink enough water to make the urine appear light in

colour at most times of the day. This is about six to eight glasses a day for most people.

■ Reduce your intake of caffeine and alcohol, which can irritate the urinary tract.

■ Stop or reduce smoking, since nicotine can irritate the urinary tract.

■ Wash your genital area with water only, to avoid irritation of the urethra.

■ Never delay urinating—the longer urine remains in the bladder, the more likely it is that bacteria will take a hold in the lining.

■ Wash your genital area before intercourse. A woman should also urinate before and after sexual intercourse.

■ Change sanitary pads or tampons frequently.

■ Avoid bath salts, bubble baths, and vaginal deodorants that may irritate the urethra.

■ Don't wear tight or thick underwear if you suffer from recurrent cystitis.

■ To help prevent bacteria from adhering to the bladder wall, drink a small glass of cranberry juice daily for its anti-inflammatory properties.

■ Eat plenty of foods rich in vitamin C and flavonoids (see p. 12) to increase your general resistance to infection.

■ Drink a cup of echinacea tea (or take a few drops of tincture each day) and have some garlic (or garlic tablets) to boost your immunity.

■ If cystitis occurs after swimming in a chlorinated pool, avoid chlorinated water, if possible, or drink plenty of water after swimming.

■ If cystitis follows sex, reduce friction between the penis and the front of the upper vagina by getting aroused before penetration, or using saliva or a commercial lubricant, and by avoiding any positions that encourage symptoms.

A grain for pain
Barley water is a traditional remedy for urinary-tract infections. Make it by boiling a teaspoon of barley in two pints of water for an hour. Strain and flavour with lemon juice and honey.

Treatment

Natural remedies are not a substitute for orthodox medical treatment. However, the measures outlined under "Prevention" and the additional steps described below can be used alongside treatments prescribed by your doctor.

Kitchen-cabinet remedies: Try one of the following barley-water remedies to reduce the burning pain on urination often associated with urinary-tract infections.

■ Have two large cups of barley water (see photograph). Then drink one large cup every 20 minutes for three hours.

■ To make the urine less acidic, every hour drink a cup of barley water with an added teaspoon of bicarbonate of soda.

Herbal remedies

■ Soothe symptoms with corn silk tea, up to three times a day.

- Benefit from buchu's antiseptic and diuretic qualities by drinking buchu tea, but no more than three times a day.
- Drink a small cup of tea made from marshmallow leaf or sea holly (eryngium) root three times daily. This is soothing and mildly diuretic.
- Uva ursi can help reduce the growth of urinary-tract bacteria.
- If burning is severe, sit in a bath of strong chamomile tea.
- For a kidney infection, drink frequent cups of lukewarm echinacea, couch grass, and buchu tea.

Caution: For safety concerns, see pp. 34–37.

Aromatherapy: Essential oils can be of benefit because of their antiseptic effect.

- Sprinkle two drops each of juniper berry, eucalyptus, and sandalwood oils into a bathtub of warm water. Kneel and swish the water over your pelvic area. Repeat several times, then sit in the water for 10 minutes. Omit juniper berry oil if you are or might be pregnant.

Hydrotherapy: Warm and cold spraying of the pelvic area—three minutes using warm water, then one minute cold, repeated three or four times—may help combat urinary-tract infections. This boosts the local circulation and so brings more white cells and other infection-fighting agents to the area. Or take alternate hot and cold sitz baths (see p. 49).

Marshmallow
This common herb is used to soothe inflammatory conditions of the urinary system, such as cystitis and urethritis.

Echinacea
The immunity-enhancing and antimicrobial properties of this herb may help urinary-tract infections.

Homeopathy
- Apis: for burning and soreness while urinating, with last drops particularly painful.
- Cantharis: for frequent urination with scalding pain, and when urine is hard to pass.
- Causticum: for burning pain while urinating.
- Sarsaparilla: for pain at the end of urination.

Spanish fly
This bug is the source of Cantharis, a homeopathic remedy for burning pains, such as those of cystitis.

When to get medical help
- You experience any of the symptoms of a urinary-tract infection (see p. 357).
- Your urine is bloody, pink, or cloudy.

Get help right away if:
- You have back pain in addition to any of the above symptoms.
- A baby or child is affected.

See also:
Abdominal Pain, Incontinence, Prostate Problems, Urinary Difficulties

Vaginal Problems

Itching or inflammation of the vagina and vulva, abnormal vaginal discharge, and pain during intercourse are the most frequent problems affecting the female genital area. Such symptoms are usually a sign of a bacterial or yeast infection. Several home remedies and other self-help measures can help prevent or relieve these conditions.

Itch relief
Geranium, tea tree, lavender, eucalyptus, and yarrow (shown from left to right) provide essential oils that can alleviate vaginal symptoms.

The vagina is normally kept free from infection by natural colourless secretions that keep the lining moist and produce a slightly acidic environment, which prevents excessive growth of any of the normal population of microorganisms. Any alteration to these conditions—for example, following a course of antibiotics—can cause an overgrowth of *Candida albicans*, a yeast-like fungus (see p. 118). Symptoms include itching, tenderness, soreness, and a thick, white discharge.

Other organisms may also multiply in the vagina. Bacterial vaginosis results from an overgrowth of one or more of the types of bacteria normally present in the vagina. It may cause a greyish discharge that has an odour. The vagina and vulva may itch, burn, and become inflamed. Bacterial vaginosis during pregnancy makes premature birth more likely. Microorganisms may also cause trichomoniasis, gonorrhoea, or a chlamydial infection, any of which can be sexually transmitted. These disorders may produce a greenish-yellow discharge.

Possible causes of redness, itching, and soreness—vaginitis—include diabetes, changing hormone levels during pregnancy, and a reaction to chemicals in soaps, bubble baths and other toiletries, or laundry detergents. After the menopause, the drop in oestrogen may make the vaginal lining drier and thinner, leading to discomfort during sexual intercourse.

Prevention

- Wash the genital area regularly, and gently pat dry with a clean towel.
- Avoid hot baths and showers, and don't remain in the bath longer than 15 minutes.
- Don't use scented soaps, bubble bath, or other bath additives.
- Don't use feminine deodorants or douches (unless prescribed by your doctor).
- Wear cotton underwear, and stockings rather than tights. Avoid trousers that are very tight.
- Avoid enzyme-containing laundry detergents.
- After using the lavatory, wipe from front to back to prevent the transfer of intestinal organisms to your vagina.
- Apply lubricating gel or cream before intercourse, if necessary to avoid discomfort.
- Consider having your partner wear a condom. This will help prevent the transmission of infecting bacteria and other organisms.
- Change tampons every three to four hours.

Treatment
Herbal remedies

- To soothe irritation, sit in a shallow bath containing a cup of herbal tea made from a combination of chamomile, chickweed, goldenseal,

Helpful herbs
If you suffer from vaginal infections, improve immunity by drinking a daily cup of tea (or a few drops of tincture in half a cup of water) made from thyme, echinacea, cleavers, and goldenseal.

marigold, and thyme. Or bathe the area with a solution of distilled witch hazel.

- Relieve itching by bathing the genital area with chickweed tea, which has a gentle anti-inflammatory effect.
- Try the teas suggested under "Helpful herbs", above. The hormone-balancing herbs black cohosh and dong quai may also be beneficial.

Caution: For safety concerns, see pp. 34–37.

Diet

- To counter infection, boost your immune system with a wholesome diet (see pp. 9–12). Include plenty of foods rich in beta-carotene, folic acid, flavonoids, iron, magnesium, zinc, essential fatty acids, and vitamins B, C, and E.
- Restrict your intake of saturated fats, sugar, meat, and caffeine (which stimulates insulin production, releasing sugar and encouraging candida). Include in your diet oils rich in monounsaturated fats, such as olive oil, which act against candida.
- If you are postmenopausal, include soybeans and soy-based foods in your regular diet. These contain plant hormones that may help counter vaginal dryness.

Aromatherapy

- Soothe inflammation and irritation by applying this mixture to the external genital area: two drops of myrrh oil and four of lavender oil added to two and a half teaspoons of unscented lotion or cold cream or jojoba oil.
- Reduce itching by wearing a panty liner on which you have sprinkled a few drops of sweet thyme or tea tree oil diluted with two teaspoons of jojoba oil.
- Lessen postmenopausal inflammation with chamomile, geranium, lavender, yarrow, tea tree, or eucalyptus oil. Add a total of six drops to a warm bath, and soak for 15 minutes.

When to get medical help

- You have an unusual vaginal discharge or unusual bleeding.
- Your symptoms last longer than three days.
- You are or might be pregnant.
- You have pain during or after intercourse.
- You have abdominal or pelvic pain.

See also:
**Candida Infections,
Menopausal Problems,
Urinary Difficulties,
Urinary-tract
Infections**

Varicose Veins

Too much pressure inside veins can damage their valves and lead to a build-up of blood that makes them twisted and swollen, or varicose. This can happen to any vein, but superficial leg veins are those most frequently affected. Changing your lifestyle and using some natural remedies can help you avoid the need for surgery.

Improving blood flow
Make a half hour's daily exercise—such as walking, swimming, or cycling—a priority. This tones the veins and boosts the circulation.

In a normal leg vein, valves stop blood from draining back down the leg under the force of gravity. When these valves no longer work efficiently, blood collects and dilates the veins, causing those just under the skin to turn blue and lumpy, especially on the backs of the calves and the insides of the legs. Some people have no additional symptoms, but others experience swelling of the feet and ankles, aching legs, itching, and in severe cases, eczema, skin discoloration, or ulceration.

One person in five develops varicose veins. Women are four times more likely to be affected than men, and the disorder tends to run in families. Other factors that increase your risk of developing varicose veins include a poor diet, too little exercise, standing or sitting still for long periods, and being overweight. Pregnancy also heightens the risk because of an increase in blood volume and because the enlarged uterus presses on the abdominal veins, causing more pressure in the leg veins.

Prevention
To reduce the risk of varicose veins:
- Whenever possible, avoid standing or sitting still for long periods.
- Lose excess weight.
- Improve your circulation with daily activity, such as walking.
- On long plane flights, get up and move around for 10 minutes every hour or two.
- When sitting, put your feet up, if possible.
- If your legs are tired or aching, rest them by raising them above hip level.
- Make sure tights, stockings, suspender belts, and socks are not too tight.

Treatment
General measures: Take the precautions listed above. In addition:
- Don't stand still for too long, especially in hot weather. If you must stand for an extended time, tighten and relax your calf muscles frequently to prevent blood from collecting in your veins and making them swell further.
- When sitting, don't let the edge of your chair obstruct your blood flow by cutting into your thighs, and don't cross your legs.
- Consider wearing support stockings.

Diet: Eat at least five daily servings of vegetables and fruit. Citrus fruits are especially rich in nutrients that help maintain the strength and elasticity of the vein walls, including vitamin C, silica, and flavonoids (plant pigments). Berries and cherries, rich in red and purple flavonoids called proanthocyanidins, and buckwheat, rich in

another vein-strengthening flavonoid called rutin, are also valuable.

Eat plenty of fibre-rich foods, such as brown rice and wholegrain cereals. These help prevent constipation and the subsequent straining that increases pressure in the leg veins.

Kitchen-cabinet remedy: Apple-cider vinegar has astringent properties and contains flavonoids,

which can ease vein swelling and inflammation. Apply over varicose veins morning and night.

From the chemist: You should take daily supplements of vitamins C and E, proanthocyanidins, rutin, silica, and zinc.

Herbal remedies: The astringent and anti-inflammatory properties of several herbs—some

Yoga to the rescue

Practising yoga exercises on a regular basis will stimulate the circulation and improve the drainage of tissue fluid and lymph which may, as a result, prevent varicose veins from getting worse. Inverted yoga postures such as these are the most beneficial for this purpose, but you should not attempt them if you have untreated high blood pressure or if there is a possibility that you may have blood clots in your legs.

Resting position
Try this position to ease your symptoms if you are new to yoga.

Lie on your back with your buttocks on a cushion about 18 inches from the wall. Rest your feet against the wall, and breathe slowly and deeply. Hold the posture for about five minutes.

The shoulder stand
This posture can help reduce the pressure inside your leg veins. Do not attempt a shoulder stand if you are new to yoga, unless you are supervised by an experienced teacher.

1 Lie on your back with your legs together and hands palms-down by your sides. As you inhale, push down on your hands and slowly swing your legs up over your head. Raise your hips from the floor and support your lower back with your hands as far down as possible. This is the half shoulder stand.

2 As you exhale, bring your legs to a vertical position and straighten your spine into a full shoulder stand. Hold the pose for about 20 breaths, breathing slowly and deeply. Release the pose by bending your knees and slowly lowering your legs to the floor.

taken internally and some applied externally—lend themselves to alleviating the symptoms of varicose veins.

■ To tone veins and reduce discomfort, bathe the affected area with one of the following: distilled witch hazel alone, four drops of calendula tincture in a tablespoon of distilled witch hazel, or cooled comfrey (if the skin is not broken) or marigold (calendula) tea.

■ Soothe aching veins with a cold compress made with two teaspoons of marigold (calendula) tincture in half a pint of cold water.

■ To strengthen your vein walls and promote circulation, take a daily dose of gotu kola and/or ginkgo biloba. Bilberry and butcher's broom may also be effective.

■ To relieve pain caused by varicose veins, apply comfrey ointment. Do not use this ointment on broken skin.

Caution: For safety concerns, see pp. 34–37.

Hydrotherapy

■ Use alternate hot and cold compresses, leaving each one in place for 30 seconds and repeating the sequence three times. Always finish with a cold compress. Carry out this treatment once a day.

■ Spray your legs and feet with cold water for two minutes twice daily to ease swelling and discomfort.

A circulatory tonic
Horse chestnut has anti-inflammatory properties and improves the tone of the vein walls. It should, however, be taken only under the supervision of a qualified practitioner.

An astringent healer
Witch hazel (*Hamamelis virginiana*) is a native North American plant. Applied externally, its distilled extract promotes constriction of blood vessels, thereby reducing swelling.

Aromatherapy: Make a toning and soothing massage oil for varicose veins by adding three drops of cypress oil, two drops of sandalwood oil, and one drop of peppermint oil to five teaspoons of sweet almond oil or calendula lotion. Omit cypress oil if you are in the first 20 weeks of pregnancy. In the morning and evening, put your legs up and smooth the mixture over your veins with upward strokes.

Homeopathy

■ Hamamelis: for veins that feel bruised and sore, with a bursting feeling. This remedy can also be used in the form of an ointment, massaged into the skin over the swollen veins.

■ Vipera: for inflamed veins (phlebitis).

When to get medical help
● Your varicose veins cause pain or bleed.
● Your leg becomes swollen, inflamed, or ulcerated.

Get help right away if:
● You have chest pain or sudden breathlessness.

See also:
Dermatitis, Haemorrhoids, Skin Problems

Voice Loss

Any disorder of the vocal cords can affect the ability to speak normally. Most temporary voice loss is caused by laryngitis—inflammation of the voice box (larynx) and the vocal cords within—as a result of infection. Another reason for the disorder is overuse of the voice. The condition is often painless and usually lasts only a few days. It responds well to home remedies.

Voice loss occurs when the vocal cords, the fibrous bands of tissue inside the larynx that vibrate when we speak, are unable to move normally. This is frequently the result of a viral infection, such as a cold. Tobacco smoke and other irritants may also cause inflammation of the voice box. Singers and others, such as teachers, who use their voices a lot are susceptible to voice loss resulting from improper or excessive use. Persistent voice loss or hoarseness may be caused by polyps (benign growths) on the vocal cords, or, in rare cases, cancer.

Treatment

- If you have an infection, limit your activities, and treat as for a cold or sore throat.
- Rest your voice, but don't whisper.
- Keep the throat lining moist by humidifying the air with a humidifier, a boiling kettle of water, or bowls of water placed by radiators.
- Drink plenty of fluids, but avoid caffeine and alcohol, since they encourage dehydration by increasing urine production.
- Don't smoke or breathe polluted air.

Aromatherapy
- To bring swift relief, inhale steam scented with a total of four to six drops of antiseptic and anti-inflammatory oils, such as lavender, sandalwood, and chamomile.
- To reduce inflammation, gargle every two hours with two drops each of sandalwood and lemon oils in half a glass of warm water.

Herbal remedies
- Make a hot compress (see p. 37), using mullein, sage, thyme, or hyssop tea. Apply it to your throat. Wrap a dry towel around your neck to keep the heat in. Apply a new compress when the previous one has cooled.
- Gargle three to six times a day with a few drops of myrrh, sage, or tea tree tincture in half a glass of warm water.
- Every two hours, drink a cup of tea made from echinacea, mullein, catnip, coltsfoot, and thyme.

Caution: For safety concerns, see pp. 34–37.

Homeopathy
- Phosphorus: for all types of laryngitis.

When to get medical help
- Voice loss is accompanied by high fever and throat pain lasting more than two days.
- Voice changes persist longer than one week.

See also:
Colds, Coughs, Influenza, Sore Throat

Warts

Contagious growths consisting of dead skin cells, warts affect only the top layer of skin. These lumps have a rough surface and can appear almost anywhere, but most often they affect the hands and face. A wart on the sole of the foot, called a verruca or plantar wart, appears relatively flat because of pressure from the weight of the body.

Help from a weed
Squeeze a drop of juice from a dandelion stalk onto a wart each day.

Warts occur when viruses, usually the papilloma type, invade skin cells. They may be unsightly, and they occasionally itch. A plantar wart on the heel or ball of the foot can cause pain when you are walking, because the weight on the foot presses it inwards. Most warts disappear within six months to two years without treatment.

Prevention

Wart viruses spread easily on moist floors, so cover your feet in public places, such as communal showers, jacuzzis, and the area around swimming pools. Dry your feet well, and don't share flannels or towels. Boost your resistance to infection with a healthy diet (see pp. 9–12), two or three cloves of raw garlic each day, and foods rich in beta-carotene and vitamins C and E.

Treatments

Few treatments are entirely dependable, but several are worth trying.

Herbal remedies
- Paint the wart, morning and evening, with a few drops of thuja or marigold (calendula) tincture.
- In addition to external treatment, drink teas or take tinctures of echinacea, burdock, dandelion root, or red clover.

Caution: For safety concerns, see pp. 34–37.

Kitchen-cabinet remedies
- Put lemon juice or the juice of crushed garlic on the wart morning and evening.
- Cover onion slices with salt, and leave overnight. Dab the juice on the wart twice daily.
- Every night, apply a bandage that has been soaked in apple-cider vinegar or that encloses a flat sliver of garlic.

Aromatherapy: Twice daily, apply lemon, tea tree, cypress, or lavender oil (for a young child, dilute one drop of oil in half a teaspoon of vegetable oil). Try not to touch the wart, and don't allow the oil to touch uninfected skin. Avoid cypress oil in the first 20 weeks of pregnancy.

Homeopathy
- Thuja, in pill form or as a cream or tincture.

When to get medical help
- Your wart hurts or looks or feels inflamed or infected.
- You have warts on the face or in the genital or rectal area.
- You develop a wart after age 45.

See also:
Skin Disorders

Weight Loss

Everyone's weight fluctuates to some extent, generally because of varying levels of food intake, exercise, and hormones, or a change in the type of food in the diet. Although losing pounds may be beneficial if you are overweight, a sudden, unexplained weight loss can indicate a serious underlying disorder, so it is important to discover the cause.

Unusual weight loss most often results from a restrictive diet or from being unable to eat—for example, after an operation or during a serious illness. Weight loss can also result from an eating disorder. However, there are other times when it is less easily explained, and your physician may need to exclude the possibility of a food sensitivity or an underlying disease, such as a peptic ulcer, inflammatory bowel disease, or irritable bowel syndrome. Depression and anxiety are also frequent causes of this problem.

Prevention

If you have a tendency to lose weight, make sure that you are eating a healthy, well-balanced diet. Opt for frequent, small, appetizing meals, and aid digestion by taking your time and chewing well. Don't skip meals, and plan the day's meals in advance. For example, if you won't be able to have a proper lunch, pack a nutritious meal and take it with you. Make a point of never going without a good breakfast. You may also want to take a multiple vitamin and mineral supplement to provide you with nutrients, such as vitamin B and zinc, that help maintain a good appetite.

Treatment

General measures: Try to increase your weight to the normal level for your height and build (see table, p. 302).

- Choosing items from all the food groups, eat more to provide more calories. Carbohydrates (such as bread and pasta) and monounsaturated fats (such as olive oil) are especially good. Liquid nutritional supplements between meals may also help.
- Get daily exercise, of both the aerobic and muscle-strengthening varieties, tailored to your own fitness level. Some people lose weight because they exercise too much, and a few seem to become addicted to excessive exercise levels, so don't overdo it.
- Make sure you get some sunlight on your bare skin on most days. Sunshine encourages your body to manufacture vitamin D, which helps keep your bones strong. This is important, because if you are underweight as a result of poor nutrition, your body "steals" calcium from the bones to maintain other vital functions, thereby weakening them. Do not overexpose your skin, however, and use a sunscreen in strong sunlight. If getting sun is a problem, ask your doctor about taking extra vitamin D.

Food sensitivity: By preventing the proper absorption of nutrients, a food sensitivity can cause weight loss. Symptoms of such a sensitivity —which may appear within minutes, hours, or even days of eating a particular food—include asthma, eczema, migraine, arthritis, depression,

diarrhoea, fluid retention, food cravings, and recurrent abdominal pain. If you think your weight loss may be the result of a reaction to a food, track down the cause with help from a doctor or dietician, who will help you correct your sensitivity while maintaining your intake of vital nutrients.

Emotional causes: If you are anxious or depressed, you may neglect your general health and lose weight. This is of particular concern if you are elderly, especially if you live alone.

It is important to address the underlying cause of anxiety or depression, to discover and use stress-management strategies (see p. 347), and to get expert advice, if necessary. The following measures may also help:

- Use lavender oil for its soothing properties. Sprinkle three or four drops into your bath-water. Also sprinkle a few drops onto a hand-kerchief so that you can inhale the vapour during moments of stress.
- Ask a friend to give you a massage using two drops each of lavender, geranium, and sandal-wood oils, and one drop of ylang ylang oil, in two tablespoons of sweet almond oil.
- Try relaxation exercises and meditation. Yoga can also be a great help for stress relief; just 10 minutes a day will help you learn to relax. Join a class or practise at home.
- Drink a cup of herbal tea when you feel particularly pressured. Choose from yarrow, rosemary, nettle, and cinnamon.
- Try taking an appropriate flower essence, such as Olive, if you are physically and emotionally depleted by a period of stress.

Acupressure for calming the mind

Regular acupressure sessions can sometimes reverse loss of weight that is linked to emotional causes. Ask a friend or partner to do the following at least once a week:

1 Using only gentle pressure, move the palm over the point four thumb-widths above the navel (CV 12). At the same time, using the other hand, apply gentle thumb pressure to the inner arm at the point two thumb-widths above the wrist crease (HP 6). Treat each side for two minutes.

2 Using the same techniques as in step 1, apply simultaneous pressure to the point eight finger-widths above the navel (CV 14) and to the point on the outside edge of the little-finger side of the wrist crease (HP 7).

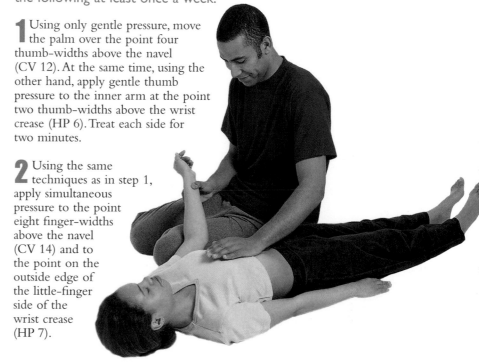

When to get medical help

- Unwanted weight loss continues for four weeks or longer.
- You have other unexplained symptoms.
- You are more than 10 percent below your ideal body weight.

See also:
Anxiety, Appetite Loss, Depression, Eating Disorders, Irritable Bowel Syndrome, Overweight

Glossary of Terms

This glossary is intended to help you understand possibly unfamiliar terms related to natural medicine that you may encounter in this book or elsewhere. Words in capitals refer to entries defined elsewhere in this section.

Acidophilus
Lactobacillus acidophilus—beneficial BACTERIA found in live yogurt and which are also available in supplement form. These bacteria can help maintain a healthy balance of microorganisms in the intestine and counter any overgrowth of the yeast-like fungus *Candida albicans*, which causes thrush.

Adaptogens
Remedies, generally from plant sources, that help regulate the body's systems and restore a healthy balance. For example, ginseng helps balance the release of STRESS HORMONES.

Aerobic
Literally, "requiring oxygen". This usually refers to exercise that makes you breathe hard and boosts your heart rate, thereby increasing the capacity of your heart and lungs.

Allergen
A normally harmless substance that triggers an allergic reaction in the IMMUNE SYSTEM of a susceptible person. Most allergens are PROTEINS, and among the most common examples are pollen, dust-mite droppings, and animal dander.

Allergy
A condition in which the IMMUNE SYSTEM responds defensively to a normally harmless substance, causing symptoms such as a skin rash, breathing difficulties, itchy eyes, and a runny nose.

Amino acids
The building blocks of PROTEIN, which the body needs to make and repair cells and to control metabolism. The body is able to manufacture some amino acids itself, but others must be obtained from food sources such as meat, fish and dairy products.

Antibodies
PROTEIN molecules that circulate in the bloodstream, latch on to potentially damaging particles (such as VIRUSES), and help destroy them. They are part of the IMMUNE SYSTEM.

Antioxidants
Compounds that help counteract the damage caused to the body by FREE RADICALS and thus offer some protection against infections and certain cancers. Vitamins A, C and E, and the minerals selenium and zinc are antioxidants, as are some phytochemicals, found in berries, green tea, and other plant foods.

Antiseptic
A substance that disables or destroys BACTERIA and other microorganisms on the skin or on objects—for example, in the home—thereby preventing the spread of infection.

Astringent
A substance that dries out body tissues, reducing any secretion, discharge, or bleeding.

Atopy
An inherited tendency to develop an allergic condition, such as asthma, allergic rhinitis, or eczema (dermatitis).

Bacteria
Microscopic organisms. Some types usually live harmlessly in the body and may even be essential to such processes as digestion, while others are capable of causing illnesses ranging from gastroenteritis to more unusual conditions, such as meningitis.

Blood sugar
Sugar that is produced when CARBOHYDRATES are digested, and that circulates in the blood so that it can be used by the body as fuel. Absorption into the cells is regulated by the HORMONE insulin, which is produced in the pancreas and released in response to an increased level of blood sugar. Eating sugary food causes a relatively rapid surge in the blood-sugar level.

Carbohydrate
The energy-producing component of plant-based foods, which consists of sugars and starches. Refined carbohydrates (from grains that have had their less digestible parts, such as the fibre, removed) are broken down and absorbed into the bloodstream quickly. Complex carbohydrates—from legumes and unrefined cereal grains, for example—take longer to digest and so release their sugar content more slowly.

Carcinogen
A substance that may promote the development of cancer. Well-known carcinogens include ultraviolet light, asbestos, and chemicals in cigarette smoke.

Cardiovascular system
The system, comprising the heart and blood vessels (arteries, veins, and capillaries), that circulates the blood to all parts of the body.

Collagen
A PROTEIN that is a major component of the connective tissue found throughout the body, for example in bones, cartilage, tendons, ligaments, and skin. It becomes less resilient with age.

Decoction
A herbal tea made by simmering the roots or the tough, woody parts of a plant in water.

Distension
Swelling or bloating, particularly of the abdomen. It can have a variety of causes, including INFLAMMATION, fluid retention, and excessive wind production.

Glossary of Terms

Diuretic
A substance that stimulates an increase in the production of urine.

Electrolyte
An element essential for normal cellular function, such as sodium or potassium. Its levels may be depleted when excessive amounts of fluid are lost in sweat, urine, or stools.

Endorphins
A group of NEUROTRANSMITTERS, released in the brain, that reduces the body's response to pain. Endorphins generate a feeling of well-being, as well as playing a role in other functions, such as temperature regulation.

Energy
For many complementary therapists, especially those influenced by Asian therapies, this is the "life force" (*chi*, *qi*, or *prana*), which is said to flow around the body in a network of channels called MERIDIANS in Chinese medicine and *nadis* in Indian medicine. Therapies such as yoga and acupressure aim to restore health by rebalancing this energy flow.

Enzymes
Chemicals whose task is to facilitate chemical reactions—for example, enzymes in the digestive system that break down food into simpler forms that can be easily absorbed by the body.

Essential fatty acids
Fatty acids (see FATS) that are needed for growth and other processes but that cannot be manufactured by the body and must therefore be included in the diet. Omega-6 fatty acids, derived from the essential fatty acid linoleic acid (found in nuts, corn and seeds and their oils, beans and avocados) are vital for healthy circulation and skin and for other bodily functions. Omega-3 fatty acids, derived from the essential fatty acid alpha-linolenic acid (found in beans, walnuts, linseeds and pumpkin seeds and their oils, rapeseed oil, broccoli, and whole grains), and also derived from oily fish, are equally important.

Essential oils
Aromatic oils derived from plants and used by aromatherapists in a variety of ways to try to heal, boost the IMMUNE SYSTEM, and rebalance the body's ENERGY.

Eustachian tubes
Fine tubes that link the air-filled middle ears with the throat, equalizing the pressure within the ear with that in the throat.

Fats
The most concentrated food source of energy. Certain fats (ESSENTIAL FATTY ACIDS) must come from the diet. Others circulate in the blood or are stored in the body. Dietary fats are divided into two broad groups: saturated and unsaturated. Saturated fats are mostly solid at room temperature and are derived mainly from animal sources. Foods containing large amounts include fatty cuts of meat, butter, and most cheeses. Excessive consumption of these fats can lead to serious health problems, such as heart disease, stroke, and obesity. Unsaturated fats are further subdivided into polyunsaturated and monounsaturated forms. Both types are usually liquid at room temperature. Most vegetable, seed, and nut oils are unsaturated. Oily fish also contain unsaturated fats. In general, eaten in moderation, unsaturated fats pose fewer health risks than saturated fats.

Fibre
Sometimes called nonstarch polysaccharides or roughage, this is the indigestible part of edible plant matter, which provides bulk in the stools. Fibre helps keep the digestive system working properly and may offer some protection against certain bowel diseases, breast cancer, and heart disease. There are two types: soluble and insoluble. Soluble fibre, which is softened by water, is found mainly in legumes, fruit, and vegetables. Insoluble fibre, found mainly in the husks of cereal grains and present in the diet in wholegrain bread, brown rice, and bran, passes through the digestive system largely unchanged.

Flavonoids
Pigments responsible for most of the bright colours of fruit and vegetables. Flavonoids (sometimes called bioflavonoids) act as important ANTIOXIDANTS. Their action is enhanced by vitamin C.

Food sensitivity
A reaction provoked by eating a food, or a constituent of food, which causes no problem for most people. The reaction may appear hours or even days afterwards.

Free radicals
Unstable particles created during oxidation reactions in the body and encouraged by a variety of physical stressors, including a diet lacking in ANTIOXIDANTS, pollutants such as cigarette smoke, and excessive exercise or sunlight. They play a role in the aging process and in triggering certain diseases, such as arterial disease and cancer.

Gluten
A PROTEIN that is present in most cereals, especially wheat, barley, rye, and oats, and thus in many foods made with cereal flour, such as bread, cakes, pies, and pasta. People who have gluten intolerance cannot digest it and must therefore follow a gluten-free diet to prevent damage to the intestines. Such damage may lead to coeliac disease, fertility problems, anaemia, and even cancer.

Hormones
Chemicals released into the bloodstream by the endocrine glands to stimulate cells or tissues to perform certain actions. They control an enormous range of bodily

processes—from growth to reproduction. Any improper hormone level or hormone imbalance can lead to disease.

Immune system
A complex armoury, including different types of white blood cells and ANTIBODIES, designed to resist and fight infection or invasion by foreign particles or—in the case of auto-immune diseases—when a person's own cells are perceived as foreign. A wide range of conditions and symptoms can result from malfunction of the immune system.

Inflammation
Swelling, redness, heat, and tenderness around a damaged or diseased area of the body. These changes are produced by the response of the IMMUNE SYSTEM to injury and infection as it acts to protect and repair affected tissues.

Infusion
A herbal tea made by steeping the leaves or flowers of a plant in boiling water.

Legumes and pulses
Peas, beans, lentils and peanuts all contain complex CARBOHYDRATES that are broken down into glucose by the digestive system. They are also sources of PROTEIN and PHYTO-OESTROGENS, and are high in soluble FIBRE.

Lymphatic system
A complex system of one-way drainage vessels (lymph vessels) and collections of cells, called lymph nodes ("glands"), that contain white blood cells, including ones which produce ANTIBODIES. This system drains fluid from the tissues, traps foreign particles, microorganisms, and cancer cells, and carries infection-fighting cells to wherever they are needed.

Mantra
A word or short phrase repeated silently during meditation to help focus the mind and eliminate distracting thoughts and anxieties.

Melatonin
A HORMONE produced by the pineal gland in the brain in response to darkness. Melatonin is made from a NEUROTRANSMITTER called serotonin. It induces drowsiness and sleep, and the cyclical fluctuation in its levels is disrupted by jet lag and night-shift work.

Meridians
The 14 channels through which ENERGY, or *qi*, is said by practitioners of Asian healing techniques to run in the body.

Metabolism
The process by which your body's cells turn food into energy. The rate at which this happens varies between individuals but is increased by regular exercise and decreased by eating a very low-calorie diet. Certain conditions, such as an underactive thyroid gland, can depress the metabolic rate.

Micronutrients
Compounds, such as some VITAMINS and MINERALS, that are needed by the body only in very small amounts and for most people are supplied by a varied diet.

Minerals
Non-organic substances that are required as NUTRIENTS for good health. Some minerals, such as calcium, are found in the body in large amounts. Others, such as chromium, are present in only trace amounts.

Monosodium glutamate (MSG)
A flavour enhancer commonly used in Chinese cuisine. It may cause bloating and headaches in some individuals.

Mucous membrane
A layer of cells that line the openings of the body to the "outside", such as the mouth, nose, urethra, and vagina. Protective secretions from these cells form part of the body's outer defences against infection.

Neurotransmitters
Chemical messengers released by nerve cells that enable an electrical signal to cross the gap between one nerve cell and the next.

Nutrients
Chemical compounds vital for the maintenance of life that are usually obtained from the diet. These include PROTEINS, CARBO-HYDRATES, FATS, VITAMINS, and MINERALS.

Oesophagus
The tube from the back of the throat into the stomach. A reflux of acidic stomach contents into the oesophagus can cause its lining to become inflamed.

Oestrogens
Sex HORMONES released by the ovaries, adrenal glands, and fatty tissue that trigger female development at puberty and promote the monthly buildup of the uterine lining. Their levels change during the menstrual cycle and tail off as the menopause begins. Men also produce small amounts of oestrogens.

Organic
Any food, whether from plants or animals, that is grown in natural conditions, without the use of pesticides, fungicides, HORMONES, and chemical fertilizers.

Phobia
An often disabling fear of things or activities that offer no genuine threat. The reasons for this fear lie in the unconscious mind. Common triggers include heights, enclosed spaces, public speaking, thunderstorms, mice, snakes, and spiders. Controlled exposure therapy, alongside cognitive therapy to aid understanding, is often successful.

Phyto-oestrogens
Compounds naturally present in plant foods that can mimic in a mild way the action of certain HORMONES (such as OESTROGENS) in

the body, or counteract the effect of an overly high level of certain hormones in the body. Good sources include soya beans and other beans, seeds, rhubarb, fennel, and celery.

Pituitary gland
Situated in the brain behind the nose cavity, the pituitary produces important HORMONES, such as growth hormone, and also controls the hormone production of most other endocrine glands.

Poultice
A paste—usually warm—of herbs or other ingredients that is spread on a cloth or pad of gauze and applied to the skin to treat conditions such as boils.

Platelets
Tiny particles manufactured in the bone marrow and carried in the bloodstream. Their role is to help seal any injury to the blood vessels and enable a blood clot to form over the site to halt the loss of blood.

Progesterone
A sex HORMONE produced by the ovaries during the second half of the menstrual cycle. It helps build and maintain the uterine lining after ovulation. Production ceases if the egg is not fertilized, triggering menstrual bleeding. Progesterone is produced by the placenta during pregnancy, to help maintain the pregnancy. Production falls after the menopause.

Proteins
A group of NUTRIENTS made of AMINO ACIDS that are essential for growth and body maintenance as well as for the manufacture of many important body parts, including cell membranes, HORMONES, NEURO-TRANSMITTERS, ENZYMES, and the blood pigment haemoglobin. Animal protein in the diet comes from meat, poultry, dairy foods, eggs, and fish. Vegetable sources include grains, LEGUMES AND PULSES, nuts, and seeds.

Sinuses
The air-filled spaces in the bones in the cheeks and forehead. The MUCOUS MEMBRANES lining them may become inflamed as a result of infection or ALLERGY, and their openings may become blocked, causing pain, as fluid accumulates inside.

Stress hormones
The HORMONES adrenaline and cortisol, which are secreted by the adrenal glands when the brain signals danger from physical or psychological stress. They shut down non-essential body processes, such as digestion, and boost those, such as respiration and heart rate, needed for "fight or flight".

Toxin
A substance that may have a harmful effect on the body. Examples in the environment include pesticides, lead, atmospheric pollutants, such as those in cigarette smoke, and industrial chemicals, such as dioxins. Certain disease-producing BACTERIA may also produce toxins that are responsible for the symptoms of infection.

Virus
An infection-causing agent that can invade living cells and make use of their structure to reproduce, causing damage and the symptoms of disease. Viral illnesses include the common cold, influenza, chickenpox, and the various types of herpes infection.

Vitamins
NUTRIENTS required in varying quantities by the body. All except vitamin D (mostly manufactured by the action of sunlight on the skin) and vitamin K (made by bacteria in the intestines) must be obtained from the diet. Vitamins are divided into two groups: water soluble and fat soluble. Water-soluble vitamins (B and C) are not stored in the body and so have to be consumed on a frequent basis. Fat-soluble vitamins (A, D, E, and K)

are stored in fatty tissue. A deficiency of any vitamin can cause disease; however, the consumption of excessive amounts, especially of fat-soluble vitamins, can also be harmful.

Whole grains
Unrefined cereal grains, which include the germ and husks and retain more of the original grain's nutritional value as well as the natural FIBRE, which is largely lost when the grain is refined. Whole grains are digested more slowly than refined cereals and therefore release their CARBOHYDRATES over a longer period, which helps prevent excessive fluctuations in BLOOD SUGAR. Common food sources of whole grains include brown rice and wholegrain bread and pasta.

Yeast
An ingredient of many bakery foods and alcoholic drinks, which may provoke a wide range of symptoms of FOOD SENSITIVITY in susceptible people. A diet high in yeast may also make a candida infection more difficult to eradicate. Brewer's yeast, used in beer-making, is a useful source of a variety of B vitamins and the MINERALS chromium and selenium.

Resource Guide

General

British Complementary Medicine Association
249 Fosse Road,
Leicester LE3 1AE
0116 282 5511

British Holistic Medical Association
179 Gloucester Place,
London NW1 6DX
020 7262 5299

Institute for Complementary Medicine
PO Box 194, Tavern Quay,
London SE16 7QZ
020 7237 5165
www.icmedicine.co.uk

NHS Direct
(Telephone information and advice about health)
0845 4647
www.nhsdirect.nhs.uk

Acupuncture/Acupressure

British Acupuncture Council
Park House, 63 Jeddo Road,
London W12 9QH
020 8735 0400
www.acupuncture.org.uk

British Medical Acupuncture Society
Newton House, Newton Lane,
Whitley, Warrington WA4 4JA
01925 730727

Alexander Technique

Society of Teachers of the Alexander Technique
20 London House,
266 Fulham Road,
London SW10 9EL
020 7284 3338
www.stat.org.uk

Aromatherapy

Aromatherapy Organisations Council
PO Box 355, Croydon,
Surrey CR9 2QP
020 8251 7912
www.aromatherapy.uk.org

Bates Method

Bates Association of Great Britain
PO Box 25, Shoreham-by-Sea,
West Sussex BN43 6ZF
01273 422090

Biofeedback

Aleph One
The Old Courthouse,
Bottisham,
Cambridge CB5 9BA
01223 811679
www.aleph1.co.uk/bio

Chiropractic

British Chiropractic Association
Blagrave House,
17 Blagrave Street, Reading,
Berkshire RG1 1QB
0118 950 5950
www.chiropractic-uk.co.uk

Craniosacral Therapy

Upledger Institute UK
2 Marshall Place,
Perth, PH2 8AH
01738 444404

Flower Essences

Dr Edward Bach Centre
Mount Vernon, Sotwell,
Wallingford,
Oxfordshire OX10 0PZ
01491 833712
www.bachcentre.com

Herbal Medicine

National Institute of Medical Herbalists
56 Longbrook Street, Exeter,
Devon EX4 6AH
01392 426022
www.btinternet.com/~nimh/

Homeopathy

Homeopathic Medical Association
01474 560336
www.the-hma.org

Society of Homeopaths
2 Artizan Road,
Northampton NN1 4HU
01604 621400
Email: info@homeopathy-soh.org

Light Therapy

SAD Association
PO Box 989, Steyning,
West Sussex BN44 3HG

Massage

British Massage Therapy Council
17 Rymes Lane,
Oxford OX4 3JU
01865 774123
www.bmtc.co.uk

Naturopathy

British College of Naturopathy and Osteopathy
Frazer House, 6 Netherhall
Gardens, London NW3 5RR
020 7435 6464
www.bcno.co.uk

General Council and Register of Naturopaths
Goswell House, Goswell Road,
Street, Somerset BA16 0JG
01458 840072

Nutrition

British Association of Nutritional Therapists.
c/o SPNT, PO Box 47,
Heathfield,
East Sussex TN21 8ZX.
0870 606 1284

British Nutrition Foundation
High Holborn House,
52-54 High Holborn,
London WC1V 6RQ
020 7404 6504.
www.nutrition.org.uk

Osteopathy

British College of Naturopathy and Osteopathy
Frazer House,
6 Netherhall Gardens,
London NW3 5RR
020 7435 6464
www.bcno.co.uk

Reflexology

Association of Reflexologists
27 Old Gloucester Street,
London WC1N 3XX
0870 567 3320
www.reflexology.org

British Reflexology Association
Monks Orchard, Whitbourne,
Worcester WR6 5RB
01886 821207
www.britreflex.co.uk

Yoga

British Wheel of Yoga
Central Office,
1 Hamilton Place,
Boston Road, Sleaford,
Lincolnshire NG34 7ES
01529 306851
www.bwy.org.uk

Index

Acknowledgments

Photographic credits
8 The Stock Market; 35 The Geffrye Museum, London;
48 Telegraph Picture Library; 81 Science Photo Library;
102 Science Photo Library; 114 Mother & Baby Picture
Library/Emap élan; 123 Mother & Baby Picture
Library/Emap élan; 129 (all) Science Photo Library;
170 Frank Murphy/Rosie McCormick; 232 Science Photo
Library; 270 The Stock Market; 288 (left) National
Medical Slide Bank; 359 NHPA.

**The publishers would also like to thank the
following individuals and organizations for their
help in the preparation of this book:**
Aleph One; The British Diabetic Association;
Amy Godfrey-Smythe; Freddie Godfrey-Smythe;
Lorie Gould; Katie Greenhalgh; Morgan Kilgallen;
Jon Meakin; Anna Moture; Neal Street East;
Kevin Owen; Mehmet Oz, M.D., New York–Presbyterian
Hospital, New York; Eljay Yildirim.
**Special thanks are also due to the Marquess
Research Trustees:** Ray Coventry, John Clutton,
William Greaves, Kil Hamilton, John Pearson,
Mickie Steinmann, Roger Stong, and Associates;
Dennis Bergkamp, Iain Carson, and Nwanko Kanu.

A. PATHOGENS

1. *Yersinia enterocolitica*

Yersinia enterocolitica is a psychrotrophic organism and unlike most other enteric pathogens it can grow at refrigeration temperatures. According to Stern and Oblinger (1980) this bacterium is one of the only potentially pathogenic microorganisms able to develop easily in refrigerated foods.

Yersinia enterocolitica and related species constitute a fairly heterogenous group of bacteria which includes both pathogens and a range of environmental strains which are ubiquitous in terrestrial and freshwater ecosystems. In general there is an association with consumption of contaminated foods, especially pork. Indeed during the last 10-15 years the reported incidences of yersiniosis have increased along with increases in reported isolations of *Y. enterocolitica* from food. This is probably due to awareness of the disease and improved methods for its detection (Schofield 1993). The growth/ survival of this organism in meat depends upon the pH of meat the temperature and the environmental conditions (vp/map) used for its storage. Studies *in vitro* (Zee *et al.* 1984) with *Y.enterocolitica* have shown that, in general high concentration of carbon dioxide ($>40\%$) inhibits the growth of this organism. In their study Eklund and Jarmund (1983) have found that, compared to the percentage growth in air, growth at 2,6 and 20° C (for 23 days) under 100% carbon dioxide atmosphere, was reduced by 100, 98 and 43% respectively. Indeed although Kleinlein and Untermann (1990) have shown that *Y. enterocolitica* fail to growth in minced beef at 4° C in 20% CO_2 / 80% O_2, while at 15°C grew equal to that of the air control. The storage at 10° C was only slightly delayed the growth of this organims. In experiments with *Y. enterocolitica* on lamb pieces (Sheridan *et al* 1992), it was established that growth was inhibited only in an atmosphere containing 100% CO_2 or 80% O_2/ 20% CO_2 at a storage temperature of 0° C. At 0° C for all other atmopshere (50% N_2/ 50% CO_2; vacuum pack; 80% O_2/ 20% CO_2), growth took place, and at 5° C there was no inhibition of growth, irrespective of gas atmopshere (Table 1).

Table 1

Growth\Survival of *Yersinia enterocolitica* ($\log_{10}$ cfu/g) on lamb pieces and minced at 42 days in different gas atmospheres at 0 and 5° C (Initial inoculum *ca* $\log_{10}$ 1.8 cfu)

	pieces				minced			
T°C	A[*]	B	C	D	A	B	C	D
0	7.0	5.0	NG[&]	NG	4.0	NG	NG[*]	NG
5	9.0	8.0	7.0	6.0	6.0	6.0	4.0	NG

[*] : A: vacuum pack; B: 50%/50% N_2/CO_2; C: 80%/20% O_2/CO_2 D: 100% CO_2

[&] : not growth

The inhibition of *Yersinia* in an atmosphere containing 80% O_2/ 20% CO_2 is difficult to explain, in view of its growth in an atmosphere where the CO_2 level was much higher (50% CO_2/ 50% N_2). The consolidation of this result in a repeat experiment suggests that this is real effect of some significance and requires further investigation. Furthermore the inhibited effect of this particular gas atmosphere on *Yesrinia enterocolitica* was confirmed from Church *et al.* 1992.

Gill and Reichel (1989) have found that the map have no advantage on the quality of poultry inoculated with this pathogen while Patterson and Grant (1989) suggested that carbon dioxide may offer a solution in retarding growth of *Y. enterocolitica* at 5° C

2. *Listeria monocytogenes*

The recent realisation that foodborne transmission of *Listeria monocytogenes* can play a role in the aetiology of epidemic human listeriosis has led to the need for information on the characteristics of this psychrotrophic pathogen in foods, in particular meat, poultry and seafood products.

While these products have not been implicated in epidemic foodborne outbreaks of human listeriosis, they remain product classes of concern for several reasons, including substantial incidence rates of *Listeria* contamination in retail-level samples and ability of the microorganism to survive or grow under conditions associated with normal processing, distribution or marketing (Buchanan *et al* 1989).

Few studies have been made to obtain a quantitative comparison between different microorganisms grown under defined atmospheric conditions, except in the case of carbon dioxide (Eklund and Jarmund 1993). This is particularly so for the "emerging pathogens" *L.monocytogenes* and the verotoxigenic *Escherichia coli*. The effect on laboratory studies of inoculum growth conditions is also a cause of increasing concern. It is reported that low temperature growth of the inoculum reduces the lag phase of *L. monocytogenes* under chill conditions (Walker *et al.* 1990; Hart *et al.* 1991). During slaughter and subsequent chilling of carcases at the abattoir, there is the potential for *L. monocytogenes* (if present) to be spread from an incoming warm carcase on to a chilled carcass and *vice versa* from contaminated chilled working surfaces and equipment on to either warm or chilled carcases. In poultry processing it is reported that contamination of carcases with *L. monocytogenes* is mainly *via* contaminated working surfaces and equipment (Genigeorgis *et al* 1989; Hudson and Mead 1989) and hence low temperature growth of the inoculum may be the more relevant situation.

Church *et al.* (1992) found that *L.monocytogenes* did not grow on beef steaks stored at 0 and 5° C (Table 2). *Listeria monocytogenes* found to be more sensitive in samples packed in 100% CO_2 in both temperatures used in their study.

Table 2
Growth\Survival of *Listeria monocytogenes* on beef steaks stored at 0° and 5°C

days of storage	vacuum pack		50%/50% N_2/CO_2		80%/20% O_2/CO_2		100% CO_2	
	0	5	0	5	0	5	0	5
0	4.94	4.94	4.94	4.94	4.94	4.94	4.94	4.94
2	-	4.80	-	4.68	-	4.61	-	4.76
7	-	4.58	-	4.71	-	4.84	-	5.20
9	4.69	-	4.57	-	4.65	-	4.68	-
10	-	4.86	-	5.14	-	4.89	-	4.63
14	-	4.52	-	4.47	-	4.72	-	4.34
20	-	4.51	-	4.49	-	4.69	-	4.39
22	4.56	-	4.60	-	4.75	-	4.72	-
24	-	-		4.63	-	4.74	-	
28	-	4.35	-	-	-	-	-	4.07
29	-	-	-	4.45	-	-	-	-
34	-	4.12	-	-	-	-	-	3.59
36	4.29	-	4.67	-	4.50	-	4.38	-
49	4.02	-	4.41	-	4.43	-	3.77	-
63	3.98	-	3.67	-	3.43	-	3.83	-
77	3.93	-	3.37	-	-	-	2.82	-

Sheridan *et al.* (1992) in their experiments with *L. monocytogenes* in lamb pieces found also that 100% CO_2 was the most effective storage condition to inhibit the growth of this pathogen (Table 3)

Table 3
Growth\Survival of *Listeria monocytogenes* on lamb pieces stored at 5° C

days of storage	vacuum pack	50%/50% N_2/CO_2	80%/20% O_2/CO_2	100% CO_2
0	2.2	2.2	2.2	2.2
7	1.8	2.3	2.6	2.0
14	3.3	2.4	3.7	1.8
21	5.9	4.1	5.8	2.0
28	6.1	4.8	6.8	1.7
35	6.3	5.3	6.9	2.9
42	5.7	6.9	7.3	2.4

In the literature there is conflicting evidence about the ability of this organims to grow on chilled beef. Grau and Vantderlinde (1988) report growth of *L. monocytogenes* on vacuum-packed beef of pH 5.6 and 6.0 at both 5.3° C and 0° C. *Listeria monocytogenes* growth was faster at the higher temperature and pH and was initiated more rapidly on fatty tissue. Kaya and Schmidt (1991) conclude that growth of *L. monocytogenes* on vacuum-packed beef is chiefly dependent on the following factors;